Mike R. Jeffrey

Hidden Dynamics

The Mathematics of Switches, Decisions and Other Discontinuous Behaviour

Springer

Mike R. Jeffrey
Department of Engineering Mathematics
University of Bristol
Bristol, UK

ISBN 978-3-030-02106-1 ISBN 978-3-030-02107-8 (eBook)
https://doi.org/10.1007/978-3-030-02107-8

Library of Congress Control Number: 2018959419

Mathematics Subject Classification: 00-02, 03H05, 34E10, 34E13, 34E15, 34N05, 37N25, 37M99, 41A60, 92B99, 70G60, 34C23

This Springer imprint is published by the registered company Springer Nature Switzerland AG
The registered company address is: Gewerbestrasse 11, 6330 Cham, Switzerland

Hidden Dynamics

For Arthur.

A smooth sea never made a skillful sailor
– African proverb

Life has no smooth road for any of us;
and in the bracing atmosphere of a high aim
the very roughness stimulates the climber to steadier steps...
– William C. Doane

Artwork by Martin Williamson (http://www.cobbybrook.co.uk/) 2017

Preface

Discontinuities are encountered when objects collide, when decisions are made, when switches are turned on or off, when light and sound refract as they pass between different media, when cells divide, or when neurons are activated; examples are to be found throughout the modern applications of dynamical systems theory.

Mathematicians and physicists have long known about the importance of discontinuities. Studying light caustics (the intense peaks that create rainbows or the bright ripples in sunlit swimming pools), George Gabriel Stokes lamented to his fiancée in a letter from 1857:

...sitting up til 3 o'clock in the morning
...I almost made myself ill, I could not get over it
...the discontinuity of arbitrary constants.

Discontinuities are not a welcome feature in dynamical or differential equations, because they introduce indeterminacy, the possibility of one problem having many possible solutions, many possible behaviours. How interesting it is then to consider the thoughts of the influential engineer Ove Arup:

Engineering is not a science ... its problems are under-defined,
there are many solutions, good, bad, or indifferent.
The art is ... to arrive at a good solution.

For Arup, 'science studies particular events to find general laws'. Many mathematical scientists would agree that the goal is to achieve generality and banish indeterminacy. But why should the two be mutually exclusive?

Unlocking the potential of discontinuities requires tackling these issues of determinacy and generality. While accepting that some parts of the world lie beyond precise expression, discontinuities nonetheless give us a way to express them, to approximate that which cannot be approximated by standard 'well-posed' equations, and to explore a world where certain things will ever remain hidden from view.

With these ideas in mind, this book attempts to ready the field of nonsmooth dynamics for turning to a wider range of applications, simultaneously moving beyond the traditional scope of, and bringing our subject closer into line with, the traditional theory of differentiable dynamical systems.

At a discontinuity, we lose access to some of the most powerful theorems of dynamical systems, and it has long been the task of nonsmooth dynamical theory to redress this. Progress has been impressive in some areas, limited in others. We suggest here that much of what has gone before constitutes a *linear* approach to discontinuities, and here, we lay the foundations for a *nonlinear* theory. Making use of advances in nonlinearity and asymptotics, once we can extend elementary methods such as linearization and stability analysis to nonsmooth systems, discontinuities stop being objects of nuisance and start becoming versatile tools to apply to modelling the real world.

Several examples of applications are studied towards the end of the book, and many more could have been included. Interest in piecewise-smooth systems has been spreading across scientific and engineering disciplines because they offer reliable models of all manner of abrupt switching processes. Our aim is to set out in this book the basic methods required to gain an in-depth understanding of discontinuities in dynamics, in whatever form they arise.

In this book, a discontinuity is *blown up* into a *switching layer*, inside which *switching multipliers* evolve infinitely fast across the discontinuity. Several concepts may be at least partly familiar in other areas of mathematics, in particular algebraic geometry, boundary layers, singularity theory, perturbation theory, and multiple timescales. The terminology used here does not exactly correspond to the usage in those fields, and attempting to refer to or resolve all of the clashes in nomenclature would not make for an easier read. Moreover, we do not use the concepts themselves in strictly the same way. For example, we use the idea of an infinitesimal ε-width of a discontinuity that we can manipulate algebraically, but we are interested solely in the limit $\varepsilon = 0$. This proves to be a sufficiently rich problem, and though it raises the question of what happens when we perturb to $\varepsilon > 0$, that is left for future work. As we discuss in Chapter 1 and Chapter 12, more so than in any smooth system, the perturbation of a discontinuity is a many faceted problem.

This work builds on the pioneering efforts particularly of Aleksei Fedorovich Filippov, Vadim I. Utkin, Marco Antonio Teixeira, and Thomas I. Seidman. I have been lucky to meet and work with all but the first of these, and much in the spirit of modern science, they represent the truly international and interdisciplinary endeavour of what was for too long a niche field of study.

In essence, the framework introduced in this book seeks to explore Filippov's world more explicitly, to make non-uniqueness itself a useful modelling tool in dynamics. Sitting somewhere between deterministic dynamics and stochastic dynamics, nonsmooth dynamics offers a third way: systems that are only piecewise-defined, rendering them *almost everywhere* deterministic.

Bristol, UK Mike R. Jeffrey

Chapter Outline

The book is roughly split into three parts: introductory material in Chapters 1 and 2, fundamental concepts at the level of the student or non-expert in Chapters 3 to 6 and Chapter 14, and advanced topics in Chapters 7 to 13.

Chapter 1 is almost a stand-alone and informal essay, surveying the reasons why discontinuities occur, what forms they take, why they matter, and how imperfect our knowledge of them is. The chapter is intended to provoke thought and discussion, not to be detailed reference on the many theoretical and applied concepts it touches on.

Chapter 2 is a stand-alone "lecture"-style outline, a crash course on the topic, and a taster of the main concepts that will be developed in the book.

Chapter 3 contains the complete foundation for everything that follows, the formalities for how we define piecewise-smooth systems in a solvable way. This chapter contains the elements necessary for the eager researcher to rediscover for themselves the contents of the remainder of the book and beyond.

Chapter 4 sets out the basic themes that dominate piecewise-smooth dynamics, the kinds of orbits, the key singularities, and the concepts of stability and bifurcation theory.

Chapter 5 defines a general prototype expression for piecewise-smooth vector fields in the form of a series expansion.

Chapter 6 describes the basic forms of contact between a flow and a discontinuity threshold.

Chapter 7 contains the most important new theoretical elements of the book, setting out the analytical methods required to understand piecewise-smooth systems.

Chapter 8 takes a step back, applying the previous chapters in the more standard setting of linear switching.

Chapter 9 begins the leap forward into nonlinear switching, revealing some of the novel phenomena of piecewise-smooth systems.

Chapter 10 focusses on the most extreme consequences of discontinuity, via determinacy breaking and loss of uniqueness.

Chapter 11 tackles how we understand large-scale behaviour, with new notions of global dynamics and associated bifurcations.

Chapter 12 asks how robust everything that has come before is. We consider how our mathematical framework can be interpreted in a practical setting and what happens to it in the face of nonideal perturbations.

Chapter 13 visits an old friend and long-term obsession of piecewise-smooth systems, the two fold singularity.

Chapter 14 is a series of case studies applying the foregoing analysis to 'real-world' models.

Exercises are provided at the back of the book to further facilitate a more in-depth reading or lecture course.

How to Use This Book

This book will look rather different to other works in the area. In Chapter 1, we start from a tour of some less quoted, wide-ranging, but fundamental, examples of how discontinuity arises. Chapter 3 presents the formalism for studying nonsmooth dynamics that forms the foundation for everything that follows and should be the starting point for any course. It is quite possible to jump from there to Chapter 12 to focus on the application and robustness of the formalism.

A proper understanding of the dynamics of nonsmooth system, or a course in it, should progress through Chapters 4 to 11, and I would suggest focussing on (and indeed extending) the analytical methods in Chapter 7. The great peculiarities of nonsmooth systems begin to be revealed in Chapter 9 and Chapter 10, and there are numerous examples therein to explore and build on. More in-depth applications are given in the form of case studies in Chapter 14. Exercises provided at the back of the book provide further insight into the various examples and theorems explored, chapter by chapter.

Prerequisites. In reading this book, it will be helpful to have a grounding in (though we give elementary introductions where possible): single and multi variable calculus, Taylor series, ordinary differential equations and elementary dynamical systems, some linear algebra (eigenvectors, etc.), and a little introductory (highschool) physics. Applications will be explained with background where they are discussed.

Contents

Chapter 1
Origins of Discontinuity

Discontinuities occur when light refracts, when neurons or electronic switches activate, and when collisions or decisions or mitosis or myriad other processes enact a change of regime. We observe them in empirical laws, in the structure of solid bodies, and also in the series expansions of certain mathematical functions.

As commonplace as they may be, discontinuities are a curious thing to try to build into dynamical models. They violate that central requirement of calculus: to be continuous. They permit determinism to collapse in fleeting bursts of non-uniqueness. They conjure up a new realm of nonlinear dynamics.

In this first chapter we start by exploring what it means for a system to be discontinuous. Some discontinuities we understand, but many, the kinds of discontinuities we encounter in engineering, the life sciences, and economics and desperately wish to develop more sophisticated models for, we hardly understand at all.

Formally, this book asks what happens to the trajectories of variables

$$\mathbf{x}(t) = (\ x_1(t),\ x_2(t),\ \ldots, x_n(t)\ ) \ \in\ \mathbb{R}^n\ , \tag{1.1}$$

as they evolve in time t, according to a set of ordinary differential equations

$$\frac{d}{dt}\mathbf{x} = \mathbf{f}(\mathbf{x})\ , \tag{1.2}$$

when the right-hand side is only piecewise-smooth, changing smoothly with respect to $\mathbf{x}$ almost everywhere, except at certain thresholds $\sigma(\mathbf{x}) = 0$ where the value of $\mathbf{f}$ jumps, i.e. is discontinuous. But this rather dry statement hints at few of the pitfalls and paradoxes of dynamics afflicted by discontinuities.

M. R. Jeffrey, *Hidden Dynamics*,
https://doi.org/10.1007/978-3-030-02107-8_1

1.1 Discontinuities and Dynamics

When Isaac Newton set down the laws of motion that form the basis of classical mechanics, he helpfully also set out the route to understand them using calculus. Yet in doing so he mischievously threw into the stirring pot some laws of motion *not* amenable to calculus. Century upon century since, a juxtaposition of continuous and discontinuous change at the heart of physics has remained, with consequences that remain only partly understood.

Collisions offer a tangible example (Figure 1.1). Newton's laws tell us the forces acting on a moving object, and from those forces, calculus provides its speed and position. Yet when that object collides with another, instead of calculus we must employ a little mathematical sleight of hand. Calculus works for the pre-collision motion, and it works for the post-collision motion, but then we must stitch the two together somewhat artificially. To disguise the conceit—the discontinuity in the laws of motion—we give the procedure a lofty title: an *impact law*.

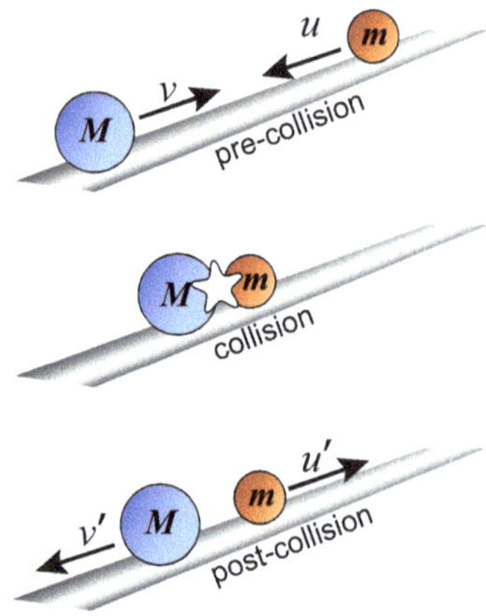

Fig. 1.1 Two objects collide and recoil. An *impact law* relates their incoming speed $u + v$ to their recoil speed $u' + v'$ by $u' + v' = e(u + v)$ for some $0 \leq e \leq 1$.

Discontinuities allow us to gloss over small details that seem to have no major effect on our large-scale view. The last century, however, has taught us that no matter how small, details can change everything.

The reason that we cannot follow motion through a collision, in the same way we can follow objects that are rolling or in free flight, is because the collision involves stepping between irreconcilable physical regimes: free motion and rigid contact. One way to understand the regime change is to step into a different modelling approach entirely, perhaps on a finer scale allowing bodies to be more compliant and less idealized. But this can bring its own problems and ambiguities, introducing much greater complexity, often probing areas where our knowledge is less complete, and ultimately being difficult to marry up with the original discontinuous model.

To serve those situations, our task in this book, and in the field of *nonsmooth* or *piecewise-smooth* dynamics more widely, is to provide a way within a given dynamical model, to follow motion across the discontinuities between irreconcilable regimes.

As science spreads its interest to new technological and sociological vistas, it increasingly encounters a world full of irreconcilable regimes, of media not behaving like steady waves rolling over the ocean, like electromagnetic waves vibrating through spacetime, or like spheres orbiting and tumbling through the vacuum of the heavens. Instead we find abrupt changes that we patch over with ad hoc rules, such as *switch from behaviour A to behaviour B*. Figure 1.2, for example, shows a discontinuity that turns up in climate models—the reflectivity of the Earth's surface jumping across the edge of an ice shelf. The mathematical implications of such switches are not obvious.

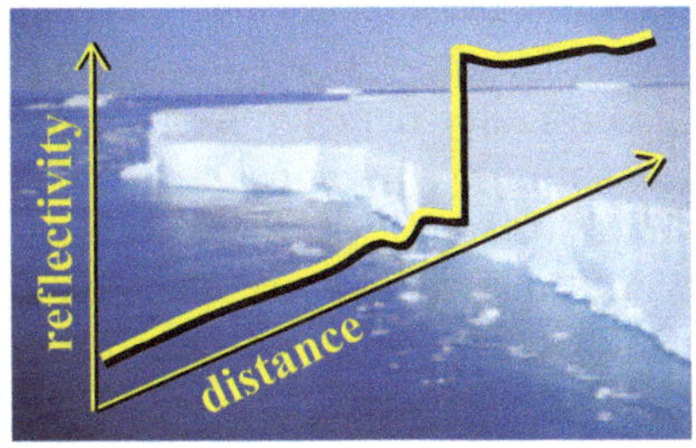

Fig. 1.2 A jump in surface reflectivity between ice and water oceans.

Discontinuities like these are what endow the world around us with structure. The boundaries of solid objects are marked by jumps in properties like density, elasticity, or reflectivity. People make decisions changing the course of their day. Storms and waves and glasses break, social regimes change, lives are stopped and started.

As with collisions, we tend to skirt around the edges of these discontinuities with a little sleight of hand and so describe *almost* everything going on in a system, glancing over the discontinuities which, after all, are but fleeting. When I choose to go left or right, when a cell chooses to grow or divide, and when a machine switches on or off—that brief moment when the choice in enacted is trivial, isn't it?

Far from it. Three centuries of calculus have left mathematicians uneasy with discontinuities and reluctant to give up the continuity that provides so many theorems concerning stability, attractors, bifurcations, and chaos, because discontinuities leave these theorems in tatters.

From the mathematical point of view, a discontinuity renders a system 'ill-posed'. A well-posed system has equations whose solutions: (i) exist, (ii) are unique, and (iii) vary continuously with initial conditions. To satisfy all three, a system must be smooth enough (meaning differentiable some number of times, and certainly anything with a discontinuity does not qualify).

It turns out that at discontinuities we will often have to give up properties (ii) and (iii), but not (i), not existence. It may seem perverse to give up uniqueness and continuous dependence on initial conditions, but that is what discontinuities are, events by which continuity and uniqueness are lost, and our task is not to judge, but to learn how those losses can be exploited to understand more about the world around us.

This book is an exploration of that idea. It is an attempt to extend the methods of nonlinear dynamics beyond the barriers that discontinuities have previously made impassable. In pushing back these boundaries, we find some intriguing behaviours. The methods, the theory behind them, and the phenomena we discover, all require deeper future study. Though we prove results where possible, not everything we do can be elevated to the level of rigour that can be achieved with smooth systems (at least not yet), so we do not claim a rigorous study here, only a development of ideas and methods.

Throughout the book we study the discontinuous system, with all of the difficulties that brings, breaking only in Chapter 12 to consider nearby 'perturbations'. There are various obvious ways that one may try to avoid discontinuity depending on context. We might, for example, smooth out a discontinuity, perhaps believing that smooth physical laws underlie it or simply to make it easily computable. Or we might blur the discontinuity with a distributive or stochastic process. An entire book mirroring this one could be written using each approach, one smooth and deterministic and one stochastic.

The discontinuous approach accepts that either of these, or numerous other *perturbations* of the discontinuous model, could be the right approach. Let us first attempt to understand the underlying discontinuity, and later we will probe a little into what happens when we perturb, in one way or another, by smoothing, randomizing, or blurring the discontinuity in other ways.

The book starts and ends with less formal chapters which set the context for our subject matter with the use of practical examples. This is one such chapter and takes us on a short tour of how discontinuities arise and some phenomena they produce. This expedition is not vital for those seeking an introduction to piecewise-smooth dynamical systems theory, nor is it a comprehensive study of the topics touched on, but I hope you will at least skim through it as motivation for what is to come.

In between those less formal chapters come more technical theory, aimed at developing methods to understand the geometry and stability of solutions, rather than focussing on proofs of solvability and universality of classes, but opening numerous avenues for future study. After the theory is established in Chapters 3 to 7 and explored at little in Chapters 8 to 11, we delve more deeply into applications and 'real-world' switches in Chapters 12 to 14.

Towards the end of the book, we return to the question of what a discontinuity is. Discontinuities allow us to model abrupt change without imposing undue structure. In a story that will unfurl as we reach Chapter 12, we will learn that the best achievable representation of reality is not always the most precise. We will see that it is sometimes unuseful, and even misleading, to model processes in finer detail than our understanding allows and that discontinuities provide not an obstacle to calculus but a new vehicle for it to traverse uneven terrain.

To rely on continuity is to overlook that discontinuities are inescapable. They arise not only in our everyday reality but within calculus itself, in the midst of divergent series and singular perturbations, leaving mathematics no less rich or rigorous for it. To rely on continuity is to risk overlooking that differentiability reaches only so far into the complexities of a real world where discordant media interact over disparate scales, and discontinuities are often the result. We visit all of these in this chapter.

So let us see why discontinuity matters, where it comes from, and what it looks like.

1.2 Discontinuities and Determinism

One issue will concern us only in limited situations, but will not go away altogether, and that is:

where there are discontinuities there is non-uniqueness.

This non-uniqueness comes in many guises, but with just two main sources that we can introduce briefly.

The first comes from a lack of knowledge of what happens inside a discontinuity. We may know that a quantity jumps between two values, but not know precisely how it does so. We then use *hidden* terms to bring this uncertainty to life, to express the different possible modes of behaviour inside the jump. We shall show these constitute a form of nonlinearity. This is one of the more subtle notions that will unfold throughout this book, and we will introduce them a little more in Section 1.3.

The second source of non-uniqueness is more obvious, more well known, and is the reason why mathematicians are taught a reluctance to study non-smooth systems. It afflicts the solutions of a differential equation at a discontinuity. A classic example is the equation

$$\frac{dx}{dt} = |x|^{\alpha} , \tag{1.3}$$

for different values of $\alpha \geq 0$. Its solutions take the form

$$x(t) = x_0 \left(1 + \frac{1-\alpha}{x_0|x_0|^{-\alpha}} t\right)^{1/(1-\alpha)} , \tag{1.4}$$

with an initial condition $x(0) = x_0$. Although we can write the solution (fairly) simply, upon closer inspection we start to find problems with it.

For $\alpha \geq 1$ solutions come in three types: those that start at $x_0 = 0$ and sit there forever, those that start at $x_0 < 0$ and tend to $x = 0$ but never quite reach it, and those that start at $x_0 > 0$ and head off towards infinity. For instance, in the special case $\alpha = 1$, we simply have $\frac{dx}{dt} = |x|$, and the solutions become $x(t) = x_0 e^{\mathrm{sign}(x_0)t}$. The solution through any x_0 is therefore unique: if we know the 'x_0' where we start, then all future (or indeed past) evolution of $x(t)$ is determined. This follows from the continuity of $|x|^{\alpha}$ for $\alpha \geq 1$ (more precisely the Lipschitz continuity of $|x|^{\alpha}$, by the so-called Picard-Lindelöf theorem [149]).

For $0 < \alpha < 1$ the situation is entirely different. The discontinuity in the derivative of (1.3) takes over. Every solution through any $x_0 < 0$ reaches $x = 0$ in a future time $t = |x_0|^{1-\alpha}/(1-\alpha)$, while every solution through any $x_0 > 0$ must have left $x = 0$ at a past time $t = -|x_0|^{1-\alpha}/(1-\alpha)$. Does this mean that we just have one solution that passes through zero? No, because the point $x = 0$ is a solution itself. So if a solution from $x < 0$ reaches $x = 0$, it can sit there arbitrarily long before setting off again towards $x > 0$. This means that an infinity of different solutions, all pausing to rest for different amounts of time at $x = 0$, all overlap at the origin and we cannot tell them apart. As a result, the history and future of the point $x_0 = 0$ are non-unique.

Non-unique histories are part of everyday experience and are one of the reasons why nonsmooth systems have such broad applications. For example, imagine an object that has been propelled along a surface and brought to rest by friction. It is subsequently impossible to reconstruct the object's motion before it came to rest or to determine how much time has elapsed since it stopped. A discontinuity in the frictional interaction between the object and the surface has destroyed this information. This is an important effect in our everyday lives. When you hit the brakes in your car, you want them to behave like $0 < \alpha < 1$ in the example above, to come to rest in finite time, not to slow interminably towards the scene of an accident.

Non-unique futures are something less comfortable. A solution can start out being unique and well behaved, but in the presence of a discontinuity, it can find itself ripped apart and endowed with infinitely many possible futures. We call these *determinacy-breaking* events.

Figure 1.3 depicts the scenario schematically. The picture shows the trajectories of a system evolving through space. Those trajectories are deterministic everywhere except at a single point, the determinacy-breaking singularity.

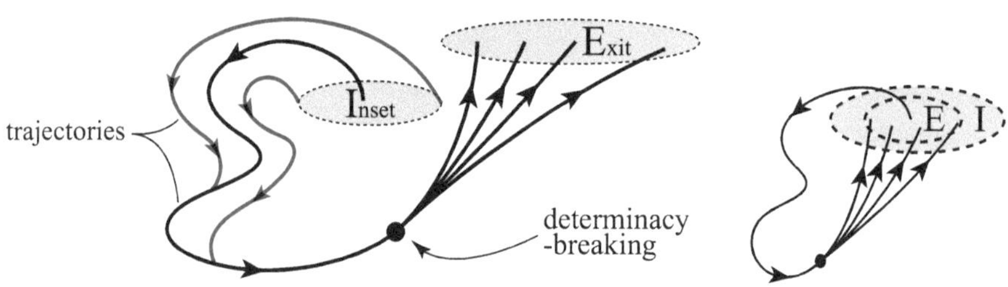

Fig. 1.3 A determinacy-breaking event. Solutions before and after the singularity are deterministic. Any trajectory starting in I hits the singularity. All trajectories in E originate at the singularity. Inset right: forming a closed set.

Such singularities are common in nonsmooth systems. They result in new kinds of nonlinear dynamics, new kinds of chaos and bifurcations, and even new kinds of attractors. Imagine in Figure 1.3, for instance, if the inset I of trajectories that are pulled into the singularity is intersected by the exit set E of trajectories leaving the singularity (shown inset right). Then trajectories will exist that make repeated yet unpredictable excursions, trapped forever to return to the singularity, despite their exit path from it being uncertain.

With its inherent ambiguities of various sources of non-uniqueness, it is easy to dismiss discontinuities from serious dynamical theory. But the non-uniqueness turns out to be useful, not to be swept under the rug or axiomatized into oblivion, and closely intwined in all its forms with nonlinearity.

1.3 Discontinuities in Approximations

How do you approximate near a discontinuity? This is what we are doing very often when we are studying discontinuous systems and their dynamics, whether in theoretical equations or in empirical models. Consider the following.

Example 1.1 (Approximating a Nonlinear Switch). Let us try to approximate a pair of functions

$$g(x) = \frac{\sin x}{|x|} \qquad \text{and} \qquad f(x) = (1 + 2g(x))^2 \ , \tag{1.5}$$

sketched in Figure 1.4. (In a strict sense we should not refer to f and g as functions if they take many values at $x = 0$, but we allow this small abuse of terminology, much as the Heaviside step 'function' or sign 'function' are so-called, with the values at $x = 0$ being, after all, our topic of interest).

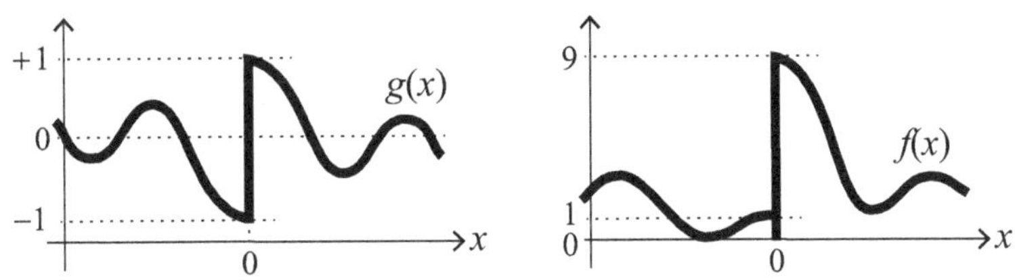

Fig. 1.4 The graph of two functions $g(x)$ and $f(x)$ with a discontinuity at $x = 0$.

These are both well behaved for x away from zero, and if we wish to approximate them near a point $c \neq 0$, we can expand them as Taylor series,

$$g(x) = \tfrac{\sin c}{|c|} + (x - c)\tfrac{c\cos c - \sin c}{c|c|} + \mathrm{O}\left((x - c)^2\right) \ , \tag{1.6a}$$

$$f(x) = \tfrac{(|c|+2\sin c)^2}{|c|^2} + 4(x - c)\tfrac{(|c|+2\sin c)(c\cos c - \sin c)}{c|c|^2} + \mathrm{O}\left((x - c)^2\right) \ . \tag{1.6b}$$

These series are unique, with successive terms telling us the values, gradients, curvature, etc. of f and g around $x = c$.

If we attempt to expand about $x = 0$, however, we obtain two different series depending on whether we consider $x > 0$ and $x < 0$. The expansion of g is $g(x) = \text{sign}(x) - \frac{1}{3!}x|x| + \frac{1}{5!}x|x|^3 - \ldots$, or to lowest order, just

$$g(x) = \text{sign}(x) + \mathrm{O}\left(x^2\right) \ . \tag{1.7a}$$

Substituting this into $f(x) = (1 + 2g(x))^2$ we have

$$f(x) = 5 + 4\,\text{sign}(x) + \mathrm{O}\left(x^2\right) \ . \tag{1.7b}$$

This result is inconsistent, however, with the definition of f. Let us assume that g lies between ± 1 at the discontinuity, that is, $-1 < g(0) < +1$. Then (1.7b) implies $1 < f(0) < 9$. This is contrary to the definition of f in (1.11), which reaches a minimum with respect to g at $g = -1/2$, where $f = 0$, and therefore implies $0 < f(0) < 9$.

We are only looking at behaviour at and near $x = 0$, so we should expect the approximations of f and g to give consistent answers. The discrepancy does not lie in the $\mathsf{O}\left(x^2\right)$ terms we have neglected, since they vanish for small x. So what has gone wrong? How can we tell unambiguously the range of values f takes as x changes sign and g jumps through the interval $[-1, +1]$?

The series expansions (1.6) to (1.7) are not strictly valid at $x = 0$ because g and f are not continuous there, but there is a more useful way of looking at what has gone wrong. The equation in (1.11) depends nonlinearly on the discontinuous quantity g. In (1.7b) we are ignoring that nonlinearity, and this, in fact, is the source of the contradictory ranges for f, not the series expansion itself.

A better way to handle this turns out to be to define a *switching multiplier*

$$\lambda = \begin{cases} +1 \text{ if } x > 0\,, \\ -1 \text{ if } x < 0\,, \end{cases} \tag{1.8}$$

and to define this as lying in $-1 < \lambda < +1$ for $x = 0$. In terms of λ we can write

$$g(x) = \lambda \frac{\sin x}{x} \qquad \text{and} \qquad f(x) = (1 + 2g(x))^2\;, \tag{1.9}$$

then expanding f gives

$$\begin{aligned} f(x) &= 1 + 4g(x) + 4g(x)^2 \\ &= 1 + 4\lambda \tfrac{\sin x}{x} + 4\lambda^2 \tfrac{\sin^2 x}{x^2}\;. \end{aligned} \tag{1.10}$$

The term λ^2 is simply unity for $x \neq 0$, and this is what went missing when we approximated for small x above. If we are careful to keep the λ^2 term, we can now approximate for small x, and we obtain

$$f(x) = 1 + 4\lambda + 4\lambda^2 + \mathsf{O}\left(x^2\right)\;. \tag{1.11}$$

For $-1 < \lambda < +1$ the function $g(0) = \lambda$ still takes values $-1 < g(0) < +1$, but we now see that f correctly takes values $0 < f(0) < 9$ (with a minimum at $\lambda = -1/2$). Note how the '5' from (1.7b) has become '$1 + 4\lambda^2$'.

■

Why would it matter what values f passes through at the discontinuity? One reason is that peaks or troughs—turning points with respect to λ—in such a function can act like potential wells at the discontinuity, whose presence or absence in a dynamic system may decide whether states can pass through the discontinuity or become trapped within it.

Let us imagine that the discontinuity in f lies not perfectly at $x = 0$, but is spread out over some $|x| < \varepsilon$, like the graphs shown on the left of Figure 1.5. As we let ε tend to zero, we recover our discontinuous system, shown on the right of Figure 1.5. Then consider a dynamical law

$$\dot{x} = -df/dx \ .$$

Figure 1.5 depicts three different scenarios. If f is monotonic (top graph on the left), then the variable x will evolve straight through the jump that occurs at $x \approx 0$. For some f with a peak or a trough around $x = 0$ (bottom two graphs on the left), the variable x will get stuck in a potential well as it tries to pass through the jump.

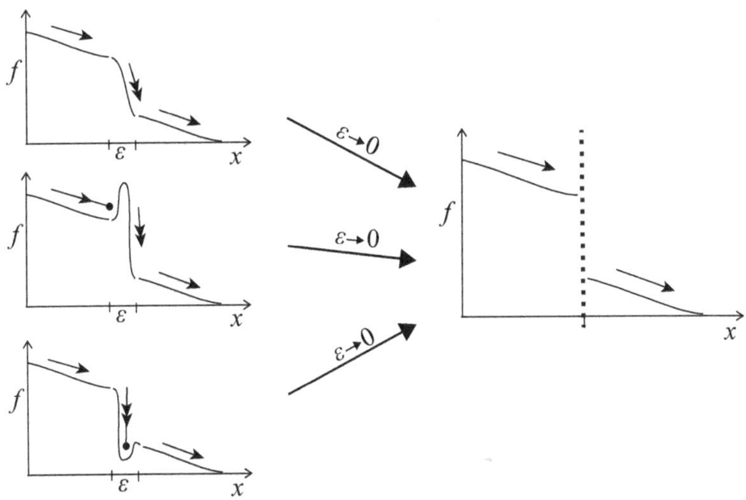

Fig. 1.5 A system 'rolls' down a potential ϕ, which has a jump over $|x| \lesssim \varepsilon$. In the limit $\varepsilon \to 0$, the shape of the potential at the jump becomes hidden inside the discontinuity.

In the limit $\varepsilon \to 0$, these potential wells become squashed into the discontinuity at $x = 0$ and indistinguishable as a function of x (right-hand graph in Figure 1.5). However, we can use nonlinear switching terms, as we used λ in the graph of f above, to resolve the difference between the three cases.

What this exercise shows us is that:

- we can use switching multipliers like λ to endow discontinuities with nontrivial structure;
- we must respect nonlinear dependence on those multipliers.

Accepting that a system can depend nonlinearly on a discontinuous quantity essentially brings nonsmooth dynamics into the era of *nonlinear switching dynamics*, into which this book is a first tentative step. Already the outlook appears to be as rich for nonsmooth systems as the era of nonlinear dynamics has been for smooth systems. While this book seeks to set out the new tools required as extensively as possible, they will necessarily be less well matured than the now standard concepts of smooth dynamic systems. Even the theory of approximating around discontinuities is not as fully developed as that of Taylor series, and the implications for the dynamics of systems would take a whole book to even begin exploring.

So here we go.

1.4 Discontinuities in Physics and Other Disciplines

Before we set out our framework for handling discontinuities in subsequent chapters, we should have some idea of their origins in mathematics and in nature and the forms they take. Some, like collisions or cellular mitosis, are part of nature. Others are imposed by us to control nature.

Switching or switch-like behaviour occurs in countless phenomena, from the mundane to the arcane. Figure 1.6 collects together a handful of examples from experimental data, showing sometimes simple monotonic jumps; some with peaks or ringing, with random perturbations, and with moving thresholds; or multipart processes. All of these are based on real experimental data or models derived from them. They are taken from varied disciplines:

(i) current and voltage in an electronic controller, based on measurement data from [28]. Large spikes are observed during switching, much larger than the small residual ringing that persists between each switch.

(ii) pressure against distance along a column of water as it pinches off to form a droplet, at different times, showing high and low peaks either side of the sharp transition, a fluid bridge where the droplet will eventually separate off; based on analytic solutions from [58].

(iii) resistivity of a superconductor against temperature in different magnetic fields, based on measurement data from [157].

(iv) particle count registered by Voyager 1 at the edge of the solar system, measurement data available online; see [164, 217].

(v) potential across a permeable cell membrane, showing a peak during transition. This heuristic is found in various textbooks; see, e.g. [128].

(vi) steady-state sodium current for an activating silicon neuron, including measurement data from a voltage clamp (dots), best fit (full curve), and a fitting by a Hill function (dashed, for Hill functions refer back to Table 1.1 in Section 1.4.4), based on data in [193].

(vii) two switches in the concentration of cells across a permeable membrane, a heuristic law found in various sources, e.g. [214].

(viii) two switches in the action potential of heart muscle, the first only showing an overshoot, another heuristic curve found in various sources, e.g. [185].

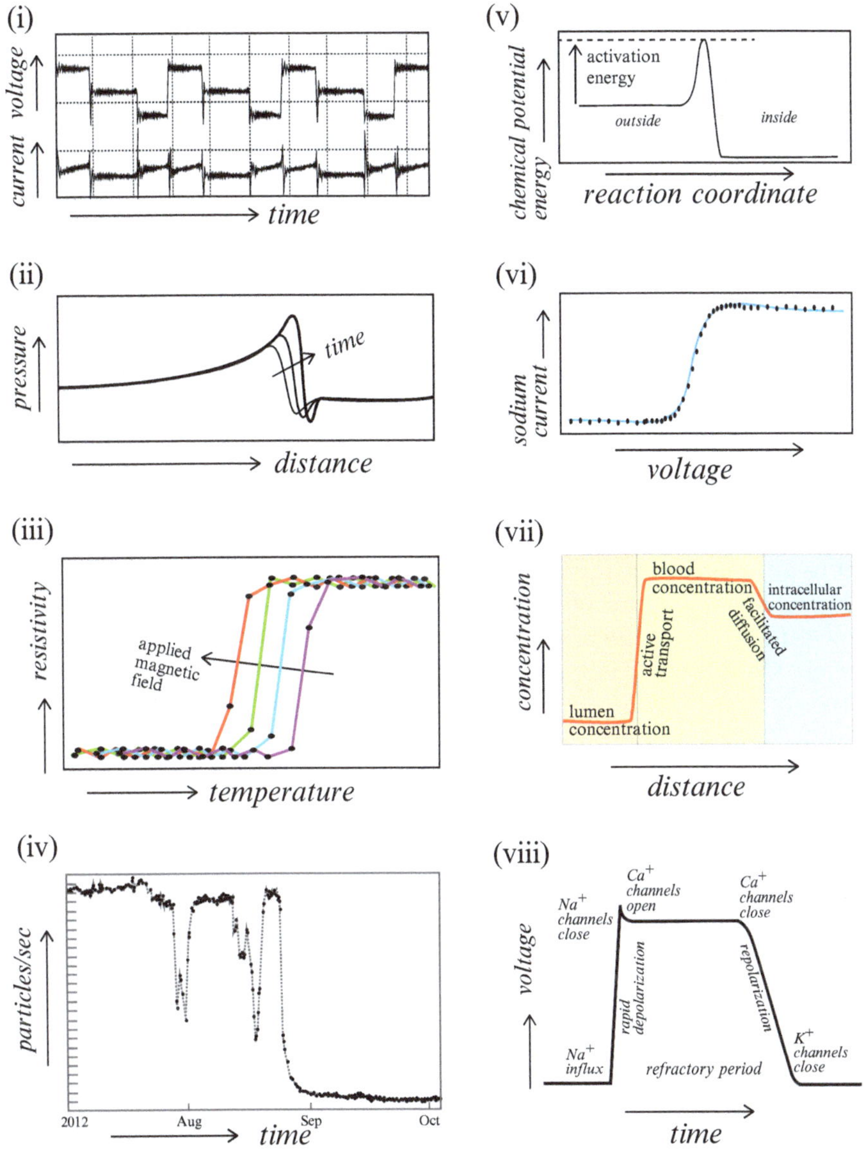

Fig. 1.6 A flavour of the many ways discontinuities manifest in experimental data and heuristic laws: (i) electronic control, (ii) fluid droplet formation, (iii) superconductors, (iv) the Voyager satellite, (v) cell membrane potentials, (vi) Neuron activation, (vii) movement across cell membranes, (viii) heart muscle potentials.

The reader may be aware of many more, such as in morphing and buckling aerodynamic surfaces, in decision-making for trades or prey preference, in gene regulation, in mitosis, in albedo jumps at the boundary of ice sheets, and on and on, with ever more to be discovered and investigated.

There is a lot going on in these graphs, leaving a choice for the scientist tasked with studying them of whether to model the jumps as discontinuous, smooth, stochastic, or otherwise. Part of our task is to understand in what sense these systems *are* discontinuous, or can be approximated by discontinuities, and what we gain by representing them as discontinuities.

But discontinuities are not just a way of handling jumps in data or empirical models. They also arise in the geometric analysis of integrals and approximation of differential equations. Over the next few sections, we trace a few physical and mathematical origins of discontinuity.

1.4.1 In Mechanics: Collisions and Contact Forces

Let us put a little more flesh on the collision problem we began with. Consider a block of mass m, with position x, driven by a force f, and colliding with a wall at position c (Figure 1.7). In motion the block satisfies Newton's second law, and in collision it satisfies an impact law. Writing $\frac{dx}{dt} \equiv \dot{x}$ and $\frac{d^2x}{dt^2} \equiv \ddot{x}$, these are given together by

$$m\ddot{x} = f \quad : \qquad \dot{x} \mapsto -r\dot{x} \quad \text{if } x = c \ \& \ \dot{x} > 0\,, \tag{1.12}$$

where r is the coefficient of restitution.

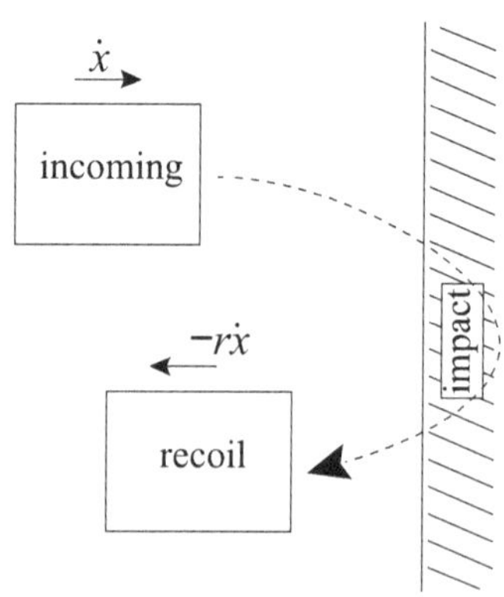

Fig. 1.7 The discontinuity in velocity resulting from impact with a hard wall.

If we assume the wall is at least slightly compliant, then the impact event is not instantaneous but represents a jump through a continuous contact phase, and instead we might write

$$m\ddot{x} = f - k\,\mathrm{step}(x-c)\ , \quad \mathrm{step}(x-c) = \begin{cases} 1 \text{ if } x > c\,, \\ 0 \text{ if } x < c\,, \end{cases} \tag{1.13}$$

introducing a large wall stiffness k. These are the most simple models of impact. We can make them more complex by trying to take account of the microscopic shapes and forces at the surface, the deformation or wear of the bodies, and the medium in which they sit, but this requires moving to other levels of description on the micro-scale. If we consider a two-ball collision like that in Figure 1.1, then we must consider deformation of both objects, and the problem becomes even more difficult. Marrying up the different scales, understanding whether micro-scale details constitute merely perturbations of (1.12) or (1.13), is not a straightforward (or indeed solved) problem.

So let us look at another kind of motion. Place the same block on a table-top, and push it in a straight line. It resists with a force due to the two surfaces rubbing against each other. That force is independent of the *speed* the object moves (at least approximately), but it always resists the direction of motion, so it opposes the object's *velocity* (Figure 1.8). Newton's second law gives us

$$m\ddot{x} \;=\; T - \mu_k R\,\mathrm{sign}(\dot{x})\ , \quad \mathrm{sign}(\dot{x}) = \begin{cases} +1 \text{ if } \dot{x} > 0, \\ -1 \text{ if } \dot{x} < 0, \end{cases} \tag{1.14}$$

where T is the force we apply to push the object, the constant R is the reaction force of the table on the object to resist its weight, and the constant μ_k is called the coefficient of friction.

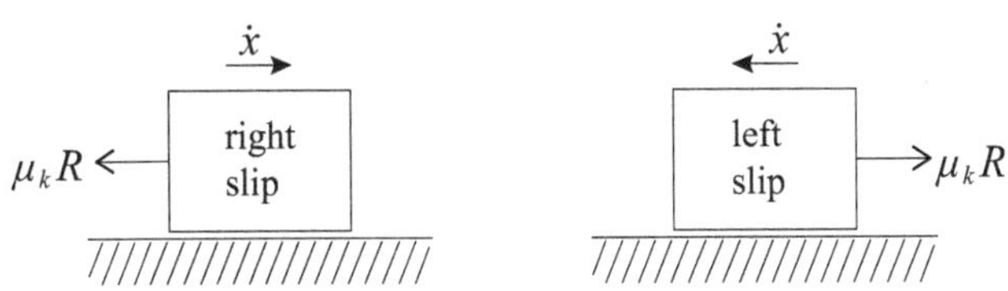

Fig. 1.8 The discontinuous force of friction.

There is an easy way to show that the friction force '$\mu_k R\,\mathrm{sign}(\dot{x})$' does something interesting at $\dot{x} = 0$.

Try pushing a few different objects along your tabletop: try your laptop, your coffee mug, your phone, your purse, and perhaps your shoe. Observe carefully how much force it takes to keep the object in motion, and how much it takes to get the object moving in the first place. For some objects you should be able to tell that the force required to instigate motion (to exit the rest state $\dot{x} = 0$) is *more* than that required to sustain motion (to maintain $\dot{x} > 0$). It may help to place the object on an inclined surface, and tilt it as much as possible before the object starts slipping (be careful not to spill your coffee, the fluid may affect the results!). Now the only force driving the object will be its weight. Tap the object gently, applying a force just for an instant, so the object starts slipping. Some of those objects will stop again almost immediately. Some of them will keep slipping (if your incline was sufficient) and even accelerate away down the slope: with the same forces as before (weight and friction), the object is now in sustained motion. So the friction force must have dropped when the object started moving.

We capture this by saying that a 'static friction coefficient', μ_s, applies at $\dot{x} = 0$, in place of the 'kinetic friction coefficient' μ_k for $\dot{x} \gtrless 0$. For some objects we have $\mu_s > \mu_k$, so the friction force μR 'peaks' at $\dot{x} = 0$. The whole experiment can be described with more detailed equations; these will come in Section 2.2.

These are ad hoc fixes, though, like our impact rules used in collisions. It is easy to pragmatically accept them as an inescapable symptom of complicated underlying physics. But a lack of understanding of the underlying mathematics is also to blame.

1.4.2 In Optics: Illuminating a Victorian Discontinuity

We might expect discontinuities in the rough and tumble of mechanical interaction, but they also plague that ideal playground of geometry: light.

As light passes from one medium to another, its path typically kinks or *refracts*. This happens because the speed of light is only piecewise-constant: it has a discontinuity in value from one medium to another.

Hence a straw in a drink appears broken (itself discontinuous!) at the fluid surface. And hence white light splits into its constituent colours as it passes through a prism or a raindrop, because the effect of the discontinuity depends on colour.

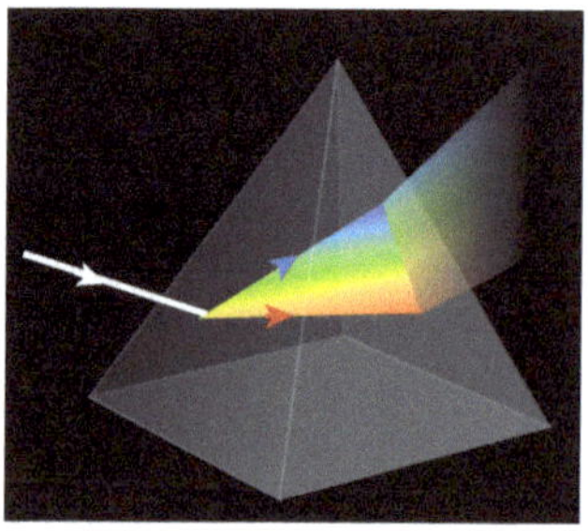

The discontinuity in the speed of light can also create a loss of uniqueness in the trajectories of light rays.

Upon entering a typical transparent crystal, any one ray of light actually splits into two rays, R_+ and R_- in Figure 1.9, of different *polarizations*. This is known as double refraction. Less well known is that along a special direction, called the optic axis, one light ray splits into an infinity of rays, R_∞ in Figure 1.9, each with a different direction. These directions form a cone, and the phenomenon is called conical refraction. It is a fine example of a discontinuity-induced singularity causing one trajectory to splinter into infinitely many, similar to Figure 1.3.

We will look more into this phenomenon in Section 14.5. Its discovery fits into an interesting period in the history of physics, as the competition between integral and geometric theories (or wave and particle theories if you prefer) of light reached its climax. William Rowan Hamilton pierced the seeming perfection of the wave theory by discovering the conical singularity responsible for this refraction phenomenon.

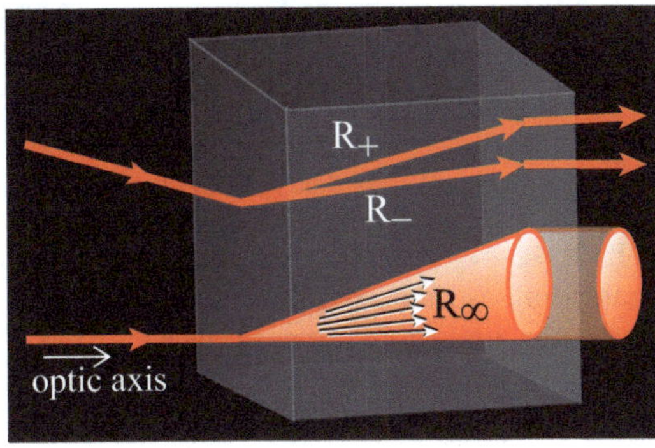

Fig. 1.9 The phenomena of double refraction and conical refraction.

In doing so Hamilton also discovered perhaps the first discontinuity-induced explosion. The loss of uniqueness in the paths of light rays was deemed implausible upon its discovery by Hamilton [97], yet it was there and was confirmed experimentally a year later [154, 153] using candlelight and imperfect crystals. The phenomenon was discovered subsequently to be hiding intricate wave effects not to be fully understood for another 170 years. See Section 14.5 for more of the story.

Conical refraction shows how discontinuity (in the refractive index) and non-uniqueness (in the paths of light) go right to the fundamentals of physics. Given the examples we will see later inspired by problems from engineering and biology, it is worth noting that these problems occur in something as fundamental and ideal as the classical optics of light.

1.4.3 In Sound: Wavefronts and Shocks

When something, be it information, a chemical, light, or sound, spreads out over space, the simplest model for its propagation is as a series of uniform fronts, travelling out from the source at a fixed speed c, Figure 1.10. As this something—let us call it the 'signal'—spreads out, its total strength is stretched out over spheres of growing area $4\pi r^2$, so its intensity decays as the inverse square of its distance r from the source, $I \propto 1/r^2$ (or as $1/r^{n-1}$ in n dimensions).

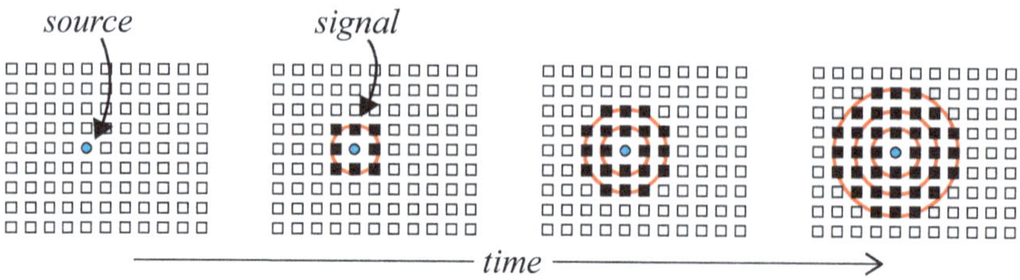

Fig. 1.10 A signal spreading out from a source, shown propagating across a grid.

If the source itself moves, say at speed v along the x direction, then each signal front is emitted from a time-dependent location $x = vt$. The signal propagates uniformly outward once emitted. This causes the fronts to bunch up in front of the source as it chases their leading edge, Figure 1.11.

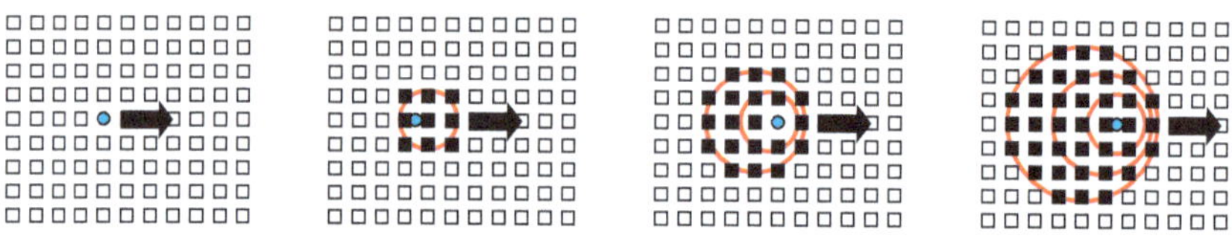

Fig. 1.11 A source moving slower than the signal it emits.

If the source moves more quickly than the signal, then it leaves the fronts behind, and they form an envelope called a *shock*, shown in Figure 1.12.

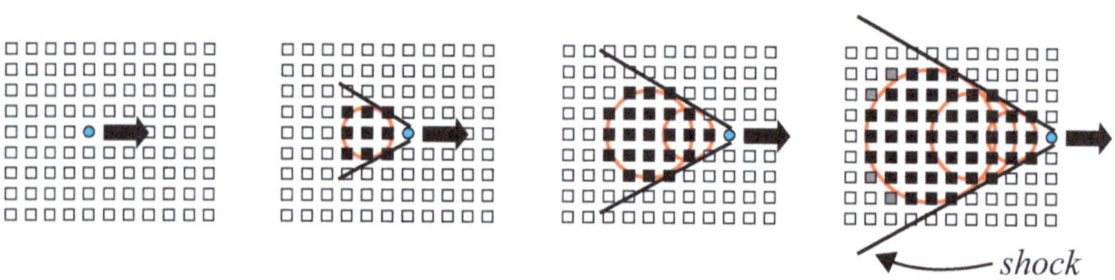

Fig. 1.12 A source moving faster than the signal it emits.

This creates a discontinuity in the intensity of the signal, $I \propto 1/r^2$. The intensity is plotted for each different speed in Figure 1.13.

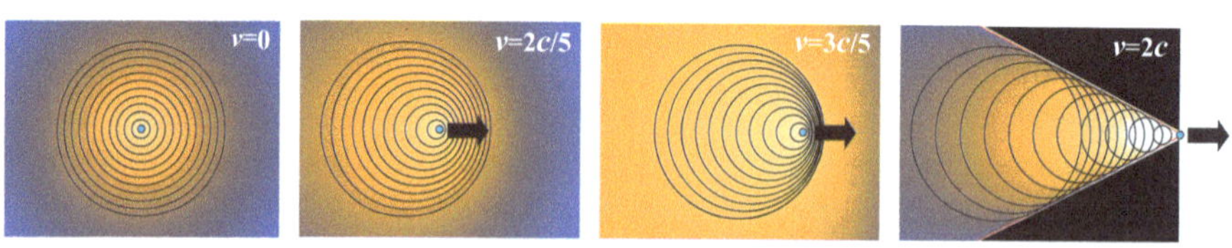

Fig. 1.13 Intensity plotted for the last figure, at speeds $v = 0, \frac{2}{5}c, \frac{3}{5}c, 2c$. Lighter shading indicates higher intensity. For $v > c$ the intensity cuts off sharply at the shock envelope.

To plot the intensity, we have to take account of the fact that the radial distance from the source is not simply $r = \sqrt{x^2 + y^2 + z^2}$. Because the source is moving, if we observe a circular front of radius r, then it came from a source whose x coordinate has since moved a distance rv/c (Figure 1.14), hence

$$r^2 = (x - rv/c)^2 + y^2 + z^2 ,$$

which can be solved for r, giving

$$r = \frac{-xv/c + \sqrt{x^2 + (1 - v^2/c^2)(y^2 + z^2)}}{1 - v^2/c^2} . \tag{1.15}$$

(There are of course two solutions to the quadratic equation, but the other one would correspond to an advanced, rather than retarded, problem, where the source absorbs rather than emits waves).

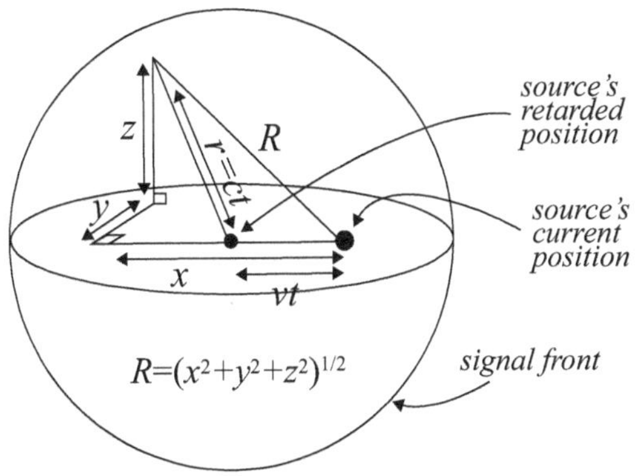

Fig. 1.14 Wavefront and retardation geometry for source moving at speed c to the right.

Plotting the intensity as $I \propto 1/r^2$ now gives the shading in Figure 1.13. This is also plotted as a graph at a fixed y and z coordinate in Figure 1.15. For $v > c$ the shock forms a discontinuity in the intensity. Inside the shock the signal peaks, outside it abruptly vanishes.

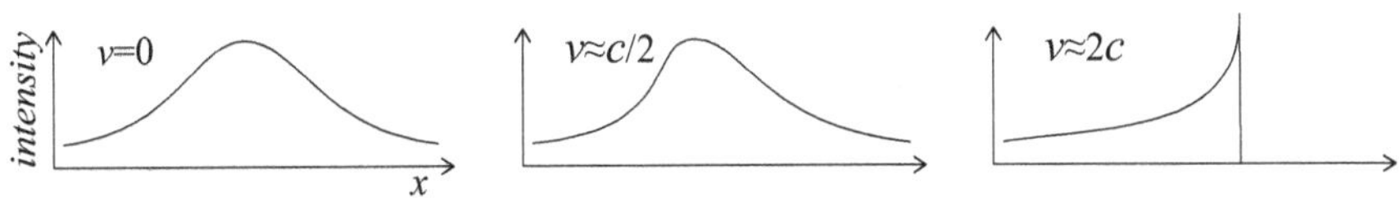

Fig. 1.15 Intensity along a horizontal line through each plot in Figure 1.13.

The vanishing intensity outside the cone $x^2 = (v^2/c^2 - 1)(y^2 + z^2)$ has different interpretations. Mathematically it arises because the solution for r

becomes complex outside the cone. Geometrically the circles with the geometry in Figure 1.14 cannot exist outside the cone. Physically the signal is moving too slowly to reach places outside the cone. In any interpretation a discontinuity in intensity is a real part of the problem's structure, marking the underpinning cone boundary.

This is a geometrical approximation. Depending on the medium, the discontinuity may be softened due to elastic effects such as viscosity of the fluid medium, as happens with sound in air. But even when softened, the effect is real, the underlying geometric discontinuity remains, and in the case of sound propagation is observed as a sonic boom.

1.4.4 In Graphs: Sigmoid Transition Functions

Many problems that involve sudden but nonetheless rapid regime changes are regulated by sigmoid functions. In modern applications to technology and the life sciences, switches are often represented by sigmoids for convenience more than scientific rationale, when their exact form is not known.

Such functions are always hiding a discontinuity which can be revealed by a series expansion. Some of the most common sigmoid functions that are found in the applied literature are listed in Table 1.1.

Name	Definition	For small ε
Inverse tangent	$\arctan(x/\varepsilon)$	$\approx \frac{\pi}{2}\,\mathrm{sign}\,x$
Hyperbolic tangent	$\tanh(x/\varepsilon)$	$\approx \mathrm{sign}\,x$
Error function	$\mathrm{Erf}(x/\varepsilon) = \frac{2}{\sqrt{\pi}}\int_0^{x/\varepsilon} du\, e^{-u^2}$	$\approx \mathrm{sign}\,x$
Hill function	$Z(z,\varepsilon) = \frac{z^{1/\varepsilon}}{z^{1/\varepsilon}+\theta^{1/\varepsilon}}$	$\approx \mathrm{step}(z-\theta)$
Smooth non-analytic	$\mathrm{sna}(x/\varepsilon)$	$\approx \mathrm{sign}\,x$

Table 1.1 A selection of sigmoid functions, defined using a small parameter $\varepsilon > 0$.

The inverse tangent is common in numerics, the hyperbolic tangent in physics, and the error function in statistics. The Hill function is commonly used in biology to represent a process that turns on or off as a threshold $z = \theta$ is passed, but derives from a more specific use in describing the binding efficiency of oxygen to haemoglobin [103]. The smooth but non-analytic functions are common in analysis as ways to regularize discontinuities, for example, by John Nash in his work on imbeddings [165], an example of which is

$$\text{sna}(s) = \begin{cases} r(-s)^{r(s)} - r(s)^{r(-s)} & \text{if } |s| < 1 \\ \text{sign}(s) & \text{if } |s| \geq 1 \end{cases} \quad \text{where } r(s) = e^{2/(s-1)}\,.$$

Typically such functions are used with a small-scale parameter $\varepsilon > 0$ (or large parameter $1/\varepsilon$, and in the case of the Hill function the parameter is a large integer $p = 1/\varepsilon$). The third column in Table 1.1 approximates the functions for small ε, resulting in a piecewise-constant function with a discontinuity. In fact as we let ε tend to zero, the third column gives the exact value of these functions but to be more precise requires a series expansion.

A Taylor series approximation about $x = 0$ is too weak to capture the jump that occurs near $x = 0$, because it consists of a series of polynomial terms that change too gently with x to represent the step-like shape indicated in the third column. Moving the expansion point to any other finite $x = c$ will not help either.

Instead we must approximate the sigmoid for large argument, $|x|/\varepsilon \gg 1$ in each case (except the Hill function which more subtly requires $(z/\theta)^{1/\varepsilon} \gg 1$ and assumes $z, \theta > 0$). Doing so we can find series expansions:

$$\tanh(x/\varepsilon) \approx \text{sign}(x)\left(1 - 2e^{-2|x|/\varepsilon} + \ldots\right) \tag{1.16a}$$

$$\arctan(x/\varepsilon) \approx \tfrac{\pi}{2}\,\text{sign}(x) - (\varepsilon/x) + \tfrac{1}{3}(\varepsilon/x)^3 - \ldots \tag{1.16b}$$

$$\text{Erf}(x/\varepsilon) \approx \text{sign}(x)\left(1 - e^{-x^2/\varepsilon^2}\left(|\varepsilon/x| - \tfrac{1}{2}|\varepsilon/x|^3 + \ldots\right)\right)/\sqrt{\pi} \tag{1.16c}$$

$$Z(\theta + x, \varepsilon) \approx \tfrac{1}{2} + \text{sign}(x)\left(\tfrac{1}{2} - e^{-|\log((\theta+x)/\theta)|/\varepsilon} + \ldots\right) \tag{1.16d}$$

$$y(x/\varepsilon) \approx \text{sign}(x)\left\{1 - \rho - (e\rho)^{1/e} + \ldots\right\}, \quad \rho = e^{2/(|x/\varepsilon|-1)} \tag{1.16e}$$

which are called *asymptotic* series (see, e.g. [3, 19]). To try this by hand, the reader may find it easier to calculate the series expansion for a sigmoid with a rational form, like $(x/\varepsilon)/\sqrt{1 + (x/\varepsilon)^2}$ for small ε/x, which will yield similar results.

What we observe is that all of the above take a form roughly like

$$y(x/\varepsilon) \sim \text{sign}(x) \;+\; \text{higher orders},$$

revealing how the sign discontinuity appears not by ad hoc approximation, but in the leading order of rigorous asymptotic expansions of typical sigmoid functions.

The series derived above are convergent only for $|x| > \varepsilon$, and remain valid even where they are divergent, right down to $x = 0$ with the exception of the point $x = 0$ itself. In the transition from one analytic domain, $x < 0$, to the other, $x > 0$, the analytic form of the series expansions jumps. The discontinuity therefore signifies a transition between analytic domains, imposed by the sign function.

Putting a little more meat on the higher order term, the asymptotic expansion of a typical sigmoid is evidently something like

$$y(x/\varepsilon) \sim \operatorname{sign}(x) + p(x/\varepsilon) \sum_{r=1}^{\infty} q_r(x/\varepsilon) \left(\frac{\varepsilon}{x}\right)^r . \tag{1.17}$$

Similar forms arise also in the solutions of differential and integral equations, examples of which are given in Appendix A. These were first discovered in none other than the physics of light again, by George Gabriel Stokes, in the decades that followed Hamilton's discovery of conical refraction (Section 1.4.2 above). Stokes found jumps in the series expansions of integrals responsible for rainbows [197] and was so troubled by their appearance that he lamented in a series of letters to the soon-to-be Lady Stokes:

I have been doing what I guess
you won't let me do when we are married,
sitting up til 3 o'clock in the morning
fighting against a hard mathematical difficulty,
... there was one difficulty about it which, though I tried
till I almost made myself ill, I could not get over it,
... the discontinuity of arbitrary constants. [141]

The discontinuity that was troubling Stokes is similar to the jumping between two forms of a series expansion that we saw in the sigmoids above. Stokes' particular problem concerned the asymptotic approximation of integrals, for which we give more details in Appendix A.

1.5 'Can't We Just Smooth It?' (And Other Fudges)

This question of whether we can just remove discontinuities by some means often arises, so it must be addressed. Perhaps discontinuities do not belong in differential equation theory at all. Perhaps the physical world is always continuous anyway under sufficient magnification. Perhaps a well-defined equation should never have jumps.

Perhaps we should just smooth it.

Given a function $f(x;\lambda)$ with a multiplier $\lambda = \operatorname{sign} x$, so that f switches between

$$f(x;\lambda) = \begin{cases} f(x;+1) \text{ if } x > 0 , \\ f(x;-1) \text{ if } x < 0 , \end{cases} \tag{1.18}$$

the common notion of a smoothing is an interpolation

$$F_\varepsilon(x) = \tfrac{1}{2}\left[1 + \phi(x/\varepsilon)\right] f(x;+1) + \tfrac{1}{2}\left[1 - \phi(x/\varepsilon)\right] f(x;-1) \tag{1.19}$$

where $\varepsilon > 0$, with ϕ being a sigmoid function such as

$$\phi(u) = \begin{cases} \operatorname{sign} u \text{ if } |u| > 1 , \\ u \qquad \text{if } |u| \leq 1 . \end{cases} \tag{1.20}$$

Then F_ε is one degree smoother than f (F_ε is continuous and f is discontinuous). By choosing different definitions of $\phi(u)$ on $|u| \leq 1$, we can make F as smooth as we wish, so that it is differentiable once, twice, and even infinitely many times if we use the smooth non-analytic function from Table 1.1.

In the limit of small ε, the function F_ε becomes

$$F_0(x) = F(x;\lambda) := \tfrac{1}{2}(1+\lambda)f(x,+1) + \tfrac{1}{2}(1-\lambda)f(x;-1)\,, \tag{1.21}$$

with $\lambda = \operatorname{sign} x$. Note that if the original function $f(x;\lambda)$ had been nonlinear in the switching multiplier λ, this would have become lost by the smoothing above, since the function $F(x;\lambda)$ is only linear in λ.

Rather than simply one smoothing, we can therefore replace $f(x;\lambda)$ with either the standard *linear smoothing* $F_\varepsilon(x) = F(x;\phi(x/\varepsilon))$ or a *nonlinear smoothing* $f(x;\phi(x/\varepsilon))$. The latter respects the dependence of f on λ, by translating it into the same dependence on ϕ.

We have already argued briefly (and will see much more of) the importance of nonlinearity, and here lies the danger of tampering with the nonsmooth description of a system as we have above. If we decide to smooth out a discontinuity, we had better not neglect any nonlinearity without due care. This also means that we must allow for nonlinearity that has been neglected from a model suddenly becoming important, as we find happens around singularities.

Smoothing is also only one way we might attempt to handle a discontinuity. When we simulate $x(t)$ in a computer, for example, we do not really compute $x(t)$, instead we approximate $x(t)$ at time intervals $t, t+\delta_1, t+\delta_2, \ldots$, which we call discretization or *time-stepping*. Moreover, outside our computer is a real world that is full of other complications. If we model the state of a system as some $x(t)$ at a time t, the true state might actually lie a short distance δx away, or suffer a time lag δt, shifting $x(t)$ to $x(t-\delta t) + \delta x(t-\delta t)$. These *perturbations* might be simple functions or stochastic processes.

When forming a dynamic model of any system, we intend it to be robust, in the sense that such perturbations alter its behaviour only slightly. The issues of non-uniqueness that arise at a discontinuity, however, mean that any perturbation might pick out only certain behaviours from the many that are possible. Smoothing might pick out one behaviour, non-monotonic smoothing another, time-stepping another, and so on. We can show with a few simple simulations how serious an issue this is at a discontinuity. The reader may like to try to reproduce the simulations in the following section as an exercise.

1.6 Discontinuities and Simulation

When turning to a computer for insight into a system's behaviour, it is vital to first understand something about the expected outcome, because the computer itself will provide no warning if the result is absurd. The key in dynamics is often to take care of nonlinearity, whether nonlinearity in the variables or nonlinearity in switching.

For solving smooth dynamic equations, there exist powerful numerical methods encoded into numerous ready-written toolkits. At the time of writing, there are toolkits that handle certain instances of discontinuity, but of those that predate this book, none is fully capable of handling all the nonlinearities, singularities, and determinacy-breaking phenomena we will describe.

Instead we can take numerical tools intended for smooth systems and adapt them with either:

(a) event detection (detecting when a discontinuity occurs and assigning a rule to handle it); or
(b) smoothing (replace the switching multipliers $\lambda = \mathrm{sign}(\sigma)$ by smooth functions approximating the sign function) and integrating using standard numerical methods.

Neither is without its difficulties. In (a) we must take care that we understand any ad hoc rules we assign to implement switching once a discontinuity is detected. In (b) we create a system that will be 'stiff' at the discontinuity. Let us carry out a few exploratory simulations.

In each we will compute solutions to a system of differential equations with linear smoothing and nonlinear smoothing. We will also ask:

- is the result sensitive to the size of the time steps taken by the computer?
- does the result change if there is some delay in switching, i.e. if the state overshoots the discontinuity threshold by some time or distance before switching takes place?
- does the result survive in a noisy environment, if we add random kicks either to the state at each time step or to the switching process by choosing the delay randomly each time the discontinuity is detected?

Let us simulate each of these (time-stepping, delay, noise) as either small perturbations or large perturbations and observe the results.

Example 1.2 (Direction Reversal). Take a system (proposed in the seminal work [71])

$$\frac{dx_1}{dt} = \tfrac{3}{10} + \lambda^3\,, \qquad \frac{dx_2}{dt} = -\tfrac{1}{2} - \lambda\,, \qquad \lambda = \mathrm{sign}\, x_2\,. \tag{1.22}$$

To simulate these equations we need to tell our computer how to implement the switch as x_2 changes sign, and the simplest way is to smooth them. First take the right-hand side of the dx_1/dt equation. The nonlinear smoothing is

$\frac{3}{10} + [\phi(x_2/\varepsilon)]^3$ using (1.20), while the linear smoothing (just interpolating between $\frac{3}{10} \pm 1$) is $\frac{3}{10} + \phi(x_2/\varepsilon)$, both equivalent for $|x_2| > \varepsilon$. For the right-hand side of the dx_2/dt equation, we have only a linear smoothing, $-\frac{1}{2} - \phi(x_2/\varepsilon)$.

Figure 1.16(i) shows the results of simulations using these nonlinear or linear smoothings (for some $\varepsilon \ll 1$). The behaviours are contradictory, as starting from the same (or similar) initial conditions, they both travel approximately along the discontinuity but in opposite directions.

To ask which of these is 'correct' is not a well-defined problem. Certainly we can form equations for each approach, simulate them, and perhaps even find their solutions analytically. Instead we can ask whether one or both of these are robust and whether they can survive in the nonideal environment of real applications. To do this let us simulate them with perturbations of time-stepping, delay, and noise (whether we apply one or all of these, the results are qualitatively the same). We obtain simulations like those in Figure 1.16(ii). The nonlinear behaviour survives under small perturbations. Under large enough perturbations—in a system with a lot of noise or long delays, for example—nonlinearity is killed off, and the system follows the linear behaviour. So which is correct? Both, under different conditions.

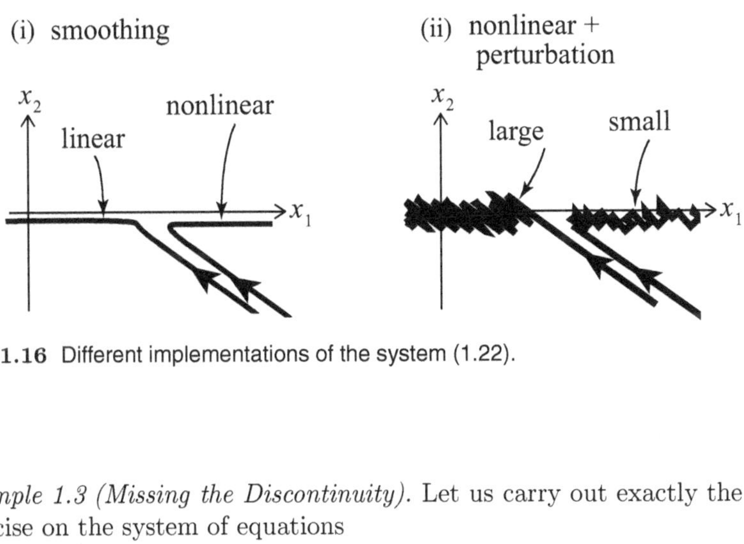

Fig. 1.16 Different implementations of the system (1.22).

■

Example 1.3 (Missing the Discontinuity). Let us carry out exactly the same exercise on the system of equations

$$\dot{x}_1 = 1 , \qquad \dot{x}_2 = 2\lambda^2 - 1 , \qquad \lambda = \operatorname{sign} x_2 . \tag{1.23}$$

Figure 1.17(i) shows the results of nonlinear or linear smoothing.

Again the behaviours are contradictory; now the nonlinear system evolves approximately along the discontinuity, while the linear system barely notices its presence. The effects of then adding small or large perturbations in the form of discretization, delays, or noise are shown in Figure 1.17(ii), and as

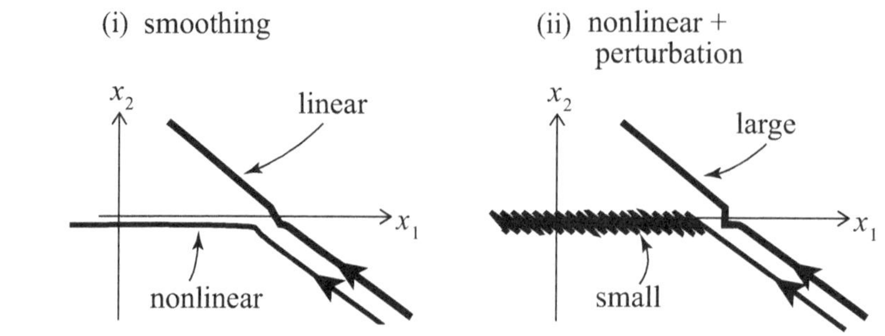

Fig. 1.17 Different implementations of the system (1.23).

before, for small perturbations the nonlinear behaviour survives and for large enough perturbations, it is killed off. Both are valid under different conditions. ■

Here is non-uniqueness playing its hand. In each example there are two possible behaviours (and more besides), captured by nonlinear terms. The behaviour of any particular model survives under perturbation, with nonlinear effects being more vulnerable to being killed off (much as we find with nonlinearity in smooth systems).

Example 1.4 (Jitter Along the Discontinuity). Even more intriguingly consider a system with two discontinuities, say

$$\tfrac{d}{dt}(x_1, x_2, x_3) = \mathbf{a} + \lambda_1\mathbf{b} + \lambda_2\mathbf{c} + \lambda_1\lambda_2\mathbf{d} \ , \tag{1.24}$$

in terms of two switching multipliers $\lambda_1 = \operatorname{sign} x_1$ and $\lambda_2 = \operatorname{sign} x_2$, for four vector fields

$$\mathbf{a} = \left(-1 - 2x_3,\ -1,\ \tfrac{5}{4}\right) \ , \qquad \mathbf{b} = \left(-\tfrac{5}{2} + 2x_3,\ -7,\ \tfrac{5}{4}\right) \ ,$$
$$\mathbf{c} = \left(\tfrac{1}{2} - 2x_3,\ 0,\ -\tfrac{27}{4}\right) \ , \qquad \mathbf{d} = \left(-7 + 2x_3,\ 1,\ -\tfrac{3}{4}\right) \ ,$$

with fairly arbitrarily chosen constants. We require only that (1.24) points towards $(x_1, x_2) = (0, 0)$ for any combination of $\lambda_1 \pm 1$ and $\lambda_2 = \pm 1$, with the simple result that whichever way we simulate (1.24), we will see x_1 and x_2 evolving towards zero and staying there. The variable x_3 then does something interesting that is highly sensitive to the method of simulation.

In this case there is no linear smoothing that can interpolate between the four vector fields in the different quadrants of the (x_1, x_2) plane, so the simplest possible is the bilinear smoothing obtained by substituting λ_1 and λ_2 with sigmoid functions $\phi(x_1/\varepsilon)$ and $\phi(x_2/\varepsilon)$. Now the simulations obtained with smoothing, time-stepping, and with time or spatial delay in enacting switching, produce markedly different results.

The upper graph in Figure 1.18 shows the rate of change of x_3 as a solution evolves along $x_1 \approx x_2 \approx 0$, at different positions along the x_3 axis. We notice two things. First, each method of simulation gives vastly different speeds of motion $\dot{x}_3$. Moreover they vary over the entire unshaded region indicated on the graphs, and this is an interesting region, because it indicates a 'convex hull' of all possible values that $\dot{x}_3$ could take. It seems that the system, as simple as it is, could hardly be more unpredictable. The lower graph shows the effect of adding noise to each of these, and rather than become more irregular, noise seems to calm the unpredictability, and the set of speeds $\dot{x}_3$ seem to collapse towards a common (if noisy) behaviour.

We return to this intricate example in Section 12.2.1, where we describe in more depth the simulation methods, but at this stage the reader may find it more instructive to explore their own ideas of how to implement them.

■

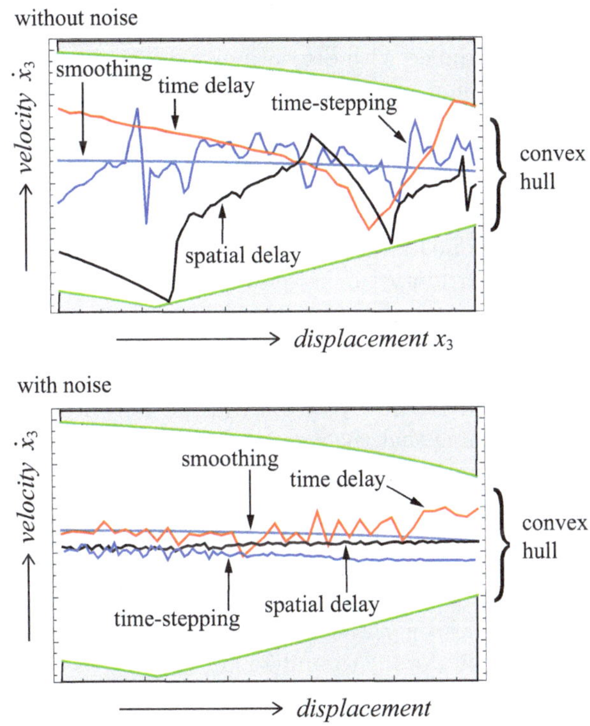

Fig. 1.18 Dynamics at an intersection of discontinuities, showing speed along the discontinuity under different simulation methods.

These examples illustrate some of the care and understanding needed when simulating a nonsmooth system and give us our first practical glimpse at the role that nonlinear switching plays.

In subsequent chapters we rely mainly on qualitative analysis and geometry, often sketching the solutions of piecewise-smooth systems. Sometimes, for illustration, it will be necessary to include computer simulations. Event detection leaves too much uncertainty in how we implement switching, running the risk of imposing unintended restrictions and unwittingly omitting the rich dynamics that is really possible. Smoothing is therefore to be our choice. It is naive but also transparent, and by the time we reach Chapter 12, we will have learnt how to ensure our simulations are faithful to the discontinuous system.

We are almost ready to begin looking at how we handle discontinuities like these when they arise inside a dynamical system. We should conclude with a short detour through the historical developments that brought us here.

1.7 Discontinuity in Dynamics: A Brief History

Many scholars have grappled with discontinuity in one discipline or another. The aphorism *Natura non facit saltus* ('Nature does not makes jumps') is sometimes held as a principle of natural philosophy. It was taken as an axiom by Gottfreid Leibniz [143], was a principle behind Charles Darwin's search for a gradual process of evolution versus abrupt development (which we might now call mutation) [41], and appeared in the early development of botany [150]. Nowadays, subatomic particle experiments and quantum field theories add new levels of uncertainty to the question of whether or not nature is continuous.

The story of discontinuities in dynamics begins with the study of collisions and of friction. Friction has been a particularly mischievous character in the evolution of modern physics. Way back in the time of Aristotle, ancient philosophers were teaching that the natural state of all bodies is to be at rest [11, 12]. The time of Descartes and Newton brought the realization that the natural state of bodies is actually to continue in their present state of motion, carried on by inertia [42, 167]. And the swindler responsible for deceiving the preceding generations is of course friction, which by resisting motion creates a non-inertial tendency towards idleness.

As we rush forward through the centuries, the manner in which friction works proves hard to pin down. Around 1500 Leonardo da Vinci shows that frictional resistance depends on load, but not on contact area. In 1699 Amontons rediscovers da Vinci's laws and describes friction as the work done to overcome—through wear and deformation—the surface roughness between two objects. In 1700 Desagulier shows that friction does *not* depend on surface roughness, seeming to contradict Amontons' theory. In 1750 Euler shows that static friction is stronger than kinetic friction, so more force is required to instigate motion of a static object than is required to keep an object in motion. In 1785 Coulomb adds clarity and depth to the former theories, especially those of da Vinci and Amontons. In 1950 Bowden and Tabor [24] clarify the role of contact area and surface roughness, showing that friction does depend on the *true* contact area, which is often less than the apparent contact area (i.e. the full contact surface). (To begin making contact with these and many other contributions to friction's history, the reader may begin with [8, 16, 24, 55, 110, 208].)

All this complexity has much to do with the presence of a discontinuity at the heart of frictional contact and the multiple scales of interaction it hides. And friction's story is still not done. The field of Tribology remains active to this day. With modern instruments we can image the contact patch between objects in tremendous detail, we can measure forces from the microscale to the macroscale, and we can model in ever more detail [94, 105, 142, 178, 174, 179, 201, 200, 221, 222]. But we have not found a set of laws that fully explains the relationship (1.14), or improve on it sufficiently to do away with the jump, without degrading its generality of application across different materials and physical arrangements [132, 221, 222, 223].

From this point of view, discontinuities represent problems of Renaissance era mechanics left yet to be married with modern developments in mathematics.

It was not until the twentieth century that serious attempts were made to formalize the study of how discontinuities affect otherwise continuous motion. Some of the earliest descriptions appear in Andronov, Khaikin, and Vitt's time [10], mostly concerning the use of switching as a control device. The reference [10] is to a book published in 1959, because the original work, from the 1930s, was the victim of the political vagaries of Stalinist Russia, which declared Vitt an 'enemy of the people' (see, e.g. [67]) and had his contribution on discontinuities stricken from the original [9], thankfully, however, only temporarily. There appears to have been a community in the area at the time, producing, for instance, Kulebakin's work on vibration control for aircraft direct-current generators [136], Nikolsky's example of boat rudder control [168], and later Tsypkin's relay control [209]. In some ways they were ahead of their time. In page 4 sec1.2 of [212] Utkin writes: *It seems that 1930s-era 'vibration control' is just the same as our contemporary sliding mode control... It is amazing that the paper published more than 60 years ago was written in the language of modern control theory: 'phase plane', 'switching line', and even sliding mode.*

Interest grew among mathematicians. According to Utkin's book, in 1954, Andronov showed that delays in switching led to equations that we now write in the form

$$\dot{x} = u(\sigma)a(x) + (1 - u(\sigma))b(x)\ , \qquad u(\sigma) \approx \begin{cases} 1 \text{ if } \sigma > 0\ , \\ 0 \text{ if } \sigma < 0\ . \end{cases} \tag{1.25}$$

Filippov, Ardne, and Seibert in 1956 showed something similar for delay and hysteresis in switching, with the result that, when u switched between 0 and

1 across a surface $\sigma = 0$, one could find values $u \in [0, 1]$ such that $\dot{\sigma} = 0$, which create *sliding* motion along a discontinuity [212, 186, 95, 96, 127].

It is with that concept that one can slide along rather than just pass through discontinuities and that our subject becomes interesting.

With the exception of a German lady, Irmgard Flügge-Lotz, who applied independent ideas to aeronautical automatic control that followed remarkably similar lines [73], it seems that only Russian research was taking discontinuities in dynamics seriously. In the 1950s other researchers, such as Yu. Kornikov, A. Popovski, and Y. Dolgolenko, all had ideas of how to handle discontinuities. Yet it seems it was again a Russian, Alexei Fedorovich Filippov, who took a simple expression like (1.25) forward, to develop the first substantive theory of *differential equations with discontinuous right-hand sides* [70, 71]. Filippov has become a towering figure of nonsmooth dynamics, his name having come to represent a whole community of Russian mathematicians who showed that equations with discontinuities can not only be solved but can succumb to rigorous analysis.

It is difficult to peer into the early Russian developments due to patchy translations and inaccessible publications, but a quite extensive description leading up to and beyond the first IFAC Congress, which was evidently a pivotal point in the early application of discontinuities in dynamics, is given on pages 1068–1069 of [5]. Certainly many authors contributed to Filippov's results, and there survive alternative approaches besides the *differential inclusions* favoured by Filippov, though many turn out to be essentially the same. For some alternatives to Filippov's formulation, see [5, 74, 224]. In [127] there is reference to the first investigations in this field being carried out by Marchaud and Zaremba in the 1930s, but sliding modes seem not to have become a central interest before Filippov and his presentation to the IFAC Congress and one more Russian who deserves mention.

The successful adoption of Filippov's work into applied mathematics owes a lot to Vadim Utkin, whose work on *variable structure control* brought Filippov's sliding mode concept to a wide engineering audience. When the two met at the First IFAC World Congress in 1960 in Moscow, along with other significant names in the subject's history like Aizerman and Pyatnitsky [4, 5], both the mathematics of discontinuities and their application for electronic control remained controversial. So intriguing is the insight given by a 1960 article by Filippov, including subsequent discussion by other Congress attendees, that we include it as an Appendix at the end of this book.

A new driving force arose amid the pioneers of structural stability theory, among names like Peixoto, Palis, and Sotomayor. Marco Antonio Teixeira began working on *divergent diagrams* [204] and *pairings of vector fields and functions* [203]. This was just piecewise-smooth systems under a pseudonym to circumvent the prevailing scepticism. Teixeira began building what would become the largest community studying the structural stability of piecewise-smooth systems.

Piecewise-smooth systems also arose independently in the 1970–1980s in a different form, as jumps in discrete time maps. They turned up in studying flows on invariant manifolds and in homoclinic bifurcations, as caricatures of nonlinear phenomena, producing the tent map [34], Lozi map [156], and robust 'routes to chaos' [75].

By this time, mathematical Study Groups with Industry were bringing academia and industry together in a new forum (the first being held in Oxford in 1968), and many of the problems that would come out of these involved nonsmooth behaviours like those in Figure 1.6. Thus began a new era of multidiscipline global interest in modern nonsmooth systems, from abstract theory to computational methods, from earthquakes to dripping taps and ocean currents and climate tipping, from electronics in vehicle control and power generation to predator prey and decision-making, from heart beats to brain spikes, from walking and rattling to wildfires and romance novels, from braking and bouncing to flocking, from squealing train breaks to morphing aerodynamic structures, etc. The breadth and excitement grow. There are now groups working in nonsmooth dynamics on six out of seven continents (while the seventh, if not home to any nonsmooth researchers, is at least the subject of their studies [173]).

There have of course been many other strands of research involving discontinuities, of no less importance, but those above are just a few that fit the narrative of this book.

One of the harbingers of changes on the horizon in piecewise-smooth systems theory has been the need to understand how widely Filippov's ideas apply to the real world. For systems with multiple switches, Alexander and Seidman considered what happened when two switches were blended [6], a concept we develop considerably here as the 'canopy' of switches. The canopy provides a series expansion for discontinuous functions, analogous to the Taylor series for analytic functions. Perhaps the first rigorous attempts to study how noise affects switching were those made by Simpson, who in [190] proved that the average of a stochastically perturbed system about a sliding region converges to Filippov's sliding mode. This is made somewhat more perplexing when we try to marry it with another insight from Alexander and Seidman, who also considered what happened when multiple switches exhibit hysteresis [7], giving not a regular sliding motion but the erratic 'jitter' seen in Example 1.4.

As we have endeavoured to apply Filippov's ideas to more and more nonideal discontinuities, so we have learnt to grapple with the things Filippov left uncertain. Still, it must be said, each advancement remains a verification of Filippov's intuition and the huge framework that now rests upon it.

Although discontinuities have earned a place in dynamical systems theory, the form made popular by Filippov and Utkin, equation (1.25), has shown particular power and generality. This is what we will call 'linear switching', i.e. depending linearly on the switching multiplier u (or λ).

Dynamical systems theory for continuous systems only came of age with the embracing of nonlinearity, and so we find for systems with discontinuities. So it is that a large part of this book is devoted to establishing the role of nonlinear dependence on switching multipliers, the role of *nonlinear switching*.

1.8 Looking Forward

With this, our dalliance among the origins of discontinuity is done. It is time to set out a unifying framework to study how dynamics is affected by discontinuity.

As you explore this book, and see some behaviours that have no place in the smooth world, you might ask whether the real-world phenomenon it represents is *really* discontinuous or just a coarse approximation of something smooth. In applied mathematics we come to realize that this is the wrong question. It doesn't matter.

There is rarely a 'right' model in practice, just a choice of approximations that are consistent with a given problem. The examples in this chapter exhibit a nature that is neither continuous nor discontinuous, where instead continuity is a matter of scale and context and of the detail of our knowledge and the questions we wish to ask. We deal not with nature directly but with our observations of and interactions with it. The aim of piecewise-smooth dynamics is to learn to employ discontinuity usefully, in the diverse applications where it is already used, and in those to which it will be applied in the future.

So the game is as follows. The presence of a discontinuity has certain effects on the dynamics of a system. Those effects are the things we want to find. And we want to pursue their mathematical consequences, no matter if they run askew of our intuition. We shall see that the behaviours we find are robust to modelling nonidealities, under somewhat counterintuitive conditions, and in intriguing fashion.

The theory of piecewise-smooth systems is still rather young; it is not our aim in this book to present a problem solved but to lay out methods that help set the way for future study.

Chapter 2
One Switch in the Plane: A Primer

This chapter presents a short course on dynamical systems with two variables and one switch,

$$(\dot{x}_1, \dot{x}_2) = (f(x_1, x_2; \lambda),\ f(x_1, x_2; \lambda)) \ , \qquad \lambda = \mathrm{sign}(\sigma(x_1, x_2)) \ .$$

These represent the simplest case of the problems we cover more generally in the rest of the book. They are the most studied and most easily understood piecewise-smooth problems, in comparison with systems on the real line which are trivial, and higher-dimensional systems which are orders more challenging. Filippov in particular covered planar systems in great detail in [71], so it is a good place to begin summarizing the state of the art and setting off in search of something more.

For more formal introductions to differentiable dynamics, see, e.g. [162], and to systems treating dynamics with discontinuities using either linear switching or inclusions, see [71]. Our study will go beyond either of these, so we must settle for introducing some basic concepts informally enough in this chapter that we can apply them more generally later.

2.1 The Elements of Piecewise-Smooth Dynamics

We start with a system

$$\dot{\mathbf{x}} = \mathbf{f}(\mathbf{x}; \lambda) \ , \tag{2.1}$$

M. R. Jeffrey, *Hidden Dynamics*,
https://doi.org/10.1007/978-3-030-02107-8_2

where $\mathbf{x} = (x_1, x_2)$ is a planar variable and $\mathbf{f} = (f_1, f_2)$ is a planar vector field that depends on $\mathbf{x}$ and some quantity λ. The dot is just a shorthand for the time derivative $\dot{\mathbf{x}} \equiv d\mathbf{x}/dt$.

The solutions of (2.1) are a family of functions $\mathbf{x}(t)$ that define paths through space whose tangent vectors are given by $\mathbf{f}$. The solution $\mathbf{x}(t)$ through an *initial point* $\mathbf{x}(0) = \mathbf{x}_0$ is called an *orbit* through $\mathbf{x}_0$. This is usually written in terms of a *flow* function $\Phi_t(\mathbf{x}_0)$ as

$$\mathbf{x}(t) = \Phi_t(\mathbf{x}_0) \quad \text{s.t.} \quad \frac{d}{dt}\Phi_t(\mathbf{x}_0) = \mathbf{f}(\Phi_t(\mathbf{x}_0); \lambda) \ , \quad \Phi_0(\mathbf{x}_0) = \mathbf{x}_0 \ .$$

The collection of all orbits makes up the *flow* whose depiction is the *phase portrait.* Many typical features of phase portraits in the plane are illustrated in Figure 2.1, and we navigate through these over the following few pages.

Let us sketch the qualitative features before meeting the necessary analytical tools.

The phase portrait contains various objects that organize its qualitative dynamics, most importantly:

- singularities—special points of the flow;
- separatrices—orbits that separate regions with different phase topology (often connected to a singularity); and
- invariant sets—sets within which an orbit or family of orbits remain for all time (often forming a separatrix in planar systems).

In a region where $\mathbf{f}$ is differentiable with respect to $\mathbf{x}$, the flow consists of orbits that are unique and nonoverlapping, as in the lower left portion of Figure 2.1. A loss of continuity or differentiability in $\mathbf{f}$ greatly adds to the richness of that geometry, as in the right portion of Figure 2.1.

Orbits can be closed or open curves. If $\mathbf{f}$ is differentiable with respect to $\mathbf{x}$ and λ is constant, the orbit through each point $\mathbf{x}$ is unique, so orbits cannot intersect but can tend asymptotically towards each other. A closed orbit that the flow tends towards is an attractor, a closed orbit that the flow tends away from is a repeller. A point $\mathbf{x}_0$ where $\mathbf{f}(\mathbf{x}_0) = 0$ forms a trivial closed orbit consisting only of the point $\mathbf{x}(t) \equiv \mathbf{x}_0$ for all t, called an *equilibrium.* This is typically an isolated point (if $d\mathbf{f}/d\mathbf{x}$ is non-singular), and its attraction or repulsion is of *nodal, focal,* or *saddle* type. A non-trivial closed orbit that is an attractor or repeller is called a *limit cycle.* These are all illustrated in the lower left portion of Figure 2.1.

The parameter λ inside $\mathbf{f}$ is a *switching multiplier* if its value jumps across a threshold $\mathcal{D} = \{\mathbf{x} : \ \sigma(\mathbf{x}) = 0\}$ for some scalar function σ. This renders the

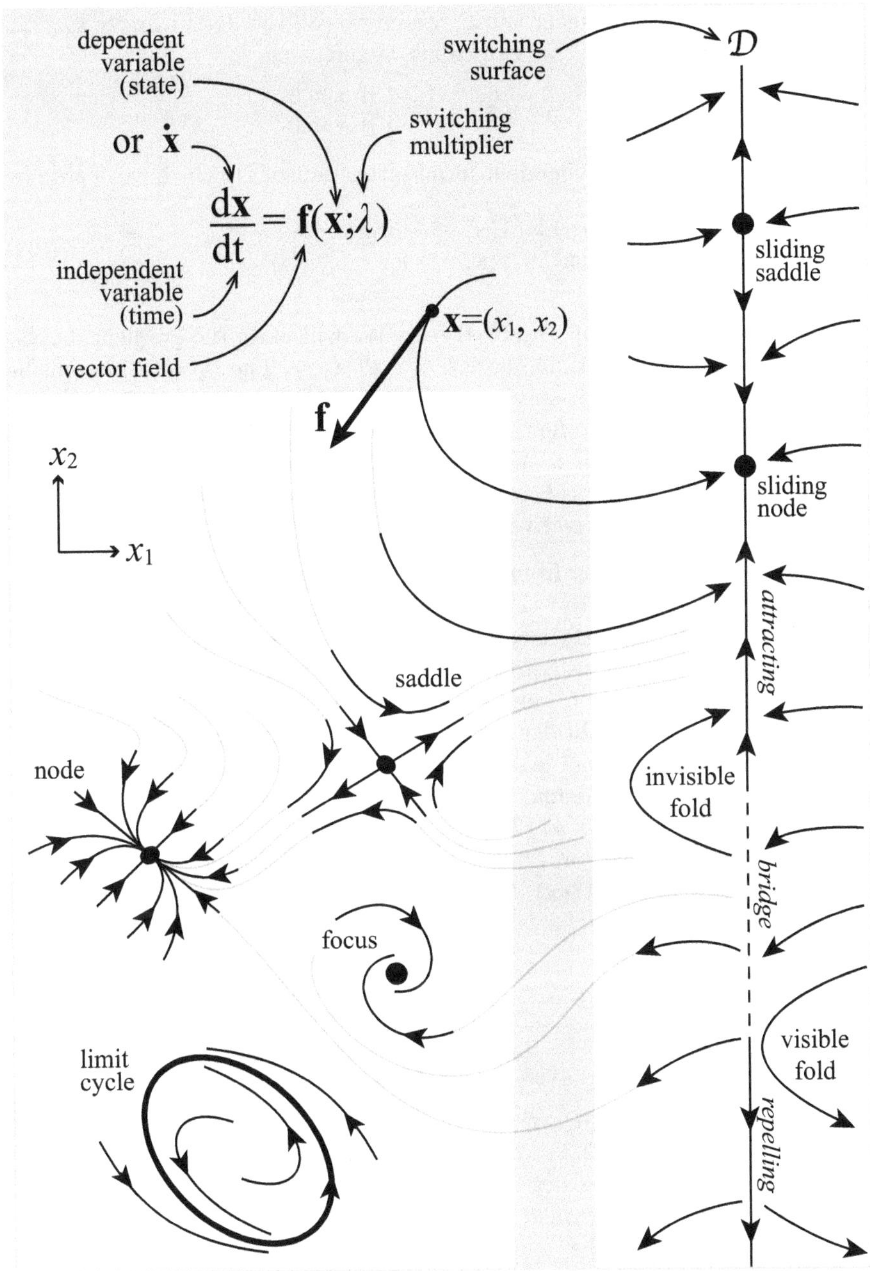

Fig. 2.1 The vector field **f** tells us the velocity with which a flow evolves through a state **x**. Some typical singularities and limit sets are shown. Lower left region: if **f** is differentiable, the flow is differentiable and unique at every point. Right region: if **f** is non-differentiable or discontinuous across a threshold $\mathcal{D}$, then the flow is continuous but may be non-differentiable and non-unique.

vector field $\mathbf{f}$ discontinuous across $\mathcal{D}$, which we call the *discontinuity surface.* Without loss of generality, we can choose λ such that

$$\lambda = \text{sign}(\sigma) = \begin{cases} +1 \text{ if } \sigma > 0 \,, \\ -1 \text{ if } \sigma < 0 \,. \end{cases}$$

The vector field takes independent forms either side of $\mathcal{D}$, which we denote by

$$\begin{aligned} \mathbf{f}^+(\mathbf{x}) &:= \mathbf{f}(\mathbf{x}; +1) \text{ for } \sigma(\mathbf{x}) > 0 \,, \\ \mathbf{f}^-(\mathbf{x}) &:= \mathbf{f}(\mathbf{x}; -1) \text{ for } \sigma(\mathbf{x}) < 0 \,. \end{aligned} \tag{2.2}$$

A Note on Symbols for Derivatives. We will write the gradient derivative $d/d\mathbf{x}$ as $\delta_{\mathbf{x}}$ and the time derivative d/dt as δ_t. The ∂_t notation will be most useful, as it is both succinct, and yet allows us to specify the time derivative operator in the flow above or below the discontinuity surface as, by the chain rule, $\mathbf{f}^{\pm} \cdot \delta_{\mathbf{x}} = \dot{\mathbf{x}} \cdot \delta_{\mathbf{x}} = \delta_{t\pm}$. We will often need the normal vector to $\mathcal{D}$, which we will write as $\delta_{\mathbf{x}}\sigma$, or the component of a vector $\mathbf{f}$ normal to $\mathcal{D}$, which is just the time derivative of σ in the flow, $\mathbf{f} \cdot \delta_{\mathbf{x}}\sigma = \dot{\mathbf{x}} \cdot \delta_{\mathbf{x}}\sigma = \partial_t\sigma = \dot{\sigma}$.

Motion towards or away from $\mathcal{D}$ typically occurs in finite time rather than asymptotically, because the vector field need not vanish as it approaches the discontinuity. The surface can be attractive or repulsive if $\mathbf{f}^{\pm}$ both point towards $\mathcal{D}$ or both point away from $\mathcal{D}$. If $\mathcal{D}$ is neither attracting nor repelling, when the components of $\mathbf{f}^+$ and $\mathbf{f}^-$ normal to $\mathcal{D}$ have the same direction, then the surface forms a bridge between the regions $\mathcal{R}_+$ and $\mathcal{R}_-$.

But a bridge, as we shall learn, cannot always be crossed.

Solutions $\mathbf{x}(t)$ can be defined in each region of space, in $\sigma > 0$, in $\sigma < 0$, and on $\sigma = 0$. If solutions to (2.1) exist that flow along $\sigma = 0$, they form a one-dimensional portrait on $\mathcal{D}$ that we call the *sliding* flow. For now let us denote its vector field as $\mathbf{f}^{\$}(\mathbf{x})$. It corresponds to a particular $\lambda = \lambda^{\$}$, and we add to (2.2) that

$$\mathbf{f}^{\$}(\mathbf{x}) := \mathbf{f}(\mathbf{x}; \lambda^{\$}) \quad \text{for} \quad \sigma(\mathbf{x}) = 0 \,, \tag{2.3a}$$

where

$$\mathbf{f}(\mathbf{x}; \lambda^{\$}) \cdot \delta_{\mathbf{x}}\sigma(\mathbf{x}) = 0 \,. \tag{2.3b}$$

In the δ_t notation, we can write the time derivative in the sliding flow as $\mathbf{f}(\mathbf{x}; \lambda^{\$}) \cdot \delta_{\mathbf{x}} = \delta_{t\$}$, so (2.3b) becomes simply $\delta_{t\$}\sigma = 0$. The sliding flow can itself tend asymptotically towards or away from points on $\mathcal{D}$ that are attracting or repelling, formed by isolated points where $\mathbf{f}^{\$}(\mathbf{x}) = 0$, called *sliding equilibria.* If the flow outside $\mathcal{D}$ is also attracted towards or repelled from $\mathcal{D}$, then we can identify these equilibria as being of saddle or node type (an extra dimension would be needed for focal type). These are all shown in the right-hand portion of Figure 2.1.

These qualitative elements of smooth and piecewise-smooth dynamics are of great assistance in gaining some intuition for the sometimes strange terrain of what we have come to call informally "*Nonsmoothland*".

There are different nomenclatures for sliding and crossing in the literature, inherited from varied applications. The attracting sliding region has been called simply 'sliding', 'stable sliding', or a 'black opaque' region, while the corresponding terms for the repelling sliding region are an 'escaping', 'unstable sliding', or 'white opaque' region, and the corresponding terms for crossing regions are 'crossing', 'sewing', or 'transparent' regions.

Example 2.1. A dry-friction oscillator (Figure 2.2). The equations

$$\ddot{x} + b\dot{x} + ax + \lambda = 0 \tag{2.4}$$

describe a simple oscillator with displacement x, speed $\dot{x}$, and acceleration $\ddot{x}$. The constants a and b typically describe influences driving and damping the motion, respectively. We will assume $a, b, v > 0$, and for $\dot{x} \neq v$ define

$$\lambda = \operatorname{sign}(\dot{x} - v) = \begin{cases} +1 \text{ if } \dot{x} > v \Leftrightarrow \text{slipping right} , \\ -1 \text{ if } \dot{x} < v \Leftrightarrow \text{slipping left} . \end{cases} \tag{2.5}$$

This models a friction-like resistance whose direction switches to oppose the oscillator's speed $\dot{x}$ relative to some threshold $v = 0$; we already mentioned such models in Chapter 1. The size of the friction force has been normalized to ± 1.

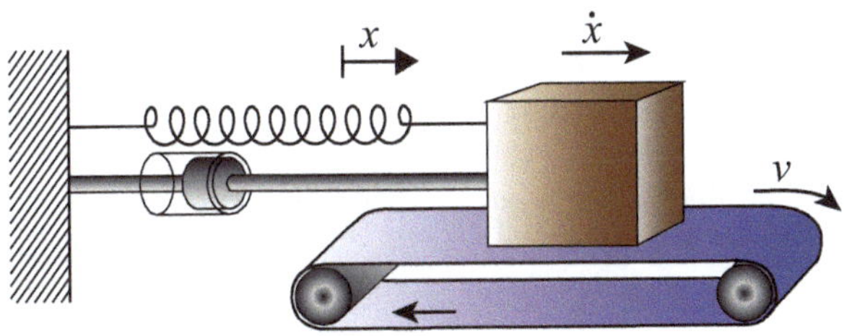

Fig. 2.2 The most obvious dry-friction oscillator is a block, connected to a spring, a damper, and resting on a moving track.

It is common to express (2.4) as a pair of first-order ordinary differential equations in two dimensions, the displacement x and speed $y = \dot{x}$, giving

$$\dot{x} = y , \qquad \dot{y} = -by - ax - \lambda , \qquad \lambda = \operatorname{sign}(y - v) ,$$

or written as a matrix equation,

$$\begin{pmatrix} \dot{x} \\ \dot{y} \end{pmatrix} = \begin{pmatrix} 0 & 1 \\ -a & -b \end{pmatrix} \begin{pmatrix} x + \lambda/a \\ y \end{pmatrix} , \qquad \lambda = \operatorname{sign}(y - v) . \tag{2.6}$$

There is an equilibrium at $(x, y) = (-\lambda/a, 0)$, where the right-hand side of the differential equation vanishes. In fact the equilibrium at $y = 0$ lies to one side of the discontinuity surface at $y = v$, and for $v > 0$, the equilibrium lies at $(x_{eq}, y_{eq}) = (1/a, 0)$.

To solve this system, we need an initial condition, so say at $t = 0$, the state is (x_0, y_0). Finding the solution in terms of a matrix exponential and then expanding in terms of sines and cosines, we have

$$\begin{aligned}\begin{pmatrix} x(t) + \lambda/a \\ y(t) \end{pmatrix} &= \exp\left\{\begin{pmatrix} 0 & 1 \\ -a & -b \end{pmatrix} t\right\} \begin{pmatrix} x_0 + \lambda/a \\ y_0 \end{pmatrix} \\ &= e^{-bt/2}\left\{\cos(\omega t) + \frac{1}{\omega}\begin{pmatrix} b/2 & 1 \\ -a & -b/2 \end{pmatrix}\sin(\omega t)\right\}\begin{pmatrix} x_0 + \lambda/a \\ y_0 \end{pmatrix},\end{aligned} \tag{2.7}$$

where $\omega = \sqrt{a - \frac{1}{4}b^2}$. Let us assume small damping, given by $b^2/4 < a$, so ω is real. These solutions oscillate clockwise around the equilibrium with a frequency ω, while decaying into it at a rate $b/2$. Taking this, we can sketch the flow of (2.6) above and below $y = v$, as in Figure 2.3.

What about dynamics on the discontinuity surface $\mathcal{D}$? If there exists motion constrained to $y = v$, then it must obey $\dot{y} = 0$. The two equations $y = v$ and $\dot{y} = 0$ imposed upon (2.6) imply

$$\begin{pmatrix} \dot{x} \\ \dot{y} \end{pmatrix} = \begin{pmatrix} v \\ 0 \end{pmatrix} \qquad \text{on } y = v\,. \tag{2.8}$$

Sketching all of these in the (x, y) plane completes Figure 2.3. Physically, solutions satisfying $y \neq v$ correspond to slipping of the oscillator relative to the threshold speed, 'rightward' if $y > v$ and 'leftward' if $y < v$. Motion along $y = v$ corresponds to sticking of the oscillator at the threshold speed. 'Sliding' describes the dynamics of the system variables—the spring displacement in this case—in the sticking phase.

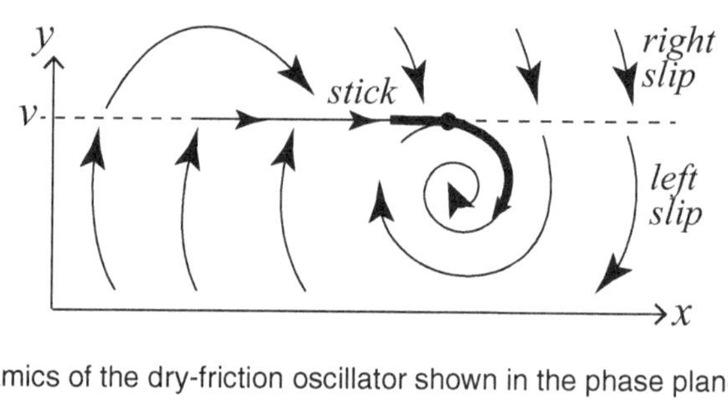

Fig. 2.3 Dynamics of the dry-friction oscillator shown in the phase plane (x, y).

■

2.2 The Value of sign(0): An Experiment

Towards the end of Section 1.4.1, we proposed an experiment to show that the friction coefficient can have 'peaks' at the discontinuity. Let us set down some equations for that experiment.

The set-up is in the spirit of the general oscillator in Example 2.1. Consider a simple block of mass m on a rough inclined table. If the table is raised up through an angle θ (with $0 < \theta < \pi/2$), then the force pulling the object down the incline is $T = mg\sin\theta$, and the reaction force from the table is $R = mg\cos\theta$. These are constant (g is the constant acceleration due to gravity, and we will keep θ fixed).

The friction force F on the block satisfies $|F| \leq \mu R$ for some coefficient of friction μ, with equality reached in slipping where $\dot{x} \neq 0$. Let us naively attempt to write this as $F = \mu_k R\,\mathrm{Sign}(\dot{x})$ for some fixed coefficient μ_k, where 'Sign($\dot{x}$)' just denotes the sign of $\dot{x}$, the specific symbol 'Sign' being used to signify that the precise character of this function is as yet unknown. Hence the equation of motion for the block's coordinate x along the slope is

$$m\ddot{x} = mg\sin\theta - \mu_k\,\mathrm{Sign}(\dot{x})mg\cos\theta\,, \tag{2.9}$$

The function 'Sign($\dot{x}$)' is not well-defined at $\dot{x} = 0$, we have to provide its definition there. Some mathematical texts select a value, perhaps the midpoint $\mathrm{Sign}(0) = 0$, just for definiteness. Instead let us appeal to our experiment to tell us the value of Sign(0) in (2.9).

If we set the incline angle θ as large as possible *without* the object slipping—the 'slipping point' angle—then we have no motion, so $\dot{x} = 0$ and $\ddot{x} = 0$. Balancing the equation of motion then implies $\mathrm{Sign}(0) = \tan\theta/\mu_k$, essentially telling us how strong the force of friction $F = \mu_k\,\mathrm{Sign}(0)$ must be to maintain equilibrium. Now there are two possibilities depending on the frictional character (crudely the 'roughness') of the block and the table.

If the 'slipping point' angle θ is small enough to satisfy $\tan\theta/\mu_k < 1$, then the rest position of the block is 'stable', meaning that if you tap the block so that it begins moving up the slope ($\dot{x} < 0$) or down the slope ($\dot{x} > 0$), the acceleration resists the motion, and the block will return to rest. This is because in motion, we have $\ddot{x} = (\tan\theta/\mu_k - \mathrm{Sign}(\dot{x}))g\mu_k\cos\theta$, in which $\tan\theta/\mu_k < 1$ implies $\ddot{x} < 0$ if $\dot{x} > 0$, and $\ddot{x} > 0$ if $\dot{x} < 0$ (or concisely $\ddot{x}\dot{x} < 0$). This is indeed the scenario encountered for some scenarios depending on what material the block and the surface are made of.

For other materials one finds the table can be inclined to a larger 'slipping point' angle such that $\tan\theta/\mu_k > 1$. Then in motion we have $\ddot{x} = (\tan\theta/\mu_k - \mathrm{Sign}(\dot{x}))g\mu_k\cos\theta > 0$, and because this is strictly positive, if you tap the block up the slope ($\dot{x} < 0$), it will return to rest, but tap it down the slope ($\dot{x} > 0$) and it will accelerate away at increasing speed. Put another way, the table is too steep for the block to be brought to rest if slipping down the slope, but friction is strong enough to overcome this *if* the block is carefully held at rest to begin with.

Because we found $\text{Sign}(0) = \tan\theta/\mu_k$ at the 'slipping point', this experiment tells us that the value of 'Sign(0)' can satisfy either $|\text{Sign}(0)| < 1$ or $|\text{Sign}(0)| > 1$. In the latter case, the $\text{Sign}(\dot{x})$ function has a peak at $\dot{x} = 0$ that does not lie between the values ± 1, so the coefficient of friction takes values outside the $\pm\mu_k$ that apply for $\dot{x} \gtrless 0$. This is actually a well-known phenomenon called 'static friction' or 'stiction'—the force of friction is greater when static than when in motion.

The way we handle this in piecewise-smooth dynamics is to work in terms of a switching multiplier, $\lambda = \text{sign}(\sigma)$, whose value we set strictly as lying in the interval $|\text{sign}(\dot{x})| < 1$. We can then let $\text{Sign}(\dot{x}) = \lambda$, or we can form other functions with a larger range, like $\text{Sign}(\dot{x}) = \lambda + (\lambda^2 - 1)\rho$ for some quantity ρ, for example, which then lies in the larger interval $|\text{Sign}(\sigma)| \leq \rho + 1/4\rho$. We put this to use in Section 2.4.

2.3 Types of Dynamics: Sliding and Crossing

Solutions in a region where $\mathbf{f}$ is differentiable consist of unique twice differentiable curves $\mathbf{x}(t)$. At a discontinuity of $\mathbf{f}$ this breaks down, but the curves $\mathbf{x}(t)$ do not terminate, they continue on by connecting to other curves $\mathbf{x}(t)$ at the discontinuity. Thus orbits are obtained by concatenating segments of paths $\mathbf{x}(t)$ in a manner that preserves the direction of time.

Crossing of the discontinuity surface is possible where it forms a bridge, that is, where $\mathbf{f}^+$ and $\mathbf{f}^-$ point in the same direction relative to $\mathcal{D}$, their normal components to $\sigma = 0$ satisfying $(\delta_{t+}\sigma)(\delta_{t-}\sigma) > 0$.

Example 2.2. Crossing in the friction oscillator. Consider the region $|bv + ax| > 1$ in (2.6). In the region $bv + ax > 1$ on $y = v$, the normal components of both slipping vector fields point downwards, since $\delta_{t+}\sigma = -bv - ax - 1 < 0$ and $\delta_{t-}\sigma = -bv - ax + 1 < 0$. Similarly in the region $bv + ax < -1$ on $y = v$, the normal components of both slipping vector fields point upwards, since $\delta_{t+}\sigma = -bv - ax - 1 > 0$ and $\delta_{t-}\sigma = -bv - ax + 1 > 0$. From (2.7) we have near a point (x_0, v) on the discontinuity surface,

$$\begin{pmatrix} x(t) + \lambda/a \\ y(t) \end{pmatrix} = \left\{ 1 + \begin{pmatrix} 0 & 1 \\ -a & -b \end{pmatrix} t + \mathsf{O}\left(t^2\right) \right\} \begin{pmatrix} x_0 + \lambda/a \\ v \end{pmatrix}$$

which rearranges to give

$$\begin{pmatrix} x(t) \\ y(t) \end{pmatrix} = \begin{pmatrix} x_0 + vt \\ v - (bv + ax_0 + \lambda)t \end{pmatrix} + \mathsf{O}\left(t^2\right) , \tag{2.10}$$

illustrated by orbits (i) and (ii) in Figure 2.4, which trace a solution over times $t_1 < 0 < t_2$. These orbits cross through $y = v$ at time $t = 0$, doing so continuously but non-differentially as $\lambda = \text{sign}(y - v)$ changes from $+1$ for $y < v$ to -1 for $y > v$.

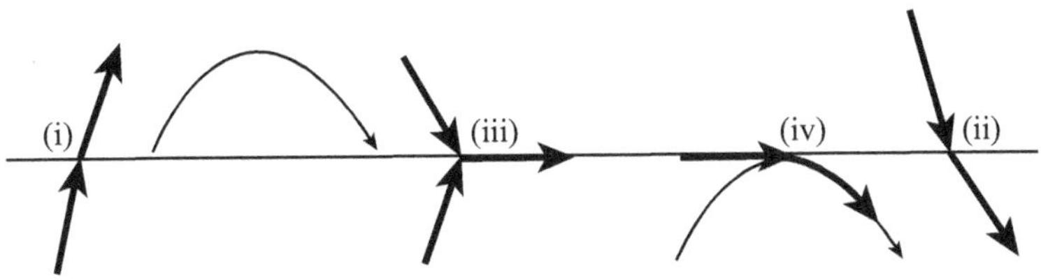

Fig. 2.4 Examples of crossing and the onset of sticking or slipping, in the friction oscillator for $v > 1$: (i)–(ii) show crossing between right/left slip given by (2.10), (iii) shows onset of sticking given by (2.12), (iv) shows onset of slipping given by (2.13).

■

Sliding along the discontinuity surface is possible if $\mathbf{f}(\mathbf{x};\lambda)$ can admit values that lie tangential to $\mathcal{D}$, given by $\delta_t\sigma = 0$ on $\sigma = 0$ (recalling $\delta_t \equiv \mathbf{f}(\mathbf{x};\lambda)\cdot\delta_{\mathbf{x}}$). Writing $\delta_{t\$} \equiv \mathbf{f}(\mathbf{x};\lambda^\$)\cdot\delta_{\mathbf{x}}$, sliding therefore occurs if there exists $\lambda = \lambda^\$$ such that

$$0 = \delta_{t\$}\sigma(\mathbf{x})\ , \qquad \lambda^\$ \in (-1,+1)\ . \tag{2.11}$$

We say $\mathcal{D}$ is attracting if $\delta_{t+}\sigma < 0 < \delta_{t-}\sigma$ or repelling if $\delta_{t-}\sigma < 0 < \delta_{t+}\sigma$. Such a region, where $\mathbf{f}^\pm$ point in different directions relative to $\mathcal{D}$, given by $(\delta_{t+}\sigma)(\delta_{t-}\sigma) < 0$, is guaranteed to admit sliding.

Lemma 2.1 (Attracting/Repelling $\mathcal{D}$ Implies Sliding). *If $\mathbf{f}(\mathbf{x};\lambda)$ is continuous in λ, and the components of $\mathbf{f}^\pm(\mathbf{x}) \equiv \mathbf{f}(\mathbf{x};\pm1)$ normal to the set $\sigma(\mathbf{x}) = 0$ are opposing, then there exist solutions to $\delta_{t\$}\sigma(\mathbf{x}) = 0$ for $-1 \le \lambda^\$ \le +1$.*

Proof. This follows from the intermediate value theorem: if $\delta_{t\pm}\sigma$ have different signs, since $\mathbf{f}^\pm(\mathbf{x}) \equiv \mathbf{f}(\mathbf{x};\pm1)$, this means $\delta_{t\pm}\sigma$ have different signs, so since $\delta_{t\pm}\sigma = \mathbf{f}(\mathbf{x};\pm1)\cdot\delta_{\mathbf{x}}\sigma(\mathbf{x})$, there exists an intermediate value $-1 \le \lambda^\$ \le +1$ such that $\delta_{t\$}\sigma = 0$.

Example 2.3. Sliding in the friction oscillator.

Consider the region $|bv + ax| < 1$ in (2.6). On $y = v$, the normal components of the vector fields have different signs, $\delta_{t\pm}\sigma = -bv - ax \mp 1 \lessgtr 0$, each pointing towards $y = v$, so the discontinuity surface is attractive. Seeking solutions that slide along $y = v$, we solve $\delta_{t\$}\sigma = \dot{y} = -bv - ax - \lambda^\$ = 0$, which implies $\lambda^\$ = -bv - ax$. The resulting dynamics is just that given by (2.8).

We can interpret this as saying that λ represents the frictional force, evolving during sticking as $\lambda = \lambda^\$ = -bv - ax$ in order to balance external forces.

If we take a point (x_0, v) on the discontinuity surface, and approximate the solutions nearby, we have $x(t) = x_0 + vt + \mathsf{O}\left(t^2\right)$, and near the onset of sticking

$$y(t) = v - \left\{ \begin{array}{ll} (bv + ax_0 + \lambda)t + \mathsf{O}\left(t^2\right) & \text{if } t < 0 \\ 0 & \text{if } t > 0 \end{array} \right\} , \tag{2.12}$$

where $\lambda = +1$ or -1 signifies that sticking initiates from rightward or leftward slip, respectively, illustrated by orbit (iii) in Figure 2.4.

Solutions can only leave the discontinuity surface where it ceases to be attracting. This occurs at $x_0 = (1 - bv)/a$, near which the onset of slipping is given by

$$y(t) = v - \left\{ \begin{array}{ll} 0 & \text{if } t < 0 \\ \frac{1}{2}avt^2 + \mathsf{O}\left(t^3\right) & \text{if } t > 0 \end{array} \right\} , \tag{2.13}$$

illustrated by orbit (iv) in Figure 2.4. To obtain this, since the first-order term $\dot{y}$ vanishes at $(x_0, 0)$, we have had to return to (2.7) and expand to the second order in t. This tells us that slipping takes the trajectory away from $y = v$ in the direction of decreasing y.

■

So we cannot make (2.1) differentiable or unique at $\mathcal{D}$, but we can make it continuous if we first permit the vector field to be set-valued, by defining the 'sign' function such that $\lambda = \text{sign}(\sigma) \in (-1, +1)$ at $\sigma = 0$, and then seeking values of λ that give valid dynamics. This was Filippov's idea. If there is a convex set $\mathcal{F} \supseteq \{\mathbf{f}(\mathbf{x}; \lambda) : \lambda \in [-1, +1]\}$, i.e. containing the values $\mathbf{f}^{\pm}$ that $\mathbf{f}$ jumps between in the neighbourhood of the switch, Filippov showed:

Theorem 2.1 (Existence of Solutions). *Solutions of* $\dot{\mathbf{x}} = \mathcal{F}$ *exist if* $\mathcal{F}(\mathbf{x})$ *is non-empty, bounded, closed, convex, and upper semicontinuous.*

The term 'upper semicontinuity' refers to an extension of the notion of continuity for sets, saying $\sup_{b \in \mathcal{F}(\mathbf{x})} \rho(b, a) \to 0$ as $p' \to p$ for $a \in \mathcal{F}(p')$ and $b \in \mathcal{F}(p)$ where $\rho(b, a) = \inf_{a \in A,\, b \in B} |a - b|$ with $|a - b|$ denoting the distance between a and b. These technicalities are not of direct use to us here but can be found in Thm1 on p77 of [71].

This has been the standard approach to piecewise-smooth systems since Filippov's work [71] (in the English-speaking world at least and much earlier (e.g. [70]) in Russian). Not everyone follows this, of course, and the other major approach is perhaps that of complementarity systems (e.g. [25]), particularly used in mechanical control. In this approach, the sticking phase is posed as a mechanical constraint, and one attempts to resolve the forces and energies that lead to a viable motion. The method is powerful, but it is unclear how it would handle the singularities and multi-valuedness we shall find later. Whatever the chosen approach, these are all lacking something so far.

By admitting sliding, we are plucking a special *sliding vector field* $\mathbf{f}(\mathbf{x}; \lambda^{\$})$ from the set $\{\mathbf{f}(\mathbf{x}; \lambda) : \lambda \in [-1, +1]\}$. We are substituting the jump between

$\lambda = \pm 1$, for a jump between $\lambda = \pm 1$ *and* this special sliding value $\lambda^{\$}$. In other words we are still ignoring the set-valuedness of λ and of $\mathbf{f}(\mathbf{x}; \lambda)$. Does this matter? Can we use it? Can the manner in which λ and $\mathbf{f}$ jump affect the dynamics? These are issues that Filippov and his contemporaries considered, and insightfully so (see Appendix B), but found no constructive resolution to, and so most of the community since have cast the persisting uncertainties aside. Advances made since Filippov's work, however, permit a new view on the problem and a reopening of the book on the elements of piecewise-smooth dynamical theory.

2.4 The Switching Layer and Hidden Dynamics

The final step is to look at the dynamics of λ itself, to ask how it switches between ± 1.

Differentiating λ with respect to time using the chain rule gives $\dot{\lambda} = \frac{d\lambda}{d\sigma}\dot{\sigma}$. This vanishes for $\sigma \neq 0$ because λ is constant there, so $d\lambda/d\sigma = 0$. But at $\sigma = 0$, the derivative $d\lambda/d\sigma$ is infinite because λ jumps across $\sigma = 0$. We need a way to map the point $\sigma = 0$ onto the interval $\lambda \in [-1, +1]$, and the simplest mapping is just $\sigma = \varepsilon\lambda$ for a constant $\varepsilon > 0$, with which the interval $\lambda \in [-1, +1]$ squashes down to $\sigma = 0$ if we let $\varepsilon \to 0$. We call $\sigma = \varepsilon\lambda$ the 'blow-up' of the point $\sigma = 0$, and elsewhere we have $\lambda = \text{sign}(\sigma)$ as usual.

This correctly gives the derivative $d\lambda/d\sigma = 1/\varepsilon$ as infinite at $\sigma = 0$ as $\varepsilon \to 0$. Then we have

$$\varepsilon\dot{\lambda} = \begin{cases} 0 & \text{if } \sigma \neq 0\,, \\ \delta_t\sigma(\mathbf{x}) & \text{if } \sigma = 0\,, \end{cases} \qquad \lambda \in (-1, +1)\;, \quad \varepsilon \to 0\,. \tag{2.14}$$

If we are studying what happens as we move ε away from zero, then an equation like this is known as a *singular perturbation* problem, a fascinating area of study for which I recommend texts such as [19, 104, 124, 135]). But for the piecewise-smooth system, our interest is solely in the limiting situation $\varepsilon \to 0$ and how it resolves the discontinuity; therefore while some of the concepts we use are similar to singular perturbation theory, they are slightly different and use different terminology.

Taking (2.14) on $\sigma = 0$, if $\delta_t\sigma$ is non-vanishing, then it carries the flow across $\mathcal{D}$ on the instantaneous timescale t/ε. If $\delta_t\sigma$ vanishes on $\sigma = 0$,

$$0 = \delta_t\sigma(\mathbf{x}) \qquad \text{at some} \quad \lambda = \lambda^{\$} \in (-1, +1)\;,$$

then $\mathbf{f}(\mathbf{x}; \lambda)$ has a vector lying in the tangent space of $\mathcal{D}$. This defines dynamics *sliding* along the surface $\mathcal{D}$ where $\sigma = 0$, and we recognize this exactly as the sliding condition we gave earlier in (2.3). Since the components $\delta_{t\pm}\sigma(\mathbf{x})$ normal to $\mathcal{D}$ have different signs where the surface is attracting or repelling, the existence of sliding in those regions is guaranteed by Lemma 2.1. Sliding is not guaranteed otherwise, but not ruled out either.

Thus we have blown the surface $\sigma = 0$ up into a layer $\lambda \in (-1, +1)$ and call $\varepsilon\dot{\lambda} = \delta_t\sigma$ the layer system. Letting $\varepsilon \to 0$ gives an instantaneous switch (instantaneous because the rate of change $\dot{\lambda} \sim 1/\varepsilon$ is then infinitely large). In fact we obtain different behaviours depending how we take the limit:

1. *The slow subsystem*: Let $\varepsilon \to 0$ in (2.14) as written. We then obtain simply an algebraic condition $\delta_t\sigma = 0$, which is the criterion for sliding along $\mathcal{D}$.
2. *The fast subsystem*: Look at (2.14) on the timescale t/ε before letting $\varepsilon \to 0$. On $\sigma = 0$, we have $\lambda \approx \lambda_0 + r_1 e^{r_2 t/\varepsilon}$ near a point λ_0, for some constants $r_{1,2}$ that depend on λ_0 and $\mathbf{x}$. Thus λ evolves through the switching interval $-1 \le \lambda \le +1$ on the fast timescale t/ε, which is infinitely fast for $\varepsilon \to 0$. If there are equilibria of the fast subsystem, where $\delta_t\sigma = 0$, then λ evolves (infinitely fast) towards or away from them, depending on the sign of $d\dot{\lambda}/d\lambda$.

These two limits provide two perspectives of a two-timescale story. The fast system tells us how λ rapidly transitions between $-1 \le \lambda \le +1$ on the infinitely fast timescale t/ε during switching. This fast system may have equilibria, and these comprise just the algebraic set inhabited by the slow subsystem, so if λ collapses onto an attractor of the fast system, then it enters the slow system $\dot{\mathbf{x}} = \mathbf{f}$, subject to the constraint $\delta_t\sigma = 0$, which means that it slides along $\sigma = 0$ on the 'slow' timescale t.

As an exercise for the reader, we suggest returning to the oscillator in Example 2.1 above and seeing how this layer system gives instantaneous transition through the regions where we found crossing and gives instantaneous attraction to the sliding modes elsewhere, permitting exit from sliding when $\lambda^{\$}$ reaches $|\lambda^{\$}| = 1$. We will apply this below once we have made the problem slightly more interesting.

We describe the function $\mathbf{f}(\mathbf{x};\lambda)$ as a **combination** of $\mathbf{f}^+(\mathbf{x})$ and $\mathbf{f}^-(\mathbf{x})$:

- E.g. Convex combination (Filippov 1988 *Def 4a p50-52* [71])

$$\dot{\mathbf{x}} = \tfrac{1}{2}\,(1+\lambda)\,\mathbf{f}^+(\mathbf{x}) + \tfrac{1}{2}\,(1-\lambda)\,\mathbf{f}^-(\mathbf{x})\ . \tag{2.15}$$

- E.g. Nonlinear combination (Jeffrey 2013 [114])

$$\dot{\mathbf{x}} = \tfrac{1}{2}\,(1+\lambda)\,\mathbf{f}^+(\mathbf{x}) + \tfrac{1}{2}\,(1-\lambda)\,\mathbf{f}^-(\mathbf{x}) + (\lambda^2-1)\mathbf{h}(\mathbf{x},\lambda)\ . \tag{2.16}$$

In both examples we have $\mathbf{f}(\mathbf{x};\pm 1) \equiv \mathbf{f}^{\pm}(\mathbf{x})$. The function $\mathbf{h}$ can be any finite-valued vector field, and we shall find a use for it in the next example below. The multiplier λ in either case satisfies $\lambda \in [-1, +1]$.

The combination $\mathbf{f}(\mathbf{x};\lambda)$ is differentiable with respect to $\mathbf{x}$ and λ (by which we mean that the partial derivatives $\frac{\partial}{\partial \mathbf{x}}\mathbf{f}$ and $\frac{\partial}{\partial \lambda}\mathbf{f}$ exist). The discontinuity is therefore encoded in the multiplier λ.

The term $(\lambda^2 - 1)\mathbf{h}(\mathbf{x}, \lambda)$ is called **hidden**, because it vanishes for $\mathbf{x} \notin \mathcal{D}$ (when $\lambda = \pm 1$). In fact it is so hidden that you won't find it in most other courses or texts on piecewise-smooth dynamics until recently.

The hidden term opens the possibility that $\mathbf{f}$ depends nonlinearly on λ and with it, the possibility that there are multiple different values of λ that satisfy the sliding conditions $\delta_t\sigma = 0$ at a given $\mathbf{x}$.

Take an extended version of the oscillator problem from Example 2.1.

Example 2.4. Consider the equations

$$\ddot{x} + b\dot{x} + ax + \mu(\lambda) = 0 \, , \qquad \mu(\lambda) = \lambda + (\lambda^2 - 1)s \, , \tag{2.17}$$

with $\lambda = \text{sign}(\dot{x} - v)$ and s some real constant. These are equivalent to (2.4) for $\dot{x} \neq v$, i.e. in slipping. The additional term in the friction force μ involving the new constant s can therefore only play a role during sticking, on $\dot{x} = v$.

A useful way to visualize the combination $\mathbf{f}(\mathbf{x}; \lambda)$, which is now the right-hand side of

$$\begin{pmatrix} \dot{x} \\ \dot{y} \end{pmatrix} = \begin{pmatrix} 0 & 1 \\ -a & -b \end{pmatrix} \begin{pmatrix} x \\ y \end{pmatrix} - \begin{pmatrix} 0 \\ \lambda + (\lambda^2 - 1)s \end{pmatrix} , \quad \lambda = \text{sign}(y - v) \, , \tag{2.18}$$

is as follows. Let us first choose a few representative points $\mathbf{x} \in \mathcal{D}$ along the discontinuity surface, labelled (i), (ii), (iii), and (iii'), in Figure 2.5, and draw on the vector field at these points. We know the vector field is $\mathbf{f}^+(\mathbf{x}) = \mathbf{f}(\mathbf{x}; +1)$ immediately above the discontinuity surface and $\mathbf{f}^-(\mathbf{x}) = \mathbf{f}(\mathbf{x}; -1)$ immediately below it.

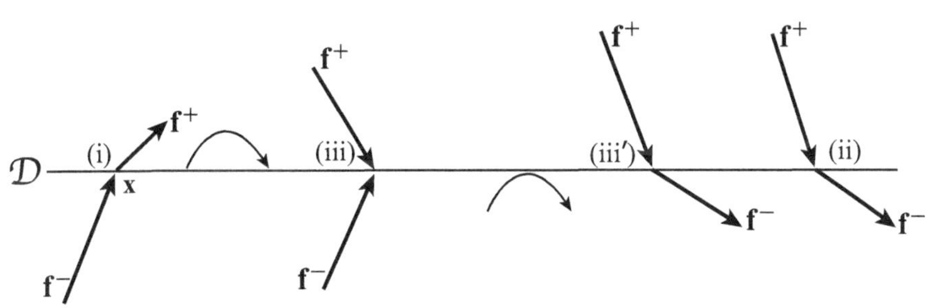

Fig. 2.5 The oscillator, showing the vector fields $\mathbf{f}^\pm$ at four points along the discontinuity surface. The labels (i-iii) correspond to those in Figure 2.4.

This tells us that, from a point $\mathbf{x} \in \mathcal{D}$, the flow wants to follow a vector $\mathbf{f}^+(\mathbf{x})$ or $\mathbf{f}^-(\mathbf{x})$. This is better indicated by drawing these vectors as extending from a point $\mathbf{x}$ on the discontinuity surface, as shown in Figure 2.6.

As the vector field jumps between $\mathbf{f}^\pm$, it traces out some family of fields in between. We can represent these by projecting the vectors $\mathbf{f}(\mathbf{x}; \lambda)$ from any point $\mathbf{x} \in \mathcal{D}$ and forming a curve $\{\mathbf{x} + \mathbf{f}(\mathbf{x}; \lambda) \ : \ \lambda \in [-1, +1]\}$ that connects

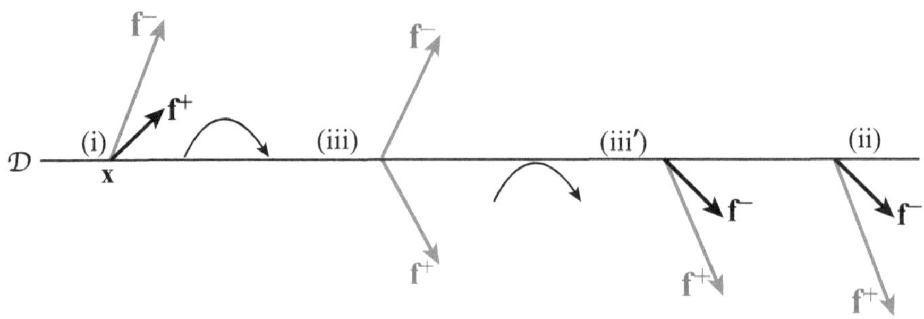

Fig. 2.6 The oscillator, with vector fields $\mathbf{f}^{\pm}$ projected from points on the discontinuity surface. The labels (i-iii') correspond to those in Figure 2.5.

$\mathbf{f}^{+}$ to $\mathbf{f}^{-}$. For the four different points in Figure 2.6, these curves are shown dashed in Figure 2.7, with $\mathbf{f}^{\pm}$ delimiting each curve's extremities.

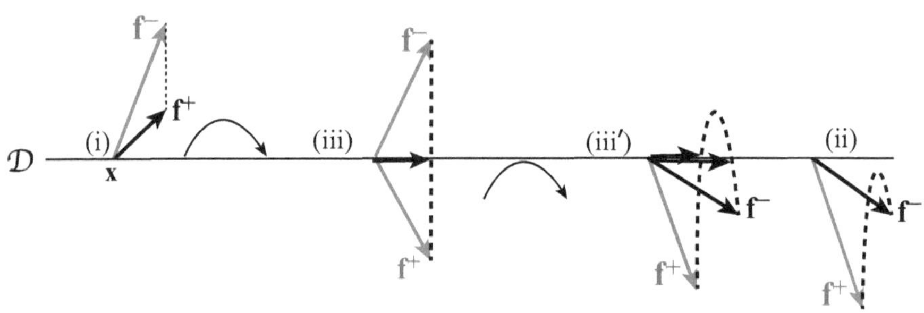

Fig. 2.7 The oscillator, with dashed curves showing all vectors $\mathbf{f}(\mathbf{x};\lambda)$ that solutions may follow from the discontinuity surface. Labels (i-iii') correspond to Figure 2.5.

We have cheated a little in drawing this picture because the vectors $\mathbf{f}^{\pm}$ have the same horizontal components, meaning the curves drawn in (iii') and (ii) are really straight lines that run up from $\mathbf{f}^{-}$ to some peak value and then run back down to $\mathbf{f}^{+}$, overlapping themselves, so we have displaced $\mathbf{f}^{\pm}$ horizontally to visualize the overlap.

What is important in Figure 2.7 is that the flow can follow any vector whose end point lies on the dashed curve. In (i) any of those vectors simply carry the flow through the discontinuity surface. In (ii) the same happens (despite the curve dipping towards $\mathcal{D}$). We can see that this dynamics makes sense considering the directions of $\mathbf{f}^{\pm}$. In (iii) and (iii'), however, the curve passes through the discontinuity surface, so there exist vectors, shown lying along $\mathcal{D}$ in Figure 2.7(iii-iii'), that carry the flow along the discontinuity surface, i.e. sliding vectors. In (iii) sliding is the only possibility, since both $\mathbf{f}^{\pm}$ push the flow onto the surface. In (iii′) it is surprising that sliding is possible, however, because $\mathbf{f}^{\pm}$ seem to point through the surface, and we might expect crossing.

The nonlinearity in (2.18) thereby means that sticking motion can occur for a greater range of displacements if $s \neq 0$. We refine this statement below but must take a closer look at the dynamics of these sticking solutions first.

There is an ambiguity in case (iii′) of Figure 2.7; it seems to be possible for solutions to follow one of the three different motions: to choose between two different sliding vectors or to cross the surface following $\mathbf{f}^-$. We resolve which happens by blowing up the discontinuity surface into a switching layer. Letting $y = \dot{x}$ as before, on $y = v$, we form the layer system,

$$\varepsilon\dot{\lambda} = -bv - ax - \lambda - (\lambda^2 - 1)s \,. \tag{2.19}$$

Since λ parameterizes the dashed curves in Figure 2.7, this equation tells us how λ varies with time on those curves, giving the double arrows in Figure 2.8.

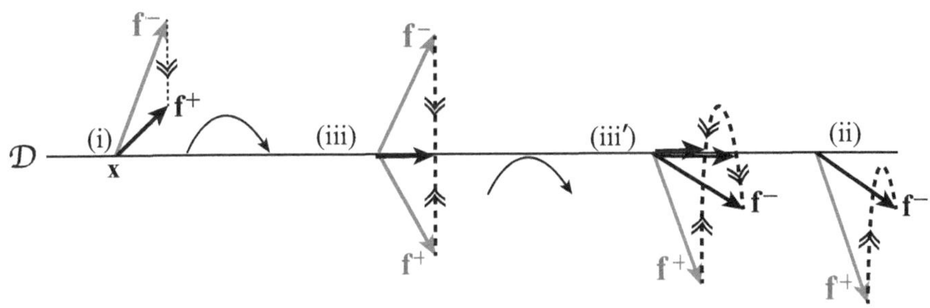

Fig. 2.8 The oscillator, with layer dynamics indicated by double arrows on the curves representing the vectors $\{\mathbf{f}(\mathbf{x};\lambda) : \lambda \in [-1,+1]\}$. Labels (i-iii′) correspond to Figure 2.5.

As $\varepsilon \to 0$ in (2.19), the multiplier λ evolves infinitely quickly across $(-1,+1)$. Unless, that is, the right-hand side of (2.19) vanishes, in which case λ has equilibria at some $\lambda^\$$ such that

$$\lambda^\$ + \left((\lambda^\$)^2 - 1\right)s = -bv - ax \quad \Rightarrow \quad \lambda^\$_\pm = -\tfrac{1}{2s} \pm R \,, \tag{2.20}$$

where $R = \sqrt{1 - (bv + ax)\frac{1}{s} + \frac{1}{4s^2}}$, and to exist these must lie in $|\lambda^\$| \leq 1$.

A little work shows that both $\lambda^\$_\pm$ exist if $s > 1/2$ and $1 < bv+ax < 1/4s+s$. Only $\lambda^\$_+$ exists in the range $|bv + ax| \leq 1$, which is where we found sliding to occur for $s = 0$. Since λ is stationary at $\lambda^\$_\pm$, we then return to the natural timescale t, where the dynamics is given (2.18), and then substituting in $\lambda = \lambda^\$_\pm$ gives (2.8) as before for the sliding dynamics but now occupying the larger region $1 < bv + ax < 1/4s + s$ if $s > 1/2$.

In this simple system, both solutions for λ result in the motion (2.8), but they are not equivalent because the friction forces $\mu(\lambda^\$_\pm)$ in these modes are different. So which of these two modes will the system follow? The layer equation tells us the attractivity of the sliding solutions.

Approximating about the points $\lambda^{\$}_{\pm}$, (2.19) gives

$$\begin{aligned}\dot{\lambda} &= -(\lambda - \lambda^{\$}_{+})(\lambda - \lambda^{\$}_{-})s/\varepsilon \\ &= \mp 2R(\lambda - \lambda^{\$}_{\pm})s/\varepsilon + \mathsf{O}\left((\lambda - \lambda^{\$}_{\pm})^2 s/\varepsilon\right)\end{aligned} \tag{2.21}$$

whose solutions behave near $\lambda = \lambda^{\$}_{\pm}$ like

$$\lambda(t) \approx (\lambda(0) - \lambda^{\$}_{\pm})e^{\mp 2Rst/\varepsilon}\,, \qquad \varepsilon \to 0\,,$$

given an initial condition $\lambda = \lambda(0)$ at $t = 0$. Therefore as λ evolves between $-1 \le \lambda \le +1$, it is attracted towards $\lambda^{\$}_{+}$ and repelled from $\lambda^{\$}_{-}$ on the fast timescale t/ε. If $|\lambda^{\$}_{+}| < 1$, then λ collapses onto $\lambda^{\$}_{+}$ and slides on $y = v$ with $\lambda = \lambda^{\$}_{+}$ according to (2.8). This motion continues until $\lambda(t)$ reaches the edge of the interval $-1 \le \lambda \le +1$, either because, as $x(t)$ varies, $\lambda^{\$}_{+}$ leaves the interval $-1 \le \lambda^{\$}_{+} \le +1$, or $\lambda^{\$}_{+}$ becomes complex and ceases to exist as an attractor for λ, which occurs at $(x, \lambda) = (s^2/4 - bvs + 1, -a/2)/as$, after which λ rapidly evolves to -1 via (2.21).

So we can return to Figure 2.8 and understand more closely what is going on. As the switch takes place, in (i) the vector $\mathbf{f}^-$ swings through $\{\mathbf{f}(\mathbf{x}; \lambda) : \lambda \in [-1, +1]\}$ to $\mathbf{f}^+$, and $\mathbf{f}^+$ carries the flow away from the surface, while in (ii), the vector swings the opposite way, and $\mathbf{f}^-$ carries the flow away. In (iii), arriving from above or below the surface, the vectors $\mathbf{f}^{\pm}$ both swing through $\{\mathbf{f}(\mathbf{x}; \lambda) : \lambda \in [-1, +1]\}$ towards a sliding vector lying in the surface, which gives sliding motion along $\mathbf{f}(\mathbf{x}; \lambda^{\$}_{+})$. The case (iii') seems more complicated, but since the flow can only arrive from above, upon arrival at $\mathcal{D}$ the vector $\mathbf{f}^+$ swings towards the sliding vector field $\mathbf{f}(\mathbf{x}; \lambda^{\$}_{+})$. Hence sliding occurs, and uniqueness is restored.

Lastly, Figure 2.9 shows the same picture for $s = 0$, when the dependence on λ is linear. The set of vectors formed by $\mathbf{f}(\mathbf{x}; \lambda)$ connecting $\mathbf{f}^{\pm}$ is now a simple line. Most notably in case (iii′), this set no longer intersects the discontinuity surface, so no sliding vectors exist, and sliding is no longer possible there.

This also implies that for $s > 1/2$ that the range of x where sticking occurs is larger than if $s = 0$. Equivalently, for $s > 1/2$, the system (2.17) supports a larger friction force $\mu(\lambda)$ as given by (2.17) during sticking than during slip, i.e. reaching $|\mu| > 1$ as λ varies over $-1 \le \lambda \le +1$. As we discussed in Section 2.2, this is a real physical phenomenon known as *static friction* or *stiction*: the force required to set an object in motion on a rough surface, i.e. to go from sticking to slipping, often exceeds the force required to keep it slipping. The model above therefore explains conceptually why stiction is to be expected in the presence of a discontinuous friction force.

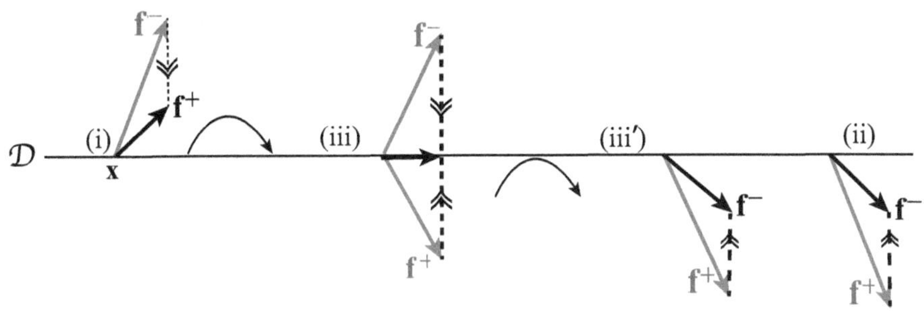

Fig. 2.9 The oscillator without stiction, corresponding to Figure 2.8 with $s = 0$.

■

2.4.1 A Note on Modelling Basic Oscillators

The oscillator example is common as a toy model of damped elastic oscillations subject to a piecewise-constant resistance, not only in mechanics.

The obvious physical realization of (2.4) or (2.17) is a block of mass m, resting on a surface that moves at velocity v, attached to a spring of stiffness k and extension x, and a motion damper with a coefficient c. There is a friction force $F = \mu R$, where $R = mg$ is the normal reaction between the surface and the block. We can rearrange Newton's second law into the form

$$m\ddot{x} = -c\dot{x} - kx - \mu R \quad \Rightarrow \quad \ddot{x} + \tfrac{c}{\sqrt{km}}\dot{x} + x + \mu = 0\,,$$

by replacing $x \mapsto xR/k$ and $t \mapsto t\sqrt{m/k}$.

This also approximates a pendulum, swinging with small angle x, with friction force $F \approx \mu R$ from a pivot which rotates with angular speed v. If the pendulum's length is l, its mass m (concentrated at its end) gives a rotating force $mg \sin x \approx mgx$, and let's say air resistance gives a damping with coefficient c, again we can rearrange Newton's second law into the form

$$ml\ddot{x} \approx -c\dot{x} - mgx - \mu R \quad \Rightarrow \quad \ddot{x} + \tfrac{c}{m\sqrt{gl}}\dot{x} + x + \mu \approx 0\,,$$

by replacing $x \mapsto xR/mg$ and $t \mapsto t\sqrt{l/g}$.

We can also use this as the basis of more exotic models of damped oscillatory behaviour overlaid with a decision. For example, μ in (2.17) could represent the tendency of an individual to perceive that 'the grass is greener on the other side' when given a choice between two situations, represented by a resistive forcing $\ddot{x} \propto -\mu$, where x is a chemical or mood indicator and terms like ax and $b\dot{x}$ represent typical regulatory processes that give damped oscillation around a desirable level $x = 0$, irrespective of the decision.

Prototypes like this allow us to begin building intuitive models of behaviour across the sciences. Our task in the coming chapters is to further our intuition to assist in the task.

2.5 Local Singularities and Bifurcations

The only pointlike singularities existing in a differentiable vector field are places where it vanishes, i.e. equilibria, where the flow is stationary. A discontinuity creates a new kind of singularity where the flow is stationary *with respect to the discontinuity surface.*

2.5.1 Equilibria and Local Stability

If all systems described by (2.1) were linear in $\mathbf{x}$ and λ was simply a constant, then qualitative dynamics would be very simple and the world rather dull.

A linear system $\dot{\mathbf{x}} = \mathbf{f}(\mathbf{x};\lambda)$ can be written as

$$\dot{\mathbf{x}} = \mathbf{a} + \underline{\underline{J}}.\mathbf{x} \tag{2.22}$$

for a constant vector $\mathbf{a} = \mathbf{f}(\mathbf{0};\lambda)$, and a constant matrix $\underline{\underline{J}} = \frac{d}{d\mathbf{x}}\mathbf{f}(\mathbf{0};\lambda)$ called the Jacobian. (For now we can take λ to denote a constant appearing in $\underline{\underline{J}}$ and/or $\mathbf{a}$). This can be solved exactly, and solutions are given by

$$\mathbf{x}(t) = \mathbf{x}_{eq} + e^{\underline{\underline{J}}t}(\mathbf{x}_0 - \mathbf{x}_{eq}) \;, \qquad \mathbf{x}_{eq} = -\underline{\underline{J}}^{-1}\mathbf{a} \;,$$

so they are deterministic, i.e. predictable. The point $\mathbf{x}(0) = \mathbf{x}_0$ is the *initial condition* of the solution $\mathbf{x}(t)$, and $\mathbf{x}_{eq}$ is an equilibrium since $\mathbf{f}(\mathbf{x}_{eq};\lambda) = 0$.

A useful way to unpack the expression for $\mathbf{x}(t)$ is to decompose $\underline{\underline{J}}$ into

$$\underline{\underline{J}} = J_0\begin{pmatrix} 1 & 0 \\ 0 & 1 \end{pmatrix} + J_1\begin{pmatrix} 1 & 0 \\ 0 & -1 \end{pmatrix} + J_2\begin{pmatrix} 0 & 1 \\ 1 & 0 \end{pmatrix} + J_3\begin{pmatrix} 0 & 1 \\ -1 & 0 \end{pmatrix} , \tag{2.23}$$

where $J_1, J_2, J_3, and\ J_4$ are scalars (a form related to *Pauli* matrices) and then the matrix exponential becomes

$$e^{\underline{\underline{J}}t} = e^{J_0 t}\left\{\begin{pmatrix} 1 & 0 \\ 0 & 1 \end{pmatrix}\cosh(\omega t) + \frac{1}{\omega}\begin{pmatrix} J_1 & J_2 + J_3 \\ J_2 - J_3 & J_1 \end{pmatrix}\sinh(\omega t)\right\} \tag{2.24}$$

with $\omega = \sqrt{J_1^2 + J_2^2 - J_3^2}$. If ω is real, then the term in curly brackets sends trajectories $\mathbf{x}(t)$ in hyperbolic paths about the point $\mathbf{x}_{eq}$; if ω is imaginary, it circulates them about the point $\mathbf{x}_{eq}$. The $e^{J_0 t}$ factor out the front simultaneously sends those solutions evolving exponentially towards or away from $\mathbf{x}_{eq}$.

As $t \to \infty$, they asymptote towards $\mathbf{x}_{eq}$ if $J_0 < 0$ (an attractor) and away from $\mathbf{x}_{eq}$ if $J_0 > 0$ (a repeller).

We can be more precise about the geometry of this attraction or repulsion. The matrix $\underline{\underline{J}}$ has eigenvectors $\mathbf{g}_i$ and eigenvalues γ_i that satisfy a characteristic equation $\underline{\underline{J}}.\mathbf{g}_i = \gamma_i \mathbf{g}_i$ for $i = 1, 2$. From the definition of the exponential $e^x = \sum_{r=0}^{\infty} x^r/r!$, this implies $e^{\underline{\underline{J}}t}\mathbf{g}_i = e^{\gamma_i t}\mathbf{g}_i$. If we take an initial condition $\mathbf{x}_0$ along either direction $\mathbf{g}_i$ from $\mathbf{x}_{eq}$, i.e. $\mathbf{x}_0 - \mathbf{x}_{eq} = \alpha \mathbf{g}_i$ for some real α, then $\mathbf{x}(t)$ gives just motion towards or away from $\mathbf{x}_{eq}$ along the direction $\mathbf{g}_i$,

$$\mathbf{x}(t) = \mathbf{x}_{eq} + e^{\underline{\underline{J}}t}\alpha \mathbf{g}_i = \mathbf{x}_{eq} + e^{\gamma_i t}\alpha \mathbf{g}_i \ . \tag{2.25}$$

From this expression, the eigenvalues tell us the exponential rates of attraction (if $\mathbb{R}\mathrm{e}\,[\gamma]_{1,2} < 0$), repulsion (if $\mathbb{R}\mathrm{e}\,[\gamma]_{1,2} > 0$), and rotation (if $\gamma_{1,2} \in \mathbb{C}$), with respect to directions $\mathbf{g}_{1,2}$ about an origin at $\mathbf{x}_{eq}$. (If there is rotation, then $\mathbf{g}_{1,2}$ may also be complex, and then we cannot take $\mathbf{x}_0 - \mathbf{x}_{eq}$ along $\mathbf{g}$, but the interpretation stands). This is linear dynamical systems theory in a nutshell. For more detail, the reader may consult any undergraduate text in dynamics (e.g. [162]) or, better, play around with rewriting the expressions above in different exponential or trigonometric forms and sketching trajectories $\mathbf{x}(t)$. Linear dynamics is entirely solvable by hand, sketchable, and forms the basis of much of the 'equilibrium physics' thinking that dominated much of scientific thought until well into the 20th century.

If a system depends nonlinearly on $\mathbf{x}$, it still behaves like a linear system in certain circumstances. We can expand about a point $\mathbf{x}_0$ as

$$\mathbf{f}(\mathbf{x};\lambda) = \mathbf{f}(\mathbf{x}_0;\lambda) + \underline{\underline{J}}.(\mathbf{x}-\mathbf{x}_0) + \mathsf{O}\left(|\mathbf{x}-\mathbf{x}_0|^2\right) \ .$$

The higher-order term is often neglectable if the Jacobian $\underline{\underline{J}} = \frac{d}{d\mathbf{x}}\mathbf{f}(\mathbf{x}_0;\lambda)$ is non-singular (if the inverse $\underline{\underline{J}}^{-1}$ exists), and locally the system can be 'linearized', meaning it can be approximated by $\dot{\mathbf{x}} \approx \mathbf{a} + \underline{\underline{J}}.\mathbf{x}$, with $\mathbf{a} = -\underline{\underline{J}}.\mathbf{x}_0$ as above. (A more precise statement is given by the *Hartman-Grobman* theorem [98].)

If $\underline{\underline{J}}$ is singular, then the inverse $\underline{\underline{J}}^{-1}$ does not exist. To rescue the analysis above, we must usually include higher-order terms (the $\mathsf{O}\left(|\mathbf{x}-\mathbf{x}_0|^2\right)$ parts we ignored), and the qualitative dynamics of the system can change if the values making up $\underline{\underline{J}}$ vary even slightly. *Local bifurcation* theory is born [137, 176], as different small perturbations of $\dot{\mathbf{x}} = \mathbf{f}(\mathbf{x})$ cause the system to bifurcate into qualitatively distinct dynamical forms. As an example, take the planar problem $(\dot{x}_1, \dot{x}_2) = (x_1^2 - a, -x_2)$, which has equilibria at $(x_1, x_2) = (\pm\sqrt{a}, 0)$ that exist only for $a > 0$ and disappear for $a < 0$. We say a *saddlenode* bifurcation has occurred at $a = 0$, because one equilibrium is a saddle and the other a node; this is verified using the Jacobian, which evaluates to $\underline{\underline{J}} = \begin{pmatrix} \pm\sqrt{a} & 0 \\ 0 & -1 \end{pmatrix}$ at $(x_1, x_2) = (\pm\sqrt{a}, 0)$, with an eigenvector $(0, 1)^{\mathsf{T}}$ associated with eigenvalue -1 (so this direction attracts the flow towards both equilibria) and an eigenvector $(1,0)^\mathsf{T}$ with eigenvalues $\pm 2\sqrt{a}$ (so along this direction the flow is attracted towards the equilibrium at $(-\sqrt{a},0)$ and repelled from the equilibrium at $(+\sqrt{a},0)$).

If $\mathbf{x}(t)$ strays far from $\mathbf{x}_0$ in a nonlinear system, then this kind of local analysis near $\mathbf{x}_0$ is no longer so useful. An orbit may undergo sequences of attraction and repulsion and rotation, towards different points or surfaces and in different directions, mixing the dynamics in space and, as the eigenvalues of the Jacobian vary, warping the flow through different timescales. These flows are rarely solvable in closed form and may exhibit complex oscillations, chaos, or mixed modes. They are the cause of much fascinating dynamics in smooth systems; see, e.g. [92, 135, 199].

For the piecewise-smooth systems of interest in this book, all of the above holds true, until the dynamics encounters a discontinuity surface. With a little ingenuity, the layer system then permits us to extend these standard ideas to the discontinuity.

2.5.2 Tangencies and Their Bifurcations

The oscillator example (2.4) exhibits two tangencies between the flow and the discontinuity surface, shown in Figure 2.3.

Generally in a system given by (2.1), a tangency between the vector field and the discontinuity surface is given by $\delta_{t\pm}\sigma = 0$ on $\sigma = 0$ for one of the vector fields $\mathbf{f}^{\pm}$. The order of the tangency is the number r of successive derivatives of σ that vanishes, so, for example, $0 = \delta_{t+}\sigma = \delta_{t+}^2\sigma = \cdots = \delta_{t+}^r\sigma$ implies that the flow of $\mathbf{x} = \mathbf{f}^+$ has $(r+1)^{th}$ order contact with $\mathcal{D}$.

In two dimensions, under typical circumstances, we expect to be able to find a point x where up to two conditions are met, e.g. $0 = \sigma = \delta_{t+}\sigma$ (but with $\delta_{t-}\sigma \neq 0$ and $\delta_{t+}^2\sigma \neq 0$), so we say *folds*—quadratic (2nd order) contact—are generic in two dimensions. They are either visible or invisible(see Figure 2.1), meaning the flow curves away from or into $\mathcal{D}$, respectively. An example is

$$(\dot{x}_1, \dot{x}_2) = (1+\lambda)(\pm 1, x_1) + (1-\lambda)(0,1)\ , \qquad \lambda = \operatorname{sign}(x_2)\ ,$$

in which the '+' gives a visible fold and the '−' gives an invisible fold.

A third-order contact or *cusp* of $\mathbf{f}^+$ obeys $0 = \sigma = \delta_{t+}\sigma = \delta_{t+}^2\sigma$, with $\delta_{t-}\sigma \neq 0$ and $\delta_{t+}^3\sigma \neq 0$. A cusp point satisfying these conditions will cease to exist under a typical small change in the system. Either the tangency disappears altogether, or it degenerates into a pair of folds at different points, each satisfying $0 = \sigma = \delta_{t+}\sigma$. An example is

$$(\dot{x}_1, \dot{x}_2) = (1+\lambda)(1, x_1^2 - c) + (1-\lambda)(0,1)\ , \qquad \lambda = \operatorname{sign}(x_2)\ ,$$

which has a cusp at $x_1 = 0$ if $c = 0$, as shown in Figure 2.10. For $c > 0$, this degenerates into a pair of folds at $x_1 = \pm\sqrt{c}$; for $c < 0$, there are no tangencies at all. The sliding region is where $x_1^2 - c < 0$, so it exists only for $c > 0$.

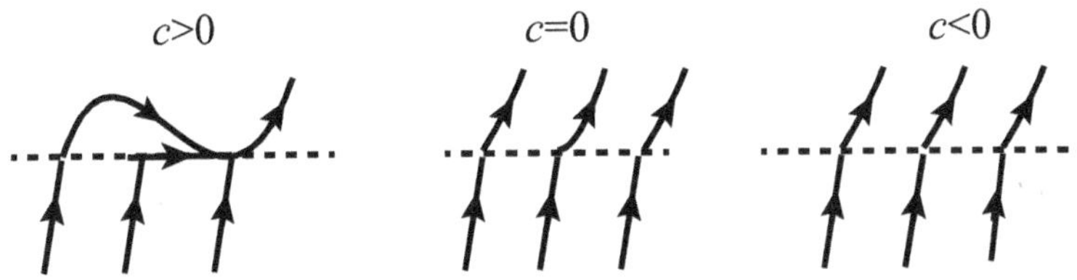

Fig. 2.10 Bifurcation of a cusp tangency.

A fold-fold obeys $0 = \sigma = \delta_{t+}\sigma = \delta_{t-}\sigma$, with $\delta_{t\pm}^2\sigma \neq 0$, but typically a small change in the system will degenerate this into a pair of folds at different points, one satisfying $0 = \sigma = \delta_{t+}\sigma$ and the other $0 = \sigma = \delta_{t-}\sigma$. An example is

$$(\dot{x}_1, \dot{x}_2) = (1+\lambda)(1, x_1 + c) - (1-\lambda)(2, x_1) , \qquad \lambda = \text{sign}(x_2) ,$$

which has a fold-fold at $x_1 = 0$ only if $c = 0$, as shown in Figure 2.11. For $c \neq 0$, this degenerates into a pair of folds at $x_1 = 0$ and $x_1 = c$. In this example one fold is visible and the other invisible. The sliding regions are where $x_1(x_1 + c) > 0$, where there is a sliding flow $(\dot{x}_1, \dot{x}_2) = -(x_1 + 2c, 0)/(2x_1 + c)$, which has a sliding equilibrium at $x_1 = -2c$.

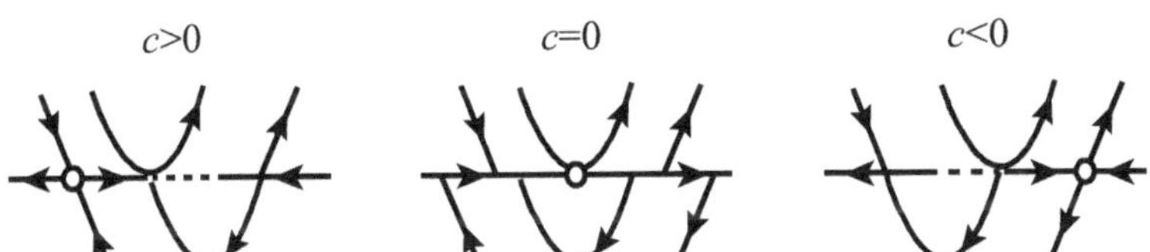

Fig. 2.11 Bifurcation of a fold-fold tangency.

The nongeneric situations of a cusp, a fold-fold, or other higher-order tangency in the plane are called *discontinuity-induced* bifurcations—under perturbation, the system degenerates into topologically different forms. Tangency bifurcations like these often also induce some kind of bifurcation in the sliding flow, where sliding equilibria are created or destroyed or move between different sliding regions. In the plane there are many different cases of one-parameter fold-fold or cusps bifurcations like those above, but we classify them all later in Section 8.6. In higher dimensions, some of the most intriguing and complex problems involve classifying tangencies and their bifurcations, and Chapter 13 is dedicated to just one particularly challenging example.

2.5.3 Equilibria, Sliding Equilibria, and Their Bifurcations

The oscillator example used throughout this chapter has one equilibrium but no sliding equilibria (see Figure 2.3).

Generally in a system given by (2.1), an equilibrium is a point where $\mathbf{f}^+(\mathbf{x}) = 0$ or $\mathbf{f}^-(\mathbf{x}) = 0$, and a sliding equilibrium is a point where $\mathbf{f}(\mathbf{x}; \lambda^\$) = 0$ when $\lambda^\$$ is the solution of $\delta_{t\$}\sigma = \sigma = 0$. When identifying either of these, we must check consistency, i.e. if $\mathbf{f}^+(\mathbf{x}) = 0$, then $\sigma(\mathbf{x}) > 0$; if $\mathbf{f}^-(\mathbf{x}) = 0$, then $\sigma(\mathbf{x}) < 0$; and if $\mathbf{f}(\mathbf{x}; \lambda^\$) = 0$, then $\mathbf{x} \in \mathcal{D}$ and $|\lambda^\$| < 1$.

Under typical circumstances, an equilibrium will have a non-singular Jacobian $d\mathbf{f}^\pm/d\mathbf{x}$ and will lie at a point $\mathbf{x}$ where $\sigma(\mathbf{x}) \neq 0$, hence away from the domain boundaries formed by $\mathcal{D}$. Similarly a sliding equilibrium will typically have a non-vanishing derivative $d\mathbf{f}^\$/d\mathbf{x}|_\mathcal{D}$, where the subscript $\mathcal{D}$ denotes that $d\mathbf{f}^\$/d\mathbf{x}$ is taken only over the components of $\mathbf{x}$ and $\mathbf{f}^\$(\mathbf{x}) \equiv \mathbf{f}(\mathbf{x}; \lambda^\$)$ lying in the tangent space of the discontinuity surface. A sliding equilibrium will typically also lie away from the boundaries of sliding, in particular away from tangencies, hence at a point $\mathbf{x}$ where $\delta_{t\pm}\sigma(\mathbf{x}) \neq 0$. An example of such a sliding equilibrium is the sliding saddle in the fold-fold bifurcation in Figure 2.11 for $c \neq 0$.

If either an equilibrium or sliding equilibrium lies on its domain boundary, this constitutes a degenerate scenario called a *boundary equilibrium bifurcation*. As we perturb the system, we may find that equilibria become sliding equilibria or vice versa or that they appear or disappear in pairs.

Example 2.5 (A Boundary Equilibrium Bifurcation). An example is given by

$$(\dot{x}_1, \dot{x}_2) = (1 + \lambda)(x_2 - c, x_1 + x_2 - c) + (1 - \lambda)(b, 1) \ , \quad \lambda = \mathrm{sign}(x_2) \ ,$$

which has an equilibrium at $(x_1, x_2) = (0, c)$ existing only if $c > 0$. A bifurcation takes place as c changes sign. The sliding region is $x_1 < c$, with a sliding flow $(\dot{x}_1, \dot{x}_2) = (bc - bx_1 - c,\ 0)/(c + 1 - x_1)$, and hence a sliding equilibrium at $x_1 = c - c/b$ existing only if $c/b > 0$. These existence conditions mean a boundary equilibrium bifurcation takes place at $c = 0$, shown in Figure 2.12. If $b > 0$, then a saddle and a sliding node collide and annihilate as c changes sign. If $b < 0$, then a saddle becomes a sliding saddle as c changes sign.
■

Whereas the tangency bifurcations in Section 2.5.2 often involve simultaneous bifurcations of sliding equilibria, boundary equilibrium bifurcations *always* involve accompanying bifurcations of tangencies. In both cases in Figure 2.12, a fold changes from invisible to visible fold as c changes sign.

There are many different cases of one-parameter boundary equilibrium bifurcations in the plane, and we classify them all later in Section 8.4.1. With nonlinear (or multilinear) dependence on $\mathbf{x}$ or λ, there may occur multiple-parameter boundary equilibrium bifurcations, with multiple equilibria and

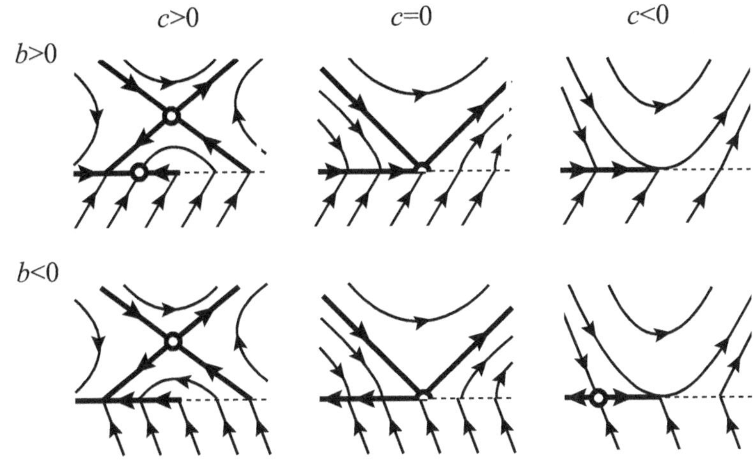

Fig. 2.12 Bifurcations of an equilibrium and sliding equilibrium.

sliding equilibria colliding an annihilating. In higher dimensions, there has been almost no general study of boundary equilibrium bifurcations, and although the forthcoming chapters focus on results applicable in any number of dimensions, this is one challenge that will remain open beyond the scope this book.

2.6 Global Bifurcations and Tangencies

Typical global bifurcations consist of changes in the orbital connections between local singularities, i.e. between equilibria, sliding equilibria, and tangencies. The latter two, of course, come into play only in piecewise-smooth systems and give rise to global discontinuity-induced bifurcations.

Global bifurcations may arise when distinguished orbits (e.g. those already connected to an equilibrium or periodic orbit) have connections to visible or invisible tangencies. In Figure 2.13, the ★ denotes the distinguished orbit. If we move this orbit slightly, then its interaction with the discontinuity surface changes qualitatively, gaining/losing segments of sliding or gaining/losing crossing points on the surface, constituting a bifurcation.

The connection need not be as simple as these and could consist of orbit segments that cross or slide on other discontinuity surfaces. Figure 2.14 shows various connections that give rise to bifurcations. Cases (i)-(iii) of Figure 2.14 show heteroclinic connections between an equilibrium outside the discontinuity surface and a tangency. Case (iv) shows heteroclinic connection between an equilibrium outside the surface and a sliding equilibrium. The

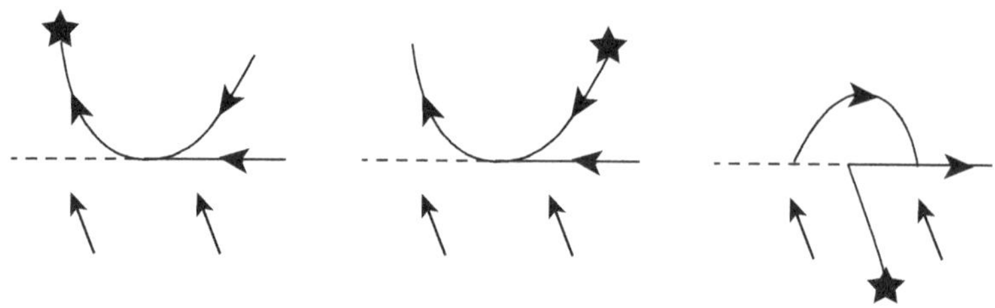

Fig. 2.13 Connections to a visible or invisible tangency lead to bifurcations.

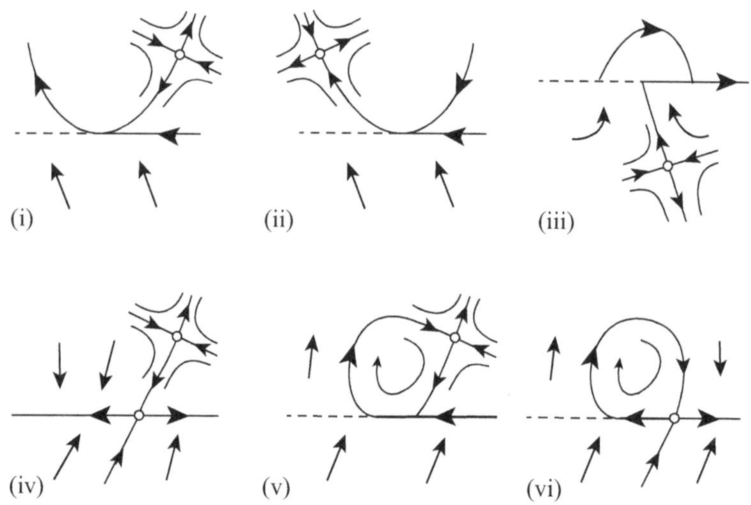

Fig. 2.14 7 A heteroclinic connection between (i–ii) a saddle and a visible tangency, (iii) a saddle and invisible tangency, and (iv) a saddle and sliding saddle. A homoclinic connection via visible tangency to (vi) a saddle or (vii) a sliding saddle.

last two cases (v)-(vi) show two ways that a visible tangency can facilitate a homoclinic connection to an equilibrium or a sliding equilibrium.

Under perturbation of any of these, the connection typically will be broken, and a bifurcation takes place. See [48, 140] for further examples. Many more examples of such connections are possible, particularly when we consider multiple dimensions or multiple switches.

A visible tangency may connect to itself to form a periodic orbit, a situation known as a *grazing-sliding* bifurcation [49, 48], shown in Figure 2.15. In (i), perturbing the system produces a smooth limit cycle on one side of the bifurcation and a 'stick-slip' oscillation on the other side. In (ii) a pair of cycles, one attracting and one repelling, exist on one side of the bifurcation, and no cycles exist on the other side.

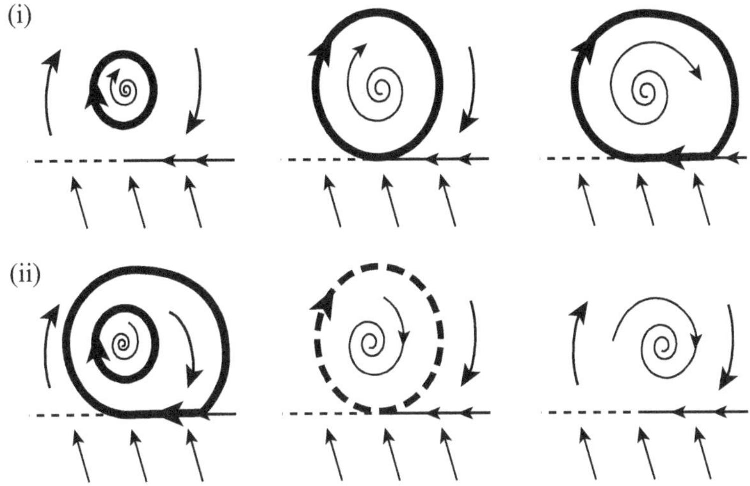

Fig. 2.15 Unfolding of a grazing-sliding bifurcation: (i) persistence of a periodic orbit, (ii) mutual annihilation of an attracting and a repelling periodic orbit.

When repelling sliding is involved, bifurcations can take another, more dramatic form shown in Figure 2.16. We call these sliding *explosions*. In the example shown, a stick-slip cycle exists on one side of the explosion but not on the other. This is not allowed in differentiable systems—an attractor (the cycle) cannot disappear without connecting (and possibly mutually annihilating) with a repeller (another cycle or a saddle equilibrium) in the process. The mechanism of the cycle's destruction lies in the central picture, where repelling sliding causes the cycle's trajectory to explode into an infinity of possible futures at the tangency. (We saw an analogous explosion of trajectories of light in Section 1.4.2). The cycle typically cannot persist through the explosion.

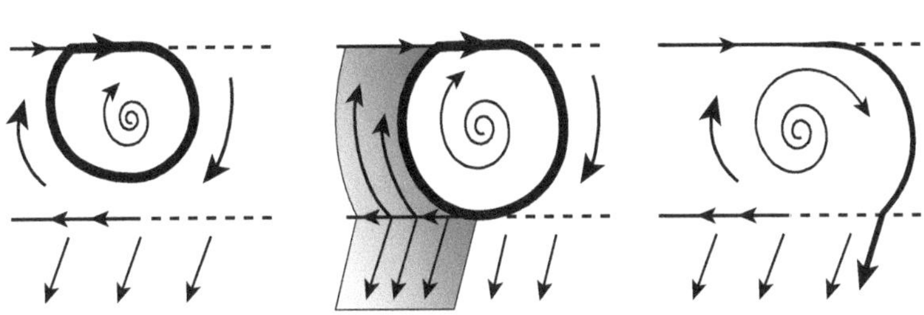

Fig. 2.16 Unfolding of a grazing-sliding explosion.

2.7 Determinacy-Breaking: A First Glimpse

An explosion is one example of *determinacy-breaking*, a general phenomenon that occurs when the flow is somehow able to enter a region of repelling sliding. This happens when a sliding region changes attractivity, so the flow passes from attracting sliding to repelling sliding. This becomes a typical phenomenon if we have more than two dimensions or more than one switch, and it can lead, for example, to novel determinacy-breaking forms of chaos (see Chapter 9).

Because of the presence of determinacy-breaking points like these, piecewise-smooth systems are *almost* deterministic, at best. The techniques of combinations and switching layers that we have explored here resolve the non-uniqueness of discontinuous systems *as far as possible* and no further, avoiding any temptation to impose artificial conditions that might select one of the many possible futures, leaving determinacy-breaking points as an essentially new fundamental phenomenon in dynamical systems theory.

2.8 Counting Limit Cycles

The study of global dynamics above brings us to the problem of oscillatory behaviour, via limit cycles. In classical (smooth) dynamics, the limit cycle is the most fundamental new object that nonlinearity in $\mathbf{x}$ makes possible, a closed oscillatory orbit in space which, like an equilibrium, can be either attracting or repelling along different directions. Their study remains an important topic of interest in piecewise-smooth dynamics. They include stick-slip cycles which contain segments of sliding (as in the top-right of Figure 2.15). They also include cycles which only cross the discontinuity transversally. Both kinds bring their own sources of intrigue.

The problem of counting how many limit cycles can occur in a system was made popular by David Hilbert who, among his many influential contributions to mathematics, posed 23 open problems in mathematics at the Paris conference of the International Congress of Mathematicians in 1900 [100, 101, 102]. The 16th and still unsolved problem was to determine the maximum number of limit cycles possessed by a planar dynamical system of a given polynomial degree. (Actually this is just the second part of a two-part problem, the first concerning the relative positions of algebraic curves.)

It is easy to construct polynomials of sufficient degree that they have arbitrarily many limit cycles. Starting with the extreme case, take a system with infinite degree,

$$\dot{\theta} = 1 \qquad \dot{r} = r\sin(\pi r) \ ,$$

with θ periodic on $[0, 2\pi)$ has infinitely many limit cycles. They are given by $r = k$ for every integer k, since at every $r = k$, we have $\dot{r} = 0$, forming a circular orbit, with odd integers giving attracting cycles and odd integers giving repulsive cycles. These are shown in Figure 2.17 in Cartesian coordinates $(x_1, x_2) = r(\cos\theta, \sin\theta)$, in which the system becomes

$$(\dot{x}_1, \dot{x}_2) = (x_1, x_2) \sin\left(\pi\sqrt{x_1^2 + x_2^2}\right) + (-x_2, x_1) \,. \tag{2.26}$$

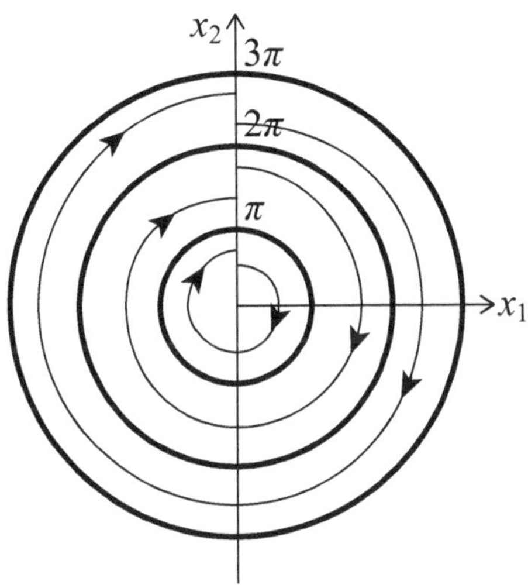

Fig. 2.17 A smooth system with infinitely many limit cycles.

As we lower the degree to polynomial functions, how many limit cycles can there be? For many unsolved problems like these, mathematicians have turned to discontinuous systems for an alternative viewpoint. The nonlinear quadratic function x^2 looks superficially like the piecewise-linear ramp function $|x|$. The circles in Figure 2.17 look superficially like pairs of parabolic arcs stitched along the vertical axis. Such associations are used to create solvable caricatures of nonlinear systems. But as we shall discover throughout this book, such comparisons require great caution.

While a smooth system that is linear in $\mathbf{x}$ cannot contain *any* limit cycles whatsoever, a piecewise-linear system can. The system

$$(\dot{x}_1, \dot{x}_2) = (x_2, -\lambda) \,, \qquad \lambda = \text{sign}\,(x_1) \,,$$

is piecewise-linear, being just $(\dot{x}_1, \dot{x}_2) = (x_2, \mp 1)$ on either side of $x_1 = 0$. Despite the linearity, *every* orbit is a closed cycle. These are not limit cycles, but if we slightly alter the discontinuity surface, the system

$$(\dot{x}_1, \dot{x}_2) = (x_2, -\lambda) \,, \qquad \lambda = \text{sign}\,(x_1 - \alpha\sin(\pi x_2)) \,, \tag{2.27}$$

is still piecewise-linear, being $(\dot{x}_1, \dot{x}_2) = (x_2, \mp 1)$ on either side of a threshold $x_1 = \alpha \sin(\pi x_2)$. This can possess infinitely many limit cycles, despite the piecewise vector fields being linear. The cycles are generated by every point where the discontinuity surface $x_1 = \alpha \sin(\pi x_2)$ intersects the line $x_1 = 0$.

The phase portrait is illustrated in Figure 2.18(left). For $0 < \alpha < 1/\pi$, the system has an infinite number of limit cycles, each with one arc $x_1 = \frac{1}{2}(k^2 - y^2)$ in $x_1 > 0$ and one arc $x_1 = \frac{1}{2}(k^2 + y^2)$ in $x_1 < 0$, where k is an integer. These cycles cross the discontinuity surface $x_1 = 0$ at every point $(0, \pm k)$ and are attractive for odd k and repulsive for even k.

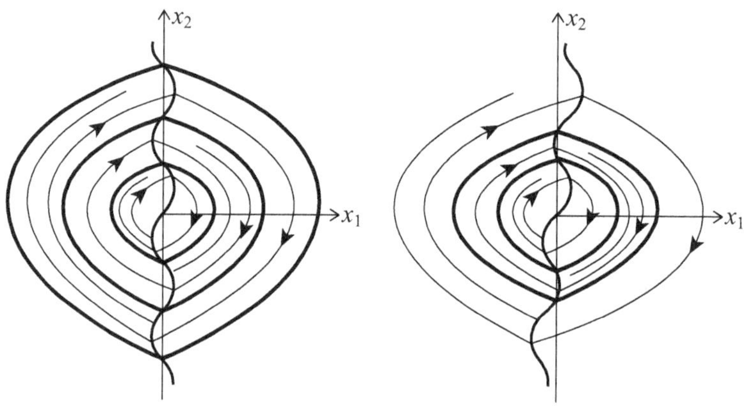

Fig. 2.18 A piecewise-linear system with infinitely many limit cycles (left) or a finite number (right, showing two cycles).

By simply tilting the discontinuity surface, letting

$$\lambda = \text{sign}\left(x_1 - \alpha \sin(\pi x_2) + \beta x_2\right) ,$$

we can obtain any finite number of limit cycles. As we move β away from zero, the discontinuity surface intersects $x_1 = 0$ at a finite number of points, which decreases as $|\beta|$ increases. An example where only two cycles remain is shown in Figure 2.18(right). Many other ways of obtaining any finite number of limit cycles in a piecewise-linear system are also conceivable, taking a lower degree discontinuity surface for example, where σ is a finite order polynomial in $\mathbf{x}$ (rather than a sine function as above). This result seems to have first been described only as recently as [170, 152].

Similar behaviour is possible in continuous but piecewise-linear systems, for example,

$$(\dot{x}_1, \dot{x}_2) = \left(x_2, -x_1 + |x_1 - \alpha \sin(\pi x_2)| /10\right) , \tag{2.28}$$

whose investigation we leave to the reader.

Hilbert's celebrated 16th problem is, it seems, particularly easy to manipulate when applied to nonsmooth dynamics.

The piecewise-linear system in (2.27) is an example of a *fused focus*, a focal point formed by fusing two inward-curving flows. Around a fused focus with a flat discontinuity surface, we can also construct systems with any number k of cycles accumulating on the focal point if the vector fields are polynomial [29], and uncountably many cycles if the vector fields are non-analytic [82].

Even if we demand a flat discontinuity surface with $\mathbf{f}^{\pm}$ being smooth in their respective regions, then nonlinear dependence on λ can cause considerable complexity. Take again the example above with the discontinuity surface $\sigma = x_1$, but add an oscillatory term with a discontinuous frequency,

$$(\dot{x}_1, \dot{x}_2) = (x_2, -\lambda) + (1,0)\sin(\tfrac{1}{2}t(3+\lambda)) \ , \quad \lambda = \mathrm{sign}\,(x_1) \ . \tag{2.29}$$

This is a nonautonomous equation, because time appears on the right-hand side, as is common when oscillatory forces are applied. The frequency of this oscillation changes between 1 and 2 as x_1 changes sign. Figure 2.19(left) shows a simulation, revealing not a simple limit cycle but a complex attractor.

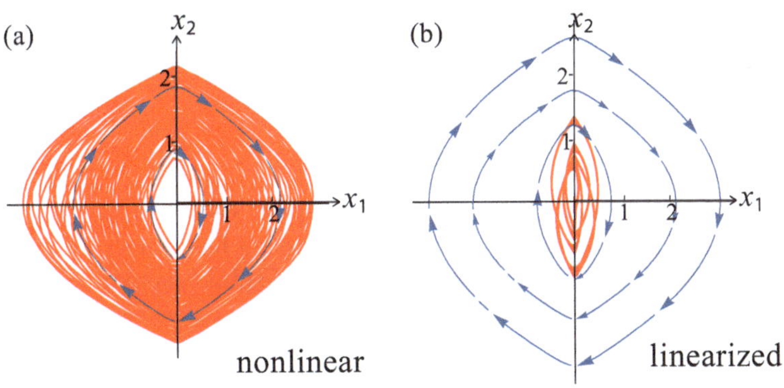

Fig. 2.19 A nonautonomous system with a complex cycle, simulated from (2.29) (left) and (2.30) (right).

Now, the equations

$$(\dot{x}_1, \dot{x}_2) = (x_2, -\lambda) + (\tfrac{1}{2},0)\left\{(1+\lambda)\sin(2t) + (1-\lambda)\sin(t)\right\} \\ \text{where} \quad \lambda = \mathrm{sign}\,(x_1) \ , \tag{2.30}$$

are identical to (2.29) for $x_1 \neq 0$, yet the simulation of them, shown in Figure 2.19(right), is completely different, revealing a simpler, smaller, limit cycle. The system has in some crude sense been 'linearized' with respect to λ (since (2.30) is linear in λ while (2.29) is nonlinear in λ), and this clearly alters the dynamics.

We can calculate the difference between the two systems (2.29) and (2.30), and it is

$$\tfrac{1}{2}(1+\lambda)\sin(2t)+\tfrac{1}{2}(1-\lambda)\sin(t) \quad - \quad \sin(\tfrac{1}{2}t(3+\lambda)) \tag{2.31}$$

which vanishes for all $x_1 \neq 0$, where $\lambda = \pm 1$. This is a *hidden term*, and it can *only* take a nonzero value at the discontinuity, on $x_1 = 0$. This is where λ somehow jumps between the values $+1$ and -1 as x_1 changes sign.

To see how λ jumps, we form the layer systems on $x_1 = 0$ for (2.29) and (2.30), differentiating $\dot{\lambda} = \frac{d\lambda}{dx_1}\dot{x}_1 = \varepsilon^{-1}\dot{x}_1$ (see Section 2.4),

$$\begin{aligned}
(2.29) \;&\mapsto\; (\varepsilon\dot{\lambda},\dot{x}_2) = (x_2,-\lambda)+(1,0)\sin(\tfrac{1}{2}t(3+\lambda))\;,\\
(2.30) \;&\mapsto\; (\varepsilon\dot{\lambda},\dot{x}_2) = (x_2,-\lambda)+(\tfrac{1}{2},0)\left\{(1+\lambda)\sin(2t)+(1-\lambda)\sin(t)\right\}\;,
\end{aligned}$$

for positive infinitesimal ε. It is this dynamics of transition over $\lambda \in (-1,+1)$, hidden inside $x_1 = 0$, that creates the different behaviours in Figure 2.19. It will take some closer inspection to really understand why.

If you attempt to simulate these, note that they are highly sensitive to the solution method. We have a lot to cover before we can sensibly discuss how to simulate such a sensitive discontinuity, but we return this in Section 9.4.3.

2.9 Looking Forward

This has been a brief informal introduction to piecewise-smooth dynamics for a single switch in the plane. A lot of the discussion has concerned equilibria and limit cycles. In smooth systems, the study of equilibria extends to higher dimensions somewhat straightforwardly in terms of eigenvectors and eigenvalues. In nonsmooth systems of higher dimension, the study of equilibria is a hard problem because their interaction with the discontinuity surface may be complex; indeed this problem remains open, but we will make important steps in this book extending the concepts of eigenvalues and eigenvectors to sliding equilibria. The study of limit cycles in higher dimensions is difficult, regardless of whether systems are smooth or nonsmooth, but the key method—taking so-called Poincaré sections through the flow—is in principle the same.

Fortunately, in nonsmooth system the interesting dynamics is not all about equilibria and limit cycles, and not even chaotic attractors. Instead, transient (i.e. short-time) dynamics takes on a far more interesting character.

It is now time for us to formalize these ideas and start putting them to use in more generality.

Chapter 3
The Vector Field: Multipliers and Combinations

Consider a set of ordinary differential equations

$$\frac{dx_1}{dt} = f_1(x_1, x_2, \ldots, x_n)\ , \quad \frac{dx_2}{dt} = f_2(x_1, x_2, \ldots, x_n)\ , \quad \ldots \quad etc.$$

or more concisely

$$\dot{\mathbf{x}} = \mathbf{f}(\mathbf{x})\ , \tag{3.1}$$

collecting the state variables x_i into an n-dimensional vector $\mathbf{x} = (x_1, x_2, \ldots, x_n)$, and the functions f_i into a vector $\mathbf{f} = (f_1, f_2, \ldots, f_n)$, with the derivative with respect to time t denoted by a dot.

The *vector field* $\mathbf{f}$ defines at every point $\mathbf{x}$ the *velocity* with which the variables x_i change. This will be assumed to vary smoothly with $\mathbf{x}$, except at certain thresholds formed by hypersurfaces in the space of $\mathbf{x}$, where the value or derivatives of $\mathbf{f}$ may jump. Our first task is to fully define the dynamical system (3.1) and its solutions under these conditions.

We will begin by expressing discontinuities in $\mathbf{f}(\mathbf{x})$ as discrete switches of certain parameters or multipliers $\boldsymbol{\lambda} = (\lambda_1, \lambda_2, \ldots)$ and write more explicitly

$$\dot{\mathbf{x}} = \mathbf{f}(\mathbf{x}; \boldsymbol{\lambda}) \tag{3.2}$$

where $\mathbf{f}(\mathbf{x}; \boldsymbol{\lambda})$ is partially differentiable in $\mathbf{x}$ and $\boldsymbol{\lambda}$.

3.1 Piecewise-Smooth Vector Fields

A vector field $\mathbf{f}$ is *piecewise smooth* if it is differentiable with respect to $\mathbf{x}$ almost everywhere, except at a hypersurface $\mathcal{D}$ called the **discontinuity surface**. We could consider $\mathbf{f}$ to be $k \geq 1$ times differentiable except at $\mathcal{D}$, but for simplicity we shall generally assume $k \to \infty$.

M. R. Jeffrey, *Hidden Dynamics*,
https://doi.org/10.1007/978-3-030-02107-8_3

Let us assume that the vector field can be expressed as a function

$$\mathbf{f} = \mathbf{f}(\mathbf{x}; \lambda_1, \lambda_2, \ldots, \lambda_m)$$

with continuous dependence on $\mathbf{x}$ and on a set of **switching multipliers** $\lambda_1, \lambda_2, \ldots, \lambda_m$, each of which jumps in value across a hypersurface $\mathcal{D}_1, \mathcal{D}_2, \ldots, \mathcal{D}_m$. We call each $\mathcal{D}_j$ a **discontinuity submanifold**. The discontinuity surface is then the union of the submanifolds $\mathcal{D} = \mathcal{D}_1 \cup \mathcal{D}_2 \cup \cdots \cup \mathcal{D}_m$.

Expressing each submanifold $\mathcal{D}_j$ as the zero contour of some **switching function** $\sigma_j(\mathbf{x})$, the switching multipliers can be chosen without loss of generality to given by $\lambda_j = \text{sign}(\sigma_j(\mathbf{x}))$ for $j = 1, 2, \ldots, m$.

A piecewise-smooth dynamical system with m independent discontinuities can then be written as a general system

$$\dot{\mathbf{x}} = \mathbf{f}(\mathbf{x}; \lambda_1, \lambda_2, \ldots, \lambda_m) \quad : \quad \lambda_j = \text{sign}(\sigma_j(\mathbf{x})) \tag{3.3}$$

for $j = 1, \ldots, m$, with $m \geq 1$, where the sign function is defined as

$$\text{sign}(\sigma) \in \begin{cases} +1 & \text{if } \sigma > 0 \, , \\ (-1, +1) & \text{if } \sigma = 0 \, , \\ -1 & \text{if } \sigma < 0 \, . \end{cases} \tag{3.4}$$

We will use this definition for the sign function throughout and assume each multiplier λ_j to take values in the interval $(-1, +1)$ on its associated discontinuity submanifold $\mathcal{D}_j$.

Note that we do not write a multiplier λ_j as a function of σ_j or $\mathbf{x}$, firstly because λ_j is (piecewise) constant outside the discontinuity surface, and secondly because no such functional dependence holds in general at $\sigma_j = 0$.

Outside the discontinuity surface the vector field is given by

$$\mathbf{f}(\mathbf{x}; \pm 1, \pm 1, \ldots, \pm 1)$$

for some combination of signs. We call these the **constituent vector fields**.

There may be parts of the expression $\mathbf{f}(\mathbf{x}; \lambda_1, \lambda_2, \ldots, \lambda_m)$ that are not present in any of the constituent vector fields $\mathbf{f}(\mathbf{x}; \pm 1, \pm 1, \ldots, \pm 1)$, i.e. they vanish everywhere outside of the discontinuity surface $\mathcal{D}$, and we call these **hidden terms**.

3.2 The Discontinuity Surface

Each discontinuity, where a multiplier λ_j switches sign, is now associated with a submanifold

$$\mathcal{D}_j = \{ \mathbf{x} \in \mathbb{R}^n \; : \; \sigma_j(\mathbf{x}) = 0 \} \tag{3.5}$$

of the discontinuity surface

$$\mathcal{D} = \bigcup_{j=1}^{m} \mathcal{D}_j = \left\{ \mathbf{x} \in \mathbb{R}^n \ : \ \sigma(\mathbf{x}) \equiv \prod_{j=1}^{m} \sigma_j(\mathbf{x}) = 0 \right\} . \tag{3.6}$$

Each σ_j is a regular scalar function, meaning its gradient vector $\delta_{\mathbf{x}}\sigma_j$ exists and is nonvanishing for all $\mathbf{x}$, with $\delta_{\mathbf{x}}$ denoting the gradient operator.

Our consideration will be almost always local, so the global topology of $\mathcal{D}$ is not restricted by this description. We typically assume that the submanifolds $\mathcal{D}_j$ are transversal, meaning the normal vectors $\delta_{\mathbf{x}}\sigma_j$ are linearly independent, so the thresholds $\sigma_j = 0$ may touch each other, but never tangentially; see e.g. Figure 3.1.

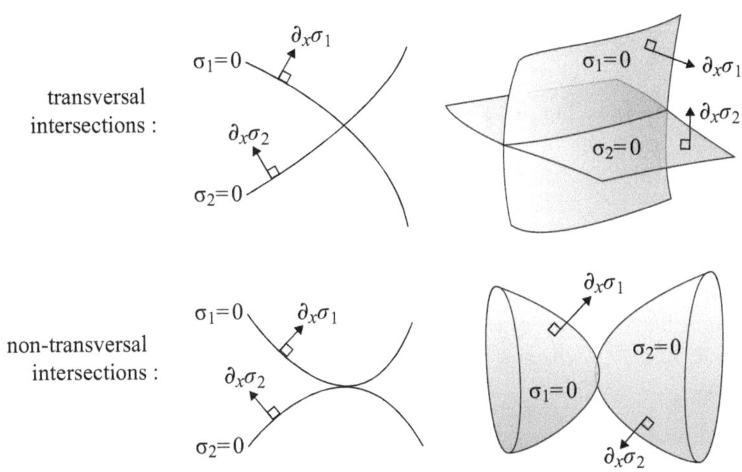

Fig. 3.1 In this book we will only consider transversally intersecting submanifolds. Our basic methods will break down at a non-transversal intersection, typically leaving the dynamics there undetermined, but we have to start somewhere!

3.3 Constituent Fields and Indexing

The discontinuity surface $\mathcal{D}$ sprawls out through phase space like a web, dividing it into open regions, say $\mathcal{R}_1$, $\mathcal{R}_2$, ..., $\mathcal{R}_N$, for some N. A region $\mathcal{R}_i$ could consist of disconnected pieces. On each region $\mathcal{R}_i$, the vector field $\mathbf{f}$ is given by a different functional form $\mathbf{f}(\mathbf{x}; \pm 1, \pm 1, \ldots, \pm 1)$ on each region $\mathcal{R}_1$, $\mathcal{R}_2$, ..., $\mathcal{R}_N$, and it is helpful to give each of these functions a different symbol, say $\mathbf{f}^1(\mathbf{x})$, $\mathbf{f}^2(\mathbf{x})$, ..., $\mathbf{f}^N(\mathbf{x})$, each corresponding to a different combination of $\pm$ signs.

These $\mathbf{f}^i$'s are the constituent vector fields, and each is assumed to be differentiable on the closure of its respective region $\mathcal{R}_i$ (i.e. on $\mathcal{R}_i$ and on its boundary). Wherever you see a vector field with a superscript like $\mathbf{f}^i(\mathbf{x})$ in this book, you can assume it is smooth with respect to $\mathbf{x}$ unless otherwise stated.

We can index the constituent vector fields $\mathbf{f}^i$ and regions $\mathcal{R}_i$ in different ways. As above we may simply number the regions $\mathcal{R}_i$ with integers $i = 1, 2, \ldots, N$, then

$$\dot{\mathbf{x}} = \mathbf{f}(\mathbf{x}) = \left\{\mathbf{f}^i(\mathbf{x}) \text{ if } \mathbf{x} \in \mathcal{R}_i\right\}_{i=1,2,\ldots,N} . \tag{3.7}$$

More convenient (and later more powerful) is to choose a binary indexing, replacing i with a string $K = \kappa_1 \ldots \kappa_m$ of $\pm$ signs, corresponding to the signs of the multipliers $\lambda_1, \ldots, \lambda_m$, such that $\lambda_j = \kappa_j 1$ for $j = 1, \ldots, m$. Then

$$\dot{\mathbf{x}} = \mathbf{f}(\mathbf{x}) = \left\{\mathbf{f}^K(\mathbf{x}) \text{ if } \mathbf{x} \in \mathcal{R}_K\right\}_{K=\kappa_1\kappa_2\ldots\kappa_m} \tag{3.8}$$

where each κ_j is either a '+' or '−' symbol. Thus the jth index of $\mathbf{f}^{\cdots\kappa_j\cdots}$ and $\mathcal{R}_{\ldots\kappa_k\ldots}$ switches between $\pm$ at the surface $\mathcal{D}_j$, and relating back to the combination in (3.3), we have

$$\mathbf{f}^{\kappa_1\kappa_2\ldots\kappa_m}(\mathbf{x}) = \mathbf{f}(\mathbf{x}; \kappa_1 1, \kappa_2 1, \ldots, \kappa_m 1) . \tag{3.9}$$

The binary indexing has the further advantage that we can let $\kappa_j = \$$ denote that $\mathbf{x}$ lies on a discontinuity submanifold $\mathcal{D}_j$. This is where κ_j switches between $\pm$ and $\lambda_j^{\$}$ switches between ± 1, so we write the vector field at such a point as

$$\mathbf{f}^{\kappa_1\ldots\$\ldots\kappa_m}(\mathbf{x}) = \mathbf{f}(\mathbf{x}; \kappa_1 1, \ldots, \lambda_j, \ldots, \kappa_m 1) , \qquad \lambda_j \in (-1, +1) .$$

Thus we can identify regions and the boundaries between them by the index κ_j, multiplier λ_j, threshold function σ_j, or submanifold $\mathcal{D}_j$, according to:

$$\begin{array}{lllllll} \kappa_j = \pm & \Leftrightarrow & \lambda_j = \pm 1 & \Leftrightarrow & \sigma_j \gtrless 0 & \Leftrightarrow & \mathbf{x} \notin \mathcal{D}_j , \\ \kappa_j = \$ & \Leftrightarrow & \lambda_j \in (-1, +1) & \Leftrightarrow & \sigma_j = 0 & \Leftrightarrow & \mathbf{x} \in \mathcal{D}_j , \end{array} \tag{3.10}$$

with $\lambda_j = \text{sign}(\sigma_j)$ as defined in (3.4).

We have chosen multipliers λ_j defined on the interval $[-1, +1]$ but could equally well choose any other interval, for instance, defining multipliers $\nu_j = (\lambda_j + 1)/2$ defined on the interval $[0, 1]$, for which everything we will do carries over directly.

In the binary indexing, the discontinuity surface $\mathcal{D}$ divides the phase space into $N = 2^m$ regions $\mathcal{R}_{\pm\cdots\pm}$. The regions $\mathcal{R}_K$ are then simply connected. More complicated geometries can be taken care of by introducing a little redundancy and by having portions of the discontinuity surface be removable so that $\mathbf{f}$ remains continuous across them. The following example shows examples of how to handle a corner in a discontinuity surface or a doubly-intersecting discontinuity surface, in the binary indexing.

Example 3.1 (Indexing of Phase Space). Consider three scenarios of a discontinuity surface formed by two intersecting submanifolds $\mathcal{D}_1$ and $\mathcal{D}_2$.

- Intersection: Figure 3.2 shows two transversally intersecting submanifolds $\mathcal{D}_1$ and $\mathcal{D}_2$ divide space into four regions, written in integer indexing as $\mathcal{R}_i$ with $i = 1,2,3,4$ or in binary indexing as $\mathcal{R}_{\kappa_1\kappa_2}$ where each $\kappa_j = \pm$ switches across $\mathcal{D}_j$. E.g.

$$(\dot{x}_1, \dot{x}_2) = (\ 2 + \lambda_1,\ 3 + \lambda_1\lambda_2\ )\ , \qquad\qquad \lambda_j = \text{sign}(x_j)\ .$$

The vector field switches between functions given in the two notations as $\mathbf{f}^1 = \mathbf{f}^{--}$, $\mathbf{f}^2 = \mathbf{f}^{-+}$, $\mathbf{f}^3 = \mathbf{f}^{+-}$, $\mathbf{f}^4 = \mathbf{f}^{++}$. We have $N = 2^m$, where $N = 4$ is the number of regions and $m = 2$ is the number of discontinuity submanifolds or switching multipliers.

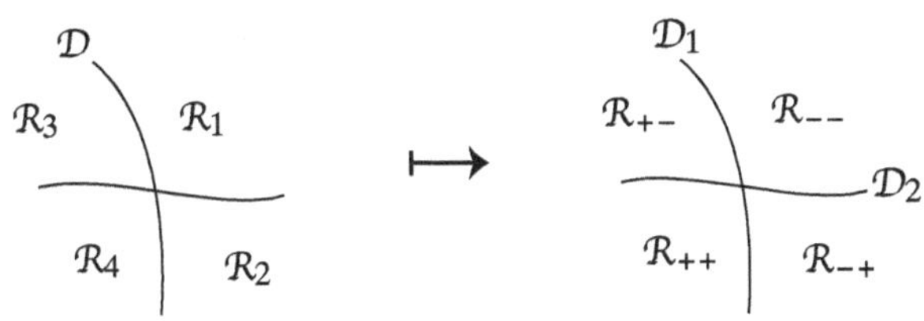

Fig. 3.2 A discontinuity surface $\mathcal{D}$ formed by intersecting discontinuity submanifolds $\mathcal{D}_1$ and $\mathcal{D}_2$. Shown using integer indexing with regions $\mathcal{R}_i$ (left) or binary indexing with regions $\mathcal{R}_K$ (right).

- Corner: Figure 3.3 shows a discontinuity surface with a sharp corner, handled in binary form by taking the previous example, but allowing $\mathcal{D}$ to have removable portions, so that three constituent vector fields are continuations of each other. E.g.

$$(\dot{x}_1, \dot{x}_2) = (\ 2,\ 3 + \lambda_1 + \lambda_2 + \lambda_1\lambda_2\ )\ , \qquad \lambda_j = \text{sign}(x_j)\ .$$

The constituent fields satisfy $\mathbf{f}^{---} = \mathbf{f}^1$ and $\mathbf{f}^{+-}(\mathbf{x}) = \mathbf{f}^{++}(\mathbf{x}) = \mathbf{f}^{-+}(\mathbf{x}) = \mathbf{f}^2(\mathbf{x})$. This gives $N = m = 2$ in integer indexing or $N = 4$ and $m = 2$ in binary indexing.

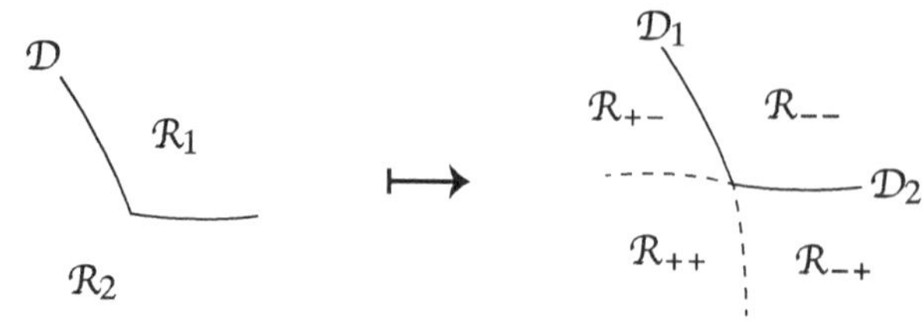

Fig. 3.3 $\mathcal{D}$ has a corner, still formed by two submanifolds $\mathcal{D}_1$ and $\mathcal{D}_2$ but forming only two distinct regions $\mathcal{R}_1 = \mathcal{R}_{--}$ and $\mathcal{R}_2 = \mathcal{R}_{+-} \cup \mathcal{R}_{++} \cup \mathcal{R}_{-+}$.

- Multiple intersections: Figure 3.4 shows a discontinuity surface intersecting itself more than once, formed by $\mathcal{D}_2$ curving back towards $\mathcal{D}_1$, creating 5 regions. A binary indexing requires a third submanifold $\mathcal{D}_3$ across which no actual switch occurs, e.g.:

$$(\dot{x}_1, \dot{x}_2) = (\, 2 + \lambda_2,\ 2 + \lambda_1 + (1 + \lambda_2)(1 + \lambda_1)\lambda_3 \,)\ ,$$
$$\lambda_1 = \mathrm{sign}(-x_1)\ , \quad \lambda_2 = \mathrm{sign}(x_1 + x_2^2 - 1)\ , \quad \lambda_3 = \mathrm{sign}(x_2)\ .$$

  The constituent fields satisfy $\mathbf{f}^{+++}(\mathbf{x}) = \mathbf{f}^1(\mathbf{x})$, $\mathbf{f}^{-++}(\mathbf{x}) = \mathbf{f}^{-+-}(\mathbf{x}) = \mathbf{f}^2(\mathbf{x})$, $\mathbf{f}^{+-+}(\mathbf{x}) = \mathbf{f}^{+--}(\mathbf{x}) = \mathbf{f}^3(\mathbf{x})$, $\mathbf{f}^{--+}(\mathbf{x}) = \mathbf{f}^{---}(\mathbf{x}) = \mathbf{f}^4(\mathbf{x})$, and $\mathbf{f}^{++-}(\mathbf{x}) = \mathbf{f}^5(\mathbf{x})$. We have $N = 5$ and $m = 2$ in integer indexing or $N = 8$ and $m = 3$ in binary indexing.

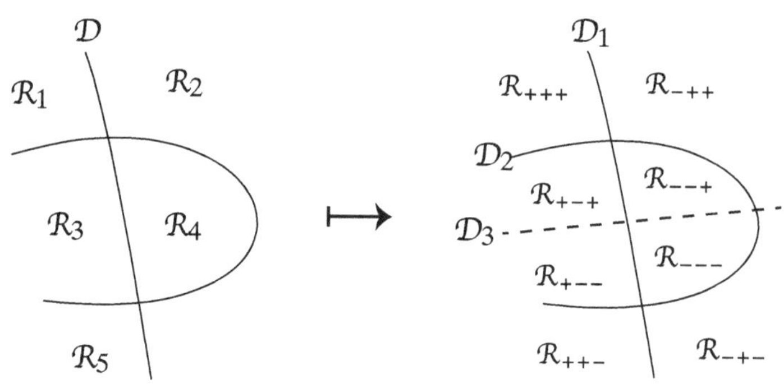

Fig. 3.4 $\mathcal{D}$ intersects itself more than once, so a binary representation requires adding a third 'removable' submanifold, creating eight regions such that $\mathcal{R}_{-++} \cup \mathcal{R}_{-+-} = \mathcal{R}_2$, $\mathcal{R}_{+-+} \cup \mathcal{R}_{+--} = \mathcal{R}_3$, $\mathcal{R}_{--+} \cup \mathcal{R}_{---} = \mathcal{R}_4$, while $\mathcal{R}_{+++} = \mathcal{R}_1$ and $\mathcal{R}_{++-} = \mathcal{R}_5$.

■

Notation for Indices: Throughout the book we will try to use i to label general integer sets, like $i = 1, \ldots, n$, usually for components such as x_i. We reserve j to label discontinuity submanifolds $\mathcal{D}_j$ or quantities associated with them, like σ_j with $j = 1, \ldots, m$. The integer m is reserved for the number of switching multipliers λ_j in a system (and hence the number of discontinuity submanifolds $\mathcal{D}_j$ or switching functions σ_j), and if we consider only a subset of them, we use $r \leq m$, e.g. $\{\mathcal{D}_1, \ldots, \mathcal{D}_r\} \subseteq \{\mathcal{D}_1, \ldots, \mathcal{D}_m\}$. The index κ_j takes values $\pm$, and K is a string $\kappa_1\kappa_2\ldots\kappa_m$; we will use this a lot for the vector fields $\mathbf{f}^K = \mathbf{f}^{\pm\pm\ldots}$.

An alternative for the binary indexing is to use 1 and 0 symbols instead of $+$ and $-$ symbols (just as the multipliers may be defined on $[0, 1]$ instead of $[-1, +1]$). Each is convenient in different contexts (we will make use of this alternative later in Section 12.1).

To solve (3.7) within each region $\mathcal{R}_i$ we have a huge arsenal at our disposal—the theory of smooth dynamical systems—which we will touch on in Chapters 7 and 8. This is sufficient for a piecewise study of $\mathbf{f}$ on the disjoint domains $\mathcal{R}_i$. It does not provide a solvable dynamical system on $\mathcal{D}$.

Having thus defined a framework to define discontinuities in the right-hand sides of ordinary differential equations, piecewise-smooth dynamical theory now faces two tasks, to examine how the right-hand side of (3.8) extends across the discontinuity surface $\mathcal{D}$ and to solve the dynamical equations that result.

The first task is sometimes casually swept aside by assuming the combination $\mathbf{f}(\mathbf{x}; \boldsymbol{\lambda})$ to take the simplest possible form on $\mathcal{D}$, namely, a linear interpolation between the constituent vector fields $\mathbf{f}^K$ (familiar to many as Filippov's convex combination for a simple discontinuity manifold [71]). In this book we will spend longer than is customary on this, instilling a little more generality and flexibility to the expression of the vector field. Only then do we progress to the more expansive second task, solving the differential equations to find their flow, their attractors and their singularities and finally applying those ideas to understanding real-world phenomena.

So to the first step: without any assumption on the form of the function $\mathbf{f}(\mathbf{x}; \boldsymbol{\lambda})$, let us see what λ_j does as it switches between ± 1.

3.4 The Switching Layer

As defined in (3.3)-(3.4), the switching multipliers λ_j vary from -1 to $+1$ across each discontinuity submanifold, so in some sense, they 'blow-up' the *surface* $\sigma_j = 0$ into a switching layer*layer* $\lambda_j \in (-1, +1)$. If we let $\sigma_j = \varepsilon_j \lambda_j$ for some non-negative infinitesimal constant ε_j, then the layer $\lambda_j \in (-1, +1)$ squashes down to the point $\sigma_j = 0$ as $\varepsilon_j \to 0$. Using this we can refine the definition of the multipliers in (3.3)-(3.4) as

$$\lambda_j = \lim_{\varepsilon_j \to 0} \begin{cases} +1 & \text{if } \sigma_j > +\varepsilon_j\,, \\ \sigma_j/\varepsilon_j & \text{if } |\sigma_j| \leq \varepsilon_j\,, \\ -1 & \text{if } \sigma_j < -\varepsilon_j\,. \end{cases} \tag{3.11}$$

Inside the layer (now $|\sigma_j| \leq \varepsilon_j$), if we differentiate with respect to time, we find $\varepsilon_j \dot{\lambda}_j = \dot{\sigma}_j = \mathbf{f} \cdot \delta_{\mathbf{x}} \sigma_j$. Thus we have the dynamics of the multiplier λ_j,

$$\dot{\lambda}_j = \begin{cases} \varepsilon_j^{-1} \mathbf{f}(\mathbf{x}; \lambda_1, \ldots, \lambda_m) \cdot \delta_{\mathbf{x}} \sigma_j(\mathbf{x}) & \text{if } \lambda_j \in (-1, +1) \ , \\ 0 & \text{if } \lambda_j = \pm 1 \ . \end{cases} \tag{3.12}$$

The top row applies when $\mathbf{x} \in \mathcal{D}_j$ for any j; the bottom row applies for $\mathbf{x} \in \mathcal{R}_K$ outside.

A more direct way to think of this is that by naively applying the chain rule, we have $\dot{\lambda}_j = \frac{d\lambda_j}{d\sigma_j} \dot{\sigma}_j$. For $\sigma_j \neq 0$, we know λ_j is constant, so $d\lambda_j / d\sigma_j$ vanishes. At $\sigma_j = 0$, however, λ_j is discontinuous with respect to σ_j, so the derivative $d\lambda_j / d\sigma_j$ is infinite, and if we represent that derivative as $\frac{d\lambda_j}{d\sigma_j} := 1/\varepsilon_j$ for infinitesimal ε_j, we obtain (3.12).

We will assume throughout this book that the quantities ε_j are constants, but the most general version of the theory would let each ε_j be a function of λ_j, bringing more complication without anything qualitatively new beyond the theory developed here; however such a treatment may become relevant in situations where λ_j and ε_j can be associated with particular physical quantities.

Take a point vx where $r \leq m$ discontinuity submanifolds intersect transversally, say where $\sigma_j(\mathbf{x}) = 0$ for $j \in \mathcal{S} \subseteq \{1, \ldots, m\}$ and $\sigma_i(\mathbf{x}) \neq 0$ for $i \notin \mathcal{S}$, $i \in \{1, \ldots, m\}$. Putting (3.12) together with (3.3) on $\sigma_j = 0$ for $j \in \mathcal{S}$, we have the **layer system**

$$\begin{cases} \varepsilon_j \dot{\lambda}_j = \mathbf{f}(\mathbf{x}; \boldsymbol{\lambda}) \cdot \delta_{\mathbf{x}} \sigma_j(\mathbf{x}) & \text{for } j \in \mathcal{S} \subseteq \{1, \ldots, m\} \ , \\ \dot{x}_i = \mathbf{f}(\mathbf{x}; \boldsymbol{\lambda}) \cdot \delta_{\mathbf{x}} x_i & \text{for } i \notin \mathcal{S}, \ i \in \{1, \ldots, n\} \ , \end{cases} \tag{3.13}$$

where each $\varepsilon_j \to 0$. This defines dynamics on the **switching layer**

$$\left\{ (y_1, \ldots, y_n) \ : \ \begin{array}{c} y_j = \lambda_j \in (-1, +1) \text{ for } j \in \mathcal{S} \ , \text{ and} \\ y_i = x_i \in \mathbb{R} \text{ for } i \notin \mathcal{S}, \ i \in \{1, \ldots, n\} \end{array} \right\} . \tag{3.14}$$

In the switching layer, the values of $\sigma_j(\mathbf{x}) = 0$ for $j \in \mathcal{S}$ and multipliers $\lambda_i = \pm 1$ for $i \notin \mathcal{S}$, $i \in \{1, \ldots, m\}$ are fixed.

This is easiest to express in coordinates lying orthogonal to the discontinuity submanifold. Assuming linear independence between the gradient vectors $\delta_{\mathbf{x}} \sigma_1, \delta_{\mathbf{x}} \sigma_2, \ldots, \delta_{\mathbf{x}} \sigma_m$, choose local coordinates in which the first m components of $\mathbf{x}$ are given by $x_j = \mathbf{x} \cdot \delta_{\mathbf{x}} \sigma_j$. The layer system is then

$$\begin{cases} (\varepsilon_1 \dot{\lambda}_1, \ldots, \varepsilon_r \dot{\lambda}_r) = (\ f_1(\mathbf{x}; \boldsymbol{\lambda}) \ , \ldots, f_r(\mathbf{x}; \boldsymbol{\lambda}) \ ) \ , \\ (\dot{x}_{r+1}, \ \ldots, \ \dot{x}_n) = (\ f_{r+1}(\mathbf{x}; \boldsymbol{\lambda}), \ldots, f_n(\mathbf{x}; \boldsymbol{\lambda}) \ ) \ , \end{cases} \tag{3.15}$$

where $\mathbf{f} = (f_1, f_2, \ldots, f_n)$. We are considering (3.15) on the switching surface, where $\mathbf{x} = (0, \ldots, 0, x_{r+1}, \ldots, x_n)$, over the switching layer now given by

$$(y_1, \ldots, y_n) = (\lambda_1, \ldots, \lambda_r, x_{r+1}, \ldots, x_n) \in (-1, +1)^r \times \mathbb{R}^{n-r} . \tag{3.16}$$

The components of $\mathbf{f}$ on the right-hand side of (3.15) are, more explicitly,

$$\begin{aligned}
&\mathbf{f}(\varepsilon_1\lambda_1, \ldots, \varepsilon_m\lambda_m, x_{m+1}, \ldots, x_n; \lambda_1, \ldots, \lambda_m) \\
&\quad = \mathbf{f}(0, \ldots, 0, x_{m+1}, \ldots, x_n; \lambda_1, \ldots, \lambda_m) + \mathsf{O}\left(\varepsilon_1, \ldots, \varepsilon_m\right) ,
\end{aligned}$$

in which we can neglect the higher-order term since all $\varepsilon_j \to 0$.

Although all of the quantities ε_j are infinitesimal, their relative values will in some cases be important.

3.5 Inclusions and Existence of a Flow

Above we introduced the *combination* $\mathbf{f}(\mathbf{x}; \boldsymbol{\lambda})$. When we let each λ_j vary over $[-1, +1]$, we obtain an *inclusion*. Let us recap by constructing them for a given set of constituent vector fields $\mathbf{f}^K$, taking (3.8) as a starting point:

Combination: Taking (3.8), associate each index $\kappa_j = \pm$ with a parameter $\lambda_j = \pm 1$, so the string $K = \kappa_1\kappa_2\ldots\kappa_m$ becomes a vector $\boldsymbol{\lambda} = (\lambda_1, \lambda_2, \ldots, \lambda_m)$ where $\lambda_j = \kappa_j 1$. Then (3.8) becomes

$$\begin{aligned}
&\dot{\mathbf{x}} = \mathbf{f}\left(\mathbf{x}; \boldsymbol{\lambda}\right) \qquad \text{where} \\
&\{\boldsymbol{\lambda} = (\kappa_1 1, \ldots, \kappa_m 1) \;\text{ if }\; \mathbf{x} \in \mathcal{R}_K\}_{K=\kappa_1\kappa_2\ldots\kappa_m} \, .
\end{aligned} \tag{3.17}$$

We fix $\lambda_j \in [-1, +1]$ and call the λ_j's *switching multipliers*. The vector field $\mathbf{f}(\mathbf{x}; \boldsymbol{\lambda})$ is called the *combination* of the constituent vector fields $\mathbf{f}^K$, which themselves are now given by

$$\mathbf{f}^{\kappa_1\ldots\kappa_m}(\mathbf{x}) = \mathbf{f}(\mathbf{x}; \lambda_1, \ldots, \lambda_m) \qquad \text{where} \quad \lambda_j = \kappa_j 1 \, .$$

Inclusion: Taking (3.8), define a continuous (simply connected) set $\mathcal{F}(\mathbf{x})$ that contains every value of $\mathbf{f}^K$ found in an infinitesimal neighbourhood of $\mathbf{x} \in \mathcal{D}$, written as

$$\dot{\mathbf{x}} \in \mathcal{F}(\mathbf{x}) = \{\mathbf{f}(\mathbf{x}'; \boldsymbol{\lambda}) \;\;\forall\;\; \mathbf{x}' \in B_\delta(\mathbf{x})\} \tag{3.18}$$

where $B_\delta(\mathbf{x})$ is an open infinitesimal δ-neighbourhood of $\mathbf{x}$.

The set $\mathcal{F}$ and the set formed by a given combination, $\mathbf{f}(\mathbf{x}; \lambda_1, \ldots, \lambda_m)$ with $\lambda_j \in [-1, +1]$, $j = 1, \ldots, m$, are now both continuous families of vector fields that contain the values of the constituent vector fields $\mathbf{f}^{\pm\pm\cdots}$. We can let the form of a given combination $\mathbf{f}(\mathbf{x}; \boldsymbol{\lambda})$ define the set $\mathcal{F}$, or vice versa, defining a function $\mathbf{f}(\mathbf{x}; \boldsymbol{\lambda})$ that will explore any values in a given set $\mathcal{F}(\mathbf{x})$ (this is possible because for any $\mathbf{v} \in \mathcal{F}$, there exists a combination $\mathbf{f}(\mathbf{x}; \boldsymbol{\lambda})$ such that $\mathbf{v} = \mathbf{f}(\mathbf{x}; \boldsymbol{\lambda})$).

The right-hand sides of the differential equations in (3.17) or (3.18) have continuous (and typically differentiable) dependence on $\mathbf{x}$ and also on the vector of switching multipliers $\boldsymbol{\lambda}$ in (3.17). They are single-valued in each $\mathcal{R}_K$ and set-valued on $\mathcal{D}$. Filippov and contemporaries used this general framework to powerful effect [71]. In particular they proved the existence of flow solutions to (3.18) and showed:

- Solutions of the problem $\dot{\mathbf{x}} = \mathcal{F}(\mathbf{x})$ (or $\dot{\mathbf{x}} = \mathcal{F}(\mathbf{x}, t)$ with time dependence) exist if $\mathcal{F}(\mathbf{x})$ is non-empty, bounded, closed, convex, and upper semicontinuous ($\sup_{b\in\mathcal{F}(x)} \rho(b, a) \to 0$ as $p' \to p$ for $a \in \mathcal{F}(p')$ and $b \in \mathcal{F}(p)$ where $\rho(b, a) = \inf_{a\in A, b\in B} |a - b|$ where $|a - b|$ is the distance between a and b). [see Filippov Thm 1 p.77].
- If $\mathcal{F}(\mathbf{x})$ is a convex set, then the limit of a uniformly convergent sequence of solutions of $\dot{\mathbf{x}} = \mathcal{F}(\mathbf{x})$ is also a solution of $\dot{\mathbf{x}} = \mathcal{F}(\mathbf{x})$.
- If $\mathcal{F}(\mathbf{x})$ is not a convex set, then the limit of a uniformly convergent sequence of solutions of $\dot{\mathbf{x}} = \mathcal{F}(\mathbf{x})$ is a solution of $\dot{\mathbf{x}} = \operatorname{co}\mathcal{F}(\mathbf{x})$, where $\operatorname{co}\mathcal{F}$ is the smallest convex set containing $\mathcal{F}$.

Seidman [183] defined a more general convex set $\mathcal{F}_\varepsilon$ that in one sweep can contain any vectors in (3.17) as well as (3.18) (Filippov's definition excludes (3.17) in general by demanding a convex set of the $\mathbf{f}^K$s) while at the same time loosening assumptions on the ideal specification of $\mathcal{D}$ and even extending existence to infinite dimensional systems for application to partial differential equations. Filippov's and Seidman's theorems guarantee the existence of a flow, and the switching layer in Section 3.4 tells us how to find it.

In n dimensions, the set $\mathcal{F}$ has up to n dimensions (where $\mathbf{x} \in \mathbb{R}^n$). The set

$$\{\mathbf{f}(\mathbf{x}; \lambda_1, \ldots, \lambda_m) : \lambda_j \in [-1, +1],\ j = 1, \ldots, m\}$$

formed directly by the combination has up to m dimensions, tracing out an m-dimensional set as m independent multipliers λ_j vary, so by using the layer system (3.13), we reduce the size of the space in which the dynamics across the discontinuity is to be studied.

This reflects the fact that, while the inclusion facilitates the proof of existence of solutions provided by Filippov [71], most of the set $\mathcal{F}$ making up the inclusion is redundant. It is comprised largely of vector fields that cannot yield continuous solution trajectories $\mathbf{x}(t)$ satisfying $\dot{\mathbf{x}}(t) \in \mathcal{F}$, so dynamically it makes no sense to consider them.

An example is shown in Figure 3.5. Outside $\mathcal{D}$ the vector field is uniquely defined at every point in space, taking functional forms $\mathbf{f}^+$ or $\mathbf{f}^-$ either side of a discontinuity surface $\mathcal{D}$. At a point $p \in \mathcal{D}$, the vector field consists of any vector in $\mathcal{F}$, represented as any vector drawn from p to an end point in the shaded region (which by definition includes $\mathbf{f}^\pm(p)$). Clearly only two scenarios are dynamically possible at p: orbits cross from $\mathcal{R}_-$ to $\mathcal{R}_+$ or remain on $\mathcal{D}$ (as the arrangement of vector fields prevents crossing in the other direction from $\mathcal{R}_+$ to $\mathcal{R}_-$). Therefore we may cut $\mathcal{F}$ into three pieces: $\mathcal{F}_A$ provides vectors that carry the flow across $\mathcal{D}$; $\mathcal{F}_B$ consists of vectors directed from $\mathcal{R}_+$ to $\mathcal{R}_-$, but these cannot be followed by the flow, and $\mathcal{F}_B$ is therefore redundant; and $\mathcal{F}_C$ consists of vectors that carry the flow along $\mathcal{D}$. Ultimately only the fact that $\mathcal{F}_A$ and/or $\mathcal{F}_C$ are non-empty are of any interest. The vectors in $\mathcal{F}_A$ apply only for an instant in carrying the flow across $\mathcal{D}$, so their length and direction are redundant, but they permit crossing of $\mathcal{D}$. Vectors in $\mathcal{F}_C$ will give orbits that travel along the discontinuity, a behaviour we call *sliding* along $\mathcal{D}$, and their size determines the velocity of this sliding.

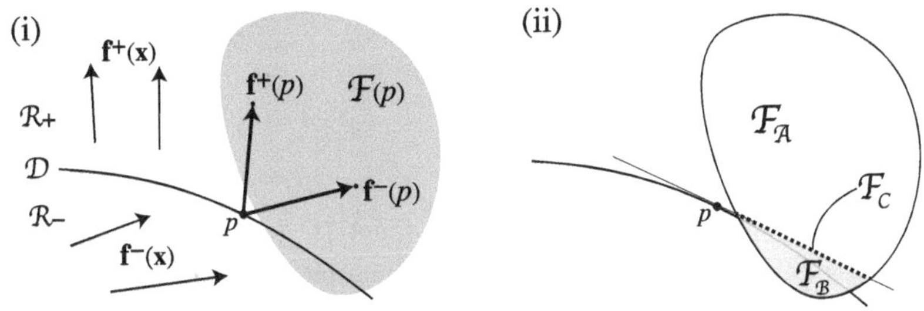

Fig. 3.5 A piecewise-defined differential equation (3.7), represented as a vector field in $\mathcal{R}_\pm$ with a discontinuity surface $\mathcal{D}$. In (i) we show a differential inclusion (3.18), where the shaded set represents the end points of the vectors in some $\mathcal{F}$ drawn from p. In (ii) this is cut into a redundant set $\mathcal{F}_\mathcal{A}$, a non-viable set $\mathcal{F}_\mathcal{B}$, and a sliding set $\mathcal{F}_\mathcal{C}$.

The combination in terms of explicit multipliers (3.17) removes some of the redundancy of $\mathcal{F}$. The set-valuedness of the vector field at the discontinuity is given a specific form, encoded in the functional dependence on the multipliers $\boldsymbol{\lambda} = (\lambda_1, \lambda_2, \ldots, \lambda_m)$ such that $\lambda_j \in [-1, +1]$. This limits the dimensionality of the inclusion, sufficiently, it turns out, to give a solvable system.

Continuing on from Figure 3.5, we show in Figure 3.6 how the combination restricts the set of motion permitted by the flow to a λ-parameterized set of vector fields, represented by a λ-parameterized curve connecting the end points of $\mathbf{f}^+$ and $\mathbf{f}^-$ drawn from p (an example is shown dotted). The only

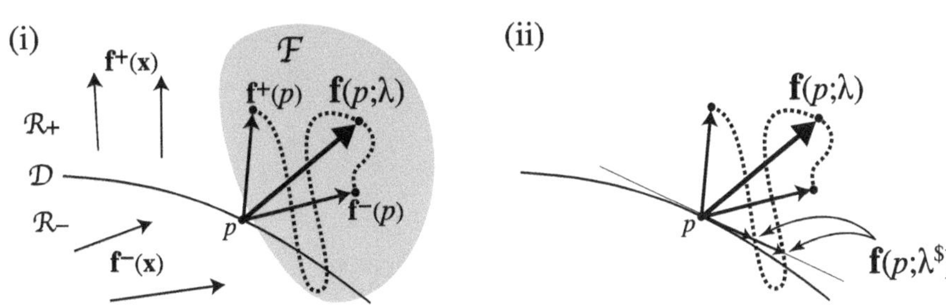

Fig. 3.6 Examples of a combination (3.17) that in (i) explores the differential inclusion via dependence on λ. As shown in (ii), the combination has isolated intersections with the tangent space to $\mathcal{D}$ at p, permitting motion along $\mathcal{D}$ following these vectors $\mathbf{f}(p;\lambda^\$)$.

vector fields in this set that give reasonable onward motion from p are then either $\mathbf{f}^+$, so that solutions cross $\mathcal{D}$, or the two vectors $\mathbf{f}(\mathbf{x}; \lambda^\$)$ that lie in the tangent space of $\mathcal{D}$ and give sliding motion along the discontinuity surface.

Until recently, the standard analytical approaches to discontinuities involved a choice between inclusions [71], complementarity constraints (e.g. [25]), and hybrids of flows with maps or rules to enact the discontinuity (e.g. [48]). Filippov's inclusions provide the most all-encompassing theory, but to fully prescribe dynamics, it is necessary to know how a system depends on switching quantities—the switching *multipliers*.

3.6 Looking Forward

The various preparations are over, the vector field has been extended across the discontinuity, and we turn now to the dynamics that results, the trajectories followed by solutions evolving through discontinuities.

Trajectories are rather different beasts at a discontinuity to those in smooth systems. The next chapter begins with a basic picture of the forms they may take. We shall follow a dynamical systems methodology, taking a differential system is defined by (3.3) and (3.15) and seeking the qualitative properties of a flow that may not be expressible in a closed form but is amenable to such things as stability analysis, numerical simulation, and various forms of perturbation.

An unfortunate consequence of discontinuity is that, without many of the sweeping statements permitted by differentiability, extending standard notions to nonsmooth systems easily degenerates into an endless enumeration of scenarios. We shall avoid such accounting where possible, instead focussing on general methodologies and on behaviours not found in smooth dynamics.

Chapter 4
The Flow: Types of Solution

This chapter sets out the basic theory of flows around a discontinuity surface that underlies the qualitative features of piecewise-smooth dynamics.

The presence of a discontinuity in the right-hand side of a differential equation $\dot{\mathbf{x}} = \mathbf{f}(\mathbf{x})$, which we express in terms of a switching multiplier $\boldsymbol{\lambda}$ as

$$\dot{\mathbf{x}} = \mathbf{f}(\mathbf{x}; \boldsymbol{\lambda}) \ ,$$

need not necessarily be an obstruction to finding a solution

$$\mathbf{x}(t) = \mathbf{x}(t_0) + \int_{t_0}^{t} d\tau \ \mathbf{f}\left(\mathbf{x}(\tau); \boldsymbol{\lambda}\right) \ .$$

The form of those solutions may, however, be drastically different to those found when $\mathbf{f}$ is differentiable. We set out some preliminaries here on the basic forms that solution trajectories can take.

4.1 Indexing

Take as a starting point the expression (3.3),

$$\dot{\mathbf{x}} = \mathbf{f}(\mathbf{x}; \boldsymbol{\lambda}) \ , \qquad \lambda_j = \text{sign}(\sigma_j(\mathbf{x})) \ ,$$

in terms of the switching multipliers $\boldsymbol{\lambda} = (\lambda_1, \lambda_2, \ldots, \lambda_m)$, with the sign function defined in (3.4).

M. R. Jeffrey, *Hidden Dynamics*,
https://doi.org/10.1007/978-3-030-02107-8_4

We will use the binary indexing introduced in (3.8), so the piecewise-smooth vector field $\mathbf{f}(\mathbf{x};\boldsymbol{\lambda})$ reduces on different regions of space to one of a family of smooth vector fields $\mathbf{f}^K(\mathbf{x})$, where $K = \kappa_1 \ldots \kappa_m$ is a string of symbols $\kappa_j = +, -,$ or \$. As given in (3.10), the symbol κ_j takes a $\pm$ sign corresponding to the sign of λ_j if $\sigma_j \neq 0$; otherwise $\kappa_j = \$$ denotes that $\mathbf{x}$ lies on the jth discontinuity submanifold $\mathcal{D}_j$, where $\sigma_j = 0$.

4.2 Types of Trajectory

From a starting condition or *initial point* $\mathbf{x} = \mathbf{x}_0$, the system (3.3) will evolve along a certain *trajectory* $\Gamma_{\mathbf{x}_0}$, which represents the set of points that evolve into or out of $\mathbf{x}_0$. Points along the trajectory $\Gamma_{\mathbf{x}_0}$ are given by

$$\mathbf{x}(t) = \Phi_t(\mathbf{x}_0) , \tag{4.1}$$

where the *flow* Φ_t is therefore a solution of

$$\frac{d}{dt}\Phi_t(\mathbf{x}_0) = \mathbf{f}\left(\Phi_t(\mathbf{x}_0);\boldsymbol{\lambda}\right) , \qquad \mathbf{x}_0 = \Phi_0(\mathbf{x}_0)) . \tag{4.2}$$

The set of all states $\mathbf{x}$ is the *phase space*, and the set of all trajectories $\{\Gamma_{\mathbf{x}}\}$ in this space is the *phase portrait.*

If $\mathbf{f}$ varies smoothly with the variable $\mathbf{x}$, then these concepts are standard and well understood. If $\mathbf{f}$ is nonsmooth, then there are subtleties to these definitions that will concern us through much of this book.

The basic kinds of trajectories in a nonsmooth system are as follows.

Definition 4.1. Regular or sliding trajectories:

(i) A **regular** trajectory

$$\mathbf{x} = \Phi_t^K(\mathbf{x}_0) \;:\; \mathbf{x} \in \mathcal{R}_K \;\text{ for }\; 0 < t < t_1 \tag{4.3}$$

is any solution segment in a region $\mathcal{R}_K$, lying therefore outside the discontinuity surface. It will sometimes be emphasized by calling it a non-sliding trajectory segment. Here $\sigma_j \neq 0$ and $\lambda_j = \kappa_j 1 = \text{sign}(\sigma_j)$ for all $j \in \{1,\ldots,m\}$, and (3.3) reduces to $\dot{\mathbf{x}} = \mathbf{f}^K(\mathbf{x})$.

(ii) A **sliding** trajectory

$$\mathbf{x} = \Phi_t^K(\mathbf{x}_0) \;:\; \mathbf{x} \in \mathcal{D} \;\text{ for }\; 0 < t < t_1 \tag{4.4}$$

where one index in $K = \kappa_1 \dots \kappa_m$ is a \$ symbol is a solution segment of (3.3) that lies on the discontinuity surface for some nonzero interval of time. Here $\sigma_j = 0$ for at least one $j \in \{1, \dots, m\}$ and $\lambda_j \in (-1, +1)$ for the corresponding multiplier.

(iii) A **codimension** r **sliding** trajectory is a solution segment

$$\mathbf{x} = \Phi_t^K(\mathbf{x}_0) \; : \; \mathbf{x} \in \mathcal{D} \;\; \text{for} \;\; 0 < t < t_1 \tag{4.5}$$

where exactly r indices in $K = \kappa_1 \dots \kappa_m$ are \$ symbols, which lies on a region of the discontinuity surface that takes the form of r transversally intersecting submanifolds $\bigcap_{j \in \mathcal{S}} \mathcal{D}_j$, for some r-dimensional set $\mathcal{S} \subseteq \{1, \dots, m\}$, then $\sigma_j = 0$ and $\lambda_j \in (-1, +1)$ for $j \in \mathcal{S}$, while $\sigma_j \neq 0$ with $\lambda_j = \text{sign}(\sigma_j)$ for $j \notin \mathcal{S}$.

Codimension $r = 1$ sliding may also be called *simple sliding.*

We have not yet discussed how to derive the sliding flow other than that it solves (3.3) and lies in $\mathcal{D}$. For now let us merely assume that such a flow does exist and gives motion along the discontinuity surface.

We can concatenate smooth regular segments

$$\left\{ \mathbf{x} = \Phi_t^K(\mathbf{x}_a) \in \mathcal{R}_K : \; 0 < t < t_a \right\}$$

and smooth sliding segments

$$\left\{ \mathbf{x} = \Phi_t^K(\mathbf{x}_b) \in \mathcal{D} : \; 0 < t < t_b \right\}$$

of different codimensions, preserving the direction of time, forming a continuous piecewise-smooth solution trajectory such as

$$\mathbf{x} = \dots \Phi_{t_3}^{j_3}\left(\Phi_{t_2}^{j_2}\left(\Phi_{t_1}^{j_1}(\mathbf{x}_1) \right) \right) \, .$$

If we label the concatenation points iteratively $\mathbf{x}_a = \Phi_{t_a}^{j_a}(\mathbf{x}_{a-1})$, then each lies on the discontinuity surface, $\mathbf{x}_1, \mathbf{x}_2, \mathbf{x}_3, \dots \in \mathcal{D}$.

Figure 4.1 shows an example of a piecewise-smooth trajectory formed by such concatenations. From an initial point $\mathbf{x}_0$, a solution $\mathbf{x}(t)$ evolves through a regular trajectory $\Phi_t^{--}(\mathbf{x}_0) \subset \mathcal{R}_{--}$, a regular trajectory $\Phi_t^{-+}(\mathbf{x}_1) \subset \mathcal{R}_{-+}$, a codimension one sliding trajectory $\Phi_t^{\$+}(\mathbf{x}_2) \subset \mathcal{D}_1$, a codimension one sliding trajectory $\Phi_t^{+\$}(\mathbf{x}_3) \subset \mathcal{D}_2$, and a codimension two sliding trajectory $\Phi_t^{\$\$}(\mathbf{x}_4) \subset \mathcal{D}_1 \cap \mathcal{D}_2$. The concatenation points consist of crossing the discontinuity surface at $\mathbf{x}_1 \in \mathcal{D}_2$, entering the surface at $\mathbf{x}_2 \in \mathcal{D}_1$, crossing between submanifolds via an intersection at $\mathbf{x}_3 \in \mathcal{D}_1 \cap \mathcal{D}_2$, and entering the intersection at $\mathbf{x}_4 \in \mathcal{D}_1 \cap \mathcal{D}_2$.

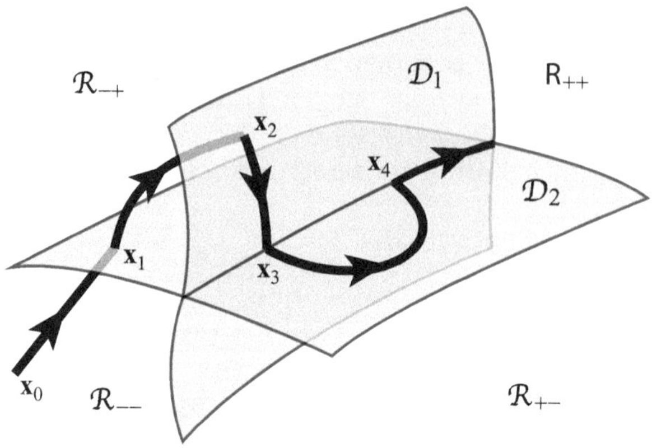

Fig. 4.1 Concatenating smooth segments to form a piecewise-smooth trajectory. The solution shown evolves through point $\mathbf{x}_0$ to $\mathbf{x}_1 = \Phi_{t_1}^{--}(\mathbf{x}_0)$, to $\mathbf{x}_2 = \Phi_{t_2}^{-+}(\mathbf{x}_1)$, to $\mathbf{x}_3 = \Phi_{t_3}^{\$+}(\mathbf{x}_2)$, to $\mathbf{x}_4 = \Phi_{t_4}^{+\$}(\mathbf{x}_3)$, to $\mathbf{x}(t) = \Phi_t^{\$\$}(\mathbf{x}_4)$, over times $t_0 < t_1 < t_2 < t_3 < t_4 \leq t$.

We then add to Definition 4.1:

Definition 4.2. An **orbit** is a concatenation of smooth segment trajectories, that is, a concatenation of segments

$$\mathbf{x} = \Phi_{t_N}^{j_N}\left(\dots \Phi_{t_2}^{j_2}\left(\Phi_{t_1}^{j_1}(\mathbf{x}_1)\right)\right) ,$$

where the duration $t_N + \cdots + t_2 + t_1$ of the orbit is the greatest that can be formed by concatenations preserving time direction.

An orbit may involve repetition of segments, where either the entire orbit repeats or only part of the orbit repeats, the latter being possible only in nonsmooth systems.

Example 4.1 (Non-unique Flow Concatenations). Figure 4.2 illustrates how just a few trajectory segments can be concatenated in different ways to create many different orbits in a piecewise-smooth system. An example exhibiting such orbits is

$$(\dot{x}_1, \dot{x}_2) = (1 - \lambda x_2, \lambda x_1) , \qquad \lambda = \text{sign}(x_2) ,$$

from which Figure 4.2 illustrates example trajectory segments, such as the arcs

$$\Phi_t^{\pm}(\mathbf{x}_b) = \left(2\cos(t - \tfrac{\pi}{6}) , \pm 1 - 2\sin(t - \tfrac{\pi}{6}) \right) ,$$

over times $0 \leq t \leq \frac{4}{3}\pi$ with $x_2 \neq 0$, and the line intervals given by $\Phi_t^{\$}(\mathbf{x}_a) = (t - \sqrt{3} , 0)$ for $0 \leq t \leq 2\sqrt{3}$ or $-\infty < t \leq 0$, and by $\Phi_t^{\$}(\mathbf{x}_b) = (t + \sqrt{3} , 0)$ for $0 \leq t < \infty$.

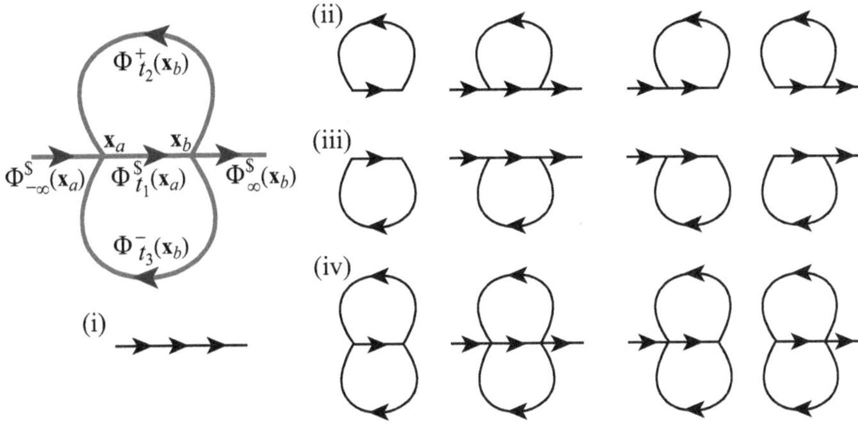

Fig. 4.2 Complex orbits formed by concatenation. The large picture shows trajectory segments through the points $\mathbf{x}_a$ and $\mathbf{x}_b$, including sliding segments $\Phi^\$_t$ and non-sliding segments $\Phi^\pm_t$. Portraits (i-iv) show the orbits that can be formed from these segments, representing infinitely many different scenarios in which loops formed by the upper and lower arcs $\Phi^+_{t_2}(\mathbf{x}_b)$ and $\Phi_{t_3}(\mathbf{x}_a)$ and the line segment $\Phi^\$_{t_1}(\mathbf{x}_b)$ may be bookended by the half-lines $\Phi_{-\infty}(\mathbf{x}_a)$ and/or $\Phi_\infty(\mathbf{x}_b)$.

■

Consider an initial point $\mathbf{x}_p \in R$ on some finite open region R. The flow through $\mathbf{x}_p$ takes one of the following forms:

1. A *stationary* point

$$\mathbf{x} = \Phi_t(\mathbf{x}_p) \ : \ \mathbf{x} = \mathbf{x}_p \quad \text{for all } t \tag{4.6}$$

   which occurs at an equilibrium $\mathbf{f}(\mathbf{x}_p;\boldsymbol{\lambda}) = 0$.
2. A *closed* orbit without self-intersections, e.g. a periodic orbit

$$\mathbf{x} = \Phi_t(\mathbf{x}_p) = \Phi_{t+T}(\mathbf{x}_p) \quad \text{for all } \ t \tag{4.7}$$

   which repeats with period $T > 0$.
3. An *open* orbit, whose locus is an open curve without self-intersections, for example:

   - a *divergent* orbit which satisfies

$$\mathbf{x} = \Phi_t(\mathbf{x}_p) \ : \ \ \mathbf{x}_p \in R\,, \ \ \Phi_t(\mathbf{x}_p) \notin R \tag{4.8}$$

     for any $t > T$ and/or $t < T$;
   - a *connection* which satisfies

$$\mathbf{x} = \Phi_t(\mathbf{x}_p) \ : \ \mathbf{x} \in R\,, \ \ \Phi_t(\mathbf{x}_p) \to \mathbf{x}_q \tag{4.9}$$

as t tends to plus or minus infinity, hence $\mathbf{x}$ asymptotes towards a stationary point $\mathbf{x}_q$ at one or both ends without reaching it (e.g. a heteroclinic or homoclinic connection);

- a *chaotic* orbit which satisfies

$$\mathbf{x} = \Phi_t(\mathbf{x}_p) \;:\; \mathbf{x} \in R \quad \text{for all } t \tag{4.10}$$

but is neither stationary nor periodic nor a connection.

4. A trajectory that has a stationary point or periodic orbit as a subset for some $t_1 \leq t \leq t_2$ or a divergent trajectory as a subset for some $[t_1 \leq t < +\infty)$ or $[-\infty < t \leq t_1)$.

The fourth class allows flows to spend intervals of time inside stationary points or periodic orbits and still have divergent tails. Such chimeric behaviour along orbits is made possible by piecewise-smooth flows.

For forward time trajectories, we also make the following distinction.

- A *nonoverlapping* orbit shares no trajectory segments with any other orbit.
- An *overlapping* orbit shares one or more trajectory segments with at least one other orbit, but the orbits are not identical (i.e. not *all* segments are shared between multiple orbits).

In a region R where the flow is non-overlapping, much of the theory of smooth dynamical systems theory can be extended, at least in principle, and the concatenation of deterministic flows remains deterministic. In a flow with overlapping orbits, the history or future of a point where two orbits overlap will typically be non-unique, and it is here that nonsmooth dynamical takes on a very different flavour.

This defines orbits and their different forms, but let us ask a slightly different question, the answer to which will provide building blocks that will be applied throughout this book. Take a typical point in phase space. What will the trajectory through that point look like?

Label the point as $\mathbf{x}_p$, and consider the trajectory through it, Γ_p, in the vicinity of $\mathbf{x}_p$. Then Γ_p can only take one of the following forms. The first three are familiar from smooth systems; the rest require discontinuities. The cases, illustrated in Figure 4.3, are:

(*a*) a smooth unbounded open curve;
(*b*) a closed curve or a point;
(*c*) a smooth curve that asymptotes (approaching as $t \to \infty$) towards a closed curve or a point;
(*d*) any of (a-c) but with kinks along the curve;
(*e*) a curve that contains a closed curve as a subset or that terminates at a point (in finite time);
(*f*) a continuum of curves that intersect at a point;
(*g*) a continuum of curves that intersect along a continuous set.

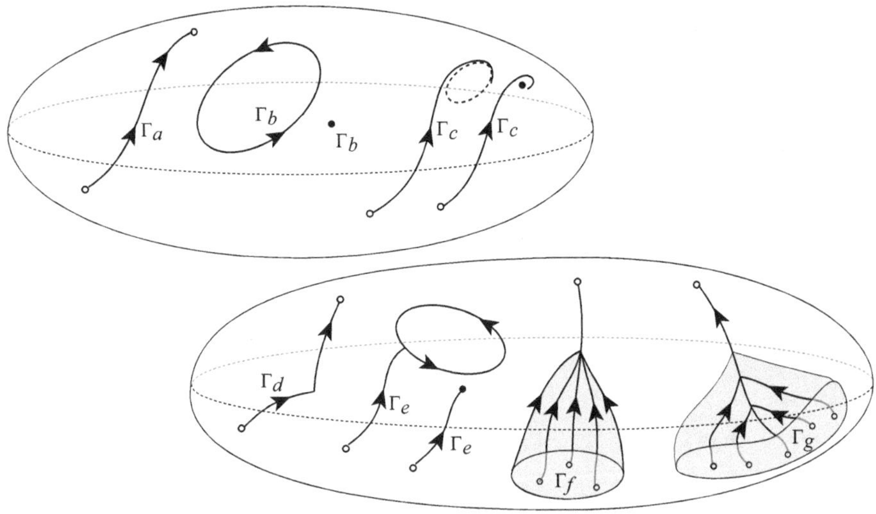

Fig. 4.3 Deterministic trajectories: smooth (top) and nonsmooth (bottom). The trajectory Γ is a curve in cases a-d, a union of two curves in e (the second curve is just a point in the second e case), and is a two-or-more-dimensional set in f-g. The trajectories labelled Γ_c are periodic orbits and stationary points; the trajectories Γ_b asymptote towards those of type Γ_c, while the trajectories Γ_e contain those of type Γ_c and enter them in finite time. The set-valued trajectories Γ_f and Γ_g may take dimension $1 < d \leq n$. The symbol $\circ$ indicates a trajectory leaving the local region illustrated.

The sets in (f) and (g) may be of any dimension $1 < d \leq n$ where n is the dimension of the space.

Full *orbits* are just sequences of such structures concatenated in a way that preserves the direction (arrow) of time. We may refer to trajectories of these types as (a) a simple or chaotic orbit, (b) a stationary point or periodic orbit, (c) an asymptotic orbit, (d) a nonsmooth orbit, (e) a collapsing orbit, (f) a funnel, and (g) sliding.

It is possible to prove these classifications, but the proof is merely a lengthy accounting of all possible changes in dimension that are possible via discontinuity surfaces. The classification above is largely as given by Filippov in [71], where it is proven that these are the only possibilities. It is more illuminating to study the mechanisms for those changes in greater detail and to discover their dynamical consequences. This we do in the forthcoming chapters, with plenty of examples. We present the types above as a preparation for what is to come, rather than claiming formally that the classification is complete.

4.3 Determinacy Breaking Events

One of the most important properties of smooth flows is the existence and uniqueness of the flow through any initial point. This is the dynamical system expression of determinism that allows the prediction of behaviour from an initial state. Smooth systems are deterministic, and while they may sometimes be sensitive to small changes in initial state, leading to chaos [199] or mixed mode oscillations [26], in principle, at least their behaviours are determined by their differential equations.

For smooth systems we have, for instance, the following.

Let there exist some family of flows $\Phi_t(\mathbf{x}_0)$ that are all valid solutions $\mathbf{x} = \Phi_t(\mathbf{x}_0)$ of the differentiable system $\dot{\mathbf{x}} = \mathbf{f}(\mathbf{x})$. Assume each member $\Gamma^i(\mathbf{x}_0)$ of the set of flows

$$\Sigma = \left\{\Gamma^1(\mathbf{x}_0), \ldots, \Gamma^k(\mathbf{x}_0)\right\} ,$$

where

$$\Gamma^a(\mathbf{x}_0) = \{\Phi_t^a(\mathbf{x}_0) \ : \ t_0 \leq t \leq t_1 \} , \tag{4.11}$$

is distinct, that is, $\nexists t : \ \Phi_t^a(\mathbf{x}_0) = \Phi_t^b(\mathbf{x}_0) \ \ \forall\, t \in [0, t_1]$ for any pair of orbits $\Gamma^a(\mathbf{x}_0)$ and $\Gamma^b(\mathbf{x}_0)$ in Σ.

Lemma 4.1. *If $\mathbf{f}(\mathbf{x})$ is differentiable in a neighbourhood of $\mathbf{x}_0$, then Σ consists of only one orbit $\Gamma^i(\mathbf{x}_0)$. The flow then exists and is unique in a neighbourhood of $\mathbf{x}_0$.*

A weaker result concerns limit sets. Let each of the orbits in the set

$$\Sigma_\infty = \left\{\Gamma_\infty^1(\mathbf{x}_1), \ldots, \Gamma_\infty^k(\mathbf{x}_k)\right\} ,$$

where

$$\Gamma_\infty^a(\mathbf{x}_a) = \{\Phi_t^a(\mathbf{x}_a) \ : \ 0 \leq t < \infty\} \tag{4.12}$$

have a common limit point $\lim_{t\to\infty} \Phi_t^a(\mathbf{x}_a) = \mathbf{x}_\infty$ on the closure $\overline{\Gamma}_\infty$ of Γ_∞.

Lemma 4.2. *If $\mathbf{f}(\mathbf{x})$ is differentiable in a neighbourhood of $\mathbf{x}_\infty$, and Σ_∞ consists of more than one trajectory $\Phi_t^a(\mathbf{x}_a)$, then Σ_∞ consists of an infinite number of distinct trajectories $\Gamma_\infty^a(\mathbf{x}_a)$, and $\mathbf{x}_\infty$ is an equilibrium, where $\mathbf{f}(\mathbf{x}_\infty) = 0$.*

These results concern point $\mathbf{x}_p$ where the vector field $\mathbf{f}$ is differentiable, and as discussed in [71], they can be strengthened to apply to weakly nonsmooth systems, namely, when $\mathbf{f}(\mathbf{x}; \boldsymbol{\lambda})$ is continuous and its derivative has a jump which is bounded by so-called Lipschitz continuity. They also therefore apply to a piecewise-smooth system $\dot{\mathbf{x}} = \mathbf{f}(\mathbf{x}; \boldsymbol{\lambda})$ restricted to a domain that excludes any discontinuities of $\mathbf{f}$.

In more general piecewise-smooth flows, the uniqueness of the flow through any point is no longer guaranteed. At best we can expect them to be *almost* deterministic—deterministic everywhere except at certain singularities where a unique trajectory segment connects to a family of trajectory segments in forward time, usually in a specific form (e.g. with certain dimensionality and time-of-flight properties and even certain bifurcations). We shall see specific phenomena later in the book; for now we simply observe the topology, based on the types of trajectories discussed so far, through which non-uniqueness arises.

Consider $\mathbf{f}(\mathbf{x};\boldsymbol{\lambda})$ to be discontinuous such that the value of $\mathbf{f}$ jumps at some point $\mathbf{x}_0$. This restricts our main interest to points $\mathbf{x}_0 \in \mathcal{D}$ on the discontinuity surface. Non-uniqueness can arise in a number of ways either forward or backward time, as accounted for in cases *e-g* above.

Reversing time in any of the cases *e-g* in Figure 4.3 (by reversing the arrows) results in a set-valued forward time flow, as shown in Figure 4.4. In some cases one trajectory is able to evolve into many; hence determinacy has been broken. We shall see that such scenarios are generic and arise so easily that they may be much more common in physical systems than previously known. When we reverse the time direction in the single-valued orbit cases *a-c* in Figure 4.3, or the nonsmooth case *d*, we obtain identical figures, only with the limit sets in *c* changed from attractive to repulsive.

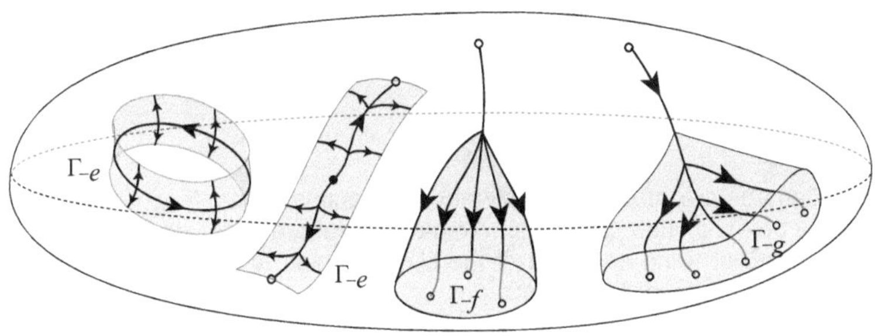

Fig. 4.4 Nondeterministic trajectories. When we reverse time in each of the cases above, we obtain set-valued forward time dynamics in cases e-f, whose orbits we denote $\Gamma_{-e}, \Gamma_{-g}, \Gamma_{-g}$. The trajectories emanating (in finite time) from the cycle or the fixed point in Γ_{-e} may take dimension $1 < d < n$ or $1 \leq d < n-1$, respectively, while Γ_{-f} and Γ_{-g} may take dimension $1 < d \leq n$.

Note that the loss of uniqueness in *e-g* does not involve stationary points; instead the trajectories involved are open. In smooth systems an open trajectory is fairly uninteresting. In nonsmooth systems open trajectories can be rather more complex, as they can form the transition between entirely distinct forms of behaviour.

If we join together the phase portraits from Figure 4.3 with those from Figure 4.4, we obtain a particular orbit of interest, called a *canard.*

Definition 4.3. A (sliding) **canard** is an orbit that evolves *from* an attracting sliding region *into* a repelling sliding region, directly via a coincidence of their boundaries.

Canards may be either *isolated*, meaning the trajectory inside the sliding region is unique, or they may be *robust*, when a continuous family of sliding trajectories are canards. Regions of robust canards may fill the sliding region, or they may fill only part of the sliding region. Some typical forms are shown in Figure 4.5. Canards will play a role later on in Section 10.3, Section 11.5.3, and Chapter 13.

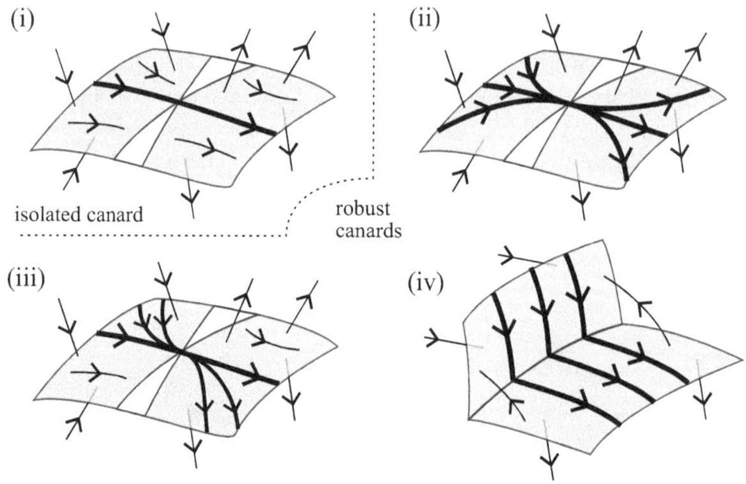

Fig. 4.5 Canards are orbits that evolve from attracting to repelling regions of sliding. Canards may be (i) isolated, (ii) robust with multiple canards passing through a common point, (iii) robust but filling only part of the sliding region, and (iv) robust with multiple canards passing through distinct points.

By reversing time we obtain **faux canards**, which evolve from repelling regions into attracting regions, and are therefore less dynamically exciting, but nonetheless useful for characterizing flows that connect repelling and attracting sliding regions.

The terminology is borrowed from the study of multiple timescale systems (see, e.g. [124, 135]). A canard describes an orbit that has the counterintuitive property of evolving *from* a region that is strongly attracting *into* a region that is strongly repelling, doing so directly without any intermediary dynamics. The term also alludes to the transitory character of canards, that they typically arise only for a tiny range of parameters or initial conditions, and that in their vicinity the flow is fast changing. (In early work on multiple timescales, the term 'canard' was partly motivated by a duck-like shape formed by canard-induced limit cycles in van der Pol's system [20], but in modern terminology (e.g. [219]), a canard is better defined locally as connecting attracting and repelling branches of invariant manifolds, rather than as a global object like a limit cycle.)

4.4 Equivalence and Stability

An analysis of a system's behaviour is not much use if that behaviour changes significantly when we slightly alter the model, so it is imperative to know under what changes the qualitative behaviour will survive. The idea of a system being robust to perturbations is vital to mathematical modelling. This leads us to ask whether two systems with the same qualitative behaviour are essentially the same system in different guises.

A family of systems with the same qualitative behaviours is called an *equivalence class*, represented by a particular member (ideally the simplest) called the *normal* form. A system is *structurally stable* to changes that move it around within that class.

It was formalizing these notions that largely allowed nonlinear dynamics to reach maturity both as a mathematical theory and as a modelling methodology (the history tracing the efforts of Poincaré, Birkhoff, Andronov, Pontrjagin, Smale, and others can be found in, e.g. [92, 177]).

Even now, these notions are not free from conceptual difficulty in the younger areas of dynamics that involve multiple timescales, emergent phenomena, or piecewise-smooth functions. Stability and equivalence should not be too casually invoked to give the appearance of rigour when their underlying assumptions and definitions are incompletely understood. We give below the best expression of them for piecewise-smooth systems based on current knowledge, and we will make some use of them, but encourage future authors to further probe the bounds of these definitions and the classes they lead to.

There are three notions of equivalence between systems that will be useful. We take $\dot{\mathbf{x}} = \mathbf{f}(\mathbf{x};\boldsymbol{\lambda})$ to denote a piecewise-smooth system as defined in (3.3). The expression as a combination $\dot{\mathbf{x}} = \mathbf{f}(\mathbf{x};\boldsymbol{\lambda})$ enables us to form definitions similar to those in smooth systems and more closely than other approaches in [48, 71, 90]. The definitions below are adapted from [92] in particular.

Definition 4.4. Orbital, differentiable, and topological equivalence:

(i) **Orbital equivalence**: If two vector fields $\mathbf{f}$ and $\mathbf{g}$ are related by $\mathbf{f}(\mathbf{x};\boldsymbol{\lambda}) = \mu(\mathbf{x};\boldsymbol{\lambda})\mathbf{g}(\mathbf{x};\boldsymbol{\lambda})$ for some continuous positive-definite scalar function $\mu(\mathbf{x};\boldsymbol{\lambda})$, the orbits of the systems $\dot{\mathbf{x}} = \mathbf{f}(\mathbf{x};\boldsymbol{\lambda})$ and $\dot{\mathbf{x}} = \mathbf{g}(\mathbf{x};\boldsymbol{\lambda})$ are identical up to a time rescaling.

(ii) $q-$**Conjugacy**: If two vector fields $\mathbf{f}$ and $\mathbf{g}$ are related by a q-times differentiable mapping $\mathbf{p}$ which takes orbits of $\dot{\mathbf{x}} = \mathbf{f}(\mathbf{x};\boldsymbol{\lambda})$ to those of $\dot{\mathbf{x}} = \mathbf{g}(\mathbf{x};\boldsymbol{\lambda})$, preserving direction but not necessarily scaling of time, then the vector fields $\mathbf{f}(\mathbf{x};\boldsymbol{\lambda})$ and $\mathbf{g}(\mathbf{x};\boldsymbol{\lambda})$ are said to be q-conjugate (or q-differentiably equivalent, sometimes called C^q equivalence).

(iii) **Topological equivalence**: If $q = 0$ in (ii) and the discontinuity surface is preserved, the vector fields $\mathbf{f}(\mathbf{x};\boldsymbol{\lambda})$ and $\mathbf{g}(\mathbf{x};\boldsymbol{\lambda})$ are said to be topologically equivalent. That is, topological equivalence between the vector fields $\mathbf{f}(\mathbf{x};\boldsymbol{\lambda})$ and $\mathbf{g}(\mathbf{x};\boldsymbol{\lambda})$ means that $\mathbf{f}$ and $\mathbf{g}$ are related by a continuous mapping $\mathbf{p}$ which takes orbits of $\dot{\mathbf{x}} = \mathbf{f}(\mathbf{x};\boldsymbol{\lambda})$ to those of $\dot{\mathbf{x}} = \mathbf{g}(\mathbf{x};\boldsymbol{\lambda})$, preserving direction but not necessarily scaling of time, and maps the discontinuity surface of one system to that of the other preserving orientation with respect to orbits.

A q-conjugacy with $q > 1$ preserves the ratios of eigenvalues associated with any equilibrium (see sect. 1.7 [92]), and if q is infinite, then the eigenvalues of the two conjugate systems are equal.

At the other end of the scale, we have $q = 0$ topological equivalence, which preserves the spatial and temporal topology of orbits themselves, but does not preserve eigenvalues. This means importantly that while the attractivity of points is preserved, the focal or nodal character of attraction, for example, may not be. So, for example, an attractive focus and an attractive node are topologically equivalent, since all orbits tend to the fixed point in infinite time, as illustrated in Figure 4.6(i).

For piecewise-smooth systems, it is necessary to include the discontinuity surface explicitly in the definition of topological equivalence, and this is achieved via the multiplier $\boldsymbol{\lambda}$. Without this, a system where orbits cross a discontinuity surface transversally would be equivalent to a smooth system with no discontinuity surface at all. This has consequences for equilibria too, for instance, while an attracting focus and attracting node are topologically equivalent, Figure 4.6(i), an attractive focus and an attractive node at a discontinuity surface are not topologically equivalent, as shown in Figure 4.6(ii). This is because while around a boundary focus all orbits hit the discontinuity surface, some orbits (shaded in (ii)) around a boundary node tend to the equilibrium without hitting the discontinuity surface.

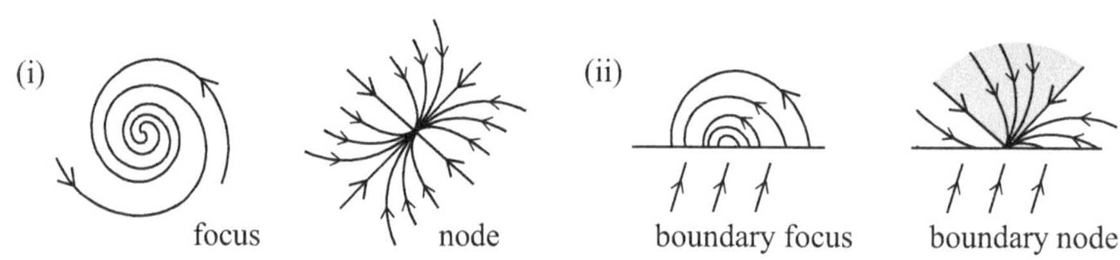

Fig. 4.6 Topological equivalence of nodes and foci. (i) A focus and a node are topologically equivalent: all trajectories tend to the fixed point. (ii) A focus and node at a discontinuity surface are not topologically equivalent.

Teixeira (see, e.g. [90]) calls this $\mathcal{D}$-equivalence[1], while the authors of [48] call it *piecewise* topological equivalence, both reasonable terminologies emphasizing that it respects the discontinuity surface, but the latter suggests an equivalence that applies on piecewise domains (which is something much weaker and would not, e.g. preserve dynamics *on* the discontinuity surface). We find that it is sufficient to characterize these simply as topological

[1] Actually M. A. Teixeira calls this Σ-equivalence, but in this book I have chosen to call the discontinuity surface $\mathcal{D}$ rather than Σ.

equivalence, recognizing that in piecewise-smooth systems, the topology of the discontinuity surface as well as the orbits is always essential.

There is a stronger (and in many ways more important) notion in dynamics, of systems that are not only equivalent but are *close*. This means that they are well approximated by similar functional expressions and exhibit equivalent behaviours. These two properties are not the same, since equivalent systems (according to the definitions above) may have very different functional expressions and two systems that appear similar in their functional expressions may in fact not be equivalent.

A system is considered robust in its behaviour—or **structurally stable**—if small changes in its expression produce equivalent systems. Intuitively, an ε-perturbation involves the addition of a term proportional to some small positive ε, which vanishes on some set K_c, and such that the rate of change of the perturbed and unperturbed systems differs by less than ε.

Definition 4.5. If $\mathbf{f} \in \mathbb{R}^n$ and $\sigma \in \mathbb{R}$ are r times differentiable (vector- and scalar-valued, respectively) functions, in which the level set $\sigma = 0$ is the discontinuity surface $\mathcal{D}$ of $\mathbf{f}$, then for some $0 < q \leq r$ and $\varepsilon > 0$, the functions $\tilde{\mathbf{f}}$ and $\tilde{\sigma}$ are a perturbation of f and σ of size ε, of differentiability class q; if there is a compact set $K \subset \mathbb{R}^{n+1}$ such that $\mathbf{f} = \tilde{\mathbf{f}}$ and $\sigma = \tilde{\sigma}$ on the complement set $K_c = \mathbb{R}^{n+1} - K$ and for all $i_1, i_2, \ldots, i_n$, with $i = i_1 + i_2 + \ldots. + i_n \leq q$, we have $\left|\left(\partial^i/\partial x_1^{i_1} \ldots \partial x_n^{i_n}\right)(\mathbf{f} - \tilde{\mathbf{f}})\right| < \varepsilon$ and $\left|\left(\partial^i/\partial x_1^{i_1} \ldots \partial x_n^{i_n}\right)(\sigma - \tilde{\sigma})\right| < \varepsilon$.

Alternatively, using the multipliers λ_j, we can write the following.

Definition 4.6. If $\mathbf{f} \in \mathbb{R}^n$ is an r times differentiable function, then for some $0 < q \leq r$ and $\varepsilon > 0$, the function $\tilde{\mathbf{f}}$ is a perturbation of f of size ε, of differentiability class q; if there is a compact set $K \subset \mathbb{R}^n$ such that $\mathbf{f} = \tilde{\mathbf{f}}$ on the complement set $K_c = \mathbb{R}^n - K$ and for all $i_1, i_2, \ldots, i_n$, with $i = i_1 + i_2 + \ldots. + i_n \leq q$ and $j_1, j_2, \ldots, j_n$, with $j = j_1 + j_2 + \ldots. + j_r \leq q$, we have $\left|\left(\partial^i/\partial x_1^{i_1} \ldots \partial x_n^{i_n}\right)(\mathbf{f} - \tilde{\mathbf{f}})\right| < \varepsilon$ and $\left|\left(\partial^j/\partial \lambda_1^{j_1} \ldots \partial \lambda_r^{j_r}\right)(\mathbf{f} - \tilde{\mathbf{f}})\right| < \varepsilon$.

Perturbations are a little subtle where piecewise-smooth systems are concerned. We permit only perturbations that are at least differentiable in $\mathbf{x}$ or $\boldsymbol{\lambda}$, but the dependence on $\boldsymbol{\lambda}$, which is itself discontinuous in $\mathbf{x}$, introduces some novelty. For example, applying a small perturbation of $+c$ for $x_1 > 0$ and $-c$ for $x_1 < 0$, to a system $\dot{x}_1 = f(x_1)$, is not allowed by the definitions above, but a perturbation $+c\lambda_1$ to a piecewise-differentiable system $\dot{x}_1 = f(x_1; \lambda_1)$ with $\lambda_1 = \text{sign}(x_1)$ as in (3.4) is allowed, because λ_1 is included in the (partially) differentiable functional form of the vector field.

Definition 4.7. A vector field $\mathbf{f} \in \mathbb{R}^n$ is **structurally stable** if there exists some $\varepsilon > 0$ such that all differentiable ($q = 1$) ε perturbations of $\mathbf{f}$ are topologically equivalent to $\mathbf{f}$.

Example 4.2 (Piecewise Perturbations).

(i) The system $\dot{x} = -\lambda$ with $\lambda = \text{sign}(x)$ is structurally stable, because a perturbation to $\dot{x} = -\lambda + c$ or $\dot{x} = -\lambda + c\lambda$, for example, gives a topologically equivalent system for small c with an attracting discontinuity surface $x = 0$.

(ii) The system $\dot{x} = -\lambda x$ with $\lambda = \text{sign}(x)$ is structurally unstable, having both a discontinuity surface and an equilibrium at $x = 0$. A perturbation to $\dot{x} = -\lambda x + c$ for any small $c \neq 0$ creates equilibria at $x = \pm c$ for $c > 0$ or none if $c < 0$. A perturbation $\lambda = \text{sign}(x - c)$ moves the discontinuity surface to $x = c$, separating it from the equilibrium at $x = 0$.

(iii) The system $\dot{x} = -\lambda^2$ with $\lambda = \text{sign}(x)$ is structurally unstable, because a perturbation to $\dot{x} = -\lambda^2 + c$ for any small c gives a non-equivalent system, where equilibrium solutions to $\dot{x} = 0$ are split into $\lambda = \pm\sqrt{c}$.

■

Filippov and Teixeira consider pseudo-orbits to be a determining factor in structural stability, but we will not concern ourselves with them. A pseudo-orbit is a concatenation of trajectories that does not preserve the direction of time, so it has no dynamical (or as far as we know physical) significance.

Example 4.3 (Stability of Foci and Centres). A fused centre consists of closed orbits crossing the discontinuity surface that form a circulating flow. This is structurally unstable, because under perturbation it degenerates into a fused focus, Figure 4.7, which is structurally stable. A pseudo fused centre and pseudo fused focus, though superficially similar to their non-pseudo counterparts, are both structurally stable and dynamically are identical to each other, with no circulation and with all trajectories starting and ending towards opposite ends of the discontinuity surface, Figure 4.7(ii).

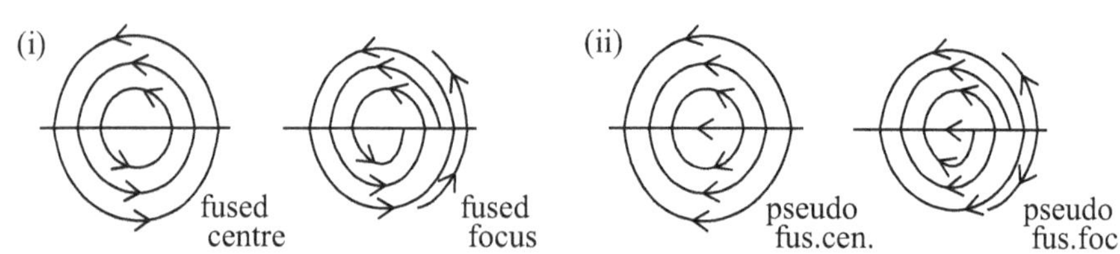

Fig. 4.7 Perturbations of orbits and pseudo-orbits, formed by fusing two parabolic systems. (i) shows a centre comprised entirely of closed orbits that cross through a discontinuity surface; this is structurally unstable and under perturbation becomes a (repelling or attracting) focus, possibly surrounded by one or more isolated closed orbits. (ii) shows a pseudo-centre, consisting of topological closed loops that are not orbits and therefore of no dynamical significance: instead all orbits begin and end on the surface; this behaviour is structurally stable and is dynamically no different from the perturbed pseudo-focus.

■

The systems we study generally depend on variables $\mathbf{x} = (x_1, x_2, \ldots, x_n)$, switching multipliers $\boldsymbol{\lambda} = (\lambda_1, \lambda_2, \ldots, \lambda_m)$, and parameters $\mathbf{p} = (a, b, c, \ldots)$, which we may include explicitly by writing $\dot{\mathbf{x}} = \mathbf{f}(\mathbf{x}; \boldsymbol{\lambda}; \mathbf{p})$. In some situations the behaviour for different parameter values can be crucial.

Definition 4.8. A **bifurcation set** is the set of parameters $\mathbf{p} = (a, b, c, \ldots)$ for which the system $\dot{\mathbf{x}} = \mathbf{f}(\mathbf{x}; \boldsymbol{\lambda}; \mathbf{p})$ is structurally unstable. Any point $\mathbf{x}$ in a neighbourhood of which the system is structurally unstable is a **singularity** (or singular point).

This is a specific kind of singularity. The term also refers to any distinguished point, e.g. an equilibrium, distinct from all other points in its neighbourhood.

In most situations we omit the explicit dependence on parameters and write $\dot{\mathbf{x}} = \mathbf{f}(\mathbf{x}; \boldsymbol{\lambda})$.

For discrete time maps (instead of continuous flows) with a piecewise-smooth character, say a system of difference equations $\mathbf{x}_n = \mathbf{f}(\mathbf{x}_{n-1}; \boldsymbol{\lambda})$ with multipliers $\boldsymbol{\lambda} = (\lambda_1, \ldots, \lambda_m)$ as in our flows above, similar notions of equivalence, perturbation, structural stability, and bifurcation follow analogously.

A bifurcation of the system—a topological change—takes place as we vary parameters through the bifurcation set. There are many bifurcations in discontinuous systems where the discontinuity plays no crucial role, in which case the behaviours observed are familiar from smooth dynamical systems. This happens if a bifurcation takes place within a region $\mathcal{R}_K$ where $\dot{\mathbf{x}} = \mathbf{f}^K$ is smoothly varying, which may be outside the discontinuity surface, or may be strictly inside the surface (if the index string K contains a '\$' index).

Then there are a whole new array of bifurcations that cannot occur in differentiable dynamical systems because they involve the discontinuity in a non-trivial way. A working definition is given as follows.

Definition 4.9. A bifurcation is said to be **discontinuity-induced** in a system $\dot{\mathbf{x}} = \mathbf{f}(\mathbf{x}; \boldsymbol{\lambda})$ if it involves a singular point on the boundary of a region of sliding on the discontinuity surface.

The boundaries of sliding are discussed in the next two chapters. This tightens a definition given in [48] which would permit any bifurcation in a discontinuous system to be considered discontinuity-induced. The theory of such bifurcations is in its infancy, as remarked in [48], and through that work and work it inspired, we have come to understand that discontinuity-induced bifurcations always involve a boundary of sliding. The definition almost certainly is still not perfect.

This definition, unlike the previous ones, is not directly applicable to discrete time-stepping maps, but for those the definition is perhaps simpler.

We would propose that a bifurcation is said to be discontinuity-induced in a system $\mathbf{x}_n = \mathbf{f}(\mathbf{x}_n; \boldsymbol{\lambda})$ if it involves a singular point at which the graph of $\mathbf{f}$ has a corner or a jump.

Classical bifurcations (i.e. of smooth systems) can interact with those that are discontinuity-induced, sometimes with one triggering the other to create novel bifurcation cascades.

4.5 Prototypes

In what follows we explore various local and global behaviours of piecewise-smooth dynamical systems by means of representative equations, which we refer to as *prototypes.* A prototype represents the leading-order terms of a series expansion

$$\dot{\mathbf{x}} = \mathbf{f}(\mathbf{x}_p) + (\mathbf{x} - \mathbf{x}_p) \cdot \frac{\partial}{\partial \mathbf{x}} \mathbf{f}(\mathbf{x}) \bigg|_{\mathbf{x}_p} + \dots \tag{4.13}$$

about a point $\mathbf{x}_p$ which is the organizing centre of a given singularity. We expect all systems satisfying certain singularity and non-degeneracy conditions to locally exhibit similar qualitative behaviour to the prototype.

A prototype is intended to satisfy two conditions, that the model:

- satisfies singularity and non-degeneracy conditions (*sufficiency*);
- represents all possible local phase portraits of the singularity (*fullness*).

The first property, sufficiency, involves the satisfaction of singularity equations that define a given singularity and of non-degeneracy inequalities that give the necessary conditions for the model to be structurally stable. It is typically a straightforward exercise of substituting into these equations and inequalities to show that a given prototype satisfies them.

The second property, fullness, can be shown in some situations, but we will generally not claim universality, i.e. that all systems of a given class can be transformed topologically into the prototype—a 'topological normal form' in standard terminology. The reason is that topological equivalence requires an intimate knowledge of the flow—the families of orbits—and much of our effort below will be an attempt to unravel the surprisingly intricate flows at seemingly simple singularities. When topological equivalence can be proven, it is often an intractably long exercise with little reward.

We will also largely rely on 'prototypes' rather than 'normal forms' because often the exact conditions, stability criteria, and possible degeneracies defining a singularity or bifurcation are insufficiently understood. To adopt the well-established term 'normal form' is premature and has led to incomplete 'normal form' classifications (see, e.g. [108]). In Chapter 13 we recount an entire saga surrounding the search for the normal form of a seemingly simple singularity. The term is therefore better reserved for a later time after due maturation of the subject, if indeed it ever becomes appropriate at all.

Difficulties of finding equivalence classes and representative normal forms are not specific to piecewise-smooth systems, having also arisen in the study of global phenomena in smooth dynamical systems theory. The added difficulty in piecewise-smooth systems is that such problems affect local phenomena too. In such situations it is more valuable to have a prototype with which to explore new phenomena than be tied to provable equivalence classes. In short, the notion of prototypes is less powerful, but more transparent.

4.6 Flow Derivatives

It will be convenient to have a concise expression for time derivatives (or so-called *Lie derivatives*) along piecewise-smooth flows. Denote the spatial derivative (gradient) operator by

$$\delta_{\mathbf{x}} := \frac{d}{d\mathbf{x}} = \nabla \ , \tag{4.14}$$

and the time derivative operator along the flow of $\mathbf{f}$ by

$$\partial_t := \frac{d}{dt} = \mathbf{f}(\mathbf{x}; \boldsymbol{\lambda}) \cdot \delta_{\mathbf{x}} \ . \tag{4.15}$$

For the constituent vector fields, where each $\mathbf{f}(\mathbf{x}; \boldsymbol{\lambda})$ equals some $\mathbf{f}^K(\mathbf{x})$ by (3.8), the time derivative ∂_t reduces to

$$\partial_{tK} := \mathbf{f}^K \cdot \delta_{\mathbf{x}} \ , \tag{4.16}$$

and applying this to $\mathbf{x}$ produces the corresponding vector field,

$$\partial_{tK}\mathbf{x} = \mathbf{f}^K(\mathbf{x}) = \mathbf{f}(\mathbf{x}; \boldsymbol{\lambda}) \tag{4.17}$$

where each $\lambda_j = \kappa_j 1$, recalling (3.8). So with constituent vector field $\mathbf{f}^{\pm}$ about a switching manifold $\mathcal{D}_1$, the time derivative δ_t reduces to $\delta_{t\pm} = \mathbf{f}^{\pm} \cdot \delta_{\mathbf{x}}$ and generates the vector field as $\delta_{t\pm}\mathbf{x} = \mathbf{f}^{\pm}(\mathbf{x}) = \mathbf{f}(\mathbf{x}; \pm 1)$. The corresponding expressions for sliding on $\mathcal{D}_1$ are $\delta_{t\$} = \mathbf{f}^{\$} \cdot \delta_{\mathbf{x}}$ and $\delta_{t\$}\mathbf{x} = \mathbf{f}^{\$}(\mathbf{x}) = \mathbf{f}(\mathbf{x}; \lambda^{\$})$.

In this notation $\delta_{\mathbf{x}}\sigma_j$ is the normal vector to a switching submanifold $\mathcal{D}_j$, and the component of the velocity field normal to the submanifold $\mathcal{D}_j$ is given equivalently by $\partial_t \sigma_j$ or $\mathbf{f} \cdot \delta_{\mathbf{x}}\sigma_j$.

4.7 Looking Forward

We will to return to all of these themes in the next ten chapters. Set-valued flows in particular will feature in Chapters 9 and 10. The singularity geometry of Section 6.1 will be of particular use in Chapter 11 when we confront global dynamics.

We have set out the basic elements of piecewise-smooth flows concerning uniqueness and structural stability, in so far as they are obtainable without close inspection of the discontinuity surface. It is to the discontinuity surface that we must turn our attention next.

Chapter 5
The Vector Field Canopy

If a problem is specified only piecewise as

$$\dot{\mathbf{x}} = \mathbf{f}(\mathbf{x}) = \left\{ \mathbf{f}^K(\mathbf{x}) \ \text{ if } \ \mathbf{x} \in \mathcal{R}_K \right\}_{K=\kappa_1\kappa_2\ldots\kappa_m} \tag{5.1}$$

in terms of constituent vector fields $\mathbf{f}^K$, we may need to extend this across the discontinuity surface to form a combination $\mathbf{f}(\mathbf{x};\boldsymbol{\lambda})$. There is an expression for $\mathbf{f}(\mathbf{x};\boldsymbol{\lambda})$ that provides a series expansion in the switching multipliers $\boldsymbol{\lambda} = (\lambda_1, \ldots, \lambda_m)$. We call this the *canopy* of the constituent vector fields.

5.1 The Vector Field Canopy

Consider a set of constituent vector fields $\mathbf{f}^K$ at a point $\mathbf{x}$, indexed by $K = \kappa_1\kappa_2\ldots\kappa_m$ where each κ_j is a $\pm$ sign such that

$$\kappa_j 1 = \operatorname{sign}(\sigma_j(\mathbf{x})) \ , \tag{5.2}$$

for switching functions σ_j, $j = 1, \ldots, m$. We then define the **canopy** combination of the vector fields $\mathbf{f}^K = \mathbf{f}^{\kappa_1\kappa_2\ldots\kappa_m}$ as

$$\mathbf{f}(\mathbf{x};\boldsymbol{\lambda}) = \sum_{\kappa_1=\pm} \cdots \sum_{\kappa_m=\pm} \lambda_1^{(\kappa_1)} \ldots \lambda_m^{(\kappa_m)} \mathbf{f}^{\kappa_1\kappa_2\ldots\kappa_m}(\mathbf{x}) \ + \ \mathbf{n}(\mathbf{x};\boldsymbol{\lambda}) \ ,$$

using shorthand

$$\lambda_j^{(\pm)} \equiv \tfrac{1}{2}(1 \pm \lambda_j) \ , \tag{5.3}$$

M. R. Jeffrey, *Hidden Dynamics*,
https://doi.org/10.1007/978-3-030-02107-8_5

where

$$\sigma(\mathbf{x})\mathbf{n}(\mathbf{x};\boldsymbol{\lambda}) = 0 \,. \tag{5.4}$$

The multipliers λ_j are defined such that $\lambda_j = \text{sign}(\sigma_j(\mathbf{x}))$; hence

$$\begin{cases} \lambda_j = \pm 1 & \text{if} \quad \mathbf{x} \notin \mathcal{D}_j \,, \\ \lambda_j \in (-1, +1) & \text{if} \quad \mathbf{x} \in \mathcal{D}_j \,. \end{cases} \tag{5.5}$$

In the general form (5.3), the notation is somewhat dense, but we unpick it below for the cases $m = 1$ and 2 and show how it builds hierarchically as m increases. Its interpretation as a series expansion is discussed in Section 5.2.

The canopy reduces to (5.1) outside $\mathcal{D}$. It has a linear part, the summation term in the first line of (5.3), which is uniquely determined by the constituent vector fields $\mathbf{f}^K$ (we prove this in Section 5.2.3 below) and which is essentially a multilinear interpolation between the constituent vector fields.

All terms nonlinear in $\boldsymbol{\lambda}$ reside in $\mathbf{n}$. The condition (5.4) defines $\mathbf{n}$ as a hidden term, so that at any $\mathbf{x}$, either the switching function $\sigma = \sigma_1 \ldots \sigma_m$ or the vector field $\mathbf{n}$ vanishes, but not necessarily both. Thus $\mathbf{n}(\mathbf{x};\boldsymbol{\lambda}) = \mathbf{n}(\mathbf{x};\lambda_1,\ldots,\lambda_m)$ vanishes when all of the λ_js takes a value ± 1, i.e. $\mathbf{n}(\mathbf{x};\pm 1,\ldots,\pm 1) \equiv 0$. The hidden term is not determined by the constituent vector fields outside $\mathcal{D}$. We derive expressions for it in Section 5.2, finding that we can typically write

$$\mathbf{n}(\mathbf{x};\boldsymbol{\lambda}) = \sum_{j=1}^{m} (\lambda_j^2 - 1)\mathbf{h}_j(\mathbf{x};\boldsymbol{\lambda}) \tag{5.6}$$

where each $\mathbf{h}_j$ is an arbitrary finite vector field in $\mathbb{R}^n$.

The vector fields $\mathbf{n}$ and $\mathbf{h}$ are known only if the vector field $\mathbf{f}(\mathbf{x};\boldsymbol{\lambda})$ is given in a modelling problem, typically when $\boldsymbol{\lambda}$ denotes a physical quantity that switches on or off, such as a current in an electrical or biochemical switch or a contact force in mechanics. If only the constituent vector fields $\mathbf{f}^K(\mathbf{x})$ are given in a model, then the canopy (5.3) can be used as a basis to infer a combination $\mathbf{f}(\mathbf{x};\boldsymbol{\lambda})$, whose nonlinear switching terms we can attempt to match to the dynamics of the system.

The expression (5.3) is clearer if we deconstruct it a little for one or two switches.

5.1.1 The Canopy for One Switch

Take a domain on which $\mathcal{D}$ consists only of the discontinuity submanifold $\mathcal{D}_1$, where the right-hand side of (5.1) switches between two values, call them $\mathbf{f}^+$ and $\mathbf{f}^-$. The canopy (5.3) gives (omitting arguments)

$$\mathbf{f} = \tfrac{1}{2}(1+\lambda_1)\mathbf{f}^+ + \tfrac{1}{2}(1-\lambda_1)\mathbf{f}^- + (\lambda_1^2 - 1)\mathbf{h} \tag{5.7}$$

with $\lambda_1 = \text{sign}(\sigma_1)$, where we have used (5.6) for the hidden term for some arbitrary vector field $\mathbf{h}$.

For $\mathbf{x} \in \mathcal{D}_1$, as λ_1 explores the interval $[-1, +1]$, the vector $\mathbf{f}$ explores a one parameter family connecting the vectors $\mathbf{f}^+$ and $\mathbf{f}^-$.

We can represent this on a vector field plot as the locus of points $\{\mathbf{x} + \mathbf{f} : -1 \leq \lambda_1 \leq +1\}$; see Figure 5.1. Without the hidden term, this would be a straight line, with $\mathbf{f}$ being just the *convex combination*

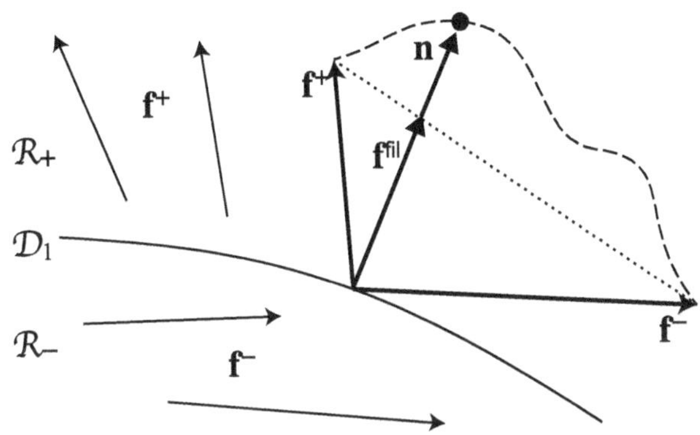

Fig. 5.1 The canopy at a point $\mathbf{x} \in \mathcal{D}_1$. As a vector field switches between $\mathbf{f}^+$ on $\mathcal{R}_+$ and $\mathbf{f}^-$ on $\mathcal{R}_-$, its value might lie on the linear interpolation $\mathbf{f}^{\text{fil}}$ (dotted), which forms a convex set between $\mathbf{f}^{\pm}$. If the vector field actually lies elsewhere, the addition of some nonzero $\mathbf{n}$ provides the necessary generalization to a curve $\mathbf{f}^{\text{fil}} + \mathbf{n}$ (dashed) which reaches more general families of vector fields joining $\mathbf{f}^+$ and $\mathbf{f}^-$.

$$\mathbf{f}^{\text{fil}} = \tfrac{1}{2}(1 + \lambda_1)\mathbf{f}^+ + \tfrac{1}{2}(1 - \lambda_1)\mathbf{f}^-$$

(the basis of most work on so-called *Filippov systems* [71]). The vector $\mathbf{n}$ deforms this line into a curve.

5.1.2 The Canopy for Two Switches

Now imagine that the vector fields $\mathbf{f}^{\pm}$ in (5.7) are themselves found to be only piecewise-smooth, switching across a second submanifold $\mathcal{D}_2$, across which $\mathbf{f}^+$ switches between $\mathbf{f}^{++}$ and $\mathbf{f}^{+-}$ and $\mathbf{f}^-$ switches between $\mathbf{f}^{-+}$ and $\mathbf{f}^{--}$. To each of the vector fields $\mathbf{f}^{\pm}$, we can apply a combination analogous to (5.7) over a switching multiplier λ_2, replacing each $\mathbf{f}^{\kappa_1}$ with

$$\mathbf{f}^{\kappa_1} = \tfrac{1}{2}(1 + \lambda_2)\mathbf{f}^{\kappa_1 +} + \tfrac{1}{2}(1 - \lambda_2)\mathbf{f}^{\kappa_1 -} + \mathbf{n}\,.$$

Substituting this into (5.7) for $\kappa_1 = \pm$ gives a bilinear combination

$$\begin{aligned} \mathbf{f} = & \tfrac{1}{2}(1+\lambda_1)\left[\tfrac{1}{2}(1+\lambda_2)\mathbf{f}^{++} + \tfrac{1}{2}(1-\lambda_1)\mathbf{f}^{+-}\right] \\ & + \tfrac{1}{2}(1-\lambda_1)\left[\tfrac{1}{2}(1+\lambda_2)\mathbf{f}^{-+} + \tfrac{1}{2}(1-\lambda_2)\mathbf{f}^{--}\right] + \mathbf{n}\,, \end{aligned} \tag{5.8}$$

which is the canopy (5.3) near a point $\mathbf{x} \in \mathcal{D}_1 \cap \mathcal{D}_2$. We can write $\mathbf{n} = (\lambda_1^2 - 1)\mathbf{h}_1 + (\lambda_2^2 - 1)\mathbf{h}_2$ for arbitrary vector fields $\mathbf{h}_1$ and $\mathbf{h}_2$. An example with bilinear dependence on the multipliers is illustrated in Figure 5.2.

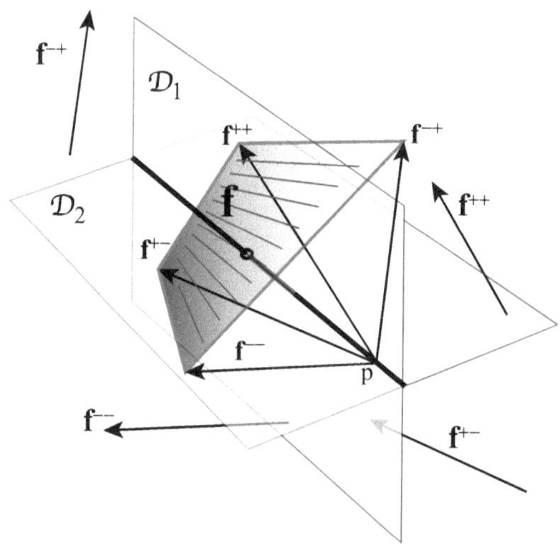

Fig. 5.2 The canopy at a point $p \in \mathcal{D}_1 \cap \mathcal{D}_2$, with $\mathbf{n} = 0$ (If we introduce a term $\mathbf{n} \neq 0$, it will alter the curvature, but not the corners, of the canopy).

5.1.3 The Canopy for m Switches

We can continue in this way iteratively, assuming each constituent vector field is only piecewise-smooth, and adding a new discontinuity submanifold, until we have m independent switches and obtain the canopy (5.3). We build up hierarchically by summing over one index at a time, ignoring $\mathbf{n}$ at first,

$$\begin{aligned} \mathbf{f}(\mathbf{x};\boldsymbol{\lambda}) &= \tfrac{1}{2}(1+\lambda_1)\,\mathbf{f}^{+}(\mathbf{x}) \;+\; \tfrac{1}{2}(1-\lambda_1)\,\mathbf{f}^{-}(\mathbf{x}) \\ \text{where}\quad \mathbf{f}^{\kappa_1}(\mathbf{x}) &= \tfrac{1}{2}(1+\lambda_2)\,\mathbf{f}^{\kappa_1+}(\mathbf{x}) \;+\; \tfrac{1}{2}(1-\lambda_2)\,\mathbf{f}^{\kappa_1-}(\mathbf{x}) \\ \text{where}\quad \mathbf{f}^{\kappa_1\kappa_2}(\mathbf{x}) &= \tfrac{1}{2}(1+\lambda_3)\mathbf{f}^{\kappa_1\kappa_2+}(\mathbf{x}) + \tfrac{1}{2}(1-\lambda_3)\mathbf{f}^{\kappa_1\kappa_2-}(\mathbf{x}) \\ \text{where}\quad \mathbf{f}^{\kappa_1\kappa_2\kappa_3}(\mathbf{x}) &= \ldots etc. \end{aligned} \tag{5.9}$$

Note that the series we obtain does not depend on the order in which we consider the switches. By the same iterative argument, we also obtain a specific form for the vector $\mathbf{n}$ in (5.3), namely,

$$\mathbf{n}(\mathbf{x};\boldsymbol{\lambda})=\sum_{j=1}^{m}\mathbf{n}_j(\mathbf{x};\boldsymbol{\lambda})\quad\text{such that}\quad\mathbf{n}_j(\mathbf{x};\boldsymbol{\lambda})=0\ \text{ if }\ \mathbf{x}\notin\mathcal{D}_j\ . \tag{5.10}$$

At a point on the intersection of r discontinuity submanifolds, as the multipliers vary over $(-1,+1)$, the canopy typically describes an r-dimensional sheet stretched out between the 2^r vector fields $\mathbf{f}^i$ of the neighbouring regions. This sheet has a particular geometry formed of curves (straight lines if $\mathbf{n}\equiv 0$) as each λ_j varies between -1 and $+1$, which we can think of as the gridlines of the λ_js coordinatizing the canopy surface.

5.2 Deriving the Canopy

There are different ways of deriving the expansion (5.3). We first give an argument in terms of local expansion in λ at the discontinuity surface $\sigma=0$ and then an argument based on assuming that $\mathbf{f}$ has a series expansion in powers of switching multipliers.

5.2.1 Joint Expansions and Matching

If we assume a system can be modelled as $\dot{\mathbf{x}}=\mathbf{f}(\mathbf{x};\lambda)$ for some $\mathbf{f}$, then we can start building a picture of what the function $\mathbf{f}$ should be by expanding it about different values. Rather than expanding about any particular $\mathbf{x}$, we are concerned with its dependence on λ and have the values $\mathbf{f}(\mathbf{x};+1)=\mathbf{f}^+(\mathbf{x})$ and $\mathbf{f}(\mathbf{x};-1)=\mathbf{f}^-(\mathbf{x})$ to expand about.

It does not make strict sense to take a Taylor series about $\lambda=\pm1$. Because these λ values lie on the boundaries of the domain $[-1,+1]$ on which λ is defined, there is no open neighbourhood about $\lambda=+1$ or $\lambda=-1$ around which a series expansion will be valid. Nevertheless we can expand formally and worry about the interpretation afterwards.

If we expand $\mathbf{f}$ about $\lambda=+1$, we have

$$\mathbf{f}(\mathbf{x};\lambda)=\mathbf{f}^+(\mathbf{x})+(\lambda-1)\mathbf{f}^+_{,\lambda}(\mathbf{x})+\tfrac{1}{2}(\lambda-1)^2\mathbf{f}^+_{,\lambda\lambda}(\mathbf{x})+\mathsf{O}\left((\lambda-1)^3\right):=P\ , \tag{5.11}$$

while expanding about $\lambda=-1$ gives

$$\mathbf{f}(\mathbf{x};\lambda)=\mathbf{f}^-(\mathbf{x})+(\lambda+1)\mathbf{f}^-_{,\lambda}(\mathbf{x})+\tfrac{1}{2}(\lambda+1)^2\mathbf{f}^-_{,\lambda\lambda}(\mathbf{x})+\mathsf{O}\left((\lambda+1)^3\right):=Q\ , \tag{5.12}$$

using the notation

$$\mathbf{f}^+_{,\lambda}(\mathbf{x})=\tfrac{\partial}{\partial\lambda}\mathbf{f}(\mathbf{x};\lambda)\big|_{\lambda=+1}\ ,\qquad \mathbf{f}^-_{,\lambda}(\mathbf{x})=\tfrac{\partial}{\partial\lambda}\mathbf{f}(\mathbf{x};\lambda)\big|_{\lambda=-1}\ . \tag{5.13}$$

The function $\mathbf{f}$ must satisfy both approximations (5.11) and (5.12), so it must be an interpolation between P and Q as λ goes from $+1$ to -1, the obvious choice being $\frac{1}{2}(1+\lambda)P + \frac{1}{2}(1-\lambda)Q$, giving

$$\dot{\mathbf{x}} = \mathbf{f}(\mathbf{x};\lambda) = \tfrac{1}{2}(1+\lambda)\,\mathbf{f}^{+}(\mathbf{x}) + \tfrac{1}{2}(1-\lambda)\,\mathbf{f}^{-}(\mathbf{x}) + (\lambda^2-1)\mathbf{h}(\mathbf{x};\lambda)\,, \quad (5.14)$$

where

$$\mathbf{h}(\mathbf{x};\lambda) = \tfrac{1}{2}\left(\mathbf{f}^{+}_{,\lambda}(\mathbf{x}) - \mathbf{f}^{-}_{,\lambda}(\mathbf{x})\right) - \tfrac{1}{4}(1+\lambda)\,\mathbf{f}^{-}_{,\lambda\lambda}(\mathbf{x}) - \tfrac{1}{4}(1-\lambda)\,\mathbf{f}^{+}_{,\lambda\lambda}(\mathbf{x}) + \dots\,. \quad (5.15)$$

Thus we have the canopy for one switch (5.7) with hidden term (5.15).

Suppose that motion occurs on the discontinuity surface, with the multiplier λ lying at some $\lambda = \lambda^{\$} \in (-1,+1)$. Then expanding in λ about $\lambda^{\$}$ gives

$$\mathbf{f}(\mathbf{x};\lambda) = \mathbf{f}^{\$}(\mathbf{x}) + (\lambda-\lambda^{\$})\mathbf{f}^{\$}_{,\lambda}(\mathbf{x}) + \tfrac{1}{2}(\lambda-\lambda^{\$})^2\mathbf{f}^{\$}_{,\lambda\lambda}(\mathbf{x}) + \mathsf{O}\left((\lambda-\lambda^{\$})^3\right)\,. \quad (5.16)$$

Here $\mathbf{f}^{\$}(\mathbf{x}) = \mathbf{f}(\mathbf{x};\lambda^{\$})$ is given, using the joint expansion above, by

$$\mathbf{f}^{\$}(\mathbf{x}) = \tfrac{1}{2}(1+\lambda^{\$})\mathbf{f}^{+}(\mathbf{x}) + \tfrac{1}{2}(1-\lambda^{\$})\mathbf{f}^{-}(\mathbf{x}) + \left((\lambda^{\$})^2-1\right)\mathbf{h}(\mathbf{x};\lambda^{\$})\,. \quad (5.17)$$

If the vector field on the discontinuity surface, $\mathbf{f}^{\$}(\mathbf{x})$, is known, then (5.17) can be used to fix $\mathbf{h}$ more precisely than (5.15) (which was an approximation about $\lambda = \pm 1$ rather than about $\lambda = \lambda^{\$}$).

Thus, by attempting to derive a vector field by joint series expansions across a discontinuity surface, we have arrived at the canopy for one switch, (5.7). The canopy (5.3) for multiple switches can be obtained by a similar exercise using multivariable expansions in the multiplier $\boldsymbol{\lambda} = (\lambda_1,\dots,\lambda_m)$.

5.2.2 Series of Signs

If we assume that a vector field depends on a quantity $\lambda = \mathrm{sign}(\sigma)$, given by $\mathbf{f}(\mathbf{x};\lambda)$ where $\mathbf{f}$ is analytic with respect to λ, then $\mathbf{f}$ can be expressed as a power series of λ,

$$\mathbf{f}(\mathbf{x};\lambda) = \sum_{p=0}^{\infty} \mathbf{a}_p(\mathbf{x})\lambda^p\,, \quad (5.18)$$

in terms of coefficients $\mathbf{a}_p$. Given that $\mathbf{f}(\mathbf{x};+1) = \mathbf{f}^{+}(\mathbf{x})$ and $\mathbf{f}(\mathbf{x};-1) = \mathbf{f}^{-}(\mathbf{x})$, we have immediately

$$\sum_{p=0}^{\infty} \mathbf{a}_p = \mathbf{f}^{+}\,, \qquad \sum_{p=0}^{\infty} (-1)^p\mathbf{a}_p = \mathbf{f}^{-}\,, \quad (5.19)$$

(omitting arguments). From the sum and difference of these, we can pull out expressions for the first two coefficients, $\mathbf{a}_0$ and $\mathbf{a}_1$, in terms of $\mathbf{f}^+$ and $\mathbf{f}^-$,

$$\begin{aligned}\mathbf{a}_0 &= \mathbf{f}^+ + \mathbf{f}^- - 2(\mathbf{a}_2 + \mathbf{a}_4 + \mathbf{a}_6 + \dots)\,,\\ \mathbf{a}_1 &= \mathbf{f}^+ - \mathbf{f}^- - 2(\mathbf{a}_3 + \mathbf{a}_5 + \mathbf{a}_7 - \dots)\,,\end{aligned}$$

and substituting these into (5.18) gives

$$\dot{\mathbf{x}} = \tfrac{1}{2}(1+\lambda)\mathbf{f}^+ + \tfrac{1}{2}(1-\lambda)\mathbf{f}^- + \mathbf{n}_1$$

where

$$\mathbf{n}_1 = \sum_{p=1}^{\infty}(\mathbf{a}_{2p} + \lambda\mathbf{a}_{2p+1})\left(\lambda^{2p}-1\right)\,.$$

Note that $\mathbf{n}_1$ vanishes when $\lambda = \pm 1$ and is therefore a hidden term.

We can take a factor $\lambda^2 - 1$ out of $\mathbf{n}$ since $\lambda^{2p} - 1 = (\lambda^2 - 1)\sum_{q=0}^{p-1}\lambda^{2q}$. Then if we rearrange slightly, we have

$$\mathbf{f} = \tfrac{1}{2}(1+\lambda)\mathbf{f}^+ + \tfrac{1}{2}(1-\lambda)\mathbf{f}^- + \left(\lambda^2-1\right)\mathbf{h}\,,$$

giving (5.7) where

$$\mathbf{h} = \sum_{p=1}^{\infty}(\mathbf{a}_{2p} + \lambda\mathbf{a}_{2p+1})\sum_{q=0}^{p-1}\lambda^{2q}\,.$$

The coefficients $\mathbf{a}_{p>1}$ of $\mathbf{h}$ are not fixed by the vector fields $\mathbf{f}^\pm$ either side of $\mathcal{D}$ and can only be fixed if we have more information to fix a model on $\mathcal{D}$, either through direct observation or through indirect dynamical consequences of $\mathbf{h}$ of the kind we shall see in Chapter 14.

We can extend this to build up a series of signs λ for m switches. First replace λ and σ above by λ_1 and σ_1, the former being the first component of $\boldsymbol{\lambda} = (\lambda_1, \dots, \lambda_m)$.

Say $\mathbf{f}^+$ is only piecewise-smooth and switches between values $\mathbf{f}^{++}$ for $\sigma_2(\mathbf{x}) > 0$ and $\mathbf{f}^{+-}$ for $\sigma_2(\mathbf{x}) < 0$, controlled by some continuous or discontinuous quantity λ_2 such that $\lambda_2 \sim \operatorname{sign}(\sigma_2)$. In a similar manner to that above, we obtain

$$\mathbf{f}^+ = \tfrac{1}{2}(\mathbf{f}^{++} + \mathbf{f}^{+-}) + \tfrac{1}{2}(\mathbf{f}^{++} - \mathbf{f}^{+-})\lambda_2 + \mathbf{n}_2\,.$$

We can do similarly for $\mathbf{f}^-$ if we assume it is also only piecewise-smooth, and we obtain

$$\mathbf{f}^{\kappa_1} = \tfrac{1}{2}(\mathbf{f}^{\kappa_1+} + \mathbf{f}^{\kappa_1-}) + \tfrac{1}{2}(\mathbf{f}^{\kappa_1+} - \mathbf{f}^{\kappa_1-})\lambda_2 + \mathbf{n}_2\,, \qquad \kappa_1 = \pm\,.$$

Now assume that the vector fields $\mathbf{f}^{\kappa_1\kappa_2}$ are only piecewise-smooth, switching between values $\mathbf{f}^{\kappa_1\kappa_2\pm}$ at a threshold $\sigma_3 = 0$, and we obtain the next iteration in a sequence

$$\mathbf{f}^{\kappa_1\kappa_2\ldots\kappa_r} = \tfrac{1}{2}(\mathbf{f}^{\kappa_1\kappa_2\ldots\kappa_{r-1}+} \\ + \mathbf{f}^{\kappa_1\kappa_2\ldots\kappa_{r-1}-}) + \tfrac{1}{2}(\mathbf{f}^{\kappa_1\kappa_2\ldots\kappa_{r-1}+} - \mathbf{f}^{\kappa_1\kappa_2\ldots\kappa_{r-1}-})\lambda_r + \mathbf{n}_2$$

where each $\kappa_j = \pm$, and we putting these together hierarchically we obtain

$$\begin{aligned}
\mathbf{f} &= \tfrac{1}{2}(1+\lambda_1)\mathbf{f}^+ + \tfrac{1}{2}(1-\lambda_1)\mathbf{f}^- + (\lambda_1^2-1)\mathbf{h}_1 \\
&= \tfrac{1}{2}(1+\lambda_1)\left\{\tfrac{1}{2}(1+\lambda_2)\mathbf{f}^{++} + \tfrac{1}{2}(1-\lambda_2)\mathbf{f}^{+-}\right\} \qquad (5.20) \\
&\quad + \tfrac{1}{2}(1-\lambda_1)\left\{\tfrac{1}{2}(1+\lambda_2)\mathbf{f}^{-+} + \tfrac{1}{2}(1-\lambda_2)\mathbf{f}^{--}\right\} \\
&\quad + (\lambda_2^2-1)\mathbf{h}_2 + (\lambda_1^2-1)\mathbf{h}_1 \\
&\ \vdots \\
&= \tfrac{1}{2}\sum_{\kappa_1=\pm}\cdots\sum_{\kappa_m=\pm}\left(\prod_{j=1}^{m}(1+\kappa_j\lambda_j)\right)\mathbf{f}^{\kappa_1\kappa_2\ldots\kappa_m}(\mathbf{x}) + \sum_{j=1}^{m}(\lambda_j^2-1)\mathbf{h}_j(\mathbf{x};\boldsymbol{\lambda}) \,, \qquad (5.21)
\end{aligned}$$

where the $\mathbf{f}^K$s are our piecewise-defined vector fields, while the $\mathbf{h}_j$s are a set of smooth vector fields particular to the switching process across each threshold $\sigma_j = 0$. Thus we arrive at the general canopy of vector fields (5.3).

5.2.3 Uniqueness of the Multilinear Term

The following two results follow from (5.3) by direct calculation. For both of these, we are concerned only with what happens to the canopy outside the discontinuity surface $\mathcal{D}$; therefore we can neglect the hidden term $\mathbf{n}$.

First we show that outside $\mathcal{D}$ the formula (5.3) reduces just to the vector fields $\mathbf{f}^K$ as it should.

Proposition 5.1. *The canopy* (5.3) *reduces to* (5.1) *for* $\mathbf{x} \notin \mathcal{D}$.

Proof. By expanding the series (5.3), we see that each of the 2^m different values $\mathbf{f}(\mathbf{x};\lambda_1,\lambda_2,\ldots)$ maps one-to-one to the constituent functions $\mathbf{f}^{\kappa_1\kappa_2\ldots}$ when we identify each index κ_j with the sign preceding each $\lambda_j = \pm 1$. Note that since the λ_js appear in pairs $\frac{1}{2}(1+\lambda_j)$ and $\frac{1}{2}(1-\lambda_j)$ whose sum is unity, the coefficients of all $\mathbf{f}^K$s automatically sum to unity. □

Secondly we show that the linear part of the canopy is uniquely determined by the constituent vector fields $\mathbf{f}^K$.

Proposition 5.2. *Given 2^m vector fields $\mathbf{f}^{\kappa_1\kappa_2\ldots\kappa_m}(\mathbf{x})$ on $\mathbb{R}^n$ where $\kappa_j = \pm$ and $n \geq m$, there is a unique function $\mathbf{f}(\mathbf{x};\boldsymbol{\lambda})$ which is multilinear in terms of m independent switching functions $\boldsymbol{\lambda} = (\lambda_1, \lambda_2, \ldots, \lambda_m)$ such that $\mathbf{f}(\mathbf{x};\boldsymbol{\lambda})$ generates each of the vector fields, i.e.*

$$\mathbf{f}(\mathbf{x};+1,+1,+1,\ldots) = \mathbf{f}^{+++\cdots}(\mathbf{x})\,, \quad \mathbf{f}(\mathbf{x};-1,+1,+1,\ldots) = \mathbf{f}^{-++\cdots}(\mathbf{x})\,,$$
$$\mathbf{f}(\mathbf{x};+1,-1,+1,\ldots) = \mathbf{f}^{+-+\cdots}(\mathbf{x})\,, \quad \ldots$$

The function $\mathbf{f}$ is given in the canopy form, using the shorthand $\lambda_j^{(\kappa_j)} = \frac{1}{2}(1+\kappa_j\lambda_j)$, as

$$\mathbf{f}(\mathbf{x};\boldsymbol{\lambda}) = \sum_{\kappa_1=\pm}\sum_{\kappa_2=\pm}\cdots\sum_{\kappa_m=\pm} \lambda_1^{(\kappa_1)}\lambda_2^{(\kappa_2)}\ldots\lambda_m^{(\kappa_m)}\mathbf{f}^{\kappa_1\kappa_2\ldots\kappa_m}(\mathbf{x})$$

for $\mathbf{x}$ outside $\mathcal{D}$ (where $\mathbf{n} \equiv 0$), or equivalently in the form

$$\mathbf{f}(\mathbf{x};\boldsymbol{\lambda}) = \sum_{\kappa_1=\pm}\cdots\sum_{\kappa_m=\pm}\;\sum_{p_1=0,1}\cdots\sum_{p_m=0,1} M^{p_1\ldots p_m}_{\kappa_1\ldots\kappa_m}\lambda_1^{p_1}\ldots\lambda_m^{p_m}\mathbf{f}^{\kappa_1\ldots\kappa_m}(\mathbf{x})$$

where

$$M^{p_1p_2\ldots p_m}_{\kappa_1\kappa_2\ldots\kappa_m} = (-1)^{\frac{1-\kappa_1 1}{2}p_1+\frac{1-\kappa_2 1}{2}p_2\ldots\frac{1-\kappa_m 1}{2}p_m}\,.$$

Proof. Express $\mathbf{f}$ as the general multilinear expression $\mathbf{f} = \mathbf{c}_0 + \mathbf{c}_1\lambda_1 + \mathbf{c}_2\lambda_2 + \ldots + \mathbf{c}_{m+1}\lambda_1\lambda_2 + \ldots$. To keep track of the indices, it is more convenient to use the binary indexing for the $\mathbf{c}_i$s and write

$$\mathbf{f}(\mathbf{x};\boldsymbol{\lambda}) = \sum_{p_1=0,1}\cdots\sum_{p_m=0,1}\lambda_1^{p_1}\lambda_2^{p_2}\ldots\lambda_m^{p_m}\,\mathbf{c}_{p_1p_2\ldots p_m}$$

where each coefficient $\mathbf{c}_{p_1\ldots p_m}$ is a vector in $\mathbb{R}^n$. We must show that the vector coefficients $\mathbf{c}_{p_1\ldots p_m}$ are uniquely determined by the vector values $\mathbf{f}^{\kappa_1\ldots\kappa_m}$, i.e. that the matrix problem

$$\begin{pmatrix}\mathbf{f}^{+++\ldots}\\ \mathbf{f}^{-++\ldots}\\ \mathbf{f}^{+-+\ldots}\\ \mathbf{f}^{--+\ldots}\\ \vdots\end{pmatrix} = \begin{pmatrix}+1 & +1 & +1 & \ldots\\ -1 & +1 & +1 & \ldots\\ +1 & -1 & +1 & \ldots\\ -1 & -1 & +1 & \ldots\\ \vdots & \vdots & \vdots & \ddots\end{pmatrix}\begin{pmatrix}\mathbf{c}_{111\ldots}\\ \mathbf{c}_{011\ldots}\\ \mathbf{c}_{101\ldots}\\ \mathbf{c}_{001\ldots}\\ \vdots\end{pmatrix}$$

can be solved for the coefficients $\mathbf{c}_{\pm\ldots}$, each of which is a vector in $\mathbb{R}^n$. Clearly there are 2^m vector coefficients $\mathbf{c}_{p_1\ldots p_m}$ and 2^m vector values $\mathbf{f}^{\kappa_1\ldots\kappa_m}$. Moreover the rows of the matrix of ±1 symbols are orthogonal, so the matrix is non-singular. Thus we can invert this expression. In index notation the expression above is

$$\mathbf{f}^{\kappa_1\ldots\kappa_m}=\sum_{p_1=0,1}\cdots\sum_{p_m=0,1}W^{\kappa_1\ldots\kappa_m}_{p_1\ldots p_m}\mathbf{c}_{p_1\ldots p_m}$$

where

$$W^{\kappa_1\kappa_2\ldots\kappa_m}_{p_1p_2\ldots p_m}=(\kappa_1 1)^{p_1}(\kappa_2 1)^{p_2}\ldots(\kappa_m 1)^{p_m}=(-1)^{\frac{1-\kappa_1 1}{2}p_1+\frac{1-\kappa_2 1}{2}p_2\ldots\frac{1-\kappa_m 1}{2}p_m}\;.$$

The matrix W has a determinant $\det W=2^m$, and the relation between $\mathbf{f}^{\kappa_1\ldots}$ and $\mathbf{c}_{p_1\ldots}$ is easily inverted to give

$$\mathbf{c}_{p_1\ldots p_m}=\sum_{\kappa_1=\pm}\cdots\sum_{\kappa_m=\pm}M^{\kappa_1\ldots\kappa_m}_{p_1\ldots p_m}\mathbf{f}^{\kappa_1\ldots\kappa_m}\;,$$

where M is the inverse of W, given by

$$M^{p_1p_2\ldots p_m}_{\kappa_1\kappa_2\ldots\kappa_m}=(-1)^{\frac{1-\kappa_1 1}{2}p_1+\frac{1-\kappa_2 1}{2}p_2\ldots\frac{1-\kappa_m 1}{2}p_m}\;.$$

□

Finally we have, rather trivially, that the multiplier canopy (5.3) is sufficient to express any member of a set $\mathcal{F}$ defining an inclusion (3.18).

Proposition 5.3. *For any member* $\boldsymbol{\nu}\in\mathcal{F}$ *of the inclusion* (3.18), *there exists a function* $\mathbf{n}$ *in the canopy* (5.3) *such that* $\boldsymbol{\nu}=\mathbf{f}(\mathbf{x};\boldsymbol{\lambda})$.

Proof. Say $\mathbf{f}\equiv\mathbf{f}^{\mathcal{D}}\in\mathcal{F}$ for $\mathbf{x}\in\mathcal{D}$, then we can write

$$\mathbf{n}=\left\{\sum_{\kappa_1=\pm}\cdots\sum_{\kappa_m=\pm}\prod_{r=1}^{m}\lambda_r^{(\kappa_r)}\mathbf{f}^{\kappa_1\kappa_2\ldots\kappa_m}-\mathbf{f}^{\mathcal{D}}\right\}\Theta(\boldsymbol{\lambda})$$

where $\Theta(\boldsymbol{\lambda})=1$ inside $\mathcal{D}$ and 0 outside, e.g.

$$\Theta(\boldsymbol{\lambda})=\lim_{\varepsilon\to 0}\prod_{r=1}^{m}\frac{2}{1+\exp(\varepsilon/(\lambda_r^2-1))}\;.$$

□

5.3 Looking Forward

The description (5.3) seems to have appeared in piecewise-smooth dynamical theory only recently, and we can take (5.3) as a prototype for nonlinear switching. Prior to the twenty-first century, attention focussed solely on the case of a single linear switch, $m=1$ with $\mathbf{h}\equiv 0$, which we might consider the age of *linear switching* theory. The nonlinear terms of $\mathbf{h}$ were introduced in [114]. The multilinear expansion $\lambda_1..\lambda_m\mathbf{f}^{\kappa_1..\kappa_m}$ appeared in [6] and later again in [113]. It also appears as the singular limit of genetic regulation models in [158] (and in related works), with some notable and fascinating dynamics that we will see later. Filippov considered a very special case of $m > 1$ which was entirely linear, a case used also in control theory, and part of the subject of Chapter 8.

A general theory of nonlinear switching—the theory of piecewise-linear flows using (5.3) with $m > 1$ and $\mathbf{h} \neq 0$—has been lacking, so in the coming pages, we take the first tentative steps into a new world of piecewise-smooth, fully nonlinear, dynamical theory.

Chapter 6
Tangencies: The Shape of the Discontinuity Surface

The presence of a discontinuity introduces a new elementary singularity to local dynamical systems theory, in the form of a tangency between the flow and the discontinuity surface.

This chapter sets out the basic geometry of flows around such tangencies. They underlie the most important qualitative features of piecewise-smooth dynamics, forming the boundaries of sliding regions, and points where the flow is poised between smooth evolution and switching.

6.1 Flow Tangencies

Figure 6.1(ii-v) shows the basic tangencies between the vector field $\mathbf{f}$ and the discontinuity surface $\mathcal{D}$ in a planar system.

Here we seek to characterize the basic forms of interaction between the discontinuity surface $\mathcal{D}$ and the flow outside it, formally defining and then generalizing the singularities in Figure 6.1. To do so we make the flow itself as near to trivial as possible. In particular we assume that the vector fields either side of the discontinuity surface are non-vanishing, i.e. contain no equilibria. The basic singularity of this interaction therefore consists of points where the vector fields are stationary with respect to the surface, meaning they lie in the tangent plane to $\mathcal{D}$. Hence in the following we seek to characterize these *tangencies* between the piecewise-smooth vector field and the discontinuity surface. These tangencies form boundaries of sliding regions.

M. R. Jeffrey, *Hidden Dynamics*,
https://doi.org/10.1007/978-3-030-02107-8_6

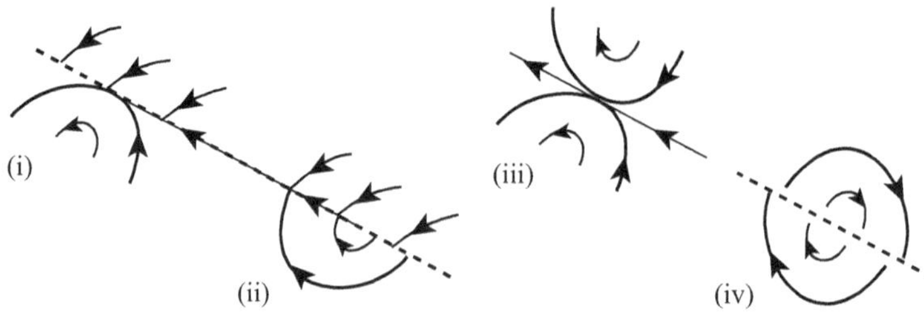

Fig. 6.1 Local singularities: (i-ii) tangencies to the discontinuity surface, (iii-iv) tangencies to both sides of the discontinuity surface.

We restrict our attention here to the flows outside the discontinuity surface. For the case of linear switching, we will derive the associated vector fields for dynamics inside the discontinuity surface in Section 8.5.

Consider a point $\mathbf{x}_p$ on the discontinuity surface $\mathcal{D}$ where it is locally a manifold, in the neighbourhood of which the system has the form

$$\dot{\mathbf{x}} = \mathbf{f}(\mathbf{x};\lambda) = \begin{cases} \mathbf{f}^+(\mathbf{x}) \text{ if } \sigma(\mathbf{x}) > 0\,, \\ \mathbf{f}^-(\mathbf{x}) \text{ if } \sigma(\mathbf{x}) < 0\,, \end{cases} \tag{6.1}$$

where $\mathbf{f}^+$ and $\mathbf{f}^-$ are linearly independent. To provide a general form for $\mathbf{f}$, since we are only interested in the flow for $\sigma \neq 0$ where $\lambda = \pm 1$, it is enough to write

$$\mathbf{f}(\mathbf{x};\lambda) = \tfrac{1}{2}(1+\lambda)\mathbf{f}^+(\mathbf{x}) + +\tfrac{1}{2}(1-\lambda)\mathbf{f}^+(\mathbf{x})\,,$$

with $\lambda = \text{sign}(\sigma(\mathbf{x}))$. We will also need the notation from (4.15), writing the time derivative along the flow as

$$\delta_t = \tfrac{1}{2}(1+\lambda)\delta_{t+} + \tfrac{1}{2}(1-\lambda)\delta_{t-} \tag{6.2}$$

in terms of the time derivatives $\delta_{t\pm}$ along the constituent flows.

A tangency between the '+' or '−' vector field and the discontinuity surface $\mathcal{D}$ is defined as a point $\mathbf{x}$ where

$$0 = \sigma(\mathbf{x}) = \partial_{t+}\sigma(\mathbf{x}) \qquad \text{or} \qquad 0 = \sigma(\mathbf{x}) = \partial_{t-}\sigma(\mathbf{x})\,. \tag{6.3}$$

Tangencies form the boundaries between different behaviours at the discontinuity surface, because they signal that $\mathbf{f}^+$ or $\mathbf{f}^-$ is changing direction with respect to the surface, altering whether $\mathcal{D}$ attracts, repels, or transmits the flow.

A smooth vector field that is locally non-vanishing is topologically equivalent to a constant vector field, known as *straightening* the flow. The presence of a discontinuity surface means that straightening is generally not possible even when $\mathbf{f}$ is piecewise-constant, as there is insufficient coordinate freedom to straighten two independent vector fields. Still, near a typical point on a discontinuity surface, it is possible to piecewise-straighten the vector field so that it becomes piecewise-constant. The discontinuity surface $\mathcal{D}$ will typically not be flat in such coordinates; see Figure 6.2. This useful device reduces study of a non-vanishing flow to that of a near-trivial piecewise-constant vector field, leaving only the non-trivial geometry of the curved level set $\sigma = 0$.

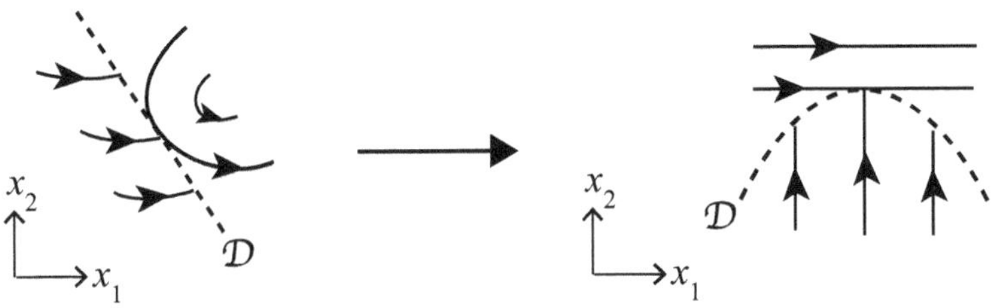

Fig. 6.2 Piecewise straightening of one of the piecewise-smooth flows from Figure 6.1.

The flow is then trivial, but $\mathcal{D}$ becomes curved. The local forms of $\mathcal{D}$, i.e. of the function σ, in these piecewise-straightened flows are found as follows.

Assuming that a point $\mathbf{x}_p$ is not an equilibrium of either of the constituent fields, and $\mathbf{f}^{\pm}$ are linearly independent, that is:

$$\mathbf{f}^{+}(\mathbf{x}_p) \neq 0\,, \quad \mathbf{f}^{-}(\mathbf{x}_p) \neq 0\,, \quad \& \quad \mathbf{f}^{+}(\mathbf{x}_p) \neq \mu \mathbf{f}^{-}(\mathbf{x}_p) \tag{6.4}$$

for any $\mu \in \mathbb{R}$, take local coordinates $\mathbf{x} = (y_1, y_2, \ldots, y_n)$ such that $\mathbf{x}_p = (0, 0, \ldots, 0)$ and

$$\delta_t = \begin{cases} \delta_{t+} = \frac{\partial}{\partial y_1} \text{ if } \sigma(\mathbf{x}) > 0\,, \\ \delta_{t-} = \frac{\partial}{\partial y_2} \text{ if } \sigma(\mathbf{x}) < 0\,. \end{cases} \tag{6.5}$$

In these coordinates the vector fields are simply

$$\begin{aligned} \mathbf{f}^{+}(\mathbf{x}) &= (1, 0, 0, \ldots)\,, \\ \mathbf{f}^{-}(\mathbf{x}) &= (0, 1, 0, \ldots)\,. \end{aligned} \tag{6.6}$$

It only remains to find an expression for the discontinuity surface $\mathcal{D}$, which we do by local expansion of σ and the use of a little singularity theory.

The shape of $\mathcal{D}$ is related to the vector fields by applying the derivative along their flows, given by (6.5), to the switching function σ. Thus if the vector field $\mathbf{f}^{+}$ is tangent to $\mathcal{D}$ at $\mathbf{x}_p = \mathbf{0}$ (where $\mathbf{0}$ denotes $(0, \ldots, 0)$), then $\partial\sigma(\mathbf{0})/\partial y_1 = 0$. If higher derivatives $\partial^2\sigma(\mathbf{0})/\partial y_1^2$ or $\partial^2\sigma(\mathbf{0})/\partial y_1 \partial y_2$ vanish, each one tells us something about how the manifold $\sigma = 0$ curves relative to the constituent flows along y_1 and y_2. If $\mathcal{D}$ simply folds as in Figure 6.2, then only one derivative vanishes, say $\partial\sigma(\mathbf{0})/\partial y_1 = 0$, while $\partial^2\sigma(\mathbf{0})/\partial^2 y_1 \neq 0$ and $\partial\sigma(\mathbf{0})/\partial y_2 \neq 0$. If either of these last two derivatives vanished, they would constitute a higher-order singularity. Similarly if the vector field $\mathbf{f}^{-}$ is tangent to $\mathcal{D}$ at $\mathbf{x}_p = (0, \ldots, 0)$, then $\partial\sigma(\mathbf{0})/\partial y_2 = 0$, and so on.

We call the vanishing derivatives *singularity conditions*, and the inequalities we call *non-degeneracy* conditions. Thus the problem of classifying tangencies of a piecewise-smooth system has been reduced to classifying the singularities of a scalar function σ, a job that has been done for us by catastrophe theory (see [13, 14, 176, 207] for good references). We apply this as follows.

It is useful to write

$$\sigma(y_1, y_2, \dots) = y_k + V(y_1, \dots, y_{k-1}, y_{k+1}, \dots) \tag{6.7}$$

for some integer k, such that $V(\mathbf{0}) = 0$ and $\delta_{\mathbf{x}} V(\mathbf{0}) = 0$. Then $\delta_{\mathbf{x}}\sigma(\mathbf{0})$ points along the coordinate y_k, i.e. y_k varies orthogonally to $\mathcal{D}$. If $\mathbf{f}^+$ is tangent to $\mathcal{D}$ but $\mathbf{f}^-$ is not, then we can choose $k = 2$ as in Figure 6.2 (and if $\mathbf{f}^-$ is tangent to $\mathcal{D}$ but $\mathbf{f}^+$ is not, then we can choose $k = 1$). If both $\mathbf{f}^+$ and $\mathbf{f}^-$ are tangent to $\mathcal{D}$, then the smallest k we can choose is $k = 3$.

The number of variables in the argument of V is the smallest possible for this expression to be structurally stable. It corresponds to the *codimension d* of the singularity at $\mathbf{x}_p = \mathbf{0}$, equal to the number of singularity conditions.

When we expand the scalar-valued function V locally as a multivariable Taylor series:

$$V(y_1, \dots) = V(0, \dots) + \mathbf{x} \cdot \delta_{\mathbf{x}} V(0, \dots) + \dots \tag{6.8}$$

the first two terms actually vanish by our choice of coordinates, so the expansion consists of higher-order terms in $y_1, y_2, \dots$, excluding y_k.

We can then build up a classification based on how many derivatives of V vanish. In the y_i coordinates, this amounts to considering different values of the codimension d and the transversal direction labelled k (Figures 6.3–6.8). We can lift the different forms from Thom's catastrophe classification (see [176]):

- ($d = 2$, $k = 2$): If there is quadratic contact in the projection of $\sigma = 0$ along the y_1 direction, i.e. along the flow of $\mathbf{f}^+$, we can find local coordinates such that $V = \frac{1}{2}y_1^2$, and take $k = 2$, then

$$\sigma = y_2 \pm \tfrac{1}{2}y_1^2 \,. \tag{6.9}$$

We call this a **fold** of $\mathcal{D}$ with respect to $\mathbf{f}^+$.

Fig. 6.3

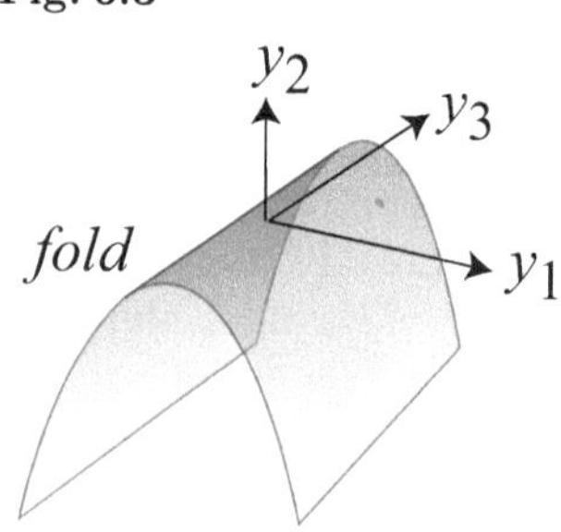

The next order of singularity could involve either a fold also along the y_2 direction, called a *two-fold*, or cubic contact along the y_1 direction, called a *cusp*. Let us take these in order.

- ($d = 3$, $k = 3$): If there are folds in the projection of $\sigma = 0$ along the y_1 and y_2 directions, i.e. along the flows of $\mathbf{f}^+$ and $\mathbf{f}^-$, we can find local coordinates such that $V = \pm\frac{1}{2}y_1^2 \pm \frac{1}{2}y_2^2$, forming a hill or saddle. We must then take $k = 3$, giving $\sigma = y_3 \pm \frac{1}{2}y_1^2 \pm \frac{1}{2}y_2^2$. This form is not structurally stable for our choice of coordinates, because the fold set in the y_1 projection (where $\partial V/\partial y_1 = 0$) lies along the y_2 axis and vice versa. A perturbation produces a $y_1 y_2$ term, and we can instead take

$$\sigma = y_3 + \tfrac{1}{2}\alpha y_1^2 - \tfrac{1}{2}\beta y_2^2 + y_1 y_2 \quad (6.10)$$

for some parameters α and β. We call this a **two-fold** of $\mathcal{D}$ with respect to $\mathbf{f}^+$ and $\mathbf{f}^-$.

Fig. 6.4

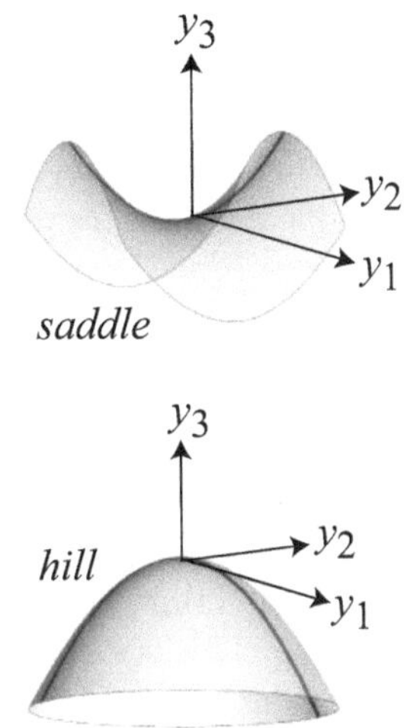

- ($d = 3$, $k = 2$): If there is a cubic contact in the projection of $\sigma = 0$ along the y_1 direction, we can find local coordinates such that $V = \pm\frac{1}{3}y_1^3$, and take $k = 2$. This is structurally unstable; a perturbation produces terms linear or quadratic in y_1, which we can reduce to $y_3 y_1$ by choosing a suitable coordinate y_3; thus

$$\sigma = y_2 \pm \left(\tfrac{1}{3}y_1^3 + y_3 y_1\right) . \quad (6.11)$$

We call this a **cusp** of $\mathcal{D}$ with respect to $\mathbf{f}^+$.

Fig. 6.5

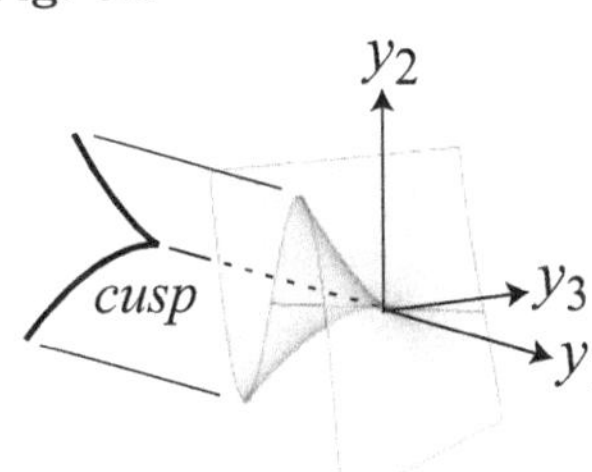

In the cusp we see a pair of folds annihilating at the singularity. The next highest orders of singularity will involve either quadratic contact along the y_1 direction where cusps annihilate with folds to form a *swallowtail*, or cusps along y_1 annihilating pairwise by approaching along another coordinate direction to form *umbilics*, or a cusp along y_1 meeting folds along y_2 to form *fold-cusps*. Let us take these in order.

- ($d = 4$, $k = 2$): If there is a quartic contact in the projection of $\sigma = 0$ along the y_1 direction, we can find local coordinates such that $V = \pm\frac{1}{4}y_1^4$, and take $k = 2$. This is structurally unstable; a perturbation produces terms of up to cubic order in y_1, but we can reduce these to $\frac{1}{2}y_3y_1^2 + y_4y_1$ by choosing suitable coordinates y_3 and y_4; thus

$$\sigma = y_2 \pm (\tfrac{1}{4}y_1^4 + \tfrac{1}{2}y_3y_1^2 + y_4y_1) \,. \quad (6.12)$$

We call this a **swallowtail** of $\mathcal{D}$ with respect to $\mathbf{f}^+$.

Fig. 6.6

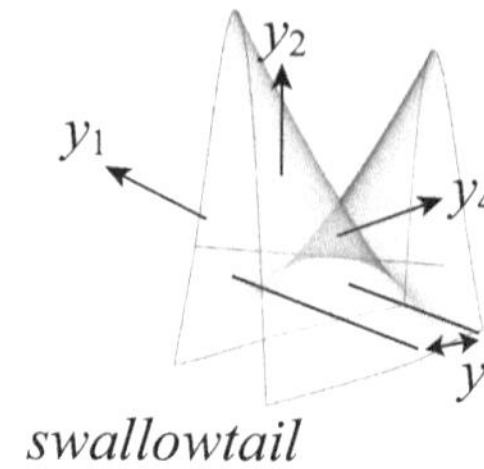

- ($d = 4$, $k = 2$): If there is a cubic contact in the projection of $\sigma = 0$ along the y_1 direction, we take $k = 2$ as in the cusp. This is structurally unstable, and under perturbation, symmetry in the y_3 direction means this can only be reduced to $\pm(y_3^2 + y_4)y_1$ by choosing suitable coordinates y_3 and y_4; thus

$$\sigma = y_2 \pm (\tfrac{1}{3}y_1^3 \pm (y_3^2 + y_4)y_1) \,. \quad (6.13)$$

We call this an **umbilic** of $\mathcal{D}$ with respect to $\mathbf{f}^+$, and the second $\pm$ sign produces two distinct cases, called the **lips** and **beak-to-beak**.

Fig. 6.7

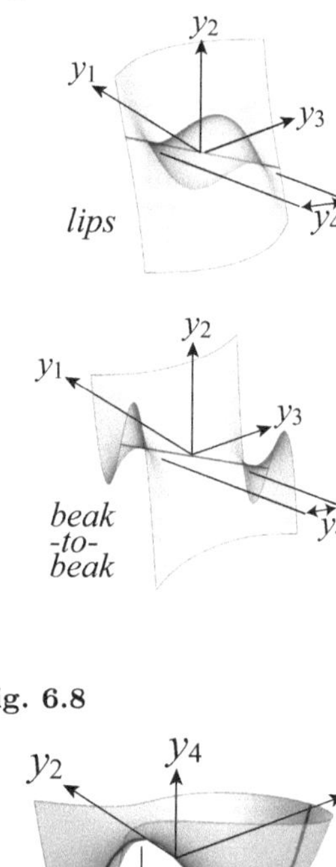

- ($d = 4$, $k = 3$): If there is a cubic contact in the projection of $\sigma = 0$ along the y_1 direction, *and* a quadratic contact in the projection along the y_2 direction, we can find local coordinates such that $V = \pm\frac{1}{3}y_1^3 \pm \frac{1}{2}y_2^2$, and take $k = 2$. This is structurally unstable; a perturbation produces linear or quadratic terms in y_1, which we can reduce to y_4y_1 by choosing a suitable coordinate y_4; thus

$$\sigma = y_3 \pm (\tfrac{1}{3}y_1^3 + y_4y_1 + y_2^2) \,. \quad (6.14)$$

Fig. 6.8

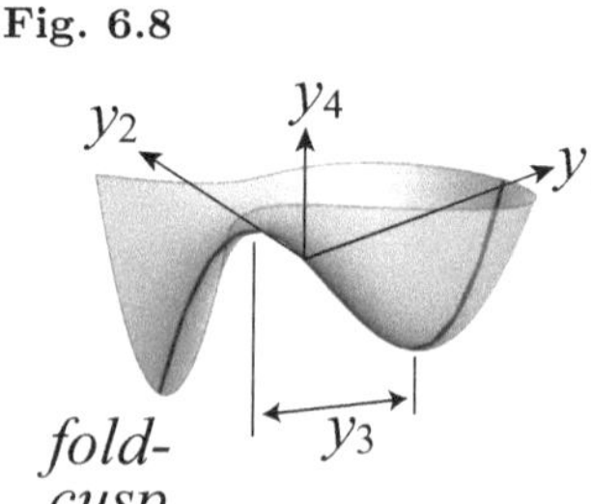

The possibilities increase with the codimension d, and the classification goes on without end. The singularities listed above will be more than sufficient for our purposes. The implications of these for the flow are worked through in Section 6.2 to Section 6.7 below.

The transversality of certain nullclines in the flow will be important for genericity, and for establishing this, we make use of a shorthand

$$\Delta(a, b, \dots) = \det |\delta_{\mathbf{x}} a, \delta_{\mathbf{x}} b, \dots| \,. \tag{6.15}$$

If $\Delta(a, b, \dots) \neq 0$, this means the gradient vectors $\delta_{\mathbf{x}} a, \delta_{\mathbf{x}} b, \dots$ are linearly independent, i.e. the matrix $(\delta_{\mathbf{x}} a, \delta_{\mathbf{x}} b, \dots)$ has full rank.

6.2 Fold ($d = 2,\ k = 2$)

Definition 6.1. A *fold* is a point $\mathbf{x}_p$ where

$$\sigma(\mathbf{x}_p) = \delta_{t+}\sigma(\mathbf{x}_p) = 0 \,,$$

with

$$\delta^2_{t+}\sigma(\mathbf{x}_p) \neq 0 \quad \text{and} \quad \delta_{t-}\sigma(\mathbf{x}_p) \neq 0 \,.$$

The fold is:

- *attracting* if $\delta_{t-}\sigma(\mathbf{x}_p) > 0$ and *repelling* if $\delta_{t-}\sigma(\mathbf{x}_p) < 0$,
- *visible* if $\delta^2_{t+}\sigma(\mathbf{x}_p) > 0$ and *invisible* if $\delta^2_{t+}\sigma(\mathbf{x}_p) < 0$.

These are shown in Figure 6.9.

Attractivity at a tangency will generally refer to the attractivity of the abutting region of sliding. Visibility refers to the concavity of a tangency: if a vector field curves towards a surface, it has an invisible tangency (since the relevant constituent field trajectory through the tangency point is not a visible part of the flow); if the field curves away, then the tangency is called visible. This means there are four distinct cases of folds: attracting-visible, attracting-invisible, repelling-visible, and repelling-invisible.

A prototype is given by taking (6.5) with (6.9), so

$$\sigma = y_2 \pm \tfrac{1}{2} y_1^2 \,.$$

In the straightened piecewise-constant vector field (6.6), this satisfies the defining conditions above, namely, at $\mathbf{x} = (0, 0, \dots, 0)$:

$$\sigma = 0 \qquad \& \qquad \frac{\partial}{\partial y_1}\sigma = \pm y_1 = 0 \,, \tag{6.16}$$

with the non-degeneracy conditions

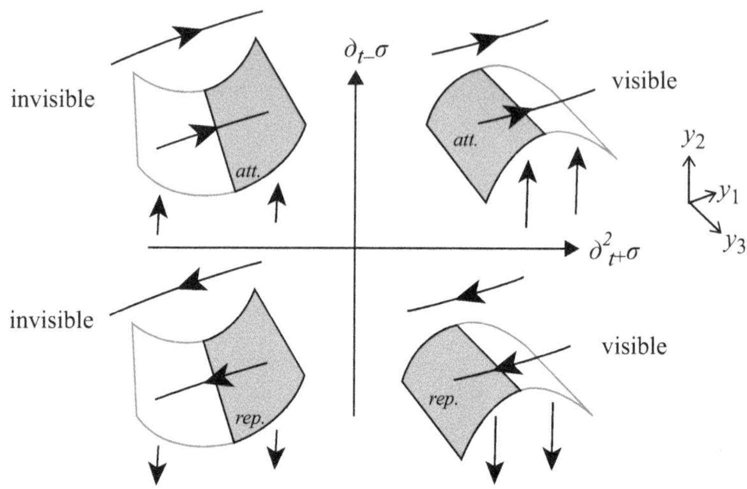

Fig. 6.9 Folds are visible or invisible and separate regions of crossing (unshaded) from attracting (*att.*) or repelling (*rep.*) sliding (shaded). The term *visible / invisible* identifies a flow curving away from/towards the discontinuity surface.

$$\frac{\partial^2}{\partial y_1^2}\sigma = \pm 1\,, \quad \frac{\partial}{\partial y_2}\sigma = 1\,, \quad \Delta = \left|\begin{pmatrix} 0 & 1 \\ \pm 1 & 0 \end{pmatrix}\right| = \mp 1\,. \tag{6.17}$$

Thus the '+' sign in (6.9) describes a visible fold and the '−' sign an invisible fold. Both are of the attracting type, with repelling cases being obtained by reversing time throughout the flow. On the discontinuity surface $\sigma = 0$, the fold lies along $y_1 = y_2 = 0$.

To see what this looks like in coordinates where the discontinuity surface is flat, we transform to $(x_1, x_2) = (y_1, \sigma)$, and then the fold vector field near a flat discontinuity surface is given by

$$(\dot{x}_1, \dot{x}_2, \dots) = \begin{cases} (1, \pm x_1, 0, 0, \dots) \text{ if } x_2 > 0\,, \\ (0, 1 \quad\;\, , 0, 0, \dots) \text{ if } x_2 < 0\,. \end{cases} \tag{6.18}$$

6.3 Two-Fold ($d = 3$, $k = 3$)

Definition 6.2. A *two-fold* is a point $\mathbf{x}_p$ where

$$\sigma(\mathbf{x}_p) = \delta_{t+}\sigma(\mathbf{x}_p) = \delta_{t-}\sigma(\mathbf{x}_p) = 0\,,$$

with

$$\delta^2_{t\pm}\sigma \neq 0\,, \quad 0 \notin \delta_{t\lambda}\mathbf{x}\,, \quad \& \quad \Delta(\sigma, \delta_{t+}\sigma, \delta_{t-}\sigma) \neq 0$$

at $\mathbf{x}_p$. The folds are:

- both *visible* if $\delta_{t+}\sigma(\mathbf{x}_p) > 0$ and $\delta_{t-}\sigma(\mathbf{x}_p) < 0$,
- both *invisible* if $\delta_{t+}\sigma(\mathbf{x}_p) < 0$ and $\delta_{t-}\sigma(\mathbf{x}_p) > 0$,
- one *visible* and one *invisible* if $\delta_{t+}\sigma(\mathbf{x}_p)\ \delta_{t-}\sigma(\mathbf{x}_p) > 0$.

These are shown in Figure 6.10.

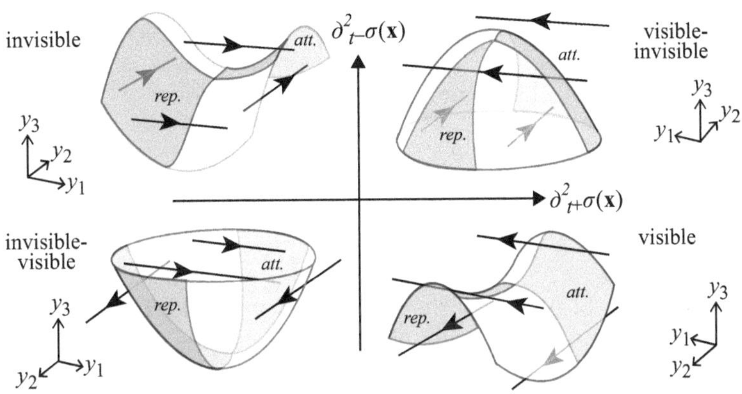

Fig. 6.10 Two-folds come in three flavours depending on whether the folds that comprise them are visible or invisible as determined by the signs of $\partial^2_{t\pm}\sigma$ (the top-right and bottom-left cases are topologically equivalent). Regions of attracting sliding (*att.*, shaded), repelling sliding (*rep.*, shaded), and crossing (unshaded) all meet at the singularity.

An important consequence of the non-degeneracy conditions is that $0 \notin \frac{d}{dt}\mathbf{x}$, so the flow is not stationary at the two-fold, yet *how* the flow traverses the singularity turns out to be highly non-trivial.

A prototype is given by taking (6.5) with (6.10), so

$$\sigma = y_3 + \tfrac{1}{2}\alpha y_1^2 - \tfrac{1}{2}\beta y_2^2 + y_1 y_2$$

with $\alpha\beta \neq -1$ and $\alpha, \beta \neq 0$. This satisfies the defining conditions since in the straightened piecewise-constant vector field (6.6) at $\mathbf{x} = (0, 0, \ldots, 0)$:

$$\sigma = 0\,, \quad \frac{\partial}{\partial y_1}\sigma = \alpha y_1 + y_2 = 0\,, \quad \frac{\partial}{\partial y_2}\sigma = y_1 - \beta y_2\,, \tag{6.19}$$

with the non-degeneracy conditions

$$\frac{\partial^2}{\partial y_1^2}\sigma = \alpha\,, \quad \frac{\partial^2}{\partial y_2^2}\sigma = -\beta\,, \quad \Delta = \begin{vmatrix} 0 & 0 & 1 \\ \alpha & 1 & 0 \\ 1 & -\beta & 0 \end{vmatrix} = -1 - \alpha\beta\,, \tag{6.20}$$

which are non-vanishing. If the Hessian determinant $\Delta = -1-\alpha\beta$ is negative, then $\mathcal{D}$ is locally a saddle; if it is positive, then $\mathcal{D}$ is locally a bowl.

The folds on $\sigma = 0$ lie along $\alpha y_1 + y_2 = 0$ and $y_1 - \beta y_2 = 0$, with the two-fold at $y_1 = y_2 = y_3 = 0$.

To obtain all cases for complex folds like the two-fold requires a little trick, redefining σ in (6.7) such that for $\sigma < 0$, we may flip the definition $\sigma \mapsto -\sigma$. This is done by introducing a new switching function $\hat{\sigma} = \gamma\sigma$ where $\gamma = 1$ for $\mathbf{x} \in \mathcal{R}_+$ and $\gamma = \pm 1$ for $\mathbf{x} \in \mathcal{R}_-$, so that in effect $\mathcal{R}_+$ and $\mathcal{R}_-$ lie on the same side of $\hat{\sigma} = 0$. It is easier to see what this means in coordinates where $\sigma = 0$ is flat.

In coordinates $(x_1, x_2, x_3) = (-s^+ y_1 + \nu^- y_2, \nu^+ y_1 + s^- y_2, \gamma\sigma)$, let $\alpha = \nu^-/s^-$, $\beta = \nu^+/s^+$, and $\gamma = \mp\nu^\mp/(s^+ s^- + \nu^+ \nu^-)$ for $\sigma \gtrless 0$; the two-fold vector field near a flat discontinuity surface is given by

$$(\dot{x}_1, \dot{x}_2, \dot{x}_3, \dots) = \begin{cases} (-s^+, \nu^+, -x_1,\ 0, 0, \dots) \text{ if } x_3 > 0\,, \\ (\ \ \nu^-, s^-, \ \ \ x_2,\ 0, 0, \dots) \text{ if } x_3 < 0\,. \end{cases} \tag{6.21}$$

The constants $s^\pm$ are the curvatures $\ddot{x}_3 = \delta^2_{t\pm} x_3 = s^\pm$ at the singularity $x_1 = x_2 = x_3 = 0$.

This is only the beginning of the story for the two-fold. It has proven to be one of the most important and illuminating singularities in piecewise-smooth dynamical theory, and we shall see that it remains one of the most intriguing. It was historically our first brush with determinacy-breaking via tangencies in piecewise-smooth flows, and its story does not end until we have studied its intricate crossing dynamics and its structural stability in the switching layer in Chapter 13.

6.4 Cusp ($d = 3,\ k = 2$)

Definition 6.3. A *cusp* is a point $\mathbf{x}_p$ where

$$\sigma(\mathbf{x}_p) = \delta_{t+}\sigma(\mathbf{x}_p) = \delta^2_{t+}\sigma(\mathbf{x}_p) = 0$$

with

$$\delta^3_{t+}\sigma \neq 0\,, \quad \delta_{t-}\sigma \neq 0 \quad \text{and} \quad \Delta\left(\sigma, \delta_{t+}\sigma, \delta^2_{t+}\sigma\right) \neq 0$$

at $\mathbf{x}_p$. The cusp is

- *attracting* if $\delta_{t-}\sigma(\mathbf{x}_p) > 0$ and *repelling* if $\delta_{t-}\sigma(\mathbf{x}_p) < 0$,
- *visible* if $(\delta^3_{t+}\sigma(\mathbf{x}_p))(\delta_{t-}\sigma(\mathbf{x}_p)) < 0$ and
  invisible if $(\delta^3_{t+}\sigma(\mathbf{x}_p))(\delta_{t-}\sigma(\mathbf{x}_p)) > 0$.

These are shown in Figure 6.11.

As in the case of the fold, attractivity here refers to whether the abutting sliding region is attracting or repelling. The visibility, on the other hand, refers here not to the vector fields either side of the discontinuity surface (the cusp being a coalescence of visible and invisible folds), but to a tangency that

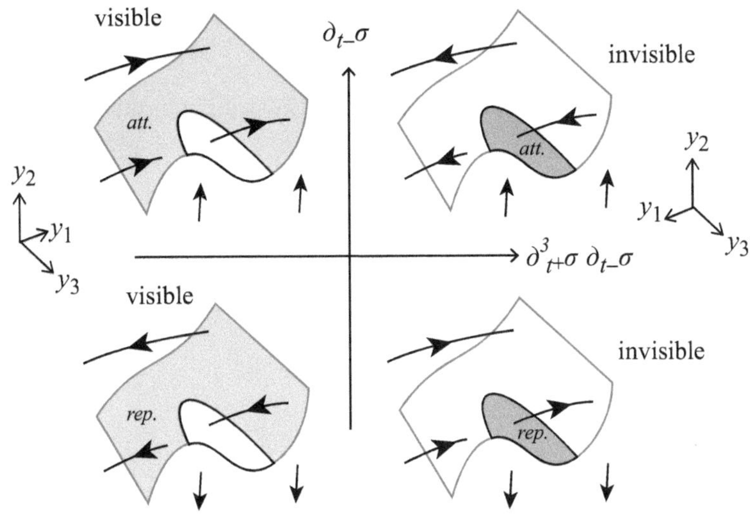

Fig. 6.11 Cusps are visible or invisible and sit at the branching of visible and invisible folds which separate regions of crossing (unshaded) from attracting (*att.*) or repelling (*rep.*) sliding (shaded). The term *visible* [or *invisible*] identifies the sliding flow curving away from [or towards] the fold curve.

arises between the sliding vector field and the sliding boundary; to really see this, we must wait until Section 8.5.3. Combining stability and visibility again gives four cases.

A prototype is given by taking (6.5) with (6.11), so

$$\sigma = y_2 \pm \left(\tfrac{1}{3}y_1^3 + y_3y_1\right) .$$

In the straightened piecewise-constant vector field (6.6), this satisfies the defining conditions above, namely, at $\mathbf{x} = (0, 0, \dots, 0)$:

$$\sigma = 0\,, \quad \frac{\partial}{\partial y_1}\sigma = \pm(y_1^2 + y_3) = 0\,, \quad \frac{\partial^2}{\partial y_1^2}\sigma = \pm 2y_1 = 0\,, \tag{6.22}$$

with the non-degeneracy conditions

$$\frac{\partial^3}{\partial y_1^3}\sigma = \pm 2, \quad \frac{\partial}{\partial y_2}\sigma = 1\,, \quad \Delta = \begin{vmatrix} 0 & 1 & 0 \\ 0 & 0 & \pm 1 \\ \pm 2 & 0 & 0 \end{vmatrix} = 2\,. \tag{6.23}$$

On $\sigma = 0$, there are folds along $y_1^2 + y_3 = 0$, one visible and one invisible branch along each of $y_1 = \pm\sqrt{3}$, meeting in the cusp at $y_1 = y_2 = y_3 = 0$.

Transforming to coordinates $(x_1, x_2, x_3) = (y_1, \sigma, y_3)$, the cusp vector field near a flat discontinuity surface is given by

$$(\dot{x}_1, \dot{x}_2, \dot{x}_3, \dots) = \begin{cases} (1, \pm(x_1^2 + x_3, 0, \dots) & \text{if } x_2 > 0, \\ (0, 1, 0, \dots) & \text{if } x_2 < 0, \end{cases} \tag{6.24}$$

and a particularly simple form arises if we also straighten out the tangency set $x_1^2 + x_3 = 0$; letting $\xi_3 = \frac{1}{2}(x_1^2 + x_3)$ and $\xi_1 = x_1$, $\xi_2 = \frac{1}{2}x_2$, then

$$(\dot{\xi}_1, \dot{\xi}_2, \dot{\xi}_3, \dots) = \begin{cases} (1, \pm\xi_3, \xi_1, \dots) & \text{if } \xi_2 > 0, \\ (0, \frac{1}{2}, 0, \dots) & \text{if } \xi_2 < 0. \end{cases} \tag{6.25}$$

6.5 Swallowtail ($d = 4,\ k = 2$)

Definition 6.4. A *swallowtail* is a point $\mathbf{x}_p$ where

$$\sigma(\mathbf{x}_p) = \delta_{t+}\sigma(\mathbf{x}_p) = \delta_{t+}^2\sigma(\mathbf{x}_p) = \delta_{t+}^3\sigma(\mathbf{x}_p) = 0$$

with

$$\delta_{t+}^4\sigma \neq 0\,, \quad \delta_{t-}\sigma \neq 0\,, \quad \text{and}$$
$$\Delta\left(\sigma, \delta_{t+}\sigma, \delta_{t+}^2\sigma, \delta_{t+}^3\sigma\right) \neq 0$$

at $\mathbf{x}_p$. This is illustrated in Figure 6.12.

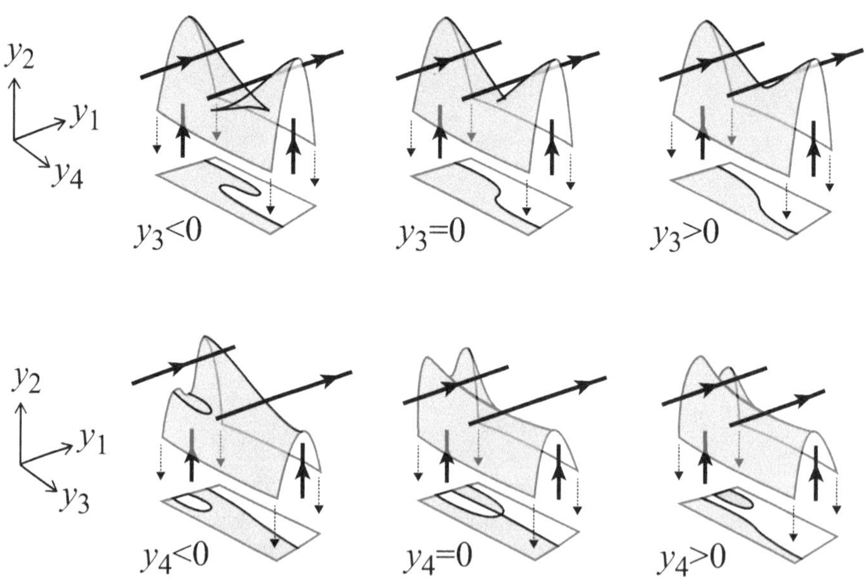

Fig. 6.12 Swallowtails require four dimensions to unfold; here we take different three-dimensional slices and two-dimensional projections showing the fold curves.

A prototype is given by taking (6.5) with (6.12), so

$$\sigma = y_2 \pm \left(\tfrac{1}{4}y_1^4 + \tfrac{1}{2}y_3y_1^2 + y_4y_1\right) .$$

This satisfies the defining conditions at $\mathbf{x} = (0, 0, \dots, 0)$:

$$\begin{aligned} &\sigma = 0 \,, \quad \tfrac{\partial}{\partial y_1}\sigma = \pm(y_1^3 + y_3y_1 + y_4) = 0 \,, \\ &\tfrac{\partial^2}{\partial y_1^2}\sigma = \pm(3y_1^2 + y_3) = 0 \,, \quad \tfrac{\partial^3}{\partial y_1^3}\sigma = \pm 6y_1 = 0 \,, \end{aligned} \tag{6.26}$$

with the non-degeneracy conditions

$$\frac{\partial^4}{\partial y_1^4}\sigma = \pm 6 \,, \quad \frac{\partial}{\partial y_2}\sigma = 1 \,, \quad \Delta = \left| \begin{pmatrix} 0 & 1 & 0 & 0 \\ 0 & 0 & 0 & \pm 1 \\ 0 & 0 & \pm 1 & 0 \\ \pm 6 & 0 & 0 & 0 \end{pmatrix} \right| = \pm 6 \,. \tag{6.27}$$

On $\sigma = 0$ there are folds along $y_1^3 + y_3y_1 + y_4 = 0$, which coalesce at cusps along $3y_1^2 + y_3 = 0$, which coalesce at the swallowtail at $y_1 = y_2 = y_3 = y_4 = 0$.

Transforming to coordinates $(x_1, x_2, x_3, x_4) = (y_1, \sigma, y_3, y_4)$, the swallowtail vector field near a flat discontinuity surface is given by

$$(\dot{x}_1, \dot{x}_2, \dot{x}_3, \dot{x}_4, \dots) = \begin{cases} (1, \pm(x_1^3 + x_3x_1 + x_4), 0, 0, \dots) & \text{if } x_2 > 0, \\ (0, 1, 0, 0, \dots) & \text{if } x_2 < 0. \end{cases} \tag{6.28}$$

6.6 Umbilic: Lips and Beaks ($d = 4,\ k = 2$)

Definition 6.5. An *umbilic* is a point $\mathbf{x}_p$ where

$$\sigma(\mathbf{x}_p) = \delta_{t+}\sigma(\mathbf{x}_p) = \delta_{t+}^2\sigma(\mathbf{x}_p) = \frac{\partial}{\partial y_3}\delta_{t+}\sigma(\mathbf{x}_p) = 0$$

where y_3 is a direction transverse to the two flow directions at $\mathbf{x}_p$, with

$$\delta_{t+}^3\sigma \neq 0 \,, \quad \delta_{t-}\sigma \neq 0 \,, \quad \frac{\partial^2}{\partial y_3^2}\delta_{t+}\sigma \neq 0 \,, \quad \text{and}$$

$$\Delta\left(\sigma, \delta_{t+}\sigma, \delta_{t+}^2\sigma, \frac{\partial}{\partial y_3}\delta_{t+}\sigma\right) \neq 0 \,,$$

at $\mathbf{x}_p$.

The condition $\frac{\partial}{\partial y_3}\delta_{t+}\sigma(\mathbf{x}_p) = 0$ with the last non-degeneracy condition means that the function $\delta_{t+}\sigma$ has a fold at $\mathbf{x}_p$ (a Morse point); we choose the coordinate y_3 such that the conditions appear as written here in terms of $\frac{\partial}{\partial y_3}\delta_{t+}\sigma$.

There are two types of umbilic.

Near a lips umbilic, a prototype is given by taking (6.5) from (6.13) with the + sign, so

$$\sigma = y_2 \pm \left(\tfrac{1}{3}y_1^3 + (y_3^2 + y_4)y_1\right) .$$

This is shown in Figure 6.13. This satisfies the defining conditions at $\mathbf{x} = (0, 0, \ldots, 0)$:

$$\begin{array}{ll} \sigma = 0 \,, & \frac{\partial}{\partial y_1}\sigma = \pm(y_1^2 + y_3^2 + y_4) = 0 \,, \\ \frac{\partial^2}{\partial y_1^2}\sigma = \pm 2y_1 = 0 \,, & \frac{\partial^2}{\partial y_3 y_1}\sigma = \pm 2y_3 = 0 \,, \end{array} \tag{6.29}$$

with the non-degeneracy conditions

$$\begin{array}{ll} \frac{\partial^3}{\partial y_1^3}\sigma = \pm 2 \,, & \frac{\partial}{\partial y_2}\sigma = \pm 1 \,, \\ \frac{\partial^3}{\partial y_3^2 y_1}\sigma = \pm 2 \,, & \Delta = \begin{vmatrix} 0 & 1 & 0 & 0 \\ 0 & 0 & 0 & \pm 1 \\ \pm 2 & 0 & 0 & 0 \\ 0 & 0 & \pm 2 & 0 \end{vmatrix} = \mp 4 \,. \end{array} \tag{6.30}$$

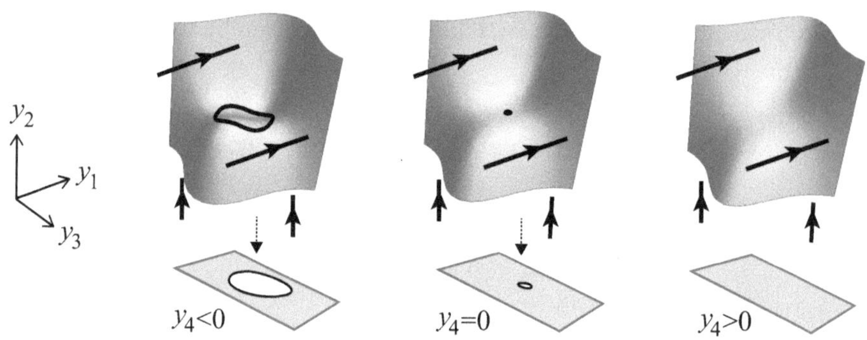

Fig. 6.13 Lips require four dimensions to unfold; here we take different three-dimensional slices and two-dimensional projections showing the fold curves.

Near a beak-to-beak umbilic, a prototype is given by taking (6.5) from (6.13) with the − sign, so

$$\sigma = y_2 \pm \left(\tfrac{1}{3}y_1^3 - (y_3^2 + y_4)y_1\right) . \tag{6.31}$$

This is shown in Figure 6.14. This also satisfies the defining conditions at $\mathbf{x} = (0, 0, \ldots, 0)$:

$$\begin{array}{ll} \sigma = 0 \,, & \frac{\partial}{\partial y_1}\sigma = \pm(y_1^2 - y_3^2 - y_4) = 0 \,, \\ \frac{\partial^2}{\partial y_1^2}\sigma = \pm 2y_1 = 0 \,, & \frac{\partial^2}{\partial y_3 y_1}\sigma = \mp 2y_3 = 0 \,, \end{array} \tag{6.32}$$

with the non-degeneracy conditions

$$\frac{\partial^3}{\partial y_1^3}\sigma = \pm 2\,, \quad \frac{\partial}{\partial y_2}\sigma = \pm 1\,,$$
$$\frac{\partial^3}{\partial y_3^2 y_1}\sigma = \pm 2\,, \quad \Delta = \begin{vmatrix} 0 & 1 & 0 & 0 \\ 0 & 0 & 0 & \mp 1 \\ \pm 2 & 0 & 0 & 0 \\ 0 & 0 & \mp 2 & 0 \end{vmatrix} = \mp 4\,. \tag{6.33}$$

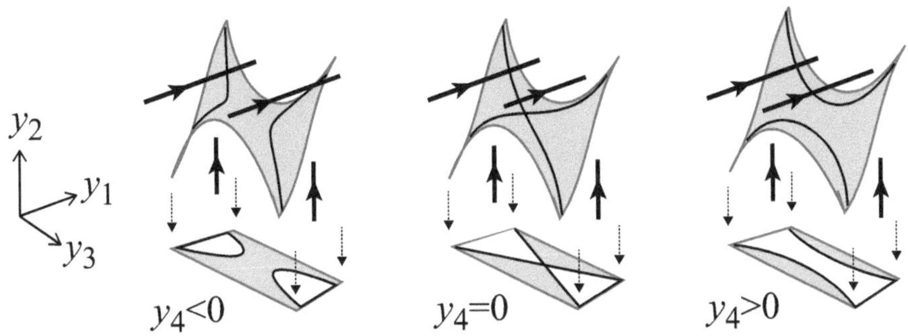

Fig. 6.14 Beak-to-beaks require four dimensions to unfold; here we take different three-dimensional slices and two-dimensional projections showing the fold curves.

In the umbilics, on $\sigma = 0$, there are folds along $y_1^2 \pm (y_3^2 + y_4) = 0$, which coalesce at cusps along $y_1 = 0$, which coalesce at the umbilic at $y_1 = y_2 = y_3 = y_4 = 0$.

Transforming to coordinates $(x_1, x_2, x_3, x_4) = (y_1, \sigma, y_3, y_4)$, the umbilic vector fields near a flat discontinuity surface are given by

$$(\dot{x}_1, \dot{x}_2, \dot{x}_3, \dot{x}_4, \dots) = \begin{cases} (1, \pm(x_1^2 + x_3^2 \pm x_4), 0, 0, \dots) & \text{if } x_2 > 0, \\ (0, 1, 0, 0, \dots) & \text{if } x_2 < 0. \end{cases} \tag{6.34}$$

6.7 Fold-Cusp ($d = 4,\ k = 3$)

Definition 6.6. A *fold-cusp* is a point $\mathbf{x}_p$ where

$$\sigma(\mathbf{x}_p) = \delta_{t+}\sigma(\mathbf{x}_p) = \delta_{t+}^2\sigma(\mathbf{x}_p) = \delta_{t-}\sigma(\mathbf{x}_p) = 0\,,$$

with

$$\delta_{t+}^3\sigma \neq 0\,, \quad \delta_{t-}^2\sigma \neq 0\,, \quad \text{and}$$
$$\Delta\left(\sigma, \delta_{t+}\sigma, \delta_{t+}^2\sigma, \delta_{t-}\sigma\right) \neq 0$$

at $\mathbf{x}_p$. These are shown in Figure 6.15.

Near a fold-cusp, a prototype is given by taking (6.5) from (6.14),

$$\sigma = y_3 \pm \left(\tfrac{1}{3}y_1^3 + y_4 y_1 + \tfrac{1}{2}y_2^2\right) .$$

This satisfies the defining conditions at $\mathbf{x} = (0, 0, \ldots, 0)$:

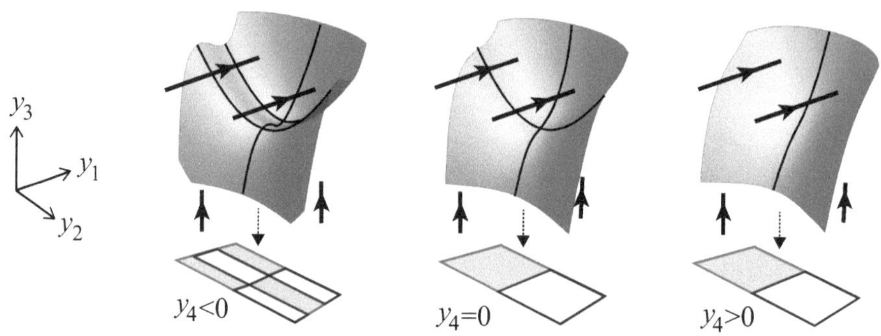

Fig. 6.15 Fold-cusps require four dimensions to unfold; here we take different three-dimensional slices and two-dimensional projections showing the fold curves.

$$\begin{array}{ll} \sigma = 0 \,, & \frac{\partial}{\partial y_1}\sigma = \pm(y_1^2 + y_4) = 0 \,, \\ \frac{\partial^2}{\partial y_1^2}\sigma = \pm 2y_1 = 0 \,, & \frac{\partial}{\partial y_2}\sigma = \pm y_2 = 0 \,, \end{array} \tag{6.35}$$

with the non-degeneracy conditions

$$\frac{\partial^3}{\partial y_1^3}\sigma = 2 \,, \quad \frac{\partial^2}{\partial y_2^2}\sigma = \pm 1 \,, \quad \Delta = \begin{vmatrix} 0 & 0 & \pm 1 & 0 \\ 0 & 0 & 0 & \pm 1 \\ \pm 2 & 0 & 0 & 0 \\ 0 & 1 & 0 & 0 \end{vmatrix} = \pm 2 \,. \tag{6.36}$$

In the fold-cusp, on $\sigma = 0$, there are folds along $y_1^2 + y_4 = 0$ which coalesce at $y_1 = 0$, and there is a cusp at $y_2 = 0$, all of which coalesce at the umbilic at $y_1 = y_2 = y_3 = y_4 = 0$.

Transforming to coordinates $(x_1, x_2, x_3, x_4) = (y_1, \sigma, y_3, y_4)$, the fold-cusp vector field near a flat discontinuity surface is given by

$$(\dot{x}_1, \dot{x}_2, \dot{x}_3, \dot{x}_4, \ldots) = \begin{cases} (1, \pm(x_1^2 + x_4), 0, 0, \ldots) & \text{if } x_2 > 0, \\ (0, 1, x_2, 0, \ldots) & \text{if } x_2 < 0. \end{cases} \tag{6.37}$$

6.8 Many-fold Singularities, Cusp-Cusps, and So On

This classification continues into higher codimensions without end, beginning with the fifth-order *butterfly* singularity (which follows directly on from the cusp and swallowtail), the wigwam, second umbilics, and so on.

We have used catastrophe theory to obtain, somewhat 'out of thin air', the prototypes above based on topological arguments for piecewise straightening of the local flow. More detailed derivations can be found in [206, 120], but to prove rigorously that a system can be locally transformed into these prototypes is not straightforward, even for the simple fold. The proofs for the fold and two-fold singularities are as follows; the proofs for the higher-order singularities are left to the reader.

6.9 Proofs of Leading-Order Expressions for the Fold and Two-Fold

The classification above provides expressions that capture the generic local topology. On a case-by-case basis, we can prove, or at least seek to prove, that any system with such singularities can locally be put into these forms. To show this can be done, we give proofs for the folds and two-fold only. We give these in coordinates where the discontinuity surface is flat.

To prove that Section 6.2 provides the leading-order behaviour of any singularity in Definition 6.1, we have the following:

Theorem 6.1. *Let the point* $\mathbf{x}_p = 0$ *be a fold on a discontinuity surface* $\sigma(\mathbf{x}) = 0$. *Then there exist coordinates* $\mathbf{x} = (x_1, x_2, \ldots, x_n)$ *in which the constituent vector fields are given by*

$$\begin{aligned} (\dot{x}_1, \dot{x}_2) &= \begin{cases} \left(1 + \mathsf{O}(|\mathbf{x}|) \,,\ \operatorname{sign}\left[\delta_{t+}^2\sigma(\mathbf{0})\right] x_1 \right) & \text{if } x_2 > 0, \\ \left(\mathsf{O}(|\mathbf{x}|) \qquad ,\ \operatorname{sign}\left[\delta_{t-}\sigma(\mathbf{0})\right] \right) & \text{if } x_2 < 0, \end{cases} \\ \dot{x}_i &= \mathsf{O}(|\mathbf{x}|) \quad \text{for } i = 3, \ldots, n. \end{aligned} \tag{6.38}$$

Essentially the proof (from [38]) involves taking coordinates $\mathbf{x} = (x_1, x_2, \ldots)$ in which $x_2 = 0$ represents the discontinuity surfaceand $x_1 = x_2 = 0$ represents the fold line. The details are a little more involved and are an example of the nontriviality of such results in nonsmooth systems.

Proof. Choose coordinates

$$x_1 = \gamma(\mathbf{x}) \left(\beta(\mathbf{x})\sigma(\mathbf{x}) + \delta_{t+}\sigma(\mathbf{x}) \right) \,, \qquad x_2 = |\alpha(\mathbf{x})|\sigma(\mathbf{x}) \,, \tag{6.39}$$

in terms of functions

$$\begin{aligned} \gamma(\mathbf{x}) &= \tfrac{1}{\delta_{t+}^2\sigma(\mathbf{x})} \,, \\ \{\alpha(\mathbf{x}), \beta(\mathbf{x})\} &= \begin{cases} \left\{ \gamma(\mathbf{x}),\ \tfrac{1}{\gamma}(\mathbf{x})\delta_{t+}\gamma(\mathbf{x}) \right\} & \text{if } x_2 > 0 \,, \\ \left\{ 1,\ -\tfrac{\delta_{t-}\delta_{t+}\sigma(\mathbf{x})}{\delta_{t-}\sigma(\mathbf{x})} \right\} & \text{if } x_2 < 0 \,. \end{cases} \end{aligned} \tag{6.40}$$

The quantities $\alpha\sigma$ and $\beta\sigma$ are continuous (since $x_2 = 0$ coincides with $\sigma = 0$); hence the right-hand sides in (6.39) are continuous but non-differentiable at $x_2 = 0$. The inequalities in Definition 6.1 guarantee that the functions α, β, γ, are finite and that α and γ are nonzero. Applying the operators $\delta_{t\pm}$ to the first two coordinates gives

$$(\dot{x}_1, \dot{x}_2) = \begin{cases} \left(1 + \mathbf{x} \cdot \boldsymbol{\varepsilon}^+(\mathbf{x})\,,\ \mathrm{sign}[\delta_{t+}^2\sigma(\mathbf{x})]x_1\right) & \text{if } x_2 > 0, \\ \left(\mathbf{x} \cdot \boldsymbol{\varepsilon}^-(\mathbf{x})\,,\ \delta_{t-}\sigma(\mathbf{x})\right) & \text{if } x_2 < 0, \end{cases} \tag{6.41}$$

in terms of functions

$$\begin{aligned} \boldsymbol{\varepsilon}^+ &= \left(2\beta,\ \mathrm{sign}(\gamma)(\delta_{t+}\beta - \beta^2),\ 0, 0, \ldots\right)\,, \\ \boldsymbol{\varepsilon}^- &= \left(\gamma^{-1}\delta_{t-}\gamma\,,\ \gamma\delta_{t-}\beta\ ,\ 0, 0, \ldots\right)\,. \end{aligned}$$

We can rescale time by $t \mapsto t/\mu$ without changing the phase portrait of the flow, provided μ is strictly positive (since rescaling time in the system (3.3) and its layer system (3.15) is an orbital equivalence, according to Definition 4.4). Let $\mu = 1$ for $x_2 > 0$ and let $\mu = |\delta_{t-}\sigma|$ for $x_2 < 0$, giving

$$(\dot{x}_1, \dot{x}_2) = \begin{cases} \left(1 + \mathbf{x} \cdot \boldsymbol{\varepsilon}^+(\mathbf{x}),\ \mathrm{sign}[\delta_{t+}^2\sigma(\mathbf{x})]x_1\right) & \text{if } x_2 > 0, \\ \left(\mathbf{x} \cdot \hat{\boldsymbol{\varepsilon}}^-(\mathbf{x}),\ \mathrm{sign}[\delta_{t-}\sigma(\mathbf{x})]\right) & \text{if } x_2 < 0, \end{cases} \tag{6.42}$$

where $\hat{\boldsymbol{\varepsilon}}^-(\mathbf{x}) = \boldsymbol{\varepsilon}^-(\mathbf{x})/|\delta_{t-}\sigma(\mathbf{x})|$. Taking a series expansion about $\mathbf{x}_p = 0$ gives the first line of the theorem

$$(\dot{x}_1, \dot{x}_2) = \begin{cases} \left(1 + \mathsf{O}\left(|\mathbf{x}|\right),\ \mathrm{sign}[\delta_{t+}^2\sigma(\mathbf{0})]x_1\right) & \text{if } x_2 > 0, \\ \left(\mathsf{O}\left(|\mathbf{x}|\right),\ \mathrm{sign}[\delta_{t-}\sigma(\mathbf{0})]\right) & \text{if } x_2 < 0. \end{cases} \tag{6.43}$$

The intersection of the hyperplane $x_1 = 0$ and the hyperplane $x_2 = 0$ is an $n-2$ dimensional surface, coordinatized by some $x_3, \ldots, x_n$, which remain to be chosen. We can choose a basis of these coordinates that is normal to the flow at $\mathbf{x}_p = 0$. Let $u_3, \ldots, u_n$, be any $n-2$ coordinates orthogonal to x_1, x_2; then let

$$x_i = u_i + \alpha_i(\mathbf{x})\sigma(\mathbf{x}) + \varphi(\mathbf{x})\delta_{t+}\sigma(\mathbf{x}) \qquad \text{for} \quad i = 3, 4, \ldots \tag{6.44}$$

where

$$\alpha_i = -\left(\delta_{t-}y_i + \varphi\delta_{t-}\delta_{t+}\sigma\right) / \left(\delta_{t-}\sigma\right)$$

and

$$\varphi = -\left(\delta_{t+}y_i\right) / \left(\delta_{t+}^2\sigma\right)\,.$$

Applying the derivatives $\partial_{t\pm}$ gives

$$\dot{x}_i = \begin{cases} \boldsymbol{\varepsilon}_i^+ \cdot \mathbf{x} \text{ if } x_2 > 0, \\ \boldsymbol{\varepsilon}_i^- \cdot \mathbf{x} \text{ if } x_2 < 0, \end{cases} + \mathsf{O}\left(|\mathbf{x}|^2\right) = \mathsf{O}\left(|\mathbf{x}|\right) \tag{6.45}$$

where

$$\begin{aligned}
\varepsilon_i^+ &= \left(\delta_{t+}\alpha_i - (\alpha_i + \delta_{t+}\varphi)\tfrac{\delta_{t+}\gamma}{\gamma},\ \alpha_i + \delta_{t+}\varphi,\ 0,\ 0, \dots \right)/\gamma, \\
\varepsilon_i^- &= \left(\delta_{t-}\alpha_i + \tfrac{(\delta_{t-}\delta_{t+}\sigma)\delta_{t-}\varphi}{\delta_{t-}\sigma},\ \tfrac{\delta_{t-}\varphi}{\gamma},\ 0,\ 0,\ \dots \right),
\end{aligned}$$

and since $\dot{x}_i = \mathsf{O}(|\mathbf{x}|)$, this completes the result (6.38). □

By the implicit function theorem, there exists an open set

$$\{(x_1, \dots, x_n) : x_1 = x_2 = 0\}$$

of fold points, in a region around $\mathbf{x}_p$ on which $\delta_{t+}^2\sigma(\mathbf{x})$ and $\delta_{t-}\sigma(\mathbf{x})$ remain nonzero. This result for n dimensions was derived in [38], building on local forms in two and three dimensions from [71, 206, 119].

To prove that Section 6.4 provides the leading-order behaviour of any singularity defined by Definition 6.2, we have the following.

Theorem 6.2. *Let the point $\mathbf{x}_p = 0$ be a two-fold singularity on a discontinuity surface $\sigma(\mathbf{x}) = 0$. Then there exist coordinates $\mathbf{x} = (x_1, \dots, x_n)$ in which the constituent vector fields are given by*

$$\left.\begin{aligned}
(\dot{x}_1, \dot{x}_2, \dot{x}_3) &= \begin{cases} \left(-\operatorname{sign}\left[\delta_{t+}^2\sigma(\mathbf{0})\right], \nu^+, -x_1\right) & \text{if } x_3 > 0, \\ \left(\nu^-, \operatorname{sign}\left[\delta_{t-}^2\sigma(\mathbf{0})\right], \ x_2\right) & \text{if } x_3 < 0, \end{cases} \\
&\quad + \begin{cases} \left(\mathsf{O}(|\mathbf{x}|), \mathsf{O}(|\mathbf{x}|), 0\right) & \text{if } x_3 > 0, \\ \left(\mathsf{O}(|\mathbf{x}|), \mathsf{O}(|\mathbf{x}|), 0\right) & \text{if } x_3 < 0. \end{cases} \\
\dot{x}_i &= \mathsf{O}(|\mathbf{x}|) \qquad \text{for } i = 4, 5, \dots, n,
\end{aligned}\right\} \tag{6.46}$$

The proof is in principle similar to that for the fold (again from [38]).

Proof. Define coordinates $\mathbf{x} = (x_1, \dots, x_n)$ in which the first three variables are given by

$$x_1 = -\beta(\mathbf{x})\delta_{t+}(\alpha(\mathbf{x})\sigma(\mathbf{x})), \quad x_2 = \frac{\delta_{t-}(\alpha(\mathbf{x})\sigma(\mathbf{x}))}{\beta(\mathbf{x})}, \quad x_3 = \alpha\sigma(\mathbf{x}), \tag{6.47}$$

where

$$\alpha = |\delta_{t+}^2\sigma\delta_{t-}^2\sigma|^{-1/2}, \qquad \beta = |\delta_{t-}^2\sigma/\delta_{t+}^2\sigma|^{1/4}. \tag{6.48}$$

The functions α and β are well-defined because the derivatives $\delta_{t\pm}^2\sigma$ are nonvanishing by Definition 6.2, and moreover the variables x_1, x_2, x_3 are linearly independent at the origin and form a valid coordinate system. The $n-3$ dimensional manifold

$$\{(x_1, \dots, x_n) :\ x_1 = x_2 = x_3 = 0\}$$

is a two-fold singularity, and the coordinates $x_4, \ldots, x_n$ remain to be chosen. Before we define those, simply rearranging (6.47) gives the dynamics on the first variable,

$$\delta_{t+} x_3 = -x_1/\beta(\mathbf{x}), \qquad \delta_{t-} x_3 = x_2 \beta(\mathbf{x}). \tag{6.49}$$

Applying the derivative operators $\delta_{t\pm}$ to the first two coordinates, x_1 and x_2, gives (omitting arguments $\mathbf{x}$)

$$\begin{aligned}(\dot{x}_1, \dot{x}_2, \dot{x}_3) = &\begin{cases} \left(-\alpha\beta^2 \delta_{t+}^2 \sigma \ , \alpha \delta_{t+} \delta_{t-} \sigma, -x_1\right)/\beta \text{ if } x_3 > 0, \\ \left(-\alpha \delta_{t-} \delta_{t+} \sigma, \alpha \beta^{-2} \delta_{t-}^2 \sigma, \ x_2\right) \beta \ \text{ if } x_3 < 0, \end{cases} \\ + &\begin{cases} \left(\mathbf{x} \cdot \boldsymbol{\varepsilon}_1^+, \ \mathbf{x} \cdot \boldsymbol{\varepsilon}_2^+, \ 0\right) \text{ if } x_3 > 0, \\ \left(\mathbf{x} \cdot \boldsymbol{\varepsilon}_1^-, \ \mathbf{x} \cdot \boldsymbol{\varepsilon}_2^-, \ 0\right) \text{ if } x_3 < 0, \end{cases}\end{aligned} \tag{6.50}$$

where some rearrangement of derivatives yields

$$\begin{aligned}
\boldsymbol{\varepsilon}_1^+ &= \left(\alpha\beta\delta_{t+}^2 \tfrac{1}{\alpha} \ , \tfrac{1}{\alpha^2\beta}\delta_{t+}(\alpha^2\beta), \quad 0 \quad , \ 0, 0, \ldots \right), \\
\boldsymbol{\varepsilon}_2^+ &= \left(-\tfrac{\alpha}{\beta}\delta_{t+}\delta_{t-}\tfrac{1}{\alpha}, -\tfrac{1}{\alpha\beta^2}\delta_{t-}\alpha \ , \ \tfrac{\beta}{\alpha}\delta_{t+}\tfrac{\alpha}{\beta} \ , \ 0, 0, \ldots \right), \\
\boldsymbol{\varepsilon}_1^- &= \left(\alpha\beta\delta_{t-}\delta_{t+}\tfrac{1}{\alpha} \ , \ \tfrac{1}{\alpha\beta}\delta_{t-}(\alpha\beta) \ , -\tfrac{\beta^2}{\alpha}\delta_{t+}\alpha, \ 0, 0, \ldots \right), \\
\boldsymbol{\varepsilon}_2^- &= \left(-\tfrac{\alpha}{\beta}\delta_{t-}^2 \tfrac{1}{\alpha} \ , \quad 0 \quad , \ \tfrac{\beta}{\alpha^2}\delta_{t-}\tfrac{\alpha^2}{\beta}, \ 0, 0, \ldots \right),
\end{aligned}$$

noting that α, β, and $\boldsymbol{\varepsilon}_i^\pm$ are all functions of $\mathbf{x}$. As in the proof of Theorem 6.1, we can apply a time rescaling $t \mapsto t/\mu$ provided that μ is strictly positive. Let $\mu = 1/\beta$ for $x_3 < 0$ and $\mu = \beta$ for $x_3 > 0$, given $\beta > 0$ by (6.48). Applying $t \mapsto t/\mu$ to (6.50), we then have

$$\begin{aligned}(\dot{x}_1, \dot{x}_2, \dot{x}_3) = &\begin{cases} bb - \operatorname{sign}[\delta_{t+}^2 \sigma(\mathbf{x})] \ , v^+(\mathbf{x}), -x_1 \text{ if } x_3 > 0, \\ (\ v^-(\mathbf{x}), \ \operatorname{sign}[\delta_{t-}^2 \sigma(\mathbf{x})], \ x_2 \ ) \text{ if } x_3 < 0, \end{cases} \\ + &\begin{cases} \left(\mathbf{x} \cdot \boldsymbol{\varepsilon}_1^+(\mathbf{x}), \ \mathbf{x} \cdot \boldsymbol{\varepsilon}_2^+(\mathbf{x}), \ 0\right) \beta(\mathbf{x}) \ \text{ if } x_3 > 0, \\ \left(\mathbf{x} \cdot \boldsymbol{\varepsilon}_1^-(\mathbf{x}), \ \mathbf{x} \cdot \boldsymbol{\varepsilon}_2^-(\mathbf{x}), \ 0\right) /\beta(\mathbf{x}) \text{ if } x_3 < 0, \end{cases}\end{aligned} \tag{6.51}$$

in terms of the functions $v^\pm$ defined as

$$\begin{aligned} v^+(\mathbf{x}) &:= \frac{(\delta_{t+})(\delta_{t-})\sigma(\mathbf{x})}{\sqrt{|\delta_{t+}^2 \sigma(\mathbf{x}) \delta_{t-}^2 \sigma(\mathbf{x})|}} \\ \& \qquad v^-(\mathbf{x}) &:= \frac{-\,(\delta_{t-})(\delta_{t+})\sigma(\mathbf{x})}{\sqrt{|\delta_{t+}^2 \sigma(\mathbf{x}) \delta_{t-}^2 \sigma(\mathbf{x})|}}. \end{aligned}$$

using the fact that $\alpha = 1/|\delta_{t+}^2 \sigma \delta_{t-}^2 \sigma|^{1/2}$, $\alpha\beta^2 = 1/|\delta_{t+}^2 \sigma|$, and $\alpha\beta^{-2} = 1/|\delta_{t-}^2 \sigma|$ by (6.48). Expanding (6.51) in powers of $x_1, x_2, \ldots, x_n$, about the point $\mathbf{x}_p = 0$, the vector field becomes

$$(\dot{x}_1, \dot{x}_2, \dot{x}_3) = \begin{cases} \left(-\operatorname{sign}\left[\delta_{t+}^2\sigma(\mathbf{0})\right], v^+(0), -x_1\right) & \text{if } x_3 > 0, \\ \left(\ v^-(0),\ \operatorname{sign}\left[\delta_{t-}^2\sigma(\mathbf{0})\right],\ x_2\ \right) & \text{if } x_3 < 0, \end{cases}$$
$$+ \begin{cases} \left(0,\ \mathsf{O}\left(|\mathbf{x}|\right),\ \mathsf{O}\left(|\mathbf{x}|\right)\right) & \text{if } x_3 > 0, \\ \left(0,\ \mathsf{O}\left(|\mathbf{x}|\right),\ \mathsf{O}\left(|\mathbf{x}|\right)\right) & \text{if } x_3 < 0. \end{cases} \tag{6.52}$$

Finally, the components $dx_{j>3}/dt$ are found as follows. At the point $\mathbf{x}_p$ inside the two-fold, the flow switches between two directions, say $\delta_{t+}\mathbf{x}$ and $\delta_{t-}\mathbf{x}$ evaluated at $\mathbf{x} = \mathbf{x}_p$, which by Definition 6.2 are linearly independent; let us call these two vectors $\mathbf{e}_+$ and $\mathbf{e}_-$. Orthogonal to $\mathbf{e}_+$, $\mathbf{e}_-$, and $\delta_{\mathbf{x}}\sigma$, there exist $n-3$ mutually orthogonal unit vectors $\mathbf{e}_i$, $i = 4, \ldots, n$, which we can take as the bases of coordinates $x_4, \ldots, x_n$. Then $dx_i/dt = 0$ at $\mathbf{x_p} = 0$, and we have

$$\dot{x}_i = \begin{cases} \mathbf{a}_i^+ \cdot \mathbf{x} \text{ if } x_3 > 0, \\ \mathbf{a}_i^- \cdot \mathbf{x} \text{ if } x_3 < 0, \end{cases} + \mathsf{O}\left(|\mathbf{x}|^2\right), \tag{6.53}$$

for $i = 4, 5, \ldots, n$, where $\mathbf{a}_i^{\pm}$ are vector constants $\mathbf{a}_i^{\pm} = \delta_{\mathbf{x}}\delta_{t\pm}x_i|_{\mathbf{x}=0}$. Since $\dot{x}_i = \mathsf{O}\left(|\mathbf{x}|\right)$, this completes (6.46). □

By expanding (6.47) to give $x_1 = -\beta\sigma\delta_{t+}\alpha - \beta\alpha\delta_{t+}\sigma$ and $x_2 = \gamma\sigma\delta_{t-}\alpha + \gamma\alpha\delta_{t-}\sigma$, we see that the role of x_1, x_2, x_3, is as follows: the $n-1$ dimensional manifold $x_3 = 0$ is the discontinuity surface where $\sigma = 0$, the $n-2$ dimensional manifold $x_3 = x_1 = 0$ is a fold where $\sigma = \delta_{t+}\sigma = 0$, and the $n-2$ dimensional manifold $x_3 = x_2 = 0$ is a fold where $\sigma = \delta_{t-}\sigma = 0$ (see Definition 6.1 for folds).

To obtain the expression of the two-fold given in (6.21), now simply permute the first three coordinates $(x_1, x_2, x_3) \mapsto (x_3, x_2, x_1)$.

6.10 A Note on Alternative Classifications

Piecewise-smooth vector fields in the plane have been the subject of several classifications, firstly in [71] and later revisited and extended in [140, 90, 48]. The forms we give for the fold (6.18), cusp (6.24), and two-fold (6.21) are equivalent to, and give unifying forms for, expressions given by previous authors. The fold, cusp, and two-folds all appear in the seminal work by Filippov [71], in similar forms to the vector fields given here for a flat discontinuity surface (in coordinates (x_1, x_2, x_3)). Kuznetsov [140] calls the cusp a 'double tangency'. The cusp follows directly from the expression for a cusp at the boundary of a manifold given in [194]. Teixeira's four normal forms for two-folds [205, 206] are obtained from equation (6.21) by a linear transformation, excluding Teixeira's case "a5" where the vector fields are not everywhere transverse. Teixeira [206] drew on results from the theory of singularities of mappings and of vector fields at the boundary of a manifold [194] to state that the fold, cusp, and two-fold are the only generic singularities in a two- or three-dimensional nonvanishing piecewise-smooth vector field.

6.11 Looking Forward

The singularity geometry encountered here will turn up throughout the following chapters, as it organizes much of the local and global dynamics of a system affected by discontinuities. The geometry of Section 6.1 will be of particular use in Chapter 11 when we confront global dynamics.

Chapter 7
Layer Analysis

In this section we discuss dynamics inside the switching layer, looking at how to apply fundamental notions like stability analysis and linearization across a discontinuity.

Local methods are given for systems with arbitrary dependence on the variable $\mathbf{x} = (x_1, \ldots, x_n)$ and the switching multipliers $\boldsymbol{\lambda} = (\lambda_1, \ldots, \lambda_m)$. We start from the general dynamical system (3.3)

$$\dot{\mathbf{x}} = \mathbf{f}(\mathbf{x}; \boldsymbol{\lambda}) \quad : \quad \lambda_j = \operatorname{sign}(\sigma_j(\mathbf{x})) , \quad j = 1, \ldots, m. \tag{7.1}$$

Alongside this we need the layer system, given generally by (3.13), but which we will typically express in coordinates where $\sigma_j = x_j$, which wherever possible we express in the form of (3.15), namely:

$$\left\{ \begin{array}{l} \left(\varepsilon_1 \dot{\lambda}_1, \ldots, \varepsilon_r \dot{\lambda}_r\right) = (f_1(\mathbf{x}; \boldsymbol{\lambda}) \quad , \ldots, f_r(\mathbf{x}; \boldsymbol{\lambda})) \ , \\ (\dot{x}_{r+1}, \ \ldots, \ \dot{x}_n) = (f_{r+1}(\mathbf{x}; \boldsymbol{\lambda}), \ldots, f_n(\mathbf{x}; \boldsymbol{\lambda})) \ , \end{array} \right. \tag{7.2}$$

on $x_1 = \cdots = x_r = 0$ for $r \leq m$, where all $\varepsilon_j \to 0$. This expression assumes that $\mathbf{x}$ lies on the intersection of r discontinuity submanifolds where $\sigma_j = x_j$ for $j = 1, \ldots, r$. The switching layer

$$(\lambda_1, \ldots, \lambda_r) \in (-1, +1)^r \ , \qquad (x_{r+1}, \ldots, x_n) \in \mathbb{R}^{n-r} \ , \tag{7.3}$$

constitutes a blow-up of the discontinuity submanifold $\mathcal{D}_1 \cap \cdots \cap \mathcal{D}_r \subset \mathcal{D}$. The sign function is defined as in (3.4) such that $\operatorname{sign}(\sigma) \in [-1, +1]$.

M. R. Jeffrey, *Hidden Dynamics*,
https://doi.org/10.1007/978-3-030-02107-8_7

7.1 The Sliding Manifold

The notion of sliding along a discontinuity surface $\mathcal{D}$ descends from the work of Filippov [71] and Utkin [211], but in recent years it has become clear that the traditional notion is incomplete. The modern notion of dynamics at the discontinuity uses the *switching layer* to peer inside the discontinuity. There is some delicacy in applying this method without losing generality, but (7.1) and (7.2) contain the necessary elements.

Consider a system with a discontinuity surface $\mathcal{D} = \mathcal{D}_1 \cup \cdots \cup \mathcal{D}_m$, and choose coordinates in which each $\mathcal{D}_j$ is a plane $x_j = 0$. Take a point on $\mathcal{D}$ at which discontinuity submanifolds $\mathcal{D}_1, \ldots, \mathcal{D}_r$ intersect, so $\lambda_j \in (-1, +1)$ for $j = 1, \ldots, r$, and $\lambda_j = \operatorname{sign}(x_j)$ for $j = r+1, \ldots, m$.

The dynamics at such a point is given by (7.2), which describes how the switching multipliers λ_j change on the timescales t/ε_j during switching. From each infinitesimal ε_j, let us take out a common infinitesimal factor ε_0 and write, in terms of constants $\upsilon_j > 0$:

$$\varepsilon_j = \varepsilon_0 \upsilon_j \quad \text{such that} \quad \varepsilon_j \xrightarrow{\varepsilon_0 \to 0} 0 \qquad \forall j = 1, \ldots, m\,.$$

Thus (7.2) becomes

$$\begin{cases} \varepsilon_0 \left(\upsilon_1 \dot{\lambda}_1, \ldots, \upsilon_r \dot{\lambda}_r \right) = (f_1(\mathbf{x}; \boldsymbol{\lambda}), \ldots, f_r(\mathbf{x}; \boldsymbol{\lambda}))\,, \\ \qquad (\dot{x}_{r+1}, \ldots, \dot{x}_n) = (f_{r+1}(\mathbf{x}; \boldsymbol{\lambda}), \ldots, f_n(\mathbf{x}; \boldsymbol{\lambda}))\,. \end{cases} \tag{7.4}$$

In the following we do not explicitly consider what happens if the ratios $\nu_j = \varepsilon_j / \varepsilon_0$ vanish, that is, if $\varepsilon_j = \mathsf{O}\left(\varepsilon_0^{p_j}\right)$ for powers $p_j > 0$; nevertheless the general analysis remains valid, and vanishing ν_j's may create more intricate multi-scale behaviour within sliding that we do not delve into.

On the fast timescale t/ε_0, denoting the derivative with respect to t/ε_0 by a prime instead of a dot, the system (7.2) on $x_1 = \cdots = x_r = 0$ becomes

$$\begin{cases} (\upsilon_1 \lambda_1', \ldots, \upsilon_r \lambda_r') = (f_1(\mathbf{x}; \boldsymbol{\lambda}), \ldots, f_r(\mathbf{x}; \boldsymbol{\lambda}))\,, \\ \left(x_{r+1}', \ldots, x_n'\right) = \varepsilon_0 \left(f_{r+1}(\mathbf{x}; \boldsymbol{\lambda}), \ldots, f_n(\mathbf{x}; \boldsymbol{\lambda})\right)\,, \end{cases}$$

which in the limit $\varepsilon_0 \to 0$ gives

$$\begin{cases} (\upsilon_1 \lambda_1', \ldots, \upsilon_r \lambda_r') = (f_1(\mathbf{x}; \boldsymbol{\lambda}), \ldots, f_r(\mathbf{x}; \boldsymbol{\lambda}))\,, \\ \left(x_{r+1}', \ldots, x_n'\right) = (0, \ldots, 0)\,. \end{cases} \tag{7.5}$$

This means that while the r-dimensional system of multipliers $(\lambda_1, \ldots, \lambda_r)$ is evolving on the fast timescale t/ε_0, the coordinates $(x_{r+1}, \ldots, x_n)$ are static.

The system (7.5) may have fixed points at points where $(f_1, \ldots, f_r) = (0, \ldots, 0)$. Solving this for the multipliers $\lambda_1, \ldots, \lambda_r$, denoting the solutions as $\lambda_1^{\$}, \ldots, \lambda_r^{\$}$, gives

$$\lambda_j^\$ = \{ \lambda_j \ : \ f_j(\mathbf{x};\lambda_1,\ldots,\lambda_m) = 0,\ x_j = 0,\ \lambda_j \in (-1,+1) \} \tag{7.6}$$

for $j = 1,\ldots,r$. We called these **sliding modes** of (7.1). (For one switch and linear dependence on λ_1, these reduce to Filippov's standard definition of sliding modes, discussed later in Chapter 8.) The sliding modes form families of fixed points of (7.5) parameterized by the quantities $(x_{r+1},\ldots,x_n)$, comprising a set

$$\mathcal{M} = \left\{ \begin{array}{l} (\lambda_1,\ldots,\lambda_r,x_{r+1},\ldots,x_n) \in (-1,+1)^r \times \mathbb{R}^{n-r} \\ \text{such that } \ \lambda_j = \lambda_j^\$ \ \text{ for } \ j = 1,\ldots,r \end{array} \right\} , \tag{7.7}$$

called the **sliding manifold** of (7.1). Because $\mathcal{M}$ consists of fixed points of (7.5), when sliding modes $(\lambda_1^\$,..,\lambda_r^\$)$ do exist, they are typically constrained to remain on $\mathcal{M}$ for some interval of time, and where this happens $\mathcal{M}$ is an *invariant manifold* of the layer system (7.2).

At a point $\mathbf{x}$ on the intersection of r discontinuity submanifolds, the sliding manifold $\mathcal{M}$ as defined in (7.7) has dimension $n-r$ (the same as the dimension of the intersection). In the full system, $\mathcal{M}$ may consist of many such submanifolds or branches, possibly of different dimensions $n-r$ inside different codimension r switching layers. Strictly speaking $\mathcal{M}$ should not be termed a 'manifold' at points where it leaves a given switching layer, as it is not defined on the entire neighbourhood surrounding such a point and so is not locally equivalent to Euclidean space, but the slight relaxation in terminology should not cause confusion and will avoid frequent obfuscating caveats.

To find the dynamics inside the sliding manifold, we must look back at the original timescale t and the system (7.4). For $\varepsilon_0 \to 0$, this becomes

$$\left\{ \begin{array}{rl} (0,\ldots,0) &= (f_1(\mathbf{x};\boldsymbol{\lambda}) \quad ,\ldots,f_r(\mathbf{x};\boldsymbol{\lambda}))\ , \\ (\dot{x}_{r+1},\ldots,\dot{x}_n) &= (f_{r+1}(\mathbf{x};\boldsymbol{\lambda}),\ldots,f_n(\mathbf{x};\boldsymbol{\lambda}))\ . \end{array} \right. \tag{7.8}$$

The first row of this differential-algebraic equation simply constrains the t-timescale motion to the sliding manifold $\mathcal{M}$. The second row gives the dynamics on $\mathcal{M}$. We call the right-hand side of (7.8) the **sliding vector field**. Recalling (from (3.10)) that we denote the constituent vector fields as $\mathbf{f}^K = \mathbf{f}^{\kappa_1\ldots\kappa_m}$, where the κ_j are indices $+$, $-$, or \$, we can write this sliding vector field as

$$\mathbf{f}^{\$\ldots\$\kappa_{r+1}\ldots\kappa_m}(\mathbf{x}) = \mathbf{f}(\mathbf{x};\lambda_1^\$,\ldots,\lambda_r^\$,\lambda_{r+1},\ldots,\lambda_m)\ , \tag{7.9}$$

where $\lambda_j^\$ \in (-1,+1)$ for $j = 1,\ldots,r$, and $\lambda_j = \operatorname{sign}(\sigma_j)$ for $j = r+1,\ldots,m$.

Example 7.1 (Sliding Manifolds). Examples are shown of sliding manifolds in two systems in Figure 7.1. The first example is

$$(\dot{x}_1,\dot{x}_2) = (\ 2,\ (x_2-1)\lambda_1 - x_2 - 1\ ) \quad \text{where} \quad \lambda_1 = \operatorname{sign} x_1\ .$$

A simple sliding manifold $\mathcal{M}$ exists where $\mathcal{D}$ is attracting, and is revealed by blowing up the switching layer with $r = 1$, giving the layer system

$$(\varepsilon\dot{\lambda}_1, \dot{x}_2) = (\ 2,\ (x_2 - 1)\lambda_1 - x_2 - 1\ ) \quad s.t. \quad \lambda_1 \in (-1, +1)\ ,\ \varepsilon \to 0\ .$$

The second example is

$$(\dot{x}_1, \dot{x}_2) = (\ 1 + 5(\lambda_1 - \lambda_2) + 3\lambda_1\lambda_2,\ -4(1 + \lambda_1 + \lambda_2)\ )\ ,$$

where $\lambda_1 = \operatorname{sign} x_1$ and $\lambda_2 = \operatorname{sign} x_2$. The sliding manifold $\mathcal{M}$ consists of multiple disconnected pieces, each having different dimension $n-r$ in regions where r switching submanifolds $\mathcal{D}_j$ intersect. The switching layer system is made up of three parts:

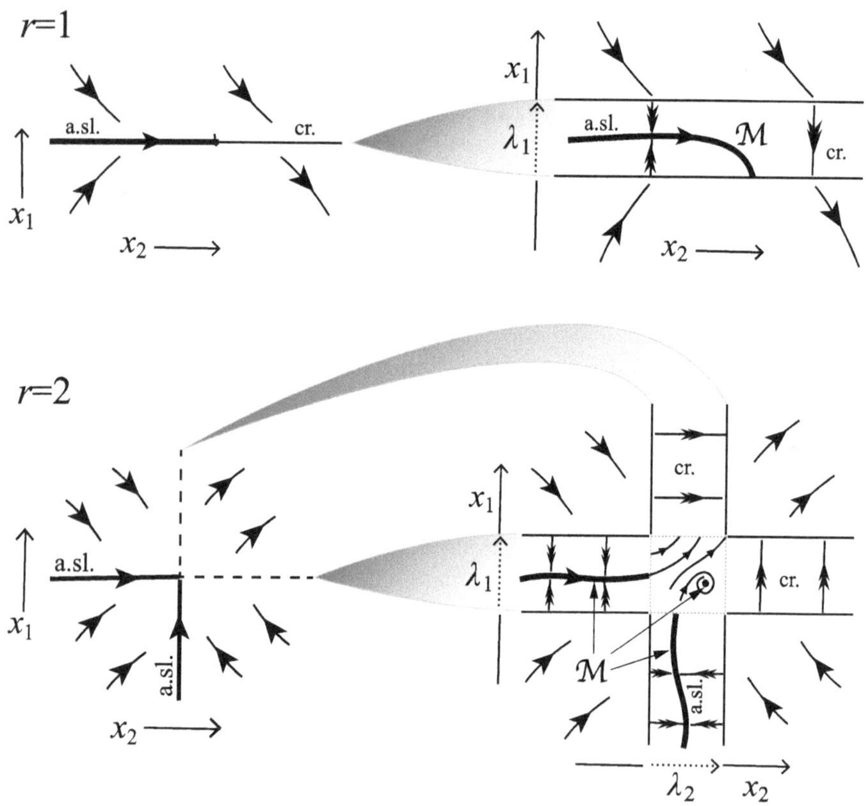

Fig. 7.1 Examples of the switching layer and sliding manifolds for one or two switches, showing crossing regions (cr.), and attracting sliding regions (a.sl.) where a sliding manifold $\mathcal{M}$ exists. At the codimension $r = 2$ intersection, there are both sliding and crossing. The left-hand pictures show the piecewise-smooth flow in the (x_1, x_2) plane; the right-hand pictures show the blow up of each discontinuity surface $\sigma_j = 0$ to reveal the switching layer $\lambda_j \in (-1, +1)$. Fast dynamics is indicated by double arrows.

- $(\varepsilon_1\dot{\lambda}_1, \dot{x}_2) = (\,1 + 5(\lambda_1 - \lambda_2) + 3\lambda_1\lambda_2,\ -4(1 + \lambda_1 + \lambda_2)\,)$ on $x_1 = 0$ such that $\lambda_1 \in (-1, +1)$ and $\lambda_2 = \operatorname{sign} x_2$;
- $(\dot{x}_1, \varepsilon_2\dot{\lambda}_2) = (\,1 + 5(\lambda_1 - \lambda_2) + 3\lambda_1\lambda_2,\ -4(1 + \lambda_1 + \lambda_2)\,)$ on $x_2 = 0$ such that $\lambda_2 \in (-1, +1)$ and $\lambda_1 = \operatorname{sign} x_2$;
- $(\varepsilon_1\dot{\lambda}_1, \varepsilon_2\dot{\lambda}_2) = (\,1 + 5(\lambda_1 - \lambda_2) + 3\lambda_1\lambda_2,\ -4(1 + \lambda_1 + \lambda_2)\,)$ on $x_1 = x_2 = 0$ such that $(\lambda_1, \lambda_2) \in (-1, +1) \times (-1, +1)$;

with $\varepsilon_1 \to 0$ and $\varepsilon_2 \to 0$ in each case.

■

If a system finds itself in a state $\mathbf{x}$ on $\mathcal{M}$ (usually having collapsed to $\mathcal{M}$ via the instantaneous system (7.5) during switching), then it evolves along a sliding mode according to the differential-algebraic system (7.8). These modes evolve inside (more precisely in the tangent space of) the discontinuity submanifolds $\mathcal{D}_1 \cap \cdots \cap \mathcal{D}_r$ (where $\sigma_1 = \cdots = \sigma_r = 0$) on the original timescale t, because $\dot{\lambda}_j = 0$ implies $\dot{\sigma}_j = 0$ (or equivalently $\delta_t\sigma_j = \mathbf{f} \cdot \nabla\sigma_j = f_j = 0$).

More generally, introduce a set $\mathcal{S}_r$ of r integers between 1 and m.

Definition 7.1. If $\mathbf{x} \in \mathcal{D}$ at a point where $\sigma_j = 0$ for all $j \in \mathcal{S}_r \subseteq \{1, \ldots, m\}$, and (7.6) has a solution $\lambda_j^\$$ for every $j \in \mathcal{S}_r$, then $\lambda_j^\$$ defines a **codimension r sliding mode** with $\lambda_j = \lambda_j^\$ \in (-1, +1)$ for $j \in \mathcal{S}_r \subseteq \{1, \ldots, m\}$, and $\lambda_i = \operatorname{sign}(\sigma_i)$ for $i \notin \mathcal{S}_r$, $i \in \{1, \ldots, m\}$, inhabiting a region where the dimension of $\mathcal{M}$ is $n - r$.

A useful convention is to take 'codimension $r = 0$ sliding' as referring to dynamics in the constituent vector fields outside $\mathcal{D}$ (i.e. not sliding).

In the absence of sliding modes, i.e. when (7.6) has no solutions within the r dimensional layer where $(-1, +1)^r$, the system (7.2) facilitates an instantaneous transition from one boundary of $\lambda_j \in (-1, +1)$ to another, and the flow is said to **cross** through the discontinuity surface. Crossing is denoted by cr. in Figure 7.1, for example.

In piecewise-smooth systems, we are concerned solely with the limit where all $\varepsilon_j \to 0$. The behaviour of systems of the form (7.2) for ε_j small but nonzero is the subject of geometric singular perturbation theory. In that context $\mathcal{M}$ is known as the critical manifold of the system (7.2), and the existence of invariant manifolds near $\mathcal{M}$ in the $\varepsilon_0 > 0$ system then follows by the theory of Fenichel under normal hyperbolicity conditions; see [68, 124]. We will briefly look at this perspective in Chapter 12 as one of a number of possible perturbations away from the class of piecewise-smooth systems.

The sliding manifold $\mathcal{M}$ is an invariant of (7.1) only where it defines a *normally hyperbolic* family of fixed points of (7.5), meaning it is attracting or repelling to first order with respect to the directions $(\lambda_1, \ldots, \lambda_r)$. To determine this, define the Jacobian matrix of the fast subsystem, evaluated at $\mathcal{M}$, as

$$\underline{\underline{B}} = \left.\frac{\partial(f_1, \ldots, f_r)}{\partial(\lambda_1, \ldots, \lambda_r)}\right|_{\mathcal{M}} = \left(\begin{array}{ccc} \frac{\partial f_1}{\partial \lambda_1} & \frac{\partial f_1}{\partial \lambda_2} & \cdots \\ \frac{\partial f_2}{\partial \lambda_1} & \frac{\partial f_2}{\partial \lambda_2} & \cdots \\ \vdots & \vdots & \ddots \end{array}\right)_{\mathcal{M}} . \tag{7.10}$$

If $\mathcal{M}$ exists, it may be comprised of many connected or disconnected branches on which the conditions (7.7) hold, and where $\mathcal{M}$ is normally hyperbolic, meaning

$$\det\left|\underline{\underline{B}}\right| \neq 0 \quad \& \quad \mathbb{R}\mathrm{e}\left[\boldsymbol{\beta}^{(j)}\right] \neq 0\,, \tag{7.11}$$

where $b^{(1)},\ldots,b^{(r)}$ are the eigenvalues of the matrix $\underline{\underline{B}}$ (such that $\underline{\underline{B}}\boldsymbol{\beta}^j = b^j\boldsymbol{\beta}^j$ for some eigenvectors $\boldsymbol{\beta}^j$.

The manifold $\mathcal{M}$ defined by (7.7) and satisfying (7.11) is invariant except at its boundaries. Those boundaries consist of points where (7.7) or (7.11) cannot be satisfied, which, respectively, give rise to:

1. *end points*: where $\mathcal{M}$ passes through the extrema of $\lambda_j \in (-1,+1)$ for some $j \in \{1,..,r\}$; or
2. *turning points*: where normal hyperbolicity of $\mathcal{M}$ is lost.

In both cases the number of roots $\lambda^\$$ of $f_j = 0$ in (7.6) changes, typically by one in case 1, because one root leaves the domain of existence, and by two in case 2 because solutions annihilate in pairs.

Example 7.2 (Boundaries of $\mathcal{M}$). Consider the system

$$(\dot{x}_1,\dot{x}_2) = (2\lambda_1^2 - \lambda_1 - x_2,\ 2\lambda_1^2 - 1)\,, \qquad \lambda_1 = \mathrm{sign}(x_1)\,.$$

with a single switch $\lambda_1 = \mathrm{sign}\, x_1$, illustrated in Figure 7.2. This shows a discontinuity surface with sliding, blown up into the layer system

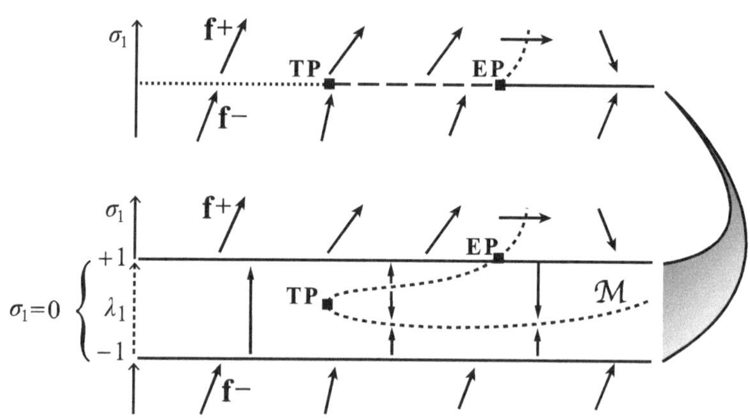

Fig. 7.2 A single switch between vector fields $\mathbf{f}^\pm$ as σ_1 changes sign, with the discontinuity surface $\sigma_1 = 0$ (shown in upper figure) blown up into the layer $\lambda_1 \in (-1,+1)$ (shown in lower figure). A sliding manifold $\mathcal{M}$ is illustrated, passing through the layer boundary $\lambda_1 = +1$ at **EP**, and becoming non-hyperbolic where it turns around at **TP**. An explicit example is given later in Example 7.6 and Figure 7.8.

$$(\varepsilon\dot\lambda_1, \dot x_2) = (2\lambda_1^2 - \lambda_1 - x_2,\ 2\lambda_1^2 - 1)\ , \qquad \lambda_1 \in (-1,+1)\ ,$$

for $\varepsilon \to 0$, revealing a sliding manifold

$$\mathcal{M} = \left\{(\lambda_1, x_2) \in (-1,+1) \times \mathbb{R}\ :\ 2\lambda_1^2 - \lambda_1 = x_2\right\}\ .$$

This has an end point at EP and a turning point at TP. At the end point, the sliding mode value of the switching multiplier $\lambda_1^\$$ passes through the upper boundary of $(-1,+1)$. At the turning point, $\mathcal{M}$ loses hyperbolicity at a point where $\partial\dot\lambda_1/\partial\lambda_1 = 0$, so (7.11) is violated because the fast λ_1 flow (7.5) is not transverse to the manifold; in this case $\partial^2\dot\lambda_1/\partial\lambda_1^2 \neq 0$, so the turning point is quadratic. ■

We take a closer look at turning points in Section 7.3 and end points in Section 7.4.

7.1.1 A Special Case: 'Higher-Order' Sliding Modes

The term 'higher-order' sliding modes is used in the literature (particularly in control electronics) to mean sliding on surfaces where

$$0 = \sigma_1 = \dot\sigma_1 = \ldots = \tfrac{d^{r-1}}{dt^{r-1}}\sigma_1\ .$$

This is not the same as what we call 'codimension r' sliding modes, on surfaces where $0 = \sigma_1 = \ldots = \sigma_r$. However, if we define switching functions

$$\sigma_j = \tfrac{d^{j-1}}{dt^{j-1}}\sigma_1 \qquad \text{for} \qquad j = 1, \ldots, r\ ,$$

then we obtain a system in the form (8.1) in which 'r^{th} higher-order' sliding modes correspond exactly to codimension r sliding.

7.2 The Sliding Region's Attractivity

The attractivity of the sliding region can be analyzed by adapting a few elementary tools of linear algebra that are usually used to understand equilibria. In essence we need to analyze the attractivity of sliding regions as fixed points of the layer's fast subsystem, from the first row of (7.8).

Points inside a codimension r sliding region satisfy (7.6) with $\lambda_j \in (-1,+1)$ on $x_j = 0$ for $j = 1, \ldots, r$. Generically $\underline{\underline{B}}$ has full rank by (7.11). (Degeneracies where $\underline{\underline{B}}$ is degenerate are discussed in Section 7.3.)

Any nondegenerate matrix $\underline{\underline{B}}$ has r linearly independent eigenvectors $\boldsymbol{\beta}^{(j)}$, with eigenvalues $b^{(j)}$, which satisfy the characteristic equation

$$\underline{\underline{B}} \cdot \boldsymbol{\beta}^{(j)} = b^{(j)}\boldsymbol{\beta}^{(j)} \quad \text{for} \quad j = 1, 2, \ldots, r. \tag{7.12}$$

This is solved by first rearranging to find the characteristic polynomial

$$\det\left|\underline{\underline{B}} - b^{(j)}\underline{\underline{I}}\right| \qquad \text{for} \quad j = 1, 2, \ldots, r, \tag{7.13}$$

whose scalar roots are the eigenvalues $b^{(j)}$. These consist of an even (or zero) number of complex eigenvalues forming complex conjugate pairings $b^{(i)} = {b^{(j)}}^*$, with the rest being real.

While the eigenvalues depend on the coordinate system chosen, the coefficients of the characteristic polynomial are coordinate invariant, the kth coefficient being a sum over all possible k-tuples of the set of eigenvalues. For example, the 0th and r^{th} coefficients are the trace and determinant

$$\mathrm{Tr}(\underline{\underline{B}}) = b^{(1)} + \cdots + b^{(r)} , \qquad \det\left|\underline{\underline{B}}\right| = b^{(1)} \ldots b^{(r)} . \tag{7.14}$$

The real part of each eigenvalue $b^{(j)}$ describes whether solutions are attracted ($\mathbb{R}\mathrm{e}[b^{(j)}] < 0$) or repelled ($\mathbb{R}\mathrm{e}[b^{(j)}] > 0$) along the direction $\boldsymbol{\beta}^{(j)}$, while imaginary parts impart oscillation in the planes spanned by their conjugate pairs.

In general the directions of the eigenvectors $\boldsymbol{\beta}^{(1)}, \ldots, \boldsymbol{\beta}^{(r)}$ bare no relation to the tangent spaces or normals to the discontinuity submanifolds $\mathcal{D}_1, \ldots, \mathcal{D}_r$. The fact that the dynamics on the sliding manifold lies inside the discontinuity submanifolds instead reveals itself in the values of the eigenvalues $b^{(1)}, \ldots, b^{(r)}$.

Example 7.3 (Eigenvectors of the Sliding Manifold). For a system with two switches, take coordinates such that $\sigma_j = x_j$ and $\lambda_j = \operatorname{sign} x_j$ for $j = 1, 2$, and consider vector fields

$$\begin{pmatrix} \dot{x}_1 \\ \dot{x}_2 \end{pmatrix} = \begin{pmatrix} a_1 \\ a_2 \end{pmatrix} + \begin{pmatrix} b_{11} & b_{12} \\ b_{21} & b_{22} \end{pmatrix} \begin{pmatrix} \lambda_1 \\ \lambda_2 \end{pmatrix} \tag{7.15}$$

where the a_i and b_{ij} may be differentiable functions of $\mathbf{x}$. Two particular examples are illustrated in Figure 7.3.

First find any codimension $r = 1$ sliding on the submanifolds $x_1 = 0$ and $x_2 = 0$. On $x_1 = 0$ the switching layer system is

$$\begin{pmatrix} \varepsilon_1 \dot{\lambda}_1 \\ \dot{x}_2 \end{pmatrix} = \begin{pmatrix} a_1 \\ a_2 \end{pmatrix} + \begin{pmatrix} b_{11} & b_{12} \\ b_{21} & b_{22} \end{pmatrix} \begin{pmatrix} \lambda_1 \\ \lambda_2 \end{pmatrix} ,$$

implying there is sliding where $\lambda_1^\$ = -\left(a_1 + b_{12} \operatorname{sign} x_2\right)/b_{11} \in (-1, +1)$. Substituting $\lambda = \lambda^\$$ back into the layer system gives the sliding dynamics

$$\dot{x}_2 = a_2 - \frac{b_{21} a_1}{b_{11}} + \frac{b_{11} b_{22} - b_{21} b_{12}}{b_{11}} \operatorname{sign} x_2 .$$

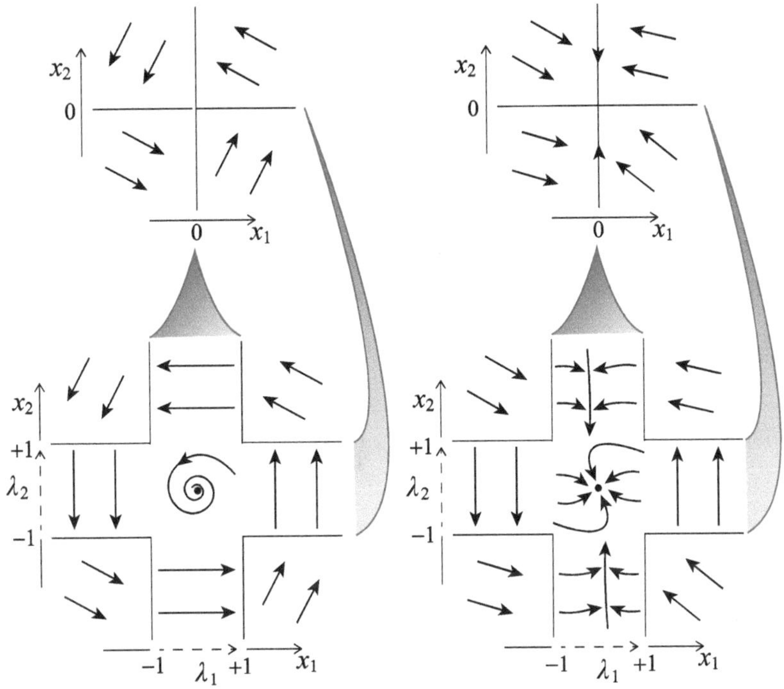

Fig. 7.3 Two illustrations of Example 7.3, with $a_1 = a_2 = 0$, $b_{21} = -b_{12} = 1$, $b_{22} = 0$, and (left) $b_{11} = -1/2$ and (right) $b_{11} = -3$. The upper figures show the piecewise-smooth system, blown up to reveal the switching layer in the lower figures. The eigenvectors of $\underline{\underline{B}}$ are (left) $\boldsymbol{\beta}^{\pm} = (-1 \pm i\sqrt{15}, 4)$ and (right) $\boldsymbol{\beta}^{\pm} = (-3 \pm \sqrt{5}, 2)$.

The derivative $\partial f_1/\partial\lambda_1 = b_{11}$ implies that the sliding manifold is attractive for $b_{11} < 0$ and repulsive for $b_{11} > 0$. On $x_2 = 0$ the switching layer system is

$$\begin{pmatrix} \dot{x}_1 \\ \varepsilon_2\dot{\lambda}_2 \end{pmatrix} = \begin{pmatrix} a_1 \\ a_2 \end{pmatrix} + \begin{pmatrix} b_{11} & b_{12} \\ b_{21} & b_{22} \end{pmatrix} \begin{pmatrix} \lambda_1 \\ \lambda_2 \end{pmatrix}$$

so there is sliding where $\lambda_2^{\$} = -\left(a_2 + b_{21}\,\mathrm{sign}\,x_1\right)/b_{22} \in (-1, +1)$. Substituting $\lambda = \lambda^{\$}$ back into the layer system gives the sliding dynamics

$$\dot{x}_1 = a_1 - \frac{b_{21}a_2}{b_{22}} + \frac{b_{11}b_{22} - b_{21}b_{21}}{b_{22}}\,\mathrm{sign}\,x_1\,.$$

The derivative $\partial f_2/\partial\lambda_2 = b_{22}$ implies that the sliding manifold is attractive for $b_{22} < 0$ and repulsive for $b_{22} > 0$. Elsewhere (where the respective multipliers $\lambda_1^{\$}$ or $\lambda_2^{\$}$ do not lie in $(-1, +1)$), the flow crosses the surfaces $x_1 = 0$

and $x_2 = 0$ transversally. The intersection of the discontinuity submanifolds $x_1 = 0$ and $x_2 = 0$ must be treated separately. On $x_1 = x_2 = 0$ the layer system is

$$\begin{pmatrix} \varepsilon_1\dot{\lambda}_1 \\ \varepsilon_2\dot{\lambda}_2 \end{pmatrix} = \begin{pmatrix} a_1 \\ a_2 \end{pmatrix} + \begin{pmatrix} b_{11} & b_{12} \\ b_{21} & b_{22} \end{pmatrix} \begin{pmatrix} \lambda_1 \\ \lambda_2 \end{pmatrix}$$

so there is codimension $r = 2$ sliding where

$$\begin{pmatrix} \lambda_1^\$ \\ \lambda_2^\$ \end{pmatrix} = \frac{1}{b_{11}b_{22} - b_{12}b_{21}} \begin{pmatrix} b_{12}a_2 - b_{22}a_1 \\ b_{21}a_1 - b_{11}a_2 \end{pmatrix} \in (-1, +1)^2$$

for which the intersection is a fixed point of the flow. Its attractivity properties can be derived either from the eigenvectors and eigenvalues of

$$\frac{d\mathbf{f}}{d\boldsymbol{\lambda}} = \begin{pmatrix} b_{11} & b_{12} \\ b_{21} & b_{22} \end{pmatrix},$$

or from the directions of the flows along the manifolds $x_1 = 0$ and $x_2 = 0$. If $b_{11}b_{22} - b_{12}b_{21} < 0$, the intersection is saddle-like, with one attractive and one repulsive direction, and if $b_{11}b_{22} - b_{12}b_{21} > 0$, then the intersection is node-like for $(b_{11} - b_{22})^2 + 4b_{12}b_{21} > 0$, focus-like for $(b_{11} - b_{22})^2 + 4b_{12}b_{21} < 0$, attractive for $b_{11} + b_{22} < 0$, and repulsive for $b_{11} + b_{22} > 0$.

Figure 7.3 shows examples with focus-like and node-like attraction to the intersection. In both figures the point at the origin is part of the sliding manifold $\mathcal{M}$; in the right-hand figure, $\mathcal{M}$ also consists of two attracting curves in the switching layer of $x_1 = 0$. The eigenvectors of $\underline{\underline{B}}$ are complex in the left-hand figure and real in the right-hand figure.

■

7.3 Singularities of the Sliding Manifold $\mathcal{M}$

In Figure 7.2 we saw a bifurcation (at the point TP) in which attracting and repelling branches of the sliding manifold $\mathcal{M}$ come together at a non-hyperbolic point, where (7.11) is violated. For the generic forms of $\mathcal{M}$ at a non-hyperbolic point, we can turn again to Thom's catastrophes, which we used in Section 6.1 to study the shape of the discontinuity surface $\mathcal{D}$.

At a simple switch ($r = 1$), the condition $\det|\underline{\underline{B}}| = 0$ from (7.11) implies a fold or higher-order catastrophe, parameterized by x_i for $i > 1$, in which potentially many attracting and repelling branches of $\mathcal{M}$ may join at non-hyperbolic points. If we consider the neighbourhood of a point where the discontinuity surface is a manifold $\mathcal{D}_1$, in coordinates where $\sigma_1(\mathbf{x}) = x_1$, and so $f_1 = \delta_t\sigma_1$, then the sliding manifold $\mathcal{M}$ is the set points where $f_1 = 0$ by (7.6). We can immediately obtain various generic forms for the normal component of the vector field, f_1, by translating the forms of σ in Section 6.1 into expressions for f_1 in the switching layer variables $(\lambda_1, x_2, x_3, \dots)$. This yields the family of sliding manifolds shown in Figure 7.4.

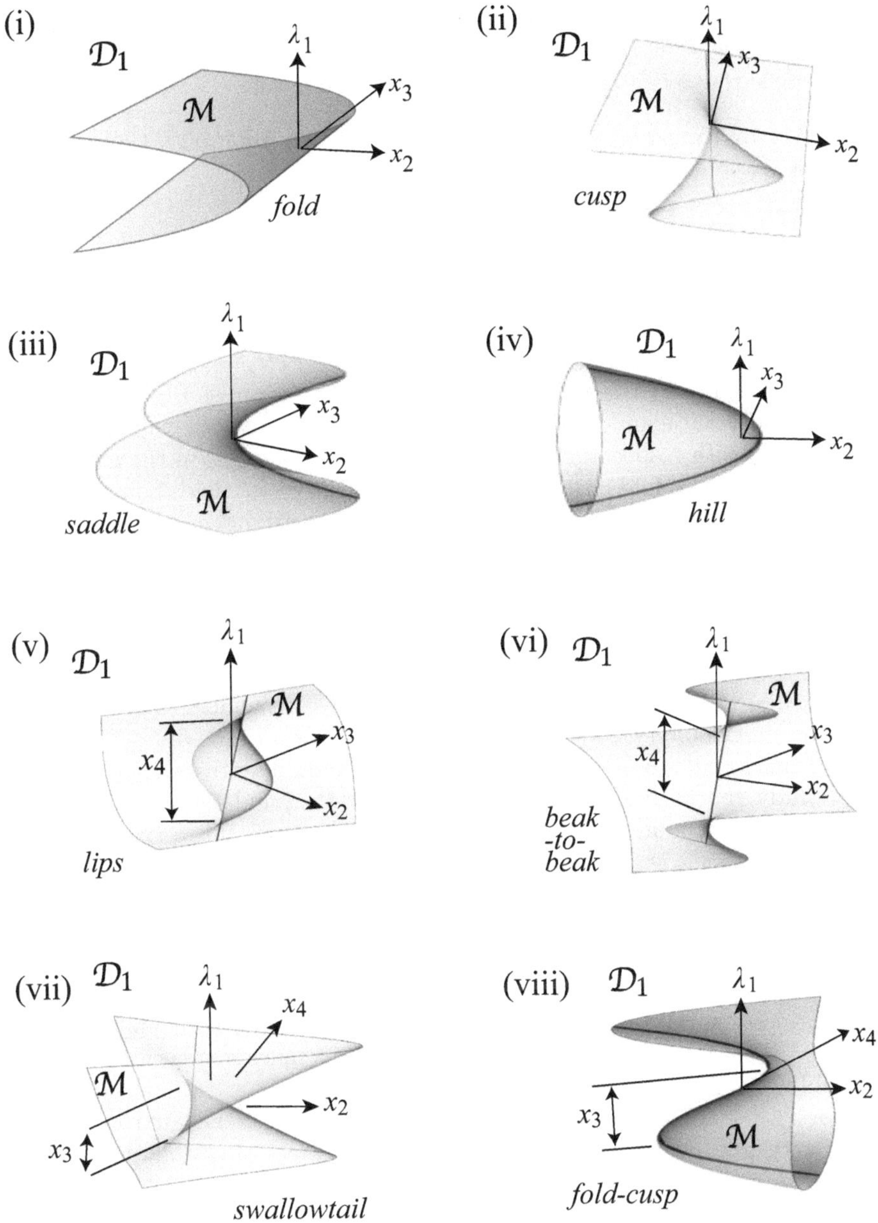

Fig. 7.4 Singularities of the sliding manifold $\mathcal{M}$. Plotting the contour $f_1 = 0$ for (i) the fold $f_1 = x_2 + \frac{1}{2}\lambda_1^2$; (ii) the cusp $f_1 = x_2 + \frac{1}{3}\lambda_1^3 + x_3\lambda_1$; the two-fold, which is either (iii) the saddle $f_1 = x_2 + \lambda_1^2 - x_3^2$ or (iv) the hill $f_1 = x_2 + \lambda_1^2 + x_3^2$; (v) the lips $f_1 = x_2 + \frac{1}{3}\lambda_1^3 + (x_3^2 + x_4)\lambda_1$; (vi) the beak-to-beak $f_1 = x_2 + \frac{1}{3}\lambda_1^3 - (x_3^2 + x_4)\lambda_1$; (vii) the swallowtail $f_1 = x_2 + \frac{1}{4}\lambda_1^4 + \frac{1}{2}x_3\lambda_1^2 + x_4\lambda_1$; (viii) the fold-cusp $f_1 = x_2 + \frac{1}{3}\lambda_1^3 + x_3\lambda_1 + \frac{1}{2}x_4^2$.

In each case depicted in Figure 7.4, attracting sheets (where $\partial f_1/\partial\lambda_1 < 0$) meet repelling sheets (where $\partial f_1/\partial\lambda_1 < 0$). The sliding dynamics that occurs on these surfaces can be studied from (7.8) with $r = 1$.

Example 7.4 (Singularities of the Nonlinear Sliding Manifold). Consider the system

$$\begin{pmatrix} \dot{x}_1 \\ \dot{x}_2 \\ \dot{x}_3 \\ \dot{x}_4 \\ \vdots \end{pmatrix} = \begin{pmatrix} f_1 \\ c_2 + x_2 \\ c_3 + x_3 \\ c_4 + x_4 \\ \vdots \end{pmatrix}, \qquad \lambda_1 = \text{sign}(x_1) . \tag{7.17}$$

where f_1 is any of the functions given in the caption of Figure 7.4. The discontinuity submanifold $x_1 = 0$ is repelling (or attracting if we replace $f_1 \mapsto -f_1$) with respect to the constituent flows in $x_1 \gtrless 0$ in the cusp, lips, beak-to-beak, and fold-cusp cases; the constituent vector fields point through the discontinuity submanifold otherwise. Taking constants $c_i \neq 0$ ensure that the origin is not an equilibrium. To obtain the sliding manifolds pictured in Figure 7.4, we form the layer system on $x_1 = 0$, given by $(\varepsilon_1\dot{\lambda}_1, \dot{x}_2, \dots) = (f_1, c_2 + x_2, \dots)$ and solve $f_1 = 0$.
■

When studying such forms, we must of course remember that $\mathcal{M}$ exists only inside the interval $\lambda_1 \in (-1, +1)$, but within that interval, a rich world of dynamics waits to be studied. To begin understanding this world, we must first study the local sliding vector fields on $\mathcal{M}$ around these singularities, the slow-fast dynamics in the switching layer, and how this connects to the dynamics outside $\mathcal{D}$. We take the first steps in this direction here.

Figure 7.4 is not complete as a classification and merely presents some obvious examples. We must leave the further study of the catastrophes of $\mathcal{M}$, and their consequences for the flow, as a challenge for the future, because it is just the beginning of a classification that rapidly expands as we consider a discontinuity surface comprised of multiple discontinuity submanifolds $\mathcal{D} = \mathcal{D}_1 \cup \dots \cup \mathcal{D}_m$. At a switching intersection for $r > 1$, much more is immediately possible.

For $r \geq 2$ the condition $\det|\underline{\underline{B}}| = 0$ where (7.11) is violated again implies a turning point of $\mathcal{M}$, creating fold or higher catastrophes in which attracting and repelling branches join at a non-hyperbolic point. For $r \geq 2$ non-hyperbolic points may also occur where the real part of any of the eigenvalues of $\underline{\underline{B}}$ vanish. For $r = 2$, for example, this implies $\text{Tr}(\underline{\underline{B}}) = 0$, which implies a Hopf bifurcation (see, e.g. [161, 138, 92]) in the two-dimensional layer fast subsystem in (λ_1, λ_2). Generically the sliding manifold then forms a parabolic (or hill-like) surface (formed by an $x_{i>2}$ parameterized family of limit cycles of the layer fast subsystem—a distributed sliding mode as we will identify it in Section 7.5.3) threaded by a line formed by $x_{i>2}$ parameterized fixed points, as illustrated in Figure 7.5.

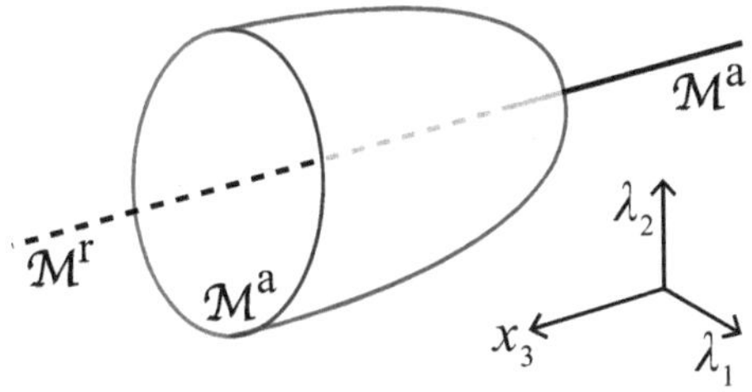

Fig. 7.5 A Hopf bifurcation of $\mathcal{M}$ in the switching layer fast subsystem on (λ_1, λ_2), parameterized by a spatial coordinate x_3. This results in a sliding manifold $\mathcal{M}$ with an attracting branch $\mathcal{M}^a$, which bifurcates into a repelling branch $\mathcal{M}^r$ surrounded by a circular attracting branch $\mathcal{M}^a$.

From standard results of smooth systems (e.g. [138]), we can immediately write down a prototype of a system with such structure.

Example 7.5 (Hopf bifurcation in a Codimension 2 Sliding Manifold). Consider the system

$$\begin{pmatrix} \dot{x}_1 \\ \dot{x}_2 \end{pmatrix} = \begin{pmatrix} \lambda_1(\mu - \lambda_1^2 - \lambda_2^2) - \lambda_2 - x_1 \\ \lambda_1 + \lambda_2(\mu - \lambda_1^2 - \lambda_2^2) - x_2 \end{pmatrix} , \qquad \lambda_j = \text{sign}(x_j) , \qquad (7.18)$$

illustrated in Figure 7.6. The discontinuity submanifolds $x_1 = 0$ and $x_2 = 0$ and their intersection $x_1 = x_2 = 0$ are all attracting sliding regions. As μ changes sign, a supercritical Hopf bifurcation occurs in the layer system

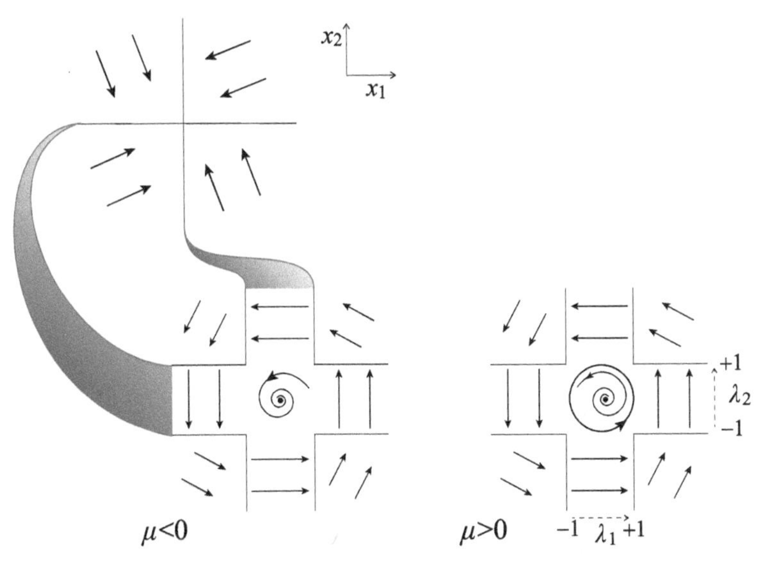

Fig. 7.6 A Hopf bifurcation in the switching layer at an intersection as μ changes sign (lower portraits) and the piecewise-smooth system in (x_1, x_2) space (upper portrait).

$$\begin{pmatrix} \varepsilon_1\dot{\lambda}_1 \\ \varepsilon_2\dot{\lambda}_2 \end{pmatrix} = \begin{pmatrix} \lambda_1(\mu - \lambda_1^2 - \lambda_2^2) - \lambda_2 - x_1 \\ \lambda_1 + \lambda_2(\mu - \lambda_1^2 - \lambda_2^2) - x_2 \end{pmatrix}, \quad \lambda_j = \text{sign}(x_j), \qquad (7.19)$$

though no qualitative change occurs outside the intersection. A practical consequence of this is discussed in Section 9.2.3.

■

In principle, any of the bifurcations possible in smooth single or multiple timescale systems (see e.g. [92, 135, 138]), of which the saddlenode and Hopf bifurcations are only the most simple, can occur inside the switching layer's fast subsystem. When they occur they enact bifurcations of the sliding manifold.

When discussing generic cases here, we interpret this only in the most general sense, but in systems with symmetries or constraints, other situations can be considered 'generic'. For example, with a single discontinuity submanifold the system

$$\begin{pmatrix} \dot{x}_1 \\ \dot{x}_2 \end{pmatrix} = \begin{pmatrix} \lambda_1(x_2 - \lambda_1^p) \\ 1 \end{pmatrix}, \qquad \lambda_1 = \text{sign}(x_1), \qquad (7.20)$$

for $p \geq 0$ describes other ways a sliding manifold can change stability without a fold. They are non-generic, as they degenerate into simple invariant manifolds and folds if we add a constant to the $\dot{x}_1$ equation, but they persist under certain symmetries (being invariant under $x_1 \mapsto -x_1$ if p is odd, and $\{x_1, x_2, t\} \mapsto -\{x_1, x_2, t\}$ if p is even). The case $p = 1$ gives a transcritical bifurcation, where attracting and repelling branches of the sliding manifold collide and exchange stability (instead of annihilating as they do at a fold. The case $p = 2$ gives a pitchfork bifurcation, where an attracting sliding manifold becomes repelling and expels two new attracting branches in the process (like taking a 'slice' through the Hopf bifurcation in Example 7.5).

We return to look at phenomena resulting from bifurcations of $\mathcal{M}$ in Chapter 9.

7.4 End Points of the Sliding Region

If $\mathcal{M}$ touches the boundary of the layer $(-1, +1)^r \times \mathbb{R}^{n-r}$, it results in a tangential contact between the flow and $\mathcal{D}$. If $\mathcal{M}$ touches the layer boundary transversally, then the contact between the flow and $\mathcal{D}$ is a fold and forms the boundary between regions where $\mathcal{D}$ is either attracting, repelling, or transverse to the external vector fields. If $\mathcal{M}$ touches the boundary of the layer $(-1, +1)^r \times \mathbb{R}^{n-r}$ tangentially, it causes bifurcations in the flow's contact with $\mathcal{D}$. The simplest example, an end point for a single switch ($r = 1$), we illustrated earlier by the point EP in Figure 7.2.

For $r = 1$ codimension sliding, the end points of $\mathcal{M}$ are places where the vector fields $\mathbf{f}(\mathbf{x}; \pm 1) \equiv \mathbf{f}^{\pm}(\mathbf{x})$ are tangent to the discontinuity surface $\sigma_1 = 0$.

At the boundary, the sliding vector field is equal to the constituent vector field which is tangent there, that is:

$$\begin{array}{lllll} \delta_{t+}\sigma_1 = 0 & \Leftrightarrow & \lambda_1^{\$} = +1 & \Leftrightarrow & \mathbf{f}(\mathbf{x};\lambda_1) = \mathbf{f}^+(\mathbf{x})\,, \\ \delta_{t-}\sigma_1 = 0 & \Leftrightarrow & \lambda_1^{\$} = -1 & \Leftrightarrow & \mathbf{f}(\mathbf{x};\lambda_1) = \mathbf{f}^-(\mathbf{x})\,. \end{array} \tag{7.21}$$

Hence the behaviour of $\mathcal{M}$ at an end point has a direct implication for the dynamics outside the switching layer. If the contact is quadratic, then the vector field has a cusp; if the contact is cubic, then the vector field has a swallowtail; if the contact is a hill or saddle, the vector field has a lips/beaks singularity, etc.

We have seen the vector fields that give rise to these already in Section 6.1. Using the simplest convex combination $\dot{\mathbf{x}} = \mathbf{f}$ with $\mathbf{f} = \frac{1}{2}(1+\lambda)\mathbf{f}^+ + \frac{1}{2}(1-\lambda)\mathbf{f}^-$, for example, the layer's fast subsystem is $\varepsilon\dot{\lambda} = \delta_t\sigma$, so the sliding manifold $\mathcal{M}$ given by $\dot{\lambda} = 0$ satisfies $\lambda^{\$} = \frac{(\delta_{t-}+\delta_{t+})\sigma}{(\delta_{t-}-\delta_{t+})\sigma} \in (-1,+1)$. Taking the vector fields from Section 6.1 (in the minimum number of dimensions necessary):

i. For the fold, the layer system from (6.18) with $\lambda = \operatorname{sign} x_2$ is

$$(\dot{x}_1, \varepsilon\dot{\lambda}) = \tfrac{1}{2}(1+\lambda)(1, \pm x_1) + \tfrac{1}{2}(1-\lambda)(0,1)\,, \tag{7.22a}$$

for which $\mathcal{M}$ is given by $\lambda^{\$} = \frac{1\pm x_1}{1\mp x_1}$. This gives the first row of Figure 7.7.

ii. For the two-fold, the layer system from (6.21) with $\lambda = \operatorname{sign} x_3$ is

$$(\dot{x}_1, \dot{x}_2, \varepsilon\dot{\lambda}) = \tfrac{1}{2}(1+\lambda)\left(-s^+, \nu^+, -x_1\right) + \tfrac{1}{2}(1-\lambda)\left(\nu^-, s^-, x_2\right)\,, \tag{7.22b}$$

for which $\mathcal{M}$ is given by $\lambda^{\$} = \frac{x_2 - x_1}{x_2 + x_1}$. This gives the second row of Figure 7.7.

iii. For the cusp, the layer system from (6.24) with $\lambda = \operatorname{sign} x_2$ is

$$(\dot{x}_1, \varepsilon\dot{\lambda}, \dot{x}_3) = \tfrac{1}{2}(1+\lambda)(1, \pm(x_1^2 + x_3), 0) + \tfrac{1}{2}(1-\lambda)(0,1,0)\,, \tag{7.22c}$$

for which $\mathcal{M}$ is given by $\lambda^{\$} = \frac{1\pm(x_1^2+x_3)}{1\mp(x_1^2+x_3)}$. This gives the third row of Figure 7.7.

iv. For the swallowtail, the layer system from (6.28) with $\lambda = \operatorname{sign} x_2$ is

$$(\dot{x}_1, \varepsilon\dot{\lambda}, \dot{x}_3, \dot{x}_4) = \tfrac{1}{2}(1+\lambda)(1, \pm(x_1^3 + x_3x_1 + x_4), 0, 0) + \tfrac{1}{2}(1-\lambda)(0,1,0,0)\,, \tag{7.22d}$$

for which $\mathcal{M}$ is given by $\lambda^{\$} = \frac{1\pm(x_1^3+x_3x_1+x_4)}{1\mp(x_1^3+x_3x_1+x_4)}$, giving the last row of Figure 7.7.

We can continue this to higher codimensions ad infinitum, as in Section 6.1. Extending it to a discontinuity surface comprised of multiple discontinuity submanifolds is more difficult, as the number of cases rapidly multiplies.

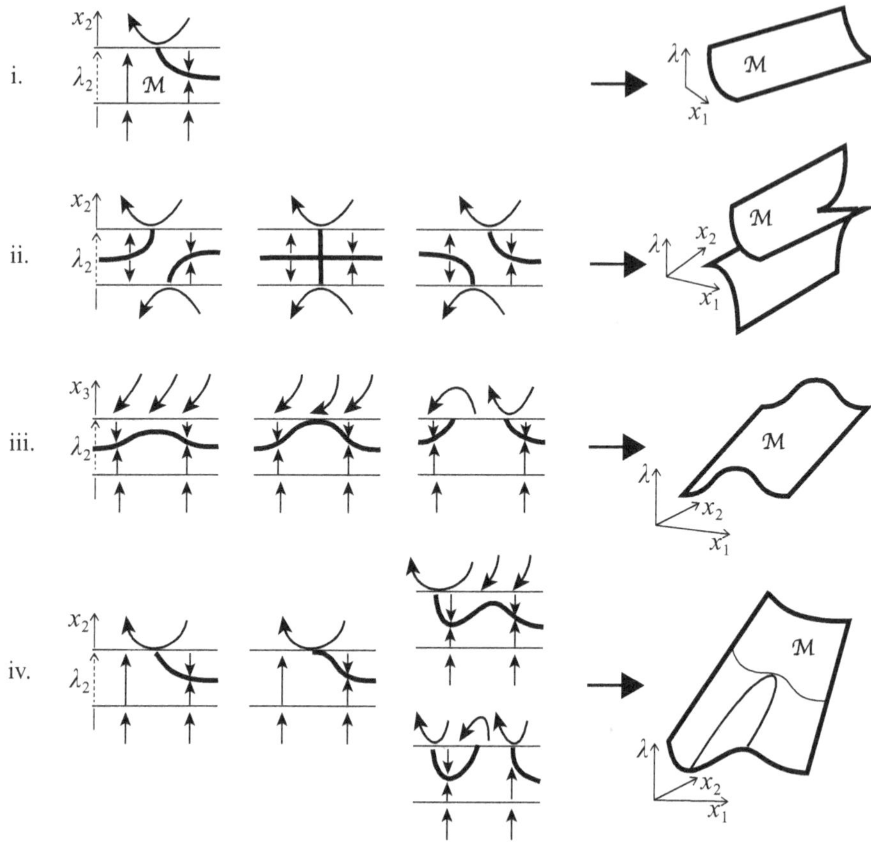

Fig. 7.7 Intersections between the sliding manifold $\mathcal{M}$ and the layer boundary. Each figure on the left shows the layer $\lambda \in (-1, +1)$, containing $\mathcal{M}$, with the fast flow inside, and surrounding vector fields $\mathbf{f}^{\pm}$ represented above and below the layer. On the left we see only sections through the manifold $\mathcal{M}$ that is illustrated on the right.

7.5 Multiplicity and Attractivity of Sliding Modes

As we have seen, the sliding manifold $\mathcal{M}$ may have multiple branches at a given point $\mathbf{x} \in \mathcal{D}$, with a different sliding mode on each. Multiplicity of sliding can occur if there is nonlinear dependence of $\mathbf{f}$ on the switching multipliers $\boldsymbol{\lambda} = (\lambda_1, \ldots, \lambda_r)$. For a single switch, this only happens if there is a hidden term. For multiple switches, the multilinearity of the canopy (5.3) naturally leads to multiple solutions of (7.6), even without hidden terms.

7.5.1 One Switch, Multiple Sliding Modes

At a point where the discontinuity surface is locally a manifold, the number of sliding modes is bounded by the order of dependence on the switching multiplier. Take the form (5.7) for the vector field

$$\dot{\mathbf{x}} = \mathbf{f}(\mathbf{x}; \lambda_1) = \tfrac{1}{2}(1+\lambda_1)\mathbf{f}^+(\mathbf{x}) + \tfrac{1}{2}(1-\lambda_1)\mathbf{f}^-(\mathbf{x}) + (\lambda_1^2 - 1)\mathbf{h}(\mathbf{x}; \lambda_1) .$$

Theorem 7.1. *If* $\mathbf{h}$ *is a polynomial of order* p *in* λ_1*, then* (7.8) *for* $r = 1$ *has up to* p *solutions, i.e. up to* p *different sliding modes.*

This is simply a consequence of the fundamental theorem of algebra applied to the equation $\delta_t \sigma_1 = 0$ as a polynomial of order p in λ_1.

Example 7.6 (Multiplicity of Sliding Modes for 1 Switch). Consider again the planar system from Example 7.2, with a discontinuity at $x_1 = 0$. Pulling out the hidden term, we write

$$(\dot{x}_1, \dot{x}_2) = (2 - \lambda_1 - x_2, 1) + 2(\lambda_1^2 - 1)(1, 1) ,$$

where $\lambda_1 = \mathrm{sign}(x_1)$, so $\mathbf{f}^{\pm} = (2 \mp 1 - x_2, 1)$ and $\mathbf{h} = 2(1, 1)$. The layer system is

$$(\varepsilon\dot{\lambda}_1, \dot{x}_2) = (2 - \lambda_1 - x_2, 1) + 2(\lambda_1^2 - 1)(1, 1) .$$

There are then regions of 0, 1, or 2 sliding modes on $x_1 = 0$ depending on the coordinate x_2, as shown in Figure 7.8. The layer's fast subsystem is

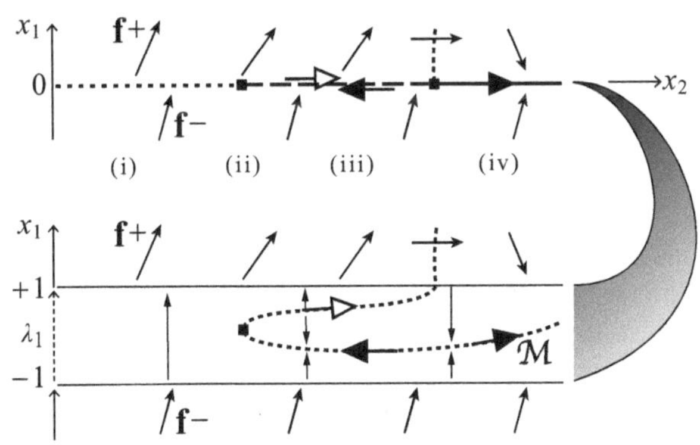

Fig. 7.8 A single switch between vector fields $\mathbf{f}^{\pm}$ as σ_1 changes sign, illustrating multiplicity of sliding modes in the nonlinear sliding region (dashed, upper figure), one mode attracting (black arrow) and one repelling (white arrow) with respect to the fast flow. The upper figure shows the piecewise-smooth system, blown up to reveal the switching layer in the lower figure. The labels (i–iv) are referred to in Figure 7.10.

$\varepsilon\dot{\lambda}_1 = 2\lambda_1^2 - \lambda_1 - x_2$, which vanishes on $\lambda_1^\$ = \frac{1}{4}(1 \pm \sqrt{1+8x_2})$, with real solutions for $x_2 > -\frac{1}{8}$ that lie in the interval $(-1, +1)$ for $1 < x_2 < 3$. In $x_2 < -\frac{1}{8}$, the flow crosses; in $-\frac{1}{8} < x_2 < 1$, there are two sliding modes, one attracting and one repelling (corresponding to the '−' and '+' roots $\lambda_1^\$$, respectively); and in $1 < x_2 < 3$, there is a single attracting sliding mode. At $x_2 = -\frac{1}{8}$, the sliding manifold has a turning point, where two sliding modes coincide. At $x_2 = 1$, as in the linear case, the manifold passes through the layer boundary $\lambda_1 = +1$, an end point of $\mathcal{M}$, coinciding with a tangency in between the vector field $\mathbf{f}^+$ and the discontinuity surface. If we continue the picture out to $x_2 = 3$, we shall see that the sliding manifold passes through the other side of the layer boundary $\lambda_1 = -1$.

As an aside, if we neglect the hidden term (since $\lambda_1^2 - 1$ vanishes for all $x_1 \neq 0$), we obtain the linear switching system depicted in Figure 7.9. The layer's fast subsystem is $\varepsilon\dot{\lambda}_1 = 2 - \lambda_1 - x_2$, and solving $\dot{\lambda}_1 = 0$ gives an attracting sliding mode on $x_1 = 0$ for $1 < x_2 < 3$ with $\lambda_1^\$ = 2 - x_2$, and crossing elsewhere, with an end point of $\mathcal{M}$ at $x_2 = 1$. Since $\varepsilon\partial\dot{\lambda}_1/\partial\lambda_1 = -1$, the sliding manifold is attracting. This is shown in Figure 7.9 and is the standard 'Filippov system' interpretation of a system switching between $(\dot{x}_1, \dot{x}_2) = (2 \mp 1 - x_2, 1)$.

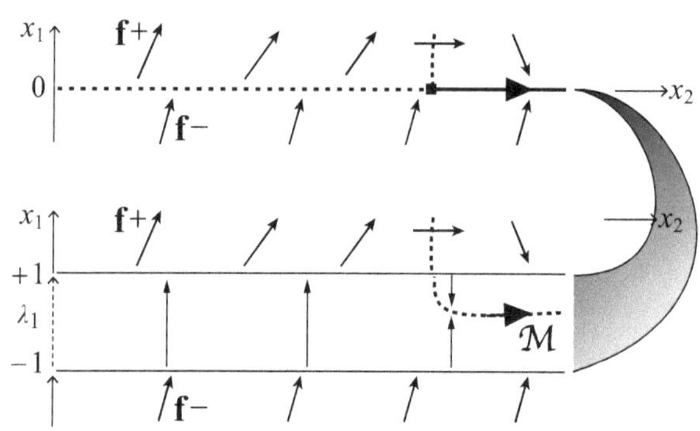

Fig. 7.9 The 'Filippov' system corresponding to Figure 7.8, neglecting the hidden term.

For the four different locations along the surface indicated by (i–iv) in Figure 7.8, the combinations $\{\mathbf{f}(x_1, x_2; \lambda_1) : \lambda_1 \in (-1, +1)\}$ are illustrated in Figure 7.10. Nonlinear sliding implies (i) no sliding modes, (ii) a turning point, and (iii) attracting and repelling sliding modes $\mathbf{f}_a^\$$ and $\mathbf{f}_r^\$$. In (iv) there are unique attracting sliding modes both in the convex and nonlinear combinations, but their vector fields (labelled $\mathbf{f}^{co}$ and $\mathbf{f}^\$$) are different. Note that in (i)–(iii) the vectors $\mathbf{f}^+$ and $\mathbf{f}^-$ do not change qualitatively. Neglecting the hidden term as in Figure 7.9 gives instead a convex combination where the values $\mathbf{f}(x_1, x_2; \lambda_1)$ for $\lambda_1 \in (-1, +1)$ form just the dotted straight line between $\mathbf{f}^\pm$ in Figure 7.10, which predicts no sliding modes for (i)–(iii), and one from $\mathbf{f}^{co}$ in (iv).

■

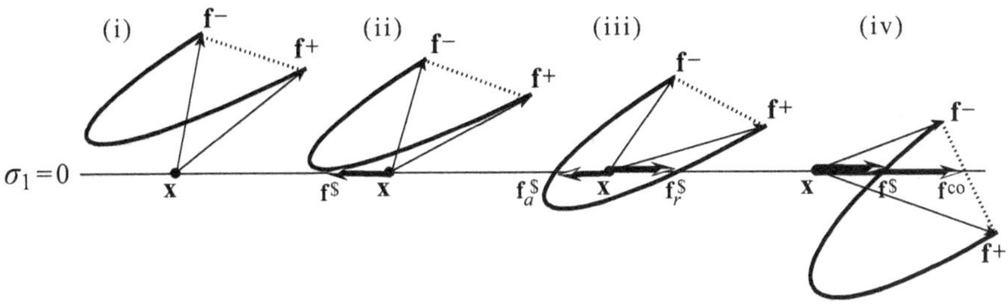

Fig. 7.10 Linear and nonlinear sliding dynamics in Example 7.6. The bold curve is the nonlinear combination $\mathbf{f}(\mathbf{x};\lambda_1)$ with $\lambda_1 \in (-1,+1)$. The dotted line joining $\mathbf{f}^{\pm}$ is Filippov's convex combination $\frac{1}{2}(1+\lambda_1)\mathbf{f}^+ + \frac{1}{2}(1-\lambda_1)\mathbf{f}^-$ with $\lambda_1 \in (-1,+1)$.

7.5.2 Multiplicity of Sliding Modes at Intersections

In systems of many switches, multiplicity of sliding modes can occur even if the dependence on each individual switching multiplier is linear. With $\mathbf{h} \equiv 0$, the combination $\mathbf{f}(\mathbf{x};\boldsymbol{\lambda})$ is a multilinear polynomial in the switching multipliers $\boldsymbol{\lambda} = (\lambda_1, \ldots, \lambda_m)$.

Theorem 7.2. *If* $\mathbf{h} = 0$*, then at a codimension* r *switching intersection, if the matrix* $\underline{\underline{B}}$ *from* (7.10) *has full rank, the layer fast subsystem* (7.8) *has up to* $r!$ *solutions and hence up to* $r!$ *different sliding modes.*

The proof of this is in [113]. Essentially, (7.6) is a system of r simultaneous equations that are multilinear in the λ_j's. Solving for each multiplier $\lambda_j \in (-1,+1)$ in turn, the solution for λ_1 gives at most one real solution in terms of the remaining $r-1$ multipliers λ_j, then two solutions for λ_2 in terms of the remaining $r-2$ multipliers λ_j, three solutions for λ_3 in terms of the remaining $r-3$ multipliers λ_j, and so on, producing $1 \times 2 \times 3 \times \ldots \times r = r!$ different solutions for the full set $\{\lambda_1^\$, \ldots, \lambda_r^\$\}$, each valid only if they are real and lie in $(-1,+1)^r$.

If $\mathbf{h} \neq 0$ then $\mathbf{f}$ is nonlinear in the λ_j's, and the number of solutions at each stage of this argument can be larger.

Example 7.7 (Multiplicity of Sliding Modes for Two Switches). Consider a system $\dot{\mathbf{x}} = \mathbf{f}$ in coordinates $\mathbf{x} = (x_1, x_2, x_3)$ where $\mathbf{f}$ switches between values

$$\mathbf{f}^{++} = (-1, 4\alpha/5, \zeta)\,, \quad \mathbf{f}^{-+} = (+1/2, -4/5, \zeta)\,,$$
$$\mathbf{f}^{--} = (+1, 4\alpha/5, \xi)\,, \quad \mathbf{f}^{+-} = (-1/2, -4/5, \xi)\,,$$

across two manifolds $\mathcal{D}_1$ and $\mathcal{D}_2$ given by the coordinate planes $x_1 = 0$ and $x_2 = 0$, respectively. The quantities ζ, ξ, are arbitrary (constants or smooth functions), while α is a constant. This is illustrated in Figure 7.11.

To find the dynamics on the discontinuity surface, form the canopy combination, using (5.8) (and assuming $\mathbf{h} = 0$ for simplicity),

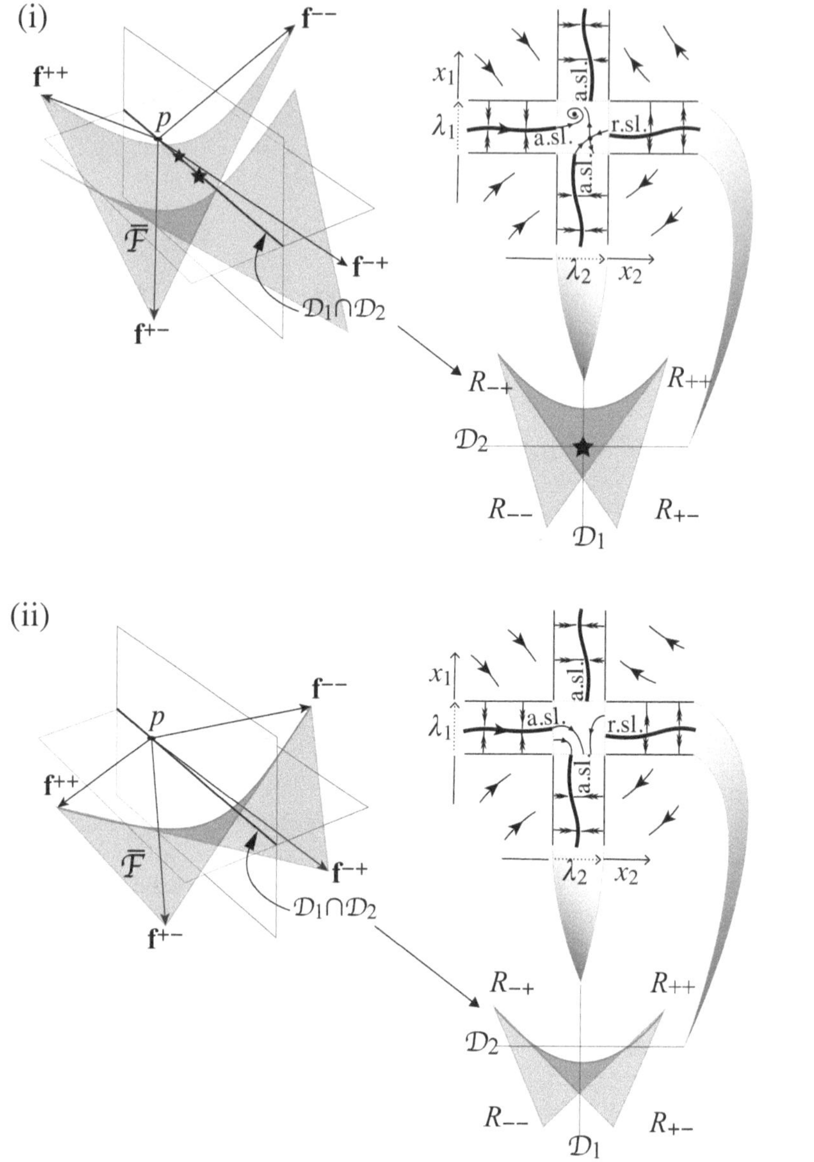

Fig. 7.11 The bifurcation of sliding modes. The canopy (5.8) stretched between the constituent vector fields $\mathbf{f}^{\pm\pm}$ is in general a curved surface (shaded). In (i) the intersection of the discontinuity submanifolds $\mathcal{D}_1 \cap \mathcal{D}_2$ pierces the canopy $\mathcal{F}$ at two points (starred) for $1 < \alpha| < 2$. In (ii) the intersection misses the canopy for $|\alpha| < 1$. This is easier to see in the end-on views below the main figures, where the regions R_κ and discontinuity submanifolds $\mathcal{D}_j$ are also labelled.

$$\dot{\mathbf{x}} = \mathbf{f} = \tfrac{1}{2}(1+\lambda_1)\left\{\tfrac{1}{2}(1+\lambda_2)\mathbf{f}^{++} + \tfrac{1}{2}(1-\lambda_2)\mathbf{f}^{+-}\right\}$$
$$+\tfrac{1}{2}(1-\lambda_1)\left\{\tfrac{1}{2}(1+\lambda_2)\mathbf{f}^{-+} + \tfrac{1}{2}(1-\lambda_2)\mathbf{f}^{--}\right\} .$$

The canopy at a point $p \in \mathcal{D}_1 \cap \mathcal{D}_2$ is shown in Figure 7.11 for (i) $\alpha = 5/4$ and (ii) $\alpha = 5/8$. Concerning ourselves only with the switching intersection $x_1 = x_2 = 0$, we have a switching layer system

$$\begin{pmatrix} \varepsilon_1\dot{\lambda}_1 \\ \varepsilon_2\dot{\lambda}_2 \\ \dot{x}_3 \end{pmatrix} = \begin{pmatrix} -\frac{1}{4}(3\lambda_1+\lambda_2) \\ \frac{2}{5}(\alpha - 1 + \lambda_1\lambda_2(\alpha+1)) \\ \frac{1}{2}(\zeta(1+\lambda_2) + \xi(1-\lambda_2)) \end{pmatrix} .$$

Looking for fixed points of the fast subsystem, i.e. solving $\dot{\lambda}_1 = \dot{\lambda}_2 = 0$, gives the codimension $r = 2$ sliding modes

$$\lambda_1^{\$} = \mp\sqrt{\frac{\alpha-1}{3(\alpha+1)}} , \qquad \lambda_2^{\$} = 1 - 3\lambda_1^{\$} .$$

Hence there exist a pair of sliding modes inside $(\lambda_1, \lambda_2) \in (-1,+1)\times(-1,+1)$ for $1 < \alpha < 2$ and no viable sliding modes otherwise. (A short cut to find these is to seek solutions for $\lambda_1, \lambda_2 \in (-1,+1)$ such that $\mathbf{f}$ lies in the tangent space of the intersection, $\dot{x}_1 = 0$ and $\dot{x}_2 = 0$, but this will not tell us about their stability.) For $|\alpha| > 1$ the sliding vector field gives motion along the intersection with two different possible speeds

$$\begin{pmatrix} \varepsilon_1\dot{\lambda}_1 \\ \varepsilon_2\dot{\lambda}_2 \\ \dot{x}_3 \end{pmatrix} = \mathbf{f}^{\$\$} = \frac{1}{2}\begin{pmatrix} 0 \\ 0 \\ \zeta + \xi \pm (\zeta - \xi)\sqrt{3 - \frac{6}{\alpha+1}} \end{pmatrix} .$$

In fact the mode with $\lambda_1^{\$} < 0$ is an attractor and that with $\lambda_1^{\$} > 0$ is of saddle type; see Figure 7.11. To prove this requires looking at the Jacobian of the layer system, but we do not yet know how should we express such a derivative: should we take

$$\frac{\partial(\dot{\lambda}_1, \dot{\lambda}_2)}{\partial(\dot{\lambda}_1, \dot{\lambda}_2)} \quad \text{or} \quad \frac{\partial(\varepsilon_1\dot{\lambda}_1, \varepsilon_2\dot{\lambda}_2)}{\partial(\varepsilon_1\dot{\lambda}_1, \varepsilon_2\dot{\lambda}_2)} \quad \text{or} \quad \frac{\partial(\varepsilon_1\dot{\lambda}_1, \varepsilon_2\dot{\lambda}_2)}{\partial(\dot{\lambda}_1, \dot{\lambda}_2)}$$

or some other scaling in terms of the ε_j infinitesimals? We learn how to do such stability analysis in the layer in Section 7.6.

■

In general there is no way to determine how many sliding modes exist directly from the constituent vector fields. One must solve the system (7.5) and investigate how many valid solutions there are satisfying (7.6), that is, belonging to both the convex canopy (5.3) and the tangent space to the discontinuity surface $\mathcal{D}$.

In the case when all of the local vector fields are directed towards the intersection (i.e. it is attractive), then unique sliding vectors are guaranteed to exist by the boundary conditions (e.g. when all $\mathbf{f}^{\pm\pm}$ point inwards at the intersection of two discontinuity submanifolds, as considered in [53, 52]). But discontinuous dynamics can be much richer than this, and we shall see some of the novel phenomena that arise from multiplicity in later chapters.

7.5.3 Classification of Sliding Modes/Equilibria

Classification in piecewise-smooth systems is a difficult game. The zoology of specific cases becomes too voluminous to contain. Instead of enumerating every individual case, we can classify general forms that capture key behaviours at singularities. In the case of sliding modes, it is useful to distinguish whether a given class of sliding mode uniquely prescribes a system's local dynamics.

Definition 7.2. We separate attracting regions of the sliding manifold $\mathcal{M}$ into four classes:

(i) **Regular**—the sliding manifold has a unique branch at a given $\mathbf{x}$, to which all trajectories that enter the switching layer are attracted.
(ii) **Weak**—the sliding manifold has at least one branch at a given $\mathbf{x}$, but not all trajectories that enter the switching layer are attracted to it (instead they may cross through or find another sliding mode). At least some trajectories depart the intersection.
(iii) **Unreachable**—the sliding manifold has an attracting branch at a given $\mathbf{x}$, but no trajectories that enter the switching layer are able to reach it in forward time.
(iv) **Distributed**—the sliding manifold has an attracting branch at a given $\mathbf{x}$ that is distributed over a set of points of the fast subsystem over $(\lambda_1, \ldots, \lambda_r) \in (-1, +1)^2$ in the switching layer, e.g. it is a limit cycle or chaotic set.
(0) **Empty**—there exist no sliding modes (i.e. crossing).

Examples inside a codimension $r = 2$ switching layer are illustrated in Figure 7.12. We have already seen an example of a distributed sliding mode in Figure 7.5 and of a weak sliding mode in Figure 7.11.

These types of sliding modes have implications for the kinds of attractors they can form. In a linear switching system (i.e. with linear dependence on $\boldsymbol{\lambda}$), a sliding attractor could only take the form of an equilibrium, while in a nonlinear switching system, a sliding attractor may be of a more complex kind, being distributed and being either regular, weak, or unreachable.

If the discontinuity surface is a local attractor, that is, if all neighbouring trajectories enter it in finite time or asymptote towards it, then a regular or distributed sliding attractor must exist. However the converse is not true; the

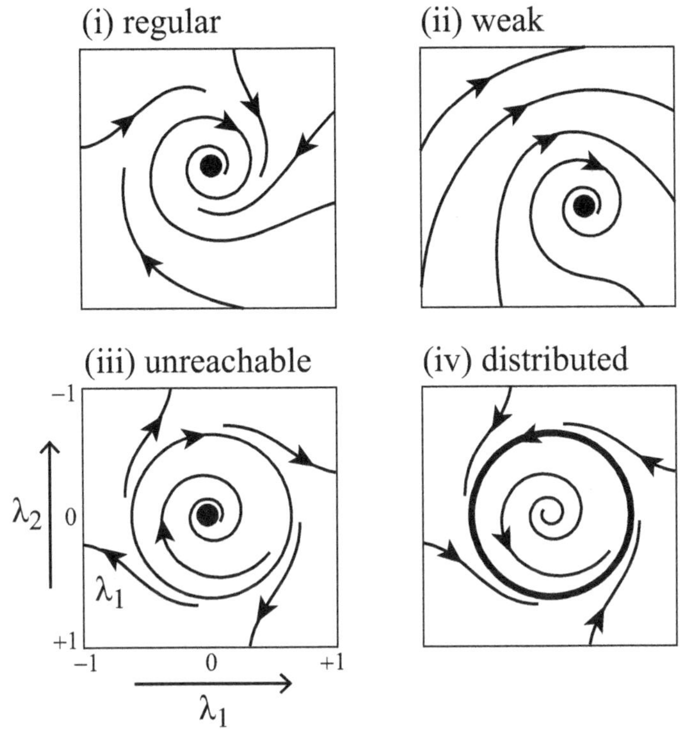

Fig. 7.12 Examples of the four kinds of sliding modes, shown in the switching layer at the intersection of discontinuity submanifolds $\sigma_1 = \sigma_2 = 0$. A regular sliding mode (the focal point in (i)) is attracting around the whole boundary. A weak sliding mode (the focal point in (ii)) attracts only some parts of the boundary and also permits crossing. An unreachable sliding mode (the focal point in (iii)) cannot be reached from any point on the boundary. A distributed sliding mode (the limit cycle in (iv)) is a set attractor. A distributed sliding mode could also be regular (as in in (iv)), weak, or unreachable.

existence of a regular or distributed sliding attractor does not guarantee that the discontinuity surface is locally an attractor, as there may exist nearby trajectories that diverge from the discontinuity surface's vicinity.

With these considerations, one of five things can happen when a non-sliding (codimension $r = 0$) trajectory arrives at a codimension 1 discontinuity submanifoldor a sliding (codimension $r \geq 1$) trajectory arrives at an intersection of discontinuity surfaces (of codimension $r+1$). We say a trajectory:

- *crosses* if it passes through the discontinuity surface with no sliding;
- *exits* if it passes to a region of equal or lower codimension sliding (including leaving the discontinuity surface altogether);
- *sticks* if it remains on the codimension $r+1$ intersection of discontinuity submanifolds and begins sliding along it;

- becomes *jammed* if no subsequent motion occurs because the trajectory has arrived at an equilibrium of codimension $r+1$ sliding; and
- *pauses* if it reaches an equilibrium of codimension $r+1$ sliding in finite time, but can be continued beyond it after an arbitrary finite time.

Example 7.8 (Exiting, Crossing, Sticking, Jamming, Pausing). The following piecewise-constant systems generate each of these behaviours at an intersection of discontinuity submanifolds $x_1 = 0$ and $x_2 = 0$:

$$\text{exiting}: \quad (\dot{x}_1, \dot{x}_2) = (1, 1 + \lambda_1 - \tfrac{1}{2}\lambda_2) \tag{7.23}$$
$$\text{crossing}: \quad (\dot{x}_1, \dot{x}_2) = (-\lambda_1, \tfrac{3}{2} + \lambda_1 - \lambda_2) \tag{7.24}$$
$$\text{jamming}: \quad (\dot{x}_1, \dot{x}_2) = (-\lambda_1, \tfrac{1}{2} + \lambda_1 - \lambda_2) \tag{7.25}$$
$$\text{pausing}: \quad (\dot{x}_1, \dot{x}_2) = (-\lambda_1, \lambda_2 + \lambda_1\lambda_2) \tag{7.26}$$

with switches $\lambda_1 = \text{sign}(x_1)$, $\lambda_2 = \text{sign}(x_2)$. Figure 7.13 shows the local portraits. If we added a dimension with, say, $\dot{x}_3 = c$, then (7.25) would still exhibit jamming if $c = 0$ but would instead show sticking at the intersection if $c \neq 0$. (Sticking or crossing at the codimension $r = 1$ discontinuity submanifolds $x_1 = 0$ or $x_2 = 0$ can also be found in each of these examples from an initial trajectory outside the discontinuity surface.)

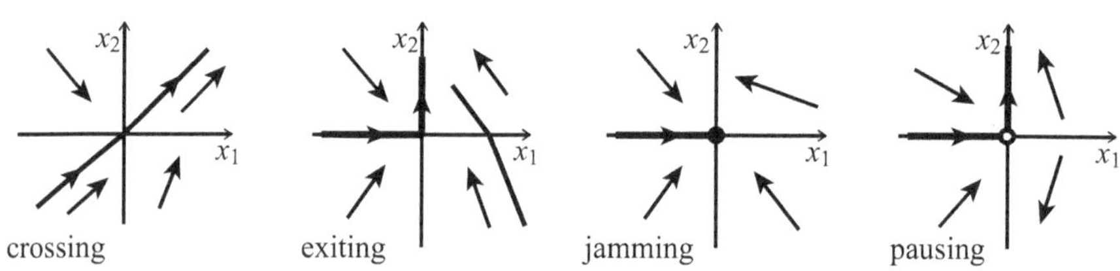

Fig. 7.13 The basic possibilities when a trajectory reaches a switching intersection in finite time: crossing, exiting, jamming, or pausing. In higher dimensions the jamming case would become sticking if there is motion along the discontinuity surface.

■

Bifurcations can occur that allow a system to transition between these behaviours. In the example in Figure 7.11, for example, a bifurcation inside the switching layer of the intersection at $\alpha = 1$ permits the system to change from sticking (or jamming if $\zeta = \xi = 0$) at the intersection to crossing through it.

7.6 Layer Variables

There is an important coordinate rescaling of the switching layer expression (3.12) that makes local analysis possible. Recall that a codimension r sliding region is a point where the discontinuity surface $\mathcal{D}$ consists of r intersecting submanifolds $\mathcal{D}_j$ and where the sliding manifold $\mathcal{M}$ exists.

Take a dynamical system $\dot{\mathbf{x}} = \mathbf{f}(\mathbf{x};\boldsymbol{\lambda})$ or more explicitly

$$\begin{pmatrix} \dot{x}_1 \\ \vdots \\ \dot{x}_n \end{pmatrix} = \begin{pmatrix} f_1(x_1,\dots,x_n;\lambda_1,\dots,\lambda_m) \\ \vdots \\ f_n(x_1,\dots,x_n;\lambda_1,\dots,\lambda_m) \end{pmatrix} , \tag{7.27}$$

with switching multipliers $\lambda_j = \text{sign}(\sigma_j)$ for $j = 1,2,\dots,m$, in terms of switching functions $\sigma_j = \sigma_j(\mathbf{x})$ whose zero level sets are transversal. As usual choose coordinates $\mathbf{x} = (x_1,\dots,x_n)$ such that $\sigma_j = x_j$ for $j = 1,\dots,m$.

At a point $\mathbf{x}$ on a codimension r sliding region where $\sigma_j = 0$ for $j \in \mathcal{S}_r \subseteq \{1,2,\dots,m\}$, the switching layer system (7.2) is then

$$\begin{pmatrix} \varepsilon_1\dot{\lambda}_1 \\ \vdots \\ \varepsilon_r\dot{\lambda}_r \\ \dot{x}_{r+1} \\ \vdots \\ \dot{x}_n \end{pmatrix} = \begin{pmatrix} f_1(0,\dots,0,x_{r+1},\dots,x_n;\lambda_1,\dots,\lambda_m) \\ \vdots \\ f_r(0,\dots,0,x_{r+1},\dots,x_n;\lambda_1,\dots,\lambda_m) \\ f_{r+1}(0,\dots,0,x_{r+1},\dots,x_n;\lambda_1,\dots,\lambda_m) \\ \vdots \\ f_n(0,\dots,0,x_{r+1},\dots,x_n;\lambda_1,\dots,\lambda_m) \end{pmatrix} , \tag{7.28}$$

where $\lambda_j \in (-1,+1)$.

Now define a *layer variable* $\boldsymbol{\xi}$ so that the left-hand side of (7.28) is $\dot{\boldsymbol{\xi}}$:

$$\boldsymbol{\xi} = (\xi_1,\dots,\xi_n) = (\varepsilon_1\lambda_1,\dots,\varepsilon_r\lambda_r,x_{r+1},\dots,x_n) . \tag{7.29}$$

The trivial case is $r = 0$ outside the discontinuity surface, where $\boldsymbol{\xi} = \mathbf{x}$.

More generally:

Definition 7.3. In coordinates where $\sigma_j = x_j$, at a codimension r-switching point where $x_j = 0$ for $j \in \mathcal{S}_r = \{1,\dots,r\}$, the **layer variable** is given by

$$\boldsymbol{\xi} = (\xi_1,\dots,\xi_n) \quad \text{s.t.} \quad \xi_j = \begin{cases} \varepsilon_j\lambda_j & \text{if } j \in \mathcal{S}_r , \\ x_j & \text{if } j \notin \mathcal{S}_r , \end{cases} \tag{7.30}$$

(This can be defined in more general coordinates, of course.)

The vector field has important properties when expressed in the layer variable, particularly concerning its Jacobian.

Relabel the vector field as

$$\mathbf{F}(\xi_1,\ldots,\xi_n;\lambda_{r+1..m}) = \mathbf{f}\Big(0,\ldots,0,x_{n-r},\ldots,x_n;\frac{\xi_1}{\varepsilon_1},\ldots,\frac{\xi_r}{\varepsilon_r},\lambda_{r+1..m}\Big)\,, \tag{7.31}$$

letting the shorthand $\lambda_{r+1..m}$ denote $\lambda_{r+1},\ldots,\lambda_m$, and where $\lambda_j = \operatorname{sign}(\sigma_j)$ for $j \geq r+1$. The switching layer system (7.28) is now

$$\dot{\boldsymbol{\xi}} := \mathbf{F}(\boldsymbol{\xi};\lambda_{r+1..m}) \tag{7.32a}$$

or in components

$$\begin{pmatrix}\dot{\xi}_1\\ \vdots \\ \dot{\xi}_n\end{pmatrix} = \begin{pmatrix}F_1(\xi_1,\ldots,\xi_n;\lambda_{r+1..m})\\ \vdots \\ F_n(\xi_1,\ldots,\xi_n;\lambda_{r+1..m})\end{pmatrix}\,, \tag{7.32b}$$

where

$$\xi_j = \begin{cases}\varepsilon_j\lambda_j & \text{if } j\in\mathcal{S}_r\,,\\ x_j & \text{if } j\notin\mathcal{S}_r\,,\end{cases}\qquad \lambda_j\in\begin{cases}(-1,+1) & \text{if } j\in\mathcal{S}_r\,,\\ \operatorname{sign}(x_j) & \text{if } j\notin\mathcal{S}_r\,.\end{cases} \tag{7.33}$$

We can carry out local differential analysis in the layer variable (something we could not do in $\mathbf{x}$, with respect to which $\mathbf{f}$ was discontinuous). Expanding $\mathbf{F}$ in a series about $\boldsymbol{\xi} = \boldsymbol{\xi}_0$, we have

$$\dot{\boldsymbol{\xi}} = \mathbf{F}(\boldsymbol{\xi}_0;\lambda_{r+1},\ldots) + \underline{\underline{J}}(\boldsymbol{\xi}_0;\lambda_{r+1..m})\cdot(\boldsymbol{\xi}-\boldsymbol{\xi}_0) + \ldots$$

where the Jacobian of $\mathbf{F}$ with respect to the layer variable is

$$\underline{\underline{J}}(\boldsymbol{\xi};\lambda_{r+1..m}) \equiv \frac{d}{d\boldsymbol{\xi}}\mathbf{F}(\boldsymbol{\xi};\lambda_{r+1..m})\,, \tag{7.34a}$$

with components

$$\underline{\underline{J}} \equiv \frac{d}{d\boldsymbol{\xi}}\mathbf{F} = \begin{pmatrix}\frac{\partial F_1}{\partial\xi_1} & \cdots & \frac{\partial F_1}{\partial\xi_n}\\ \vdots & \ddots & \vdots\\ \frac{\partial F_n}{\partial\xi_1} & \cdots & \frac{\partial F_n}{\partial\xi_n}\end{pmatrix}\,. \tag{7.34b}$$

At $\boldsymbol{\xi} = \boldsymbol{\xi}_0$, letting the vector field be $\mathbf{F}_0 = \mathbf{F}(\boldsymbol{\xi}_0;\lambda_{r+1..m})$ and the Jacobian be $\underline{\underline{J}}_0 = \underline{\underline{J}}(\boldsymbol{\xi}_0;\lambda_{r+1..m})$, solutions of (7.32) near $\boldsymbol{\xi}_0$ behave like

$$\boldsymbol{\xi}(t) \approx \boldsymbol{\xi}_0 + (t-t_0)\mathbf{F}_0 + \ldots \tag{7.35a}$$

to leading order in $t-t_0$, or

$$\boldsymbol{\xi}(t) \approx \boldsymbol{\xi}_0 + (e^{(t-t_0)\underline{\underline{J}}_0} - 1)(\boldsymbol{\xi}_0 + \underline{\underline{J}}_0^{-1}\mathbf{F}_0) + \ldots \tag{7.35b}$$

to linear order in $\boldsymbol{\xi}-\boldsymbol{\xi}_0$. The local behaviour, approximated to linear order, is therefore described by $\underline{\underline{J}}$, specifying any exponential contraction or expansion or rotation close to $\boldsymbol{\xi}_0$. We can now extend standard local stability analysis as follows.

The Jacobian $\underline{\underline{J}}$ obeys a characteristic equation

$$\underline{\underline{J}}.\gamma = c\gamma \; , \tag{7.36}$$

where γ and c are the eigenvectors and eigenvalues of $\underline{\underline{J}}$, of which there are n, written $c^{(i)}$ for $i = 1, \ldots, n$. Rearranging gives $0 = (\underline{\underline{J}} - c\underline{\underline{\mathcal{I}}}).\gamma$, which implies

$$\det\left|\underline{\underline{J}} - c\underline{\underline{\mathcal{I}}}\right| = 0 \; , \tag{7.37}$$

giving the *characteristic polynomial* of $\underline{\underline{J}}$, whose n roots are the eigenvalues $c^{(i)}$. While the eigenvalues change with coordinates, the coefficients of the characteristic polynomial are coordinate invariant, the kth coefficient being a sum over all possible k-tuples of the set of eigenvalues, for example, the coefficients of the 0th and n^{th} power of $c^{(i)}$ are the trace and determinant

$$\mathsf{T}(\underline{\underline{J}}) = c^{(1)} + \cdots + c^{(n)} \; , \qquad \det\left|\underline{\underline{J}}\right| = c^{(1)} \ldots c^{(n)} \; . \tag{7.38}$$

If the Jacobian has full rank, so $\det\left|\underline{\underline{J}}\right| \neq 0$, then it has n independent eigenvectors γ_i which satisfy the characteristic polynomial. The scalar roots of this equation, $c^{(i)}$, are the *eigenvalues*, consisting of an even number (or zero) of complex eigenvalues forming complex conjugate pairings ($c^{(i)} = c^{(j)*}$ for some i and j), and the remainder being real. The real part of each eigenvalue $c^{(i)}$ describes whether solutions are contracting $\mathbb{R}\mathrm{e}\left[c^{(i)}\right] < 0$ or expanding $\mathbb{R}\mathrm{e}\left[c^{(i)}\right] > 0$ along the direction $\gamma^{(i)}$, while imaginary parts impart an oscillation in the planes spanned by their conjugate pairs.

These are standard linear algebra concepts (which we already saw in Section 7.2) applied to $\underline{\underline{J}}$. What follows, however, is not standard, as we seek to decompose $\underline{\underline{J}}$ into components associated with dynamics inside and outside the discontinuity surface.

The Jacobian $\underline{\underline{J}}$ in layer variables typically has:

- $n - r$ eigenvalues associated with $n - r$ linearly independent eigenvectors spanning the tangent plane to the codimension r intersection $\sigma_1 = \ldots = \sigma_r = 0$, these eigenvalues remain finite as the parameters $\varepsilon_1, \ldots, \varepsilon_r \to 0$;
- r eigenvalues associated with r linearly independent eigenvectors spanning the normal directions to the codimension r intersection, these eigenvalues satisfy $c^{(j)} = \mathsf{O}(1/\varepsilon_j)$ and are therefore infinite as $\varepsilon_j \to 0$, providing finite time contraction/expansion in directions transverse to the discontinuity surface.

The eigenvalues may be real-valued or occur in complex conjugate pairs, as usual. We can find them as follows. First note that

$$\underline{\underline{J}} = \begin{pmatrix} \frac{1}{\varepsilon_1}\frac{\partial f_1}{\partial \lambda_1} & \cdots & \frac{1}{\varepsilon_r}\frac{\partial f_1}{\partial \lambda_r} & \frac{\partial f_1}{\partial x_{r+1}} & \cdots & \frac{\partial f_1}{\partial x_n} \\ \vdots & \ddots & \vdots & \vdots & \ddots & \vdots \\ \frac{1}{\varepsilon_1}\frac{\partial f_n}{\partial \lambda_1} & \cdots & \frac{1}{\varepsilon_r}\frac{\partial f_n}{\partial \lambda_r} & \frac{\partial f_n}{\partial x_{r+1}} & \cdots & \frac{\partial f_n}{\partial x_n} \end{pmatrix}$$

$$= \underline{\underline{A}}.\underline{\underline{E}} \tag{7.39}$$

in terms of $n \times n$ square matrices

$$\underline{\underline{A}} = \begin{pmatrix} \frac{\partial f_1}{\partial \lambda_1} & \cdots & \frac{\partial f_1}{\partial \lambda_r} & \frac{\partial f_1}{\partial x_{r+1}} & \cdots & \frac{\partial f_1}{\partial x_n} \\ \vdots & \ddots & \vdots & \vdots & \ddots & \vdots \\ \frac{\partial f_n}{\partial \lambda_1} & \cdots & \frac{\partial f_n}{\partial \lambda_r} & \frac{\partial f_n}{\partial x_{r+1}} & \cdots & \frac{\partial f_n}{\partial x_n} \end{pmatrix},$$

$$\underline{\underline{E}} = \begin{pmatrix} \frac{1}{\varepsilon_1} & \cdots & 0 & 0 & \cdots & 0 \\ \vdots & \ddots & \vdots & \vdots & \ddots & \vdots \\ 0 & \cdots & \frac{1}{\varepsilon_r} & 0 & \cdots & 0 \\ 0 & \cdots & 0 & 1 & \cdots & 0 \\ \vdots & \ddots & \vdots & \vdots & \ddots & \vdots \\ 0 & \cdots & 0 & 0 & \cdots & 1 \end{pmatrix}. \tag{7.40}$$

The matrix $\underline{\underline{A}}$ is the Jacobian $\left(\frac{\partial(f_1,\ldots,f_n)}{\partial(\lambda_1,\ldots,\lambda_r)}, \frac{\partial(f_1,\ldots,f_n)}{\partial(x_{r+1},\ldots,x_n)}\right)$, which is ε_j-independent and typically non-singular, while $\underline{\underline{E}}$ is a diagonal matrix containing the factors $1/\varepsilon_j$ which are singular as $\varepsilon_j \to 0$.

The two groups of eigenvalues come from two different ways that we can absorb $\underline{\underline{E}}$ into c and/or $\boldsymbol{\gamma}$.

Denoting the ith eigenvector as $\boldsymbol{\gamma}^{(i)} = (\gamma_1^{(i)}, \ldots, \gamma_n^{(i)})$, we can rewrite (7.36) as

$$\begin{pmatrix} \frac{\partial(f_1,\ldots,f_n)}{\partial(\lambda_1,\ldots,\lambda_r)} & \frac{\partial(f_1,\ldots,f_n)}{\partial(x_{r+1},\ldots,x_n)} \end{pmatrix} \begin{pmatrix} \frac{1}{\varepsilon_1}\gamma_1^{(i)} \\ \vdots \\ \frac{1}{\varepsilon_r}\gamma_r^{(i)} \\ \gamma_{r+1}^{(i)} \\ \vdots \\ \gamma_n^{(i)} \end{pmatrix} = c \begin{pmatrix} \gamma_1^{(i)} \\ \vdots \\ \gamma_r^{(i)} \\ \gamma_{r+1}^{(i)} \\ \vdots \\ \gamma_n^{(i)} \end{pmatrix}, \tag{7.41a}$$

which will produce finite eigenvalues and eigenvectors if the eigenvector components $\gamma_1^{(i)}, \ldots, \gamma_r^{(i)}$, scale with $1/\varepsilon_1, \ldots, 1/\varepsilon_r$, so let $\gamma_j^{(i)} = \varepsilon_j \hat{\gamma}^{(i)}$ for $j = 1, \ldots, r$, and then for $\varepsilon_j \to 0$ we have

$$\begin{pmatrix} \frac{\partial(f_1,\ldots,f_n)}{\partial(\lambda_1,\ldots,\lambda_r)} & \frac{\partial(f_1,\ldots,f_n)}{\partial(x_{r+1},\ldots,x_n)} \end{pmatrix} \begin{pmatrix} \hat{\gamma}_1^{(i)} \\ \vdots \\ \hat{\gamma}_r^{(i)} \\ \gamma_{r+1}^{(i)} \\ \vdots \\ \gamma_n^{(i)} \end{pmatrix} = c \begin{pmatrix} 0 \\ \vdots \\ 0 \\ \gamma_{r+1}^{(i)} \\ \vdots \\ \gamma_n^{(i)} \end{pmatrix} . \tag{7.41b}$$

The first r rows of (7.41b) are trivially only satisfied if $c \equiv 0$, and we discard these as they are not solutions of (7.36). The lower $n-r$ rows provide $n-r$ eigenvalues c, and $n-r$ associated eigenvectors $(\varepsilon_1\hat{\gamma}_1^{(i)},\ldots,\varepsilon_r\hat{\gamma}_r^{(i)},\gamma_{r+1}^{(i)},\ldots,\gamma_n^{(i)})$ whose first r components vanish as $\varepsilon_j \to 0$; therefore, they span the last $n-r$ directions, which are linearly independent to the planes $\sigma_1 = 0, \ldots, \sigma_r = 0$.

Alternatively we can let $\varepsilon_j = \varepsilon_0 \upsilon_j$ such that all $\varepsilon_j \to 0$ as $\varepsilon_0 \to 0$, and then rewrite (7.36) as

$$\begin{pmatrix} \frac{\partial(f_1,\ldots,f_n)}{\partial(\upsilon_1\lambda_1,\ldots,\upsilon_r\lambda_r)} & \varepsilon_0\frac{\partial(f_1,\ldots,f_n)}{\partial(x_{r+1},\ldots,x_n)} \end{pmatrix} \begin{pmatrix} \gamma_1^{(i)} \\ \vdots \\ \gamma_n^{(i)} \end{pmatrix} = \varepsilon_0 c \begin{pmatrix} \gamma_1^{(i)} \\ \vdots \\ \gamma_n^{(i)} \end{pmatrix} . \tag{7.42a}$$

If we assume the eigenvalues are of order $\mathsf{O}(1/\varepsilon_0)$, we let $\hat{c} = \varepsilon_0 c$ and obtain

$$\begin{pmatrix} \frac{\partial(f_1,\ldots,f_n)}{\partial(\upsilon_1\lambda_1,\ldots,\upsilon_r\lambda_r)} & \underline{\underline{0}} \end{pmatrix} \begin{pmatrix} \gamma_1^{(i)} \\ \vdots \\ \gamma_n^{(i)} \end{pmatrix} = \hat{c} \begin{pmatrix} \gamma_1^{(i)} \\ \vdots \\ \gamma_n^{(i)} \end{pmatrix} . \tag{7.42b}$$

This produces $n-r$ eigenvectors spanning the last $n-r$ directions, but their eigenvalues must be trivially zero to balance the $\mathsf{O}(\varepsilon_0)$ scaling, so we discard these as, again, they are not solutions of (7.36). Equation (7.42b) also produces r eigenvectors spanning the first r directions in the tangent space of the switching intersection, associated with eigenvalues $\hat{c}$. As we let $\varepsilon_0 \to 0$ so that all $\varepsilon_j \to 0$, the eigenvalues of (7.36) given by $c = \hat{c}/\varepsilon_0$ diverge to infinity.

These methods will come in useful throughout the rest of this chapter.

The eigenvalues that are infinite as $\varepsilon_j \to 0$ describe the rate of attraction to or from the discontinuity surface from outside the sliding region. This means that, as $\varepsilon_j \to 0$, the flow reaches or departs along these directions in finite time. The finite eigenvalues (those of order ε_j^p for $p \geq 0$) describe the finite rate of asymptotic attraction/repulsion along the sliding vector field inside the surface. Within the inner or outer parts of the eigen-decomposition, the finite or infinite eigenvalues may occur in complex conjugate pairs, implying circulation inside or about the sliding region, respectively.

7.7 Equilibria and Sliding Equilibria

An equilibrium point of a dynamical system $\dot{\mathbf{x}} = \mathbf{f}(\mathbf{x})$ is a point where $\mathbf{x}$ is stationary, since $\mathbf{f}(\mathbf{x}) = \mathbf{0}$. Equilibria in sliding vector fields are similar beasts, except they occupy the reduced dimensions of the discontinuity surface. In some texts they are called 'pseudoequilibria', but in our context they *are* equilibria, not something artificial as the term 'pseudo' might suggest, and this should be clearer from the way we study them below. Instead to emphasize their confinement to the discontinuity surface as part of the sliding dynamics, we call them *sliding equilibria.*

Definition 7.4. An **equilibrium** of the system (7.27) is a point $\mathbf{x}_{eq}$ where the right-hand side of (7.27) vanishes. A **sliding equilibrium** (of the codimension r sliding vector field) of the system (7.27) is a point where the right-hand side of (7.28) vanishes.

Generically an equilibrium lies away from the discontinuity surface, where all σ_j are nonzero, and a sliding equilibrium lies away from the boundaries of the switching layer, where $\lambda_j \in (-1, +1)$ for $j = 1, \dots, r$.

Consider a point $\mathbf{x}_p$ at a codimension r switching intersection point for $0 \leq r \leq m$. Let $\mathbf{x}_p$ lie on the intersection of manifolds $\sigma_j(\mathbf{x}_p) = 0$ for $j \in \mathcal{S}_r \subseteq \{1, 2, \dots, m\}$. So $\mathcal{S}_r$ contains the integers 1 to r with $r \leq m$. Then

$$\mathbf{x}_p = (x_1, \dots, x_n) \quad \text{such that} \quad x_j = 0 \ \text{ if } \ j \in \mathcal{S}_r \,. \tag{7.43}$$

An equilibrium is a point $\boldsymbol{\xi} = \boldsymbol{\xi}_{eq}$ where the right-hand $\mathbf{F}(\boldsymbol{\xi}; \lambda_{r+1}, \dots, \lambda_m)$ of (7.32) vanishes. Expanding in a series about $\boldsymbol{\xi} = \boldsymbol{\xi}_{eq}$ gives

$$\dot{\boldsymbol{\xi}} = \underline{\underline{J}}(\boldsymbol{\xi}_{eq}) \cdot (\boldsymbol{\xi} - \boldsymbol{\xi}_{eq}) + \dots$$

where $\underline{\underline{J}}(\boldsymbol{\xi}) \equiv \frac{d}{d\boldsymbol{\xi}} \mathbf{F}(\boldsymbol{\xi}; \lambda_{r+1..m})$ is the Jacobian matrix (7.34b). Since the constant term $\mathbf{F}(\boldsymbol{\xi}_{eq}; \lambda_{r+1..m})$ vanishes, the linear term describes the local flow to leading order.

These are the elementary tools used to understand simple equilibria in the space of $\boldsymbol{\xi}$, and they apply equally well in this case when $\boldsymbol{\xi}$ is a layer variable, of which some components are the spacial variables x_i and some are the scaled switching multipliers λ_j.

Example 7.9 (Eigenvectors of an Equilibrium in cod. 1 Sliding, Part I). Letting $\lambda = \text{sign}(x_1)$, the planar system

$$\begin{pmatrix} \dot{x}_1 \\ \dot{x}_2 \end{pmatrix} = \begin{pmatrix} -\lambda \\ c - x_2 - \lambda(c + x_2) \end{pmatrix} = \begin{cases} \begin{pmatrix} -1 \\ -2x_2 \end{pmatrix} & \text{if } x_1 > 0 \,, \\ \begin{pmatrix} 1 \\ 2c \end{pmatrix} & \text{if } x_1 < 0 \,, \end{cases}$$

has no equilibria for $x_1 \neq 0$. The switching layer system is

$$\begin{pmatrix} \varepsilon\dot{\lambda} \\ \dot{x}_2 \end{pmatrix} = \begin{pmatrix} -\lambda \\ c - x_2 - \lambda(c + x_2) \end{pmatrix} .$$

Sliding modes are given by $\lambda^{\$} = 0$ and exist for all x_2.

There is a sliding equilibrium at $(\lambda, x_2) = (0, c)$. To find out if it is an attractor, we need its eigenvalues and eigenvectors.

The Jacobian in the layer variables $\boldsymbol{\xi} = (\xi_1, \xi_2) = (\varepsilon\lambda, x_2)$, evaluated at the equilibrium, is

$$\underline{\underline{J}} = \begin{pmatrix} \frac{\partial \dot{\xi}_1}{\partial \xi_1} & \frac{\partial \dot{\xi}_1}{\partial x_2} \\ \frac{\partial \dot{x}_2}{\partial \xi_1} & \frac{\partial \dot{x}_2}{\partial \dot{x}_2} \end{pmatrix} = -\frac{1}{\varepsilon} \begin{pmatrix} 1 & 0 \\ c + x_2 & \xi_1 + \varepsilon \end{pmatrix} = -\frac{1}{\varepsilon} \begin{pmatrix} 1 & 0 \\ 2c & \varepsilon \end{pmatrix}$$

with eigenvalues $\gamma^{(i)}$ and eigenvectors $\boldsymbol{\gamma}^{(i)}$ as $\varepsilon \to 0$ given by

$$\begin{aligned} \gamma^{(1)} &= -1/\varepsilon \to -\infty &&: \quad \boldsymbol{\gamma}^{(1)} \to (1, 2c) \ , \\ \gamma^{(2)} &= -1 &&: \quad \boldsymbol{\gamma}^{(2)} = (0,\ 1\ ) \ . \end{aligned}$$

The second eigenvector lies along the direction $(0, 1)$, giving asymptotic attraction along the tangent space of $\mathcal{D}$ at unit rate $x_2 - c \sim (x_{20} - c)e^{-t}$. The first eigenvector lies along the direction $(1, 2c)$ out of $\mathcal{D}$, along which attraction is infinitely fast, $\xi_1 \sim e^{-t/\varepsilon}$ for $\varepsilon \to 0$, after reaching the discontinuity surface in finite time. An example is shown in Figure 7.14(i).

■

Example 7.10 (Eigenvectors of an Equilibrium in cod. 1 *Sliding, Part II).* Letting $\lambda = \text{sign}(x_1)$, the planar system

$$\begin{pmatrix} \dot{x}_1 \\ \dot{x}_2 \end{pmatrix} = \begin{pmatrix} 1 - x_2 - \lambda(1 + x_2) \\ c - 1 - \lambda(c + 1) \end{pmatrix} = \begin{cases} 2\begin{pmatrix} -x_2 \\ -1 \end{pmatrix} & \text{if } x_1 > 0 \ , \\ 2\begin{pmatrix} 1 \\ c \end{pmatrix} & \text{if } x_1 < 0 \ , \end{cases}$$

has no equilibria for $x_1 \neq 0$. The switching layer system is

$$\begin{pmatrix} \varepsilon\dot{\lambda} \\ \dot{x}_2 \end{pmatrix} = \begin{pmatrix} 1 - x_2 - \lambda(1 + x_2) \\ c - 1 - \lambda(c + 1) \end{pmatrix} .$$

This has a sliding equilibrium at $(\lambda, x_2) = (\frac{c-1}{c+1}, \frac{1}{c})$. However, sliding modes are given by $\lambda^{\$} = (1 - x_2)/(1 + x_2)$ and therefore exist only for $x_2 > 0$. The sliding equilibrium only exists when lying in this region and hence when $c > 0$.

To find out if it is an attractor, we find the Jacobian in the layer variables $\boldsymbol{\xi} = (\xi_1, \xi_2) = (\varepsilon\lambda, x_2)$, evaluated at the equilibrium:

$$\underline{\underline{J}} = \begin{pmatrix} \frac{\partial\dot{\xi}_1}{\partial\xi_1} & \frac{\partial\dot{\xi}_1}{\partial x_2} \\ \frac{\partial\dot{x}_2}{\partial\xi_1} & \frac{\partial\dot{x}_2}{\partial x_2} \end{pmatrix} = -\frac{1}{\varepsilon}\begin{pmatrix} 1+x_2 & \xi_1+\varepsilon \\ c+1 & 0 \end{pmatrix} = -\frac{1}{\varepsilon}\begin{pmatrix} 1+\frac{1}{c} & \frac{2c\varepsilon}{c+1} \\ c+1 & 0 \end{pmatrix}$$

with eigenvalues $\gamma^{(i)}$ and eigenvectors $\boldsymbol{\gamma}^{(i)}$ as $\varepsilon \to 0$ given by

$$\begin{aligned} \gamma^{(1)} &= \tfrac{c+1+\sqrt{R}}{-2c\varepsilon} \to -\infty &&: \quad \boldsymbol{\gamma}^{(1)} \to (1, c)^{\mathsf{T}} \,, \\ \gamma^{(2)} &= \tfrac{c+1-\sqrt{R}}{-2c\varepsilon} \to \tfrac{2c^2}{c+1} &&: \quad \boldsymbol{\gamma}^{(2)} = (0,\ 1\ )^{\mathsf{T}} \,, \end{aligned}$$

where $R = (c+1)^2 + 8c^3\varepsilon$. The first eigenvector lies along the $(0,1)$ direction, giving asymptotic repulsion for $c > 0$ along the tangent space of $\mathcal{D}$ as $x_2 - \frac{1}{c} \sim (x_{20} - \frac{1}{c})e^{-2tc^2/(c+1)}$. The second eigenvector lies along the $(1, c)$ direction, giving infinitely fast attraction for $c > 0$ out of $\mathcal{D}$ as $\xi_1 - \frac{c-1}{c+1} \sim (\xi_{10} - \frac{c-1}{c+1})e^{-t(c+1)/2c\varepsilon}$, so the contraction to $\xi_1 = 0$ is in finite time, after reaching the discontinuity surface in finite time. An example is shown in Figure 7.14(ii).

■

Example 7.11 (Eigenvectors of an Equilibrium in cod. 1 Sliding, Part III). Letting $\lambda = \operatorname{sign}(x_1)$, the three-dimensional system

$$\begin{pmatrix} \dot{x}_1 \\ \dot{x}_2 \\ \dot{x}_3 \end{pmatrix} = \frac{1}{2}\begin{pmatrix} -2\lambda \\ (1+\lambda)(ax_2+bx_3)+(1-\lambda)e \\ (1+\lambda)(cx_2+dx_3) \end{pmatrix} = \begin{cases} \begin{pmatrix} -1 \\ ax_2+bx_3 \\ cx_2+dx_3 \end{pmatrix} \text{ if } x_1 > 0\,, \quad \begin{pmatrix} 1 \\ e \\ 0 \end{pmatrix} \text{ if } x_1 < 0\,, \end{cases}$$

has no equilibria. The switching layer system ((7.2) for $r = 1$) is

$$\begin{pmatrix} \varepsilon\dot{\lambda} \\ \dot{x}_2 \\ \dot{x}_3 \end{pmatrix} = \tfrac{1}{2}\begin{pmatrix} -2\lambda \\ (1+\lambda)(ax_2+bx_3)+(1-\lambda)e \\ (1+\lambda)(cx_2+dx_3) \end{pmatrix}.$$

Sliding modes with $\lambda^{\$} = 0$ exist everywhere on $x_1 = 0$. There are sliding equilibria at $(\lambda, x_2, x_3) = (0, -d, c)e/(ad-bc)$. The Jacobian at this point in the layer variables $(\xi_1, \xi_2, \xi_3) = (\varepsilon\lambda_1, x_2, x_3)$ is

$$\underline{\underline{J}} = \frac{\partial(\dot{\xi}_1, \dot{\xi}_2, \dot{\xi}_3)}{\partial(\xi_1, \xi_2, \xi_3)} = \tfrac{1}{2}\begin{pmatrix} -2/\varepsilon & 0 & 0 \\ -2e/\varepsilon & a & b \\ 0 & c & d \end{pmatrix}$$

with eigenvalues $\gamma^{(i)}$ and eigenvectors $\boldsymbol{\gamma}^{(i)}$ as $\varepsilon \to 0$ given by

$$\begin{array}{lcl} \gamma^{(1)} = -1/\varepsilon \to -\infty & : & \boldsymbol{\gamma}^{(1)} \to (1, e, 0)^{\mathsf{T}} \\ \gamma^{(2)} = -\frac{a+d}{4} - \sqrt{R} & : & \boldsymbol{\gamma}^{(2)} = (0, 2\gamma^{(2)} - d, c)^{\mathsf{T}} \\ \gamma^{(3)} = -\frac{a+d}{4} + \sqrt{R} & : & \boldsymbol{\gamma}^{(3)} = (0, 2\gamma^{(3)} - d, c)^{\mathsf{T}} \end{array}$$

where $R = \left(\frac{a+d}{4}\right)^2 + bc - ad$. The first row says that the sliding region $x_1 = 0$ is attracting with an infinite rate (i.e. attracting in finite time) along the direction $(1/e, 1, 0)$, which necessarily lies transverse to the discontinuity surface, and in fact lies along the directions of the two composite vector fields at the equilibrium, as shown in Figure 7.14(iii).

The other two eigenvalues are finite and provide the attractivity of the fixed point of the (x_2, x_3) subsystem in the sliding vector field in the usual way for planar systems: a focus if $R < 0$, a node or saddle if $R > 0$, a saddle if $ad - bc < 0$, and focus or node if $ad - bc > 0$, attracting if $a + d < 0$ and repelling if $a + d > 0$.

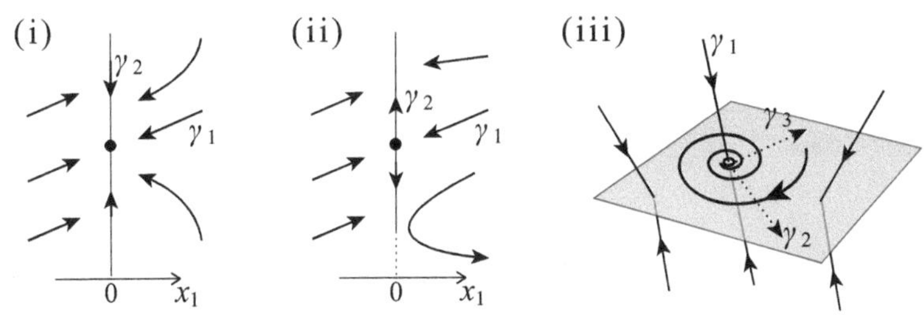

Fig. 7.14 The sliding equilibria and their eigenvectors, for example, Example 7.9 (left), Example 7.10 (middle), and Example 7.11 (right). The right figure is for a set of constants that gives a sliding focus; the eigenvector $\gamma_{2,3}$ span the surface $\mathcal{D}$ but in this case are complex.

■

Example 7.12 (Eigenvectors of an Equilibrium in Codimension 2 Sliding). Consider the three-dimensional system

$$\begin{pmatrix} \dot{x}_1 \\ \dot{x}_2 \\ \dot{x}_3 \end{pmatrix} = \begin{pmatrix} a_1 + c_1 x_1 \\ a_2 + c_2 x_2 \\ -x_3 \end{pmatrix} + \begin{pmatrix} b_{11} & b_{12} & 0 \\ b_{21} & b_{22} & 0 \\ 0 & 0 & 0 \end{pmatrix} \begin{pmatrix} \lambda_1 \\ \lambda_2 \\ \lambda_3 \end{pmatrix}$$

where $\lambda_i = \operatorname{sign} x_i$, for nonzero constants a_i, b_{ij}, c_i. Three examples are shown in Figure 7.15. The system is piecewise constant, so there are no equilibria outside the discontinuity surface. Taking the intersection first, the switching layer system ((7.2) for $r = 2$) is

$$\begin{pmatrix} \varepsilon_1 \dot{\lambda}_1 \\ \varepsilon_2 \dot{\lambda}_2 \\ \dot{x}_3 \end{pmatrix} = \begin{pmatrix} a_1 + b_{11}\lambda_1 + b_{12}\lambda_2 \\ a_2 + b_{21}\lambda_1 + b_{22}\lambda_2 \\ -x_3 \end{pmatrix} \qquad \text{on } x_1 = x_2 = 0\,.$$

This has a unique equilibrium at

$$(\lambda_1, \lambda_2, x_3) = \frac{(a_2 b_{12} - a_1 b_{22},\ a_1 b_{21} - a_2 b_{11},\ 0)}{b_{11} b_{22} - b_{12} b_{21}},$$

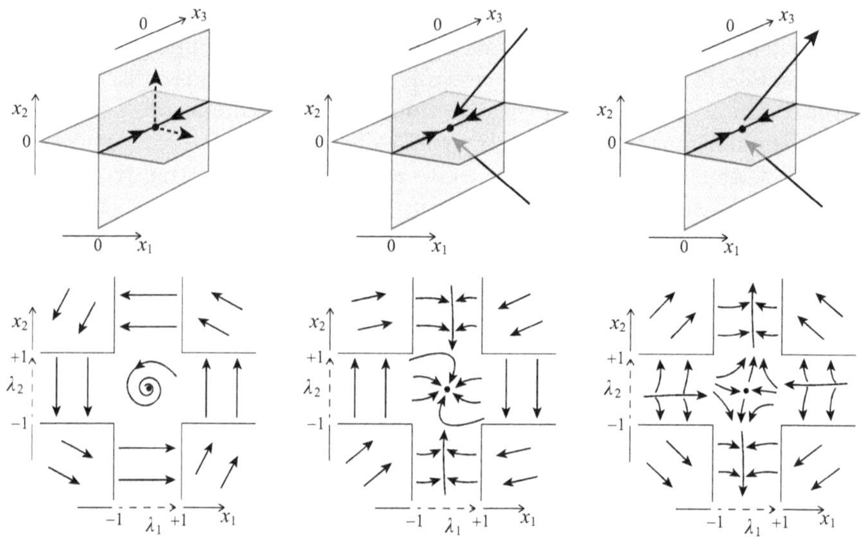

Fig. 7.15 Three illustrations of Example 7.12, with $a_1 = a_2 = 0$ and: (left) $b_{21} = -b_{12} = 1$, $b_{22} = 0$, $b_{11} = -1/2$; (middle) $b_{21} = -b_{12} = -1$, $b_{22} = 0$, $b_{11} = -3$; (right) $b_{21} = b_{12} = 1$, $b_{22} = 2$, $b_{11} = -3$. The upper figures show the directions of the eigenvectors in the layer variables (one vector along the intersection and two vectors in the x_1-x_2 plane in each case); the lower figures show the piecewise-smooth system in the x_1-x_2 plane, including the switching layer.

the first two components of which must lie inside $(-1, +1) \times (-1, +1)$; otherwise the equilibrium ceases to exist. (Where does the sliding equilibrium go if the parameters leave this range? We return to this question in Example 7.15.) The Jacobian of the equilibrium by (7.34b) in the layer variables $(\xi_1, \xi_2, \xi_3) = (\varepsilon_1 \lambda_1, \varepsilon_2 \lambda_2, x_3)$ is

$$\underline{\underline{J}} = \frac{\partial(\dot{\xi}_1, \dot{\xi}_2, \dot{\xi}_3)}{\partial(\xi_1, \xi_2, \xi_3)} = \begin{pmatrix} \varepsilon_1^{-1} b_{11} & \varepsilon_2^{-1} b_{12} & 0 \\ \varepsilon_1^{-1} b_{21} & \varepsilon_2^{-1} b_{22} & 0 \\ 0 & 0 & -1 \end{pmatrix}.$$

with eigenvalues $\gamma^{(i)}$ and eigenvectors $\boldsymbol{\gamma}^{(i)}$ as each $\varepsilon_j \to 0$ given by

$$\begin{aligned} \gamma^{(1)} &= \left(\tfrac{b_{11} + \upsilon b_{22}}{2} + \sqrt{R}\right)/\varepsilon_1 &&: \quad \boldsymbol{\gamma}^{(1)} = \left(\gamma^{(1)} - \upsilon b_{22}, b_{21}, 0\right)^{\mathsf{T}} \\ \gamma^{(2)} &= \left(\tfrac{b_{11} + \upsilon b_{22}}{2} - \sqrt{R}\right)/\varepsilon_1 &&: \quad \boldsymbol{\gamma}^{(2)} = \left(\gamma^{(2)} - \upsilon b_{22}, b_{21}, 0\right)^{\mathsf{T}} \\ \gamma^{(3)} &= -1 &&: \quad \boldsymbol{\gamma}^{(3)} = (0, 0, 1)^{\mathsf{T}} \end{aligned}$$

where $\upsilon = \varepsilon_1/\varepsilon_2$ and $R = \left(\frac{b_{11}+\upsilon b_{22}}{2}\right)^2 + \upsilon b_{12}b_{21} - \upsilon b_{11}b_{22}$. The eigenvector $\gamma^{(3)}$ gives a finite rate of attraction along the intersection. The eigenvalues $\gamma_{1,2}$ have a magnitude that is infinite as $\varepsilon_{1,2} \to 0$, since they describe dynamics in the plane transverse to the intersection, i.e. in the directions out of the codimension 2 sliding region. Their eigenvectors, however, are finite, assuming that the ratio $\upsilon = \varepsilon_1/\varepsilon_2$ is finite and nonzero as $\varepsilon_{1,2} \to 0$. Thus the parameters b_{ij} and the ratio υ determine whether the equilibrium is a node, focus, or saddle in the $\lambda_{1,2}$ plane. This closely mirrors what happens outside the intersection, and if $\upsilon = 1$ it corresponds directly.

■

7.8 Invariant Subspaces

Having extended the elements of linearization to the discontinuity via the switching layer and via rescaling to the *layer variable*, we can also extend certain elements of the theory of stable manifolds and centre manifolds.

If the Jacobian at $\boldsymbol{\xi}_{eq}$ is nondegenerate, we can then approximate $\dot{\boldsymbol{\xi}} = \mathbf{F}(\boldsymbol{\xi};\lambda_{r+1..m})$ from (7.32) by its *linearization*

$$\dot{\boldsymbol{\xi}} = \underline{\underline{J}}_{eq}\cdot(\boldsymbol{\xi} - \boldsymbol{\xi}_{eq}) \,. \tag{7.44}$$

We showed in the previous section that the eigenvectors of $\underline{\underline{J}}_{eq}$ are well-defined and associated with eigenvalues that may be finite or infinite (as $\varepsilon_j \to 0$) in a well-defined manner, with well-defined signs of the real and imaginary parts. We therefore have the following.

Provided there exists an open neighbourhood $\mathcal{U}$ around $\boldsymbol{\xi}_{eq}$ that lies strictly inside the switching layer, so $\boldsymbol{\xi}_{eq} \in \mathcal{U} \subset (-1,+1)^r \times \mathbb{R}^{n-r}$, then the spaces spanned by the eigenvectors of $\underline{\underline{J}}_{eq}$ (the eigenspaces of $\underline{\underline{J}}_{eq}$) are invariant subspaces for the flow. These consist of:

- the *stable subspace* $\mathcal{S}^s(\boldsymbol{\xi}_{eq}) = \text{span}\{\gamma_s^{(1)},\ldots,\gamma_s^{(i_s)}\}$,
- the *unstable subspace* $\mathcal{S}^u(\boldsymbol{\xi}_{eq}) = \text{span}\{\gamma_u^{(1)},\ldots,\gamma_u^{(i_u)}\}$,
- the *centre subspace* $\mathcal{S}^c(\boldsymbol{\xi}_{eq}) = \text{span}\{\gamma_c^{(1)},\ldots,\gamma_c^{(i_c)}\}$,

where $i_s+i_u+i_c = n$, with each eigenvector $\gamma_s^{(i)}$, $\gamma_u^{(i)}$, or $\gamma_c^{(i)}$, having a corresponding eigenvalue whose real part is negative, positive, or zero, respectively. The terms stable/unstable refer to the exponential attraction to/repulsion from $\boldsymbol{\xi}_{eq}$ implied by the eigenvalue in the solution of (7.44) (given by (7.35b) with $\mathbf{F}_{eq} = 0$):

$$\boldsymbol{\xi}(t) \approx \boldsymbol{\xi}_{eq} + e^{(t-t_0)\underline{\underline{J}}_{eq}}\boldsymbol{\xi}_0 = \boldsymbol{\xi}_{eq} + \sum_{i=1}^{n} e^{(t-t_0)c^{(i)}}\alpha_i\gamma^{(i)} \tag{7.45}$$

where $\boldsymbol{\xi}_0 = \sum_{i=1}^{n} \alpha_i \boldsymbol{\gamma}^{(i)}$ is some initial condition expressed in terms of the eigenvectors $\boldsymbol{\gamma}^{(i)}$.

The linearization is important because of the following results from standard dynamical systems theory.

Theorem 7.3 (Hartman-Grobman Applied to Switching Layers). *If $\underline{\underline{J}}_{eq}$ has no eigenvalues with zero real part and $\boldsymbol{\xi}_{eq} \in \mathcal{U} \subset (-1,+1)^r \times \mathbb{R}^{n-r}$, then the flow of $\dot{\boldsymbol{\xi}} = \mathbf{F}(\boldsymbol{\xi}; \lambda_{r+1..m})$ is homeomorphic to the flow of the linearization $\dot{\boldsymbol{\xi}} = \underline{\underline{J}}_{eq} . (\boldsymbol{\xi} - \boldsymbol{\xi}_{eq})$.*

The extension of this to a diffeomorphism between the flows (Sternberg's theorem) applies on $\mathcal{U}$ given the standard non-resonance conditions between the eigenvalues of $\underline{\underline{J}}_{eq}$ (see [98, 92]).

The invariant *subspaces* of the linearization are local approximations of invariant *manifolds* of a nonlinear system near equilibrium. The stable manifold $\mathcal{W}^s(\boldsymbol{\xi}_{eq})$ and unstable manifold $\mathcal{W}^u(\boldsymbol{\xi}_{eq})$ are the sets of trajectories that tend to $\boldsymbol{\xi}_{eq}$ as $t \to +\infty$ and $t \to -\infty$, respectively, and their tangent spaces are given by $\mathcal{S}^s(\boldsymbol{\xi}_{eq})$ and $\mathcal{S}^u(\boldsymbol{\xi}_{eq})$ at $\boldsymbol{\xi}_{eq}$. This is the extension of the standard *stable manifold theorem* for switching layers.

If any of the eigenvalues of the Jacobian at $\boldsymbol{\xi}_{eq}$ have vanishing real parts, there exists also a centre subspace. The *centre manifold theorem* then applies inside $\mathcal{U}$, that is, there exists a *centre manifold* $\mathcal{W}^c(\boldsymbol{\xi}_{eq})$ tangent to $\mathcal{S}^c(\boldsymbol{\xi}_{eq})$, invariant under the local flow. Unlike the stable and unstable manifolds, the centre manifold need not be unique.

For a discussion of these elements in the context of nonlinear flows differentiable in a variable $\boldsymbol{\xi}$, see [92]; for proofs of the Hartman-Grobman and stable manifold theorems, see [98, 30]; and for the centre manifold theorem, see [161, 30, 187, 126]; note these only apply provided the flow in the neighbourhood $\mathcal{U}$ lies strictly inside the switching layer; we shall see below situations where $\mathcal{U}$ extends outside the layer, namely, in discontinuity-induced bifurcations.

7.9 Bifurcations of Equilibria and Sliding Equilibria

Proposition 7.1. *A generic equilibrium of a codimension r sliding vector field has a Jacobian matrix with full rank and satisfies*

$$\lambda_j \in (-1,+1) \;\text{ for }\; j = 1, \ldots, r, \qquad x_i \neq 0 \;\text{ for }\; i = r+1, \ldots, n. \tag{7.46}$$

For $r = 0$ this means that a generic non-sliding equilibrium, where the constituent fields $\mathbf{f}^i(\mathbf{x})$ vanish, lies outside the discontinuity surface, i.e. $\mathbf{x}_{eq} \notin \mathcal{D}$ (or equivalently $\sigma(\mathbf{x}_{eq}) \neq 0$), for which the qualitative local theory can be found in any textbook on local dynamical theory for differentiable systems.

For $r \geq 1$ this means that a sliding equilibrium will generically lie inside the layer $(-1,+1)^r \times \mathbb{R}^{n-r}$ and not on its boundary or equivalently lies on the interior (i.e. not the boundary) of a sliding region. This condition translates into the following theorem, which forms the starting point for so-called *boundary equilibrium* bifurcations.

Theorem 7.4. *If the codimension r sliding vector field* (7.28) *has an equilibrium at some point $\mathbf{x}_{eq}$, at which $\lambda_j = \pm 1$ for $1 \leq j \leq r$, then the neighbouring codimension $r-1$ sliding vector field also has an equilibrium at $\mathbf{x}_{eq}$.*

Proof. Say without loss of generality in (7.28) that $\lambda_r = \pm 1$. Let us write the vector field $\mathbf{f}(\mathbf{x}_{eq}; \lambda_1^\$, \dots, \lambda_r^\$)$ as a combination (5.3) of the codimension $r-1$ sliding vector fields $\mathbf{f}(\mathbf{x}_{eq}; \lambda_1^\$, \dots, \lambda_{r-1}^\$, \pm 1)$, writing '$\$_{r-1}$' as shorthand for '$\lambda_1^\$, \dots, \lambda_{r-1}^\$$', so

$$\mathbf{f}(\mathbf{x}_{eq}; \$_{r-1}, \lambda_r) = \tfrac{1}{2}(1+\lambda_r)\mathbf{f}(\mathbf{x}_{eq}; \$_{r-1}, +1) + \tfrac{1}{2}(1-\lambda_r)\mathbf{f}(\mathbf{x}_{eq}; \$_{r-1}, -1)\ ,$$

then $\lambda_r = \pm 1$ implies $\mathbf{f}(\mathbf{x}_{eq}; \$_{r-1}, \lambda_r) = \mathbf{f}(\mathbf{x}_{eq}; \$_{r-1}, \pm 1)$, respectively. Hence if $\lambda_r = \pm 1$ and $\mathbf{f}(\mathbf{x}_{eq}; \$_{r-1}, \lambda_r)$ vanishes by hypothesis, hence $\mathbf{f}(\mathbf{x}_{eq}; \$_{r-1}, \pm 1)$ also vanishes. □

The implication of Theorem 7.4 is that the scenario therein described is a bifurcation point: since both the codimension r and $r-1$ sliding vector field contain an equilibrium, under perturbation there may exist equilibria in one, both, or neither vector fields.

Definition 7.5. A **boundary equilibrium** is a (non-sliding) equilibrium at some $\mathbf{x}_{eq}$ such that $\sigma_j(\mathbf{x}_{eq}) = 0$ for some $j \in 1, \dots, m$. If $\sigma_j(\mathbf{x}_{eq}) = 0$ for d different indices j, this constitutes at least a codimension d boundary equilibrium. A **boundary sliding equilibrium** is a sliding equilibrium at some $\mathbf{x}_{eq}$ such $\lambda_j = +1$ or -1 for some $j \in 1, \dots, r$. If $\lambda_j = \pm 1$ for d different indices j from the set $\{1, \dots, r\}$, this constitutes at least a codimension d boundary sliding equilibrium.

Definition 7.6. A discontinuity-induced bifurcation of an equilibrium may be called a **boundary equilibrium bifurcation**; it occurs under perturbation of a boundary equilibrium.

According to Definition 4.8 and Proposition 7.1, an equilibrium will undergo a boundary equilibrium bifurcation when it lies on the boundary of a sliding region. The forms that such boundaries themselves may take are considered in the following sections. We prepare with a few examples of boundary equilibrium bifurcations and their layer analysis.

Let us recall briefly the simple saddlenode bifurcation in a smooth system.

Example 7.13 (Saddlenode Bifurcations in the Plane, Part I). Consider the smooth system

$$\begin{pmatrix} \dot{x}_1 \\ \dot{x}_2 \end{pmatrix} = \begin{pmatrix} x_1^2 - c \\ -x_2 \end{pmatrix}$$

which has equilibria at $(x_1, x_2) = (\pm 2\sqrt{c}, 0)$ that exist only for $c > 0$. The Jacobian is $\underline{\underline{J}} = \begin{pmatrix} \pm\sqrt{c} & 0 \\ 0 & -1 \end{pmatrix}$ at $(x_1, x_2) = (\pm\sqrt{c}, 0)$, with eigenvectors $(0, 1)^{\mathsf{T}}$ and $(1, 0)^{\mathsf{T}}$ associated with eigenvalues -1 and $\pm 2\sqrt{c}$. These denote a saddle and an attractive node, respectively. A saddlenode bifurcation occurs as c passes through zero: as c becomes negative, the two equilibria annihilate each other and leave a nonvanishing flow (Figure 7.16).

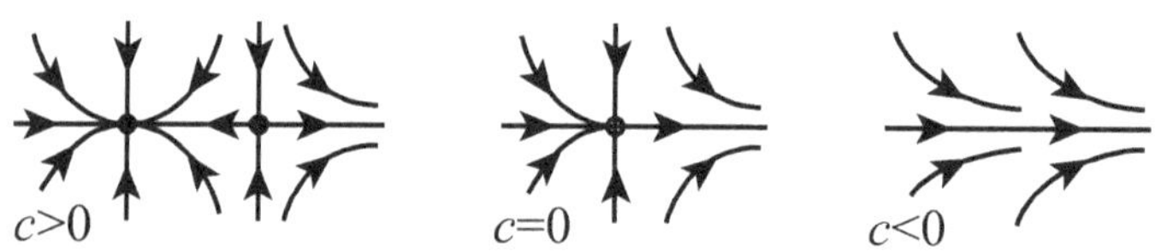

Fig. 7.16 A saddlenode bifurcation in a smooth system.

■

Let us contrast this with a similar boundary equilibrium bifurcation. Take a system with the same local form at the node, which we place at the origin, so $(\dot{x}_1, \dot{x}_2) = (-\frac{1}{2}x_1, -x_2)$. Let this apply for $\sigma < 0$ where $\sigma = x_2 - x_1 - c$, and for $\sigma > 0$ take a constant rightward drift $(\dot{x}_1, \dot{x}_2) = (1, -b)$ for some fixed $|b| < 1$. Hence consider the following.

Example 7.14 (Saddlenode Bifurcations in the Plane, Part II). Take a piecewise-smooth system

$$\begin{pmatrix} \dot{x}_1 \\ \dot{x}_2 \end{pmatrix} = \tfrac{1}{2}(1+\lambda)\begin{pmatrix} 1 \\ -b \end{pmatrix} + \tfrac{1}{2}(1-\lambda)\begin{pmatrix} -\frac{1}{2}x_1 \\ -x_2 \end{pmatrix}$$

where $\lambda = \operatorname{sign}\sigma$ and $\sigma = x_1 - x_2 - c$. The bifurcation parameter is c, and b is a small nominal constant. To write the layer system, it is easiest to rotate coordinates, so let $(y_1, y_2) = (x_1 - x_2 - c, x_1 + x_2)/4$, giving

$$\begin{pmatrix} \dot{y}_1 \\ \dot{y}_2 \end{pmatrix} = \tfrac{1}{2}(1+\lambda)\begin{pmatrix} 1-b \\ 1+b \end{pmatrix} + \tfrac{1}{2}(1-\lambda)\begin{pmatrix} y_2 - 3y_1 - \frac{3}{4}c \\ y_1 - 3y_2 + \frac{1}{4}c \end{pmatrix}$$

in which the node lies at $(x_1, x_2) = (-\frac{c}{4}, 0)$, at which the Jacobian is

$$\underline{\underline{J}} = \frac{\partial(\dot{x}_1, \dot{x}_2)}{\partial(x_1, x_2)} = \begin{pmatrix} -3 & 1 \\ 1 & -3 \end{pmatrix}$$

with eigenvectors $(-1,1)$ and $(1,1)$ (rotated 45° from the vertical and horizontal in x-coordinates) and eigenvalues $-4, -2$. The layer system on $y_1 = 0$ is

$$\begin{pmatrix} \varepsilon\dot{\lambda}_1 \\ \dot{y}_2 \end{pmatrix} = \tfrac{1}{2}(1+\lambda)\begin{pmatrix} 1-b \\ 1+b \end{pmatrix} + \tfrac{1}{2}(1-\lambda)\begin{pmatrix} y_2 - \frac{3}{4}c \\ -3y_2 + \frac{1}{4}c \end{pmatrix}$$

with a fixed point at $(\lambda, y_2) = \left(\frac{c+b-2}{c-b+2}, \frac{c(2+b)}{4(2-b)}\right)$, where the Jacobian in layer variables $(\xi_1, \xi_2) = (\varepsilon\lambda_1, x_2)$ is

$$\underline{\underline{J}} = \frac{\partial(\dot{\xi}_1, \dot{\xi}_2)}{\partial(\xi_1, \xi_2)} = \begin{pmatrix} \frac{(b-1)(c-b+2)}{2\varepsilon(b-2)} & \frac{2-b}{c-b+2} \\ \frac{2+c+b+cb-b^2}{2\varepsilon(2-b)} & \frac{3(2-b)}{b-c-2} \end{pmatrix}$$

with eigenvalues and corresponding eigenvectors as $\varepsilon \to 0$

$$\begin{aligned} \gamma^{(1)} &= \tfrac{1}{\varepsilon}\tfrac{(1-b)(c-b+2)}{2(2-b)} \to +\infty \quad &: \quad \gamma^{(1)} &\to (1-b, 1+b)^{\mathsf{T}} \\ \gamma^{(2)} &= \tfrac{2(2-b)^2}{(b-1)(c-b+2)} < 0 \quad &: \quad \gamma^{(2)} &= (0,1)^{\mathsf{T}} \end{aligned}$$

assuming small c and b. Hence this equilibrium is a saddle, with an infinite rate of repulsion along the $\pm(1-b, 1+b)$ directions out of the discontinuity surfaceand asymptotic attraction along the vertical inside the surface.

As in the saddlenode Example 7.13, the two equilibria exist only for $c > 0$. At $c = 0$ the equilibria coalesce. As c becomes negative, the two equilibria leave their domains of definition: the node leaves the region $\sigma < 0$; the saddle leaves the layer $\lambda \in (-1, +1)$ as its $\lambda^{\$}$ value becomes less than -1. A saddlenode boundary equilibrium bifurcation has occurred at $c = 0$ (Figure 7.17).

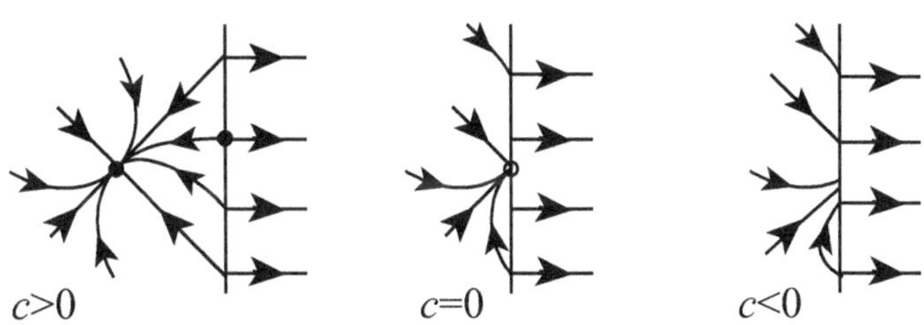

Fig. 7.17 A saddlenode bifurcation in a nonsmooth system.

■

Example 7.15 (Boundary Equilibrium Bifurcations Between Codimension 1 and 2 Sliding). Let us return to the three-dimensional system in Example 7.12. There are no equilibria outside the discontinuity surface, and we saw that a sliding equilibrium exists inside the intersection $x_1 = x_2 = 0$ at

$$(\lambda_1, \lambda_2, x_3) = \frac{(a_2b_{12} - a_1b_{22},\ a_1b_{21} - a_2b_{11},\ 0)}{b_{11}b_{22} - b_{12}b_{21}},$$

but this exists only for

$$\left|\frac{a_2 b_{12} - a_1 b_{22}}{b_{11} b_{22} - b_{12} b_{21}}\right| < 1\,, \qquad \left|\frac{a_1 b_{21} - a_2 b_{11}}{b_{11} b_{22} - b_{12} b_{21}}\right| < 1\,.$$

When these conditions are violated, there no longer exists a sliding equilibrium inside the intersection. Bifurcations therefore occur at

$$\left|\frac{a_2 b_{12} - a_1 b_{22}}{b_{11} b_{22} - b_{12} b_{21}}\right| = 1 \quad \& \quad \left|\frac{a_1 b_{21} - a_2 b_{11}}{b_{11} b_{22} - b_{12} b_{21}}\right| = 1\,.$$

Where does the equilibrium go?

We did not consider sliding on the switching submanifolds $x_1 = 0 \neq x_2$ and $x_2 = 0 \neq x_1$ in Example 7.12. Let us look at them now. On the discontinuity submanifolds $x_1 = 0$ or x_2 outside the intersection, the layer systems are

$$\begin{pmatrix} \varepsilon_1 \dot\lambda_1 \\ \dot x_2 \\ \dot x_3 \end{pmatrix} = \begin{pmatrix} a_1 + b_{11}\lambda_1 + b_{12}\,\mathrm{sign}(x_2) \\ a_2 + c_2 x_2 + b_{21}\lambda_1 + b_{22}\,\mathrm{sign}(x_2) \\ -x_3 \end{pmatrix} \quad \text{on } x_1 = 0 \neq x_2\,,$$
$$\begin{pmatrix} \dot x_1 \\ \varepsilon_2 \dot\lambda_2 \\ \dot x_3 \end{pmatrix} = \begin{pmatrix} a_1 + c_1 x_1 + b_{11}\,\mathrm{sign}(x_1) + b_{12}\lambda_2 \\ a_2 + b_{21}\,\mathrm{sign}(x_1) + b_{22}\lambda_2 \\ -x_3 \end{pmatrix} \quad \text{on } x_2 = 0 \neq x_1\,.$$

Sliding (where the $\dot\lambda_j$ subsystems vanish) occurs for $\lambda_1^\$ = -\frac{a_1 + b_{12}\,\mathrm{sign}(x_2)}{b_{11}}$ and $\lambda_2^\$ = -\frac{a_2 + b_{21}\,\mathrm{sign}(x_1)}{b_{22}}$, respectively, provided $\lambda_1^\$$ and $\lambda_2^\$$ lie in $(-1, +1)$, giving dynamics

$$\begin{aligned} \dot x_2 &= c_2 x_2 + \tfrac{a_2 b_{11} - a_1 b_{21}}{b_{11}} - \tfrac{b_{21} b_{12} - b_{11} b_{22}}{b_{11}}\,\mathrm{sign}(x_2) \quad \text{on } x_1 = 0 \neq x_2\,, \\ \dot x_1 &= c_1 x_1 + \tfrac{a_1 b_{22} - a_2 b_{12}}{b_{22}} - \tfrac{b_{12} b_{21} - b_{11} b_{22}}{b_{22}}\,\mathrm{sign}(x_1) \quad \text{on } x_2 = 0 \neq x_1\,. \end{aligned}$$

Thus there exists a sliding equilibrium:

- on $x_1 = 0 < x_2$, at $x_2 = \frac{-a_2 b_{11} + a_1 b_{21} + b_{21} b_{12} - b_{11} b_{22}}{b_{11} c_2}$, if $\left|\frac{a_1 + b_{12}}{b_{11}}\right| < 1$,
- on $x_1 = 0 > x_2$, at $x_2 = \frac{-a_2 b_{11} + a_1 b_{21} - b_{21} b_{12} + b_{11} b_{22}}{b_{11} c_2}$, if $\left|\frac{a_1 - b_{12}}{b_{11}}\right| < 1$,
- on $x_2 = 0 < x_1$, at $x_1 = \frac{-a_1 b_{22} + a_2 b_{12} + b_{12} b_{21} - b_{11} b_{22}}{b_{22} c_1}$, if $\left|\frac{a_2 + b_{21}}{b_{22}}\right| < 1$,
- on $x_2 = 0 > x_1$, at $x_1 = \frac{-a_1 b_{22} + a_2 b_{12} - b_{12} b_{21} + b_{11} b_{22}}{b_{22} c_1}$, if $\left|\frac{a_2 - b_{21}}{b_{22}}\right| < 1$.

These tell us that when $\left|\frac{a_1 b_{21} - a_2 b_{11}}{b_{11} b_{22} - b_{12} b_{21}}\right| = 1$, the sliding equilibrium on $x_1 = 0$ lies at $x_2 = 0$ with $|\lambda_1^\$| = 1$, and when $\left|\frac{a_2 b_{12} - a_1 b_{22}}{b_{11} b_{22} - b_{12} b_{21}}\right| = 1$, the sliding equilibrium on $x_2 = 0$ lies at $x_1 = 0$ with $|\lambda_2^\$| = 1$.

This means that at the bifurcation values, the equilibrium vanishes from codimension two sliding on the intersection, and either passes into one of these codimension $r = 1$ systems or collides with another equilibrium from one of the codimension $r = 1$ systems with the two annihilating in another example of a discontinuity-induced saddlenode bifurcation.

We leave it to the reader to explore the many different scenarios for choices of the various constants, such as those in Figure 7.18.

The upper panel of Figure 7.18 shows an example of a persistence bifurcation, where a sliding equilibrium passes from the intersection $x_1 = x_2 = 0$ to the submanifold $x_1 = 0 < x_2$ as parameter values change. The lower panel of Figure 7.18 shows an example of a saddlenode bifurcation, where two sliding

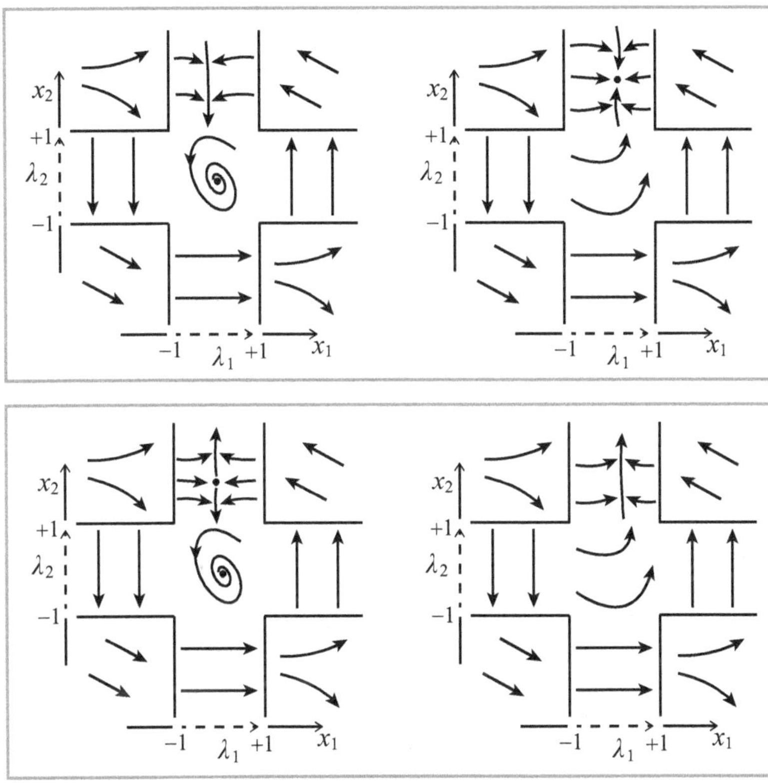

Fig. 7.18 Bifurcations of a sliding equilibrium from Example 7.15. Top panel: persistence of a sliding equilibrium, passing from the intersection to the upper branch of sliding on $x_1 = 0$. Bottom panel: saddlenode bifurcation of a sliding equilibrium, in which a sliding focus on the intersection and a sliding saddle on the upper branch of sliding on $x_1 = 0$ annihilate. With parameter values $a_2 = 0$, $c_1 = 0$, $b_{11} = b_{12} = -1$, $b_{21} = 2$, $b_{22} = 0$, and (top) $c_2 = -1$ or (bottom) $c_2 = 1$. The bifurcation takes place as a_1 passes through unity, showing $a_1 = 1/2$ (left) and $a_2 = 2/3$ (right).

equilibria annihilate, a sliding focus inside the intersection $x_1 = x_2 = 0$ and a sliding saddle on the submanifold $x_1 = 0 < x_2$.

■

The extension of *linearization*—analyzing the local properties of an equilibrium from its Jacobian—is thus extended to piecewise systems in a very general way using the switching layer. The fact that we can generalize local stability theory to piecewise-smooth dynamical systems is important and opens a significant window into characterizing equilibria in higher dimensions. Equilibria are locally characterized by their linearization, even in high-dimensional systems with many switches. A significant challenge remains in characterizing the curvature of the flow surrounding the equilibrium with respect to the discontinuity surface, something that is not an issue in smooth systems. Clearly this is of greatest interest in the neighbourhood of a boundary equilibrium bifurcation, because by Proposition 7.1 these occur at the boundaries of sliding regions.

There is some interesting geometry of the vector fields that gives insight into this.

Theorem 7.5. *If an equilibrium in a codimension r sliding region lies on the sliding boundary in a system* (7.1), *then either:*

(i) it coincides with an equilibrium of one of the neighbouring codimension $r-1$ vector fields; or
(ii) both of the neighbouring codimension $r-1$ vector fields are tangent to the codimension r switching intersection.

Proof. For $r = 1$ the theorem says that a simple sliding equilibrium at the boundary of its sliding region either (i) coincides with an equilibrium of $\mathbf{f}^+$ or $\mathbf{f}^-$ or (ii) coincides with tangencies of both $\mathbf{f}^+$ and $\mathbf{f}^-$ to the discontinuity surface. So take first a single switching multiplier, with

$$\dot{\mathbf{x}} = \mathbf{f}(\mathbf{x};\lambda) = \tfrac{1}{2}(1+\lambda)\mathbf{f}(\mathbf{x};+1) + \tfrac{1}{2}(1-\lambda)\mathbf{f}(\mathbf{x};-1)$$

and $\mathbf{f}(\mathbf{x}_{eq};\lambda^{\$}) = 0$, and then $\mathbf{x}_{eq}$ on sliding boundary implies $\lambda = \pm 1$, so either:

(i) if $\lambda^{\$} = +1$ then $\mathbf{f}^+(\mathbf{x}_{eq}) = \mathbf{f}^{\$}(\mathbf{x}_{eq}) = 0$, if $\lambda^{\$} = -1$ then $\mathbf{f}^-(\mathbf{x}_{eq}) = \mathbf{f}^{\$}(\mathbf{x}_{eq}) = 0$;
(ii) if $\mathbf{f}^{\pm}(\mathbf{x}_{eq}) \neq 0$, the sliding condition implies $\lambda^{\$} = \frac{(\delta_{t-}+\delta_{t+})\sigma(\mathbf{x}_{eq})}{(\delta_{t-}-\delta_{t+})\sigma(\mathbf{x}_{eq})}$, then either of $\delta_{t\pm}\sigma(\mathbf{x}_{eq}) = 0$ imply $\lambda = \pm 1$ and give case (i), the only other case is both $\delta_{t+}\sigma(\mathbf{x}_{eq}) = \delta_{t-}\sigma(\mathbf{x}_{eq}) = 0$, in which case $\lambda^{\$}$ is undefined, but this means both $\mathbf{f}^{\pm}(\mathbf{x}_{eq})$ lie tangent to $\sigma = 0$.

The proof for higher codimension sliding follows similarly. □

To describe the bifurcation that occurs in such a situation, we then have

Theorem 7.6. *At a boundary equilibrium bifurcation in $\mathbb{R}^n$, in which an equilibrium $\mathbf{x}_{eq}$ of a codimension r sliding vector field lies on a codimension $r+1$ switching intersection, under perturbation there exist in a neighbourhood of $\mathbf{x}_{eq}$*

(i) an equilibrium of the codimension r sliding vector field on one side of the bifurcation;
(ii) an equilibrium of the neighbouring codimension $r+1$ sliding vector field on one side of the bifurcation;
(iii) tangencies of order $k = 2, 3, \ldots, n-1$, between the codimension r sliding vector field and the codimension $r+1$ intersection, on both sides of the bifurcation.

Proof. We assume that $\mathbf{f}^{\pm}(\mathbf{x}) \equiv \mathbf{f}(\mathbf{x}; \pm 1)$ are defined for all $\mathbf{x}$, although they form part of the dynamical system (7.1) only on regions $\mathcal{R}_{\pm}$ either side of $\mathcal{D}$. Part (i) is obvious. Then, from an equilibrium $\mathbf{x}_{eq}$ for which $\text{sign}(\sigma(\mathbf{x}_{eq})) = +1$ without loss of generality, there extends a curve

$$\mathcal{T} = \left\{\mathbf{x} \in \mathbb{R}^n : 0 = \delta_{t+}\sigma = \cdots = \delta_{t+}^{n-1}\sigma\right\}$$

on which $\mathbf{f}^+$ has an $(n-1)^{th}$ order tangency to level sets of σ, and there extends a curve

$$\mathcal{A} = \left\{\mathbf{x} \in \mathbb{R}^n : \mathbf{f}^+(\mathbf{x}) = \alpha\mathbf{f}^-(\mathbf{x}),\ \alpha \in \mathbb{R}\backslash 0\right\}$$

on which $\mathbf{f}^{\pm}$ are (anti-)collinear. This gives, under a perturbation of a point $\mathbf{x}_{eq} \in \mathcal{D}$ such that it lies just outside $\mathcal{D}$:

ii. generically $\mathcal{A}$ will intersect $\mathcal{D}$ transversally; if it intersects in the sliding region, then the intersection point is a sliding equilibrium (see Theorem 8.1 later);
iii. generically $\mathcal{T}$ will intersect $\mathcal{D}$ transversally, at which point $\mathbf{f}^+$ has an $(n-1)^{th}$ order tangency to $\mathcal{D}$.

The proof for higher codimension follows similarly. □

When a boundary equilibrium is perturbed, if the layer Jacobian $\underline{\underline{J}}$ of any equilibrium remains non-degenerate, we may therefore find that an equilibrium passes from a region with a given sliding codimension, to a region whose sliding codimension is one higher/lower, or that a pair of equilibria from some codimension r sliding region annihilates with one from a codimension $r+1$ sliding region in the form of saddlenode bifurcation. If $\underline{\underline{J}}$ is degenerate or the equilibria are non-hyperbolic, meaning either $\det \underline{\underline{J}} = 0$ or any of its eigenvalues have zero real part, then higher codimension bifurcations will occur.

7.10 A Saddlenode/Persistence Criterion

When a boundary equilibrium bifurcation occurs, there is a simple criterion to determine whether it takes the form of an equilibrium persisting from one sliding codimension to another (a *persistence*) or a folding of two equilibria of different sliding codimensions (a *saddlenode*). This is a more general alternative to a similar criterion derived in [48] at a single discontinuity surface.

If the point $(\mathbf{x}, \boldsymbol{\lambda}) = (\mathbf{x}_{eq}, \boldsymbol{\lambda}^{\$})$ is a (sliding) equilibrium lying on the boundary between codimension r and $r-1$ sliding, let $\underline{\underline{J}}_r$ and $\underline{\underline{J}}_{r-1}$ denote the Jacobian matrices in the layer variables

$$\underline{\underline{J}} = \frac{\partial(\dot{\xi}_1, \dot{\xi}_2, \ldots, \dot{\xi}_n)}{\partial(\xi_1, \xi_2, \ldots, \xi_n)}$$

associated with each region and evaluated at $\mathbf{x}_{eq}$. Then a generic perturbation that pushes $\mathbf{x}$ away from $\mathbf{x}_{eq}$ will result in:

- persistence if $\det|\underline{\underline{J}}_r| \det|\underline{\underline{J}}_{r-1}| > 0$,
- a saddlenode if $\det|\underline{\underline{J}}_r| \det|\underline{\underline{J}}_{r-1}| < 0$.

This follows as a consequence of the linearization methods we have set out in this chapter. In a persistence bifurcation, the eigenvalues of the Jacobian in the switching layer variables identify the topological type of the (sliding) equilibrium, and the number of them that have positive or negative real parts is preserved through the bifurcation (the number of them that are complex may change pairwise as new ε_j multipliers are introduced between different sets of layer variables). This means the determinants of the Jacobians associated with the two sets of layer variables, in codimension r and $r-1$ sliding, will have the same sign. If their signs are opposing, then by the continuity of the layer equations, a persistence cannot occur, and the equilibria are of different topological type, generically with one eigenvalue having a real part of differing sign. The two equilibria will generically exist then only on one side of the bifurcation, meaning a saddlenode bifurcation has occurred, because for them both to exist on one side of the bifurcation, they would have to coexist in the same codimension r (or $r-1$) sliding field and their coincidence with the sliding boundary would be a higher codimension event.

7.11 Looking Forward

We have seen some general tools and principles for analysing the local sliding modes, singularities, and bifurcations induced by switching. It quickly becomes clear that everything possible in nonlinear dynamics is possible inside the switching layer, but also by patching together different sliding regions of different sliding codimension, all manner of new bifurcations are brought into being. These are largely analogous to bifurcations in smooth systems, and while the invariant surfaces formed by the sliding manifolds add to the complexity, they also provide the necessary insight, along with switching layer analysis, to study all behaviours that seemingly might arise.

We shall illustrate some more novel local phenomena in Chapter 9 before moving on to global behaviour. The discontinuity, while giving us a world of new phenomena to classify, also provides a powerful means to facilitate such classification. We will use the local geometry developed so far to give a general classification of global dynamics, looking into exit from codimension r sliding by focussing on exit points from $\mathcal{M}$ in Chapter 10 and returning to concentrate mostly on a single switch in Chapter 11.

Before all that, armed with some new methods, we must revisit the kinds of systems that have dominated piecewise-smooth research until recently, those restricted to linear switching or what are sometimes called *Filippov* systems.

Chapter 8
Linear Switching (Local Theory)

In this section we take a look at systems that depend only linearly on the switching multipliers $\boldsymbol{\lambda} = (\lambda_1, \ldots, \lambda_m)$ and are therefore expressible in the form

$$\dot{\mathbf{x}} = \mathbf{f}(\mathbf{x}; \boldsymbol{\lambda}) = \mathbf{a}(\mathbf{x}) + \underline{\underline{B}}(\mathbf{x})\boldsymbol{\lambda} \,, \tag{8.1}$$

where $\underline{\underline{B}}$ is an $n \times m$ matrix. In relation to (5.3), the quantities in (8.1) are

$$\mathbf{a} = \tfrac{1}{2}\sum_{j=1}^{m}\sum_{\kappa_j=\pm} \mathbf{f}^{\kappa_1\ldots\kappa_j\ldots\kappa_m} \,, \qquad \mathbf{B}_j = \tfrac{1}{2}\sum_{j=1}^{m}\sum_{\kappa_j=\pm} \kappa_j \mathbf{f}^{\kappa_1\ldots\kappa_j\ldots\kappa_m} \,,$$

where $\mathbf{B}_j$ is the jth column of $\underline{\underline{B}}$.

These are the systems predominantly studied in the piecewise-smooth dynamics literature to date, under the methods of either *Filippov systems* or *Utkin's equivalent control*; see, e.g. [71, 211, 212, 48, 36]. The hidden terms $\mathbf{h}$ or $\mathbf{n}$ from Section 3.1 are zero everywhere.

The reliable application of linear switching models require that the switching layer dynamics is in some sense trivial—having unique structurally stable attractors—reducible to just the sliding dynamics on invariant manifolds inside the layer. We will see the conditions under which this holds. We also provide some key advances on standard theory and see how the layer (3.15) simplifies analysis of switching, even for systems as simple as (8.1).

Following on from the previous section, our interest will be local. Our main concern will be to establish the typical shapes of the boundaries of sliding regions and to establish typical forms of equilibria on or off the discontinuity surface, along with some elementary bifurcations thereof.

M. R. Jeffrey, *Hidden Dynamics*,
https://doi.org/10.1007/978-3-030-02107-8_8

8.1 The Sliding Manifold

In a linear system, it is possible to give general and explicit expressions for sliding modes. Taking the system (8.1) at a point where

$$\sigma_1 = \cdots = \sigma_m = 0 \,,$$

to find the sliding manifold $\mathcal{M}$, we first define a matrix $\underline{\underline{S}} = (\delta_{\mathbf{x}}\sigma_1, \ldots, \delta_{\mathbf{x}}\sigma_m)^{\mathsf{T}}$ whose rows are the gradient vectors to the discontinuity submanifolds $\mathcal{D}_j$. Then solving (7.6) gives

$$\begin{aligned} 0 &= \underline{\underline{S}}(\mathbf{x}).\underline{\underline{B}}(\mathbf{x})\boldsymbol{\lambda}^{\$} + \underline{\underline{S}}(\mathbf{x}).\mathbf{a}(\mathbf{x}) \\ \Rightarrow \quad \boldsymbol{\lambda}^{\$}(\mathbf{x}) &= -[\underline{\underline{S}}(\mathbf{x}).\underline{\underline{B}}(\mathbf{x})]^{-1}\underline{\underline{S}}(\mathbf{x}).\mathbf{a}(\mathbf{x}) \ \in (-1,+1)^m \end{aligned} \tag{8.2}$$

where $\underline{\underline{S}}(\mathbf{x}).\underline{\underline{B}}(\mathbf{x})$ is an $m \times m$ square matrix. (If $m = n$ such that $\underline{\underline{B}}$ and $\underline{\underline{S}}$ have inverses, then this simplifies to $\boldsymbol{\lambda}^{\$}(\mathbf{x}) = -[\underline{\underline{B}}(\mathbf{x})]^{-1}\mathbf{a}(\mathbf{x})$.) This inhabits the switching layer

$$(\lambda_1, \ldots, \lambda_m, x_{m+1}, \ldots, x_n) \in (-1,+1)^m \times \mathbb{R}^{n-m} \,.$$

The matrix that describes the attractivity of $\mathcal{M}$ becomes

$$\left.\frac{\partial(f_1, \ldots, f_M)}{\partial(\lambda_1, \ldots, \lambda_M)}\right|_{\mathcal{M}} = \underline{\underline{B}} \,, \tag{8.3}$$

so by (7.11) the normal hyperbolicity of $\mathcal{M}$ requires that $\underline{\underline{B}}$ has nonzero determinant and no eigenvalues on the imaginary line.

The sliding dynamics on $\mathcal{M}$ is given by

$$\dot{\mathbf{x}} = \mathbf{f}(\mathbf{x}; \boldsymbol{\lambda}^{\$}) = \mathbf{a}(\mathbf{x}) - \underline{\underline{B}}(\mathbf{x})[\underline{\underline{S}}(\mathbf{x}).\underline{\underline{B}}(\mathbf{x})]^{-1}\underline{\underline{S}}(\mathbf{x}).\mathbf{a}(\mathbf{x}) \,. \tag{8.4}$$

At a point where the discontinuity surface is locally a manifold given by $\sigma_j = 0$ for some j, writing $\dot{\mathbf{x}} = \mathbf{a} + \mathbf{B}_j\lambda_j$, omitting the argument $\mathbf{x}$, (8.2) becomes

$$\begin{aligned} 0 &= (\delta_{\mathbf{x}}\sigma_j) \cdot (\mathbf{a} + \mathbf{B}_j\, \lambda_j^{\$}) \\ \Rightarrow \quad \lambda_j^{\$} &= -\frac{(\delta_{\mathbf{x}}\sigma_j) \cdot \mathbf{a}}{(\delta_{\mathbf{x}}\sigma_j) \cdot \mathbf{B}_j} = \frac{(\delta_{t-} + \delta_{t+})\sigma_j}{(\delta_{t-} - \delta_{t+})\sigma_j} \ \in (-1,+1) \end{aligned} \tag{8.5}$$

where $\mathbf{f}^{\pm}$ denote the vector fields in $\sigma_j \gtrless 0$, and (8.4) becomes

$$\begin{aligned} \dot{\mathbf{x}} = \mathbf{f}(\mathbf{x}; \lambda_j^{\$}) &= \frac{(\mathbf{B}_j \cdot \delta_{\mathbf{x}}\sigma_j)\mathbf{a} - (\mathbf{a} \cdot \delta_{\mathbf{x}}\sigma_j)\mathbf{B}_j}{\mathbf{B}_j \cdot \delta_{\mathbf{x}}\sigma_j} \\ &= \frac{(\delta_{t-}\sigma_j)\mathbf{f}^{+} - (\delta_{t+}\sigma_j)\mathbf{f}^{-}}{(\delta_{t-} - \delta_{t+})\sigma_j} \,. \end{aligned} \tag{8.6}$$

The numerator can be compactly written as $(\delta_{\mathbf{x}}\sigma_j) \times (\mathbf{f}^+ \times \mathbf{f}^-)$ where '$\times$' denotes the generalized vector (curl) product. It is useful to observe, from (8.6), that

$$\begin{array}{lcl} \delta_{t+}\sigma_j = 0 & \Rightarrow & \mathbf{f}^\$ = \mathbf{f}^+ \quad \& \quad \lambda_j^\$ = +1 \,, \\ \delta_{t-}\sigma_j = 0 & \Rightarrow & \mathbf{f}^\$ = \mathbf{f}^- \quad \& \quad \lambda_j^\$ = -1 \,. \end{array} \tag{8.7}$$

This constitutes a boundary condition which connects tangencies of the constituent vector fields $\mathbf{f}^\pm$ (where $\delta_{t\pm}\sigma = 0$) to the sliding vector field $\mathbf{f}^\$$ at the end points $\lambda^\$ = \pm 1$ of the sliding manifold.

8.2 The Convex Combination

The combination (8.1) is convex, in the sense that every line joining two points in the combination $\mathbf{f}(\mathbf{x};\boldsymbol{\lambda})$ lies inside the combination. Examples at points of crossing or sliding on a single discontinuity submanifold or an intersection of two discontinuity submanifolds are shown in Figure 8.1. In symbols, convexity means that given two points $\mathbf{z}_p = \mathbf{f}(\mathbf{x};\boldsymbol{\lambda}_p)$ and $\mathbf{z}_q = \mathbf{f}(\mathbf{x};\boldsymbol{\lambda}_q)$ for some $\boldsymbol{\lambda}_p \in (-1,+1)^m$ and $\boldsymbol{\lambda}_q \in (-1,+1)^m$, we have

$$\alpha\mathbf{z}_p + (1-\alpha)\mathbf{z}_q \;:\in\; \{\mathbf{f}(\mathbf{x};\boldsymbol{\lambda}) \,:\, \boldsymbol{\lambda} \in (-1,+1)^m\} \quad \text{for } \alpha \in (0,1) \,.$$

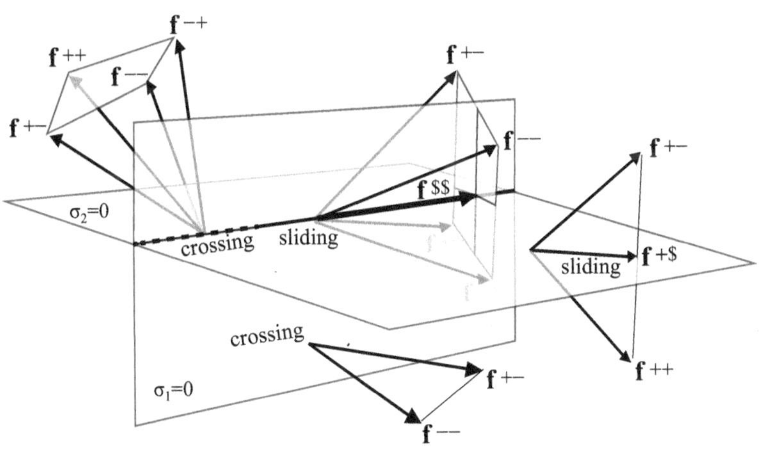

Fig. 8.1 Convex combinations. A system that switches between vector fields $\mathbf{f}^{\pm\pm}$ across two submanifolds $\sigma_1 = 0$ and $\sigma_2 = 0$ is shown. Convex combinations are shown at points on the individual submanifolds, and on their intersection, forming sliding vectors if the convex combination lies in the tangent space of the discontinuity surface.

Convexity has two appealing implications. The more obvious one is that the sliding manifold is uniquely given by (8.2). The more subtle is that the limit of a convergent sequence of solutions of (8.1) is also a solution of (8.1), as discussed in [71]. Yet convexity does not prevent multiplicity of solutions or guarantee determinacy. Motion in forward time is still set-valued at repelling sliding regions and determinacy-breaking singularities (which we will meet in Chapter 10). Equilibria can undergo bifurcations that appear counterintuitive and which turn out to involve degeneracies that can only be broken by adding nonlinear dependence on $\boldsymbol{\lambda}$. Points may also arise where deciding whether the flow slides along or crosses through the discontinuity surface cannot be decided without inspection of the switching layer.

8.3 Equilibria and Sliding Equilibria

In systems of the form (8.1), equilibria in a sliding region are induced when the constituent vector fields outside the discontinuity surface are anti-collinear (i.e. having the same direction with opposite orientation). Intuitively, the vector fields either side of the switch 'cancel out', creating a sliding equilibrium. Figure 8.2 illustrates this in two or three dimensions, for equilibria in codimension one or two sliding.

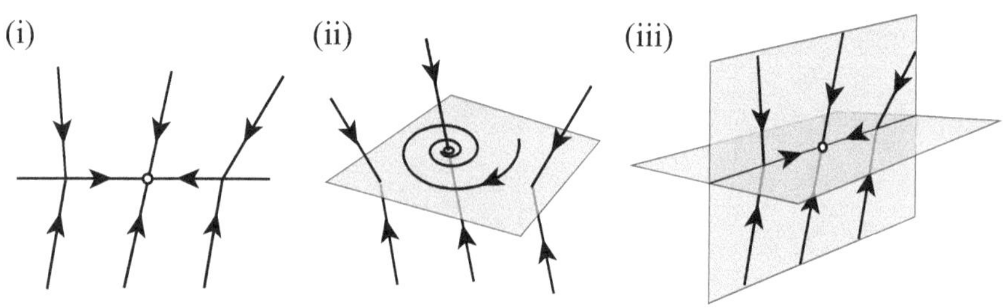

Fig. 8.2 Sliding equilibria of a linear switching system occur where the adjacent vector fields are anti-collinear. The figure shows (i) a sliding node in the plane in codimension one sliding, (ii) a sliding focus in three dimensions in codimension one sliding, and (iii) a sliding node in codimension two sliding.

The principle extends to higher codimension sliding.

Theorem 8.1. *If an equilibrium of a codimension r sliding vector field exists at a point $\mathbf{x}_{eq} \in \mathcal{D}$, then the neighbouring codimension $r-1$ (sliding) vector fields are anti-colinear at $\mathbf{x}_{eq}$.*

Proof. Consider without loss of generality a point where $\sigma_1 = \cdots = \sigma_r = 0$. The codimension r vector field canopy, in terms of the codimension $r-1$ fields $\mathbf{f}(\mathbf{x}; \$_{r-1}, \pm 1)$, where $\$_{r-1}$ denotes "$\lambda_1^\$, \ldots, \lambda_{r-1}^\$$", is

$$\dot{\mathbf{x}} = \mathbf{f}(\mathbf{x}; \$_{r-1}, \lambda_r) = \tfrac{1}{2}(1+\lambda_r)\mathbf{f}(\mathbf{x}; \$_{r-1}, +1) + \tfrac{1}{2}(1-\lambda_r)\mathbf{f}(\mathbf{x}; \$_{r-1}, -1) \,.$$

If this vanishes at a point $\mathbf{x}_p$ for some λ_r, then $\mathbf{f}(\mathbf{x}; \$_{r-1}, +1)$ and $\mathbf{f}(\mathbf{x}; \$_{r-1}, -1)$ are linearly dependent at $\mathbf{x}_p$, and if $\mathbf{x}_p$ lies in a sliding region, then $\lambda_r \in (-1, +1)$, therefore $\mathbf{f}(\mathbf{x}; \$_{r-1}, +1) = \frac{\lambda_r - 1}{\lambda_r + 1}\mathbf{f}(\mathbf{x}; \$_{r-1}, -1)$ where $\frac{\lambda_r - 1}{\lambda_r + 1} < 0$, so $\mathbf{f}(\mathbf{x}; \$_{r-1}, +1)$ and $\mathbf{f}(\mathbf{x}; \$_{r-1}, -1)$ are anti-collinear. □

The examples we looked at in Section 7.9 were linear in $\boldsymbol{\lambda}$, and so they were of the form (8.1) of interest here. There we used layer analysis to find the eigenvectors of the equilibria, which for a linear switching system coincide with the anti-collinear directions of the adjacent constituent vector fields.

8.4 Boundary Equilibrium Bifurcations

In Section 7.9 we saw some examples of boundary equilibrium bifurcations. The layer analysis of Chapter 7 permits equilibria and sliding equilibria to be characterized by their eigenvectors and eigenvalues, much as in smooth systems. The bifurcations that such equilibria can undergo are somewhat more complicated to classify for two reasons. First is that not only the equilibria themselves but their interaction with sliding boundaries must be characterized. Second is that linear dependence on the switching multipliers λ_j does not mean the layer system is entirely linear, with the possibility of nonlinear products between the variables x_i and the multipliers λ_j.

As a result, there are too many different scenarios for all boundary equilibrium bifurcation to be classified, even in low dimensions. As an introduction we provide the classification for the very simplest boundary equilibrium bifurcations, namely, those in planar systems with a single switch. We also provide prototypes for further study in higher dimensions.

8.4.1 One-Parameter BEBs in the Plane

A complete classification of one-parameter boundary equilibrium bifurcations in the plane is quite simple to obtain based on topological arguments. Only the type of equilibrium and the direction of the sliding vector field are needed to find all possible cases. Take a system of the form

$$\dot{\mathbf{x}} = \mathbf{f}(\mathbf{x}; \lambda, \mu) = \tfrac{1}{2}(1+\lambda)\mathbf{f}^{+}(\mathbf{x}; \mu) + \tfrac{1}{2}(1-\lambda)\mathbf{f}^{-}(\mathbf{x}; \mu) \,, \quad \lambda = \mathrm{sign}(\sigma) \,, \quad (8.8)$$

where μ is a bifurcation parameter. Assume that a boundary equilibrium exists at $\mathbf{x}_{eq}$ when $\mu = 0$, such that $\mathbf{f}^{+}(\mathbf{x}_{eq}; 0) = 0$ and $\sigma(\mathbf{x}_{eq}) = 0$.

As non-degeneracy conditions, assume that the Jacobian $d\mathbf{f}^{+}/d\mathbf{x}$ is non-singular at $\mathbf{x}_{eq}$ and has eigenvectors transverse to the discontinuity surface, while the other vector field $\mathbf{f}^-$ is neither vanishing nor tangent to the discontinuity surface, given by $\delta_{t-}\sigma(\mathbf{x}_{eq}) \neq 0$. If any of these are violated, then more than one parameter is needed to unfold the bifurcation.

The non-singular Jacobian implies that $\mathbf{x}_{eq}$ is either a saddle, node, or focus of $\mathbf{f}^+(\mathbf{x})$. Its location on $\sigma = 0$ means that, solving the sliding condition $\delta_t\sigma(\mathbf{x}_{eq}) = 0$ for $\lambda = \lambda^\$$, we find $\lambda^\$ = +1$. Therefore $\mathbf{x}_{eq}$ is also a sliding equilibrium, as $\mathbf{f}^\$(\mathbf{x}_{eq}; 0) = \mathbf{f}(\mathbf{x}_{eq}; \lambda^\$, 0) = 0$. Since $\lambda^\$ = +1$, the point $\mathbf{x}_{eq}$ is a boundary of sliding; therefore to one side of it, there is crossing, to the other there is sliding, and the sliding vector field $\mathbf{f}^\$(\mathbf{x}; 0)$ may point either towards or away from $\mathbf{x}_{eq}$.

We can now build up a classification from two properties: the type of equilibrium of $\mathbf{f}^+$ and whether the sliding vector field $\mathbf{f}^\$$ is inwards or outwards with respect to it. We proceed qualitatively and provide a prototype that confirms the classification afterwards.

Building piece by piece from column 1 to column 3 in Figures 8.3 to 8.5, we first place the equilibrium in a general orientation with respect to the discontinuity surface, neglecting the direction of time at first, giving column 1. We then consider the different permutations of sliding dynamics to obtain column 2, in which sliding lies to one side or the other of the equilibrium and is directed inwards or outwards. Then we apply the arrows that indicate the flow of time, discarding scenarios that are either duplicates up to topological equivalence or are dynamically inconsistent, giving the middle portrait in column 3. Dynamical consistency now determines the angle of the lower vector field $\mathbf{f}^-$, namely, that its direction in convex combination with $\mathbf{f}^+$ must yield the appropriate direction of $\mathbf{f}^\$$. Finally the bifurcation is unfolded by perturbing the equilibrium, so that it rises above the discontinuity surface or disappears into it, with sliding equilibria appearing wherever the upper and lower vector fields are anti-collinear. A proof of the classification, for example, in [71], involves showing exhaustively that all possible permutations of these topologies have been considered.

Take first a boundary saddle, as shown in the left column of Figure 8.3. When we apply sliding inwards or outwards with respect to the equilibrium, we obtain the two cases in the second column. Adding time to the entire phase portrait results in three different possible phase portraits with the saddle resting just on the discontinuity surface, shown in the middle image on each row of the third column. Lastly we unfold the bifurcation, obtaining the three portraits on each row of the third column. We obtain either a persistence bifurcation (upper row) or a saddlenode bifurcation (lower rows). The saddlenode has two cases, where the saddle separatrix hits the sliding region to one side of the sliding node or the other. Because the sliding equilibrium appears where the vector fields are anti-collinear, the two cases of saddlenode occur when the lower vector field is either steeper or shallower than the unstable manifold of the saddle on $\mathcal{D}$.

Secondly, take a boundary node as shown in the left column of Figure 8.4. When we add sliding inwards or outwards, we obtain the four possible cases

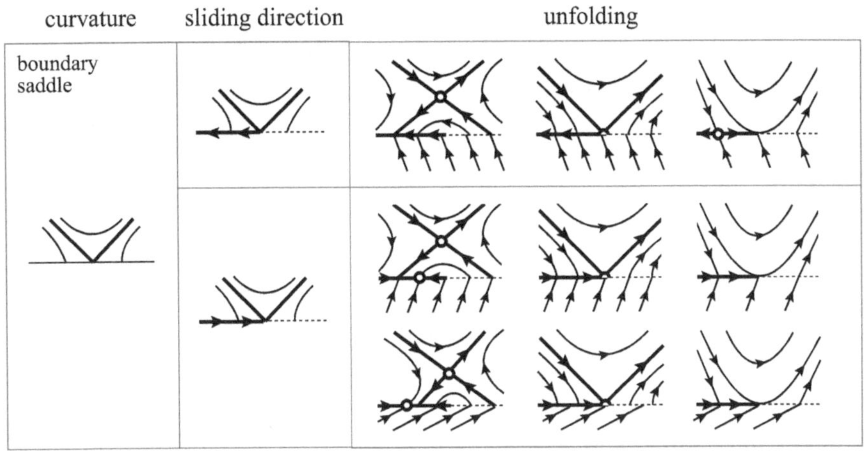

Fig. 8.3 Boundary equilibrium bifurcations of a saddle in the plane: a persistence (upper row) and a saddlenode (lower row).

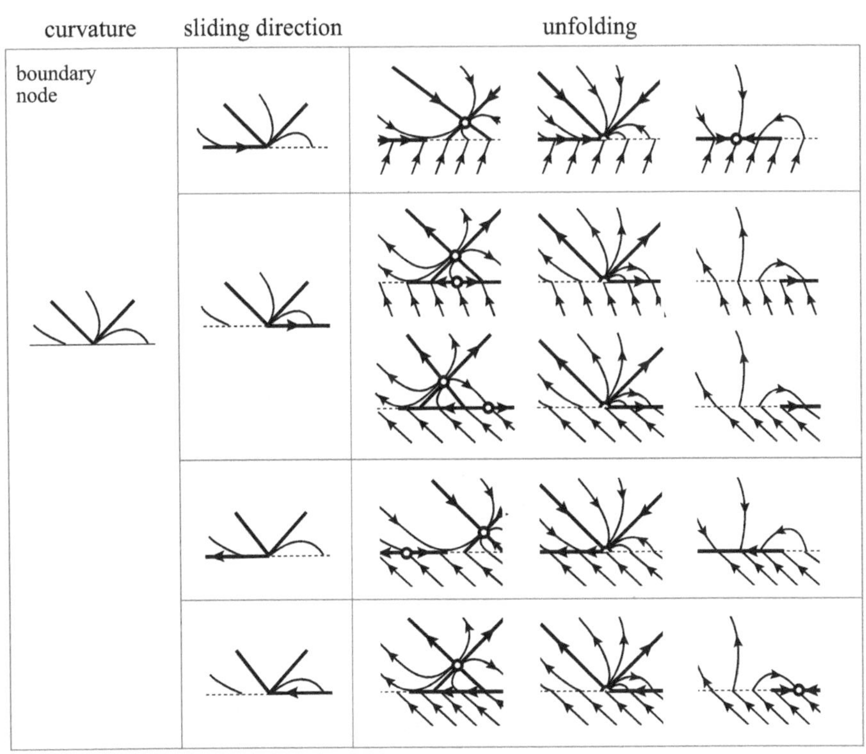

Fig. 8.4 Boundary equilibrium bifurcations of a node in the plane: a persistence (first and fourth rows), a saddlenode (second and third rows).

in the middle column. The positioning of the node's weak and strong unstable manifolds mean that it matters (unlike for the saddle) whether the sliding region is to the left or right of the node, and we immediately obtain four cases. Each results either in a persistence (first and last rows) or a saddlenode (the remaining rows), shown in the third column, when the arrows of time are added and the bifurcation unfolded. In one saddlenode case, the strong manifold to the node may lie to one side of the sliding saddle or the other, creating two cases corresponding to whether the lower vector field is steeper or shallower than the strong manifold. In the first two rows of column three, the lower vector field can take any angle. To obtain the lower two rows, the lower vector field must be shallower than the strong eigendirection of the equilibrium. The difficulty of piecewise-smooth classifications is exemplified by the lower two cases of Figure 8.4, as these were missing from classifications in [140, 48, 90] due to insufficient criteria, despite having appeared in [71] (further discussion of this can be found in [108]).

Thirdly, take a boundary focus, shown in the left column of Figure 8.5. When we add sliding inwards or outwards, we obtain the two cases in the second column. Similarly to the saddle, only two cases are evident until time is added. Again each case leads either to a saddlenode (upper rows) or persistence (lower rows). The added complication of the focus is the circulation

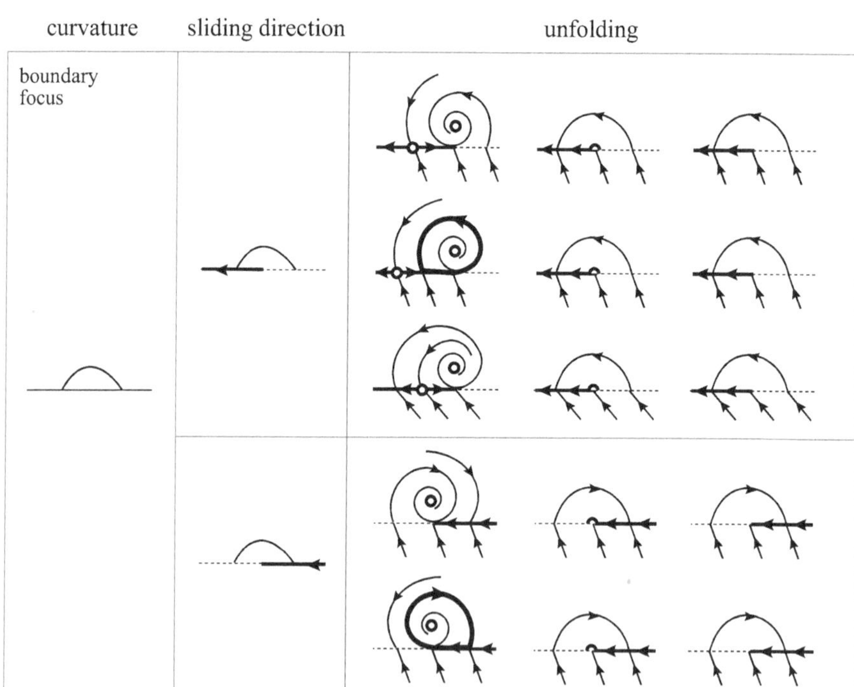

Fig. 8.5 Boundary equilibrium bifurcations of a focus in the plane: a saddlenode (top row) and a persistence (bottom row).

of the flow, which brings an orbit back around to intersect the discontinuity surface, either in the crossing region (if the focus is attracting) or in the sliding region (if the focus is repelling), and if the latter case is a saddlenode, then the return orbit may hit to one side of the sliding saddle or the other. In two cases (the second and last rows of column three), an attracting stick-slip cycle is formed.

All possible portraits for a generic one-parameter boundary equilibrium bifurcation in the plane are found to be topologically equivalent to these. The cases involving a repelling rather than attracting sliding region, for example, are obtained by reversing the direction of time. The qualitative scheme above is formalized rigorously in [71] at considerable length and some cost in transparency, so we shall not repeat it here, especially as it is clear that as a methodology it is highly specific and does not generalize easily to other kinds of bifurcations. A more explicit attempt to proceed by series expansions of the constituent vector fields is made in [80], and while not providing a proof of completeness, this approach is more instructive.

A prototype for these one-parameter boundary equilibrium bifurcations in the plane is given by

$$\begin{pmatrix} \dot{x}_1 \\ \dot{x}_2 \end{pmatrix} = \tfrac{1}{2}(1+\lambda)\begin{pmatrix} a(x_1-\mu)+bx_2 \\ c(x_1-\mu) \end{pmatrix} + \tfrac{1}{2}(1-\lambda)\begin{pmatrix} 1 \\ d \end{pmatrix} \tag{8.9}$$

with $\lambda = \text{sign}(x_1)$, through which one can reproduce all of the cases in Figures 8.3 to 8.5 and verify that no others occur. The $x_1 > 0$ vector field has an equilibrium at $(x_1, x_2) = (\mu, 0)$ which exists for $\mu > 0$ and at which the Jacobian is

$$\frac{d\mathbf{f}^+}{d\mathbf{x}} = \begin{pmatrix} a \; b \\ c \; 0 \end{pmatrix},$$

implying a saddle for $bc > 0$, node for $-a^2 < 4bc < 0$, and focus for $4bc < -a^2$, the latter two being attracting for $a < 0$ and repelling for $a > 0$. The eigenvectors $(a \pm \sqrt{a^2+4bc}, 2c)$ are in a general position with respect to the discontinuity surface provided $c \neq 0$. Sliding modes satisfy $\lambda^\$ = (1 - a\mu + bx_2)/(1 + a\mu - bx_2)$, giving a sliding vector field (by (8.6))

$$\dot{\mathbf{x}} = \mathbf{f}(\mathbf{x}; \lambda^\$) = \frac{(ad-c)\mu - bdx_2}{1 + a\mu - bx_2}$$

and for $\mu = 0$ this is directed inwards to the point $x_2 = 0$ if $bd > 0$ and outwards if $bd < 0$. The sliding region is where $bx_2 > a\mu$, and the sliding equilibrium at $x_2 = (ad-c)\mu/bd$ exists if $\mu c/d < 0$. Therefore if $c/d > 0$, the sliding equilibrium exists for $\mu < 0$ so there is persistence; if $c/d < 0$, the sliding equilibrium exists for $\mu < 0$ so there is a saddlenode bifurcation. The slope of the lower vector field is controlled by d.

The obvious extension to study the analogous bifurcation in higher dimensions is to take as a prototype

$$\dot{\mathbf{x}} = \mathbf{f}(\mathbf{x};\lambda) = \tfrac{1}{2}(1+\lambda)\underline{\underline{A}}.(\mathbf{x}-\boldsymbol{\mu}) + \tfrac{1}{2}(1-\lambda)\mathbf{d}\;, \tag{8.10}$$

with $\lambda = \text{sign}(x_1)$, $\mathbf{d} = (1, d, 0, 0, \dots)^{\mathsf{T}}$, $\boldsymbol{\mu} = (\mu, 0, 0, \dots)^{\mathsf{T}}$, and $\underline{\underline{A}}$ a constant $n \times n$ matrix. As in the plane, care must be taken in identifying the locations of invariant sets and separatrices in relation to the discontinuity surface. Alas this means that the topological scheme for classification in the plane does not extend immediately to higher dimensions or multiple switches. For those one must turn to the local methods established in Chapter 7, studying any boundary equilibria encountered on a case-by-case basis, at least until new profound insights into the problem are discovered.

8.5 Boundaries of Sliding: For a Single Switch

When the vector field $\mathbf{f}$ depends only linearly on the switching multipliers λ_j, the attractivity of the discontinuity surface is determined solely by the rate of change of $\mathbf{f}$ with $\boldsymbol{\lambda}$. At a simple intersection where $r = 1$, by (8.5) sliding modes satisfy the equation $\delta_t \sigma_1 = 0$. Then:

- $\sigma_1 = 0$ is repelling if

$$\frac{\partial}{\partial \lambda_1}\mathbf{f}(\mathbf{x};\lambda_1)\cdot \delta_{\mathbf{x}}\sigma_1 > 0 \quad \Leftrightarrow \quad \delta_{t+}\sigma_1 < 0 < \delta_{t-}\sigma_1\;, \tag{8.11a}$$

- $\sigma_1 = 0$ is attracting if

$$\frac{\partial}{\partial \lambda_1}\mathbf{f}(\mathbf{x};\lambda_1)\cdot \delta_{\mathbf{x}}\sigma_1 < 0 \quad \Leftrightarrow \quad \delta_{t-}\sigma_1 < 0 < \delta_{t+}\sigma_1\;. \tag{8.11b}$$

If there exists no λ_1 such that $\delta_t \sigma_1 = 0$, then the flow crosses $\sigma_1 = 0$ transversally. The conditions (8.11) thus define regions on $\sigma_1 = 0$ where sliding occurs. The sliding region (8.11a) repels the surrounding flow outside the discontinuity surface; the sliding region (8.11b) attracts it.

For $r > 1$ things are not so simple, and we will come to this in Section 8.7.

By (8.7), the boundaries between sliding and crossing regions are places where $\delta_{t\pm}\sigma_1 = 0$, that is, where one of the vector fields $\mathbf{f}^{\pm}$ lies tangent to the discontinuity surface $\sigma_1 = 0$.

Here we describe the local classification of such boundaries. Since we classified flow tangencies already in Section 6.1, all that remains is to derive the sliding vector fields for each case, restricted to systems with linear switching.

Due to (8.7) the sliding dynamics is quite trivial near a simple fold, while near cusps or multiple folds it can be anything but.

We assume there are no local equilibria, that is,

$$\mathbf{f}^+(\mathbf{x}_p) \neq 0\,, \quad \mathbf{f}^-(\mathbf{x}_p) \neq 0\,, \quad \& \quad \mathbf{f}^+(\mathbf{x}_p) \neq \mu \mathbf{f}^-(\mathbf{x}_p) \tag{8.12}$$

for any $\mu \in \mathbb{R}$ (where the third condition prohibits sliding equilibria).

We shall first derive the sliding vector field $\mathbf{f}^{\$}$ for the straightened flows in Section 6.1 that is in local coordinates $\mathbf{x} = (y_1, y_2, \ldots, y_n)$ such that $\mathbf{x}_p = (0, 0, \ldots, 0)$, where the flow is given by (6.5) and (6.6), hence

$$\begin{aligned} \mathbf{f}^+(\mathbf{x}) &= (1, 0, 0, \ldots)\,, \\ \mathbf{f}^-(\mathbf{x}) &= (0, 1, 0, \ldots)\,. \end{aligned} \tag{8.13}$$

The switching function is given by $\sigma_1 = y_k + V(\mathbf{x})$, in terms of the potential function $V(\mathbf{x})$ defined in Section 6.1, where k is the smallest index such that $\partial\sigma_1(0)/\partial y_k \neq 0$.

Denote the unit vector along the x_i axis as $\mathbf{e}_i$, then $\mathbf{f}^+(\mathbf{x}) = \mathbf{e}_1$, $\mathbf{f}^-(\mathbf{x}) = \mathbf{e}_2$, $\mathbf{f}(\mathbf{x}; \lambda_1) = \frac{1}{2}(1+\lambda_1)\mathbf{e}_1 + \frac{1}{2}(1-\lambda_1)\mathbf{e}_2$, and $\delta_{\mathbf{x}}\sigma_1 = \mathbf{e}_k + \delta_{\mathbf{x}} V(\mathbf{x})$. The sliding condition now gives

$$0 = \delta_t \sigma_1(\mathbf{x}) = \tfrac{1}{2}\left((1+\lambda^{\$})\mathbf{e}_1 + (1-\lambda^{\$})\mathbf{e}_2\right) \cdot (\mathbf{e}_k + \delta_{\mathbf{x}} V(\mathbf{x}))$$

which we can solve to find

$$\lambda^{\$}(\mathbf{x}) = \frac{(\mathbf{e}_2 + \mathbf{e}_1) \cdot (\mathbf{e}_k + \delta_{\mathbf{x}} V(\mathbf{x}))}{(\mathbf{e}_2 - \mathbf{e}_1) \cdot (\mathbf{e}_k + \delta_{\mathbf{x}} V(\mathbf{x}))}\,, \tag{8.14}$$

and this must lie in the interval $(-1, +1)$ for sliding modes to exist, which requires

$$\frac{\mathbf{e}_1 \cdot (\mathbf{e}_k + \delta_{\mathbf{x}} V(\mathbf{x}))}{\mathbf{e}_2 \cdot (\mathbf{e}_k + \delta_{\mathbf{x}} V(\mathbf{x}))} < 0\,. \tag{8.15}$$

The sliding vector field $\mathbf{f}^{\$}(\mathbf{x}) \equiv \mathbf{f}(\mathbf{x}; \lambda^{\$})$ is then

$$\mathbf{f}^{\$}(\mathbf{x}) = \frac{(\mathbf{e}_2 \cdot (\mathbf{e}_k + \delta_{\mathbf{x}} V(\mathbf{x}))\,,\ -\mathbf{e}_1 \cdot (\mathbf{e}_k + \delta_{\mathbf{x}} V(\mathbf{x}))\,,\ 0,\ 0,\ \ldots)}{(\mathbf{e}_2 - \mathbf{e}_1) \cdot (\mathbf{e}_k + \delta_{\mathbf{x}} V(\mathbf{x}))}\,. \tag{8.16}$$

The piecewise-straightened flows were useful in Section 6.1 to classify local tangencies, but in applications it is more common to find coordinates where the discontinuity surface is flat and the flow is curved. The vector fields for each tangency in such coordinates were given throughout Section 6.1; we shall refer to these and use (8.6) to derive their sliding vector fields.

8.5.1 Fold

Take the system (6.18),

$$(\dot{x}_1, \dot{x}_2, \dots) = \begin{cases} (1, \pm x_1, 0, 0, \dots) \text{ if } x_2 > 0, \\ (0, 1 \quad , 0, 0, \dots) \text{ if } x_2 < 0. \end{cases}$$

Sliding regions are given correspondingly by $\pm x_1 < 0$ on the discontinuity surface $x_2 = 0$, and sliding modes satisfy

$$\lambda^{\$} = \frac{1 \pm x_1}{1 \mp x_1} \in (-1, +1) \quad \text{for} \quad \pm x_1 < 0 \,, \tag{8.17}$$

giving a sliding vector field on $\mathcal{M} = \big\{(x_1, \lambda) : \lambda = \lambda^{\$},\ \pm x_1 < 0\big\}$ of

$$(\dot{x}_1, \dot{x}_2, \dots) = \mathbf{f}^{\$} = (1/(1 \mp x_1), 0, 0, \dots)\,. \tag{8.18}$$

The different cases are illustrated in Figure 8.6. The sliding dynamics is somewhat simple, being directed transversally inwards or outwards at the sliding boundary, in a manner consistent with the $\mathbf{f}^+$ vector field there. At least two dimensions are needed to observe nondegenerate folds. The attracting cases come from the '+' (visible) and '−' (invisible) signs in (6.18); the repelling cases are obtained by reversing the direction of time.

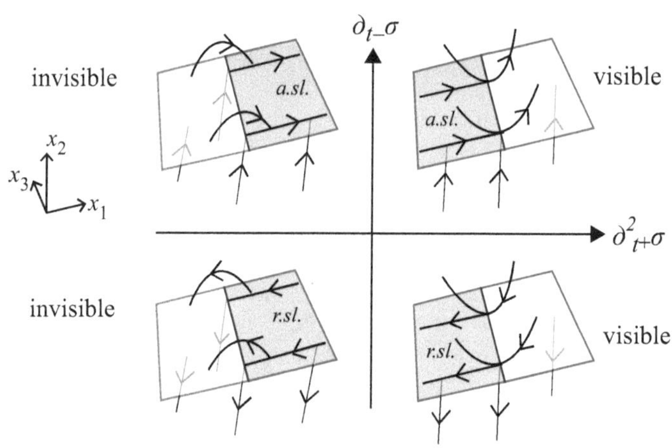

Fig. 8.6 Folds are visible or invisible and separate regions of crossing (unshaded) from attracting ($a.sl$) or repelling ($r.sl.$) sliding (shaded). Linear sliding dynamics is shown.

8.5.2 Two-Fold

Taking the system (6.21),

$$(\dot{x}_1, \dot{x}_2, \dot{x}_3, \dots) = \begin{cases} (-s^+, \nu^+, -x_1, 0, \dots) \text{ if } x_3 > 0\,, \\ (\ \nu^-, s^-, \quad x_2,\ 0, \dots) \text{ if } x_3 < 0\,, \end{cases}$$

the sliding regions on the discontinuity surface $x_3 = 0$ are where $x_1 x_2 > 0$. At least three dimensions are needed to observe nondegenerate two-folds.

The sliding modes are given by

$$\lambda^{\$} = \frac{x_2 - x_1}{x_2 + x_1} \tag{8.19}$$

and the sliding vector field is

$$(\dot{x}_1, \dot{x}_2, \dots) = \mathbf{f}^{\$} = \frac{(\nu^- x_1 - s^+ x_2, s^- x_1 + \nu^+ x_2, 0, 0, \dots)}{x_1 + x_2}\,. \tag{8.20}$$

The different cases are illustrated in Figure 8.7. The field $\mathbf{f}^{\$}$ is directed transversally inwards at an invisible fold and outwards at a visible fold, as dictated by (8.7) and consistent with (8.18), except at the two-fold singularity itself. There are two sliding regions, one attractive where $x_1, x_2 > 0$ and one repelling where $x_1, x_2 < 0$, plus two crossing regions where $x_1 x_2 < 0$, and their boundaries all meet at the singularity.

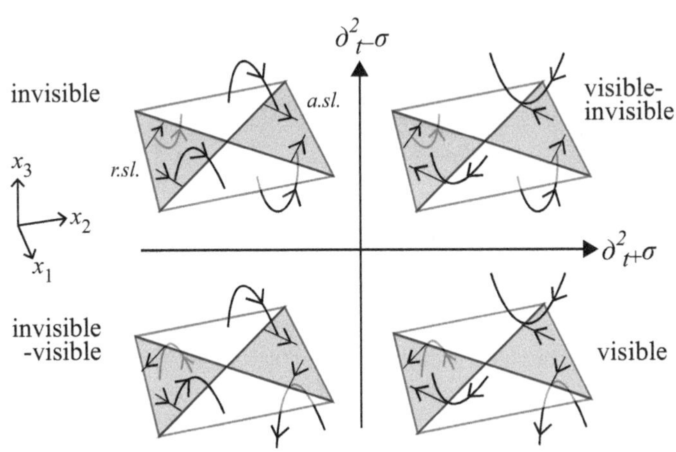

Fig. 8.7 Two-folds are visible or invisible and sit at the branching of visible and invisible folds which separate regions of crossing (unshaded) from attracting (*a.sl.*) or repelling (*r.sl.*) sliding (shaded). The term *visible* [or *invisible*] identifies the sliding flow curving away from [or towards] the fold curve.

The different sliding vector fields this generates are illustrated in Figure 8.8. Near the singularity, the vector field $\mathbf{f}^{\$}$ is able to take any of the generic local topological forms associated with equilibria, as the numerator of (8.20) has the form of a focus, node, or saddle, in the (x_1, x_2) plane for different $s^{\pm}, \nu^{\pm}$. There are no actual equilibria in the local sliding vector field (8.20), however, as the denominator provides a singular scaling: it is positive in the attracting sliding region and negative in the repelling sliding region and vanishes at the singularity. This 'folds up' the phase portrait to resemble an attracting equilibrium in one sliding region and a repelling equilibrium in the other.

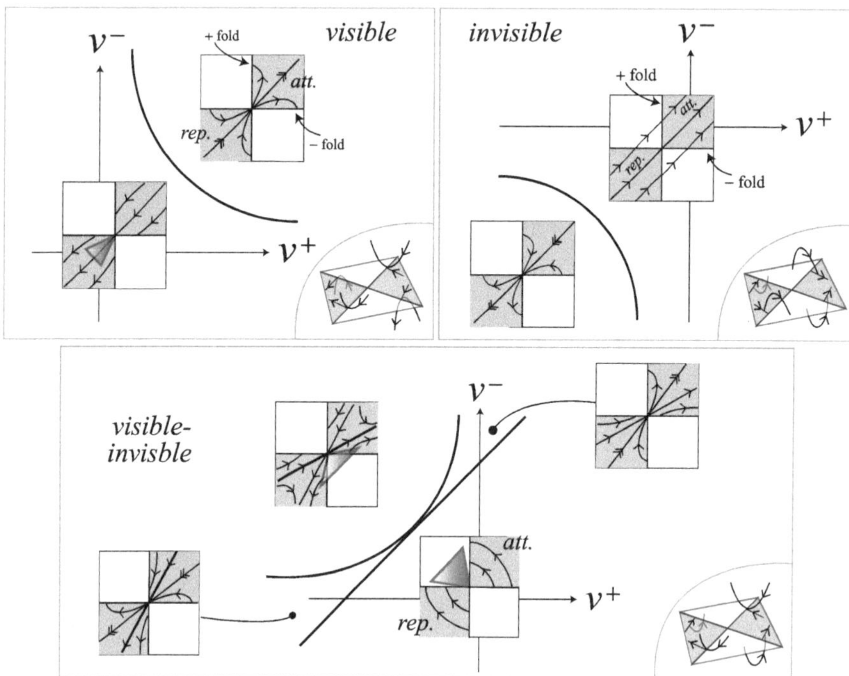

Fig. 8.8 Sliding vector fields of the two-folds.

One way to obtain the sliding phase portraits in Figure 8.8 is therefore to sketch the equilibria in the numerator of (8.20), using standard linear stability theory, and then simply reverse the directions of arrows in the repelling sliding region. For this one requires the eigenvectors of the numerator,

$$\left(\nu^- - \nu^+ \pm \sqrt{(\nu^- - \nu^+)^2 - 4s^+s^-}, 2\nu^-\right) ,$$

along which the flow tends inwards/outwards with respect to the singularity, if the associated eigenvalues $\nu^- + \nu^+ \pm \sqrt{(\nu^- - \nu^+)^2 - 4s^+s^-}$ are negative/positive. The eigenvectors are of interest only if they lie in the sliding regions, i.e. if they point into the upper right or lower left quadrants of the (x_1, x_2) plane, when

$$\left(\nu^- - \nu^+ \pm \sqrt{(\nu^- - \nu^+)^2 - 4s^+s^-}\right)/2\nu^- > 0\,,$$

otherwise they lie outside of the domain of (8.20), and topologically the sliding flow has no eigendirections and circulates from one fold to the other. If the eigenvalues are complex, then the sliding flow circulates around the singularity and has no real eigendirections. Thus a focus of the numerator of (8.20), with no real eigenvectors (when $(\nu^- - \nu^+)^2 - 4s^+s^-$), yields a sliding vector field topologically equivalent to a node or saddle of the numerator of (8.20) whose eigenvectors lie outside the sliding regions (when $(\nu^- - \nu^+ \pm \sqrt{(\nu^- - \nu^+)^2 - 4s^+s^-})/2\nu^- < 0)$. These features are responsible for there being so few different portraits for each type of two-fold in Figure 8.8, despite the numerator of (8.20) having focus, node, and saddle forms in each case.

These phase portraits beg two important questions, concerning firstly whether the sliding flow is able to traverse the two-fold in cases where there are both inward and outward trajectories and secondly how the sliding flow interacts with the flow outside the discontinuity surface. The two questions lead to surprisingly intricate dynamics which we return to in Chapter 13.

8.5.3 Cusp

Taking the system (6.24),

$$(\dot{x}_1, \dot{x}_2, \dot{x}_3, \dots) = \begin{cases} (1, \pm(x_1^2 + x_3), \dots) & \text{if } x_2 > 0, \\ (0, 1, 0, \dots) & \text{if } x_2 < 0, \end{cases}$$

the sliding regions are found to be $\pm(x_3 + x_1^2) < 0$. The different cases are illustrated in Figure 8.9.

The sliding modes are given by

$$\lambda^\$ = \frac{1 \pm (x_3 + x_1^2)}{1 \mp (x_3 + x_1^2)} \in (-1, +1) \tag{8.21}$$

on which the sliding vector field is

$$(\dot{x}_1, \dot{x}_2, \dots) = \mathbf{f}^\$ = \frac{(1, 0, 0, \dots)}{\pm 1 - x_3}\,. \tag{8.22}$$

At least three dimensions are needed to observe nondegenerate cusps. The sliding dynamics is again rather simple, being transverse to the folds as dictated by Figure 8.6, except at the cusp point where the sliding vector field is tangent to the boundary of the sliding region.

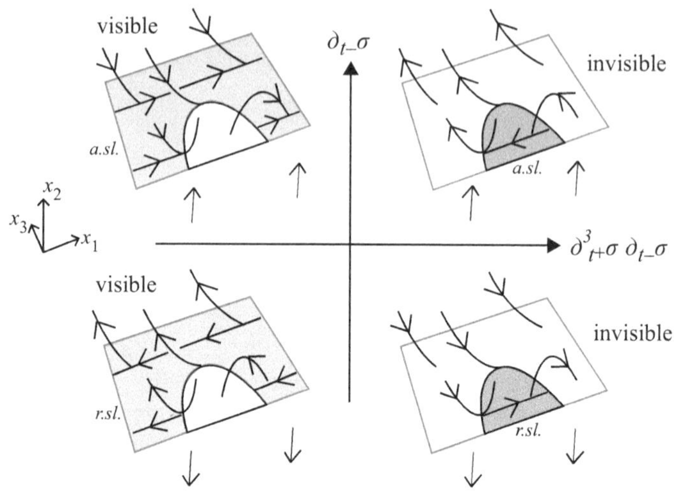

Fig. 8.9 Cusps are visible or invisible and sit at the branching of visible and invisible folds which separate regions of crossing (unshaded) from attracting ($a.sl.$) or repelling ($r.sl.$) sliding (shaded). Linear sliding dynamics is shown.

This explains our terminology for cusps in Section 6.4 that they are *visible* or *invisible* according to whether the sliding flow curves away from or towards the sliding boundary (formed by the curve of folds).

8.5.4 Swallowtail

Taking the system (6.28),

$$(\dot{x}_1, \dot{x}_2, \dot{x}_3, \dot{x}_4, \dots) = \begin{cases} (1, \pm(x_1^3 + x_3x_1 + x_4), 0, 0, \dots) & \text{if } x_2 > 0, \\ (0, 1, 0, 0, \dots) & \text{if } x_2 < 0, \end{cases}$$

the sliding regions are found to be $\pm(x_4 + x_3x_1 + x_1^3) < 0$. The vector fields are represented in (x_1, x_2, x_4) space for $x_3 > 0$ in Figure 8.10, showing the tangencies along the folds and cusps that emanate from the swallowtail.

The sliding modes are given by

$$\lambda^{\$} = \frac{\pm 1 + x_4 + x_3x_1 + x_1^3}{\pm 1 - x_4 - x_3x_1 - x_1^3} \tag{8.23}$$

giving a sliding vector field

$$\mathbf{f}^{\$} = \frac{(1, 0, 0, \dots)}{\pm 1 - x_4 - x_3x_1 - x_1^3} . \tag{8.24}$$

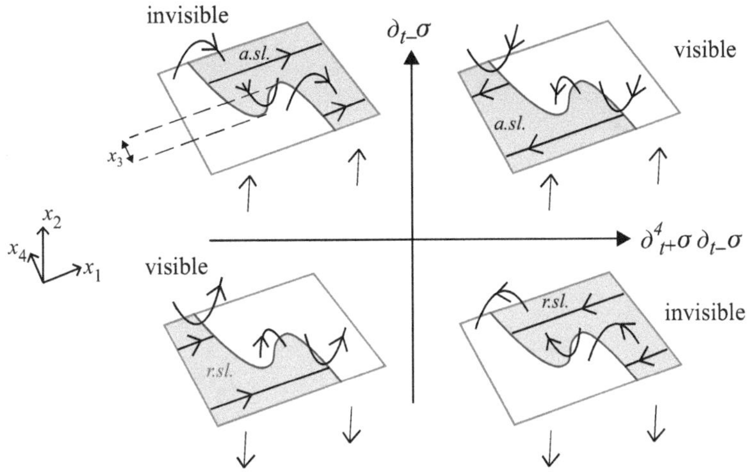

Fig. 8.10 Flow topologies around a swallowtail. The swallowtail occurs here when $x_3 = 0$, when the two cusps coincide. Swallowtails are visible or invisible as determined by the flow outside the discontinuity surface. Linear sliding dynamics is shown.

At least four dimensions are needed to observe nondegenerate swallowtails, and we give different representations of the swallowtail below that show how it unfolds through them. The sliding vector field is tangent to the sliding boundary at the cusps, two branches of which (one of visible and one of invisible type) collide to form the swallowtail, shown in Figure 8.11.

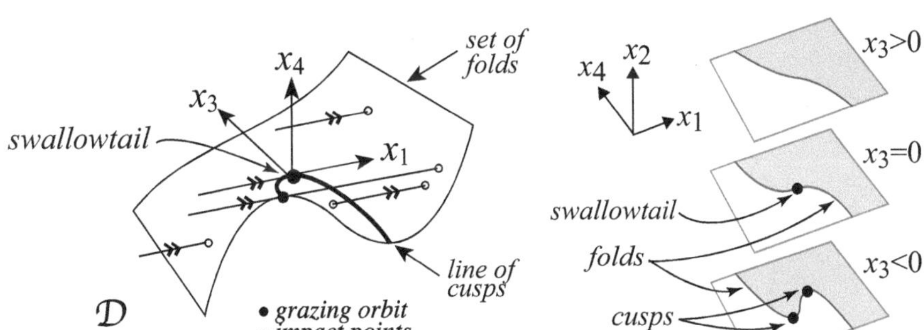

Fig. 8.11 Singular sets on the boundary of sliding near a swallowtail. The left figure is shown in the space of the discontinuity surface $\mathcal{D}$; the right shows the singularity sets on the discontinuity surface at fixed values of x_3.

8.5.5 Umbilic

Taking the system (6.34),

$$(\dot{x}_1, \dot{x}_2, \dot{x}_3, \dot{x}_4, \dots) = \begin{cases} (1, \pm(x_1^2 \pm x_3^2 \pm x_4), 0, 0, \dots) & \text{if } x_2 > 0, \\ (0, 1, 0, 0, \dots) & \text{if } x_2 < 0, \end{cases}$$

the sliding regions are found to be $\pm\left(x_1^2 \pm (x_3^2 + x_4)\right) < 0$ on which the switching multiplier is

$$\lambda^{\$} = \frac{\pm 1 + x_1^2 \pm (x_3^2 + x_4)}{\pm 1 - x_1^2 \mp (x_3^2 + x_4)} \in (-1, +1) \;, \tag{8.25}$$

giving a sliding vector field

$$\mathbf{f}^{\$}(\mathbf{x}) = \frac{(1,\ 0,\ 0,\ \dots)}{\pm 1 - x_1^2 \mp (x_3^2 + x_4)} \;, \tag{8.26}$$

for the lips (upper signs) and beak-to-beak (lower signs).

The lips involves a parabolic fold surface, with two branches of cusps, which meet at the lips singularity, which unfold in a minimum of four dimensions, shown in Figure 8.12.

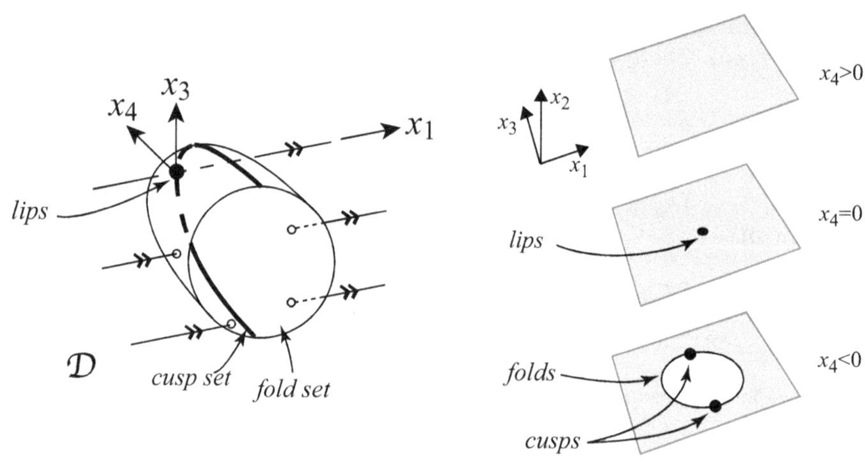

Fig. 8.12 Singular sets on the boundary of sliding near a lips singularity. The left figure is shown in the space of the discontinuity surface $\mathcal{D}$; the right shows the singularity sets on the discontinuity surface at fixed values of x_4.

The beaks involves a hyperbolic fold surface, with two branches of cusps that meet at the beak-to-beak singularity, again unfolding in a minimum of four dimensions, shown in Figure 8.13.

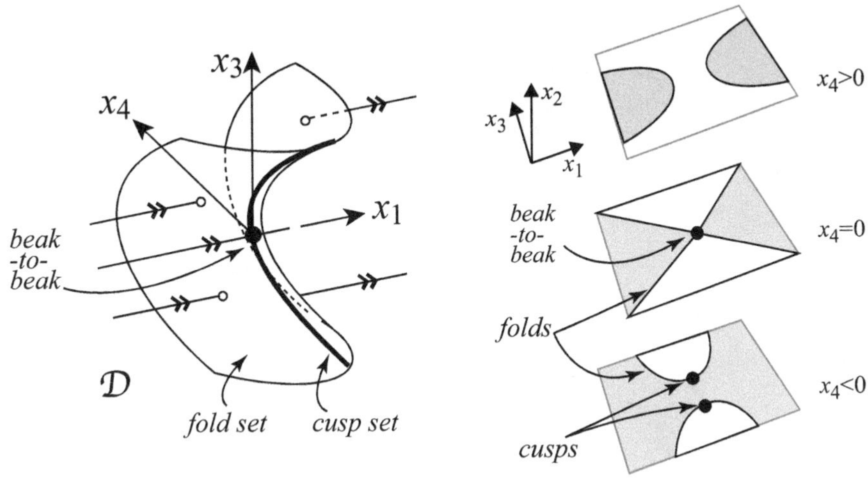

Fig. 8.13 Singular sets on the boundary of sliding near a beak-to-beak singularity. The left figure is shown in the space of the discontinuity surface $\mathcal{D}$; the right shows the singularity sets on the discontinuity surface at fixed values of x_4.

Since at least four dimensions are needed to observe nondegenerate umbilics, the vector fields are represented in (x_1, x_2, x_3) space for $x_4 < 0$ in Figures 8.14 and 8.15, showing the tangencies along the folds and cusps that emanate from the lips and beaks singularities.

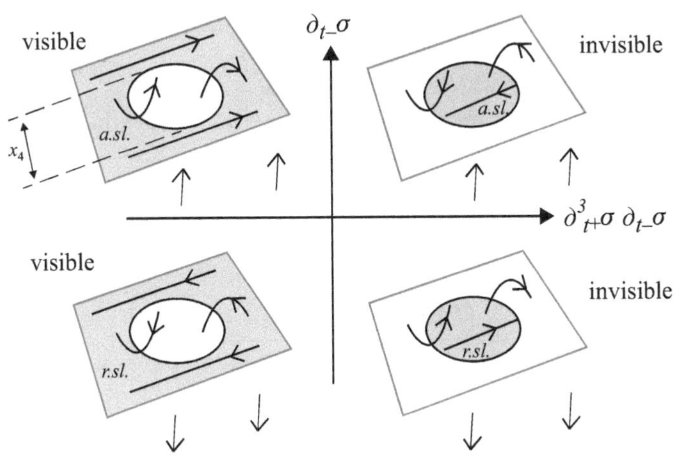

Fig. 8.14 Flow topologies around a lips singularity. The lips occurs here when $x_4 = 0$, when the two cusps coincide. Lips are visible or invisible as determined by the flow outside the discontinuity surface. Linear sliding dynamics is shown.

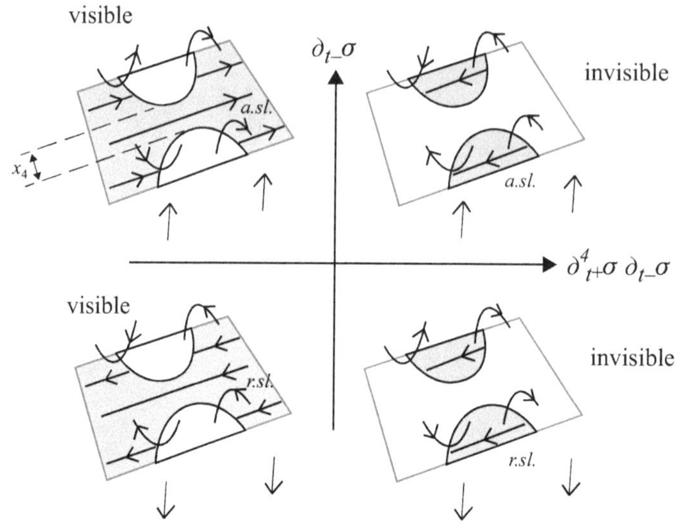

Fig. 8.15 Flow topologies around a beak-to-beak. The beak-to-beak occurs here when $x_4 = 0$, when the two cusps coincide. Beak-to-beaks are visible or invisible as determined by the flow outside the discontinuity surface. Linear sliding dynamics is shown.

8.5.6 Fold-Cusp

Taking the system (6.37),

$$(\dot{x}_1, \dot{x}_2, \dot{x}_3, \dot{x}_4, \dots) = \begin{cases} (1, \pm 1)x_1^2 + x_4), 0, 0, \dots) & \text{if } x_2 > 0, \\ (0, 1, x_2, 0, \dots) & \text{if } x_2 < 0, \end{cases}$$

the sliding regions are found to be $\pm(x_1^2 + x_4)/x_2 < 0$, on which the switching multiplier is

$$\lambda^{\$} = \frac{\pm 1 + x_1^2 + x_4}{\pm 1 - x_1^2 - x_4} \in (-1, +1) \ , \tag{8.27}$$

giving a sliding vector field

$$\mathbf{f}^{\$}(\mathbf{x}) = \frac{(1,\ 0,\ -x_2(x_1^2 + x_4),\ 0,\ 0,\ \dots)}{\pm 1 - x_1^2 - x_4} \ . \tag{8.28}$$

The fold-cusp involves a simple fold set of one vector field, intersecting the parabolic fold surface of the other vector field in which two fold branches are joined by a cusp. These unfold in a minimum of four dimensions, shown in Figure 8.16.

The vector fields are represented in (x_1, x_2, x_4) space for $x_3 > 0$ in Figure 8.17, showing the tangencies along the folds and cusps that emanate from the fold-cusp singularity.

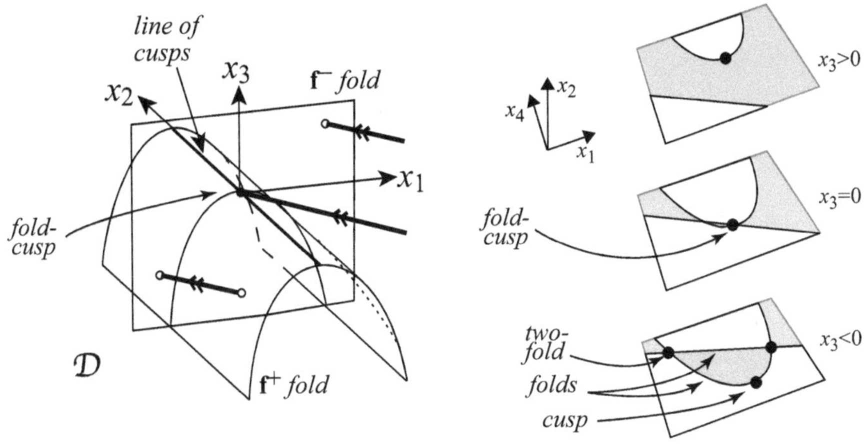

Fig. 8.16 Singular sets on the boundary of sliding near a fold-cusp. The left figure is shown in the space of the discontinuity surface $\mathcal{D}$; the right shows the singularity sets on the discontinuity surface at fixed values of x_3.

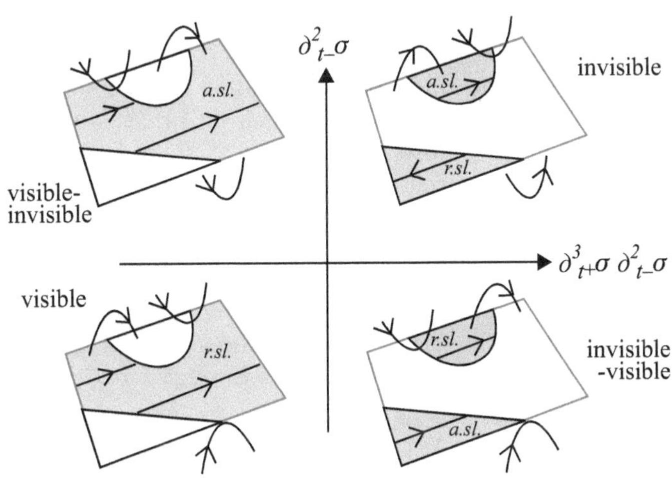

Fig. 8.17 A fold-cusp occurs here when $x_3 = 0$, when the two cusps coincide. Fold-cusps are visible or invisible. Linear sliding dynamics is shown.

One can continue such a classification into higher dimensions and codimensions without end.

If instead we increase the codimension of the singularity but restrict the dimension, taking, for example, a cusp in two dimensions, we obtain bifurcations of the sliding boundary.

8.6 Bifurcations of Sliding Boundaries in the Plane

Similarly to Section 8.4 where we classified one-parameter boundary equilibrium bifurcations in the plane, it ought to be possible to classify all one-parameter bifurcations of sliding boundaries in planar systems. In essence this involves taking generic two-dimensional slices of the boundary singularities in Section 8.5. In restricting to lower dimensions, however, sliding equilibria are created and complicate the classification considerably.

A reliable classification scheme follows topological principles similar to those in Section 8.4, and again, choosing sufficient criteria is all important.

A simple fold, where one of the constituent vector fields $\mathbf{f}^+$ or $\mathbf{f}^-$ is tangent to the discontinuity surface and the other is transverse to it, forms a structurally stable sliding boundary. One-parameter bifurcations in planar systems are obtained when two folds collide, forming either a cusp if both folds are associated with the same constituent vector field or a two-fold if one fold belongs to each constitutive vector field.

The topological properties that must be considered are the curvature (i.e. fold or cusp) of the vector fields $\mathbf{f}^\pm$ at the boundary, with the directions of $\mathbf{f}^\pm$ and the sliding vector field $\mathbf{f}^\$$.

As we noted in (8.7), the direction of the sliding vector field is determined by the constituent vector fields $\mathbf{f}^\pm$, unless their directions are in opposition, in which case the different possible directions of $\mathbf{f}^\$$ lead to different cases (though sliding might only occur at the bifurcation point itself). The choice of sliding directions is only important at the visible-invisible fold-fold.

The resulting classification is shown in Figure 8.18. The singularity classification appeared in [71], with rigorous (and again lengthy case-by-case) proofs of its completeness, and was later reiterated with unfoldings in [140].

A prototype that allows all of the fold-fold cases from Figure 8.18 to be explored, as a bifurcation parameter μ passes through zero, is

$$\begin{pmatrix} \dot{x}_1 \\ \dot{x}_2 \end{pmatrix} = \tfrac{1}{2}(1+\lambda)\begin{pmatrix} a(x_2-\mu) \\ b+cx_2 \end{pmatrix} + \tfrac{1}{2}(1-\lambda)\begin{pmatrix} x_2 \\ d \end{pmatrix} \tag{8.29}$$

with $\lambda = \operatorname{sign}(x_1)$, for different values of the constants a, b, c, d. There is a tangency of the $x_1 > 0$ vector field at $x_2 = \mu$, with curvature $\ddot{x}_1 = ab$ (implying a visible fold if $ab > 0$ and invisible fold if $ab < 0$), and a tangency of the $x_1 < 0$ vector field at $x_2 = 0$, with curvature $\ddot{x}_1 = d$ (implying a visible fold if $d < 0$ and invisible fold if $d > 0$). Sliding modes satisfy

$$\lambda^\$ = \frac{x_2 + ax_2 - a\mu}{x_2 - ax_2 + a\mu} \in (-1, +1)\ , \tag{8.30}$$

implying that sliding occurs on the region where $a(1-\mu/x_2) > 0$ on $x_1 = 0$, with dynamics

$$\dot{x}_2 = \frac{x_2(b - ad + cx_2) + ad\mu}{x_2 - ax_2 + a\mu}\ . \tag{8.31}$$

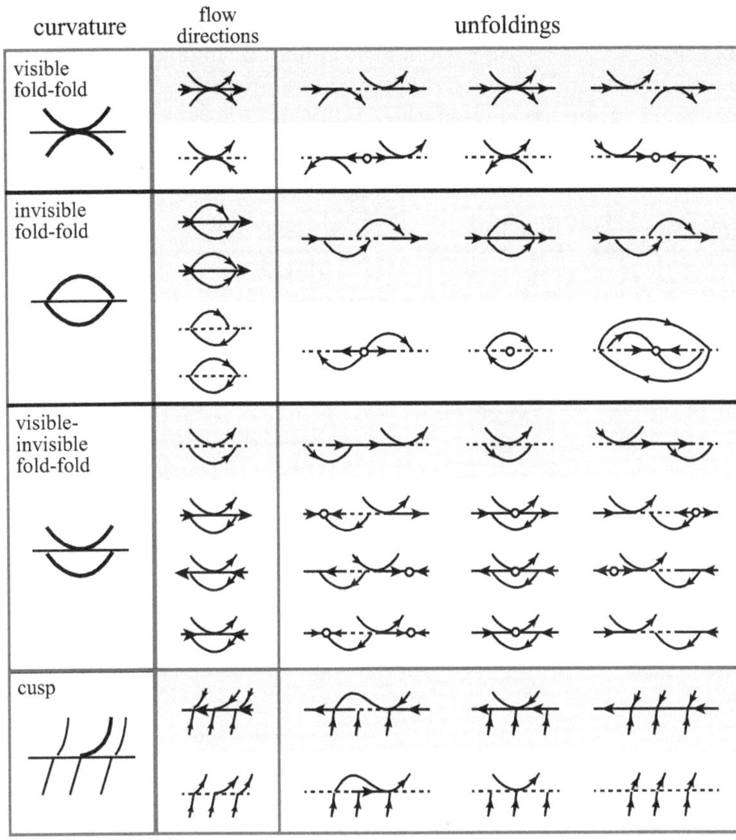

Fig. 8.18 Phase portraits of one-parameter sliding boundary bifurcations in the plane.

A sliding equilibrium exists at $x_2 = (ad-b)/c$ if $a(1-\mu c/(ad-b)) > 0$. Using (8.29) to (8.31), all of the one-parameter sliding boundary bifurcations in the plane can be found as given in Figure 8.18. (See the start of Section 9.6.2 to be led more closely through the second visible-invisible fold case in Figure 8.18.)

To unfold the cusp bifurcation from Figure 8.18, we can take a prototype

$$\begin{pmatrix} \dot{x}_1 \\ \dot{x}_2 \end{pmatrix} = \tfrac{1}{2}(1+\lambda)\begin{pmatrix} \frac{1}{2}x_2^2 - \mu \\ b \end{pmatrix} + \tfrac{1}{2}(1-\lambda)\begin{pmatrix} a \\ 0 \end{pmatrix} \tag{8.32}$$

with $\lambda = \text{sign}(x_1)$, for constants a and b. We leave its investigation to the reader.

The extension of these to higher dimensions is given by the generic two-folds and cusps from Section 8.5.

These can also be extended to two or more parameter bifurcations—changes that require more than one parameter to fully unfold. Two-parameter bifurcations in the plane were studied in [90]. Filippov preferred an alternative and more far-reaching classification, classifying all singularities that involved a tangency above and below the discontinuity surface, of any order. Unfortunately in [71] Filippov quoted the number of classes without identifying them explicitly. The classification is given in Figure 8.19 and is the subject

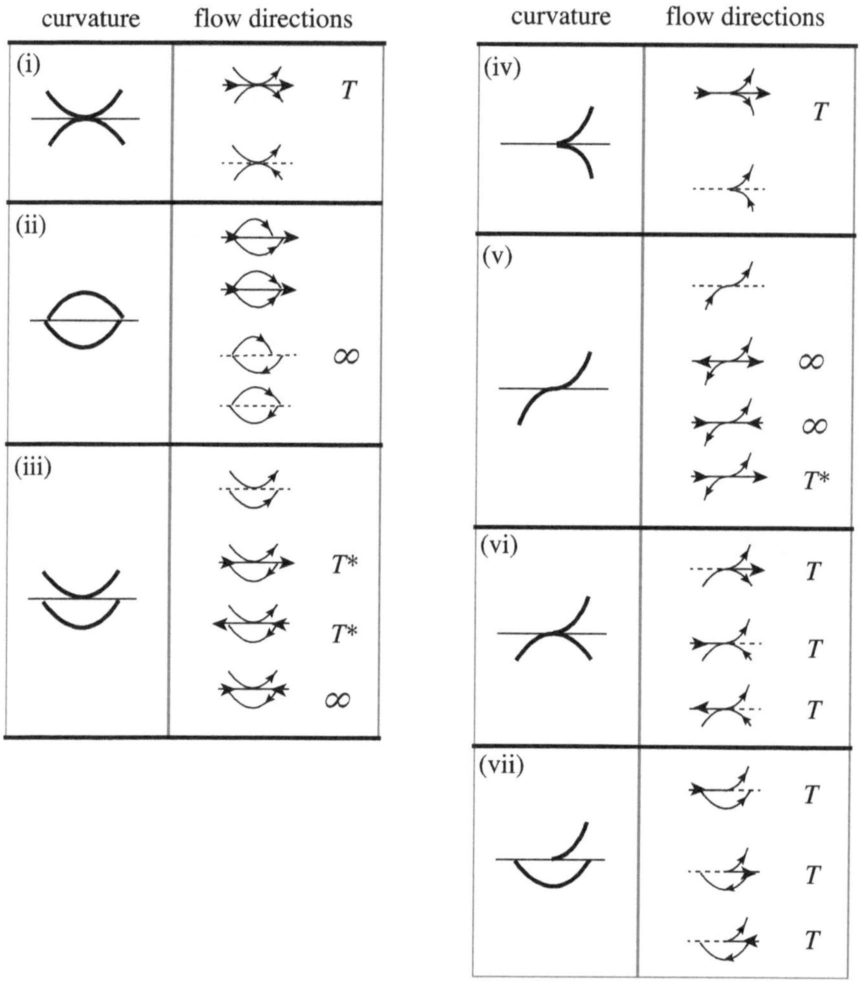

Fig. 8.19 Phase portraits of double tangency bifurcations in the plane. T or ∞ denotes that convergence to the singularity via the sliding mode is in finite or infinite time, respectively, and the cases T^* can take infinite time if the folds are higher order than folds or cusp.

of more extensive study in [83].

The cases (i-iii) in Figure 8.19 are not simply those in Figure 8.18. They illustrate any flow with order $2k$ contact to both sides of the discontinuity surface. Similarly in (iv-v) the flow has order $2k+1$ contact to both sides of the discontinuity surface, and in (vi-vii) there is order $2k+1$ contact on one side and order $2k'$ contact on the other, for positive integers $k \neq k'$. The symbols T or ∞ indicate that convergence to the singularity via the sliding mode takes place in finite or infinite time, respectively, and the cases T^* can take infinite time if the folds are higher order than folds or cusp.

It is impossible to give a complete unfolding of the singularities in Figure 8.19 if the tangencies are not simply folds or cusps, as they require many parameters to fully unfold (at least $p+q-3$ parameters if contact is of order p and q on either side of the discontinuity surface).

8.7 Boundaries of Sliding: For r Switches

At a codimension r switching intersection (where $\sigma_1 = \sigma_2 = \cdots = \sigma_r = 0$), the constituent flows in the regions $\mathcal{R}_\kappa$ can generically be tangent to k of the submanifolds $\sigma_j = 0$ (satisfying conditions, e.g. $\delta_{ti}\sigma_1 = \delta_{ti}\sigma_2 = \cdots = \delta_{ti}\sigma_k = 0$) in a system of $n \geq k+r$ dimensions. Alternatively, tangency to k submanifolds $\sigma_j = 0$ can occur as a codimension $k+r-n$ bifurcation in a system of $n < k+r$ dimensions.

To begin to understand the structure of these boundaries, we have the following result, which can be considered as an extension of (8.7) to multiple switches. We take first the case of two switches, illustrated in Figure 8.20.

Lemma 8.1. *Tangency between a vector field* $\mathbf{f}^{++}$ *and a switching intersection* $\sigma_1 = \sigma_2 = 0$ *implies tangency between the neighbouring sliding vector fields* $\mathbf{f}^{\$+}$ *and* $\mathbf{f}^{+\$}$ *and the intersection.*

Proof. If $\mathbf{f}^{++}$ is tangent to both $\sigma_1 = 0$ and $\sigma_2 = 0$ at a point $\mathbf{x}$, then at that point it satisfies

$$0 = \delta_{t++}\sigma_1 = \delta_{t++}\sigma_2 \,. \tag{8.33}$$

The neighbouring sliding vector field on $\sigma_1 = 0$ is

$$\mathbf{f}^{\$+} = \tfrac{1}{2}(1+\lambda_1)\mathbf{f}^{++} + \tfrac{1}{2}(1-\lambda_1)\mathbf{f}^{-+} \tag{8.34}$$

in which sliding modes, if they exist, satisfy

$$0 = \delta_{t\$+}\sigma_1 \,. \tag{8.35}$$

Since $\delta_{t++}\sigma_1 = 0$ by (8.33), the condition (8.35) with (8.34) reduces to

$$0 = \delta_{t\$+}\sigma_1 = \tfrac{1}{2}(1-\lambda_1)\delta_{t-+}\sigma_1 \tag{8.36}$$

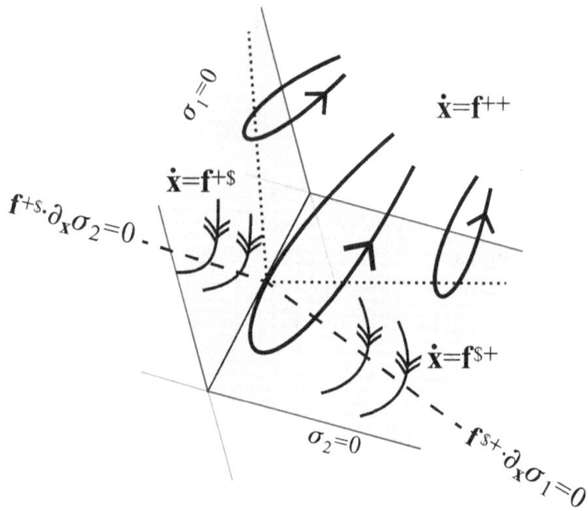

Fig. 8.20 $\mathbf{f}^{++}$ is tangent to the intersection curve $\sigma_1 = \sigma_2 = 0$. The theorem concerns the dashed curves, on which the neighbouring sliding vector fields $\mathbf{f}^{\$+}$ and $\mathbf{f}^{+\$}$ are tangent to the intersection. These will not typically be the same set (dotted) on which $\mathbf{f}^{++}$ is tangent to the individual discontinuity submanifold.

implying either (i) $\lambda_1 = 1$, hence $\mathbf{f}^{\$+} = \mathbf{f}^{++}$ by (8.34), which by assumption is tangent to the intersection, or (ii) $\delta_{t-+}\sigma_1 = 0$; hence $\mathbf{f}^{-+}$ lies tangent to the intersection, as does $\mathbf{f}^{++}$ by assumption, and since $\mathbf{f}^{\$+}$ is a convex combination of $\mathbf{f}^{\pm+}$, therefore it also lies tangent to the intersection; i.e. in either case

$$0 = \delta_{t\$+}\sigma_1 = \delta_{t\$+}\sigma_2 \,. \tag{8.37}$$

By an analogous argument following (8.34) to (8.37) on $\sigma_2 = 0$,

$$0 = \delta_{t+\$}\sigma_1 = \delta_{t+\$}\sigma_2 \,. \tag{8.38}$$

Hence tangency of $\mathbf{f}^{++}$ to $\sigma_1 = 0$ and $\sigma_2 = 0$ implies tangency of $\mathbf{f}^{\$+}$ to $\sigma_2 = 0$ and of $\mathbf{f}^{+\$}$ to $\sigma_1 = 0$. $\square$

This can be generalized substantially.

Theorem 8.2. *Tangency between a codimension r sliding vector field and a codimension $d > r + 1$ discontinuity surface (intersection of d manifolds) implies tangency between the neighbouring codimension $r+1, \ldots, d-1$, sliding vector fields, and the codimension d intersection (where $r = 0$ indicates a constituent vector field in one of the regions $\mathcal{R}_K$ off the discontinuity surface).*

Proof. Take without loss of generality a codimension r sliding vector field $\mathbf{f}^{\$_r+\mathcal{I}}$, given by

$$\mathbf{f}^{\$_r+\mathcal{I}} = \sum_{\kappa_1=\pm} \cdots \sum_{\kappa_r=\pm} \prod_{j=1}^{r} \lambda_j^{(\kappa_j)} \mathbf{f}^{\kappa_1\kappa_2\ldots\kappa_r+\mathcal{I}}(\mathbf{x}) \tag{8.39}$$

where $\$_r$ is a string of r symbols $\$\ldots\$$ and $\mathcal{I}$ is a string of $d-r-2$ fixed '+' or '−' symbols $\mathcal{I} = \kappa_{r+2}\ldots\kappa_{d-1}$, such that

$$0 = \delta_{t\$_r+\mathcal{I}}\sigma_1 = \cdots = \delta_{t\$_r+\mathcal{I}}\sigma_r \,.$$

If this is tangent to a codimension $d > r$ discontinuity surface then, without loss of generality, it satisfies in addition

$$0 = \delta_{t\$_r+\mathcal{I}}\sigma_{r+1} = \cdots = \delta_{t\$_r+\mathcal{I}}\sigma_d \,. \tag{8.40}$$

The neighbouring codimension $r+1$ sliding vector field on $\sigma_{r+1} = 0$ is

$$\mathbf{f}^{\$_r\$\mathcal{I}} = \tfrac{1}{2}(1+\lambda_{r+1})\mathbf{f}^{\$_r+\mathcal{I}} + \tfrac{1}{2}(1-\lambda_{r+1})\mathbf{f}^{\$_r-\mathcal{I}} \tag{8.41}$$

with the sliding conditions

$$0 = \delta_{t\$_r\$\mathcal{I}}\sigma_1 = \delta_{t\$_r\$\mathcal{I}}\sigma_2 = \cdots = \delta_{t\$_r\$\mathcal{I}}\sigma_{r+1} \,. \tag{8.42}$$

Since $\delta_{t\$_r+\mathcal{I}}\sigma_{r+1} = 0$ by (8.40), the last condition in (8.42) with (8.41) reduces to

$$0 = \delta_{t\$_r\$\mathcal{I}}\sigma_{r+1} = \tfrac{1}{2}(1-\lambda_{r+1})\delta_{t\$_r-\mathcal{I}}\sigma_{r+1} \tag{8.43}$$

implying either (i) $\lambda_{r+1} = 1$, hence $\mathbf{f}^{\$_r\$\mathcal{I}} = \mathbf{f}^{\$_r+\mathcal{I}}$ by (8.41), which by assumption is tangent to the intersection, or (ii) $\delta_{t\$_r-\mathcal{I}}\sigma_{r+1} = 0$; hence $\mathbf{f}^{\$_r-\mathcal{I}}$ lies tangent to the intersection, as does $\mathbf{f}^{\$_r+\mathcal{I}}$ by assumption, and since $\mathbf{f}^{\$_r\$\mathcal{I}}$ is the convex combination of $\mathbf{f}^{\$_r\pm\mathcal{I}}$, it therefore lies tangent to the intersection; i.e. in either case

$$0 = \delta_{t\$_r\$\mathcal{I}}\sigma_1 = \cdots = \delta_{t\$_r\$\mathcal{I}}\sigma_d \,. \tag{8.44}$$

By analogous argument following (8.41) to (8.44) on $\sigma_m = 0$ for $m = r+1,\ldots,d$,

$$0 = \delta_{t\$_r+\mathcal{I}_a\$\mathcal{I}_b}\sigma_1 = \cdots = \delta_{t\$_r+\$}\sigma_d \tag{8.45}$$

where $\mathcal{I}_a\$\mathcal{I}_b$ is the string $\mathcal{I}$ with \$ replacing one index.

Continuing iteratively, the neighbouring codimension $r+2$ sliding vector field on $\sigma_{r+2} = 0$ is

$$\mathbf{f}^{\$_r\$\$\mathcal{I}'} = \tfrac{1}{2}(1+\lambda_{r+2})\mathbf{f}^{\$_r\$+\mathcal{I}'} + \tfrac{1}{2}(1-\lambda_{r+2})\mathbf{f}^{\$_r\$-\mathcal{I}'} \tag{8.46}$$

analogously to (8.41) with $\mathcal{I}'$ being $\mathcal{I}$ with the first component removed, where the first term is tangent to $\sigma_d = 0$ by the result above, and then proceeding analogously to (8.42) to (8.44), we find that

$$0 = \delta_{t\$_1\ldots\$_m}\sigma_1 = \cdots = \delta_{t\$_1\ldots\$_m}\sigma_d \tag{8.47}$$

for all $r < m \leq d-1$. □

In short this means that a tangency between a given vector field and a high codimension intersection implies that certain of the neighbouring sliding vector fields will also be tangent. The converse does not hold: tangency between the codimension r vector field and the intersection does not imply tangency of the surrounding vector fields of lower codimension with the intersection.

8.8 The Hidden Degeneracy of Linear Switching

We should really have delved deeper into the classifications in Section 8.4 and Section 8.6, using switching layer analysis to closely study the attractivity of the sliding manifolds and equilibria involved. Alas, we cannot do this within the present chapter, because we would find that the switching layer system for these bifurcations is degenerate. To fix the degeneracy will require a look a *nonlinear* switching terms.

A first glance at the degeneracy can be seen in (8.30) and less obviously in (8.9). The sliding mode solution that defines $\mathcal{M}$ in (8.30) is singular at $x_2 = \mu = 0$, i.e. at the fold-fold point of the sliding boundary bifurcations, so $\mathcal{M}$ is degenerate there. When the switching layer is derived from (8.9) at the boundary focus (with $4bc < -a^2$), the resulting system is a centre when $\mu = 0$, meaning it is a degenerate system consisting entirely of closed cycles.

A symptom of these is more readily observable in their unfoldings. Close inspection reveals something peculiar in certain of the boundary equilibrium bifurcations in Figures 8.3 to 8.5 and sliding boundary bifurcations in Figure 8.18. In the last row of Figure 8.4, a *repelling* node becomes an *attracting* sliding node. In the second row of visible-invisible fold-folds in Figure 8.18, a sliding node changes from attracting to repelling. These are collected in Figure 8.21.

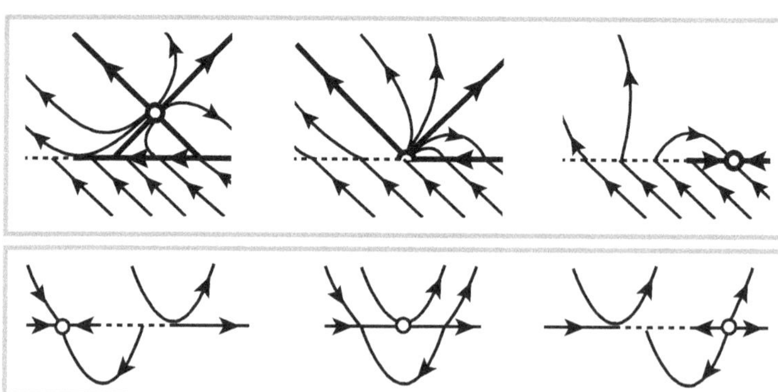

Fig. 8.21 A boundary equilibrium bifurcation (top) and a sliding boundary fold-fold bifurcation (bottom). Notice the flip in attractivity of the equilibrium.

This kind of flip in attractivity cannot happen in smooth systems without some other object being born to balance it, like a limit cycle balancing out the change in attractivity of an equilibrium in a Hopf bifurcation. Indeed we find that it cannot happen in piecewise-smooth systems either, and inspection of the dynamics inside the switching layer *should* reveal how the change has occurred. When we form the switching layer system ((7.2) with $r = 1$) for these bifurcations, we will find that the sliding manifold is degenerate and typically, therefore, structurally unstable. We will see why in Section 9.5.3, where we will also see how nonlinear dependence on the switching multipliers $\boldsymbol{\lambda}$ is needed to restore structural stability. Only then can we revisit the specific bifurcations Figure 8.21, which we do in Section 9.6.

8.9 Piecewise-Smooth Time Rescaling

It is a simple but useful observation that the phase portrait of a smooth system remains unchanged if we rescale time by a strictly positive function. This remains true if the timescaling is piecewise-smooth. This particular type of orbital equivalence (Definition 4.4) can be proven explicitly if $\mathbf{f}$ depends linearly on just one switching multiplier λ_1.

Let us show explicitly why applying different, strictly positive, time rescalings either side of the discontinuity surfacedoes not alter the phase portrait at the discontinuity surface itself. We used this fact earlier in Theorem 6.1 and Theorem 6.2.

Taking the system

$$\dot{\mathbf{x}} = \tfrac{1}{2}(1+\lambda_1)\mathbf{f}^+(\mathbf{x}) + \tfrac{1}{2}(1-\lambda_1)\mathbf{f}^-(\mathbf{x}) \,, \tag{8.48}$$

where $\lambda_1 = \operatorname{sign}(\sigma_1)$, with sliding modes by (8.5) satisfying

$$\lambda_1^\$ = \frac{(\delta_{t-}+\delta_{t+})\sigma_1}{(\delta_{t-}-\delta_{t+})\sigma_1} \,. \tag{8.49}$$

Let us ask the effect of making a piecewise-smooth time rescaling $t \mapsto t/\mu(\mathbf{x})$, where

$$\mu(\mathbf{x}) = \begin{cases} \mu^+(\mathbf{x}) \text{ if } \sigma_1(\mathbf{x}) > 0, \\ \mu^-(\mathbf{x}) \text{ if } \sigma_1(\mathbf{x}) < 0, \end{cases} \tag{8.50}$$

with $\mu^\pm(\mathbf{x})$ strictly positive. The timescaling (8.50) does not change the phase portrait of the individual subsystems in $\sigma_1(\mathbf{x}) > 0$ or $\sigma_1(\mathbf{x}) < 0$, since they are smooth in those regions, so the only possible change in phase portrait would be on $\sigma_1(\mathbf{x}) = 0$. Firstly, the signs of $\delta_{t\pm}\sigma_1$ are unchanged (multiplied by $\mu_\pm > 0$) so the attractivity or repulsivity of $\mathcal{D}_1$ with respect to the flow is preserved. Secondly, on $\mathcal{D}_1$, substituting $t \mapsto t/\mu$ into (8.6), the Lie derivative along the flow (denoted $\delta_{t\$}$ or $\mathbf{f}^\$ \cdot \delta_{\mathbf{x}}$) transforms as $\delta_{t\$} \mapsto \mu' \delta_{t\$}$ where

$$\mu' = \frac{\mu^+\mu^-(\partial_{t-}-\partial_{t+})\sigma_1}{(\mu^-\partial_{t-}-\mu^+\partial_{t+})\sigma_1} = \frac{\mu^+\mu^-(|\partial_{t-}\sigma_1|+|\partial_{t+}\sigma_1|)}{|\mu^-\partial_{t-}\sigma_1|+|\mu^+\partial_{t+}\sigma_1|} > 0.$$

The second equality follows because $\partial_{t^{\pm}}\sigma_1$ have opposite signs. Thus (8.50) scales time on the discontinuity surface by a strictly positive constant μ', so the transformation $t \mapsto t/\mu$ preserves the phase portrait of the flow.

8.10 Looking Forward

For more about systems with linear dependence on a single switching multiplier λ, often called *Filippov systems*, the reader may refer to any of the classic texts on piecewise-smooth systems, especially [71, 140, 48] for classifications of local singularities. We will have a little more to say about them when we discuss global bifurcations in Chapter 11. For now we must move on.

We must move on because we wish to avoid a trap. There has been a tendency to apply subtle concepts from smooth dynamics, such as normal forms and equivalence classes, perhaps a little too early for their application in piecewise-smooth dynamics to be well understood. As a consequence their usage has not always been as reliable as it might have been. Classifications in particular are troublesome. Despite their importance in smooth systems, classifications have produced some of the most baffling schemes for systematizing the study of piecewise-smooth systems, in which the completeness of results or assumptions are often unclear. Even local singularities require the consideration of various intersections of separatrices and the careful search for unanticipated sliding equilibria, such that a proliferation of cases with increasing dimension or codimension happens much faster than in smooth systems. Thus piecewise-smooth systems suffer a form of the *curse of dimensionality* in the number of bifurcations they can exhibit (as remarked in [84]). The zoology of classes will certainly continue to generate years of work for those pursuing it, and there are doubtless many interesting scenarios to be discovered.

Some works have also attempt to generalize bifurcations that appear in smooth systems to those that are piecewise-smooth, for example, Hopf bifurcations at a discontinuity boundary, of which we have already seen one example in Figure 8.18. We will not cover these in detail, because none are able to say with any generality how the results relate to systems with more dimensions, more switches, or nonlinear dependence on the switching multipliers.

As such we will avoid setting out many classifications in this book. Let us instead turn towards phenomena that happen in piecewise-smooth systems that *does not* happen in differentiable systems. Making use of the intuition gained from Chapter 7 and Chapter 8, the next chapter will begin to reveal just how much unprecedented behaviour there is.

Chapter 9
Nonlinear Switching (Local Theory): The Phenomena of Hidden Dynamics

We now resume the study of piecewise-smooth systems with general dependence on the variable $\mathbf{x}$ and the switching multiplier $\boldsymbol{\lambda}$.

We begin by exploring some of the more exotic behaviours created by dynamics hidden inside the switching layer.

To see how much of our knowledge of linear systems—the content of the last chapter—applies in general, we also establish which features of linear switching are robust to small perturbations that are nonlinear in the switching multipliers and which features are not.

9.1 Nonlinear Sliding

This chapter will focus on the dynamical phenomena possible in systems

$$\dot{\mathbf{x}} = \mathbf{f}(\mathbf{x}; \lambda_1, \ldots, \lambda_m) , \qquad \lambda_j = \operatorname{sign}(\sigma_j(\mathbf{x})) ,$$

when there is nonlinear dependence on the switching multipliers $\lambda_1, \ldots, \lambda_m$, in particular phenomena that are not possible in linear switching systems. We look at novel attractors, oscillations, and bifurcations of such systems. We also probe the frontier between linear and nonlinear behaviour by considering small nonlinear perturbations of linear systems.

Nonlinear dependence on $\boldsymbol{\lambda}$ can produce dynamics that cannot be anticipated outside the discontinuity surface. This is *hidden dynamics*. Perhaps the simplest demonstration of it is the following.

Example 9.1 (Nonlinear Sliding Behaviour). Take the two systems

$$\text{(a)} \quad (\dot{x}_1, \dot{x}_2) = (-\lambda_1, 2\lambda_1^2 - 1) , \qquad \text{(b)} \quad (\dot{x}_1, \dot{x}_2) = (-\lambda_1, 1) ,$$

M. R. Jeffrey, *Hidden Dynamics*,
https://doi.org/10.1007/978-3-030-02107-8_9

with $\lambda_1 = \text{sign}(x_1)$. Outside the discontinuity surface these are identical, just $(\dot{x}_1, \dot{x}_2) = (-\text{sign}(x_1), 1)$ for $x_1 \neq 0$. On the discontinuity surface it is a different story. Both vector fields point towards $x_1 = 0$ as shown in Figure 9.1, so sliding is guaranteed, but in blowing up $x_1 = 0$ into the switching layer $\lambda_1 \in (-1, +1)$ (applying (7.2) with $r = 1$), we find

$$\text{(a)} \quad (\varepsilon_1 \dot{\lambda}_1, \dot{x}_2) = (-\lambda_1, 2\lambda_1^2 - 1) , \qquad \text{(b)} \quad (\varepsilon_1 \dot{\lambda}_1, \dot{x}_2) = (-\lambda_1, 1) .$$

The fixed points of the fast $\dot{\lambda}_1$ subsystem, lying at $\lambda_1^\$ = 0$ in both (a) and (b), define the sliding modes (see Section 7.1). The dynamics in these modes is in direct contradiction, however, as substituting $\lambda_1 = \lambda_1^\$ = 0$ into the layer system gives sliding dynamics

$$\text{(a)} \quad (\varepsilon_1 \dot{\lambda}_1, \dot{x}_2) = (0, -1) , \qquad \text{(b)} \quad (\varepsilon_1 \dot{\lambda}_1, \dot{x}_2) = (0, +1) ,$$

as illustrated in Figure 9.1. Both provide unique forward time dynamics.

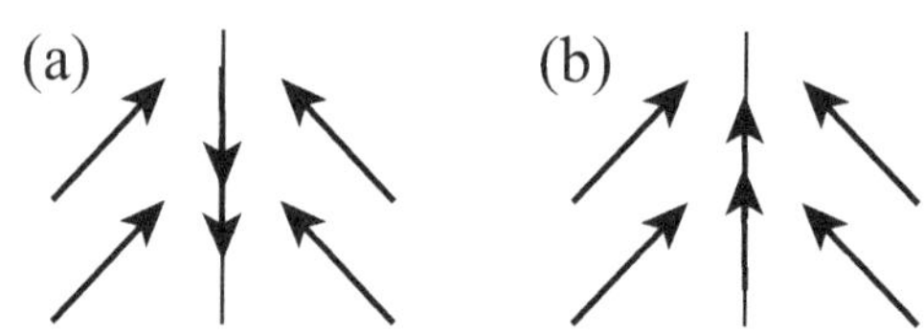

Fig. 9.1 Sketch of Example 9.1. The portrait (a) includes nonlinear switching terms and exhibits sliding downwards, due to hidden terms not evident outside the discontinuity surface. The portrait (b) excludes hidden terms, which results in sliding upwards.

■

This shows how nonlinear dependence of the vector fields on the switching multipliers $\boldsymbol{\lambda} = (\lambda_1, \lambda_2, \ldots)$ can lead to systems with the same behaviour outside the discontinuity surface having distinct, and even directly opposing, dynamics on the surface.

If $(\delta_{t+}\sigma_1)(\delta_{t-}\sigma_1) > 0$ at $\sigma_1 = 0$, both vector fields point the same way through the discontinuity surface, so the flow may cross it. If $\mathbf{h} \equiv 0$, in fact, the flow *will* cross the surface, because $\delta_{t+}\sigma_1$ and $\delta_{t-}\sigma_1$ have the same sign as each other, so the linear interpolation, given by (9.24) as λ_1, varies between ± 1 and cannot pass through zero, and there can be no sliding modes (solutions of (7.8) for $\lambda_1^\$$). If $\mathbf{h}$ is nonzero then this is not the case, and sliding modes may exist and are found using (7.8).

Example 9.2 (Nonlinear Barrier to Crossing). Consider two systems

$$\text{(a)} \quad (\dot{x}_1, \dot{x}_2) = (2\lambda_1^2 - 1, 1 + c\lambda_1) , \qquad \text{(b)} \quad (\dot{x}_1, \dot{x}_2) = (1, 1 + c\lambda_1) ,$$

with $\lambda_1 = \text{sign}(x_1)$, for some constant c. These both appear the same, $(\dot{x}_1, \dot{x}_2) = (1, 1 \pm c)$, for $x_1 \neq 0$. The layer systems are

$$\text{(a)}\quad (\varepsilon_1\dot{\lambda},\dot{x}_2)=(2\lambda_1^2-1,1+c\lambda_1)\;,\qquad \text{(b)}\quad (\dot{x}_1,\dot{x}_2)=(1,1+c\lambda_1)\;.$$

In (a), solving $\dot{\lambda}_1 = 0$ gives sliding modes $\lambda_1^\$ = \pm\sqrt{2}$, while (b) has no solutions, so the dynamics on $x_1 = 0$ collapses to

$$\text{(a)}\quad (\varepsilon_1\dot{\lambda},\dot{x}_2)=(0,1\pm c\sqrt{2})\;,\qquad \text{(b)}\quad \textit{crossing}\;,$$

as illustrated in Figure 9.2. Despite (a) having two sliding modes $\lambda_1^\$ = -\sqrt{2}$ and $\lambda_1^\$ = +\sqrt{2}$, it is easy to see that these are attracting and repelling, respectively, such that the flow sticks to $\lambda_1 = \lambda_1^\$$ and follows sliding dynamics $(\varepsilon_1\dot{\lambda}_1,\dot{x}_2) = (0,1-c\sqrt{2})$.

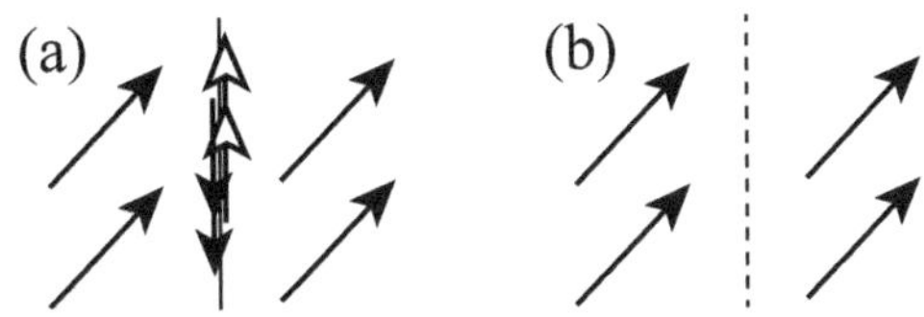

Fig. 9.2 Sketch of Example 9.1, with (a) including nonlinear switching terms and (b) excluding them. The hidden dynamics in (a) gives an attracting upwards sliding solution and repelling downwards sliding solution. Linear switching in (b) instead results in crossing.

■

The disagreement in Example 9.2 appears almost absurd when we set $c = 0$, for which the vector field is just $(1,1)$ on both sides of the discontinuity surface $\mathcal{D}$. This does not mean the systems are continuous, however, as they are defined in a piecewise fashion on disjoint domains $x_1 > 0$ and $x_1 < 0$. Despite having the same value either side of the boundary $x_1 = 0$, there is freedom for hidden terms proportional to the quantity $\lambda_1^2 - 1$ to appear.

In applications we must be aware that the presence or exclusion of hidden terms is a modelling assumption. Later in this chapter, we will see situations where hidden terms are indispensable for more fundamental reasons, either to explain observed behaviour or to achieve structural stability.

We have here an example of multiple sliding modes existing at the same place on the discontinuity surface, something we showed was possible in Section 7.5, corresponding to the sliding manifold $\mathcal{M}$ having multiple branches. This phenomenon opens up a vast array of possibilities for novel sliding dynamics that will feature repeatedly in the forthcoming chapters. To handle this one, further piece of terminology will be useful.

Definition 9.1. An **overlapping** sliding region is one where there exists more than one sliding mode for the same value of $\mathbf{x}$. **Non-overlapping** sliding regions may be called **simple**.

9.2 Hidden Attractors

Anything that can happen in a bounded smooth system can happen inside the switching layer and can do so almost independently of what happens outside the discontinuity surface. This gives rise to unprecedented phenomena associated with a discontinuity, for instance, causing variables, for reasons hidden from view, to enter into bifurcations, chaos, relaxation oscillations, and so on. We illustrate here with a few examples. As well as showcasing some novel dynamics, these suggest how observations of such dynamics at a discontinuity surface could be used to infer the nonlinear model of a system.

9.2.1 A Hidden van der Pol Oscillator

Take a planar system with a single discontinuity,

$$(\dot{x}_1, \dot{x}_2) = \left(\tfrac{1}{10}x_2 + \lambda - 2\lambda^3,\ -\lambda \right) \qquad \text{where} \quad \lambda = \operatorname{sign}(x_1)\,. \tag{9.1}$$

This is deceptively simple for $x_1 \neq 0$, where

$$(\dot{x}_1, \dot{x}_2) = \begin{cases} \left(\tfrac{1}{10}x_2 - 1,\ -1 \right) \text{ if } x_1 > 0\,, \\ \left(\tfrac{1}{10}x_2 + 1,\ +1 \right) \text{ if } x_1 < 0\,. \end{cases} \tag{9.2}$$

The discontinuity surface $x_1 = 0$ is therefore attracting, as shown in Figure 9.3(i).

The switching layer system for (9.1) reveals something equivalent to a van der Pol oscillator (see [213]), illustrated in Figure 9.3(ii) and given by

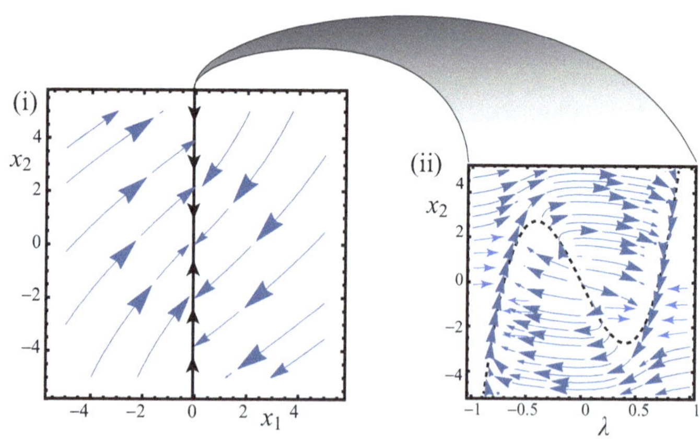

Fig. 9.3 Simulations of (9.1) showing (i) the flow in the (x_1, x_2) plane, (ii) the flow inside $x_1 = 0$ given by (9.3), showing the sliding manifold $\mathcal{M}$ (dotted).

$$(\varepsilon\dot{\lambda}, \dot{x}_2) = \left(\tfrac{1}{10}x_2 + \lambda - 2\lambda^3, \ -\lambda \right) \qquad \text{for } x_1 = 0 \,, \quad \lambda \in (-1, +1) \,, \tag{9.3}$$

The sliding manifold $\mathcal{M}$ is the cubic curve $\frac{1}{10}x_2 = 2\lambda^3 - \lambda$ inside the switching layer and is invariant except at two turning points at $(\lambda, x_2) = \pm(1, -2/3)/\sqrt{6}$. On $|x_2| < 2/3\sqrt{6}$, we have an overlapping sliding region with three sliding modes; outside this we have a simple sliding region. The branches either side of the turning points in $|\lambda| > 1/\sqrt{6}$ are attracting; the branch connecting them is repelling.

As a result the switching parameter λ cycles through relaxation oscillations inside the layer $\lambda \in (-1, +1)$, and these cause x_2 to fluctuate between the folds of the curve $\frac{1}{10}x_2 = 2\lambda^3 - \lambda$. These are both shown in Figure 9.4, where in (i) we plot the graphs of x_2 and λ against time (a brief transient is seen at time $t \lesssim 2$ before settling to the relaxation oscillation), and in (ii) the same trajectory is simulated in the space of x_1, x_2, λ. The oscillations in x_2 are on the natural timescale of the system (t rather than $\tau = t/\varepsilon$), but their origin is hidden inside $x_1 = 0$.

If we were to instead take (9.2) and form a convex combination of the two constituent vector fields, we would obtain a linear switching system

$$(\dot{x}_1, \dot{x}_2) = \left(\tfrac{1}{10}x_2 - \lambda, \ -\lambda \right) \,.$$

To relate this to (9.1), notice that we can write the nonlinear term on the right-hand side of (9.1) as $\lambda - 2\lambda^3 = -\lambda + 2\lambda(1 - \lambda^2)$, the first term on the right-hand side constituting the linear switching part, the latter term constituting the hidden part. Omitting the hidden terms leaves the linear switching ('Filippov') system. The linear system does not exhibit oscillations, and instead has a sliding mode with $\lambda^{\$} = x_2/10$ and dynamics $(\dot{x}_1, \dot{x}_2) = (0, -x_2/10)$, making the point $x_1 = x_2 = 0$ a global attractor.

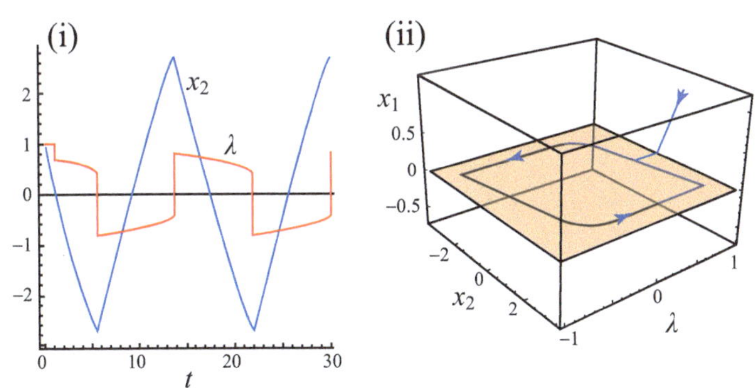

Fig. 9.4 Simulations revealing the hidden dynamics of (9.1): (i) graphs of the variable x_2 and λ, (ii) the corresponding orbit in the space of (x_1, x_2, λ), with the discontinuity surface at $x_1 = 0$.

So we see how nonlinear dependence on the switching multiplier λ can create hidden oscillations inside the discontinuity surface; in this case a hidden van der Pol oscillator that through overlapping sliding modes affects the sliding dynamics (see [213] for the origin of the van der Pol oscillator). These oscillations can also be chaotic if the system has more dimensions or time dependence.

9.2.2 Hidden Duffing Oscillator and Ueda chaos

Take the time-dependent planar system

$$(\dot{x}_1, \dot{x}_2) = \left(x_2 - cx_1,\ -\lambda^3 - bx_2 + a\cos t\right) , \tag{9.4}$$

where $\lambda = \text{sign}(x_1)$, for constants a, b, c. The phase portrait, shown Figure 9.5(i), is that of a fused focus. There are invisible tangencies of the constituent vector fields to the discontinuity surface at $x_2 = 0$. For the parameters in Figure 9.5, this creates focus-like attraction towards $x_1 = x_2 = 0$, obtainable from just the piecewise-defined system outside the discontinuity surface:

$$(\dot{x}_1, \dot{x}_2) = \begin{cases} (x_2 - cx_1,\ -1 - bx_2 + a\cos t) & \text{if } x_1 > 0\ , \\ (x_2 - cx_1,\ +1 - bx_2 + a\cos t) & \text{if } x_1 < 0\ . \end{cases} \tag{9.5}$$

The layer system on $x_1 = 0$ is

$$(\varepsilon\dot{\lambda}, \dot{x}_2) = \left(x_2,\ -\lambda^3 - bx_2 + a\cos t\right) , \qquad \lambda \in (-1, +1)\ . \tag{9.6}$$

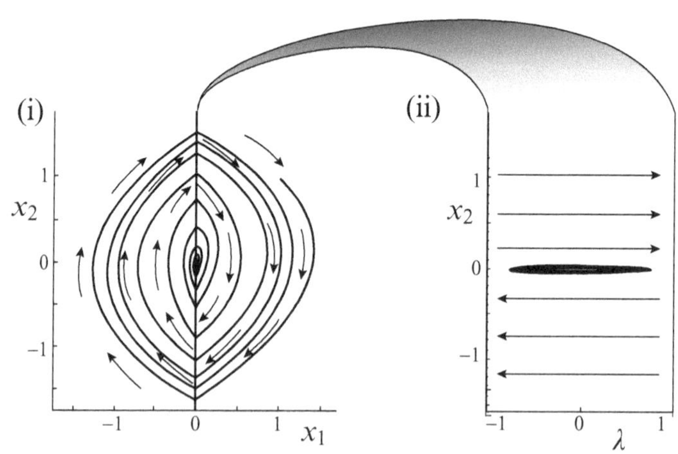

Fig. 9.5 The flow of (i) the fused focus (9.4) and (ii) its switching layer (9.6), shown at a fixed time, for constants $c = 0.1$, $b = 0.05$, and $a = 0.15$. An orbit is simulated over a time t from 0 to 500.

There are not strictly any sliding modes, the condition $\dot{\lambda} = 0$ being met only at $x_1 = x_2 = 0$; therefore by conventional ideas, the dynamics in the discontinuity would seem to be of little interest. Figure 9.5(ii) shows the discontinuity surface $x_1 = 0$ blown up into the switching layer $\lambda \in (-1, +1)$ on $x_1 = 0$ and shows the orbit from (i) continued for later times. The variable x_2 varies only slightly, while λ oscillates in an irregular pattern between values of around $|\lambda| \approx a^{1/3}$.

This oscillation is a hint of something more interesting going on inside the discontinuity. In fact the system (9.6) is a form of Duffing oscillator (see e.g. [92]), and far from being trivial, this can lead to rich dynamics inside the transition layer $\lambda \in (-1, +1)$.

Although the orbit in Figure 9.5(ii) may appear to be insignificant because it is negligible in the state space of (x_1, x_2), the oscillation of the switching multiplier λ could have a noticeable effect on a system in more variables. As an example we can introduce a third variable satisfying some simple equation involving the switch, say

$$\dot{x}_3 = \mu(\lambda - x_3) , \tag{9.7}$$

where μ is a large positive constant. Then x_3 will simply tend towards values ± 1 outside the discontinuity surface, but inside the surface, it will track the oscillatory dynamics of λ.

To visualize the dynamics through some simulations, it is helpful to take a nonzero value of ε and consider what happens as $\varepsilon \to 0$. We will take μ large enough so that x_3 tracks λ closely enough for them to be almost indistinguishable. Figure 9.6(a) then shows a plot of $x_3(t)$. After transients, $x_3(t)$ settles onto an attractor with complex oscillatory or chaotic dynamics for different parameters. For some parameters this is Ueda's chaotic attractor [210] as expected from a Duffing oscillator. The curves (i) and (ii) show $x_3(t) \approx \lambda(t)$ for $\varepsilon = 10^{-2}$ and $\varepsilon = 10^{-5}$, respectively. We see complex smaller amplitude oscillations around a regular oscillation with amplitude $\Delta x_3 = a^{1/3}$ (and period 2π of the cosine in (9.4)), and as ε shrinks, the smaller amplitude oscillations shrink with it.

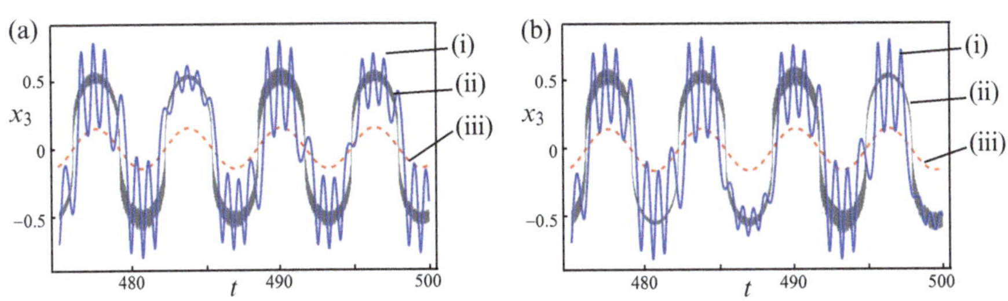

Fig. 9.6 Simulations of hidden dynamics, plotting $z(t)$ (or $\lambda(t)$ for: (i) the nonlinear system with $\varepsilon = 10^{-2}$, (ii) the nonlinear system with $\varepsilon = 10^{-5}$, and (i) the linear system with $\varepsilon = 10^{-2}$. Graph (a) is a simulation of the switching layer system (9.6), and graph (b) is a simulation of the full system (9.4) smoothed out using $\lambda = \varphi_\varepsilon(x_1)$ given in the text. All simulations are with $c = 0.1$, $b = 0.05$, and $a = 0.15$.

Let us contrast this with what would be seen if we ignored hidden terms. For $x_1 \neq 0$, (9.4) is indistinguishable from the linear switching system

$$(\dot{x}_1, \dot{x}_2) = (x_2 - cx_1,\ -\lambda - bx_2 + a\cos t)\ , \tag{9.8}$$

where $\lambda = \text{sign}(x_1)$, as both (9.4) and (9.8) satisfy (9.5). The curve (iii) in Figure 9.6(a) shows the graph of $x_3(t)$ for the linear system (9.8), which exhibits always a steady oscillation (which does not vary appreciably with ε), but with an amplitude a, rather than the $a^{1/3}$ of the nonlinear system.

Preparing ourselves for later discussion in Chapter 12 of what kinds of behaviour survive if we smooth out the discontinuity, let us simulate this system again by smoothing the discontinuity in (9.4) and (9.8), replacing λ in those expressions with a continuous function:

$$\varphi_\varepsilon(x_1) = \begin{cases} \text{sign}(x_1) & \text{if } |x_1| \geq \varepsilon\ , \\ x_1/\varepsilon & \text{if } |x_1| < \varepsilon\ . \end{cases} \tag{9.9}$$

A numerical simulation then yields the corresponding graphs Figure 9.6(b), showing dynamics similar to Figure 9.6(a) with only minor quantitative differences in the precise form of smaller oscillations. Thus the differences between the linear and nonlinear models survive under smoothing. Simulations using alternative smoothing functions such as $\varphi_\varepsilon(x_1) = \frac{2}{\pi}\arctan(x_1/\varepsilon)$ or $\varphi_\varepsilon(x_1) = \tanh(x_1/\varepsilon)$, or a smooth non-analytic function equal to exactly $\text{sign}(x_1)$ outside $|x_1| > \varepsilon$, yield results with no significant difference to those using (9.9).

9.2.3 Cross-Talk Oscillations

Hidden oscillations can arise not only from nonlinear dependence on a single switching multiplier but through multilinear dependence on two or more switching multipliers. A rather clear example was noted by Guglielmi and Hairer in [93] (although studied there by smoothing the system rather than resolving the switching layer directly). They presented a planar piecewise-constant system

$$(\dot{x}_1, \dot{x}_2) = \left(\ \lambda_2 - \lambda_1 + \lambda_1\lambda_2,\ 3(\tfrac{1}{2}\lambda_2 - 2\lambda_1 + \lambda_1\lambda_2)\ \right)\ , \tag{9.10}$$

with $\lambda_1 = \text{sign}(x_1)$, $\lambda_2 = \text{sign}(x_2)$. On the discontinuity submanifolds $x_2 = 0$ for $x_1 \gtrless 0$, the switching layer system is

$$(\dot{x}_1, \varepsilon_2\dot{\lambda}_2) = \left(\ \lambda_2 \pm 1 \pm \lambda_2,\ 3(\tfrac{1}{2}\lambda_2 \mp 2 \pm \lambda_2)\right)\ ,$$

which has no sliding modes and thus gives crossing through $x_2 = 0$. On the discontinuity submanifolds $x_1 = 0$ for $x_2 \gtrless 0$, the switching layer system is

$$(\varepsilon_1\dot{\lambda}_1, \dot{x}_2) = \left(\pm 1 - \lambda_1 - \pm\lambda_1,\ 3(\pm\tfrac{1}{2} - 2\lambda_1 \pm \lambda_1)\right) ,$$

in which the $x_2 > 0$ system gives crossing, while the $x_2 < 0$ system has sliding modes with $\lambda_1^{\$} = -1/2$.

The result is that the origin is a simple attractor in the (x_1, x_2) plane. The flow spirals around, crossing through the discontinuity submanifolds $x_1 = 0$ and $x_2 = 0$ until they reach the half-line $\{x_1 = 0 > x_2\}$, upon which they slide in towards the intersection.

The switching layer system of the intersection at $x_1 = x_2 = 0$ is rather more interesting,

$$(\varepsilon_1\dot{\lambda}_1, \varepsilon_2\dot{\lambda}_2) = \left(\ \lambda_2 - \lambda_1 + \lambda_1\lambda_2,\ 3(\tfrac{1}{2}\lambda_2 - 2\lambda_1 + \lambda_1\lambda_2)\ \right) , \tag{9.11}$$

for infinitesimal positive ε_1 and ε_2. The phase portraits containing all of these elements, with the discontinuity submanifolds blown up, are shown in Figure 9.7. They differ qualitatively for different values of the ratio $\varepsilon_2/\varepsilon_1$, showing a focal attractor in (i), a focal repeller in (ii), and a nodal repeller in (iii). In the repelling cases, inspection of the flow reveals the existence of a limit cycle as illustrated, and we will confirm its existence with simulations below.

The switching layer system has an equilibrium at $(\lambda_1, \lambda_2) = (0, 0)$, where the Jacobian (7.34b) evaluates in layer coordinates $(\xi_1, \xi_2) = (\varepsilon_1\lambda_1, \varepsilon_2\lambda_2)$ as

$$\begin{pmatrix} \frac{\partial\dot{\xi}_1}{\partial\xi_1} & \frac{\partial\dot{\xi}_1}{\partial\xi_2} \\ \frac{\partial\dot{\xi}_2}{\partial\xi_1} & \frac{\partial\dot{\xi}_2}{\partial\xi_2} \end{pmatrix} = \begin{pmatrix} -\varepsilon_1^{-1} & \varepsilon_2^{-1} \\ -6\varepsilon_1^{-1} & \frac{3}{2}\varepsilon_2^{-1} \end{pmatrix}$$

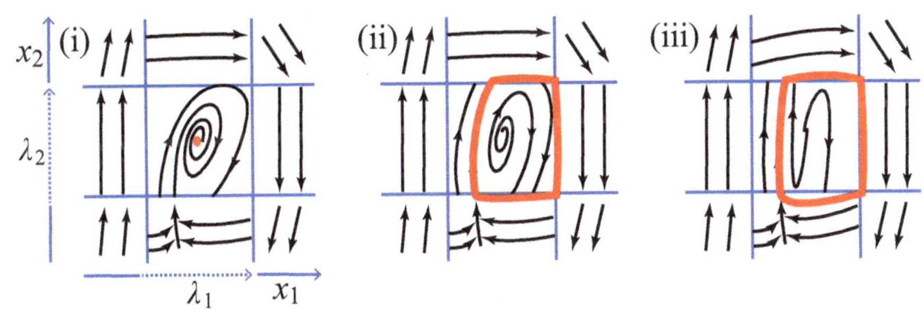

Fig. 9.7 Sketch of the piecewise-smooth system, including the switching layers, where $\{\varepsilon_1, \varepsilon_2\}$ are (i) $\{0.02.0.04\}$, (ii) $\{0.02, 0.02\}$, and (iii) $\{0.08, 0.02\}$.

with determinant $9/2\varepsilon_1\varepsilon_2 > 0$ implying that the equilibrium is a focus or node, and trace $\frac{3}{2}\varepsilon_2^{-1} - \varepsilon_1^{-1}$ implying the origin is attracting for $\varepsilon_2/\varepsilon_1 > 3/2$, repelling otherwise, with a Hopf bifurcation when $\varepsilon_2/\varepsilon_1 = 3/2$. The Hopf bifurcation is supercritical. As $\varepsilon_2/\varepsilon_1$ decreases through $3/2$, an attracting limit cycle grows out from the equilibrium.

This cycle cannot grow outside of the switching layer (since no change occurs outside the discontinuity surface as $\varepsilon_2/\varepsilon_1$ changes), so it is constrained to lie inside or on the edges of the layer $\lambda_{1,2} \in (-1, +1)$, as seen in Figure 9.7(ii)-(iii). This means the limit cycle can inhabit only the intersection where the variables x_1 and x_2 are fixed at zero, so it may seem that these oscillations are of little practical consequence, but we can easily illustrate two simple ways that they can significantly impact a system.

Similar to the previous section, the oscillations can reveal themselves in additional variables whose dynamics involves the switching parameters λ_1 or λ_2 in some way, for example

$$\dot{x}_3 = \mu(\lambda_1 - x_3) \ . \tag{9.12}$$

If the positive constant μ is large enough, then x_3 tracks the value of λ_1 very closely, and the hidden oscillations become visible. Figure 9.8 (bold curve) shows a simulation of the values of $x_i(t)$ for $i = 1, 2, 3$, showing x_3 collapsing to a steady state in (i) but forming oscillations in (ii) and (iii). These oscillations are infinitely fast (they occur at the timescale $t/\varepsilon_{1,2}$ where $\varepsilon_{1,2}$ are infinitesimal), so we represent them as a filled band of values (the shaded rectangle) in Figure 9.8.

What this means in practice depends on the precise form of the coupling, but in the example (9.12), we can expect $x_3(t)$ to average over the infinitely fast oscillations of $\lambda_1(t)$. Some insight is given by considering ε_1 and ε_2 to be nonzero. To simulate, as we have done before, it is easiest to then approximate the discontinuous system by replacing each $\lambda_j = \text{sign}(x_j)$ with a steep sigmoid function, say $\lambda_j = \tanh(x_j/\varepsilon_j)$. The 'hidden' oscillations are immediately brought to life, and we see their significance for numerical simulation. The limit cycles become cycles of order ε_j in their respective directions (which is weaker than confining them strictly to the regions $|x_j| < \varepsilon_j$). The simulations in Figure 9.9 show, in fact, the limit cycles reaching as far as $|x_2| = 4\varepsilon_j$, creating significant and observable oscillations. In (i) a trajectory crosses through $x_2 = 0$ before hitting $x_1 = 0$, sliding towards the intersection

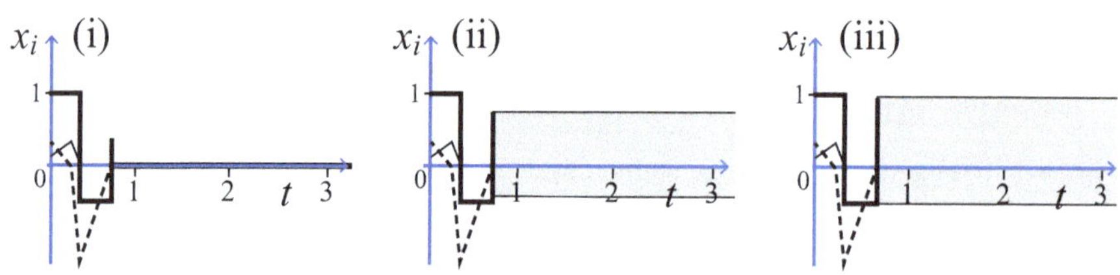

Fig. 9.8 Hidden oscillations become visible when coupled to a third variable; (i)-(iii) correspond to Figure 9.7. The thin curve shows $x_1(t)$ settling to zero at $t \approx 0.4$, and the dashed curve shows $x_2(t)$ settling to zero at $t \approx 0.8$. The bold curve is $x_3(t)$, which at $t \approx 0.8$ starts following the infinitely fast oscillations of $\lambda_1(t)$, indicated by the shaded region.

and then reaching a fixed point as implied by the switching layer dynamics above. In (ii) and (iii), the intersection is repelling and creates an oscillation, which is larger in (iii) when the switching layer is nodally, rather than focally, repelling.

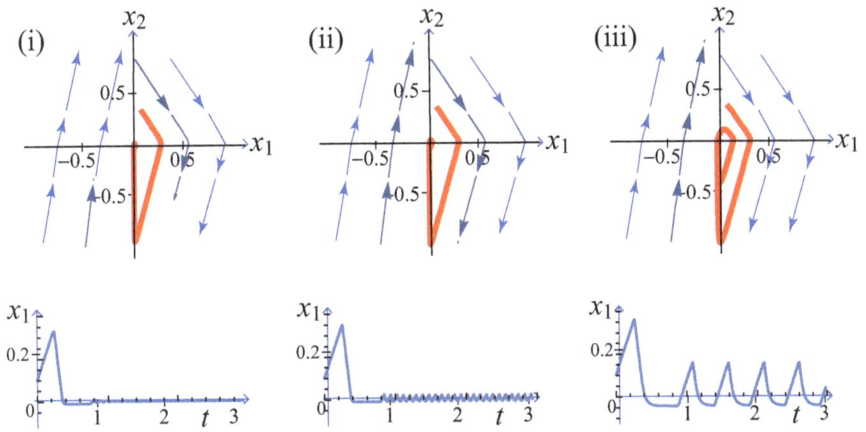

Fig. 9.9 Smoothing of (9.10) makes hidden oscillations observable. Simulations are shown in the (x_1, x_2) plane (top) and plotting $x_1(t)$ (bottom) corresponding to Figure 9.7.

The cross-talk oscillation is an example of a distributed sliding mode (Definition 7.2).

9.2.4 Hidden Lorenz Attractor

A situation in which the switching multipliers enter into hidden chaos was conceived of in [159] in the form

$$\begin{aligned} \dot{x}_1 &= 5(\lambda_2 - \lambda_1) - 75x_1 \; , \\ \dot{x}_2 &= -\lambda_1 - 15\lambda_1\lambda_3 - \tfrac{1}{2}\lambda_2 - 75x_2 \; , \\ \dot{x}_3 &= 15\lambda_1\lambda_2 - \tfrac{4}{3} - \tfrac{4}{3}\lambda_3 - 75x_3 \; , \end{aligned} \tag{9.13}$$

where $\lambda_j = \text{sign}(x_j)$ for $j = 1, 2, 3$. This is inspired by gene regulatory networks [159], as a model of gene concentrations $y_j = x_j + \frac{1}{4}$, where the multipliers λ_j arise in the limits of "on/off" Hill functions $Z_j = y_j^k/(y_j^k + \theta_j^k) \xrightarrow{k\to\infty} \frac{1}{2} + \frac{1}{2}\lambda_j$ (see Table 1.1 or [103]), with threshold values given here by $\theta_j = \frac{1}{4}$.

Similar to our previous examples in this section, the dynamics outside the discontinuity surface appears trivial, with the triple intersection point $x_1 = x_2 = x_3 = 0$ in this case being a simple nodal attractor in (x_1, x_2, x_3)

space, illustrated in Figure 9.10(left). We can analyse the system in terms of x_i or y_i and λ_j or Z_j, we take x_i and λ_j.

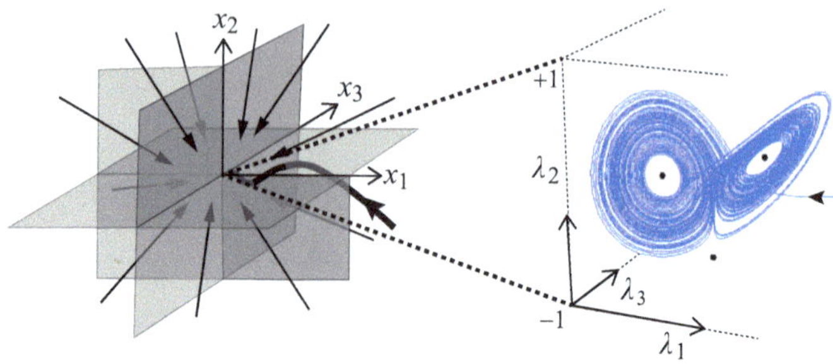

Fig. 9.10 Global attractor in x_1, x_2, x_3 space and a Lorenz attractor in the switching layer inside $(x_1, x_2, x_3) = (\theta_1, \theta_2, \theta_3)$.

The dynamics on the discontinuity submanifolds is not particularly interesting, so proceeding directly to their intersection applying (7.2) with $r = 3$ to (9.13), we obtain the switching layer dynamics

$$\begin{aligned} \varepsilon_1 \dot{\lambda}_1 &= 5(\lambda_2 - \lambda_1)\,, \\ \varepsilon_2 \dot{\lambda}_2 &= -\lambda_1 - 15\lambda_1\lambda_3 - \tfrac{1}{2}\lambda_2\,, \\ \varepsilon_3 \dot{\lambda}_3 &= 15\lambda_1\lambda_2 - \tfrac{4}{3} - \tfrac{4}{3}\lambda_3\,, \end{aligned} \tag{9.14}$$

on $x_1 = x_2 = x_3 = 0$. The authors of [159] assume that the three small parameters are the same, $\varepsilon_1 = \varepsilon_2 = \varepsilon_3$. Then the layer system has a Lorenz attractor inside $(\lambda_1, \lambda_2, \lambda_3) \in (-1, +1)^3$. As the previous example, this is a distributed sliding mode (see Definition 7.2). So while each x_j is attracted to the intersection point and then remains fixed, the switching multipliers λ_j enter into chaos.

The study in [159] was actually of the singularly perturbed ($\varepsilon_j > 0$) system using the smooth Hill function, but by comparing the reader will see that the results are consistent.

Again, the variation in the switching parameters (oscillatory in the cross-talk example, chaotic here) is not directly observable through the variables $x_{1,2,3}$, which are attracted to the origin and then remain there. Again, though, they can reveal themselves in additional variables, for example, adding

$$\dot{x}_4 = \mu(\lambda_1 - x_4)\,, \tag{9.15}$$

for a sufficiently large positive constant μ. The variable x_4 will track the chaotic value λ_1, while the variables $x_1 = x_2 = x_3 = 0$ remain fixed. The solution for x_4 coupled with (9.13) is simulated in Figure 9.11 and exhibits chaotic spiking behaviour, while x_1, x_2, x_3, remain at zero.

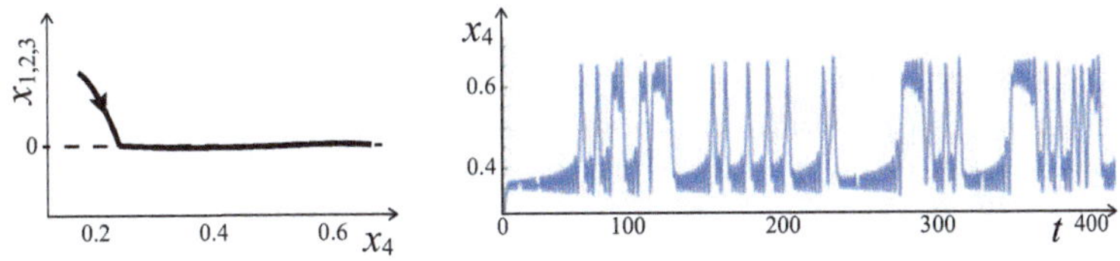

Fig. 9.11 Hidden chaos made visible the dynamics of the system (9.13) with (9.15), showing a representation of the dynamics in (x_1, x_2, x_3, x_4) space where the trajectory collapses onto the line $\{x_1 = x_2 = x_3 = 0,\ x_4 \in \mathbb{R}\}$ and a simulation of the chaotic solution $x_4(t)$.

The examples we have seen in Section 9.2.1 to Section 9.2.3 demonstrate that hidden attractors of switching can take almost any of the interesting forms known in nonlinear dynamics, induced by any number of switches. We have considered only the most simple such oscillations, namely, those constrained inside the layer, whereas more complex scenarios could involve a mixture of the dynamics inside the layer with that outside the discontinuity surface.

Such examples are evidently easy to construct, so our aim here is merely to highlight the ease with which such phenomena arise, and the visible dynamics by which hidden oscillation may be recognized, hinting at how they may appear in observations of physical data or in spurious behaviour of simulations. Given the ease with which they appear, these and countless other examples may have a significant role in physical and biological switching processes, opening an exciting world of new possibilities for modelling applications.

9.3 Hidden Bifurcations

When bifurcations occur due to nonlinear dependence on $\boldsymbol{\lambda}$, then structural changes occur inside the switching layer that are not directly observable outside the discontinuity surface. These changes can nevertheless lead to qualitative changes in the sliding or crossing dynamics that *are* observable, affecting both local and global dynamics.

9.3.1 Cross or Not at an Intersection

We saw in Example 9.2 that the question of whether or not a solution will cross the discontinuity surface cannot be determined purely from the vector fields outside it. Hidden dynamics may cause solutions to instead stick to the discontinuity surface and slide along it. The previous example we saw involved a single switch with nonlinear dependence on the switching multiplier.

Examples of this have been observed previously in models of gene regulatory networks, due to multilinear dependence on two or more switching multipliers. For certain parameters an attracting nonlinear (specifically multilinear) sliding mode exists. At the discontinuity surface we must then analyse the sliding mode to determine whether sliding or crossing will occur, according to whether it is a regular, weak, or unreachable mode (Definition 7.2). At some parameter value, the mode bifurcates. If it annihilates through collision with a coexisting repelling mode, such that for certain parameters no attracting sliding modes exist, then crossing will be permitted.

Consider the system

$$\begin{pmatrix} \dot{x}_1 \\ \dot{x}_2 \end{pmatrix} = \begin{pmatrix} \frac{1}{2}(1-\lambda_1\lambda_2) - \gamma_1(x_1+\theta_1) \\ \frac{1}{4}(3-\lambda_1-\lambda_2-\lambda_1\lambda_2) - \gamma_2(x_2+\theta_2) \end{pmatrix} \tag{9.16}$$

where $\lambda_j = \text{sign}(x_j)$ for some constants $\theta_1, \theta_2, \gamma_1, \gamma_2$. This example is taken from a model of protein product concentrations $y_j = x_j + \theta_j$ in a two gene regulatory system [175, 57]. The phase portrait of (9.16) is sketched in the main picture in Figure 9.12.

In the quadrants $x_1x_2 \gtrless 0$, the dynamics is attracted onto the discontinuity submanifolds $x_1 = 0$ and $x_2 = 0$. We find the dynamics on $x_1 = 0$ and $x_2 = 0$ by applying (7.2) with $r = 1$ to each of the switching submanifolds $x_1 = 0$ and $x_2 = 0$ independently, giving switching layer systems

$$\begin{aligned} \begin{pmatrix} \varepsilon_1\dot{\lambda}_1 \\ \dot{x}_2 \end{pmatrix} &= \begin{pmatrix} \frac{1}{2}(1-\lambda_1\lambda_2) - \gamma_1\theta_1 \\ \frac{1}{4}(3-\lambda_1-\lambda_2-\lambda_1\lambda_2) - \gamma_2(x_2+\theta_2) \end{pmatrix} \quad \text{on } x_1 = 0\,, \\ \begin{pmatrix} \dot{x}_1 \\ \varepsilon_2\dot{\lambda}_2 \end{pmatrix} &= \begin{pmatrix} \frac{1}{2}(1-\lambda_1\lambda_2) - \gamma_1(x_1+\theta_1) \\ \frac{1}{4}(3-\lambda_1-\lambda_2-\lambda_1\lambda_2) - \gamma_2\theta_2 \end{pmatrix} \quad \text{on } x_2 = 0\,. \end{aligned} \tag{9.17}$$

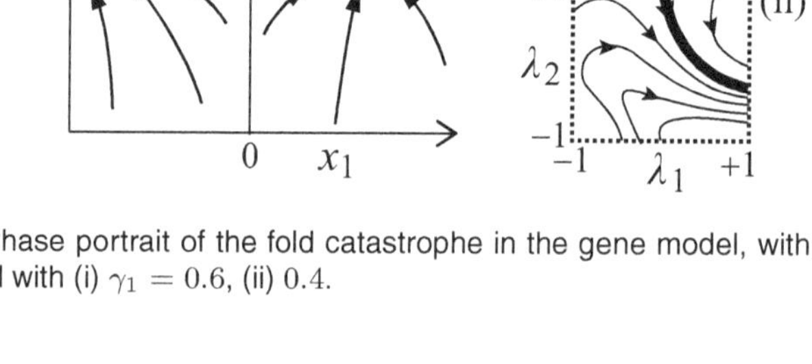

Fig. 9.12 Phase portrait of the fold catastrophe in the gene model, with $\theta_1 = \theta_2 = 1$, $\gamma_2 = 0.9$, and with (i) $\gamma_1 = 0.6$, (ii) 0.4.

By looking for fixed points of the fast $\dot{\lambda}_1$ and $\dot{\lambda}_2$ subsystems, we can use these to show that the flow crosses the discontinuity submanifolds on $x_2 = 0$ with $x_1 < 0$ and slides on $x_1 = 0$ towards $x_2 = 0$. On $x_2 = 0$ with small $x_1 > 0$, there is a sliding flow away from the origin. Thus there are sliding flows into the intersection point $x_1 = x_2 = 0$ along $x_1 = 0$ and a sliding flow out of the intersection along $x_2 = 0 < x_1$. Are these two connected: does the flow round the corner from one sliding branch to another?

To answer this we analyse the switching layer inside the intersection point, by applying (7.2) with $r = 2$, giving

$$\begin{pmatrix} \varepsilon_1 \dot{\lambda}_1 \\ \varepsilon_2 \dot{\lambda}_2 \end{pmatrix} = \begin{pmatrix} \frac{1}{2}(1 - \lambda_1 \lambda_2) - \gamma_1 \theta_1 \\ \frac{1}{4}(3 - \lambda_1 - \lambda_2 - \lambda_1 \lambda_2) - \gamma_2 \theta_2 \end{pmatrix} \tag{9.18}$$

on $x_1 = x_2 = 0$. This is sketched in the two portraits (i-ii) in Figure 9.12. It has two equilibria, a focus and a saddle at

$$\{\lambda_1^{\pm}, \lambda_2^{\pm}\} = (1 + \gamma_1 \theta_1 - 2\gamma_2 \theta_2)\{1, 1\} \pm \sqrt{d}\{-1, 1\}$$

where $d = (\frac{1}{2}\gamma_1 \theta_1 + \gamma_2 \theta_2)^2 - \gamma_1 \theta_1 - \gamma_2 \theta_2$, which exist only for $d > 0$, disappearing in a saddlenode bifurcation as $\{\lambda_1^{\pm}, \lambda_2^{\pm}\}$ become complex.

The outcome is that for $d > 0$ the variables $x_{1,2}$ become fixed at the origin, while for $d < 0$, they evolve through the intersection point $(0, 0)$ and continue along $x_2 = 0$ with increasing x_1, as shown in the main part of Figure 9.12.

This example highlights another important issue when multiple switches occur on different surfaces $\sigma_1 = 0$, $\sigma_2 = 0$, ..., $\sigma_r = 0$, and those surfaces intersect. There may be numerous vector fields pointing into *and* out of an intersection simultaneously, and whether or not sliding occurs is less obvious than at a single discontinuity surface. The piecewise-smooth vector fields outside the discontinuity surface can be deceptive, and analysis of the switching layer is vital.

9.4 The Illusion of Noise

We have tried in this chapter to highlight novel kinds of dynamics that are induced by discontinuities and are not seen in differentiable systems. While hidden dynamics holds fascinating possibilities, perhaps the most profound discontinuity-induced phenomenon, particularly with a view towards applications, is one that we have barely begun to explore: when discontinuity mimics noise by giving the deceitful appearance of randomness.

In certain situations, seemingly almost trivial local dynamics can give rise to behaviour that would easily be misinterpreted as noise. When situated in a complex environment, one would have no idea that these features have a simple geometrical origin.

To our knowledge at present, there are several sources of such discontinuity-induced *illusions of noise*: jitter, cascades, and time dependence.

9.4.1 Jitter

If the sliding manifold $\mathcal{M}$ undergoes a bifurcation, the sliding dynamics can change abruptly. We have already seen an example in Section 9.2.1, where the bifurcation points between different branches of $\mathcal{M}$ allowed a sliding mode to jump between different attracting branches, ultimately in that case forming a relaxation oscillation.

Imagine the same mechanism, but with multiple bifurcation points along $\mathcal{M}$ being traversed as a sliding mode evolves along the discontinuity surface. The sliding dynamics would undergo a sequence of abrupt jumps in its vector field $\mathbf{f}(\mathbf{x}; \boldsymbol{\lambda}_i^\$)$, between different branches $\boldsymbol{\lambda}_i^\$$ of the function $\boldsymbol{\lambda}^\$$ on different attracting branches of $\mathcal{M}$, for some $i = 1, 2, \ldots$. When numerous jumps occur, and result in irregular abrupt jumps in the sliding vector field, they are said to induce **jitter** in the sliding motion.

Example 9.3 (Jitter for One Switch). A simple illustration of jitter with one switch is given by the planar system

$$\begin{pmatrix} \dot{x}_1 \\ \dot{x}_2 \end{pmatrix} = \begin{pmatrix} x_2 - 8\lambda + \sin(10\pi\lambda) \\ 1 + \frac{4}{5}\cos(4\pi\lambda) \end{pmatrix}, \qquad \lambda = \operatorname{sign}(x_1) , \tag{9.19}$$

shown in Figure 9.13. The switching layer system is

$$\begin{pmatrix} \varepsilon\dot{\lambda} \\ \dot{x}_2 \end{pmatrix} = \begin{pmatrix} x_2 - 8\lambda + \sin(10\pi\lambda) \\ 1 + \frac{4}{5}\cos(4\pi\lambda) \end{pmatrix}, \qquad \varepsilon \to 0 . \tag{9.20}$$

As depicted in Figure 9.13, the sliding manifold $\mathcal{M}$ is the set where

$$x_2 - 8\lambda + \sin(10\pi\lambda) = 0 \qquad \text{for} \qquad \lambda \in (-1, +1) ,$$

which has turning points (making up disjoint branches of a non-hyperbolic set $\mathcal{L}$) between attracting and repelling branches, at every λ such that $\cos(10\pi\lambda) = 4/5\pi$. The sliding modes are therefore the solutions of $0 = x_2 - 8\lambda^\$ + \sin(10\pi\lambda^\$)$, with sliding dynamics $\dot{x}_2 = 1 + \frac{4}{5}\cos(4\pi\lambda^\$)$, for $\lambda^\$ \in (-1, +1)$. At each turning point, a jump occurs to the next attracting branch of $\mathcal{M}$, as shown in Figure 9.13(right).

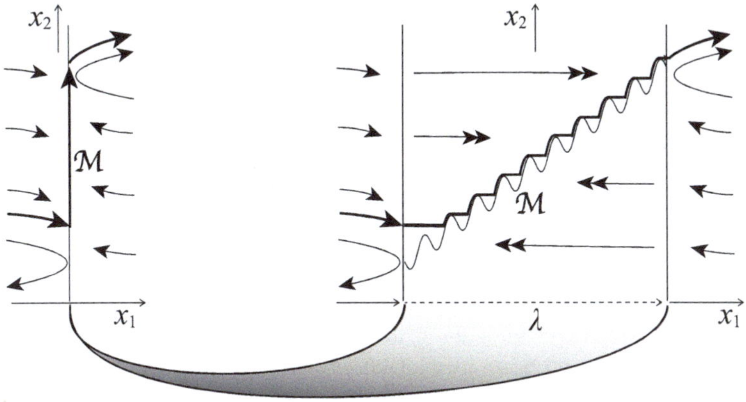

Fig. 9.13 Jitter in sliding motion, due to hidden jumps between branches of $\mathcal{M}$. The piecewise-smooth flow is sketched along with the layer system.

Because the $\dot{x}_2$ component depends on λ, this incurs an accompanying jump in $\dot{x}_2$ simulated in Figure 9.14(left). Thus the sliding solution evolves in bursts or 'jitters' along the discontinuity surface, its vector field undergoing repeated jumps that lead to an irregular trajectory, simulated in Figure 9.14(right).

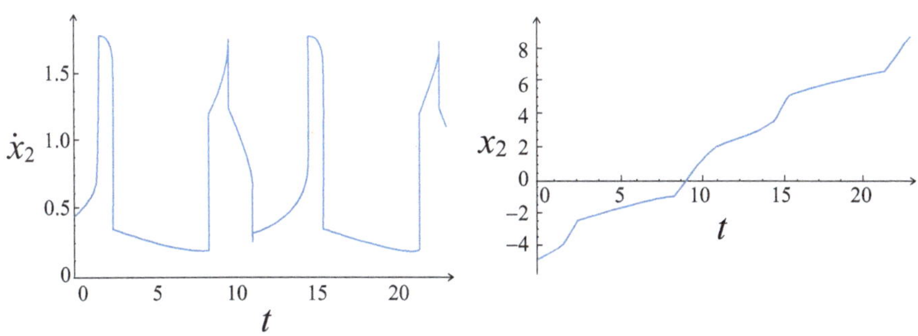

Fig. 9.14 Simulations of the effect of jitter on dynamics are shown by the vector field component $\dot{x}_2(t)$ (plotted left), which shows irregular jumps in sliding speed along the discontinuity surface, which result in jitter the trajectory $x_2(t)$ (plotted right).

■

9.4.2 Slip Cascades

In systems with many switches, complex behaviour arises from the transition between sliding and non-sliding behaviour, via the kind of high codimension tangencies described by Theorem 8.2. Consider the oscillator system

$$\ddot{x}_i + M^{\alpha}_{ij}\dot{x}_j + M^{\beta}_{ij}x_j + \mu\lambda_i = 0\ , \qquad \lambda_i = \text{sign}\,(\dot{x}_i - v) \tag{9.21}$$

for $i = 1, \ldots, n$, where M^{α} and M^{β} are $n \times n$ square matrices, and we sum over the index $j = 1, \ldots, n$. The matrix M^{α} represents dampings and is diagonal; the matrix M^{β} represents couplings between states x_j and has a diagonal part (of self-couplings) and an antisymmetric part (of relations between states). A force $\mu\lambda_i$ resists motion relative to some reference rate $\dot{x}_i = v$.

Let $y_i = \dot{x}_i$ to obtain a $2n$-dimensional set of first-order ordinary differential equations in (x_i, y_i), $i = 1, \ldots, n$. Each oscillator is associated with one of n switching thresholds $\dot{x}_i = v$. On each threshold there is typically a sliding region, with two boundaries (where $\lambda_i = +1$ and $\lambda_i = -1$ systems are tangent to $y_i = 0$), at which the oscillators can exit from sliding. If r of the n oscillators are sliding, the dynamics occupies a $2n - r$ dimensional phase space, upon which k sliding boundaries may intersect if $n + k \leq 2n - r$, and there will be up to 2^k such points. Thus there exist up to 2^k points where up to $k \leq n$ of $r \leq k$ oscillators may exit from sliding simultaneously, nearby which cascades of up to 2^k oscillators may be seen exiting from the switching thresholds, one after another in rapid succession.

The central feature of interest is the codimension r of the sliding motion at any instant, giving the number of oscillators stuck to the switching threshold. Exit points are then easily observed as incremental decreases in the sliding codimension. If cascades of exit points occur as expected, they will be seen as cascades of decreasing sliding codimension r, or *slip cascades*.

The following plots are taken from a simulation of the system (9.21) with $v = -0.2$, with all $\mu_i = 1$, and with initial conditions $x_i = -0.7$, $\dot{x}_i = -0.4$, for all oscillators. For simplicity we approximate each sign function $\mu_i(u)$ by a sigmoid $\mu(u) = \tanh(u/\epsilon)$ with $\epsilon = 0.03$, permitting us to simulate using a standard ODE package (in this case the Mathematica routine NDSolve). The damping constants are taken as random values in the range $M^{\alpha}_{ij} \in [\alpha, 2\alpha]$ with $\alpha = 3.5$, the spring constants as random values in the range $M^{\beta}_{ij} \in [0, \beta]$ with $\beta = 0.2$. These various details have little impact on the qualitative outcomes of the simulations, the key parameters which provide the behaviour below being v, α, and β.

Figure 9.15(i) shows the sliding codimension for a system of 20 oscillators. Four oscillators enter sliding almost immediately, followed by a succession of exit points through which all cease sliding at around $t \approx 24$, and after a short return to sliding all oscillators eventually exit. The behaviour becomes more interesting as we increase the system's size. For 100 oscillators, in Figure 9.15(ii), numerous cascades of exit points, and countering returns to sliding, occur over a time period of around 100 or so increments (but after $t \gtrsim 150$ all oscillators are eventually found to leave sliding). For 200 oscillators, in Figure 9.15(iii), these complex stick-slip transitions become self-sustaining, with cascades of exit points and collapse back to sliding mediating each other over long times.

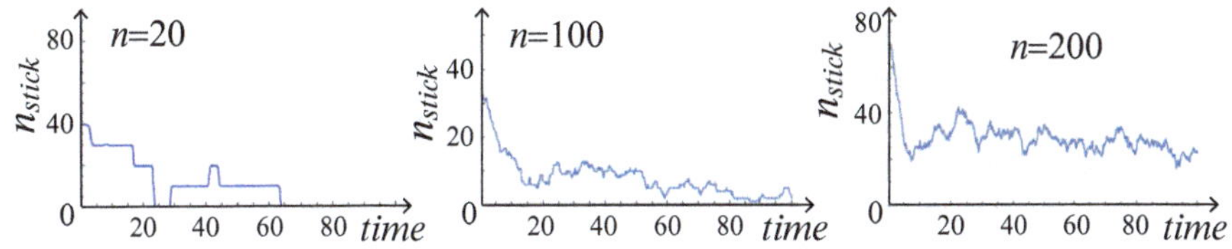

Fig. 9.15 Plots of the sliding codimension r for systems of n oscillators.

Let us explore this long-term dynamics a little further. Again we fix all $R_i = 1$ with $\epsilon = 0.03$ for the simulation. Now we shall perform simulations on a system of 200 oscillators, with different random matrices of damping and coupling constants M^α and M^β. We simulate these for a significantly longer period of time and observe in most cases (as in the three shown) that the complex stick-slip cascades persist over time in a self-sustaining manner, though occasional instances of constants can be found that collapse to $r = 0$. The plots of sliding codimension for three sample cases are shown in Figure 9.16(i).

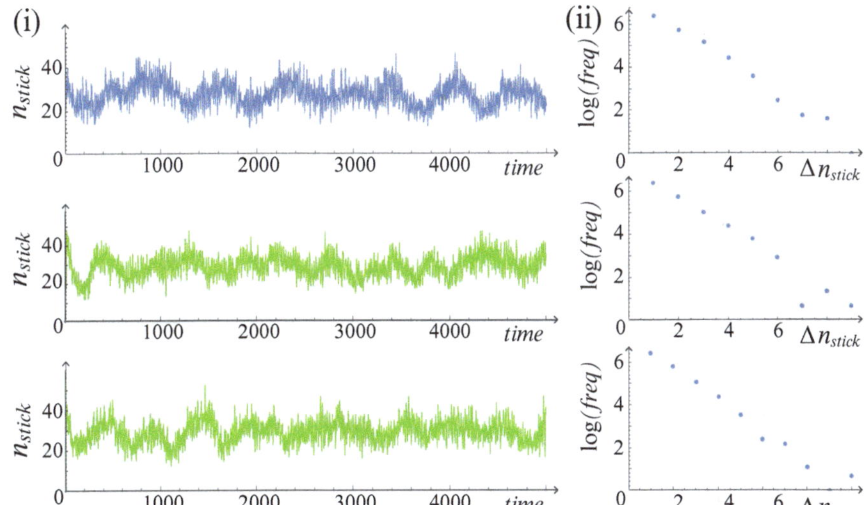

Fig. 9.16 Plots of the sliding codimension r for systems of 200 oscillators for three different sets of coupling and damping coefficients in (i), and the frequency of stick-slip cascades of size Δr in (ii) shows a logarithmic pattern for each system in (i).

With so many oscillators and such long timescales, further insight is provided by considering the statistics of these self-sustaining stick-slip events. We plot in Figure 9.16(ii) the size of cascades Δr (defined as the number of successive time increments in which the sliding codimension decreases), against the frequency of events of each size, showing that the frequency ν

of cascades of size Δr fits a relation $\nu \propto e^{-\gamma\Delta r}$ with $\gamma \approx 0.74$, that is, the expected exponential decay of frequency for events of greater size.

We expect a system to be very sensitive to perturbation around high codimension exit points, due to the different exit routes available in close proximity, and such behaviour should therefore be sensitive to modelling assumptions. In Figure 9.17 we investigate the system's robustness, first to perturbations of the switching model and then to perturbations of the parameters. In Figure 9.17(i)-(ii), each switching function $\mu_i(u)$ is replaced by a sigmoid $\mu(u) = \tanh(u/R_i\epsilon)$, each with a different stiffness $R_i\epsilon$ where the R_i's are a random numbers between 0 and 1. The result of four such simulations for 20 and 200 oscillators is shown. As expected from the sensitivity of high codimension exit points to initial conditions, the short-time dynamics is sensitive to these small changes in the modelling stiffness. Surprisingly, however, the long-time behaviour is largely unaffected, even exhibiting similar large-scale structures in the sustained oscillations of 200 oscillators. That is, we observe short-term sensitivity but long-term robustness, despite the complexity of the long-term dynamics.

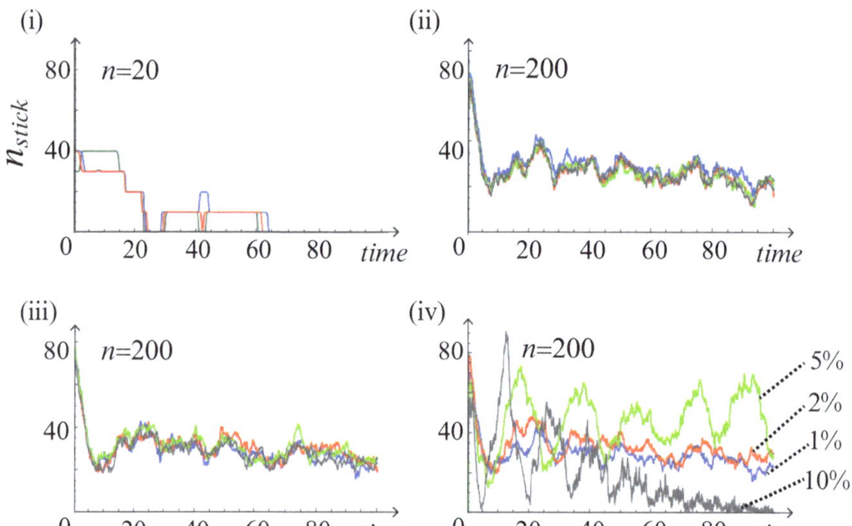

Fig. 9.17 Plots of the sliding codimension r for systems of n oscillators, simulated: (i-ii) for four random perturbations of the stiffness parameter $R_i\epsilon$ with $R_i \in [0,1]$; (iii) for four different random sets of perturbations of the mechanical constants up to 1% of β and α; (iv) for random sets of perturbations of the mechanical constants up to 1%, 2%, 5%, 10%, of β and α.

In Figure 9.17(iii)-(iv), we probe the robustness of the physical system rather than the simulation itself. We return to modelling each switch as $\mu(u) = \tanh(u/\epsilon)$, but vary the damping and spring stiffness constants (the components of M^α and M^β), by adding random values on the order of 1% in (iii), for three different sets of random perturbations, and of orders 1%, 2%, 5%, and 10%, in (iv). For small perturbations in (iii), the results are similar to before, namely, that short-time behaviour is sensitive and unpredictable, but long-term behaviour is robust, and large-scale structures are preserved. In Figure 9.17(iv), we verify that sufficiently large changes in the constants do yield a different system, with perturbations on the order of a few percent or more giving a very distinct system with large (period ~ 20) oscillations at 5%, and total collapse such that all oscillators have left sliding at 10%.

9.4.3 Switching with Time Dependence

Nonlinear switching can also make dynamics at the discontinuity surface so sensitive that it becomes impossible to simulate reliably. Consider a very simple oscillator where a control is applied with a frequency that switches between two different values $\pi/2$ and $3\pi/2$:

$$(\dot{x}_1, \dot{x}_2) = (-\tfrac{x_1}{100} - x_2, x_1) - (1,0)\sin\left[(1+\tfrac{\lambda}{2})\pi t\right] , \tag{9.22a}$$

where $\lambda = \mathrm{sign}(x_1)$. This clearly depends nonlinearly on the switching multiplier λ. (We simulated a very similar example in Figure 2.19 of Section 2.8, which can be analysed by steps similar to those below. This particular example comes from [117].)

We can extract the hidden term by expanding

$$\sin\left((1+\tfrac{1}{2}\lambda)\pi t\right) = \tfrac{1+\lambda}{2}\sin\tfrac{3\pi t}{2} + \tfrac{1-\lambda}{2}\cos\tfrac{\pi t}{2} + \left(\lambda^2-1\right)h(t;\lambda)$$

where some lengthy algebra yields

$$h(t;\lambda) = \tfrac{1}{4}\left(\sin\tfrac{3\pi t}{2} + \sin\tfrac{\pi t}{2} - 2\pi t\right) + \tfrac{1}{4}\sum_{i=1}^{\infty}\left(\tfrac{\pi t}{2}\right)^{2i+1}\left(\tfrac{a_i(t;\lambda)}{(2i+1)!} - \tfrac{b_i(t;\lambda)}{(2i)!}\right) ,$$

$$a_i(t;\lambda) = \sum_{j=1}^{2i-1}\left\{\left(\tfrac{1+\lambda}{2}\right)^{2j}3^{2i+1} + \left(\tfrac{1-\lambda}{2}\right)^{2j}\right\} ,$$

$$b_i(t;\lambda) = \sum_{j=0}^{\infty}\left(\tfrac{\pi t}{2}\right)^{2j}\frac{(1-\lambda)^{2i-1}(1+\lambda)^{2j}3^{2j+1} + (1+\lambda)^{2i-1}(1-\lambda)^{2j}3^{2i}}{2^{2i+2j-1}(2j+1)!} .$$

All we really need to understand from this complicated expression is that the function h is finite. If we neglect the hidden term altogether, we are left with

$$(\dot{x}_1, \dot{x}_2) = (-\tfrac{x_1}{100} - x_2, x_1) - (1,0)\left\{\tfrac{1+\lambda}{2}\sin\tfrac{3\pi t}{2} + \tfrac{1-\lambda}{2}\sin\tfrac{\pi t}{2}\right\} , \tag{9.22b}$$

a linear switching (or Filippov) system. The systems (9.22a) and (9.22b) are identical for all $x_1 \neq 0$, namely

$$(\dot{x}_1, \dot{x}_2) = (-\tfrac{x_1}{100} - x_2 - \sin(\pi\omega t), x_1) \quad \text{with} \quad \omega = \begin{cases} 3/2 \text{ if } x_1 > 0\,, \\ 1/2 \text{ if } x_1 < 0\,. \end{cases} \tag{9.23}$$

Figure 9.18 shows a simulation of each system. The nonlinear system has a complex oscillation, shown in (a.i). This is not seen in a simulation of the Filippov system in (b). A coarse simulation of the nonlinear system misses the fine features that create instability, shown in (a.ii), and we shall discuss why this happens in Chapter 12. This is the first hint that the level of coarseness in a simulation—coarseness being introduced by large noise, large delay, or large hysteresis, for example—affects how a discontinuity is observed. If irregularity in the system is small, it will exhibit the irregular behaviour in (a.i); if irregularity is high, it will wash out nonlinear effects and, counterintuitively, result in the more regular behaviour in (a.ii)-(b).

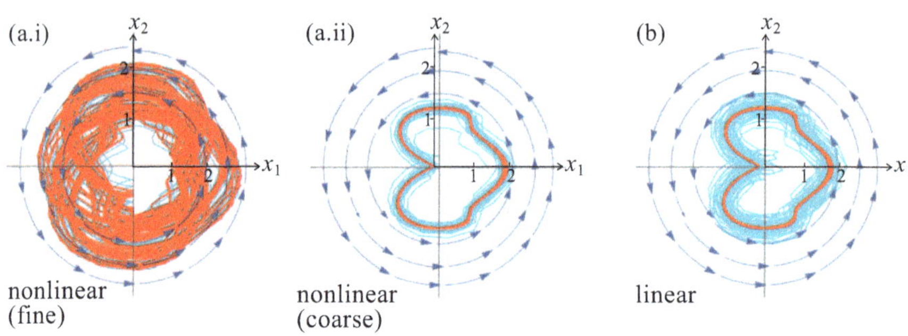

Fig. 9.18 A solution simulated from an initial point $(1, 0)$ until time $t = 2000$ (times $0 < t < 1000$ are shown in cyan colour as transients). The nonlinear system (9.22a) is simulated with fine discretization in (a.i) and coarse discretization in (a.ii). The linear system (9.22b) is simulated in (b). Simulations are made by smoothing out the discontinuity with $\lambda = \tanh(x_1/\varepsilon)$ and $\varepsilon = 10^{-3}$, calculated using explicit Euler discretization with step size s: (a.i) fine integration $s = 10^{-5}$; (a.ii) coarse integration $s = 10^{-4}$; (b) fine integration $s = 10^{-5}$ (though $s = 10^{-4}$ gives similar).

We gain more insight by plotting the trajectories $x_1(t)$ or $x_2(t)$ as graphs against time, as done in Figure 9.19 for $x_1(t)$. The nonlinear system, when finely simulated in (a.i), is delayed at the discontinuity surface by incidents of sliding. The coarse simulation in (a.ii), and linear switching system (b), both miss the sliding behaviour, crossing the discontinuity surface transversally and resulting in simpler cycles.

The method of simulation matters, of course, given such sensitive behaviour. The simulations here are made from an initial point $(1, 0)$, evolved up to a time $t = 2000$ ($t < 1000$ is shown lightly as transients). The discontinuity is smoothed by replacing the sign function with $\lambda = \tanh(x_1/\varepsilon)$ for

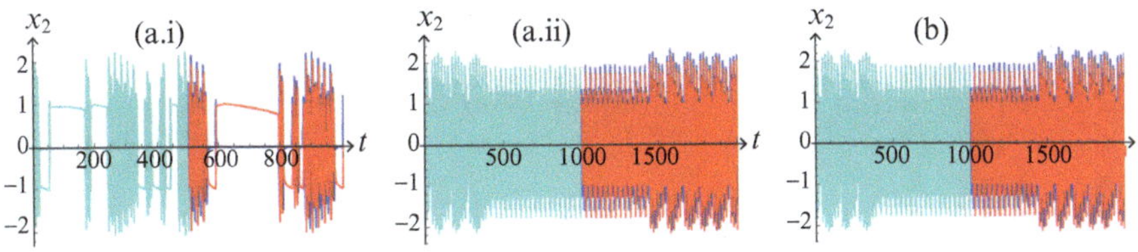

Fig. 9.19 Plots of x_1 and x_2 against time for the simulations in Figure 9.18.

small positive ε, and using explicit Euler discretization in fixed time steps described in the figure, representing (a.i) precise simulation of the full system (9.22a); (a.ii) coarse simulation of the full system (9.22a); (b.i) precise or coarse simulation of the linearized system (9.22b).

Essentially, in this system, the nonlinear term, however complicated, has a fairly benign effect: it merely delays the flow as it crosses the discontinuity surface. Given the time-dependent sinusoidal control, this is enough to alter the connection between trajectories either side of the switch sufficiently to significantly destabilise the oscillation.

9.5 Nonlinear Switching as a Small Perturbation

We must complete this chapter with a little due diligence. Having encountered new phenomena in moving to systems with nonlinear dependence on $\boldsymbol{\lambda}$, it is important to understand what features of linear switching systems remain qualitatively *un*changed. We will show here that crossing and sliding are robust to perturbation by small terms nonlinear in $\boldsymbol{\lambda}$ and that the boundaries between them are robust except at certain singularities we have encountered already at the end of Chapter 8.

A dynamical system that depends linearly on a switching multiplier $\boldsymbol{\lambda}$ is what we called a *linear switching system* in the last chapter, taking a form $\dot{\mathbf{x}} = \mathbf{a}(\mathbf{x}) + \underline{\underline{B}}.\boldsymbol{\lambda}$, or for one switch simply

$$\dot{\mathbf{x}} = \tfrac{1}{2}(1+\lambda_1)\mathbf{f}^+(\mathbf{x}) + \tfrac{1}{2}(1-\lambda_1)\mathbf{f}^-(\mathbf{x}) \;, \tag{9.24}$$

with arbitrary dependence on $\mathbf{x}$. In the remainder of this chapter, we will consider nonlinear switching systems of the form

$$\dot{\mathbf{x}} = \tfrac{1}{2}(1+\lambda_1)\mathbf{f}^+(\mathbf{x}) + \tfrac{1}{2}(1-\lambda_1)\mathbf{f}^-(\mathbf{x}) + (\lambda_1^2-1)\mathbf{h}(\mathbf{x};\lambda_1) \;, \tag{9.25}$$

where $\mathbf{h}$ constitutes a small perturbation of (9.24). We will also consider generalisations to multiple switches.

Firstly, the Examples 9.1–9.2 indicate that some restriction exists on the effect of hidden terms: they can turn crossing into sliding, but not the reverse.

Lemma 9.1. *If (9.24) has a sliding mode at a point* $\mathbf{x}$*, there will be at least one sliding mode in the nonlinear switching system (9.25) with* $\mathbf{h} \neq 0$*. If (9.24) has no sliding modes (i.e. crossing) at a point* $\mathbf{x}$*, there may or may not exist sliding modes in the nonlinear switching system (9.25) with* $\mathbf{h} \neq 0$ *at* $\mathbf{x}$*.*

Proof. This is a straightforward consequence of the boundary conditions either side of $\sigma_1 = 0$. If (9.24) has a sliding mode at a point $\mathbf{x}$, then (7.8) for $r = 1$ and $\mathbf{h} = 0$ has one solution, implying that the normal components $\delta_{t\pm}\sigma_1$ have opposite signs. Therefore $\delta_t\sigma_1$ passes through zero at least once in the range $-1 < \lambda_1 < +1$, implying that if $\mathbf{f}$ depends nonlinearly on λ_1, then (7.8) has at least one solution giving at least one sliding mode in (9.25). On the other hand, if there are no solutions to the linear system (hence there is crossing), this implies that the normal components $\delta_{t\pm}\sigma_1$ must have the same signs, but this does not prevent $\delta_t\sigma_1$ passing through zero in the range $-1 < \lambda_1 < +1$, in which case (7.8) has solutions and defines sliding modes of (9.25) for $\mathbf{h} \neq 0$. □

A simple extension of this is that crossing/sliding in a linear switching system typically implies an even/odd number of sliding modes if hidden terms are added, respectively. The fast $\varepsilon_1\dot{\lambda}_1$ component of the layer system implies also that multiple sliding modes alternate, when ordered by their λ_1 values, between attracting and repelling, corresponding to whether $\delta_t\sigma$ as a function of λ is decreasing or increasing, respectively, as it passes through zero.

More generally we have:

Lemma 9.2. *Different numbers of sliding modes can exist on subsets of* $\sigma_1 = 0$*, and the boundaries of these subsets consist of points where* $\delta_{t+}\sigma_1 = 0$*, or* $\delta_{t-}\sigma_1 = 0$*, or* $\frac{d}{d\lambda}\delta_t\sigma_1 = 0$*.*

Proof. If $\lambda_1^\$$ lies on the edges of the interval $(-1, +1)$, each condition $\lambda_1^\$ = -1$ or $\lambda_1^\$ = +1$ generically defines a codimension one subset of the discontinuity surface, given by $\{\mathbf{x} : \ \sigma_1 = 0, \ |\lambda_1^\$| = 1\}$ where $\lambda_1^\$$ is a solution of (7.8). Across this subset, $\lambda_1^\$$ enters or leaves the interval $(-1, +1)$ (an end point from Section 7.1); therefore the number of sliding modes changes by one. At $\lambda_1^\$ = -1$, we have $\delta_{t-}\sigma_1 = 0$. At $\lambda_1^\$ = +1$, we have $\delta_{t+}\sigma_1 = 0$. Assuming generically that $\mathbf{f}^\pm \neq 0$ at $\sigma_1 = 0$, these constitute tangencies between the vector fields $\mathbf{f}^\pm$ and the discontinuity surface $\sigma_1 = 0$. In the remaining case, $\lambda_1^\$$ is a degenerate solution of (7.8) when

$$\frac{d}{d\lambda_1}[\delta_t\sigma_1(\mathbf{x})] = 0 \qquad \text{at} \quad \lambda_1 = \lambda_1^\$ \,. \tag{9.26}$$

This condition generically defines a codimension one subset of the discontinuity surface, on which two or more solutions $\lambda_1^\$$ coalesce (a turning point from Section 7.1), so the number of solutions $\lambda_1^\$$ changes by at least two across this subset. □

These are the 'end points' and 'turning points' identified in Section 7.1 as boundaries of the sliding manifold $\mathcal{M}$.

9.5.1 Crossing

The structural stability of crossing regions is given by the following.

Lemma 9.3. *If sliding does not occur at a point* $\mathbf{x}$ *where* $\sigma_1(\mathbf{x}) = 0$ *in* (9.24) *(i.e.with* $\mathbf{h} = 0$*), then it does not occur for sufficiently small* $\mathbf{h} \neq 0$ *in* (9.25).

Proof. It is sufficient to consider a two-dimensional system $\mathbf{x} = (x_1, x_2)$ such that $\sigma = x_1$ (and we may simply replace x_2 by a vector of $n-1$ variables $x_2, \ldots, x_n$, to obtain the general case). If $f_1^+(0,x_2)f_1^-(0,x_2) > 0$ then the normal components of $\mathbf{f}^\pm$ point in the same direction at $x_1 = 0$, so if a trajectory arrives at $x_1 = 0$ at some point via $\mathbf{f}^\pm$, it will depart from the same point via $\mathbf{f}^\mp$. Thus crossing regions form submanifolds of the discontinuity surface given by

$$\left\{(x_1, x_2) \in \mathbb{R}^2 :\ x_1 = 0,\ f_1^+(0,x_2)f_1^-(0,x_2) > 0\right\} . \tag{9.27}$$

A sliding manifold $\mathcal{M}$ in the unperturbed system (9.24) exists only if there are solutions to (7.8). For some $\lambda_1 = \lambda_1^*$, we thus solve

$$\begin{aligned} 0 &= f_1\left(0, x_2; \lambda_1^*\right) \\ &= \frac{1+\lambda_1^*}{2} f_1^+\left(0, x_2\right) + \frac{1-\lambda_1^*}{2} f_1^-\left(0, x_2\right) \\ \Rightarrow \lambda_1^* &= \frac{f_1^-\left(0, x_2\right) + f_1^+\left(0, x_2\right)}{f_1^-\left(0, x_2\right) - f_1^+\left(0, x_2\right)} , \end{aligned}$$

but then $|\lambda_1^*| > 1$ because the signs of $f_1^\pm(0,x_2)$ are the same, so there are no solutions $\lambda_1^\$$ to (7.6). Hence no sliding manifold $\mathcal{M}$ exists, and since f_1 does not change sign at the discontinuity surface , the flow must cross it.

Now let us consider this for a perturbed system (9.25) with small $\mathbf{h} \neq 0$, seeking solutions such that

$$0 = \frac{1+\lambda_1^\$}{2} f_1^+\left(0, x_2\right) + \frac{1-\lambda_1^\$}{2} f_1^-\left(0, x_2\right) + \left((\lambda_1^\$)^2 - 1\right) h_1\left(0, x_2\right) ,$$

which upon substituting $\lambda_1^\$ = \lambda_1^* + \delta\lambda$ for small $\delta\lambda$ rearranges to

$$0 = (\lambda_1^*)^2 - 1 + 2\left(P + \lambda_1^*\right)\delta\lambda + \delta\lambda^2 , \qquad P = \frac{f_1^+\left(0, x_2\right) - f_1^-\left(0, x_2\right)}{4h_1\left(0, x_2\right)} ,$$

with roots $\lambda_1 = \lambda_1^* + \delta\lambda = -P \pm \sqrt{1 + P(2\lambda_1^* + P)}$. Then $|h_1| \ll 1$ implies $|P| \gg 1$, in which case $\lambda_1 = \lambda_1^* + \mathsf{O}(P^{-1})$ or $\lambda_1 = -2P - \lambda_1^* + \mathsf{O}(P^{-1})$, either of which give $|\lambda_1| > 1$ since $|\lambda_1^*| > 1$ and $|P| \gg 1$. □

Therefore the sliding mode is equivalent for the perturbed and unperturbed systems, in that $\lambda_1^\$$ does not exist, and so there is no sliding manifold $\mathcal{M}$. Hence there are no sliding modes, so transversal crossing between the systems $\dot{\mathbf{x}} = \mathbf{f}^{\pm}$ occurs in both the perturbed and unperturbed systems. Thus the system (9.24) is stable to perturbations of the form (9.25) near points of crossing where the discontinuity surface is locally a manifold $\sigma_1(\mathbf{x}) = 0$.

While our interest here is in small perturbations, and this result holds for small $\mathbf{h}$, it does not hold for large values of $\mathbf{h}$. We dealt with systems where $\mathbf{h}$ is of significant size—not a small perturbation—in Chapter 7 and return to them later in this chapter.

The extension to show that if no codimension r sliding modes exist in the absence of hidden terms, then no codimension r sliding modes exist in the presence of small hidden terms $(\lambda_j^2 - 1)\mathbf{h}(\mathbf{x}; \boldsymbol{\lambda})$, is a straightforward exercise applying the above to the components of $\boldsymbol{\lambda} = (\lambda_1, \ldots, \lambda_r)$.

9.5.2 Sliding

To show that sliding regions are structurally stable, we have the following.

Lemma 9.4. *If sliding occurs at a point* $\mathbf{x}$ *where* $\sigma_1(\mathbf{x}) = 0$ *in* (9.24), *then the sliding manifold* $\mathcal{M}$ *of* (9.25) *is well-defined and has a unique solution nearby for small* $\mathbf{h} \neq 0$.

Proof. As in Lemma 9.3, it is sufficient to consider a two-dimensional system $\mathbf{x} = (x_1, x_2)$ such that $\sigma = x_1$. If $f_1^+(0, x_2; \lambda_1) f_1^-(0, x_2; \lambda_1) < 0$, then the normal components of $\mathbf{f}^{\pm}$ point in opposite directions, so any point on $x_1 = 0$ marks the arrival point of trajectories via both $\mathbf{f}^{\pm}$ or the exit point of trajectories via both $\mathbf{f}^{\pm}$. Thus sliding regions form submanifolds of the discontinuity surface given by

$$\left\{\mathbf{x} = (x_1, x_2) \in \mathbb{R}^2 : \; x_1 = 0, \; f_1^+(0, x_2) f_1^-(0, x_2) < 0\right\} . \tag{9.28}$$

On these we seek a sliding manifold $\mathcal{M}$ of the unperturbed system (9.24) as before. The conditions in (7.8) give that there exists λ_1^* satisfying

$$\begin{aligned} 0 &= f\left(0, x_2; \lambda_1^*\right) \\ &= \frac{1 + \lambda_1^*}{2} f_1^+\left(0, x_2\right) + \frac{1 - \lambda_1^*}{2} f_1^-\left(0, x_2\right) \\ \Rightarrow \lambda_1^* &= \frac{f_1^-\left(0, x_2\right) + f_1^+\left(0, x_2\right)}{f_1^-\left(0, x_2\right) - f_1^+\left(0, x_2\right)} \quad := \; \lambda_1^\$ \end{aligned}$$

where $|\lambda_1^*| < 1$ because the signs of $f_1^{\pm}(0, x_2)$ are opposing, so these give valid solutions $\lambda_1^{\$}$ defining a unique sliding manifold $\mathcal{M}$.

Considering the same problem for a perturbed system of the form (9.25) with small $\mathbf{h} \neq 0$, calculations similar to the proof of Lemma 9.3 yield again $\lambda_1 = \lambda_1^* + \mathsf{O}\left(P^{-1}\right)$ or $\lambda_1 = -2P - \lambda_1^* + \mathsf{O}\left(P^{-1}\right)$ with $P = \frac{f_1^+(0,x_2) - f_1^-(0,x_2)}{4h_1(0,x_2)}$, and for $|h_1| \ll 1$, we have $|P| \gg 1$. Then the first of these solutions lies inside $(-1, +1)$ since $|\lambda_1^*| < 1$, while the second lies outside $(-1, +1)$ since $|P| \gg 1$. Thus there exists a unique sliding manifold $\mathcal{M}$ on which $\lambda_1^{\$} = \lambda_1^* + \mathsf{O}\left(P^{-1}\right) = \lambda_1^* + \mathsf{O}\left(h_1\right)$, a regular perturbation of the unperturbed sliding mode λ_1^*.

Having shown that the manifold $\mathcal{M}$ exists, it remains to show that the manifold is invariant and hence represents sliding motion. The invariance of $\mathcal{M}$ holds in the neighbourhood of points where $\mathcal{M}$ is normally hyperbolic, meaning $\partial\dot{\lambda}_1/\partial\lambda_1$ is nonzero. Normal hyperbolicity holds in a sliding region (9.28) of the unperturbed system (9.24), because

$$\varepsilon_1 \left.\frac{\partial\dot{\lambda}_1}{\partial\lambda_1}\right|_{\mathcal{M}} = \frac{\partial}{\partial\lambda_1} f_1\left(0, x_2; \lambda_1^{\$}\right) = \tfrac{1}{2}\left(f_1^+(0, x_2) - f_1^-(0, x_2)\right) \neq 0\,. \tag{9.29}$$

The inequality follows since $f_1\left(0, x_2; +1\right)$ and $f_1\left(0, x_2; -1\right)$ have opposing signs by (9.28). In the perturbed system (9.25), we have

$$\varepsilon_1 \left.\frac{\partial\dot{\lambda}_1}{\partial\lambda_1}\right|_{\mathcal{M}} = \frac{\partial}{\partial\lambda_1} f_1\left(0, x_2; \lambda_1^{\$}\right) = \tfrac{1}{2}\left(f_1^+(0, x_2) - f_1^-(0, x_2)\right) + E \neq 0 \tag{9.30}$$

where $E = \mathsf{O}\left(h_1(0, x_2; \lambda_1^*), \frac{\partial}{\partial\lambda_1} h_1(0, x_2; \lambda_1^*)\right)$. If $\mathbf{h}$ is small in a neighbourhood of $\lambda_1 = \lambda_1^{\$}$, then E is small and normal hyperbolicity is maintained. Hence for small perturbations $\mathbf{h}$ the manifold $\mathcal{M}$ persists, is an invariant of the flow (7.2), and the dynamics on it is a regular $\mathsf{O}\left(|\mathbf{h}|\right)$ perturbation of the unperturbed sliding motion (7.8). □

Again it is straightforward to extend this to show that if a codimension r sliding mode exists in the absence of hidden terms, then a slightly perturbed codimension r sliding mode exists in the presence of small hidden terms $(\lambda_j^2 - 1)\mathbf{h}(\mathbf{x}; \boldsymbol{\lambda})$.

9.5.3 Structural Stability of the Sliding Manifold

Provided $\boldsymbol{\lambda}^{\$}$ in (7.6) is well defined up to the boundary, the arguments on structural stability of crossing and of linear sliding (Sections 9.5.1 to 9.5.2) extend to the boundary between them. This applies to end points of the sliding manifold $\mathcal{M}$, where one of the switching multipliers $\boldsymbol{\lambda}_j^{\$}$ is equal to $+1$ or -1, meaning $\mathcal{M}$ touches the boundary of the switching layer.

In Chapter 8, however, we have already seen a boundary of sliding where $\boldsymbol{\lambda}^{\$}$ is not well-defined. The sliding solution near a two-fold singularity is given by (8.19), which diverges precisely at the two-fold singularity. That implies the sliding manifold $\mathcal{M}$ is degenerate and hence not structurally stable. This happens in general when two sliding boundaries intersect, hence at complex tangencies such as the two-fold or fold-cusp.

This is therefore a commonplace problem in linear switching systems, so it is important to understand its implications.

The degeneracy arises as follows. Take coordinates $\mathbf{x} = (x_1, \ldots, x_n)$ in which $\sigma_j = x_j$ for $j = 1, \ldots, r$. Normally hyperbolicity of the sliding manifold (7.7) with respect to the system (7.2) is lost where (7.11) is violated, including the set of points

$$\mathcal{L} = \left\{ (\lambda_1, \ldots, \lambda_r, x_{r+1}, \ldots, x_n) \in \mathcal{M} \ : \ \det \left| \frac{\partial \mathbf{f}}{\partial \boldsymbol{\lambda}} \right| = 0 \right\} . \tag{9.31}$$

It is this set $\mathcal{L}$ that becomes degenerate in linear switching systems.

Proposition 9.1. *If the set $\mathcal{L} \subset \mathcal{M}$ exists inside the switching layer system* (7.2) *of the linear switching system* (9.24)*, then $\mathcal{L}$ is degenerate.*

Proof. Take the case of a single switch ($r = 1$) first, in coordinates where $\sigma_1 = x_1$, letting $\mathbf{f} = (f_1, \ldots, f_n)$. With respect to the fast timescale $\tau = t/\varepsilon_1$, the layer system is

$$\begin{aligned} \lambda_1' &= f_1(0, x_2, \ldots, x_n; \lambda_1) , \\ x_i' &= \varepsilon_1 f_i(0, x_2, \ldots, x_n; \lambda_1) , \qquad i = 2, \ldots, n , \end{aligned} \tag{9.32}$$

with the prime denoting differentiation with respect to the fast time τ, and this becomes a one-dimensional fast subsystem in the limit $\varepsilon_1 \to 0$,

$$\begin{aligned} \lambda_1' &= f_1(0, x_2, \ldots, x_n; \lambda_1) , \\ x_i' &= 0 , \qquad i = 2, \ldots, n . \end{aligned} \tag{9.33}$$

This describes how λ_1 either transitions directly between ± 1 or else encounters a zero of $f_1(0, x_2, \ldots, x_n; \lambda_1)$ which defines the sliding set $\mathcal{M}$ as in (7.7), in the limit as $\varepsilon_1 \to 0$.

Expand the first component of (9.33) as a series in λ_1 about some $\lambda_1^* \in \mathcal{M}$,

$$f_1(0, x_2, \ldots, x_n; \lambda) = (\lambda_1 - \lambda_1^*) \tfrac{\partial}{\partial \lambda_1} f_1(0, x_2, \ldots, x_n; \lambda_1^*) + \mathsf{O}\left(|\lambda_1 - \lambda_1^*|^2\right) . \tag{9.34}$$

On the set $\mathcal{L}$, the lowest-order term vanishes, leaving

$$f_1(0, x_2, \ldots, x_n; \lambda) = \mathsf{O}\left((\lambda_1 - \lambda_1^*)^2\right) . \tag{9.35}$$

This vanishes identically for a system where $\mathbf{f}$ depends only linearly on λ_1, in which case the first line of the layer system (9.33) is trivially zero on $\mathcal{L}$, and the value of $\lambda_1^{\$}$ defining the sliding manifold $\mathcal{M}$ is undetermined.

Geometrically this means that $\mathcal{L}$ lies along the λ_1 coordinate direction if $\mathbf{f}$ depends linearly on λ_1, which is the fast direction of the corresponding perturbed system (9.32), giving a line of zeros of the fast subsystem (9.33).

This extends easily to $r > 1$. Assume for some infinitesimal ε_0 that the ratios $\upsilon_j = \varepsilon_j/\varepsilon_0$ are finite and nonzero as $\varepsilon_0 \to 0$ and each $\varepsilon_j \to 0$. With respect to the fast timescale $\tau = t/\varepsilon_0$, the layer system is

$$\begin{aligned} \upsilon_j\lambda_j' &= f_j(0,\dots,0,x_{r+1},\dots,x_n;\lambda_1,\dots,\lambda_r)\,, \qquad & j &= 1,\dots,r\,,\\ x_i' &= \varepsilon_0 f_i(0,\dots,0,x_{r+1},\dots,x_n;\lambda_1,\dots,\lambda_r)\,, & i &= r+1,\dots,n\,. \end{aligned} \tag{9.36}$$

with the prime denoting differentiation with respect to the fast time τ, and this becomes an r dimensional fast subsystem in the limit $\varepsilon_0 \to 0$,

$$\begin{aligned} \upsilon_j\lambda_j' &= f_j(0,\dots,0,x_{r+1},\dots,x_n;\lambda_1,\dots,\lambda_r)\,, \qquad & j &= 1,\dots,r\,,\\ x_i' &= 0\,, & i &= r+1,\dots,n\,. \end{aligned} \tag{9.37}$$

Expanding the first line of (9.37) as a series in $\boldsymbol{\lambda}$ about some $\boldsymbol{\lambda}_* \in (-1,+1)^r$,

$$\begin{aligned} f_j(0,\dots,0,x_{r+1},\dots,x_n;\lambda) = (\boldsymbol{\lambda}-\boldsymbol{\lambda}_*)\cdot\tfrac{\partial}{\partial\boldsymbol{\lambda}}f_j(0,\dots,0,x_{r+1},\dots,x_n;\boldsymbol{\lambda}_*)\\ +\mathsf{O}\left(|\boldsymbol{\lambda}-\boldsymbol{\lambda}_*|^2\right)\,. \end{aligned} \tag{9.38}$$

On the set $\mathcal{L}$, the $r\times r$ square matrix $\frac{\partial(f_1,\dots,f_r)}{\partial(\lambda_1,\dots,\lambda_r)}$ is degenerate by (9.31), so this expression vanishes to lowest order and cannot be solved for $\boldsymbol{\lambda}$. Again, geometrically this means that $\mathcal{L}$ lies in the $\boldsymbol{\lambda}$ coordinate space and represents a degenerate set of zeros of the fast subsystem (9.37). □

The condition for $\mathcal{L}$ to lie transversal to the fast direction is simply the existence of non-trivial higher-order terms, for example, for one switch the next derivative of f_1 must be nonzero, i.e. $\partial^2 f_1/\partial\lambda_1^2 \neq 0$. Perturbing a linear switching system with a term nonlinear in $\boldsymbol{\lambda}$ will therefore yield a topologically non-equivalent system.

By definition, on the set $\mathcal{L}$, the lowest-order term of (9.34) vanishes, leaving (9.35). While this is trivial if f_1 depends only linearly on λ_1, with nonlinear dependence it becomes

$$\begin{aligned} f_1(0,\dots,0,x_{r+1},\dots,x_n;\boldsymbol{\lambda}) &= \\ &\tfrac{1}{2}\sum_{ij=1}^{r}(\lambda_i-\lambda_{i0})(\lambda_j-\lambda_{j0})\partial_{\lambda_i\lambda_j}f_1(0,\dots,0,x_{r+1},\dots,x_n;\boldsymbol{\lambda}_0)\\ &+\mathsf{O}\left(|\boldsymbol{\lambda}-\boldsymbol{\lambda}_0|^3\right)\,. \end{aligned} \tag{9.39}$$

Thus for a boundary of codimension r sliding to be structurally stable at a point $\mathbf{x}\in\mathbb{R}^n$, it must satisfy:

- For $r = 1$ on $\sigma_1(\mathbf{x}) = 0$,

$$\delta_{t+}\sigma_1(\mathbf{x}) = 0 \quad \text{or} \quad \delta_{t-}\sigma_1(\mathbf{x}) = 0 \quad \text{or both,} \tag{9.40}$$

such that

$$\frac{\partial}{\partial \lambda_1}\delta_t \sigma_1(\mathbf{x}) \neq 0 \,. \tag{9.41}$$

- For $r > 1$ on $\sigma_1(\mathbf{x}) = \ldots = \sigma_r(\mathbf{x}) = 0$,

$$\frac{\partial}{\partial \lambda_j}\mathbf{f}(\mathbf{x}; \lambda_1, \ldots, \lambda_r) = 0 \quad \text{at} \quad \lambda_j = \pm 1 \quad \text{for any } j = 1, \ldots, r \,, \tag{9.42}$$

  such that (7.11) holds.

This tells us that structural stability in the switching layer requires either that the constituent vector fields are tangent to the discontinuity surface from at most one side or if tangent from two sides (a 'complex' tangency), then we require $\mathbf{h} \neq 0$. For example, in a region where the discontinuity surface is a simple manifold, by differentiating the general expression

$$\mathbf{f}(\mathbf{x}; \boldsymbol{\lambda}) = \tfrac{1}{2}(1 + \lambda_1)\mathbf{f}(\mathbf{x}; +1) + \tfrac{1}{2}(1 - \lambda_1)\mathbf{f}(\mathbf{x}; -1) + (\lambda_1^2 - 1)\mathbf{h}(\mathbf{x}; \lambda_1)$$

we have the following conditions at the sliding boundaries from Section 6.1.

- **Simple sliding boundaries**:
  if $\sigma_1 = f_1(\mathbf{x}; \pm 1) = 0$ but $f_1(\mathbf{x}; \mp 1) \neq 0$ at a point $\mathbf{x}$, then

$$\left.\frac{\partial}{\partial \lambda_1} f_1(\mathbf{x}; \lambda_1)\right|_{\lambda_1 = \pm 1} = \tfrac{1}{2}\left(f_1(\mathbf{x}; +1) - f_1(\mathbf{x}; -1)\right) \pm 2h_1(\mathbf{x}; \pm 1) \neq 0 \,,$$

  so such a point is structurally stable for typical

$$h_1 \neq \pm\tfrac{1}{4}\left(f_1(\mathbf{x}; -1) - f_1(\mathbf{x}; +1)\right)$$

  (including if $\mathbf{h} = 0$).

- **Cusp or higher order sliding boundaries**:
  if $\sigma_1 = f_1(\mathbf{x}; \pm 1) = \ldots = \delta_{t\pm}^k f_1(\mathbf{x}; \pm 1) = 0$ for $k \geq 1$ ($k = 1$ is a cusp, $k = 2$ a swallowtail, etc.), but $f_1(\mathbf{x}; \mp 1) \neq 0$, at a point $\mathbf{x}$, then

$$\left.\frac{\partial}{\partial \lambda_1} f_1(\mathbf{x}; \lambda_1)\right|_{\lambda_1 = \pm 1} = \mp\tfrac{1}{2} f_1(\mathbf{x}; \mp 1) \pm 2h_1(\mathbf{x}; \pm 1) \neq 0 \,,$$

  so such a point is structurally stable for typical $h_1 \neq \frac{1}{4} f_1(\mathbf{x}; \mp 1)$ (including if $\mathbf{h} = 0$).

- **Two-fold sliding boundaries**:
  if $\sigma_1 = f_1(\mathbf{x}; +1) = f_1(\mathbf{x}; -1) = 0$, then

$$\left.\frac{\partial}{\partial \lambda_1} f_1(\mathbf{x}; \lambda_1)\right|_{\lambda_1 = \pm 1} = \pm 2h_1(\mathbf{x}; \pm 1) \neq 0 \,,$$

  so for such a point to be structurally stable we must have $\mathbf{h} \neq 0$.

An example of the degeneracy of $\mathcal{M}$ on the non-hyperbolic set $\mathcal{L}$ is given for a planar system in Section 9.6.2 below and later for a similar system in higher dimensions in Section 13.5.

9.6 Hidden Degeneracy at Local Bifurcations

Local classifications are intended to tell us all of the ways a system may behave near certain singularities, and structural stability in the switching layer plays an important role, particularly when studying bifurcations. In Section 8.8 we saw two bifurcations that involved seemingly unexplained changes of attractivity of an equilibrium. It turns out that both of these are structurally unstable inside the switching layer, but for different reasons.

9.6.1 Boundary Node Flip

Let us revisit the boundary equilibrium bifurcation from Figure 8.21(top), by considering the planar system

$$(\dot{x}_1, \dot{x}_2) = \tfrac{1}{2}(1+\lambda)(x_1 + 2x_2 - \alpha, x_2) + \tfrac{1}{2}(1-\lambda)(1,3) \;, \tag{9.43}$$

where $\lambda = \text{sign}(x_1)$ for $x_1 \neq 0$, and α is a bifurcation parameter. From the arrangements of the vector fields, it is simple to sketch the phase portrait, shown in Figure 9.20. The system has a unique equilibrium, a repelling node at $(x_1, x_2) = (\alpha, 0)$ that exists for $\alpha \geq 0$, hitting the discontinuity surface as α decreases through zero to become an attracting sliding equilibrium for $\alpha \leq 0$.

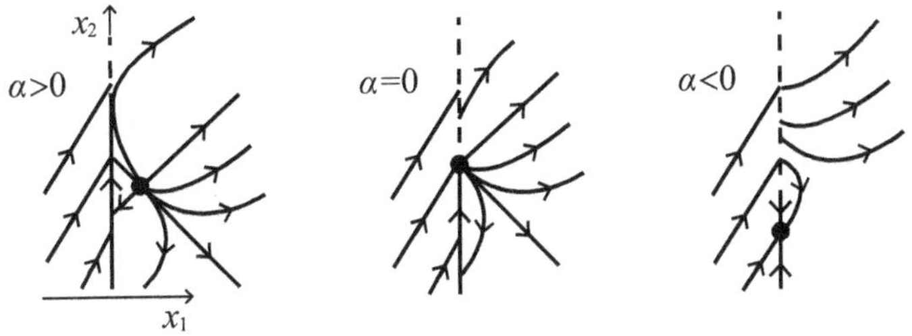

Fig. 9.20 As α changes sign, a repelling node hits the discontinuity surfaceand becomes an attracting sliding node.

The curiosity of this boundary equilibrium bifurcation is that the attractivity of the node flips during the bifurcation. To see how this happens requires closer inspection of the layer dynamics and its structural stability.

The switching layer system is

$$(\varepsilon\dot{\lambda}, \dot{x}_2) = \tfrac{1}{2}(1+\lambda)(2x_2-\alpha, x_2) + \tfrac{1}{2}(1-\lambda)(1,3) \quad \text{on} \quad x_1 = 0\,, \tag{9.44}$$

which possesses a sliding manifold $\mathcal{M}$ given by

$$\lambda^{\$} = \frac{1+2x_2-\alpha}{1-2x_2+\alpha}\,.$$

This exists on $(\lambda^{\$}, x_2) \in (-1,+1) \times \mathbb{R}$ only for $x_2 < \alpha/2$. Substituting $\lambda^{\$}$ into (9.44) gives sliding dynamics

$$\dot{x}_2 = \frac{3\alpha - 5x_2}{1-2x_2+\alpha} \quad \text{on} \quad x_1 = 0\,.$$

The denominator of $\dot{x}_2$ is positive since $x_2 < \alpha/2 < (1+\alpha)/2$. The numerator implies that there exists an attracting sliding node in the sliding dynamics at $(\lambda^*, x_2^*) = ((5+\alpha)/(5-\alpha), 3\alpha/5)$, and since this can exist only for $x_2^* < \alpha/2$, the sliding equilibrium exists only for $\alpha < 0$, precisely when the repelling node in the $\lambda = +1$ system vanishes.

So far this only confirms the description we gave of the phase portrait in Figure 9.20. We must look a little closer to understand the flip in stability that befalls the equilibrium at $\alpha = 0$.

We can understand the node as entering the switching layer at $\lambda = +1$ when α becomes negative. The Jacobian (given by (7.34b) for $r = 1$) at this equilibrium in the layer variables $(\xi_1, \xi_2) = (\varepsilon\lambda_1, x_2)$ is

$$\begin{pmatrix} \frac{\partial\dot{\xi}_1}{\partial\xi_1} & \frac{\partial\dot{\xi}_1}{\partial\xi_2} \\ \frac{\partial\dot{\xi}_2}{\partial\xi_1} & \frac{\partial\dot{\xi}_2}{\partial\xi_2} \end{pmatrix} = \begin{pmatrix} \frac{\alpha-5}{10\varepsilon} & \frac{10}{5-\alpha} \\ \frac{3(\alpha-5)}{10\varepsilon} & \frac{5}{5-\alpha} \end{pmatrix},$$

with determinant and trace

$$det = \frac{5}{2\varepsilon}\,, \qquad tr = \frac{5}{5-\alpha} - \frac{5-\alpha}{10\varepsilon} \to -\frac{5-\alpha}{10\varepsilon}$$

as $\varepsilon \to 0$. These confirm that the equilibrium is an attractor for $\alpha < 5(1-\sqrt{2\varepsilon}) \to 5$; hence for small α (with $\varepsilon \to 0$ by definition), it is always an attractor. The eigenvalues of the Jacobian are $\frac{25}{\alpha-5} + \mathrm{O}(\varepsilon)$ and $\frac{\alpha-5}{10\varepsilon} + \mathrm{O}(1)$, both negative for small α, consistent with this being a sliding node.

At $\alpha = 0$ we therefore have a matching of two opposing systems with an equilibrium at $x_1 = x_2 = 0$, that of a repelling node outside the switching layer and an attracting node inside the layer, yielding the phase portraits in Figure 9.21. We can describe this as a *folded node*, the stitching together of phase portraits of nodes that are attracting and repelling either side of some curve. This is peculiar in itself, and occurs in smooth systems only in the midst of bifurcations, when multiple equilibria of different kinds collide.

This coincides with another structural instability when $\alpha = 0$. Figure 9.21 reveals that there exist a whole family of cycles through the folded equilibrium, in the region bounded by $\mathcal{M}$ and the strong unstable manifold of the repelling node (labelled $\mathcal{U}$).

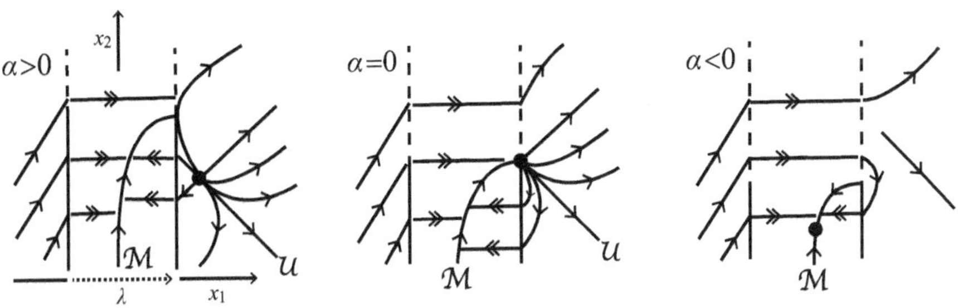

Fig. 9.21 The bifurcation in Figure 9.20 with the discontinuity surface $x_1 = 0$ blown up into a layer $\lambda \in (-1, +1)$.

Together these elements show the phase portrait at $\alpha = 0$ to be degenerate, possessing both a folded node and an infinite family of cycles. The system (9.43) is evidently deficient as a prototype of this boundary equilibrium bifurcation. Its degeneracy is not broken by adding functions of x_1 or x_2 to (9.43) and can only be broken by adding a term nonlinear in λ.

It is sufficient to consider simple perturbation of (9.43) in the form (9.25)

$$(\dot{x}_1, \dot{x}_2) = \tfrac{1}{2}(1+\lambda)(x_1 + 2x_2 - \alpha, x_2) + \tfrac{1}{2}(1-\lambda)(1,3) + (\lambda^2 - 1)(\beta, 0), \tag{9.45}$$

again with $\lambda = \text{sign}(x_1)$ and introducing a constant $\beta > 1/4$ (following [116]). In this system, the degeneracy that occurred at $\alpha = 0$ is broken and revealed as a Hopf bifurcation. We show this as follows.

The perturbation in (9.45) does not change the system outside $x_1 = 0$ because β is a hidden term. There is still a unique equilibrium in the switching layer, which now lies at $\lambda^* = \frac{1}{4\beta}(\alpha - 5 + R)$, $x_2^* = \frac{3}{10}(\alpha + 5 + 4\beta + R)$, where $R = \sqrt{(4\beta+5)^2 + \alpha^2 + 8\alpha\beta}$, and this lies at $\lambda^* = 1$ when $\alpha = 0$. The associated Jacobian (7.34b) in the layer variables becomes

$$\begin{pmatrix} \frac{\partial \dot{\xi}_1}{\partial \xi_1} & \frac{\partial \dot{\xi}_1}{\partial \xi_2} \\ \frac{\partial \dot{\xi}_2}{\partial \xi_1} & \frac{\partial \dot{\xi}_2}{\partial \xi_2} \end{pmatrix} = \begin{pmatrix} \frac{3r+2R}{10\varepsilon} & \frac{r+R}{4\beta}) \\ \frac{3(r-R)}{20\varepsilon} & \frac{r+R}{8\beta} \end{pmatrix}$$

where $r = \alpha - 5 + 4\beta$. Its eigenvalues are $\frac{5R(r+R)}{8\beta(3r+2R)} + \mathrm{O}(\varepsilon)$ and $\frac{3r+2R}{10\varepsilon} + \mathrm{O}(1)$. Its trace is

$$tr = \frac{r+R}{8\beta} + \frac{3r+2R}{10\varepsilon} \to \frac{3r+2R}{10\varepsilon}. \tag{9.46}$$

At the bifurcation value $\alpha = 0$, this simplifies to $tr = (4\beta - 1)/2\varepsilon$, which is positive for $\beta > 1/4$, making the sliding equilibrium a repelling node, which now matches with the repelling node outside the switching layer, and therefore breaking the degeneracy of the node's 'folded stability'.

This is only a perturbation of the original system (9.43), so the sliding equilibrium must return to being attracting somewhere nearby as α decreases. Indeed, a flip in attractivity of the sliding node now occurs via a Hopf bifurcation inside the switching layer, happening when the trace (9.46) vanishes, at an α value

$$\alpha_h = 5 - 8\sqrt{\beta} - 4\beta + O(\varepsilon) \ . \tag{9.47}$$

Just before this happens, the sliding equilibrium changes from a node into a focus, when the eigenvalues of the Jacobian change from real to complex. This transition is squashed into an order $\sqrt{\varepsilon}$ neighbourhood of the bifurcation, and to find it, we must expand the eigenvalues not solely in ε as above, but jointly in α and ε, to find that the focus-node transition occurs at an α value

$$\alpha_f = 5 - 8\sqrt{\beta} - 4\beta + 12\sqrt{3\varepsilon} + O(\varepsilon) \ . \tag{9.48}$$

These different bifurcation values satisfy $0 < \alpha_f < \alpha_h$, giving an overall phase portrait through the boundary-node bifurcation as shown in Figure 9.22.

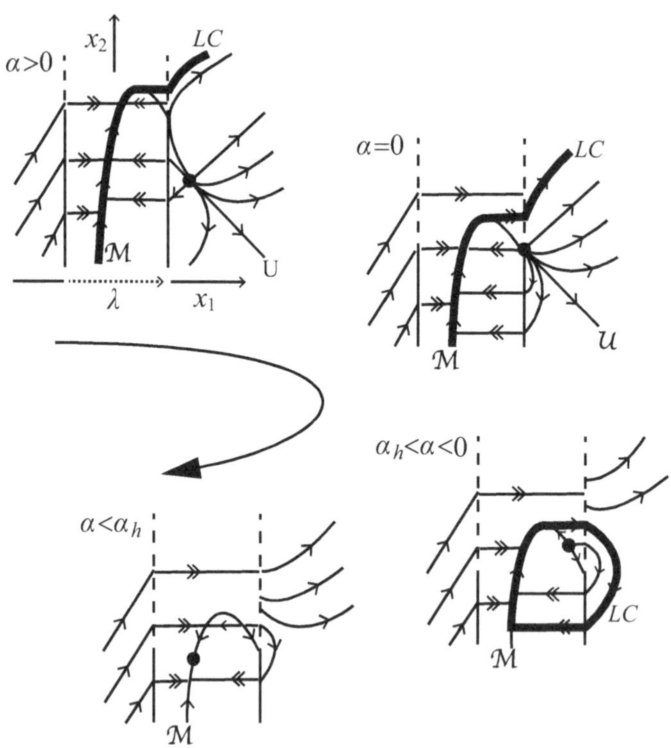

Fig. 9.22 The bifurcation in the perturbed system as α passes through $\alpha = 0$ and α_h. A limit cycle labelled LC in (ii-iii) shrinks until a supercritical Hopf bifurcation occurs, giving the change in the equilibrium's attractivity from (iii) to (iv).

The bifurcation is revealed to be actually a sequence of bifurcations facilitating the node's change from repelling for $\alpha > 0$ to attracting for $\alpha \lesssim 0$.

For $\alpha > 0$, there exists a repelling node outside the discontinuity surface (the first portrait in Figure 9.22), from which all trajectories in the system evolve towards infinity. The orbit labelled LC, formed by continuing the flow from the sliding manifold $\mathcal{M}$, is a kind of 'limit cycle at infinity'; we will see the role it plays in the subsequent bifurcation.

At $\alpha = 0$, the repelling node reaches the discontinuity surface, and becomes a repelling node of the switching layer system, but in terms of stability, the phase portrait is not critically different from $\alpha > 0$. This node turns into a focus at $\alpha = \alpha_f$ (not shown), resulting in the third portrait in Figure 9.22 for $\alpha < \alpha_f$, and this brings the orbit LC into play. At some $0 < \alpha < \alpha_h$ between the second and third portraits of Figure 9.22, the trajectory LC emitted from $\mathcal{M}$ intersects the strong manifold of the vanished equilibrium of the $x_1 > 0$ system (labelled $\mathcal{U}$), after which LC is seen to form an attracting limit cycle surrounding the focus.

The change in stability we have sought now occurs at $\alpha = \alpha_h$, when the limit cycle LC shrinks to zero and the focus's attractivity changes in a supercritical Hopf bifurcation, giving the last portrait in Figure 9.22. At some further $\alpha = \alpha_f < \alpha_h$, not shown, the equilibrium will become a node again, now attracting.

This analysis is a mixture of geometrical arguments and flow sketching and could be made more rigorous by careful matching of the systems from inside and outside the layer. We leave these as an open challenge. To verify our analysis, let us instead simulate (9.45) in Figure 9.23, taking ε nonzero for ease of inspection. The flow exhibits the predicted behaviour, showing

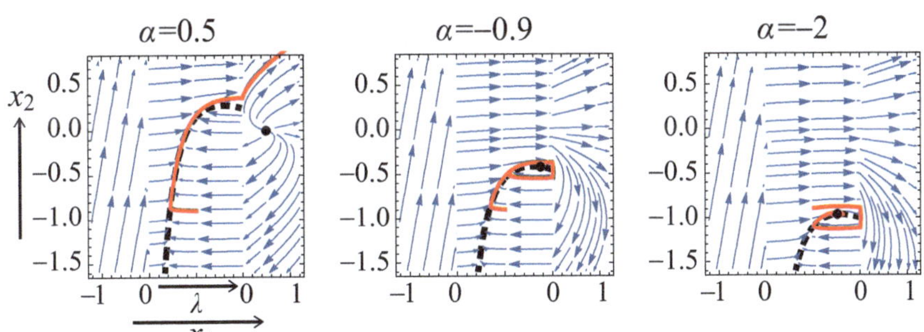

Fig. 9.23 Simulation of the boundary node bifurcation for $\beta = 0.4, \varepsilon = 0.01$, and different α values as shown. The discontinuity surface $x_1 = 0$ is blown up into a layer $\lambda \in (-1, +1)$. A trajectory is simulated from an initial point $(0, -0.9)$. For $\alpha = 0.5$, a repelling node exists in $x_1 > 0$; for $\alpha = -0.9$, a repelling sliding focus exists inside the switching layer, surrounded by an attracting stick-slip limit cycle; and for $\alpha = -2$, and attracting sliding focus exists. For these parameters $\alpha_f = 0.28$ and $\alpha = -1.7$. The dotted curve indicates the sliding manifold $\mathcal{M}$.

a repelling node entering the switching layer at $\alpha = 0$, becomes a repelling focus surrounded by an attracting limit cycle, which is then annihilated in a Hopf bifurcation as the focus changes stability.

9.6.2 Fold-Fold and Flip

Now let us revisit the bifurcation from Section 8.8(bottom), by considering the planar system with linear switching given by

$$(\dot{x}_1, \dot{x}_2) = \tfrac{1}{2}(1+\lambda)(x_2 - 2\alpha, 2) - \tfrac{1}{2}(1-\lambda)(x_2, 1) \ , \qquad (9.49)$$

where α is the bifurcation parameter. The phase portrait, shown in Figure 9.24, appears at first to unfold a simple bifurcation in which the relative position of two tangencies along the discontinuity surface is exchanged.

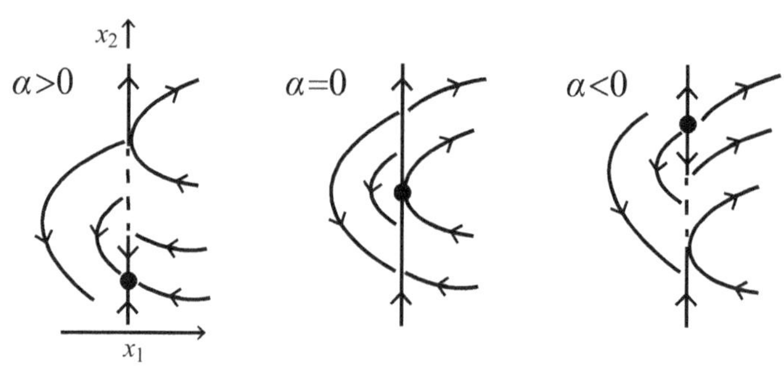

Fig. 9.24 A bifurcation in which visible and invisible folds in the flow exchange ordering is accompanied by a sliding node changing from attracting to repelling.

The discontinuity surface is attracting for $x_2 < \min(0, 2\alpha)$ and repelling for $x_2 > \max(0, 2\alpha)$. The switching layer system is

$$(\varepsilon\dot{\lambda}, \dot{x}_2) = \tfrac{1}{2}(1+\lambda)(x_2 - 2\alpha, 2) - \tfrac{1}{2}(1-\lambda)(x_2, 1) \quad \text{on } x_1 = 0 \ . \qquad (9.50)$$

Sliding modes exist for $|\alpha/(x_2 - \alpha)| < 1$ with $\lambda^{\$} = \alpha/(x_2 - \alpha)$, with dynamics $\dot{x}_2 = (x_2 + 2\alpha)/2(x_2 - \alpha)$. There is therefore a sliding node at $x_2^* = -2\alpha$, $\lambda^* = -1/3$. More simply than finding the Jacobian, near this equilibrium we can expand $\dot{x}_2 = -(x_2 + 2\alpha)/6\alpha + \mathsf{O}\left((x_2 + 2\alpha)^2\right)$, implying that the node is attracting for $\alpha > 0$ and repelling for $\alpha < 0$.

So, as sketched in Figure 9.25, an attracting node in negative x_2 moves to become a repelling node in positive x_2 as α changes sign.

Some explanation is now required for how the flip in attractivity occurs, as in the study of the boundary equilibrium bifurcation in the previous section. And as in the previous section, the flip is associated with a structural instabil-

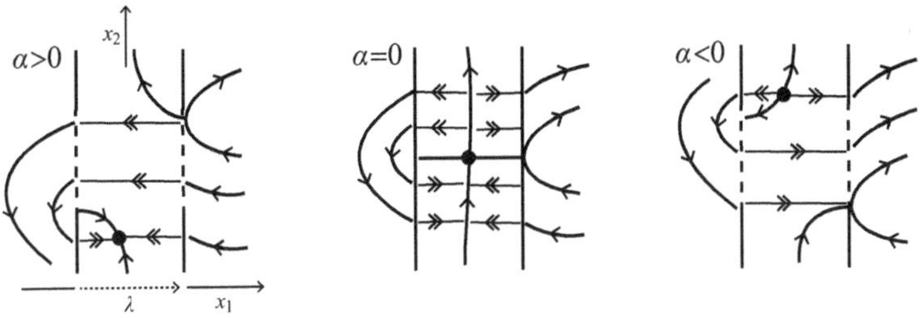

Fig. 9.25 The bifurcation in Figure 9.24, with $x_1 = 0$ blown up into the layer $\lambda \in (-1, +1)$.

ity in the switching layer. In fact the switching layer system (9.50) contains two topological degeneracies which coincide with the flip in the node's attractivity. Firstly, the sliding manifold $\mathcal{M}$ is degenerate at such a double tangency, because the nullcline $\dot{\lambda} = 0$ has segments lying parallel to the fast (λ) direction, as illustrated in Figure 9.25 at $\alpha = 0$ and as predicted in Section 9.5.3. Secondly, the layer system (9.50) becomes a centre when $\alpha = 0$. The centre degeneracy can be broken by adding an x_2 term to the $\dot{x}_2$ equation, but this does not break the sliding manifold degeneracy, which indeed can only be broken by a term nonlinear in λ. In fact, a term nonlinear in λ turns out to be sufficient to break both the sliding manifold and centre degeneracies.

Therefore we introduce a perturbed system in the form of (9.25) as

$$(\dot{x}_1, \dot{x}_2) = \tfrac{1}{2}(1+\lambda)(x_2 - 2\alpha, 2) - \tfrac{1}{2}(1-\lambda)(x_2, 1) + (\lambda^2 - 1)(\beta, 0) \ . \quad (9.51)$$

The switching layer system on $x_1 = 0$ is

$$(\varepsilon\dot{\lambda}, \dot{x}_2) = \tfrac{1}{2}(1+\lambda)(x_2 - 2\alpha, 2) - \tfrac{1}{2}(1-\lambda)(x_2, 1) + (\lambda^2 - 1)(\beta, 0) \ , \quad (9.52)$$

on $x_1 = 0$. For small β the boundaries of the sliding region change only slightly from (9.50). The fixed points of the fast $\dot{\lambda}$ subsystem give the sliding modes as satisfying $\beta\lambda^2 + (x_2 - \alpha)\lambda - \beta - \alpha = 0$, with solutions

$$\lambda^{\$} = \frac{\alpha - x_2}{2\beta} \pm \sqrt{\frac{(\alpha - x_2)^2}{4\beta^2} + 1 + \frac{\alpha}{\beta}} \ ,$$

and to lie in the layer $|\lambda^{\$}| < 1$, these must satisfy $(x_2 - \alpha)^2 > -4\beta^2(1 + \frac{\alpha}{\beta})$. There is a sliding equilibrium at $(\lambda^*, x_2^*) = (-1/3, -2\alpha - 8\beta/3)$, with Jacobian in the layer variables $(\xi_1, \xi_2) = (\varepsilon\lambda, x_2)$, given by

$$\begin{pmatrix} \frac{\partial \dot{\xi}_1}{\partial \xi_1} & \frac{\partial \dot{\xi}_1}{\partial \xi_2} \\ \frac{\partial \dot{\xi}_2}{\partial \xi_1} & \frac{\partial \dot{\xi}_2}{\partial \xi_2} \end{pmatrix} = \begin{pmatrix} \frac{10\beta + 9\alpha}{-3\varepsilon} & -\frac{1}{3} \\ \frac{3}{2\varepsilon} & 0 \end{pmatrix} ,$$

whose trace changes sign at $\alpha = \alpha_h = -\frac{10}{9}\beta$. This implies that the change in attractivity of the equilibrium is again facilitated by a Hopf bifurcation occurs in the switching layer system.

The perturbation ensures that the trace of the Jacobian when $\alpha = 0$ is $-10\beta/3$, so the equilibrium is nondegenerate. When the Hopf bifurcation occurs at $\alpha = \alpha_h$, the switching layer system is given by

$$(\varepsilon\dot{\lambda}, \dot{x}_2) = \tfrac{1}{2}(1+\lambda)(x_2 - 2\alpha_h, 2) - \tfrac{1}{2}(1-\lambda)(x_2, 1) + (\lambda^2 - 1)(\beta, 0) ,$$

which is not a centre provided $\beta \neq 0$, unlike in the unperturbed system. A nondegenerate Hopf bifurcation therefore takes place at $\alpha = \alpha_h$. Depending on the sign of β, this occurs on one side or the other of the bifurcation of tangencies which occurs at $\alpha = 0$.

Taking β negative or positive yields supercritical and subcritical cases of the Hopf bifurcation, respectively, shown in Figures 9.26 and 9.27. A single degenerate bifurcation now becomes a short sequence of bifurcations that unfold as we vary α, described qualitatively as follows.

First we take negative β in Figure 9.26.

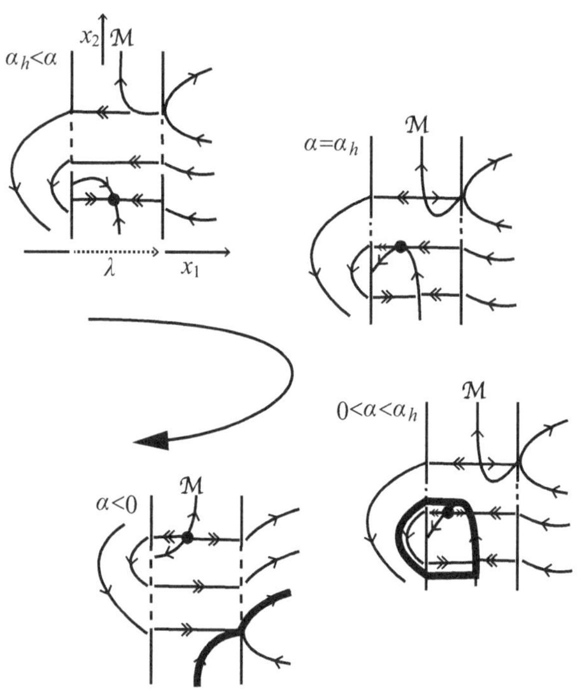

Fig. 9.26 The bifurcation for $\beta < 0$ (so $\alpha_h > 0$) gives a supercritical Hopf bifurcation.

An attracting equilibrium exists inside the switching layer for α positive and greater than α_h. This undergoes a supercritical Hopf bifurcation at $\alpha = \alpha_h$, changing stability to become a repelling equilibrium, emitting an attracting limit cycle. The cycle is formed by the continuation of the sliding manifold $\mathcal{M}$ under the flow and grows as α decreases. These form the first three portraits in Figure 9.26.

No bifurcation is evident in the vector field outside the switching layer as these changes take place, with the repelling equilibrium and cycle constituting a hidden repeller and attractor, respectively. As α decreases further, the sliding manifold $\mathcal{M}$ itself bifurcates, its two branches rearranging to form the last portrait of Figure 9.26. At the branches join between the third and fourth portraits, the repelling equilibrium passes from one branch to the other, and the limit cycle formed by the continuation of $\mathcal{M}$ tails off to infinity. The two tangencies exchange ordering along the discontinuity surface as α passes through zero, so no further qualitative changes occur inside the switching layer as α decreases further and the sequence is complete.

The process for positive β is slightly different, shown in Figure 9.27.

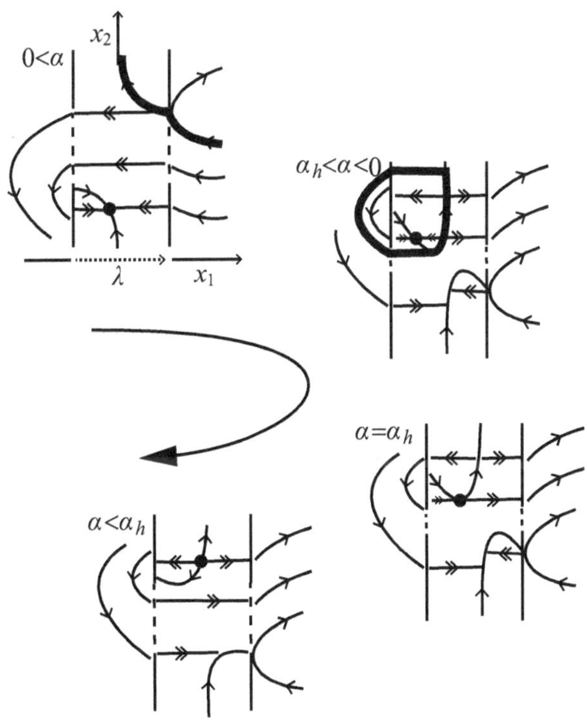

Fig. 9.27 The bifurcation for $\beta > 0$ (so $\alpha_h < 0$) gives a subcritical Hopf bifurcation.

An attracting equilibrium again exists inside the switching layer for α positive and greater than α_h, but now the continuation of the sliding manifold $\mathcal{M}$ under the flow immediately plays a role, highlighted as a bold curve, a 'limit cycle at infinity'. The tangencies exchange ordering along the discontinuity surface as α passes through zero, causing the two branches of $\mathcal{M}$ to bifurcate and driving the attracting equilibrium from one branch of $\mathcal{M}$ to the other. The continuation of $\mathcal{M}$ then forms a limit cycle, shown in the second portrait. The equilibrium and the cycle are a hidden attractor and repeller, respectively. Inspection of the vector field outside the discontinuity surface might suggest that the equilibrium is repelling, but in fact it only becomes repelling after a subcritical Hopf bifurcation at $\alpha = \alpha_h$ in the third portrait, when it enters the repelling branch of the sliding manifold $\mathcal{M}$, as in the fourth portrait of Figure 9.27.

These geometrical arguments can be made more rigorous, but let us conclude only with a simulation, Figure 9.28, taking ε nonzero as in the last section for ease of inspection. For different signs of β, we see the supercrit-

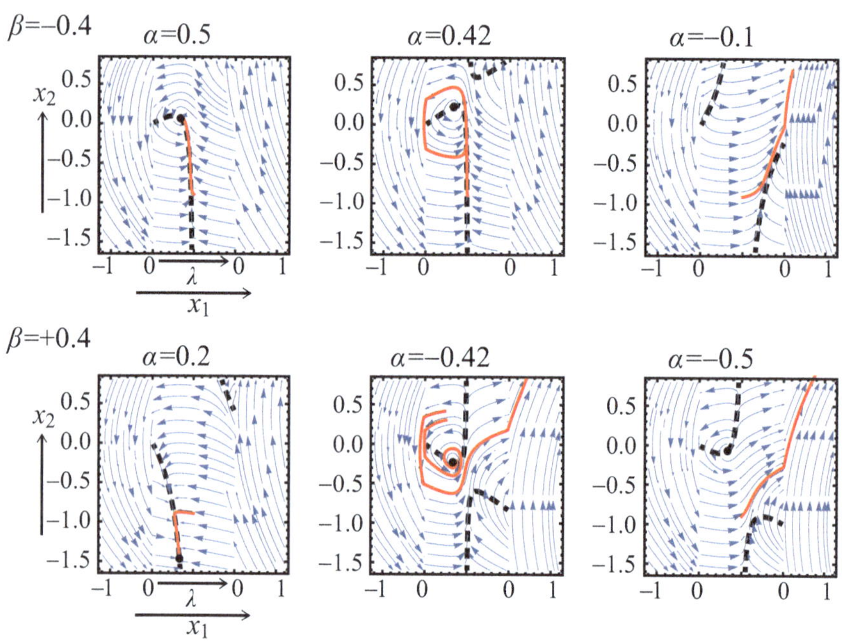

Fig. 9.28 Simulation of the two cases of visible-invisible fold-fold bifurcation, in which a sliding node changes attractivity, giving birth to a limit cycle. The dotted curve is the sliding manifold $\mathcal{M}$. In each case a trajectory is simulated from initial point $(0, -0.9)$, except in the lower middle portrait, where two trajectories are taken from initial points $(-0.5, 0.3)$ and $(-0.5, 0.5)$, one converging to the focus and the other diverging to infinity, indicating the existence of a repelling limit cycle between them (we can of course find this more accurately if we wish). Simulated with $\varepsilon = 0.1$.

ical and subcritical bifurcations, indicated by the birth of an attracting or repelling limit cycle associated with the change in attractivity of the sliding node.

The structurally stable form of a double tangency in three or more dimensions also requires hidden dynamics to be structurally stable. The simplest example is the two-fold singularity, which we study in Section 13.5.

9.7 Looking Forward

The huge possibilities for the different manifestations of hidden dynamics, and their role in local and global bifurcations, are merely hinted at by the examples in this chapter. How many others are possible or lie waiting to be recognised in applications?

A study which effectively uses hidden dynamics to classify the typical ways a trajectory may behave when it enters an intersection of two switches in a planar system has been made in [93] (though relying on smoothing the discontinuity rather than forming a layer system). For higher dimensions a full classification is generally impossible, the permutations evidently vast. Our aim here has been to identify typical behaviours that may signify the presence of hidden dynamics in general applications.

This chapter has introduced the elements and surprises of local nonlinear switching. There remains much more to discover, but for us it is time to move on to global dynamics. Section 9.4.2 in particular has introduced rich dynamics induced by motion repeatedly entering into and exiting from sliding. It is here that the issue of determinism comes to the fore, and it is to this that we now turn.

Chapter 10
Breaking Determinacy

The organizing centres of local and global behaviour in smooth systems are most commonly equilibria—fixed points of a flow—or invariant manifolds emanating from equilibria. In nonsmooth systems, it turns out to be transitional points—singularities the flow passes through in finite time—that create the most interesting dynamics.

Although it is not possible to classify all singularities in higher dimensions or at many switches, there is a particular kind of singularity, whose local geometry we *can* classify, that turns out to lie at the heart of both local events like determinacy-breaking and global events like sliding bifurcations or explosions. This singularity is called an *exit point*. In this chapter we concentrate on its local properties.

Here we come face to face with the issue that piecewise-smooth dynamics is *almost* deterministic. As we follow a flow, it will evolve deterministically, perhaps collapsing to lower dimensions in places so the history is not unique, but with determinism nevertheless intact for almost all times.

Except, that is, when the flow encounters determinacy-breaking singularities. Typically these are exit points from sliding.

10.1 Exit Points

Having made a classification of sliding modes in Chapter 9, we can now turn to what happens when a flow exits from sliding.

Here we introduce general types of exit point, with a classification that establishes their key dynamical features (e.g. deterministic or not) and the conditions for existence and methods of detection.

M. R. Jeffrey, *Hidden Dynamics*,
https://doi.org/10.1007/978-3-030-02107-8_10

Definition 10.1. An **exit point** is a point on a codimension r discontinuity surface where the flow exits from attracting sliding into a region of equal or lower sliding codimension on a different branch of the sliding manifold $\mathcal{M}$ and/or different discontinuity submanifold.

In the simplest case, for example, the flow exits from codimension 1 sliding and thereby leaves the discontinuity surface $\mathcal{D}$. Attracting sliding manifolds are invariant by definition; therefore exit points can arise only at their boundaries. We have seen the basic mechanism already in Section 7.1, namely, boundaries of the sliding manifold $\mathcal{M}$ (and hence the boundaries of sliding regions), which are either *end points*, where a sliding mode multiplier $\lambda_j^{\$}$ leaves the interval $(-1, +1)$ (and where the constituent vector fields are tangent to the discontinuity surface), or *turning points* where attracting and repelling sliding modes annihilate as $\mathcal{M}$ loses normal hyperbolicity.

Exit points may be either deterministic, so that the onward flow through the exit point is unique, or determinacy-breaking, so that the onward flow through the exit point is set-valued. Similarly the history of the flow into the exit point may be unique or set-valued. Many of these appeared for the first time (in the theoretical piecewise-smooth literature at least) in [115], and it is worth surveying those preliminary analyses here.

10.2 Exit Points: Deterministic

The routes available for a flow to exit from sliding (or decrease its sliding codimension) deterministically are somewhat limited, involving end points formed from simple tangencies or involving transversal crossing of an intersection of discontinuity submanifolds, along with certain local stability conditions.

10.2.1 Exit via a Simple Tangency

The simplest way a flow exits from sliding on a single discontinuity submanifold is via an end point of $\mathcal{M}$. From the definition of the sliding manifold (7.7) for $r = 1$, an end point of $\mathcal{M}$ occurs when $f_1^{\pm}(\mathbf{x}; \pm 1) \equiv f_1^{\pm}(\mathbf{x}) = 0$, which constitutes a tangency between the respective vector field $\mathbf{f}^{\pm}$ and the discontinuity surface $\sigma_1(\mathbf{x}) = 0$.

Thus for the simplest kind of exit point, we have to go right back to the simple folds in Figure 6.9, Section 6.2. They come in four types determined by their visibility and the attractivity of the sliding region they bound. Only one of these, a visible fold at the boundary of attracting sliding, is an exit point. (An invisible fold at the boundary of repelling sliding is not an exit point because the flow escapes the discontinuity surface at every point of the sliding region, so no reduction of codimension along the flow occurs at the fold, and the remaining cases are entry, not exit, points of the sliding region.)

At a simple fold where $\delta_{t\pm}\sigma_1 = 0$ but $\delta_{t\mp}\sigma_1 \neq 0$, by Section 9.5.3 the vector field with no hidden term is structurally stable, so we can consider linear dependence on λ_1. The sliding vector $\mathbf{f}^{\$}(\mathbf{x})$ given by the right-hand side of (7.8) is then equal to the tangential constituent vector field $\mathbf{f}^{\pm}$, and therefore the flow passes differentiably through the fold, typically (provided $\mathbf{f}^{\pm} \neq 0$) in finite time. Moreover the right-hand sides of the discontinuous system (7.1), the switching layer system (7.2), and the sliding system (7.8) are equal precisely at such points where $\lambda_1^{\$} = +1$ or -1.

The dynamics at a nondegenerate tangency, i.e. a quadratic tangency of one flow only, where only one of the conditions $\delta_{t\pm}\sigma_1 = 0$ hold, is therefore locally very simple. The flow actually transitions differentiably from sliding on the discontinuity surface into smooth motion outside it, and by implication, such a flow is deterministic.

Simple tangencies have been well studied, particularly in [48, 71, 140]. They are most interesting for their role in global dynamics as the instigators of so-called sliding bifurcations (see Chapter 11), whereby limit cycles or stable/unstable manifolds lose or gain connections to the discontinuity surface.

10.2.2 Exit Transverse to an Intersection

Consider a codimension r sliding trajectory evolving along an intersection of r discontinuity submanifolds that then encounters an intersection with another discontinuity submanifolds. Exit can then occur into sliding of equal or lower codimension along a different branch of the discontinuity submanifolds or out of the discontinuity surface altogether. The exit could be either deterministic or determinacy-breaking. We investigate the typical behaviour that characterizes deterministic exit in this section and cover determinacy-breaking exit later in Section 10.3.1.

In the simplest case, a trajectory sliding along one discontinuity submanifold $\mathcal{D}_1$ encounters an intersection with another discontinuity submanifold $\mathcal{D}_2$. Let us imagine, without loss of generality, that a trajectory is sliding along the region $x_2 < 0$ of a discontinuity submanifold $x_1 = 0$ and encounters a second discontinuity submanifold $x_2 = 0$. Assuming that the trajectory exits as in Definition 10.1, and does not instead enter into higher codimension sliding along the intersection, then it must evolve into one of the two quadrants of the plane with $x_2 > 0$ and either $x_1 < 0$ or $x_1 > 0$ or into one of the three discontinuity surface regions $x_1 = 0 < x_2$, $x_2 = 0 < x_1$, and $x_2 = 0 > x_1$. Exit into $x_1 < 0$ is impossible because the flow is attracting towards $x_1 = 0 > x_2$ by assumption. These scenarios are illustrated in Figure 10.1, showing the three scenarios in which exit occurs to lower codimension in (i) or to the same codimension of sliding within the same discontinuity submanifold in (ii) or a different discontinuity submanifold in (iii).

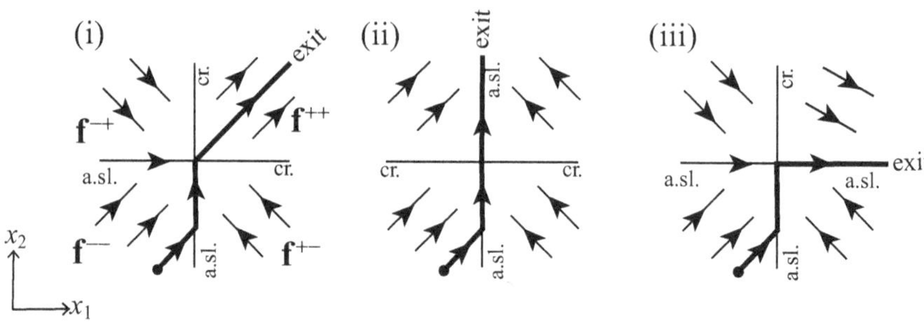

Fig. 10.1 Exit from a region of sliding via an intersection, illustrating the three systems in (10.1a)(i-iii). In (i) exit occurs from a region of codimension 1 sliding to codimension 0 sliding (i.e. exit from the discontinuity surface). In (ii)–(iii) exit occurs from one region of codimension 1 sliding to another, directly across the intersection in (ii), and around a corner in (iii).

To ensure that the exit is deterministic, we need to assume that exit is possible into only one of these regions at $x_1 = x_2 = 0$; otherwise determinacy is broken (which we leave to Section 10.3.1). If exit is not possible into any region, then higher codimension sliding occurs.

As prototypes of deterministic exit at an intersection, consider the piecewise-constant systems

$$\text{(i)} \qquad (\dot{x}_1, \dot{x}_2) = (1 + \lambda_2 + \lambda_1\lambda_2,\ 1 - \lambda_2 + \lambda_1\lambda_2)\,, \tag{10.1a}$$
$$\text{(ii)} \qquad (\dot{x}_1, \dot{x}_2) = (-\lambda_1,\ 2 - \lambda_2)\,, \tag{10.1b}$$
$$\text{(iii)} \qquad (\dot{x}_1, \dot{x}_2) = (2 - \lambda_1 + 2\lambda_1\lambda_2,\ -\lambda_2)\,, \tag{10.1c}$$

where $\lambda_1 = \operatorname{sign} x_1$ and $\lambda_2 = \operatorname{sign} x_2$, consistent with those shown in Figure 10.1. These have constituent vector fields

$$\text{(i)} \qquad \begin{cases} \mathbf{f}^{++} = (3,\ 1)\,, & \mathbf{f}^{--} = (1,\ 3)\,, \\ \mathbf{f}^{-+} = (1, -1)\,, & \mathbf{f}^{+-} = (-1, 1)\,, \end{cases} \tag{10.2a}$$
$$\text{(ii)} \qquad \begin{cases} \mathbf{f}^{++} = (-1,\ 1)\,, & \mathbf{f}^{--} = (1,\ 3)\,, \\ \mathbf{f}^{-+} = (1, 1)\,, & \mathbf{f}^{+-} = (-1, 3)\,, \end{cases} \tag{10.2b}$$
$$\text{(iii)} \qquad \begin{cases} \mathbf{f}^{++} = (3,\ -1)\,, & \mathbf{f}^{--} = (5,\ 1)\,, \\ \mathbf{f}^{-+} = (1, -1)\,, & \mathbf{f}^{+-} = (-1, 1)\,. \end{cases} \tag{10.2c}$$

Take system (i) for instance. The layer system on $x_1 = 0$ is $(\varepsilon_1\dot{\lambda}_1, \dot{x}_2) = (2+\lambda_1\lambda_2,\ \lambda_1)$ for $x_2 > 0$ and $(\varepsilon_1\dot{\lambda}_1, \dot{x}_2) = (-\lambda_1\lambda_2,\ 2-\lambda_1)$ for $x_2 < 0$ and on $x_2 = 0$ is $(\dot{x}_1, \varepsilon_2\dot{\lambda}_2) = (1+2\lambda_2,\ 1)$ for $x_1 > 0$ and $(\dot{x}_1, \varepsilon_2\dot{\lambda}_2) = (1,\ 1-2\lambda_2)$ for $x_1 < 0$. From these it is quite easy to see that sliding occurs in the regions $x_2 = 0 > x_1$ and $x_1 = 0 > x_2$ and flows towards the switching intersection $x_1 = x_2 = 0$. Crossing occurs on $x_2 = 0 < x_1$ and $x_1 = 0 < x_2$. The switching layer system at the intersection $x_1 = x_2 = 0$ is

$$(\varepsilon_1\dot{\lambda}_1, \varepsilon_2\dot{\lambda}_2) = (1+\lambda_2+\lambda_1\lambda_2,\ 1-\lambda_2+\lambda_1\lambda_2)\ ,$$

which flows towards the upper right corner of the layer $(\lambda_1, \lambda_2) \in (-1, +1) \times (-1, +1)$. The result is that all orbits flow eventually into the region $x_1, x_2 > 0$. Orbit that slide initially, then exit at the intersection, do so along a common trajectory $\{x_1(t), x_2(t)\} = \{3t, t\}$ for $t \geq 0$.

This is rather simple because the flow is single-valued. Various other scenarios may be studied, but they generate little of interest for deeper study here, and the analysis of (ii)–(iii) is similar to that of (i) above. In each case an attracting branch of a sliding manifold $\mathcal{M}$ exists in each sliding region, and the different branches are connected by trajectories passing deterministically through the intersection in finite time.

10.2.3 Exit via Tangency to an Intersection

Exit from codimension r to $r-1$ sliding for $r \geq 1$ can occur where the codimension $r-1$ sliding vector field is tangent to the codimension r sliding region, on an intersection of r discontinuity submanifolds.

To study exit from sliding via a simple tangency of the flow to an intersection, take as a prototype

$$(\dot{x}_1, \dot{x}_2, \dot{x}_3) = \tfrac{1+\lambda_2}{2}\left(\tfrac{1+\lambda_1}{2}x_3 + 1, -1, \tfrac{1+\lambda_1}{2}\right) + \tfrac{1-\lambda_2}{2}(-\lambda_1, 1, 0) \tag{10.3}$$

with $\lambda_1 = \operatorname{sign} x_1$ and $\lambda_2 = \operatorname{sign} x_2$, restricting to $x_3 > -1$, whose geometry is sketched in Figure 10.2. This has constituent vector fields

$$(\dot{x}_1, \dot{x}_2, \dot{x}_3) = \begin{cases} \mathbf{f}^{++} = (x_3+1, -1, 1) \text{ if } 0 < x_1, x_2, \\ \mathbf{f}^{-+} = (+1\ ,\ -1\ ,\ 0\ ) \text{ if } x_1 < 0 < x_2, \\ \mathbf{f}^{+-} = (-1\ ,\ +1\ ,\ 0\ ) \text{ if } x_2 < 0 < x_1, \\ \mathbf{f}^{--} = (+1\ ,\ +1\ ,\ 0\ ) \text{ if } x_1, x_2 < 0. \end{cases} \tag{10.4}$$

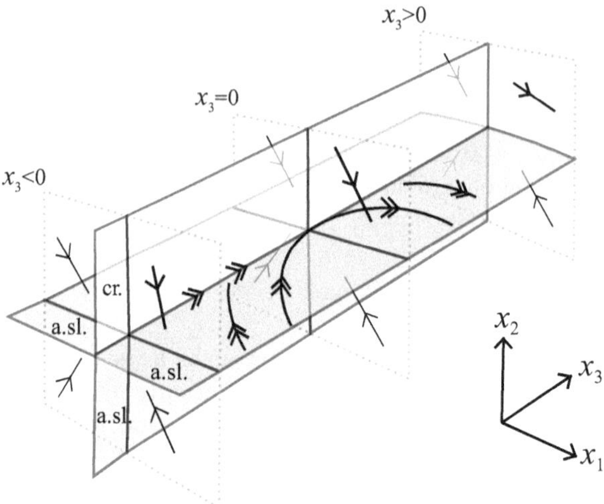

Fig. 10.2 Sketch of the system (10.3), showing projections of the vector fields into three constant x_3 planes, and the sliding dynamics on three half-planes and on the intersection.

Let us first look at the dynamics on the discontinuity submanifolds, seeking sliding modes using (7.8) for $r = 1$ on $x_1 = 0$ and $x_2 = 0$ separately. We find:

- $x_1 = 0 < x_2$ is a crossing region since $f_1^{++} f_1^{-+} = 1 + x_3 > 0$;
- $x_1 = 0 > x_2$ is a sliding region since $f_1^{+-} f_1^{--} = -1 < 0$, the sliding modes satisfy $\lambda_1 = \lambda_1^{\$} = 0$, giving a sliding system $(\dot{x}_2, \dot{x}_3) = (1, 0)$;
- $x_2 = 0 \neq x_1$ is a sliding region since $f_2^{++} f_2^{+-} = -1 < 0$ on $x_2 = 0 < x_1$ and $f_2^{-+} f_2^{--} = -1 < 0$ on $x_2 = 0 > x_1$, the sliding modes in both regions satisfy $\lambda_2^{\$} = 0$, giving sliding systems $(\dot{x}_1, \dot{x}_3) = (x_3, 1)$ and $(\dot{x}_1, \dot{x}_3) = (1, 0)$, respectively.

At the intersection $x_1 = x_2 = 0$, applying (7.8) for $r = 2$, we find sliding modes with $(\lambda_1^{\$}, \lambda_2^{\$}) = ((x_3 + 2)/(x_3 - 2), 0)$, giving one-dimensional dynamics $\dot{x}_3 = 1/(2 - x_3)$, existing only for $x_3 < 0$ where $|\lambda_1^{\$}| < 1$.

This implies that trajectories in $x_3 < 0$ are attracted onto sliding modes in $x_1 = 0 > x_2$ and $x_2 = 0 \neq x_1$ and thence attracted onto the intersection $x_1 = x_2 = 0$ whereupon they travel towards the origin. At the origin the intersection ceases to admit sliding (since $|\lambda_1^{\$}| > 1$ for $x_3 > 0$), so orbits must exit, and do so on $x_2 = 0 < x_1$ along the sliding system $(\dot{x}_1, \dot{x}_3) = (x_3, 1)$, which at the origin is tangent to the intersection as sketched in Figure 10.2.

As for the visible tangency in Section 10.2.2, the exit is deterministic, (here we have a tangency of the codimension 1 sliding flow to the switching intersection, in Section 10.2.2 we had a tangency of the codimension 0 [non-]sliding flow to the discontinuity surface). We should inspect the dynamics inside the switching layer more closely to understand how the exit occurs.

On $x_1 = 0$ the switching layer system is

$$(\varepsilon_1\dot{\lambda}_1, \dot{x}_2, \dot{x}_3) = \begin{cases} \left(\frac{1}{2}(1+\lambda_1)x_3 + 1, -1, \frac{1}{2}(1+\lambda_1)\right) & \text{if } x_2 > 0\,, \\ (-\lambda_1, 1, 0) & \text{if } x_2 < 0\,, \end{cases} \tag{10.5}$$

with $\lambda_1 \in (-1, +1)$, illustrated in Figure 10.3. A sliding manifold $\mathcal{M}$ exists for $x_2 < 0$ with mode $\lambda_1^{\$} = 0$, giving a sliding vector field $(\dot{x}_2, \dot{x}_3) = (1, 0)$, so all trajectories flow into the intersection in finite time. There are no sliding modes for $x_2 > 0$, and instead the fast subsystem $\varepsilon_1\dot{\lambda}_1 = 1 + \mathsf{O}(x_3)$ carries the flow across the discontinuity surface in the direction of increasing x_2, at least for small x_3.

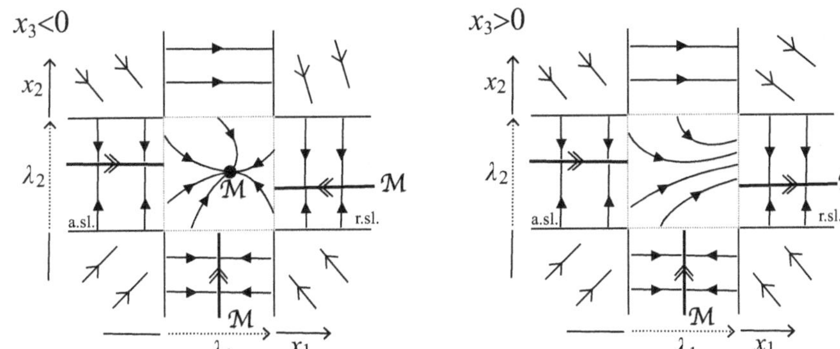

Fig. 10.3 Sketch of the layer dynamics of the system (10.3), showing an attracting sliding manifold $\mathcal{M}$ consisting of curves in the regions $x_2 = 0 < x_1$, $x_2 = 0 > x_1$, and $x_1 = 0 > x_2$ and a point inside $x_1 = x_2 = 0$ for $x_3 < 0$ (this refers to curves and points in $\mathbb{R}^2$, which are of course surfaces and curves, respectively, in the full $\mathbb{R}^3$).

The more interesting switching layer system is the one on $x_2 = 0$,

$$(\dot{x}_1, \varepsilon_2\dot{\lambda}_2, \dot{x}_3) = \begin{cases} \left(\frac{1}{2}(1+\lambda_2)x_3 + \lambda_2, -\lambda_2, \frac{1}{2}(1+\lambda_2)\right) & \text{if } x_1 > 0\,, \\ (1, -\lambda_2, 0) & \text{if } x_1 < 0\,, \end{cases} \tag{10.6}$$

with $\lambda_2 \in (-1, +1)$. This has a sliding manifold $\mathcal{M}$ for all $x_1 \neq 0$ with modes $\lambda_2^{\$} = 0$, giving sliding dynamics $(\dot{x}_1, \dot{x}_3) = (x_3, 1)/2$ for $x_1 > 0$ and $(1, 0, 0)$ for $x_1 < 0$. The $\dot{x}_1$ component implies that the sliding flow is attracted onto the intersection in $x_3 < 0$ but crosses the intersection in the direction of increasing $\dot{x}_1$ for $x_3 > 0$.

We already found the sliding modes $(\lambda_1^{\$}, \lambda_2^{\$}) = ((2+x_3)/(2-x_3),\ 0)$ that exist at the intersection $x_1 = x_2 = 0$ for $x_3 < 0$, and these should form a sliding manifold $\mathcal{M}$. This is confirmed by the switching layer system, which is

$$(\varepsilon_1\dot{\lambda}_1, \varepsilon_2\dot{\lambda}_2, \dot{x}_3) = \tfrac{1+\lambda_2}{2}\left(\tfrac{1+\lambda_1}{2}x_3 + 1, -1, \tfrac{1+\lambda_1}{2}\right) + \tfrac{1-\lambda_2}{2}(-\lambda_1, 1, 0) \quad (10.7)$$

with $(\lambda_1, \lambda_2) \in (-1, +1) \times (-1, +1)$. For $x_3 < 0$ this has an attracting sliding manifold $\mathcal{M}$ as given by the sliding modes, on which the sliding dynamics is $\dot{x}_3 = 1/(2 - x_3)$, and hence in a positive direction in the region of interest. When the flow enters the intersection in the region $x_3 < 0$, it collapses onto $\mathcal{M}$ and travels towards $x_3 = 0$, where $\mathcal{M}$ leaves the region $(\lambda_1, \lambda_2) \in (-1, +1) \times (-1, +1)$. Inside the intersection, the flow is still attracted towards the line $\lambda_2 = 0$, on which $\varepsilon_1\dot{\lambda}_1 = \frac{1}{2}(1-\lambda_1)+\frac{1}{4}x_3(1+\lambda_1)$ is strictly positive for $x_3 > 0$. This directs the flow out of the intersection into sliding on the discontinuity submanifold $x_2 = 0 < x_1$.

One may construct other examples that exhibit similar behaviour and in particular build up examples that form a hierarchy of intersections and sliding modes of successively higher codimension r, giving exit points from intersections via tangency of the codimension $r - 1$ sliding vector field. If a trajectory passed through a series of such points, it would cascade down to successively lower codimensions of sliding, eventually releasing from the discontinuity surface altogether. Each of these exit events should behave similar to that above, that is, deterministically, with each decreasing the sliding codimension by one. Coincidences of many such events could decrease the codimension of sliding by more than one in a single event, however, but this would be accompanied by determinacy-breaking, as in the following section.

10.3 Exit Points: Determinacy-Breaking

In the neighbourhood of a determinacy-breaking exit point, the local flow turns around such that there are regions of both attracting and repelling sliding, usually also with regions of crossing. This brings the possibility of the flow transitioning between the three types, in particular between attractive and repulsive regions of sliding.

If the determinacy-breaking point lies in a c-dimensional set of determinacy-breaking points in n-dimensions, the outset can be of dimension anything from $\max(2, c + 1)$ to n, depending on the local sliding flows. The inset and outset through these points are not necessarily of the same dimension as each other. As far as we know the inset and outset of the flow through a determinacy-breaking point are the same for complex tangencies, but need not be so at intersections of discontinuity submanifolds.

10.3.1 Exit Transverse to an Intersection

An intersection of discontinuity submanifolds permits changes in the direction of the constituent flows with respect to the discontinuity surfaceand with much greater freedom and genericity than a complex tangency. They therefore have the potential to play a more significant role in applications. We will show here that when determinacy-breaking occurs, it can be partially resolved by layer analysis.

As a prototype of determinacy-breaking exit from codimension $r = 1$ sliding at a codimension $r = 2$ intersection, consider

$$(\dot{x}_1, \dot{x}_2, \dot{x}_3) = (1,\ x_3 + \lambda_1\lambda_2,\ 0)\ , \qquad \lambda_j = \operatorname{sign}(x_j)\ , \tag{10.8}$$

for $|x_3| < 1$, or in terms of the constituent vector fields,

$$(\dot{x}_1, \dot{x}_2, \dot{x}_3) = \begin{cases} \mathbf{f}^{++} = \mathbf{f}^{--} = (1, x_3 + 1, 0) & \text{if } x_1 x_2 > 0, \\ \mathbf{f}^{-+} = \mathbf{f}^{+-} = (1, x_3 - 1, 0) & \text{if } x_1 x_2 < 0. \end{cases} \tag{10.9}$$

The equality between diagonally opposite vector fields in (10.9) is a convenience here and has no bearing on the results; indeed we may add small constant, linear, or nonlinear terms to any of the four vector fields without significant effect.

From inspecting of the constituent vector fields, it appears that orbits cross through the discontinuity submanifold $x_1 = 0$ everywhere and slide along the discontinuity submanifold $x_2 = 0$ everywhere. The sliding region $x_2 = 0 \geq x_1$ is attracting, while $x_2 = 0 \leq x_1$ is repelling. Moreover the sliding along $x_2 = 0$ is everywhere to the right, evolving through the intersection $x_1 = x_2 = 0$ in finite time from the attracting to the repelling region. Of course to confirm these we must inspect the switching layer systems, which we do in detail below.

The line $x_1 = x_2 = 0$ is a determinacy-breaking singularity. All trajectories in the region $x_3 - 1 \leq x_2/x_1 \leq x_3 + 1$ pass through the intersection $x_1 = x_2 = 0$ (see Figure 10.4(i)), forming a continuum of trajectories all flowing into and out of the intersection in finite time. Any point in this set with $x_1 < 0$ is connected via the flow to any point in this set with $x_1 > 0$ with the same x_3 value. Outside of the region $x_3 - 1 \leq x_2/x_1 \leq x_3 + 1$, the system is deterministic.

We have to inspect the layer dynamics to verify whether the orbits described above actually pass through the intersection. The switching layer system on $x_1 = 0$ for $x_2 \neq 0$ is

$$(\varepsilon_1\dot{\lambda}_1, \dot{x}_2, \dot{x}_3) = (1,\ x_3 + \lambda_1 \operatorname{sign}(x_2),\ 0)\ , \tag{10.10}$$

with $\lambda_1 \in (-1, +1)$. The fast $\dot{\lambda}_1$ subsystem is constant and thus provides a simple transition between the surfaces $\lambda_1 = -1$ and $\lambda_1 = +1$ on the t/ε_1 timescale.

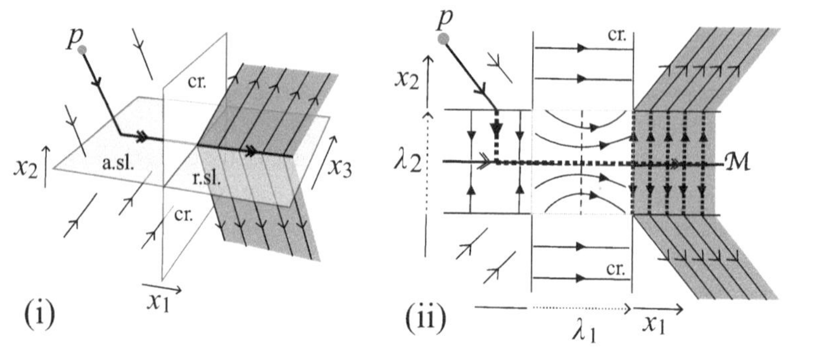

Fig. 10.4 Determinacy-breaking at a switching intersection (i) shows the discontinuous system and (ii) shows the blow-up system in the plane $x_3 = 0$. The trajectory of any point p in $x_1 < 0$ becomes multivalued as it exits the intersection, identifiable as the set $x_{1,2} = 0$ in (a) and $(\lambda_1, \lambda_2) \in (-1,+1) \times (-1,+1)$ in (b).

On $x_2 = 0$ for $x_1 \neq 0$ the switching layer system is

$$(\dot{x}_1, \varepsilon_2\dot{\lambda}_2, \dot{x}_3) = (1,\ x_3 + \lambda_2 \operatorname{sign}(x_1),\ 0)\ , \tag{10.11}$$

with $\lambda_2 \in (-1,+1)$. The fast $\dot{\lambda}_2$ subsystem has a set of x_3-parameterized fixed points forming sliding modes $\lambda_2^{\$} = -x_3 \operatorname{sign}(x_1)$, forming planar sliding manifolds

$$\mathcal{M} = \left\{ (x_1, \lambda_2, x_3) : \begin{array}{c} x_1 \neq 0,\ |x_3| < 1, \\ \lambda_2 = -x_3\ \operatorname{sign}(x_1) \end{array} \right\} \tag{10.12}$$

which are normally hyperbolic since $\partial\varepsilon_2\dot{\lambda}_2/\partial\lambda_2 = \operatorname{sign}(x_1)$ and are attracting for $x_1 < 0$ and repelling for $x_1 > 0$. On $\mathcal{M}$ the sliding dynamics is

$$(\dot{x}_1, 0, \dot{x}_3) = (1,\ x_3 + \lambda_2 \operatorname{sign}(x_1),\ 0)\ . \tag{10.13}$$

This gives a constant drift in the positive x_1 direction on $\mathcal{M}$ inside $x_2 = 0$, with $\lambda_2 = -x_3 \operatorname{sign}(x_1)$.

Lastly we have the switching layer system on the intersection $x_1 = x_2 = 0$,

$$(\varepsilon_1\dot{\lambda}_1, \varepsilon_2\dot{\lambda}_2, \dot{x}_3) = (1,\ x_3 + \lambda_1\lambda_2,\ 0)\ , \tag{10.14}$$

for $(\lambda_1, \lambda_2) \in (-1,+1) \times (-1,+1)$. There are no sliding modes in the layer system; instead orbits pass through in a positive direction without stopping. Solutions of (10.14) can be expressed as graphs

$$\lambda_2(\lambda_1) = e^{\nu^2\lambda_1^2/2}\left(\lambda_{20}e^{-\nu^2\lambda_{10}^2/2} + x_3\sqrt{\tfrac{\pi}{2}}\left\{\operatorname{Erf}\left[\tfrac{\nu\lambda_1}{\sqrt{2}}\right] - \operatorname{Erf}\left[\tfrac{\nu\lambda_{10}}{\sqrt{2}}\right]\right\}\right)\ ,$$

parameterized by initial conditions $(\lambda_1(0), \lambda_2(0)) = (\lambda_{10}, \lambda_{20})$, where $\nu = \varepsilon_1/\varepsilon_2$, and Erf is the standard error function [3]. These solutions allow us to continue the attracting and repelling planes of $\mathcal{M}$ (given by $\lambda_2 = \pm x_3$ in $x_1 \lesseqgtr \mp 1$) into the region $(\lambda_1, \lambda_2) \in (-1, +1) \times (-1, +1)$, as

$$\lambda_2(\lambda_1) = x_3 e^{\nu^2\lambda_1^2/2}\left(\pm e^{-\nu^2/2} + \sqrt{\tfrac{\pi}{2}}\left\{\mathrm{Erf}\left[\tfrac{\nu\lambda_1}{\sqrt{2}}\right] \pm \mathrm{Erf}\left[\tfrac{\nu}{\sqrt{2}}\right]\right\}\right) . \qquad (10.15)$$

Plotting these solutions, we find that the flow from the attracting plane of $\mathcal{M}$ curves towards negative λ_2 in $x_3 < 0$ whereby it exits into the region $x_1, x_2 > 0$ and curves towards positive λ_2 in $x_3 > 0$ whereby it exits either into the region $x_2 < 0 < x_1$. To confirm this, upon reaching either $\lambda_1 = +1$ or $|\lambda_2| = 1$, the $\dot{\lambda}_2$ and $\dot{\lambda}_1$ dynamics, respectively, given by (10.11) and (10.10) drive the flow into the corners where $\lambda_1 = +1$ and $\lambda_2 = \mathrm{sign}(x_3)$. This dynamics is illustrated in Figure 10.5 for $x_3 < 0$.

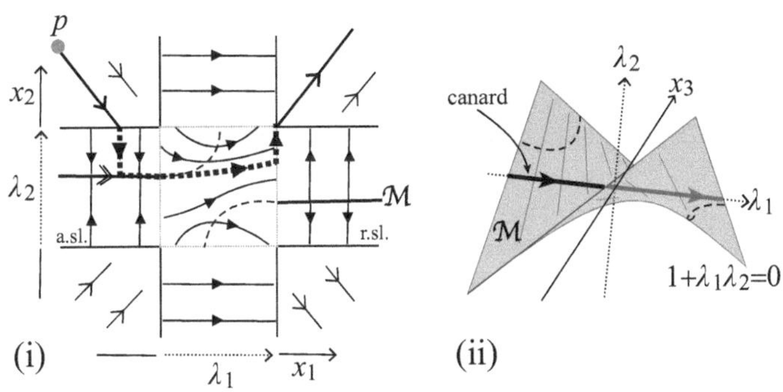

Fig. 10.5 Sketch of the switching layer system for simple exit from a crossing of discontinuity submanifolds. In (ii) we show the layer of the intersection, as well as the switching layers along $x_1 = 0$ (for $x_2 \neq 0$) and $x_2 = 0$ (for $x_1 \neq 0$), and the dynamics outside the discontinuity submanifolds; the case shown is in a coordinate plane with constant $x_3 < 0$; and in (i) we show the sliding manifold $\mathcal{M}$ inside the switching layer of the intersection point $x_1 = x_2 = 0$.

The splitting between the continuations of the branches of $\mathcal{M}$ in the x_2 direction inside the intersection, i.e. the difference between the solutions (10.15), depends linearly on x_3, given by

$$\Delta\lambda_2(\lambda_1) = x_3 e^{\nu^2\lambda_1^2/2}\left(2e^{-\nu^2/2} + \sqrt{2\pi}\,\mathrm{Erf}\left[\frac{\nu^2}{\sqrt{2}}\right]\right) .$$

Where this splitting vanishes, the attracting and repelling branches of $\mathcal{M}$ are connected by a *canard* trajectory, given by

$$\lambda_1(t) = t\,, \qquad \lambda_2(t) = 0\,, \qquad x_3(t) = 0\,, \qquad (10.16)$$

in the region $\lambda_2 \in (-1, +1)$, valid for all t and hence running along the λ_1 coordinate axis. The canard passes from the attracting plane $\lambda_2 = x_3$ for

$\lambda_1 < -1$ to the repelling plane $\lambda_2 = x_3$ for $\lambda_1 > +1$. There is only one such trajectory, and it is structurally stable, because the attracting and repelling manifolds generated by $\mathcal{M}$ intersect transversally at $\lambda_1 = \lambda_2 = x_3 = 0$.

It is important to note that the existence of one unique canard trajectory connecting the attracting and repelling branches of $\mathcal{M}$, rather than every trajectory on $x_2 = 0$ being a canard, is evident only from this layer analysis (Figure 10.5) and cannot be seen by inspecting the dynamics outside the discontinuity surface (Figure 10.4(i)) alone.

Thus a unique trajectory exists that passes through the intersection and remains asymptotic to $x_2 = 0$ as $x_1 \to \pm\infty$. All other trajectories that enter the intersection are expelled via the point $\lambda_1 = +1$, $\lambda_2 = \text{sign}(x_3)$, depending on the value of x_3 along them.

Viewed another way, treating x_3 as a parameter of a system in the (x_1, x_2) plane, the prototype above shows how hidden dynamics can lead to quantitatively different behaviour for different values of x_3, with exit upwards or downwards depending on the sign of x_3, and determinacy-breaking occurring only at $x_3 = 0$. Taken in three dimensions, the different scenarios unfold to create a structurally stable singularity and, at its heart, a canard trajectory (10.16) through the intersection, hidden inside the switching layer.

Note that, in this case, the timescale ratio $\nu = \varepsilon_1/\varepsilon_2$ plays no part in the qualitative behaviour of the system.

As in the previous sections, we could construct numerous other scenarios that exhibit similar behaviour and yield to similar analysis. For example, one may consider constituent vector fields $\mathbf{f}^{--} = (1, x_3 + 1, 0)$, $\mathbf{f}^{+-} = (1, -1, 0)$, and $\mathbf{f}^{-+} = (-1, -2, 0)$, with either $\mathbf{f}^{++} = (1, 3, 0)$ or $\mathbf{f}^{++} = (-1, 2, 0)$. In both of these, a similar determinacy-breaking passage through the intersection can be resolved by layer analysis, except at a special value of x_3 where a canard produces determinacy-breaking. In these examples, there is also reinjection of the set-valued flow back into the singularity, resulting in complex oscillatory dynamics in the neighbourhood of the intersection.

Determinacy-breaking is a local phenomenon, but it should obviously have global consequences for any flow passing through it. To illustrate we conclude with a simulation of a system that has the same qualitative phase portrait as (10.9) near the intersection $x_1 = x_2 = 0$,

$$\begin{pmatrix} \dot{x}_1 \\ \dot{x}_2 \\ \dot{x}_3 \end{pmatrix} = \begin{pmatrix} 1 - \lambda_2 x_2 + \frac{1}{5}\lambda_1 \\ x_3 + \lambda_1\lambda_2 - c\lambda_2 \\ -\frac{1}{10}x_3 - \frac{1}{5}x_2 \end{pmatrix} \tag{10.17}$$

where c is a constant in the range $0 < c < 1$.

First, observe that the constant c has little qualitative effect on the phase portraits of (10.17), shown in Figure 10.6. Nevertheless, the simulations below show the dynamics to depend highly on c due to sensitivity in the flow's exit from the intersection.

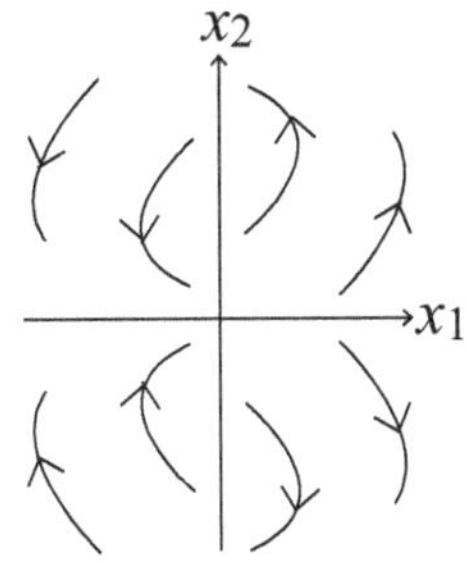

Fig. 10.6 The flow of (10.17) in a plane x_3 =constant for small x_3. The phase portrait does not change qualitatively outside the discontinuity surface as we change c.

Let us simulate the global dynamics of this system by replacing $\lambda_j = \text{sign}(x_j)$ with $\lambda_i \mapsto \phi(x_j/\varepsilon) = \tanh(x_j/\varepsilon)$ for $j = 1, 2$, with $\varepsilon = 10^{-4}$. The result, shown in Figure 10.7, is a simple periodic orbit in (i), and as we vary c the period of this attractor changes rapidly, becoming eventually the complex attractor in (ii). Closer inspection of the simulations (not shown) confirms that the flow enters the origin by sliding along $x_1 < 0 = x_2$, then exits into

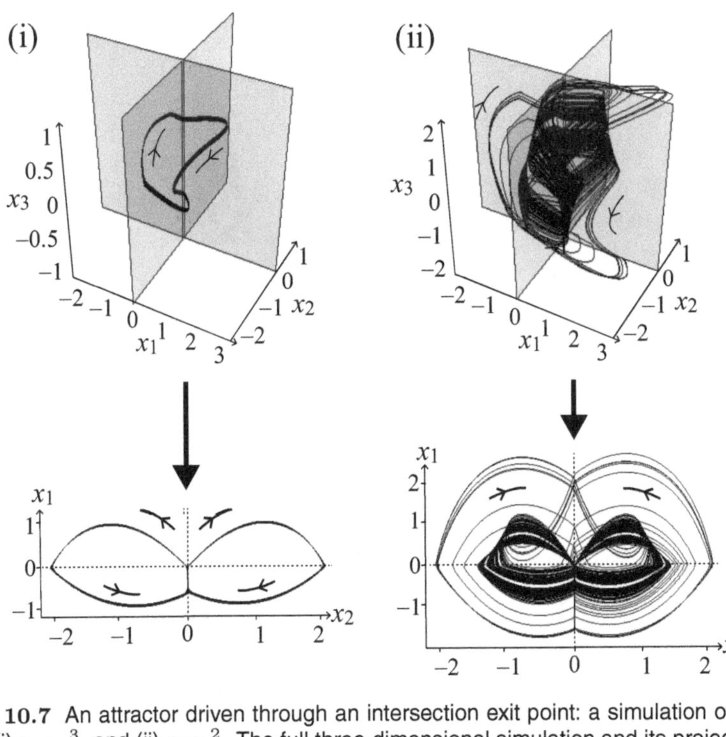

Fig. 10.7 An attractor driven through an intersection exit point: a simulation of (10.17) with (i) $c = \frac{3}{10}$ and (ii) $c = \frac{2}{5}$. The full three-dimensional simulation and its projection into the (x_1, x_2) plane are shown.

positive x_2 for $x_3 > 0$ and into negative x_2 for $x_3 < 0$, as predicted from the layer analysis above. Any trajectories that pass through $x_1 = 0$ cross it transversally (this happens in the positive x_1 direction near the intersection, but also in the negative x_1 direction at large x_3 values which allows the flow in (ii) to loop around more intricately). Any trajectories that hit the half-plane $x_2 = 0 > x_1$ do so at small enough x_3 that they then slide along $x_2 = 0$ into the singularity. In (ii-b) trajectories are also seen that cross the half-plane $x_2 = 0 < x_1$, which is allowed since these have strayed to large enough x_3 that $x_2 = 0$ is no longer a sliding region, so the flow crosses through transversally.

This dynamics is a result of the singularity geometry, and is not acutely sensitive to the manner of smoothing (at least no more than will be described in Section 12.3). To verify this we simulate the same system for the same parameters but approximate the switch by different sigmoid functions. In Figure 10.8 we repeat the simulation with the smooth rational function $\phi(x_j/\varepsilon) = (x_j/\varepsilon)/\sqrt{1+(x_j/\varepsilon)^2}$ in (i–ii.a) and the continuous but non-differentiable ramp function $\phi(x_j/\varepsilon) = \text{sign}(x_j)$ for $|x_j| > \varepsilon$ and $\phi(x_j/\varepsilon) = x_j/\varepsilon$ for $|x_j| \le \varepsilon$ in (i–ii.b). These demonstrate that the choice of smoothing has no significant effect upon the dynamics, and is not responsible for the complex dynamics observed. We could also take different values of $0 < \varepsilon \ll 1$ or introduce hysteresis, delay, or noise, with similar results.

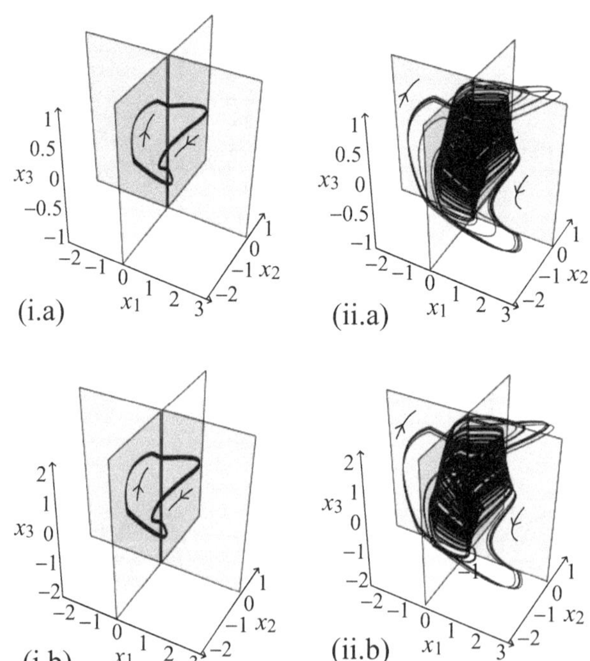

Fig. 10.8 The attractors in Figure 10.7 for the smooth rational (a) and ramp (b) smoothings described in the text, with parameters and initial conditions corresponding to those in Figure 10.7.

This is essentially the only way that a codimension $r = 1$ sliding trajectory can exit with determinacy-breaking upon encountering an intersection of codimension $r = 2$, since such a trajectory will not generically run into a codimension $r \geq 3$ intersection (i.e. where three or more discontinuity submanifolds intersect). To investigate determinacy-breaking further, we can therefore proceed to considering exit from codimension $r = 2$ sliding.

10.3.2 Exit via a Complex Tangency

A complex tangency is a place where two vector fields lie tangent to $\mathcal{D}$, both turning around such that they lie on the boundaries of attracting sliding, repelling sliding, and crossing regions. We saw the most basic examples in the two-fold of Section 6.3 and Section 8.5.2. Typically the vector field need not vanish at such a point, so the flow will traverse it in finite time

When multiple flows are tangent to a discontinuity surface $\mathcal{D}$, or multiple codimension $r - 1$ (or lower) sliding flows are tangent to a codimension r intersection of submanifolds $\mathcal{D}_j$, the conditions can exist for determinacy-breaking. They create the conditions for a flow to be conveyed *from* attractive *to* repelling sliding, via the singularity, constituting a *canard* trajectory (see Definition 4.3).

Recall that a point in the attracting sliding region can be reached from many points in the piecewise-smooth flow, consisting of a sliding trajectory $\mathbf{x}_s(t)$ and a continuum of trajectories that impact the discontinuity surface along $\mathbf{x}_s(t)$ (Figure 10.9, left). So such a point has many possible histories in the flow. Similarly, a point in the repelling sliding region has many possible futures, consisting of a sliding trajectory $\mathbf{x}_s(t)$ and a continuum of trajectories that depart the discontinuity surface from points along $\mathbf{x}_s(t)$.

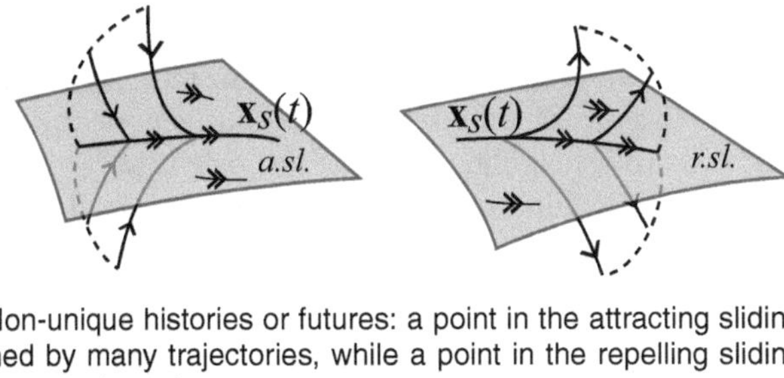

Fig. 10.9 Non-unique histories or futures: a point in the attracting sliding region ($a.sl.$) can be reached by many trajectories, while a point in the repelling sliding region ($r.sl.$) evolves into many trajectories.

Complex tangencies like the two-fold singularity compound this multivaluedness because they lie on the boundary of both attracting and repelling sliding regions. If sliding modes are able to traverse those singularities as canards, the singularity will have both a non-unique history (many orbits arriving at the singularity in forward time) and non-unique future (many possible onward orbits from the singularity). As illustrated in Figure 10.10 (shown in three dimensions), these set-valued histories and futures may have dimension between 2 and n, that is, a p-parameter family of orbits where $1 < p < n - 1$.

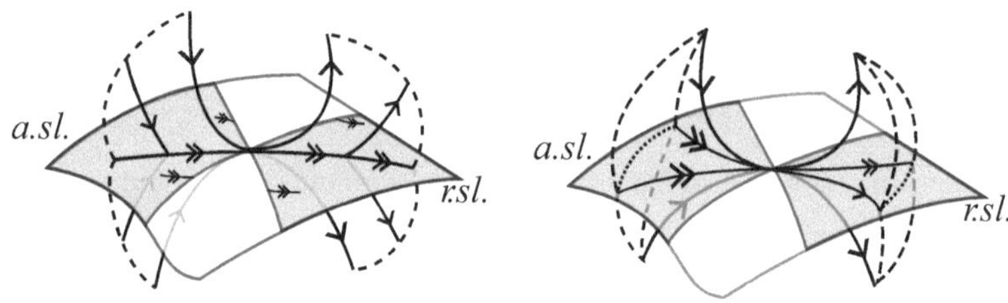

Fig. 10.10 Simple canards and robust canards from attracting sliding ($a.sl.$) to repelling sliding ($r.sl.$). Simple canards (left) form an $n - 1$ dimensional set, and robust canards (right) form an n dimensional set.

We deal with the two-fold singularity at length in Chapter 13 (as well as glimpses in Section 6.4 and Section 8.5.2), so we do not revisit it here. Below, instead, we move on to considering exit from codimension $r = 2$ sliding via a complex tangency, and in doing so, we shall find that the two-fold forms a model for higher codimension exits.

To study exit from codimension $r = 2$ sliding via a complex tangency, consider the prototype

$$\begin{pmatrix} \dot{x}_1 \\ \dot{x}_2 \\ \dot{x}_3 \\ \dot{x}_4 \end{pmatrix} = \frac{(1+\lambda_1)(1+\lambda_2)}{4} \begin{pmatrix} x_3 + 1 \\ -1 \\ a_1 \\ b_1 \end{pmatrix} + \frac{(1-\lambda_1)(1-\lambda_2)}{4} \begin{pmatrix} x_4 - d \\ d \\ -b_2 \\ -a_2 \end{pmatrix} + \frac{\lambda_1-\lambda_2}{2} \begin{pmatrix} -1 \\ +1 \\ 0 \\ 0 \end{pmatrix} \tag{10.18}$$

where $\lambda_1 = \operatorname{sign} x_1$ and $\lambda_2 = \operatorname{sign} x_2$, and in terms of constants $d = \pm 1$, $a_i = \pm 1$ and $b_i \in \mathbb{R}$. It is necessary here to consider four dimensions, as multiple tangencies to a switching intersection occur generically only in $\mathbb{R}^{n\geq 3}$. We restrict attention to a neighbourhood of the origin $|x_3| < 1$, $|x_4| < 1$.

Figures 10.11 and 10.12 illustrate the basic dynamics in the (x_1, x_2) plane in different regions of (x_3, x_4) space. The constituent vector fields are

$$(\dot{x}_1, \dot{x}_2, \dot{x}_3, \dot{x}_4) = \begin{cases} \mathbf{f}^{++} = (1 + x_3,\ -1,\ a_1,\ b_1) & \text{if} \quad 0 < x_1, x_2, \\ \mathbf{f}^{-+} = (+1,\ -1,\ 0,\ 0\ ) & \text{if} \quad x_1 < 0 < x_2, \\ \mathbf{f}^{+-} = (-1,\ +1,\ 0,\ 0\ ) & \text{if} \quad x_2 < 0 < x_1, \\ \mathbf{f}^{--} = (d - x_4, -d,\ b_2,\ a_2\ ) & \text{if} \quad x_1, x_2 < 0. \end{cases} \tag{10.19}$$

We will write components of the constituent vector fields as

$$\mathbf{f}^K = (f_1^K, f_2^K, f_3^K, f_4^K) \qquad K = \pm\pm \ .$$

Of the four regions of the discontinuity submanifolds $\{x_1 = 0 < x_2\}$, $\{x_1 = 0 > x_2\}$, $\{x_2 = 0 < x_1\}$, and $\{x_2 = 0 > x_1\}$, two exhibit crossing, and two exhibit sliding. For $d = -1$ the two sliding regions are coplanar, lying on the same discontinuity submanifold $x_1 = 0$, and for $d = +1$ the sliding regions are orthogonal, lying on different discontinuity submanifolds.

Take first the coplanar case $d = -1$, shown in the (x_1, x_2) plane in Figure 10.11.

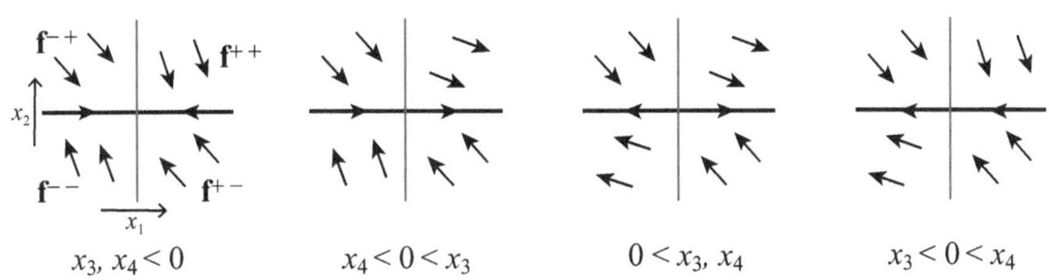

Fig. 10.11 Dynamics in the (x_1, x_2) plane for $d = -1$. As x_3 and x_4 change sign, the fields $\mathbf{f}^{00}$ and $\mathbf{f}^{--}$ rotate, and their directions relative to $\mathbf{f}^{+-}$ and $\mathbf{f}^{-+}$ change whether the sliding vector fields point towards or away from the intersection $x_1 = x_2 = 0$.

The flow crosses the discontinuity submanifold $x_1 = 0$, since $f_1^{++} f_1^{-+} = 1 + x_3 > 0$ in $x_2 > 0$ and $f_1^{+-} f_1^{--} = 1 + x_4 > 0$ in $x_2 < 0$. The discontinuity submanifold $x_2 = 0$ is an attracting sliding region for all $x_1 \neq 0$ since $f_2^{++} f_2^{+-} = -1 < 0$ in $x_1 > 0$ and $f_2^{-+} f_2^{--} = -1 < 0$ in $x_1 < 0$. By (7.6) the sliding modes are given by $\lambda_2^{\$} = 0$ and give sliding dynamics

$$(\dot{x}_1, \dot{x}_3, \dot{x}_4) = \begin{cases} (\ x_3, a_1, b_1)/2 & \text{if} \quad x_1 > 0\ , \\ (-x_4, b_2, a_2)/2 & \text{if} \quad x_1 < 0\ , \end{cases} \tag{10.20}$$

on $x_2 = 0$.

The intersection exhibits sliding for $x_3 x_4 > 0$. By (7.6) the sliding modes are given by $\lambda_1^{\$} = \frac{x_4 - x_3}{x_4 + x_3}$ and $\lambda_2^{\$} = 0$, and since they must lie inside $(-1, +1) \times (-1, +1)$, they exist only for $x_3 x_4 > 0$, giving dynamics

$$(\dot{x}_3, \dot{x}_4) = \frac{(a_1 x_4 + b_2 x_3, a_2 x_3 + b_1 x_4)}{j(x_3, x_4)} \tag{10.21}$$

on $x_1 = x_2 = 0$, where $j(x_3, x_4) = 2(x_3 + x_4)$. For $x_3 x_4 < 0$ the flow crosses through the intersection from one sliding region on the discontinuity submanifold $x_1 = 0$ to another. We will look more closely at what (10.21) means for the sliding dynamics at the intersection below.

Before that, let us look also at the orthogonal case $d = +1$, shown in the (x_1, x_2) plane in Figure 10.12.

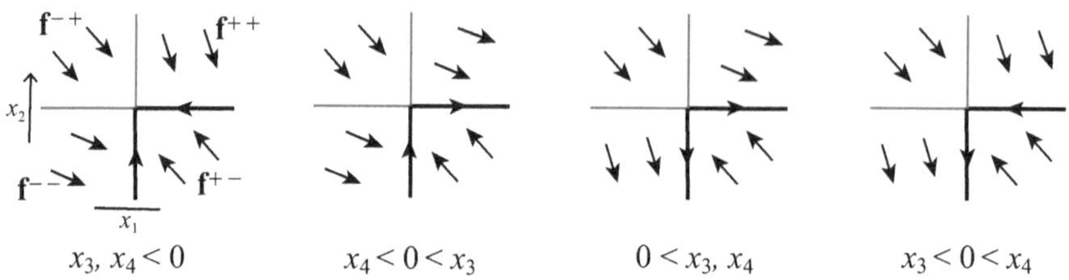

Fig. 10.12 Dynamics in the (x_1, x_2) plane for $d = +1$. As x_3 and x_4 change sign, the fields $\mathbf{f}^{00}$ and $\mathbf{f}^{--}$ rotate, and their directions relative to $\mathbf{f}^{+-}$ and $\mathbf{f}^{-+}$ change whether the sliding vector fields point towards or away from the intersection $x_1 = x_2 = 0$.

On the discontinuity submanifold $x_1 = 0$ for $x_2 > 0$ the flow crosses since $f_1^{++}f_1^{-+} = 1 + x_3 > 0$. On the discontinuity submanifold $x_2 = 0$ for $x_1 < 0$ the flow crosses since $f_2^{-+}f_2^{--} = 1 > 0$. (These simple arguments work on the discontinuity submanifolds because dependence on the switching multipliers is linear). Elsewhere on the discontinuity surface there is sliding. Sliding exists on the discontinuity submanifold $x_1 = 0$ for $x_2 < 0$ since $f_1^{+-}f_1^{--} = x_4 - 1 < 0$, and on the discontinuity submanifold $x_2 = 0$ for $x_1 > 0$ since $f_2^{++}f_2^{+-} = -1 < 0$. Applying (7.6) these sliding modes satisfy $\lambda_1^\$ = \frac{x_4}{x_4-2}$ and $\lambda_2^\$ = 0$, respectively, giving sliding dynamics

$$
\begin{aligned}
(\dot{x}_2, \dot{x}_3, \dot{x}_4) &= \frac{(-x_4, b_2, a_2)}{2 - x_4} && \text{on } x_1 = 0 > x_2 && (10.22\text{a})\\
(\dot{x}_1, \dot{x}_3, \dot{x}_4) &= (x_3, a_1, b_1)/2 && \text{on } x_2 = 0 < x_1 \,. && (10.22\text{b})
\end{aligned}
$$

Both of these sliding regions are attracting.

The intersection exhibits sliding for $x_3x_4 > 0$, where the modes satisfy $\lambda_1^\$ = 2x_4/\jmath(x_3, x_4)$ and $\lambda_2^\$ = -2x_3/\jmath(x_3, x_4)$, giving sliding dynamics

$$
(\dot{x}_3, \dot{x}_4) = \frac{(a_1x_4 + b_2x_3, a_2x_3 + b_1x_4)}{\jmath(x_3, x_4)} \tag{10.23}
$$

on $x_1 = x_2 = 0 < x_3x_4$, where

$$
\jmath(x_3, x_4) = \frac{-4x_3x_4}{x_3 + x_4 + \sqrt{(x_3 + x_4)^2 + 4x_3x_4}}
$$

and $x_3x_4 > 0 \Rightarrow \jmath(x_3, x_4) > 0$. For $x_3x_4 < 0$ the flow crosses through the intersection from one discontinuity submanifold to the other.

In both cases $d = \pm 1$, the sliding dynamics at the intersection given by (10.21) and (10.23) takes a familiar and illuminating form. The numerator is simply the vector field of a general equilibrium in the (x_3, x_4) plane, allowing us to sketch its phase portrait as a focus, node, or saddle, for different values of the constants a_i and b_i. The phase portrait is an accurate topological portrayal of the flow, but the point $x_3 = x_4 = 0$ is not actually an equilibrium because the denominator $j(x_3, x_4)$ provides a singular time rescaling, giving a nonzero speed at $x_3 = x_4 = 0$. The sign of $j(x_3, x_4)$ is then crucial.

For $d = +1$, the time scaling by $j(x_3, x_4)$ is positive in the attracting sliding region and negative in the repelling sliding region. This time scaling becomes zero at $x_3 = x_4 = 0$, such that the vector field remains finite and nonzero there, and the phase portrait of the equilibrium 'folds up' such that it appears to be attracting from one region $x_3, x_4 \gtrless 0$ and repelling from the other. This permits the flow to pass in finite time from one sliding region to another. For $d = -1$ the time scaling is strictly negative in both sliding regions, becoming zero at $x_3 = x_4 = 0$ such that the vector field remains finite and nonzero, so if the flow is attracted to/repelled from the singularity, it reaches/departs it in finite time.

There exists a singularity at $x_3 = x_4 = 0$ where these sliding modes are undefined.

A little more insight into what is happening around this singularity is given by looking at how the sliding dynamics on the discontinuity submanifolds curves relative to the intersection. The curvature of the flow towards or away from the intersection is characterized by $\ddot{x}_1$ on $x_2 = 0$ or $\ddot{x}_2$ on $x_1 = 0$. Along the set $x_3 = 0$, we have $\ddot{x}_1 = a_1$ for $x_2 = 0 < x_1$. Along the set $x_4 = 0$, we have $\ddot{x}_1 = -a_2$ on $x_2 = 0 > x_1$ and $\ddot{x}_2 = -a_2$ on $x_1 = 0 > x_2$. The result is that both tangencies of the sliding dynamics to the intersection are of visible type for $a_1 = a_2 = +1$, invisible type for $a_1 = a_2 = -1$, and of mixed type (one visible and one invisible) for $a_1 a_2 = -1$ (one curves towards and one away from the intersection in either sliding region). This curvature also implies, as seen in Figures 10.11 and 10.12, that the intersection is attracting with respect to the sliding dynamics for $x_3, x_4 < 0$, repelling for $x_3, x_4 < 0$, while (as already stated above) the flow crosses between sliding regions at the intersection for $x_3 x_4 < 0$.

The passage between the two sliding regions, each of dimension three on (x_1, x_3, x_4) or (x_2, x_3, x_4) space, closely mimics the flow between two regions on (x_1, x_2, x_3) space in the two-fold of Section 8.5.2. Taking the example of two visible folds, we sketch the situation in Figure 10.13. We should compare this to the portraits of the two-fold from Figures 8.7 and 8.8. The sliding vector fields on the intersection, given by (10.20) and (10.22) on (x_3, x_4) space, are both expressible as

$$(\varepsilon_1 \dot{\lambda}_1, \varepsilon_2 \dot{\lambda}_2, \dot{x}_3, \dot{x}_4) \propto \frac{(0, 0, b_2 x_3 + a_1 x_4,\; a_2 x_3 + b_1 x_4)}{j(x_3, x_4)}, \tag{10.24}$$

and are equivalent up to time scaling to (8.20), i.e. the canonical form sliding vector field of a two-fold singularity on the discontinuity surface $x_3 = 0$ of a system in (x_1, x_2, x_3) space.

The system of sliding resulting from (10.19) differs from the two-fold in one important aspect, the sign of the time scaling $j(x_3, x_4)$ which, as described above, crucially changes the character of the singularity at $x_3 = x_4 = 0$. The singularity for the 'coplanar' case $d = +1$ may be called a *bridge point*, forming a bridge between attracting and repelling sliding regions, similar to a two-fold singularity, while for the 'orthogonal' case $d = -1$, it may be called a *jamming* point, an equilibrium that the flow may reach or depart in *finite* time.

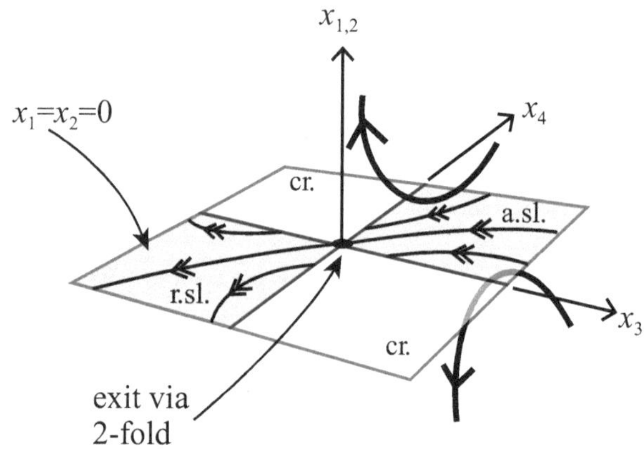

Fig. 10.13 Sketch of x_3-x_4 sliding dynamics. The visible-visible case of folds is shown with a saddle-like case of sliding dynamics. A complete catalogue of the possible two-dimensional sliding topologies in this codimension $r = 2$ surface is equivalent to the two-dimensional sliding phase portraits for the codimension $r = 1$ two-folds in [38].

The comparison to the two-fold singularity gives not only insight into the local dynamics, revealing that the singularity exhibits determinacy-breaking, but also suggest that the prototype (10.18) is actually degenerate. Because this is a complex tangency, as discussed in Section 9.5.3, the sliding manifold will have a degeneracy, and to obtain structural stability requires the addition of a nonlinear term such as $(1-\lambda_1^2)(\alpha, 0, 0, 0)$ for some small α. In Section 13.5 we shall see the straightforward steps to obtain the switching layer on the intersection for the system (10.24), obtaining the invariant sliding manifolds $\mathcal{M}$ that connect at $x_3 = x_4 = 0$.

More intriguingly, this prototype and its relation to the two-fold open the way to considering k-fold singularities for $k \geq 2$. In an n-dimensional system, at the intersection of r discontinuity submanifolds, there will generically occur sets of dimension $d = n - k$ where $k \leq 2r$ codimension $r - 1$ sliding flows are tangent to the intersection. Theorem 8.2 in Section 8.7 gives a general statement about their topology, but only extending analysis like that above can reveal their dynamics, a task we must leave open.

10.3.3 Zeno Exit from an Intersection

It is also possible for a codimension $r \geq 2$ sliding trajectory, evolving along an intersection of $r \geq 2$ discontinuity submanifolds, to exit directly out of the discontinuity surface altogether without meeting another discontinuity submanifold or a complex tangency.

The simplest case is exit from codimension $r = 2$ sliding without tangency of the neighbouring constituent or sliding vector fields and without any segments of codimension $r = 1$ sliding.

Taking a system in three dimensions and with two discontinuity submanifolds, consider the prototype

$$\begin{pmatrix} \dot{x}_1 \\ \dot{x}_2 \\ \dot{x}_3 \end{pmatrix} = \begin{pmatrix} \lambda_2 + \frac{1}{4}x_3(1+\lambda_1)(1+\lambda_2) \\ -\lambda_1 \\ \frac{1}{4}(1+\lambda_1)(1+\lambda_2) \end{pmatrix} , \tag{10.25}$$

restricted to $x_3 > -1$, where $\lambda_1 = \operatorname{sign} x_1$ and $\lambda_2 = \operatorname{sign} x_2$. The constituent vector fields are

$$(\dot{x}_1, \dot{x}_2, \dot{x}_3) = \begin{cases} \mathbf{f}^{++} = (x_3 + 1, -1, 1) \text{ if } 0 < x_1, x_2 \;, \\ \mathbf{f}^{-+} = (+1 \quad, +1, 0) \text{ if } x_1 < 0 < x_2 \;, \\ \mathbf{f}^{+-} = (-1 \quad, -1, 0) \text{ if } x_2 < 0 < x_1 \;, \\ \mathbf{f}^{--} = (-1 \quad, +1, 0) \text{ if } x_1, x_2 < 0 \;. \end{cases} \tag{10.26}$$

The precise values of the constants ± 1 in the vector fields have no qualitative bearing on the results but greatly simplify the calculations.

As we will show, this has a flow which crosses through the discontinuity submanifolds $x_1 = 0$ and $x_2 = 0$, spiralling around as it is attracted in towards the intersection $x_1 = x_2 = 0$ in $x < 3 = 0$ and then sliding along the intersection until exiting somewhere in $x > 3 > 0$ and spiralling back outward, as shown in Figure 10.14.

First we show that there is no sliding on the discontinuity submanifolds outside the intersection. On $x_1 = 0$ the fast subsystem of the switching layer is

$$\varepsilon_1 \dot{\lambda}_1 = \pm(1 + \tfrac{1}{2}x_3(1+\lambda_1)) \gtrless 0 \quad \text{for} \quad x_2 \gtrless 0 \;,$$

and on $x_2 = 0$ the fast subsystem of the switching layer is

$$\varepsilon_2 \dot{\lambda}_2 = \mp 1 \lessgtr 0 \quad \text{for} \quad x_1 \gtrless 0 \;,$$

implying the flow crosses the discontinuity submanifolds in a clockwise direction. Very often the most interesting dynamics in a system comes from sliding rather than crossing, and though this intuition will prove to be false here, let us follow it and go in search of sliding at the intersection.

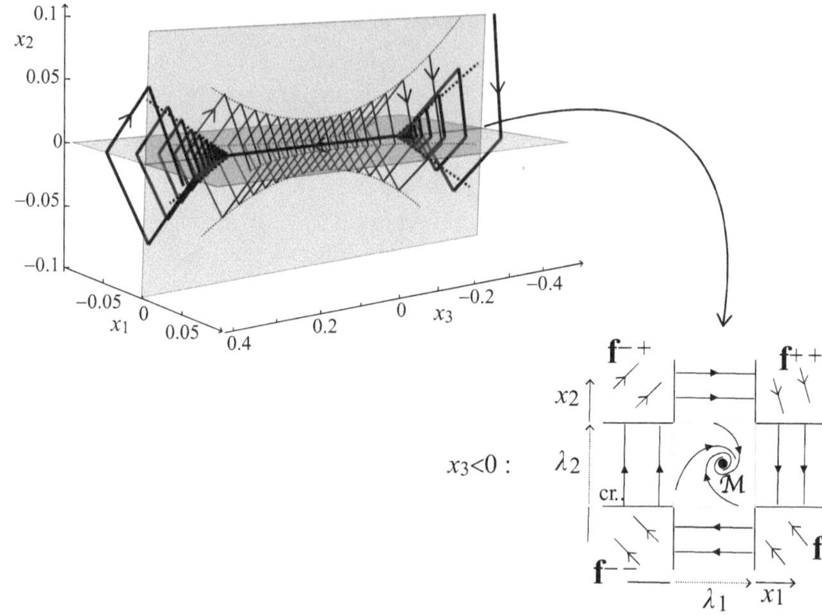

Fig. 10.14 Sketch of a Zeno trajectory (bold), taking infinitely many steps in finite time to spiral in towards the intersection in $x_3 < 0$ and out from the intersection in $x_3 > 0$. The inward and outward trajectories are connected by a codimension $r = 2$ sliding trajectory along the intersection. A simple trajectory which never reaches the intersection is also shown. The switching layer system, including a sliding manifold $\mathcal{M}$ with focal attraction, is shown inset (lower right).

The switching layer system at the intersection is

$$\begin{pmatrix} \varepsilon_1\dot{\lambda}_1 \\ \varepsilon_2\dot{\lambda}_2 \\ \dot{x}_3 \end{pmatrix} = \begin{pmatrix} \lambda_2 + \frac{1}{4}x_3(1+\lambda_1)(1+\lambda_2) \\ -\lambda_1 \\ \frac{1}{4}(1+\lambda_1)(1+\lambda_2) \end{pmatrix} . \tag{10.27}$$

The fixed points of the fast (λ_1, λ_2) subsystem give us sliding modes at $(\lambda_1^\$, \lambda_2^\$) = (0, -\frac{x_3}{x_3+4})$, forming a sliding manifold $\mathcal{M}$ on which the sliding dynamics is given by $\dot{x}_3 = 1/(4+x_3)$. The two-dimensional Jacobian derivative of $\mathcal{M}$ in layer variables $(\xi_1, \xi_2) = (\varepsilon_1\lambda_1, \varepsilon_2\lambda_2)$ is

$$\frac{\partial(\dot{\xi}_1, \dot{\xi}_2)}{\partial(\xi_1, \xi_2)} = \begin{pmatrix} \frac{x_3}{(x_3+4)\varepsilon_1} & \frac{4+x_3}{4\varepsilon_2} \\ -1/\varepsilon_1 & 0 \end{pmatrix} ,$$

which for $x_3 > -1$ has complex eigenvalues, giving focus-like rotation around $\mathcal{M}$ in the fast subsystem. The eigenvalues of the Jacobian have real part $x_3/2\varepsilon_1(4+x_3)$ whose sign is just that of x_3 (for $x_3 > -1$ at least), therefore $\mathcal{M}$ is attracting in $x_3 < 0$ and repelling in $x_3 > 0$. On $\mathcal{M}$ itself there is a drift along in the positive x_3 direction.

Thus when a trajectory enters the intersection $x_1 = x_2 = 0$ for $x_3 < 0$, it does so with a unique value of (λ_1, λ_2) lying on the set

$$\mathcal{B} = \left\{(\lambda_1, \lambda_2) \in [-1, +1]^2 : \ |\lambda_1| = 1 \text{ or } |\lambda_2| = 1\right\} .$$

It must enter the intersection in $-1 < x_3 < 0$, whereupon it will spiral around in the (λ_1, λ_2) coordinates of the switching layer system, initially decaying towards

$$\mathcal{M} = \{(\lambda_1, \lambda_2, x_3) : \lambda_1 = 0, \lambda_2 = -x_3/(x_3 + 4)\}$$

until it passes into $x_3 > 0$, where it then begins spiralling outward until it reaches $\mathcal{B}$ again and then exits. We can integrate (10.27) to find that λ_1 and λ_2 indeed evolve through the region $(\lambda_1, \lambda_2) \in (-1, +1) \times (-1, +1)$, starting and ending on the bounding box $\mathcal{B}$ on which they enter the intersection in $x_3 < 0$ and exit it in $x_3 > 0$.

The dynamics *inside* the intersection is therefore well defined, but the coordinates of entry and exit on $\mathcal{B}$, and the trajectories they connect to in $x_1 x_2 \neq 0$, may not be. Thus in this case it is the crossing dynamics *outside* the intersection that we must inspect more closely.

We said that the flow outside the discontinuity surface rotates around the intersection by continually traversing the discontinuity surfaces $x_1 = 0$ and $x_2 = 0$. To study this more closely, we construct a return map to the discontinuity submanifolds. Filippov used such maps to powerful effect in [71], and this particular map was derived in [115].

We will construct a map that describes how any trajectory on the discontinuity submanifold $x_1 = 0$ on $x_2 > 0$ returns to the same half-plane. Taking a starting point $(x_1, x_2, x_3) = (0, \eta, \zeta)$ at time $t = 0$, with $\eta > 0$ and $-1 < \zeta < 0$, we integrate the flow through crossings at times $t = t_1, t_2, t_3, t_4$. In $x_1 > 0$, $x_2 > 0$, we have $\dot{x}_2 = -1$, so to reach $x_2 = 0$ takes a time $t_1 = \eta$, arriving at $x_1(t_1) = \int_0^\eta (x_3 + 1)dt = \int_\zeta^{\eta+\zeta} (x_3 + 1)dx_3 = (\frac{1}{2}\eta + \zeta + 1)\eta$ and $x_3(t_1) = \eta + \zeta$. The next two sectors are a reflection so we arrive at $(-(\frac{1}{2}\eta + \zeta + 1)\eta, 0, \eta + \zeta)$ in time $t_3 - t_1 = 2(\frac{1}{2}\eta + \zeta + 1)\eta$. The last sector is a rotation to $(0, (\frac{1}{2}\eta + \zeta + 1)\eta, \eta + \zeta)$ in time $t_4 - t_3 = (\frac{1}{2}\eta + \zeta + 1)\eta$.

Taking the start and end coordinates gives the overall rotation map $(\eta_n, \zeta_n) \mapsto (\eta_{n+1}, \zeta_{n+1})$ on $\{x_1 = 0 < x_2\}$, namely

$$\begin{aligned} \eta_n &= (1 + \zeta_{n-1} + \tfrac{1}{2}\eta_{n-1})\eta_{n-1} , \\ \zeta_n &= \zeta_{n-1} + \eta_{n-1} . \end{aligned} \tag{10.28}$$

While we cannot solve this map, we observe that it has an invariant

$$\eta_n - \tfrac{1}{2}\zeta_n^2 = \eta_{n-1} - \tfrac{1}{2}\zeta_{n-1}^2 = \ldots = \eta_0 - \tfrac{1}{2}\zeta_0^2 ,$$

implying that its trajectories lie on the parabolic contours of the function

$$\psi(\eta, \zeta) = \eta - \tfrac{1}{2}\zeta^2 . \tag{10.29}$$

This tells us that an orbit that reaches a point $\eta_n = 0$ does so with $\zeta_n = \sqrt{\zeta_0^2 - 2\eta_0}$ and can do so only if it starts on a curve such that $\zeta_0^2 - 2\eta_0 > 0$.

We can then show that the rotation map (10.28) exhibits the Zeno phenomenon, that is, its attraction to/repulsion from the intersection occurs in finite time over infinitely many crossings.

Proposition 10.1. *An orbit starting at* (η_0, ζ_0) *such that* $\sqrt{\zeta_0^2 - 2\eta_0} > 0$ *and* $-1 < \zeta_0 < 0$ *converges to* $\eta_n = 0$ *as* $n \to \infty$ *in finite time* $\sqrt{\zeta_0^2 - 2\eta_0} - \zeta_0$. *(From [115]).*

Proof. An orbit starting at (η_0, ζ_0) such that $\sqrt{\zeta_0^2 - 2\eta_0} > 0$ and $-1 < \zeta_0 < 0$ will hit $\eta_n = 0$ when $\zeta_n = \sqrt{\zeta_0^2 - 2\eta_0}$. Since the speed of travel of the flow along the x_3 direction is unity, the time taken is $\Delta T_n = \zeta_n - \zeta_0 = \sqrt{\zeta_0^2 - 2\eta_0} - \zeta_0$, which is clearly finite. We must then show that this orbit takes infinitely many steps, i.e. $\eta_n = 0$ implies $n \to \infty$. Note that $\eta_n = 0$ is a fixed point of the map (10.28) for any ζ_n. Then by the η_n part of (10.28), we have $\frac{\eta_{n+1}}{\eta_n} = 1 + \zeta_n + \frac{1}{2}\eta_n$, and using the ζ_n part of (10.28), we can rewrite this as $\frac{\eta_{n+1}}{\eta_n} = 1 + \zeta_n + \frac{1}{2}(\zeta_{n+1} - \zeta_n) = 1 + \frac{1}{2}(\zeta_n + \zeta_{n+1})$, which is negative since $\zeta_n, \zeta_{n+1} < 0$. This implies that η_n is strictly decreasing towards 0, and therefore cannot terminate at the fixed point 0 in finitely many steps, and thus η_n asymptotes towards 0 as $n \to \infty$. □

This gives the result for attracting in $x_3 < 0$, and a similar argument for $x_3 > 0$ tells us that an orbit starting at the intersection takes infinitely many steps but finite time to exit from the intersection via the rotation map.

The Zeno phenomenon now presents us with a problem, actually, two problems, as both the history and future through the origin are non-unique. Because an orbit takes infinitely many rotations to reach the intersection in finite time, its entry point cannot be determined uniquely. Then, even though we can then integrate the layer system (10.27) along the intersection until exit occurs, even given a set points of exit, the ensuing trajectory cannot be determined uniquely as the exit takes infinitely many steps in finite time.

What this means practically is that given a point in $x_3 < 0$ outside the intersection, we can be certain it will evolve into sliding on $x_1 = x_2 = 0$, and conversely, given a point in $x_3 > 0$ outside the intersection, we can be certain it will evolve from sliding on $x_1 = x_2 = 0$, but given *any* two such points, we can find a regularization of the discontinuity or simulation method or physical model such that those points are connected.

Merely to help us form a picture of the local dynamics, let us perform a few forward time simulations by smoothing out the discontinuity in (10.26). One finds, as predicted, that the exit point along the intersection is very sensitive to numerical imprecision. We plot simulations in Figure 10.15 that replace $\lambda_i = \text{sign}(x_i)$ with $\lambda_i \approx \phi(x_i/\varepsilon) = \tanh(x_i/\varepsilon)$ for $i = 1, 2$, with (i) $\varepsilon = 10^{-4}$, (i) $\varepsilon = 10^{-3}$, (i) $\varepsilon = 10^{-2}$. The value of the smoothing stiffness ε determines how narrow (order ε) the funnelling along the intersection is, so of course in the limit $\varepsilon \to 0$ where this approximates the discontinuous system, the

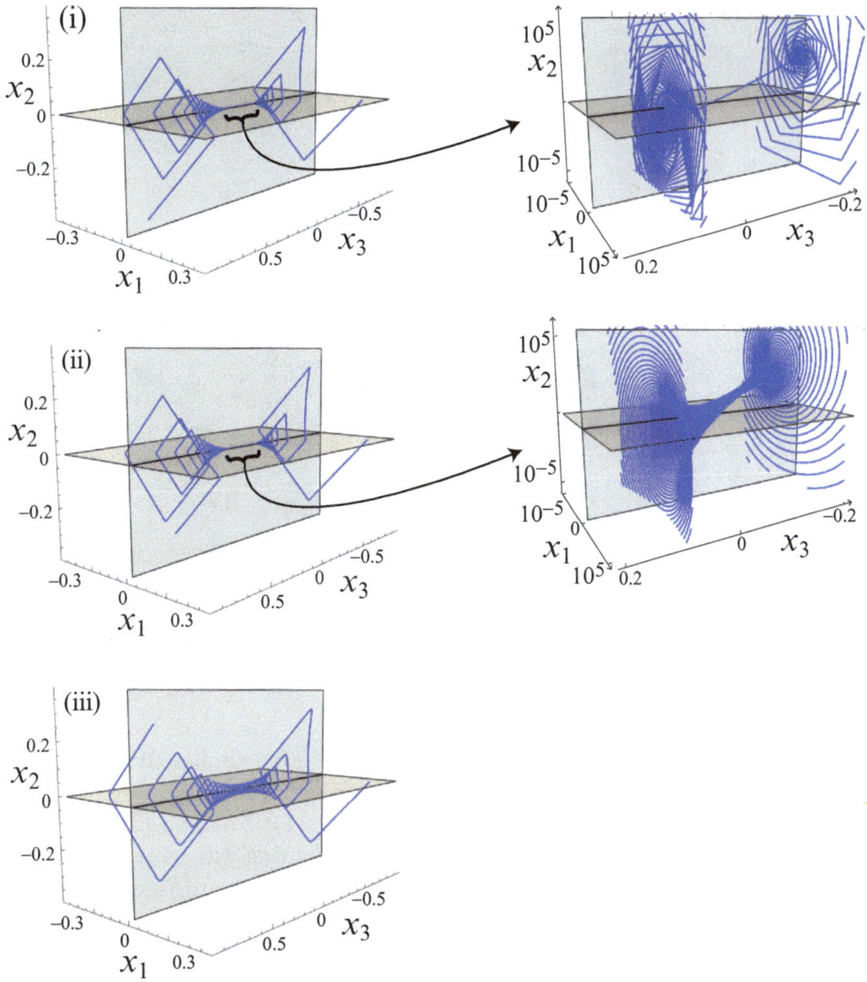

Fig. 10.15 Simulations using a tanh smoothing with ε values (i) 10^{-4}, (ii) 10^{-3}, and (iii) 10^{-2}. For (i)–(ii) a magnification is shown of the funnel along the intersection.

funnels will reach zero width. The results of the different simulations are all consistent; however, given a common initial point, they give exit occurring at similar coordinates $x_3 \approx 0.2$.

The consistency of these results is further verified by using different smoothings of the sign function, taking the rational function $\phi(x_i/\varepsilon) = (x_i/\varepsilon)/\sqrt{1+(x_i/\varepsilon)^2}$ in (i–ii.a) or the ramp function $\phi(x_i/\varepsilon) = \text{sign}(x_i)$ for $|x_i| > \varepsilon$ and $\phi(x_i/\varepsilon) = x_i/\varepsilon$ for $|x_i| \leq \varepsilon$ in (i–ii.b). The results for these rational and ramp smoothings are qualitatively indistinguishable from the tanh smoothing, with some difference in the thickness of the funnel visible for $\varepsilon = 10^{-3}$, but with similar exit points around $x_3 \approx 0.2$ (Figure 10.16).

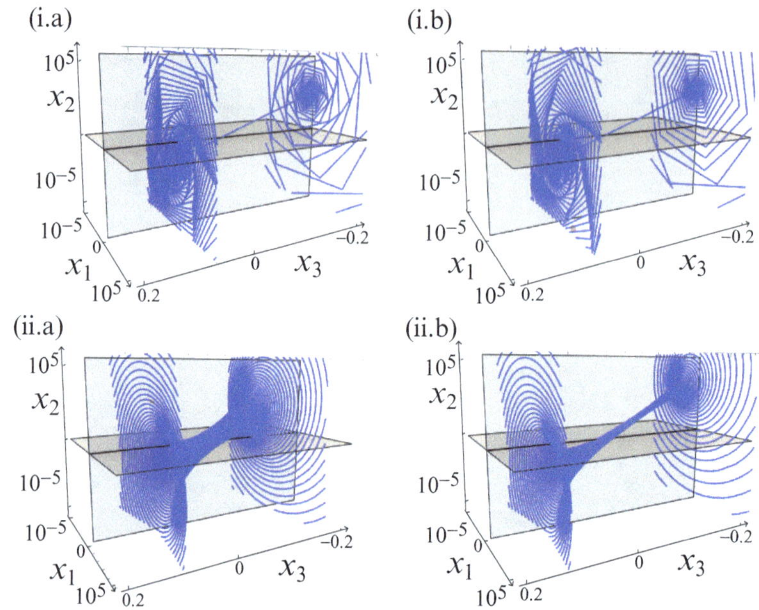

Fig. 10.16 The attractors in Figure 10.15(i–ii) for the rational (a) and ramp (b) smoothings, with the same initial conditions, for $\varepsilon = 10^{-3}$.

This suggests that similar sigmoid smoothings give similar dynamics. To obtain a wider range of behaviours, such as longer or shorter segments of sliding before exit, one may add hidden terms to (10.26) and consider not just smoothing of the discontinuity but implementing the switch between constituent vector fields hysteretically, stochastically, or with time delay. We leave these as open challenges, but as stated above, with the right combination of these ingredients in the simulation method and by varying their relative magnitudes, it should be possible to connect any point in $x_3 < 0$ to any point in $x_3 > 0$ via the flow. Thus the singularity at the origin acts as a phase randomizer, making the angle of rotation around the intersection enormously sensitive to modelling parameters, despite simple underlying geometry.

The scenario above has been studied by other authors, but usually with the focus of the problem on how orbits can spiral into an intersection without sliding. Filippov discussed a planar piecewise-constant example in [71], stating that it exhibited geometric convergence or the so-called *Zeno* phenomenon, reaching the intersection in finite time via infinitely many crossings of the local discontinuity submanifolds. Filippov also noted that this constituted a form of determinacy-breaking when the intersection is repelling. The system is so simple yet compelling that it may well have appeared elsewhere in the literature in instances that this author is unaware of, and another example is in [51], where the scenario was studied for perhaps the first time in three dimensions, including highlighting the computational problem raised by spiralling outward from an intersection.

10.3.4 Exit from a Sliding Fold

All of the examples of determinacy-breaking above occur in systems with linear dependence on λ. Most of these are structurally stable and persist if we add small nonlinear terms $+(\lambda^2-1)\mathbf{h}$.

Nonlinear dependence on λ can create other routes to determinacy-breaking. The simplest is due to a fold in the sliding manifold $\mathcal{M}$.

Consider the system

$$(\dot{x}_1,\dot{x}_2)=\tfrac{1}{2}(1+\lambda)(x_2,-1)+\tfrac{1}{2}(1-\lambda)(1,1+c)+(\lambda^2-1)(1,0)\ , \quad (10.30)$$

as c changes sign.

The sliding manifold $\mathcal{M}$ has two branches,

$$\mathcal{M}_{\pm}=\left\{(\lambda^{\$},x_2)\in(-1,+1)\times\mathbb{R}\ :\ \lambda^{\$}=\tfrac{1}{4}\left(1-x_2\pm\sqrt{(x_2-3)^2-4x_2}\right)\right\} \quad (10.31)$$

with the branches $\mathcal{M}_+$ and $\mathcal{M}_-$ being repulsive and attractive, respectively, meeting at a non-hyperbolic point $(\lambda,x_2)=(0,1)$. The flow on $\mathcal{M}$ (and everywhere else) is upward for $\lambda<c/(2+c)$ and downward for $\lambda>c/(2+c)$, between which lies a sliding equilibrium at $(\lambda,x_2)=(c,(2+3c)/(1+c))/(2+c)$.

This results in the flow illustrated in Figure 10.17.

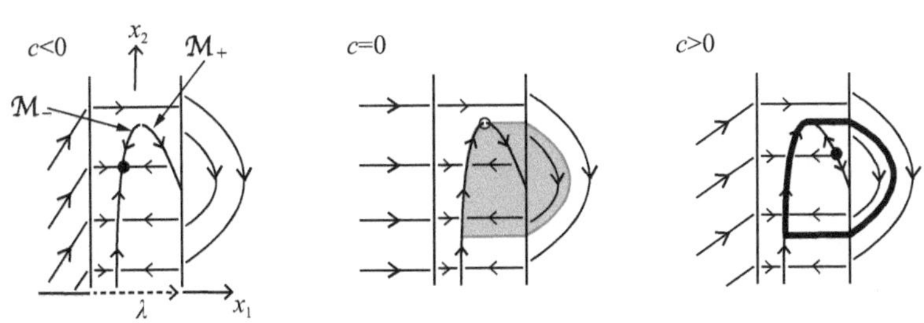

Fig. 10.17 Exit from nonlinear sliding in the flow of (10.30), showing (i) a sliding attractor, (ii) determinacy-breaking exit from a non-hyperbolic point along $\mathcal{M}$, (iii) a sliding repeller.

For $c<0$ there is no exit point, and the situation is similar to that we would find without linear switching at an invisible fold: the sliding equilibrium is an attractor, so the entire flow is attracted to this point in the nonlinear sliding region $x_1=0$, $x_2<1$. This is shown in Figure 10.17(i).

For $c > 0$ there is a deterministic exit point at $(0, 1)$ (in either (x_1, x_2) or (λ, x_2) space). After exiting, however, the flow returns to sliding, creating a periodic orbit with a branch of sliding on $\mathcal{M}_-$ and enveloping the branch $\mathcal{M}_+$ along with the repelling sliding equilibrium. This is shown in Figure 10.17(iii).

For $c = 0$ the equilibrium lies exactly on the boundary between $\mathcal{M}_\pm$, creating a canard point, where trajectories can flow from $\mathcal{M}_-$ to $\mathcal{M}_+$, generating a set-valued flow in forward time occupying the shaded region in Figure 10.17(ii). Any trajectory, from an arbitrary initial condition, will evolve eventually onto $\mathcal{M}_-$, continuing deterministically (i.e. uniquely) until reaching the exit point $(0, 1)$, whereupon the onward flow is the entire set generated by trajectories leaving from any point along $\mathcal{M}_+$.

In two dimensions this canard phenomenon is structurally unstable, occurring only for $c = 0$. In three or more dimensions, we can easily construct examples where this becomes a robust phenomenon, which we leave as an exercise for the reader.

10.4 Stranger Attractors

Strange attractors have a fractal structure, and in chaotic dynamics their topology permits global stability despite local instability, so that two trajectories initially nearby may be arbitrarily far apart after sufficient times while remaining within the attractor.

An economical term for the attracting sets formed by determinacy-breaking might be *stranger attractors*. Locally, the dynamics at any point may appear stable or unstable, except at one or more determinacy-breaking points which are unstable in the extreme, despite which an attractor is formed. Two trajectories initially nearby in the attractor, including two trajectories starting from *the same initial point*, may be arbitrary far apart after sufficient times.

A strange attractor is not necessarily chaotic. Likewise, a stranger attractor should not necessarily be interpreted as representing any kind of chaos, despite having the property (unlike a nonchaotic strange attractor) that nearby trajectories diverge arbitrarily.

Let us attempt to define the general topological principles of a stranger attractor, before considering their link to chaos *and* other behaviours.

A formal definition of stranger attractors was attempted in [37] (under the name of 'nondeterministic chaotic attractors' at the time), but this was tied strongly to extending the idea of chaos by applying notions of sensitive dependence and transitivity through the determinacy-breaking point. The full consequences of this are not entirely clear, however, and a less laboured geometrical definition lends more insight into their meaning.

Definition 10.2. Each point in a **stranger attractor** visits each other point in the set (including itself) via an infinite number of distinct orbits.

This is illustrated in Figure 10.18 for examples with one or two switches, where we choose any two points in the shaded attractors and show one of the infinitely many orbits connecting them.

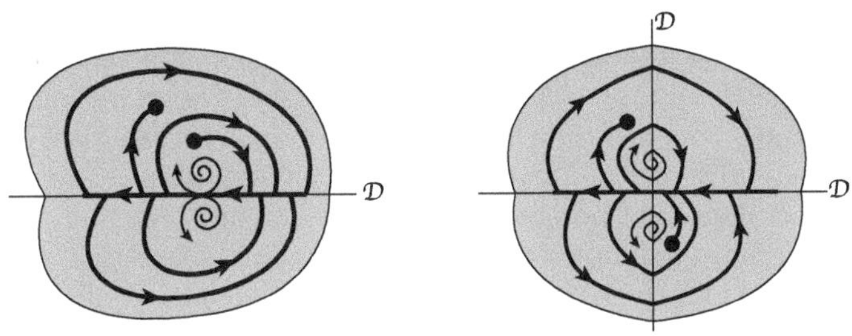

Fig. 10.18 Topology of a stranger attractor. Any two points in the attractor (shaded) are connected by infinitely many distinct orbits, filling the entire set. Here the flow is generated by two repelling foci either side of a discontinuity surface (left) or two repelling fused foci and two discontinuity submanifolds.

The dynamical objects necessary appear to be the existence of at least one of each of the following:

- a determinacy-breaking point,
- a collapsing phase (e.g. an attracting sliding region),
- an explosive phase (e.g. a repelling sliding region),
- a fully causal phase (i.e. a differentiable flow).

In the *collapsing* phase the forward time flow from any point is a unique orbit, but the history is set-valued; in the *explosive* phase, the history from any point is a unique orbit, but the forward time flow is set-valued; in the fully causal phase, both forward and backward time evolution are unique. These are illustrated schematically in Figure 10.19.

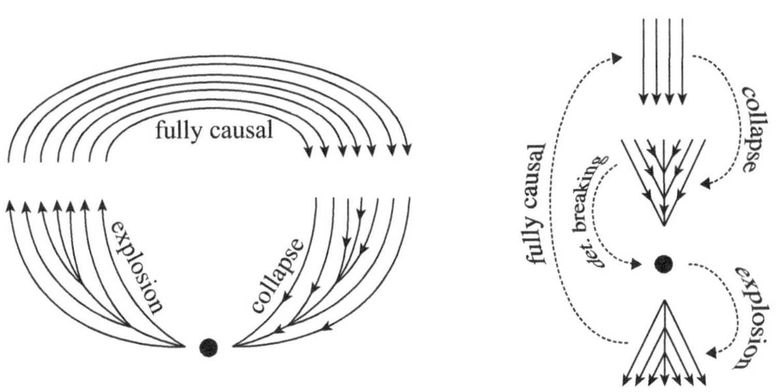

Fig. 10.19 The ingredients of a stranger attractor. Phases of fully causal, collapse, determinacy-breaking, and explosive motion follow each other. A system may be made up of multiple such phases in different regions of a flow, as the more abstract right-hand picture is supposed to imply.

Applying this to the examples in Figure 10.18, in the left picture, the visible-visible fold-fold on the discontinuity surface is the determinacy-breaking point, the half-line to the right is the collapsing region, and the half-line to the left is the explosive region, with everything else being fully deterministic. In the right picture, the same is true, but the determinacy-breaking point is an intersection of discontinuity submanifolds.

We can equivalently say that a stranger set consists of a region, or several regions, of unique forward time dynamics and unique reverse time dynamics, with regions of fully causal dynamics where the two types overlap and points lying on the boundary of both sets where determinacy-breaking occurs.

We have detailed examples of such attractors in Figure 10.6 and will see more in Figures 11.5 and 11.10, and it is easy to conceive of others. More sophisticated definitions and analyses of such objects may be possible as new examples come to light in the future.

10.5 Looking Forward

We have studied local attractors and local bifurcations and now the local portraits around exit point as organizing centres of the global flow. It is time to turn to global bifurcations and their counterpart where determinacy-breaking is involved: global explosions.

Chapter 11
Global Bifurcations and Explosions

Nonsmooth systems suffer a curse of dimensionality. Every higher dimension brings the possibility of fundamentally new local and global dynamics, as a simple object classified in low dimensions takes on unknown new complications in higher dimensions. Classifying these conventionally, by forming an accounting of all singularities and their unfoldings, can therefore make little impact on our studies in higher dimensions. As with the exit points which we used to organize local dynamics in Chapter 10, we need another approach.

Fortunately, what nature takes with one hand it gives with another. The cause of our struggles is that whenever we classify patterns in the flow, we then have to worry about how they interact with the discontinuity surface. But the peculiarities of the way the flow and the discontinuity surface interact *are* something we can classify; in fact we covered these back in Section 6.1. By focussing our attention there, we discover the mechanisms that underlay global bifurcations in higher dimensions.

11.1 Local Classification of Global Phenomena

The notion of discontinuity-induced phenomena is one of the most important to have arisen in dynamical systems in the last 30 years. From Chapter 8 onwards, we have seen a number of *local* discontinuity-induced bifurcations. One of the kickstarters of the modern resurgence of interest in nonsmooth dynamics was the realization that *global* discontinuity-induced bifurcations can also be studied with some generality.

The key idea goes back to the work of Feigin [63, 65, 66], who first began classifying the limited family of topological forms displayed by a periodic orbit when it contacts a discontinuity surface. Attention continued to centre around periodic orbits up until [48], but this is an unnecessary

M. R. Jeffrey, *Hidden Dynamics*,
https://doi.org/10.1007/978-3-030-02107-8_11

restriction, and we will consider instead *distinguished orbits*. From the qualitative dynamics viewpoint, a distinguished orbit is either an isolated periodic orbit or a separatrix, such as an orbit lying in the stable or unstable manifold to a saddle or node. From a practical dynamical viewpoint, however, a distinguished orbit may simply be the solution through a special initial condition (e.g. an electrical circuit whose current and voltage are charged from zero).

In Definition 4.9 we described a discontinuity-induced bifurcation as occurring at the boundary of sliding. In Chapter 8 onwards, we have seen how this applies to local bifurcations, and we shall now see that it applies also to global bifurcations. A distinguished orbit undergoes a discontinuity-induced bifurcation at a certain point if it touches the discontinuity surface such that an arbitrarily small perturbation gives a topologically nonequivalent orbit. We will refer to a one-parameter family of perturbed orbits as an *unfolding* of the bifurcation.

Through [49, 50, 130, 131, 48, 84, 86], study centred around so-called *sliding bifurcations*, which occur at the boundary of attracting sliding regions. After a tentative introduction in [48], in [120] the concept of a new kind of sliding bifurcation was introduced, that of *sliding explosions*, which occur at the boundary of repelling sliding and involve determinacy-breaking.

Definition 11.1. An orbit in a piecewise-smooth system $\dot{\mathbf{x}} = \mathbf{f}(\mathbf{x}; \boldsymbol{\lambda})$ undergoes a **sliding bifurcation** or **sliding explosion** as a parameter varies if the orbit's intersection with a discontinuity surface changes topologically, at the boundary of a sliding region. If the boundary abuts an attracting sliding region only, then the event is a bifurcation; if the boundary abuts a repelling sliding region, then the event is an explosion.

This definition develops on a definition in [120], which preceded the introduction of hidden dynamics and nonlinear switching to piecewise-smooth systems. If we omit nonlinearity and consider only linear dependence on a single switching multiplier λ at a manifold $\sigma = 0$, then sliding boundaries occur only where $\mathbf{f}(\mathbf{x}; \pm 1)$ lies tangent to $\sigma = 0$.

Intuitively, if an orbit gains or loses a segment on the discontinuity, then it undergoes a topological change. Because the switch is a 'hard' event (by virtue of being a *dis*continuity), the change is abrupt, not accompanied by incipient changes outside the neighbourhood of the discontinuity surface. Such events can therefore be characterized purely by the change that occurs local to the discontinuity surface, so to classify sliding bifurcations and explosions, we begin by classifying their local geometry near the discontinuity surface.

Another way to say this is that the topological instability constituting a sliding bifurcation or explosion is localized to the neighbourhood of the discontinuity surface, as illustrated in Figure 11.1.

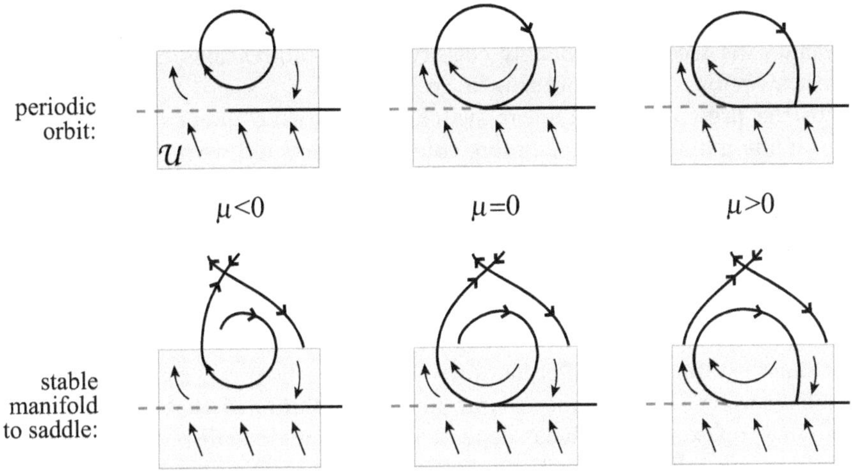

Fig. 11.1 Sliding bifurcations: a local mechanism for global bifurcations. As μ changes, a segment of sliding appears along a periodic orbit (top) and the stable manifold to a saddle (bottom). The phase portrait only changes qualitatively inside the grey box $\mathcal{U}$, so the topological change is entirely local and is the same in the top and bottom rows.

A way of formalizing this was given in [120]. Let $\mathbf{x}(t,\mu) = \phi_t(\mu)$ denote a distinguished orbit of the system $\dot{\mathbf{x}} = \mathbf{f}(\mathbf{x};\lambda,\mu)$, where λ is a switching multiplier and μ is a parameter. Begin by assuming that $\phi_t(0)$ passes through a point $\mathbf{x} = \phi_t(0) = 0$ on the boundary of a sliding region at some t, in a structurally unstable manner such that $\mathbf{x}(t;\mu)$ is topologically nonequivalent for $\mu < 0$ and $\mu > 0$, e.g. passing through the sliding boundary at some t for $\mu > 0$ and not touching the same boundary or even the discontinuity surface for $\mu < 0$. Define a map $T_\mu : \mathbb{R}^n \mapsto \mathbb{R}^n$, given in a neighbourhood $\mathcal{U}$ of $\mathbf{x} = 0$ by $T_\mu(\mathbf{f}(\mathbf{x};\lambda,0)) = \mathbf{f}(\mathbf{x};\lambda,\mu)$, and assume that T_μ does not induce any topological change in the vector field $\mathbf{f}(\mathbf{x};\lambda,\mu)$ on $\mathcal{U}$. As μ changes, we can apply the inverse map T_μ^{-1} to the vector field $\mathbf{f}(\mathbf{x};\lambda,\mu)$, so that on $\mathcal{U}$, it takes the form $\mathbf{f}(\mathbf{x};\lambda,0)$, but in doing so, the distinguished orbit $\phi_t(\mu)$ maps to some new orbit $\tilde{\phi}_t(\mu)$ (with $\tilde{\phi}_t(0) = \phi_t(0)$). The sliding bifurcation then unfolds locally as a μ-family of orbits $\tilde{\phi}_(\mu)$ in the unchanging vector field given on $\mathcal{U}$ by $\mathbf{f}(\mathbf{x};\lambda,0)$. Those μ-families or orbits in an unchanging local vector field provide an unfolding of a sliding bifurcation or explosion.

So, although a distinguished orbit is selected out of the flow by global conditions, the changes it can undergo relative to the discontinuity surface depend only on the configurations of orbits that are possible in the neighbourhood of the discontinuity surface. As the orbit explores these different configurations, it undergoes sliding bifurcations or explosions.

Fortunately we have already classified the local forms of vector fields near sliding boundaries. For systems with linear switching, sliding boundaries occur at tangencies between the vector fields $\mathbf{f}(\mathbf{x};\pm 1)$ and the discontinuity surface $\sigma = 0$. We studied these in detail Section 6.1, and they provide everything we need to completely classify sliding bifurcations/explosions with linear switching; we explore this in Section 11.2.

In the presence of nonlinear switching, things are more open-ended, as the sliding manifold can take more complex forms, and we make preliminary investigations at the end of Section 11.2.

11.2 The Sliding Eight

We must establish the different ways in which a flow can contact a discontinuity surface. Let us first assume linear switching, so that we consider a system $\dot{\mathbf{x}} = \mathbf{f}(\mathbf{x}; \lambda)$ with linear dependence on a single multiplier $\lambda = \text{sign}(\sigma)$. Then we can infer from Section 6.1 that:

Lemma 11.1. *In an $n \geq 3$ dimensional dynamical system with linear switching, the following are generic: an orbit forms a quadratic or cubic tangency to one side of a discontinuity surface, or an orbit forms a quadratic tangency to both sides of a discontinuity surface. These are the fold, cusp, and two-fold tangencies, respectively.*

This result, proven in [206], follows because each tangency requires at most three conditions to arise: a fold requires $\sigma = \delta_{t+}\sigma = 0$, a cusp requires $\sigma = \delta_{t+}\sigma = \delta_{t+}^2\sigma = 0$, and a two-fold requires $\sigma = \delta_{t+}\sigma = \delta_{t-}\sigma = 0$.

The result is important because we can then show the key result in the classification of sliding bifurcations and explosions.

Theorem 11.1. *A generic one-parameter sliding bifurcation /explosion in $\mathbb{R}^n$ is a bifurcation/explosion at a fold, cusp, or two-fold or a combination thereof.*

We will prove this in Section 11.5. Each case is classified by whether the tangency is visible or invisible and whether it occurs at the boundary of attracting or repelling sliding. By considering the generic ways a flow can encounter such tangencies as a parameter varies, we derive the necessary and sufficient conditions to find *all* the one-parameter sliding bifurcations/explosions of systems with linear switching at a simple discontinuity submanifold. The list is short.

Corollary 11.1. *One-parameter sliding bifurcations/explosions consist of the eight cases in Tables 11.1 and 11.2.*

Tables 11.1 and 11.2 give, along with the bifurcations and explosions promised by Corollary 11.1, the common names for the events, the phase portraits of their unfoldings illustrated in three dimensions, the type of singularity responsible for each event, and the conditions that define the necessary singularity. We derive these details in Sections 11.5.2 to 11.5.3.

The four sliding bifurcations are mechanisms by which an orbit gains or loses a segment of sliding. Importantly the sliding involved is attracting, and this ensures that the flow remains deterministic.

Name/portrait	Bifurcation at an ...	Conditions
grazing sliding 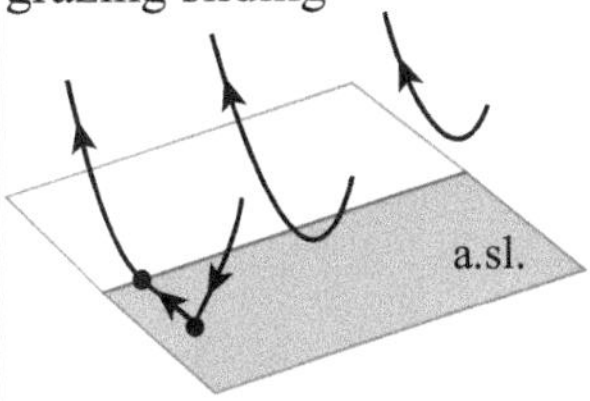	Attracting Visible fold	$\delta_{t+}\sigma = 0$ with $\delta^2_{t+}\sigma > 0,\ \delta_{t-}\sigma > 0$ or $\delta_{t-}\sigma = 0$ with $\delta^2_{t-}\sigma < 0,\ \delta_{t+}\sigma < 0$
crossing sliding 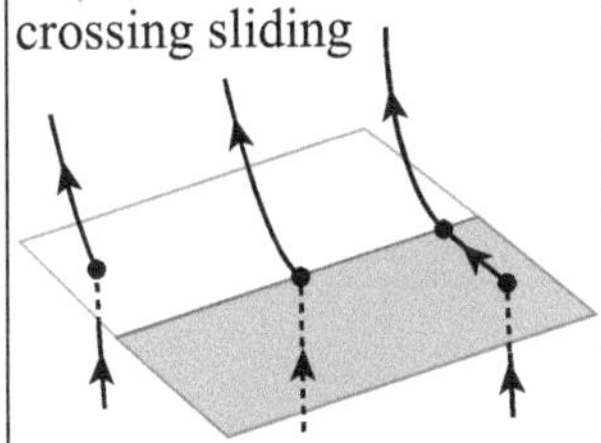	Attracting Visible fold	$\delta_{t+}\sigma = 0$ with $\delta^2_{t+}\sigma > 0,\ \delta_{t-}\sigma > 0$ or $\delta_{t-}\sigma = 0$ with $\delta^2_{t-}\sigma < 0,\ \delta_{t+}\sigma < 0$
switching sliding 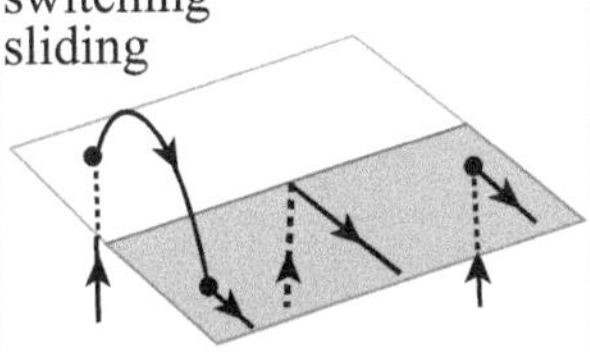	Attracting Invisible fold	$\delta_{t+}\sigma = 0$ with $\delta^2_{t+}\sigma < 0,\ \delta_{t-}\sigma > 0$ or $\delta_{t-}\sigma = 0$ with $\delta^2_{t-}\sigma > 0,\ \delta_{t+}\sigma < 0$
adding sliding 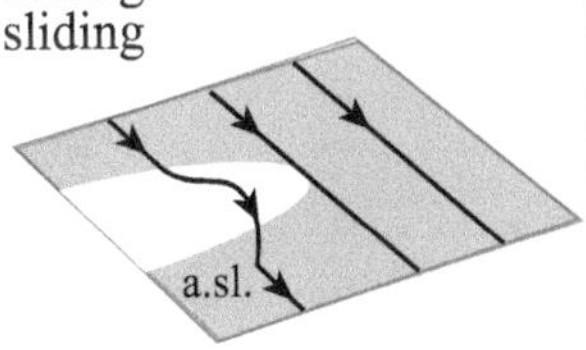	Attracting Visible cusp	$\delta_{t+}\sigma = \delta^2_{t+}\sigma = 0$ with $\delta^3_{t+}\sigma(\delta_{t-})\sigma > 0$ or $\delta_{t-}\sigma = \delta^2_{t-}\sigma = 0$ with $\delta^3_{t-}\sigma(\delta_{t+})\sigma > 0$

Table 11.1 The four sliding bifurcations and their underlying singularities.

The four sliding explosions are mechanisms by which an orbit gains or loses a segment of repelling sliding, so that determinacy-breaking occurs. In three cases, determinacy-breaking occurs only at the explosion point itself, i.e. at a special parameter value, but it causes a total separation of the orbit geometry either side of the bifurcation. In the 'robust' case, determinacy-breaking persists for all parameter values on one side of the explosion.

Name/portrait	Explosion at a ...	Conditions
grazing sliding 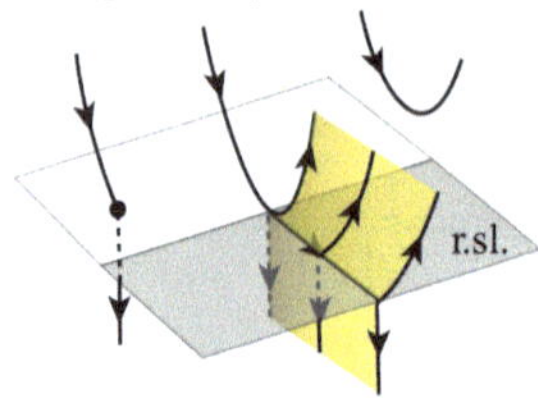	Repelling Visible fold	$\delta_{t+}\sigma = 0$ with $\delta_{t-}\sigma < 0 < \delta^2_{t+}\sigma$ or $\delta_{t-}\sigma = 0$ with $\delta^2_{t-}\sigma < 0 < \delta_{t+}\sigma$
visible canard 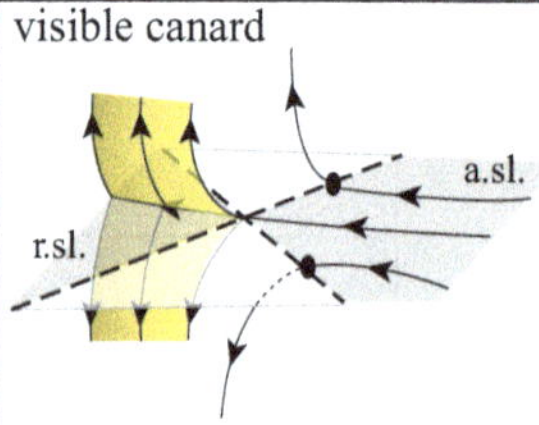	Visible Two-fold	$\delta_{t\pm}\sigma = 0$ with $\delta^2_{t-}\sigma < 0 < \delta^2_{t+}\sigma$
invisible canard 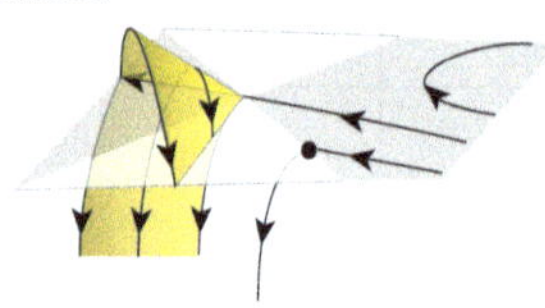	Visible-invisible Two-fold	$\delta_{t\pm}\sigma = 0$ with $\delta^2_{t+}\sigma\delta^2_{t-}\sigma > 0$
robust canard 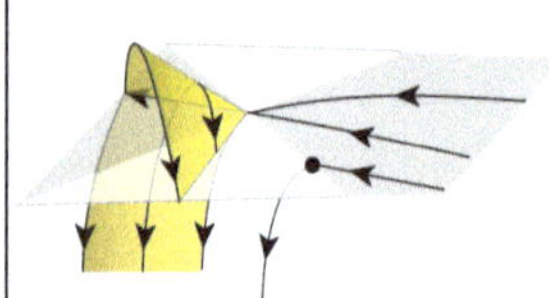	Visible-invisible Two-fold	$\delta_{t\pm}\sigma = 0$ with $\delta^2_{t+}\sigma\delta^2_{t-}\sigma > 0$

Table 11.2 The four sliding explosions and their underlying singularities.

Before we prove the theorem, let us gain some intuition for what these bifurcations and explosions represent.

11.3 Sliding Bifurcations/Explosions: The Global Picture

The classification in Tables 11.1 and 11.2 provides the local mechanisms for generic one-parameter global bifurcations/explosions of a 'distinguished orbit'. To understand their significance, we must apply them to systems with some global dynamics that identify such an orbit.

Below we illustrate examples where the distinguished orbit is the unstable manifold to a saddle (Figures 11.2 and 11.3) or a periodic orbit (Figures 11.4 and 11.5). These are shown in the minimum number of dimensions that such an event can occur as a parameter varies. The parameter μ can be interpreted as varying an initial point of the orbit, for example, the coordinate at which it pierces a hypersurface Π (an example is shown dotted in the first portrait in each figure).

In Figures 11.2 and 11.3, as μ varies, a saddle equilibrium moves in phase space, such that at $\mu = 0$, its unstable manifold passes through the singularity.

In the cases of sliding bifurcations, Figure 11.2, as μ changes from positive to negative, the unstable manifold undergoes a continuous deformation that adds a sliding segment (in grazing/crossing-sliding) or adds a non-sliding segment (in switching/adding-sliding).

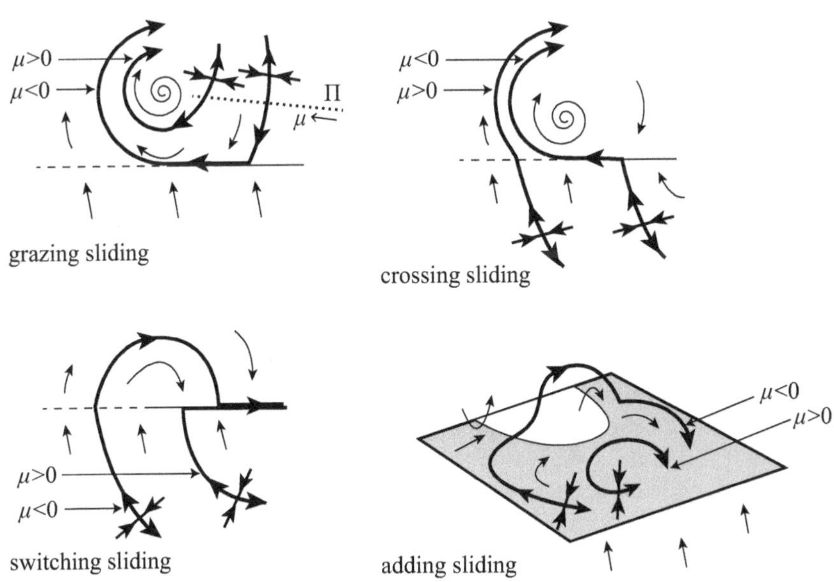

Fig. 11.2 Examples of sliding bifurcations of the unstable manifold to a saddle.

In the cases of sliding explosions, Figure 11.3, as μ passes through zero, the unstable manifold undergoes a sudden jump, its outset lying in a different region of space for different signs of μ. In the case of a robust canard explosion, the outset for $\mu > 0$ is set-valued and remains so for a range of positive μ values (hence it is 'robust'). In the other cases, the outset is set-valued only fleetingly, at $\mu = 0$.

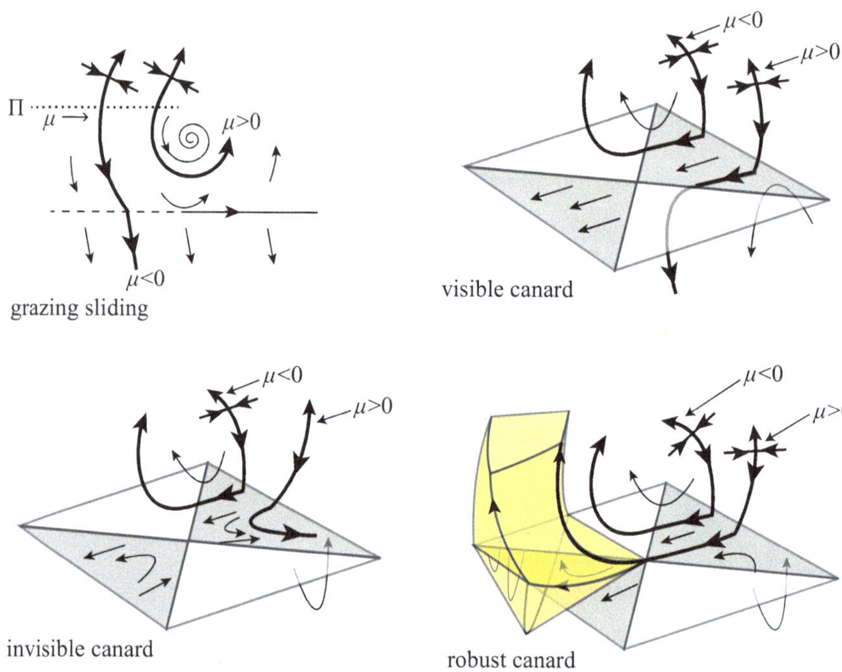

Fig. 11.3 Examples of sliding explosions of the unstable manifold to a saddle.

In Figures 11.4 and 11.5, as μ varies, a periodic orbit grows such that at $\mu = 0$, it passes through the singularity. Periodicity exists for both signs of μ through the bifurcations in Figure 11.4, but only for one sign of μ through the explosions in Figure 11.5.

For a periodic orbit undergoing a sliding bifurcation as in Figure 11.4, the cases $\mu > 0$ and $\mu < 0$ could label the same orbit undergoing a transition as μ changes sign, called a *persistence* of the periodic orbit. Alternatively the cases $\mu > 0$ and $\mu < 0$ could label two co-existing orbits with different μ-dependent initial conditions, which coalesce at $\mu = 0$. Intuitively, the latter should constitute a bifurcation of saddle-node type, meaning one of the periodic orbits should be attracting and the other repelling, at least along some centre-manifold direction along which coalescence occurs. An inspection of the phase portraits reveals this to be the case (at least in all examples known to the author); however, these notions—particularly of centre manifolds—are not well developed yet for piecewise-smooth systems beyond the scope discussed in Section 7.8. In fact, whereas centre manifolds are one of the most important concepts for making high dimension smooth systems tractable, sliding bifurcations and explosions appear to do the same for piecewise-smooth systems, and they are one of the few such concepts available, but more remains to be learnt about how to apply them.

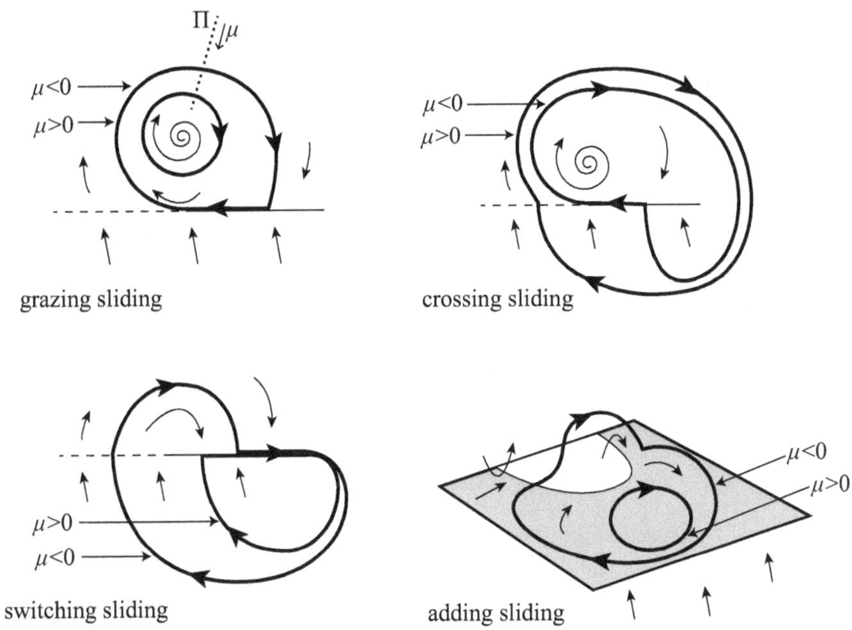

Fig. 11.4 Examples of sliding bifurcations of a periodic orbit.

A periodic orbit undergoing a sliding explosion will exist on only one side of the explosion, disappearing at $\mu = 0$ via a determinacy-breaking event. In the cases of grazing-sliding explosions and visible/invisible canard explosions, the set-valued flow encountered at $\mu = 0$ facilitates a continuous but abrupt jump to an orbit whose outset lies elsewhere in the phase space, breaking the periodic orbit. Depending on global conditions, the orbit may diverge or evolve towards some other global attractor. Only in the case of a robust canard does some form of attractor persist nearby for $\mu > 0$, and in fact the set-valued flow through the singularity is reinjected into the singularity, forming a *stranger attractor* (Section 10.4).

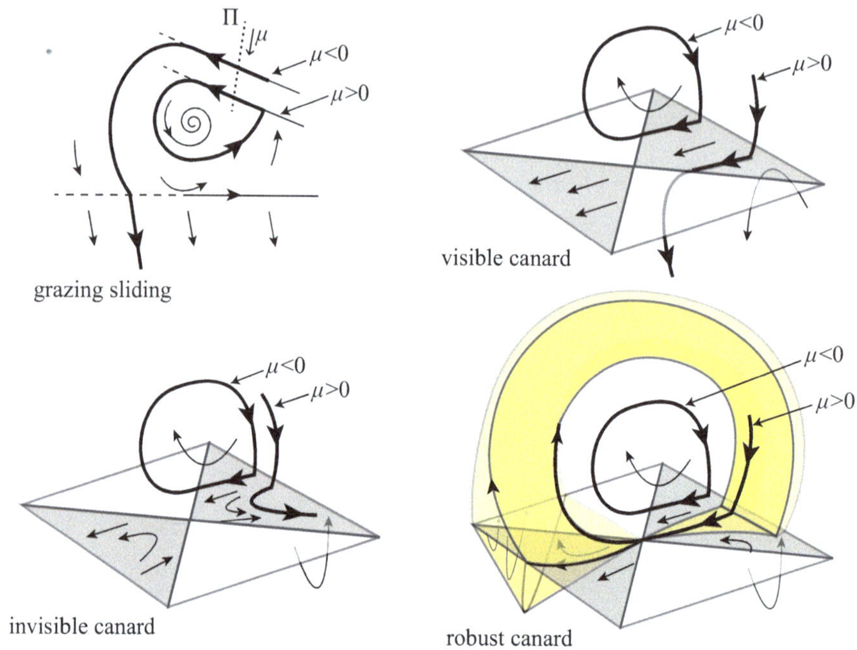

Fig. 11.5 Examples of sliding explosions of a periodic orbit.

In any of these scenarios, the classification of sliding bifurcations or explosions merely provides the local mechanism of an event, and not, obviously, its full global unfolding. As an example, let us take a saddle undergoing a grazing event at an attracting visible fold and see how it fits in to the classification.

First consider a planar system, in which the unstable manifold to a saddle passes through grazing as a parameter μ varies, Figure 11.6 (i.e. revisiting the first portrait of Figure 11.2 in more detail). The unstable manifold to the saddle undergoes the bifurcation by losing a segment of sliding as μ increases via a grazing event at $\mu = 0$. The stable manifold undergoes no topological change, always being connected to a focus. There seems to be nothing to add to the first portrait of Figure 11.2, but the other grazing example from

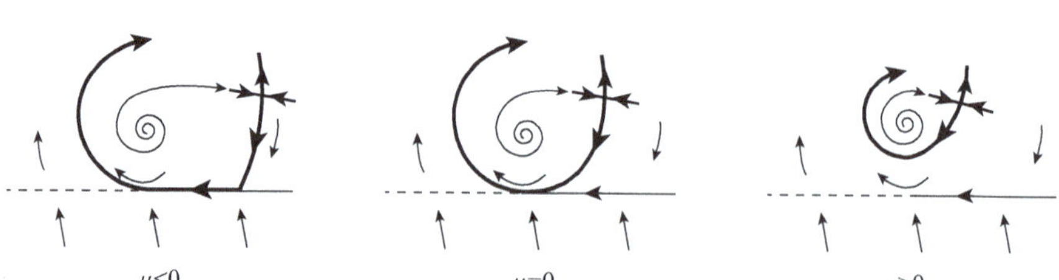

Fig. 11.6 A grazing-sliding bifurcation of the unstable manifold to a saddle.

the same figure does require closer inspection, as the following example will reveal.

Consider a similar system to that above, but let the stable manifold to the saddle pass through grazing as μ varies, Figure 11.7. The unstable manifold undergoes no local topological change, being connected to the sliding region throughout. The stable manifold does undergo a topological change, but not a sliding bifurcation, because it jumps as μ changes sign, and does not have a sliding segment on either side of $\mu = 0$. A careful inspection reveals also the existence of an attracting periodic orbit that is destroyed as μ increases. How does this fit into our classification?

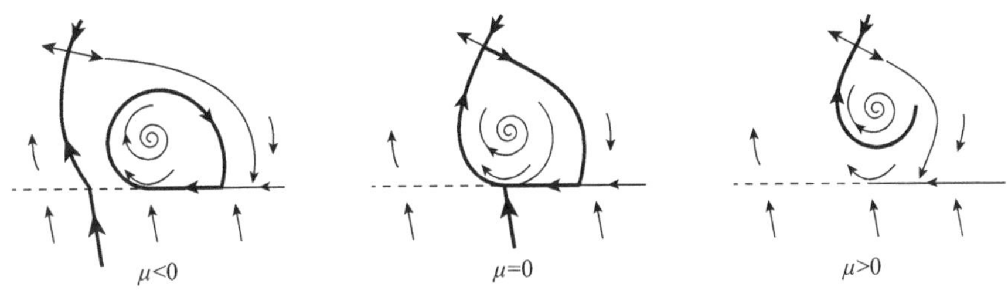

Fig. 11.7 A grazing-sliding explosion of the stable manifold to a saddle.

In fact the stable manifold has undergone a grazing-sliding explosion, because at $\mu = 0$, the history of the stable manifold through the grazing point becomes set-valued, and this facilitates the jump it undergoes as μ changes sign. Rather than describing the change in terms of reverse-time dynamics, we can simply reverse time for the entire portrait, obtaining Figure 11.8. This is just the portrait of the grazing-sliding explosion from Figure 11.2, and we now see that, if the system is planar, a (now repelling) periodic orbit must exist for $\mu < 0$ and be destroyed in the explosion.

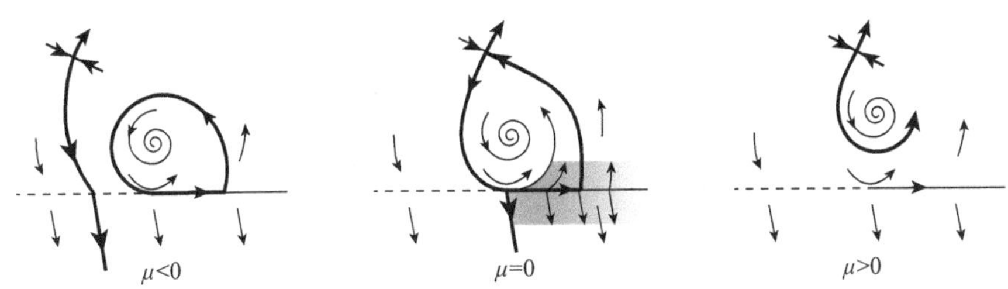

Fig. 11.8 A grazing-sliding explosion of the unstable manifold to a saddle.

In three dimensions or higher, the existence of a periodic orbit, at least such a simple one, is not so obvious, and the invariant manifolds to the saddle will have higher dimension; nevertheless the local mechanism, that of a grazing-sliding explosion, will be the same.

Thus the sliding bifurcations and explosions are a local representation of a wide class of global bifurcations, a way of reducing infinite global possibilities to a finite classification, and lending some insight on the topological forms to look for that organize the dynamics. Of course, more is needed for a complete dynamical study of any given system. One tool for using the local geometry of sliding bifurcations and explosions to quantitatively study global dynamics is described in Section 11.7.

Sliding explosions arise in singular perturbation problems such as the van der Pol system with relation to canards, bursting, and mixed-mode oscillations [46, 91, 134], where the two-fold appears in the 'reduced' system. They have also been proposed as the mechanism for sudden loss of oscillations in superconducting resonators [181, 118] and have been shown to be generic in switched feedback controllers [35]. A physical example of an explosion was identified shortly after their mathematical prediction, described as a case study in Section 14.4. The following toy mechanical example was proposed in [111].

Example 11.1 (Explosion in a Mechanical Oscillator). Consider an object of unit mass, whose displacement x satisfies a Newtonian force law $\ddot{x} = (\dot{x} - v)b - x + g(\dot{x}, t)$, where $\dot{x} = dx/dt$. This includes a spring force $-x$, a negative damping proportional to the speed relative to some reference v, plus an additional forcing g. For $\dot{x} < v$ let g grow linearly in time, say as $g = r_1 t$. For $\dot{x} > v$ let g have speed-dependent dynamics, setting $g = r_2 z$ where $\dot{z} = a + (\dot{x} - v)c$.

Letting $u = \dot{x} - v$, we obtain a first-order system

$$\begin{aligned} \tfrac{d}{dt}x &= u + v \\ \tfrac{d}{dt}u &= -x + bu + r_1 z + (r_2 - r_1) z \operatorname{step} u, \\ \tfrac{d}{dt}z &= 1 + (a - 1 + cu) \operatorname{step} u, \end{aligned} \tag{11.1}$$

where $\operatorname{step} u = 1$ for $u > 0$ and $\operatorname{step} u = 0$ for $u < 0$. The flows are tangent to the discontinuity surface, $u = 0$, along the lines $l_{1,2}$ where $x = r_{1,2} z$ (labelled in Figure 11.10), which intersect at the origin, where there is a visible-invisible two-fold.

For some parameter values, an attracting periodic orbit exists, as in Figure 11.9(top). The orbit has a segment in the attracting sliding region, leaves via the visible fold into the upper flow, and then winds around making three crossings of $u = 0$ before returning to attracting sliding. As parameters vary the sliding segment can vary until it passes through a two-fold singularity, as in Figure 11.9(middle). A stranger attractor is then formed, as a set-valued invariant set through the two-fold (see Section 10.4). The determinacy-breaking at the two-fold means this cannot be simulated directly; instead in Figure 11.9(bottom), this is partially represented by simulating forwards and backwards from a sample of points in the repelling region (actually slightly above or below it)—both their forward and backward trajectories intersect the singularity because they lie in the invariant set.

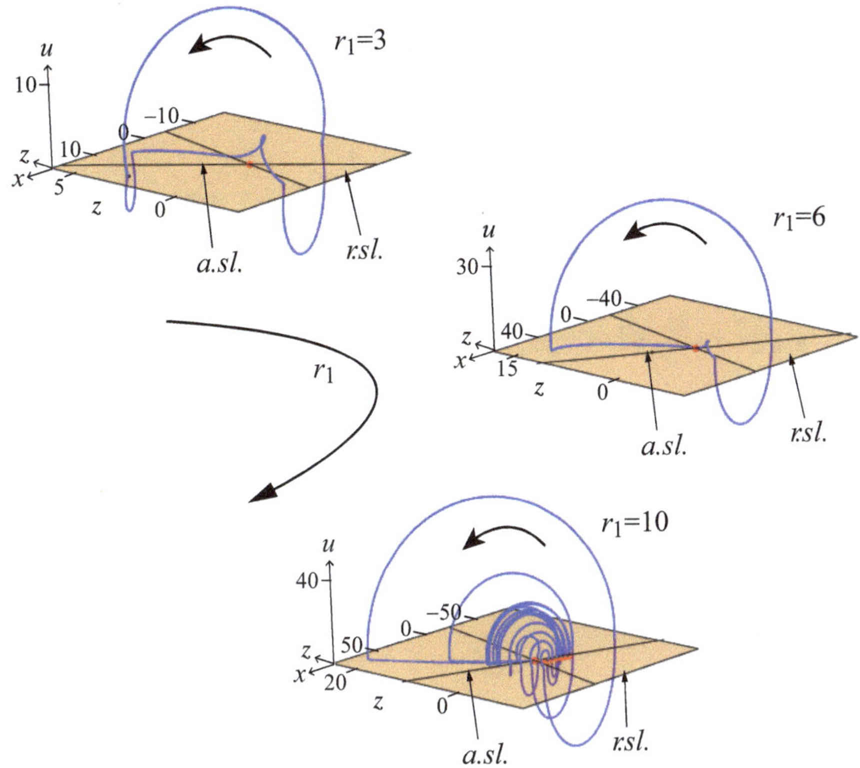

Fig. 11.9 Simulation of (11.1), for values of r_1 shown in figure, and $a = -1.3$, $b = 0.1$, $c = 0.2$, $v = -1$, $r_2 = -1$. In the last picture, to trace out part of the attractor through the singularity, we simulate the flow forwards and backwards through a sample of six points taken in the repelling sliding region, $z = -2$, $x = -\{20, 16.4, 12.8, 9.2, 5.6, 2\}$, perturbed to $u = \pm 1$ to obtain trajectories repelled above/below $u = 0$.

For some range of parameter values, as in Figure 11.9(top), part of the flow through the origin detaches from $u = 0$, winds around through both $u > 0$ and $u < 0$, returns to $u = 0$, then sticks (sliding dynamics corresponds to mechanical sticking), and returns to the origin, local to which the phase portrait resembles Figure 10.10(right). This generates a stranger attractor (see Section 10.4), on which the entire flow meets the double tangency both in forward and backward time, which is highly attracting with respect to the surrounding flow.

We can define this set precisely. The flows in $u < 0$ and $u > 0$ are tangent along the lines $(x, u, z) = l_1 := (r_1 z, 0, z)$ and $l_2 := (r_2 z, 0, z)$, respectively. The sliding flow has a straight line solution $\gamma(t) := (0.71, 0, 0.70)t$. Between γ and l_1, one finds that every sticking orbit traverses the double tangency, and moreover Figure 11.10 shows that the flows through l_1 (green) and through γ (red and blue) return to $u = 0$ and thence to the origin. Interspersing these, the flow returns to the origin recurrently yet unpredictably. Using this, the attractor's shape is traced more precisely in Figure 11.10.

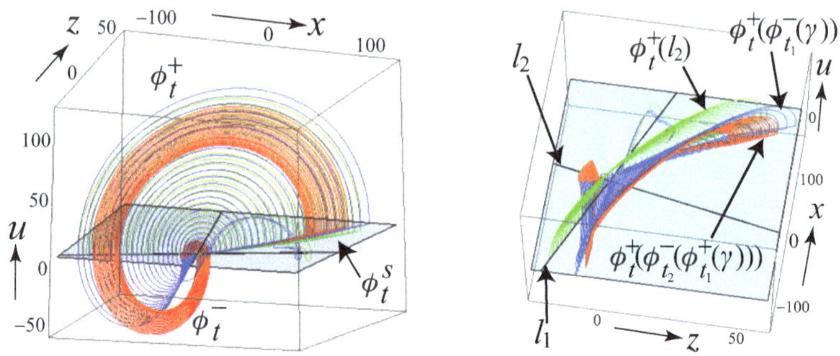

Fig. 11.10 Stranger attractor in a simulation of (11.1). The two views show the discontinuity surface $u = 0$ and tangency lines $l_{1,2}$. The flows in $u > 0$ and $u < 0$, and the sticking flow on $u = 0$, are indicated by ϕ_t^+, ϕ_t^-, ϕ_t^s. Several orbits are shown on the boundaries of the stranger attractor. Simulated for $a = -1.3$, $b = 0.1$, $c = 0.2$, $v = -1$, $r_1 = 12$, $r_2 = -1$.

11.4 Sliding Bifurcations/Explosions in Nonlinear Switching

We know from Section 9.5 that the boundaries of linear sliding formed by fold or cusp tangencies are structurally stable, so they, and the events we have unfolded above, survive perturbation by nonlinear switching terms. Below we shall ask what happens at the boundary of overlapping sliding, when nonlinear switching terms are not small and give rise to new topological instabilities.

In systems with nonlinear switching, topological instabilities also occur at the boundaries of overlapping sliding regions. The instabilities arise similarly to the folds and cusps considered above but require layer analysis to study fully. A full classification of nonlinear boundaries is not known (and indeed may not be possible), so without attempting a full classification of the bifurcations or explosions they give rise to, let us at least outline the scenarios analogous to those above.

Building on the local forms of folds and cusps, let us add a nonlinear term in the vector field component normal to the discontinuity surface. For the folds we consider a prototype system

$$(\dot{x}_1, \dot{x}_2) = \tfrac{1}{2}(1+\lambda)(-1, \pm x_1) + \tfrac{1}{2}(1-\lambda)(0,1) + (\lambda^2 - 1)(0,1) \qquad (11.2)$$

and its reverse time equivalent, and for the cusp

$$(\dot{x}_1, \dot{x}_2, \dot{x}_3) = \tfrac{1}{2}(1+\lambda)(1, x_1^2+x_3, 0)+\tfrac{1}{2}(1-\lambda)(0,1,0)+(\lambda^2-1)(0,1)\ , \quad (11.3)$$

each with $\lambda = \operatorname{sign} x_2$. In each case the sliding manifold $\mathcal{M}$ has a turning point connecting attractive and repulsive branches, which co-exist in an overlapping sliding region, leading to certain new sliding bifurcations and explosions.

To study these will require blowing up the discontinuity surface into a switching layer. In preparation, in Figure 11.11 the grazing-sliding bifurcation is shown again, now along with its blow-up to reveal the switching layer.

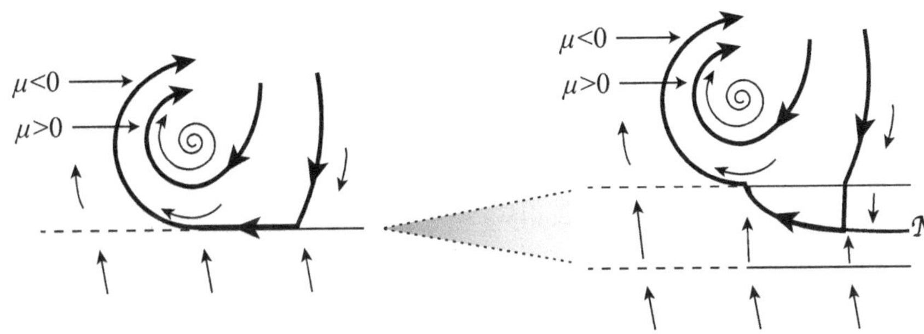

Fig. 11.11 The grazing-sliding bifurcation (left) blown-up to show its switching layer (right).

In Figures 11.12 to 11.15, the blow-up is performed on several new cases—one new bifurcation and three new explosions—that arise at a fold or cusp.

- The first is a new grazing-sliding bifurcation, Figure 11.12, at a visible fold on the boundary of a branch of attracting sliding.

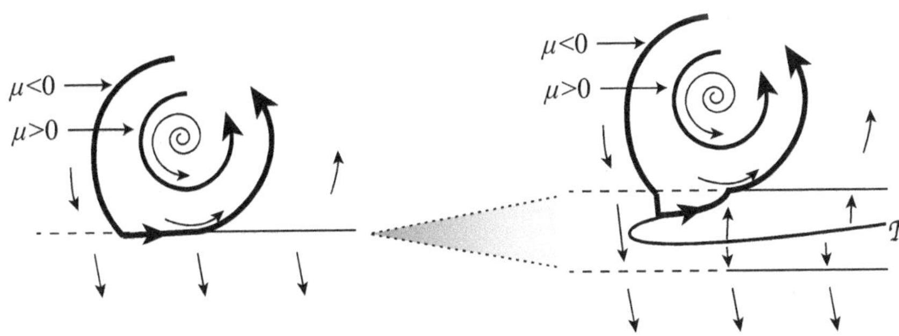

Fig. 11.12 Four new sliding bifurcations and explosions (left) arise in the presence of nonlinear switching. They are, firstly here, a nonlinear grazing-sliding bifurcation (found from (11.2) with the '–' sign and reversing time), ...

- The second is a new grazing-sliding explosion, Figure 11.13, at a visible fold on the boundary of repelling sliding, giving a similar event to the grazing-sliding bifurcation but with a jump in the orbit as μ changes sign.

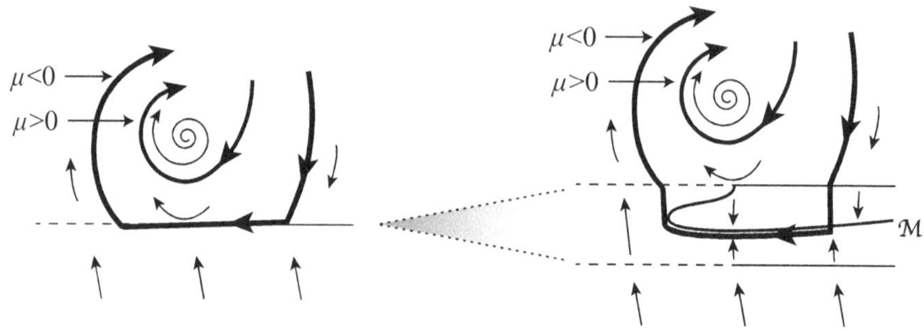

Fig. 11.13 ...a nonlinear grazing sliding explosion (from (11.2) with the '−' sign), ...

- The third is a new switching-sliding explosion, Figure 11.14, at an invisible fold on the boundary of repelling sliding, giving a similar event to the switching-sliding bifurcation but with a jump in the orbit as μ changes sign.

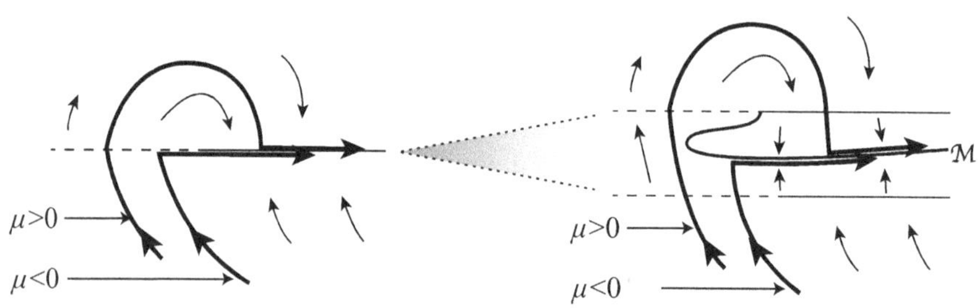

Fig. 11.14 ...a nonlinear switching sliding explosion (from (11.2) with the '+' sign), ...

- The fourth is a new adding-sliding explosion, Figure 11.15, at a visible cusp on the boundary of repelling sliding, giving a similar event to the adding-sliding bifurcation but with a jump in the orbit as μ changes sign.

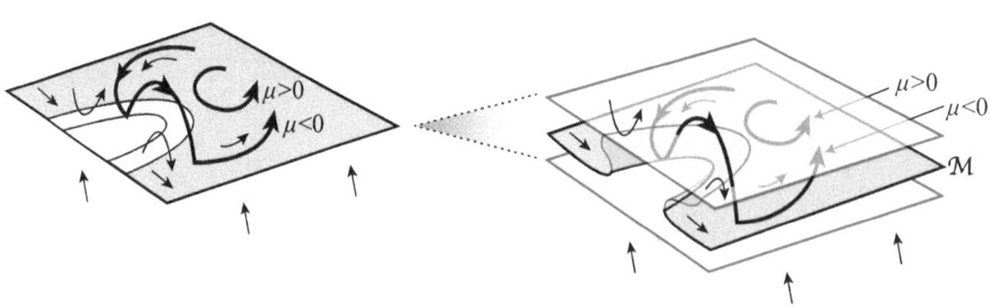

Fig. 11.15 ...and a nonlinear adding sliding explosion (from (11.3)).

The same dimensionality arguments used for the linear case suggest these are likely to be the only new generic one-parameter sliding bifurcations and explosions brought by nonlinear switching, at a single discontinuity surface, *if* we exclude the two-fold. Studying the two-fold singularity with large nonlinear terms (rather than only small perturbations of a linear switching system as in Section 13.5) is a somewhat more complex problem, and given the size of Chapter 13 on the two-fold already, the classification of sliding bifurcations and explosions in nonlinear switching systems must for now remain open.

11.5 The Classification and Its Completeness

The classification of sliding bifurcations and explosions centres around two principles (set out in [120]): (i) in the neighbourhood of a point where $\dot{\mathbf{x}} \neq 0$, the only singularities of the flow are tangency points, and (ii) for two general simply connected open sets $V \subset \mathbb{R}^m$ and $U \subset \mathbb{R}^p$ that are unbounded on some subspace of $\mathbb{R}^n$, generically there exists a stable intersection $V \cap U$ with dimension $\dim = m + p - n$ provided that $\dim \geq 0$.

Let Ξ^m be the set of surfaces of dimension m. Let $S_r \subset \Xi^{n-r-1}$ denote a singularity set of the discontinuity surface in $\mathbb{R}^n$, such that S_1 is a fold and S_2 is a cusp or two-fold. We assume implicitly in both Lemma 11.2 and Lemma 11.3 that $\dot{\mathbf{x}} \neq 0$ at the bifurcation point.

Lemma 11.2. *In a piecewise-smooth system* $\dot{\mathbf{x}} = \mathbf{f}(\mathbf{x}; \lambda)$, *a generic one-parameter bifurcation/explosion that occurs along a non-sliding segment of an orbit occurs at a fold or at a degenerate combination of folds.*

Proof. Let $\Phi(t; \mu)$ denote a non-sliding orbit segment of the system $\dot{\mathbf{x}} = \mathbf{f}(\mathbf{x}; \lambda)$ parameterised by μ. A one-parameter unfolding is a surface $U = \{\Phi(t; \mu) : \mu \in \mathbb{R}\} \subset \Xi^2$. Given a set $S_r \subset \Xi^{n-r-1}$, an intersection $S_r \cap U \subset \Xi^{(n-r-1)+2-n}$ is generic provided that $r \leq 1$. Assuming $n \geq 2$ and $\dot{\mathbf{x}} \neq 0$, the only such singularity set $S_{r \leq 1}$ is the fold.

Now assume that there exists a bifurcation/explosion at a higher codimension singularity, that is, at a singular set $S_{r \geq 2}$ in $\mathbb{R}^{n \geq 3}$. The singularity can be constructed as the intersection of fold singularities, for each of which the bifurcations are those of the fold. Let $U_k = \left\{\Phi(\mathbf{x}; \mu) : \mu \in \mathbb{R}^k\right\} \subset \Xi^{k+1}$ denote a k-parameter unfolding of non-sliding orbits. An intersection between U_k and a singularity S_r with dimension $n - r - 1 \geq n - 1 - k$ is generic, hence $r \leq k$. But every orbit in U_k belongs to the unfolding $U_1 = U$ for some μ, so an orbit $\Psi(\mathbf{x}; \mu)$ will undergo k *bifurcations/explosions at a fold* degenerately as the components of μ are varied.

Thus the bifurcations/explosions at a fold form a basis for all bifurcations/explosions that occur along non-sliding segment of orbits. □

Lemma 11.3. *In a piecewise-smooth system* $\dot{\mathbf{x}} = \mathbf{f}(\mathbf{x};\lambda)$, *a generic one-parameter bifurcation/explosion that occurs along a sliding segment of an orbit occurs either at a cusp, at a two-fold, or at a degenerate combination of thereof.*

Proof. Sliding orbits are confined to the discontinuity surface $\mathcal{D} \subset \Xi^{n-1}$, and do not belong to the set U in Lemma 11.2. Denote a sliding segment of an orbit by $\tilde{\Psi}(\mathbf{x};\mu)$, with an unfolding $\tilde{U}$. Then $\tilde{U} \subset \mathcal{D}$. Note also that $S_r \subset \mathcal{D}$.

Given a set $S_r \subset \Xi^{n-r-1}$, an intersection $S_r \cap \tilde{U} \subset \Xi^{n-r-1+2-(n-1)}$ is generic provided that $n - r - 1 + 2 - (n-1) \geq 0$, hence $r \leq 2$. Assuming $n \geq 3$, this includes only the fold set S_1 and the cusp or two-fold set S_2.

Now assume that there exists a bifurcation/explosion at a singular set $S_{r\geq 3}$ in $\mathbb{R}^{n\geq 3}$, which as in the previous lemma can be constructed as the intersection of fold, cusp, and two-fold singularities. With k bifurcation parameters, a sliding orbit may be made to intersect a singularity with dimension at least $n-2-k$, that is, an unfolding $\tilde{U}_k = \left\{\Psi(\mathbf{x};\mu) : \mu \in \mathbb{R}^k\right\}$ may generically intersect a set $S_{r\leq k+1}$. But every orbit in $\tilde{U}_k$ belongs to the unfolding $\tilde{U}_1 = \tilde{U}$ for some μ, so an orbit $\Psi(\mathbf{x};\mu)$ will undergo k *bifurcations/explosions at a cusp or two-fold* degenerately.

Thus the one-parameter bifurcations/explosions at a cusp and two-fold form a basis for all bifurcations/explosions that occur along sliding segments of orbits. □

We can then combine the two and show that no higher codimension singularities play a role. Restating the main theorem:

Theorem 11.1. In a piecewise-smooth vector field of the form $\dot{\mathbf{x}} = \mathbf{f}(\mathbf{x};\lambda)$ with linear dependence on λ, a generic one-parameter sliding bifurcation/explosion is either one of the *bifurcations/explosions at a fold, cusp, or two-fold* or a degenerate combination of them.

Proof. Singularities of the discontinuity surface consist of intersections of fold, cusp, and two-fold singularities. By Lemma 11.2 and Lemma 11.3, a generic unfolding of orbits in the neighbourhood of these singularities is comprised of the sliding bifurcations/explosions associated with them.

Assume that this classification is incomplete. Then with one bifurcation parameter in $\mathbb{R}^{n\geq 4}$, an orbit may be made to intersect a singular set $S_{r>2}$ through sliding. This is possible only if the sliding orbit is confined to an $n - r + 1$ dimensional set that contains S_r. The only generic set satisfying this is the singularity set S_{r-1}, but the sliding vector field is transverse to S_{r-1} everywhere except at S_r. Since the vector field points through S_r, no orbits are confined to S_r.

Thus there are no one-parameter unfoldings of orbits that intersect a singularity other than $S_{r\leq 2}$, which includes only a fold, cusp, or two-fold. □

11.5.1 Unfoldings

To derive the unfoldings in Tables 11.1 and 11.2 systematically, we vary an initial condition along a path generically oriented with respect to the flow and the discontinuity surfaceand then study the one-parameter family of orbits it generates—the unfolding.

Let V denote the potential function defined in Section 6.1. Define a hyperplane Ψ given by the zero set of a scalar function

$$\psi(\mathbf{x};\nu) = \nu + p(\mathbf{x}\backslash\{x_1,x_2\}) + V(x\backslash x_k) - q(\mathbf{x})\sigma(\mathbf{x}) \,, \tag{11.4}$$

where $q = \delta_t V/\delta_t\sigma$, $p = b_3x_3 + b_4x_4 + \ldots$, and ν, b_j, are parameters, and where k is the smallest index such that $\partial\sigma(0)/\partial x_k \neq 0$. The expression $\mathbf{x}\backslash x_a$ or $\mathbf{x}\backslash\{x_a,\ldots\}$ denotes dependence on the component of $\mathbf{x} = \{x_1,\ldots,x_n\}$ excluding x_a or $\{x_a,\ldots\}$. Direct calculation shows that $\delta_t\psi = 0$, meaning that Ψ is comprised of solutions of the system

$$\dot{\mathbf{x}} = \mathbf{f}(\mathbf{x};\lambda) = \tfrac{1}{2}(1+\lambda)\mathbf{f}^+(\mathbf{x}) + \tfrac{1}{2}(1-\lambda)\mathbf{f}^-(\mathbf{x}) \,, \qquad \lambda = \operatorname{sign}\sigma \,. \tag{11.5}$$

The surface Ψ is continuous but nonsmooth at $\sigma = 0$, where q is discontinuous. It intersects the discontinuity surface transversally for general values of the constants $p, q, r, \ldots$. The intersection consists of *impact coordinates* $\mathbf{x}_0$, which are solutions of $0 = \psi(\mathbf{x}_0) = p(\mathbf{x}_0\backslash\{x_1,x_2\}) + V(\mathbf{x}_0\backslash x_k)$ and $\sigma(\mathbf{x}_0) = 0$.

At the origin, $\psi(0,\nu) = \nu$. Therefore the set of conditions $\psi(\mathbf{x};0) = 0$ with $x_3 = x_4 = \cdots = 0$ define a trajectory that intersects the origin. If we take σ as given in Section 6.1, then the origin is a singularity, where the trajectory $\psi(\mathbf{x};0) = x_3 = x_4 = \cdots = 0$ is tangent to the discontinuity surface, and therefore Ψ is an unfolding of a sliding bifurcation/explosion.

We work in the $\mathbf{x} = (x_1,\ldots)$ coordinates from Section 6.1, where $\sigma = 0$ is a coordinate plane $x_k = 0$, for $k = 1$, 2, or 3, rather than the $\mathbf{y}$ coordinates where the flow is straightened and $\sigma = 0$ is curved. The hyperplane

$$\Psi = \{\mathbf{x} \in \mathcal{U} : \psi(\mathbf{x};\nu) = 0\}$$

can then be defined via the function

$$\psi(\mathbf{x};\nu) = \nu + p(\mathbf{x}\backslash\{x_1,x_2\}) + V(\mathbf{x}\backslash x_k) - \gamma q(\mathbf{x})x_k. \tag{11.6}$$

We can write q, which is 1 in a flow that grazes and 0 otherwise, as

$$q(\mathbf{x}) = \begin{cases} 1 \text{ if } \delta_t(V - x_k) = 0, \\ 0 \text{ if } \delta_t V = 0. \end{cases} \tag{11.7}$$

The quantity γ represents a sign that arises in handling complex tangencies, so here it will be $+1$, except for the two-fold where the transformation to the y-coordinates in (6.5) gives $\gamma = 1$ for $x_k > 0$ and $\gamma = \pm 1$ for $x_k < 0$.

11.5.2 Classes of Sliding Bifurcation . . .

That there are only four generic one-parameter sliding bifurcations is a remarkable result for piecewise-smooth systems. It depends firstly on Theorem 11.1, which tells us that to find all generic one-parameter sliding bifurcations, we must explore the possible flow topologies only at a fold, cusp, or two-fold. The local geometry at these singularities is described in Sections 6.2 to 6.4, with their linear sliding flows given in Sections 8.5.1 to 8.5.3. Any given singularity may generate one or more bifurcations or none at all. Below, Sections 11.5.2.1 to 11.5.2.4 list all bifurcations found by such an exhaustive exploration using the procedure outlined in Section 11.5.1.

Sliding bifurcations occur only at the boundaries of attracting sliding regions, where the flow is deterministic. They therefore involve only folds or cusps at the boundaries of attracting sliding, which we call 'attracting folds or cusps' for short. The four sliding bifurcations were discovered and classified through the works [49, 50, 63, 64, 65], where they were given the names in Table 11.1. The classification is collated in Table 11.1 and described in full below.

11.5.2.1 . . . at an Attracting Visible Fold: Grazing-Sliding

Consider first the piecewise-smooth vector field (6.18) with the '+' sign, which fixes $\delta^2_{t+}\sigma > 0$ so the fold is visible. For convenience we add a trivial third dimension (or parameter) x_3, which we can set as $x_3 = \nu + p$ in (11.6). Then Ψ is an unfolding of a *sliding bifurcation at an attracting visible fold*, given by

$$0 = x_3 + \tfrac{1}{2}x_1^2 - x_2\,\mathrm{step}\,(x_2)\ , \tag{11.8}$$

where step denotes the unit step function.

The discontinuity surface $x_2 = 0$ is partitioned into a crossing region over $x_1 < 0$ and a sliding region over $x_1 > 0$ where sliding orbits flow towards the fold (as implied by (8.18)). They leave the sliding section and enter $x_2 > 0$ along the surface

$$0 = \tfrac{1}{2}x_1^2 - x_2, \qquad x_1 < 0, \tag{11.9}$$

which is the separatrix between sliding and crossing dynamics. We make the choice not to explicitly include this separatrix or the sliding orbits in the surface (11.6), but for completeness they must be added to the unfolding.

Impact points in the unfolding are the curves $0 = \frac{1}{2}x_{10}^2 + x_3$. First consider a family of orbits parameterised by x_3. If $x_3 > 0$ then x_{10} is imaginary, so an orbit is a smooth trajectory in $x_2 > 0$. As we decrease x_3, the orbit approaches the manifold $x_2 = 0$ and grazes when $x_3 = 0$. For $x_3 < 0$ the orbit impacts and takes on a sliding segment of length $\sqrt{-2x_3}$, see Figure 11.16. This unfolds a *grazing-sliding* bifurcation.

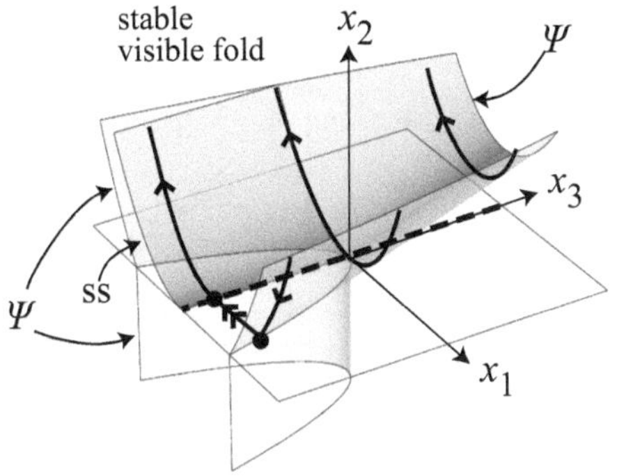

Fig. 11.16 The *grazing-sliding* bifurcation takes place at an attracting visible fold. The discontinuity surface is $x_2 = 0$ with sliding in $x_1 > 0$. The unfolding Ψ of a nongeneric orbit through the origin is shown, with the sliding separatrix ss. The nongeneric orbit visibly grazes at the boundary of attracting sliding. (Arrows can be reversed in each case to obtain the same phase portrait with the opposite stability of sliding).

Examples of a grazing-sliding bifurcation of a periodic orbit or the unstable manifold to a saddle are shown in Figures 11.2 and 11.4.

11.5.2.2 ...at an Attracting Visible Fold: Crossing-Sliding

Orbits belonging to (11.8) behave somewhat differently if they originate in $x_2 < 0$. They can be parameterized by the impact coordinate x_{10}. An orbit with $x_{10} < 0$ crosses the manifold, but as x_{10} changes sign, the orbit gains a sliding segment of length x_{10}, see Figure 11.17. A second type of *sliding bifurcation at an attracting visible fold* takes place, called a *crossing-sliding* bifurcation.

Examples of a crossing-sliding bifurcation of a periodic orbit or the unstable manifold to a saddle are shown in Figures 11.2 and 11.4.

11.5.2.3 ...at an Attracting Invisible Fold: Switching-Sliding

An invisible fold arises from the '−' case of the system (6.18), since $\delta_{t+}^2\sigma < 0$. Similarly to the visible fold, we find that an unfolding of the *sliding bifurcation at an attracting invisible fold* is given by

$$0 = x_3 + \tfrac{1}{2}x_1^2 + x_2\,\mathrm{step}\,(x_2) \tag{11.10}$$

This is the attracting case, with the repelling case found by reversing time in (6.18). In contrast to the visible fold, the phase portraits of attracting

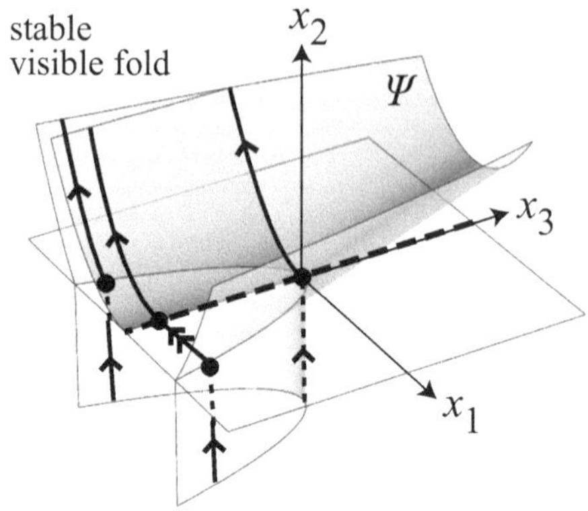

Fig. 11.17 The *crossing-sliding* bifurcation takes place at an attracting visible fold. The discontinuity surface is $x_2 = 0$ with sliding in $x_1 > 0$. The unfolding Ψ of a nongeneric orbit through the origin is shown, with the sliding separatrix ss. The nongeneric orbit is transverse to the manifold on one side but grazes visibly on the other.

and repelling *in*visible folds are equivalent up to time reversal. We therefore regard them as one case up to time direction.

The sliding and crossing regions are as for the visible case, but now all sliding orbits flow away from the fold. Orbits impact the manifold $x_2 = 0$ at $0 = x_3 + \frac{1}{2}x_1^2$, but the unfolding exists only for $x_3 < 0$. We can parameterize orbits by the impact coordinate x_{10}, and then as it passes from $x_{10} > 0$ to $x_{10} < 0$, an orbit gains a switching segment, that is, a non-sliding trajectory that turns around the fold line back onto $\sigma = 0$; see Figure 11.18. This unfolds a *switching-sliding* bifurcation.

Examples of a switching-sliding bifurcation of a periodic orbit or the unstable manifold to a saddle are shown in Figures 11.2 and 11.4.

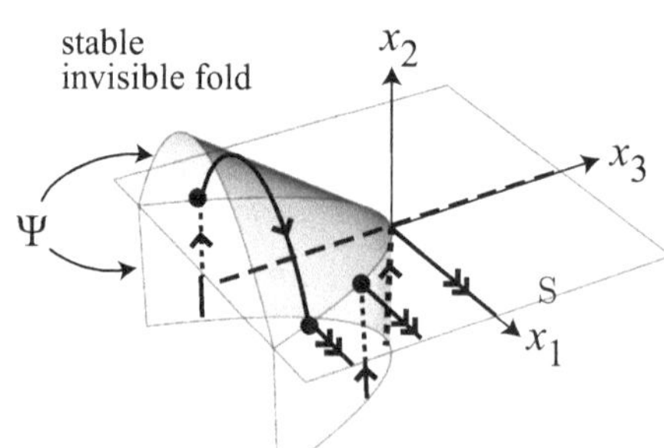

Fig. 11.18 The *switching-sliding* bifurcation takes place at an attracting visible fold. The discontinuity surface is $x_2 = 0$ with sliding in $x_1 > 0$. The unfolding Ψ of a nongeneric orbit through the origin is shown, with the sliding separatrix ss. The nongeneric orbit intersects an invisible tangency.

11.5.2.4 ... at a Visible Cusp: Adding-Sliding

Now consider the system (6.24), which has a cusp at the origin. The unfolding Ψ as given by (11.6) has the form

$$0 = \nu + px_3 + \frac{1}{3}x_1^3 + x_3x_1 + x_2 \operatorname{step}(x_2) \ . \tag{11.11}$$

We cannot remove the parameters ν and p by any transformation as for the fold, since x_3 appears in the term $V = \frac{1}{3}x_1^3 + x_3x_1$, but it can be written more conveniently by using the impact coordinates $(x_{10}, 0, x_3)$, which satisfy $0 = \frac{1}{3}x_{10}^3 + x_3x_{10} + px_3 + \nu$, to eliminate ν and p. The unfolding becomes

$$0 = \frac{1}{3}(x_1 - x_{10})(x_1^2 + x_1x_{10} + x_{10}^2 + 3x_3) + x_2 \operatorname{step}(x_2) \ . \tag{11.12}$$

This can be interpreted as a family of orbits with two parameters, namely, the impact coordinates x_{10} and x_3. That the cusp has a two-parameter unfolding indicates that a one-parameter family of orbits does not generically intersect the cusp from $\sigma > 0$ or $\sigma < 0$.

Curves of fold points emanate from the origin on the discontinuity surface $x_2 = 0$: a visible fold along $x_1 = -\sqrt{-x_3}$ and an invisible fold along $x_1 = \sqrt{-x_3}$, in the region $x_3 < 0$. The separatrix between sliding and crossing dynamics is the particular surface whose impact coordinates lie along a visible fold, that is

$$0 = \frac{1}{3}(x_1 + \sqrt{-x_3})^2(x_1 - 2\sqrt{-x_3}) + x_2 \operatorname{step}(x_2) \ . \tag{11.13}$$

Sliding orbits near an attracting cusp leave along the separatrix. The two cases are shown in Figure 11.19. The system (6.24) with the '−' sign gives

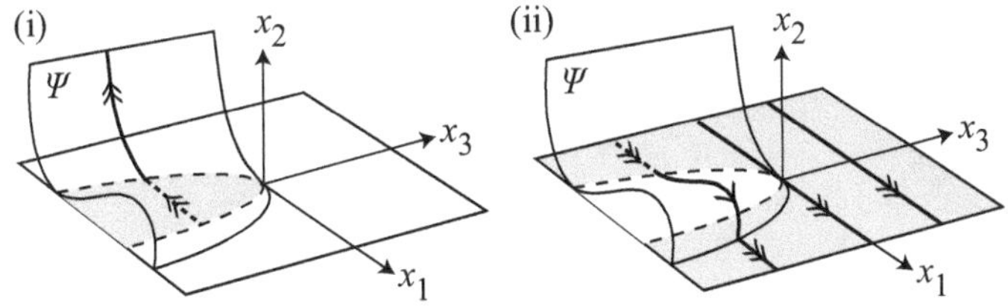

Fig. 11.19 The sliding separatrix (11.13) near an attracting cusp, with sliding orbits shown. (i) At the invisible cusp, the sliding region is $x_2 = 0$, $x_3 < -x_1^2$. (ii) At the visible cusp, the sliding region is $x_2 = 0$, $x_3 > -x_1^2$.

Figure 11.19(i), the attracting invisible cusp, which has sliding orbits only for $x_3 < 0$, making sliding bifurcations impossible. Taking the '+' sign from (6.24), we have the attracting visible cusp shown in Figure 11.19(ii). The visible case contains a nongeneric sliding orbit that is (visibly) tangent to the cusp and (11.13) represents its unfolding. But (11.13) has only one-parameter, x_3, and therefore (11.13) is the one-parameter unfolding for the *sliding bifurcation at an attracting visible cusp*, known as an *adding-sliding* bifurcation.

Reversing the time direction at the attracting cusps, we obtain the repelling visible and repelling invisible cusps, but their phase portraits are the same as the attracting cases up to time direction. Therefore the only distinct one-parameter bifurcation at a cusp is Figure 11.19(ii), at a visible cusp.

An example of this is adding-sliding of a limit cycle (Figure 11.4). The bifurcation is often depicted with the fold curve $x_3 + x_1^2 = 0$ straightened into a line as shown in Figure 11.20.

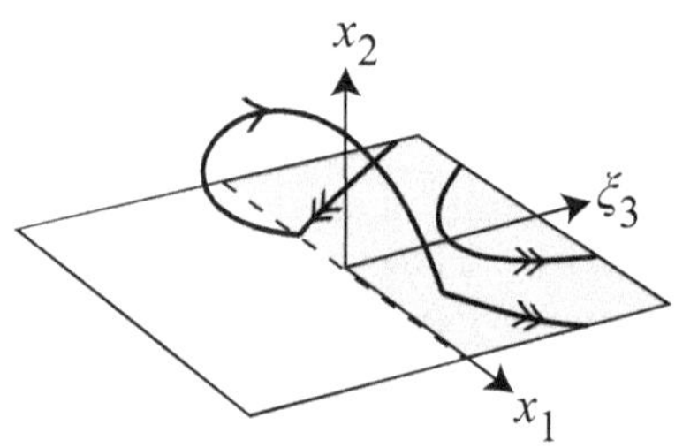

Fig. 11.20 The bifurcation at an attracting visible cusp in coordinates (x_1, x_2, ξ_3), obtained from Figure 11.19(ii) by substituting $\xi_3 = x_3 + x_1^2$.

The forms in Figures 11.19(ii) and 11.20 are both common in the literature, the latter in control applications in the form $\dot{\mathbf{z}} = \underline{\underline{A}}\mathbf{z} + \mathbf{b}\,\mathrm{sign}(\sigma(\mathbf{z}))$, with

$$\underline{\underline{A}} = \begin{pmatrix} a_1 & 1 & 0 \\ a_2 & 0 & 1 \\ a_3 & 0 & 0 \end{pmatrix}, \quad \mathbf{b} = \begin{pmatrix} b_1 \\ b_2 \\ b_3 \end{pmatrix}, \quad \sigma = z_2 \,,$$

where the a_i, b_i, are constants. There are cusps at $(z_1, z_2, z_3) = \pm(0, b_1, b_2)$. Letting $(x_1, x_2, \xi_3) = (z_2 - b_1, z_1, z_3 - b_2)/b_3$, then approximating near the '+' cusp for $|b_1/b_3| \gg |b_2/b_3|$ and $|b_1/b_3| \gg 1$, gives $(\dot{x}_1, \dot{x}_2, \dot{\xi}_3) \approx (1, \xi_3, x_1)$ for $\sigma > 0$ and $(0, 0, 1)$ for $\sigma < 0$ (up to a time rescaling in $\sigma < 0$). This is precisely the cusp form with $\xi_3 = p + \frac{1}{2}x_1^2$ shown in Figure 11.20.

The system in Figure 11.19(ii) is common in mechanics, for example, as a dry-friction oscillator $\ddot{u} + a\dot{u} + u = \cos t + F\,\mathrm{sign}(\dot{u})$ where a and F are constants. There are cusps at $t = (\frac{1}{2} + n)\pi$, $n \in \mathbb{Z}$. Letting $(x_1, x_2, x_3) = (t, -\dot{u}, u - 1 - F)$ and then approximating near the cusp at $(0, 0, 0)$, we have the system $(\dot{x}_1, \dot{x}_2, \dot{x}_3) \approx (1, x_3 + x_1^2, 0)$ for $x_2 > 0$ and $(1, 2F, 0)$ for $x_2 < 0$. This is topologically equivalent to the structural model of the cusp in (6.24). The oscillator was analysed in [89, 131] and shown to contain periodic orbits that underwent the bifurcation at an attracting cusp (or adding-sliding), and local analysis revealed curves of bifurcations at folds nearby.

A general form of the bifurcation diagram at a cusp is provided by the two-parameter unfolding (11.12), shown in Figure 11.21 in the parameter space of the impact coordinates x_{10} and x_3. The bifurcation curve has two branches: the visible fold V along $x_{10} = -\sqrt{-x_3}$ and the invisible fold I along $x_{10} = +\sqrt{-x_3}$. It also contains the curve $x_{10} = +2\sqrt{-x_3}$ where the separatrix (11.13) impacts the discontinuity surface, labelled R. Bifurcations at a fold take place where the number of solutions of (11.12) jumps from one to three. The bifurcations at the folds are those described in Sections 11.5.2.1 to 11.5.2.3. Degenerate combinations of them occur at the cusp. At an attracting cusp, the bifurcation in Figure 11.19(ii) also occurs.

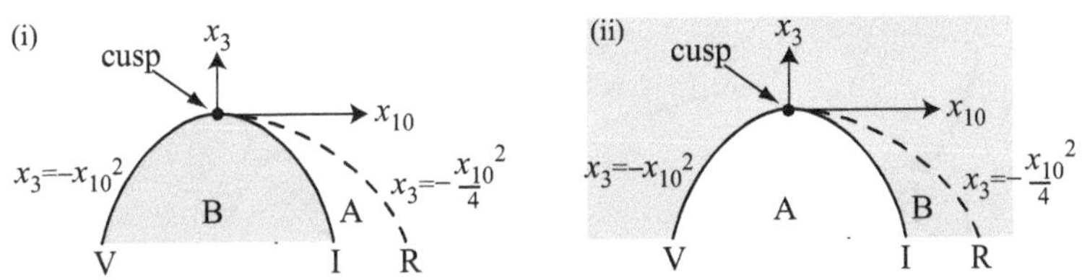

Fig. 11.21 Bifurcation curves in parameter space (x_{10}, x_3) for the (i) invisible and (ii) visible cusp. Branches of visible (V) and invisible (I) folds separate regions of sliding (shaded) and crossing (unshaded). Grazing orbits impact along R. The flow in $x_2 > 0$ maps points from region A to B.

If we return to (11.11), we can interpret ν and p as specifying the height and angle, with respect to the discontinuity surface, of a family of orbits parameterized by x_3. As shown in Figure 11.22,

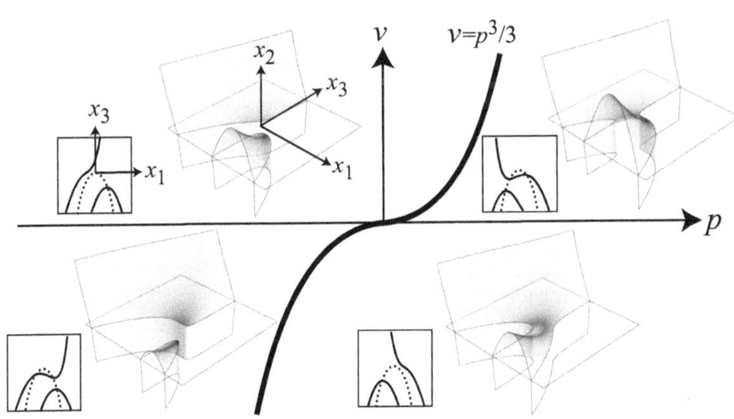

Fig. 11.22 The family of unfoldings (11.11), partitioned by the lines $\nu = 0$ and $\nu = p^3/3$. Their intersections with the discontinuity surface are inset and have turning points along the folds (dotted).

such a surface has four topologically different forms, depending on the signs of ν and $\nu - \frac{1}{3}p^3$, which determines the kind of intersection that exists between the unfolding's contact with the discontinuity surface (full curves inset) and the fold sets (dashed curves inset). If we fix the signs of these two quantities, we obtain an unfolding surface that contains all of the possible topologies of orbits that can be found as we continuously change an inifital condition of a distinguished orbit, hence unfolding all possible local bifurcations. An example of the unfolding is shown in Figure 11.23. The same surface unfolds orbits around an attracting visible or invisible cusp, (the repelling cases are obtained by reversing arrows).

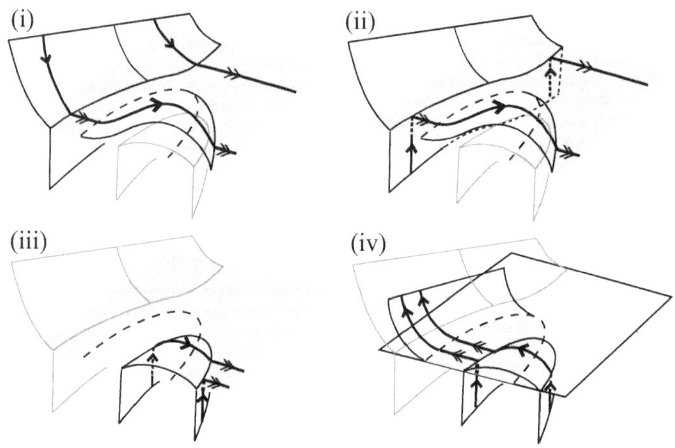

Fig. 11.23 An example of the unfolding (11.11), with $\nu > 0$ and $\nu > p^3/3$, and parts of the separatrix from Figure 11.19, exhibiting bifurcations: (i–ii) at an attracting visible cusp (Figure 11.19), (iii–iv) at an attracting invisible fold (switching sliding from Figure 11.18). The sliding region is shaded.

Examples of an adding-sliding bifurcation of a periodic orbit or the unstable manifold to a saddle are shown in Figures 11.2 and 11.4.

11.5.3 Classes of Sliding Explosion ...

That there are only four generic one-parameter sliding explosions again stems from Theorem 11.1, which tells us they can only occur at a fold, cusp, or two-fold. Any given singularity may generate one or more explosions, or none at all, and below we list all explosions found by such an exhaustive exploration using the procedure outlined in Section 11.5.1.

Explosions involve determinacy-breaking, so they occur at the boundaries of repelling sliding regions, hence at folds or cusps at the boundaries of repelling sliding, which we call for short 'repelling folds or cusps'. They can also occur at two-folds, which lie, of course, at the boundary of both repelling and attracting sliding. The fold and cusp flow topologies are described in Sections 6.2 to 6.3, with their linear sliding flows given in Sections 8.5.1 and 8.5.2. The results are collated in Table 11.2.

The first sliding explosion was conjectured in [48], the full classification emerging only in [120]. Those early works used the term ‘catastrophic bifurcation’ instead of ‘explosion’. The modern terminology is more accurate in its description, as the flow ‘explodes’ into multivaluedness as it encounters a determinacy-breaking point, with a more extreme effect on the system than a standard bifurcation.

11.5.3.1 . . . at a Repelling Visible Fold: Grazing-Sliding

Orbits belonging to (11.8) and originating in $x_2 > 0$ may also undergo a *sliding explosion at a repelling visible fold.* The unfolding surface is the same as the attracting fold, (11.8), but this case is obtained by reversing the time direction of the system (6.18). The phase portrait, in Figure 11.24, is not equivalent to the attracting case because this explosion does not involve the sliding separatrix (11.9).

Along the unfolding, a continuous change of initial conditions leads to a discontinuous change in an orbit leaving the fold. We can parameterize the orbit with the coordinate x_3. As x_3 passes from positive to negative, the impact coordinates become real. In this case, a smooth orbit in $\sigma > 0$ makes a quadratic tangency with the discontinuity surface ,but, instead of gaining a sliding segment, it crosses a distance $\sqrt{-2x_3}$ from the fold, and its outset jumps to $x_2 < 0$, shown in Figure 11.24. This unfolds a *grazing-sliding* explosion.

Examples of a grazing-sliding explosion of a periodic orbit or the unstable manifold to a saddle are shown in Figures 11.3 and 11.5. For a case study in a real physical device, see Section 14.4.

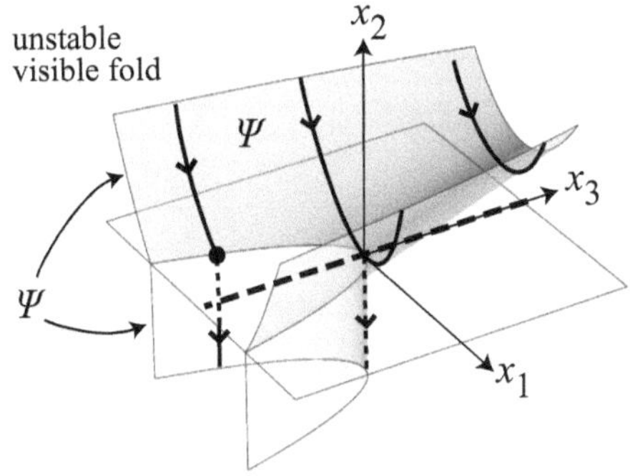

Fig. 11.24 The *grazing-sliding* explosion takes place at a repelling visible fold. The discontinuity surface is $x_2 = 0$ with sliding in $x_1 > 0$. The unfolding Ψ of a nongeneric orbit through the origin is shown, with the sliding separatrix ss. The nongeneric orbit visibly grazes the manifold at the boundary of repelling sliding.

11.5.3.2 ...at a Two-Fold: Canards

The last singularity remaining to be considered is the system (6.21). The unfolding is most simply expressed in the form (11.4), which gives $0 = \nu + (p-1)x_3$. So in the $\mathbf{y}$ coordinate system, the unfolding is a family of planes $y_3 = \nu/(p-1)$. Since ν and p are general constants, we can replace the right-hand side of this expression simply by ν. Then in the x_1, x_2, x_3, coordinates, the unfolding becomes

$$0 = \nu + \tfrac{1}{2}\alpha x_1^2 - \tfrac{1}{2}\beta x_2^2 + x_1x_2 - \gamma x_3, \tag{11.14}$$

recalling from (11.6) that γ is a piecewise constant given by

$$\gamma = \left\{ \begin{matrix} 1 \text{ if } x_3 \geq 0 \\ \omega \text{ if } x_3 < 0 \end{matrix} \right\}, \qquad \omega = \pm 1. \tag{11.15}$$

This describes a family of nonsmooth paraboloids which crease at $x_3 = 0$ shown in Figure 11.25. They consist of pieces of hills (if $\alpha\beta < -1$) and saddles (if $\alpha\beta > -1$).

As for the cusp, (11.14) is a two-parameter unfolding—we can parameterize orbits by, say, ν which controls where the surfaces in Figure 11.25 cut the x_3 axis and an impact coordinate such as $(x_{10}, x_{20}, 0)$, obtained by solving (11.14) on $x_3 = 0$.

Similarly to the cusp, the need for two parameters means an orbit will not generically intersect the two-fold point except via a sliding segment. There are regions of both attracting and repelling sliding, bounded by the lines of folds given on $x_3 = 0$ by $\alpha x_1 + x_2 = 0$ and $x_1 - \beta x_2 = 0$. Depending on the signs of $\delta_{t+}\sigma = \alpha$ and $\delta_{t-}\sigma = -\beta$, the folds are visible or invisible. At a visible fold, orbits depart the discontinuity surface along sliding separatrices,

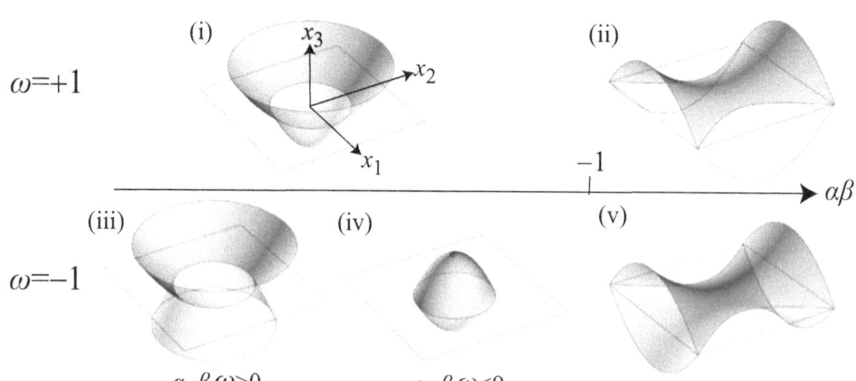

Fig. 11.25 Unfoldings near a two-fold singularity. If $\alpha\beta > -1$ they consist of saddles, with the same orientation either side of $x_3 = 0$ if $\omega > 0$. If $\alpha\beta < -1$, they consist of hills, similarly oriented if $\omega > 0$ and similarly oriented if $\omega < 0$ to form either the nonsmooth hourglass when $\alpha, \beta\omega > 0$ or the nonsmooth ball if $\alpha, \beta\omega < 0$.

$$\sigma = \frac{1}{2}\begin{cases} \frac{1}{\alpha}(\alpha x_1 + x_2)^2 & \text{if } x_3 \geq 0, \\ \frac{-\omega}{\beta}(x_1 - \beta x_2)^2 & \text{if } x_3 \leq 0, \end{cases} \tag{11.16}$$

give the full unfolding of the dynamics near a two-fold.

The unfoldings of sliding orbits reveal three one-parameter explosions which we now describe. The different sliding vector fields are given by (8.20) (and are set out in Section 13.4); from the many topologies, we show here only those that generate explosions.

If $\alpha > 0$ and $\beta\omega > 0$, then the two-fold point is the intersection of two visible folds—the visible-visible two-fold. A sliding orbit intersecting the two-fold undergoes the *explosion at a visible-visible two-fold*, in which its outset jumps between $x_3 < 0$ and $x_3 > 0$ along the sliding separatrices (11.16) as shown in Figure 11.26(i). We call this a *visible canard explosion.*

If $\alpha < 0 < \beta\omega$ then the two-fold is the intersection of one visible and one invisible fold. A sliding orbit intersecting the two-fold undergoes the *explosion at a visible-invisible fold.* There are two possible scenarios permitted by the sliding vector field's boundary conditions (8.7). On one side of the explosion, a sliding orbit intersects the visible fold and has a segment in $x_3 < 0$ on the sliding separatrix, but on the other side, it may either remain locally in the sliding region, resulting in an *invisible canard explosion* as shown in Figure 11.26(ii), or the intersection with the two-fold may persist, resulting in a *robust canard explosion* as shown in Figure 11.26(iii).

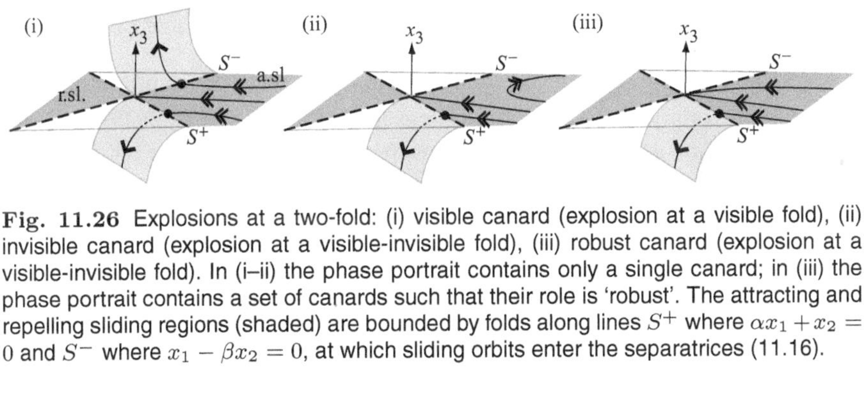

Fig. 11.26 Explosions at a two-fold: (i) visible canard (explosion at a visible fold), (ii) invisible canard (explosion at a visible-invisible fold), (iii) robust canard (explosion at a visible-invisible fold). In (i–ii) the phase portrait contains only a single canard; in (iii) the phase portrait contains a set of canards such that their role is 'robust'. The attracting and repelling sliding regions (shaded) are bounded by folds along lines S^+ where $\alpha x_1 + x_2 = 0$ and S^- where $x_1 - \beta x_2 = 0$, at which sliding orbits enter the separatrices (11.16).

In the neighbourhood of an invisible-invisible two-fold, given by $\alpha < 0$ and $\beta\omega < 0$, orbits map repeatedly back onto the discontinuity surface. The relevant unfoldings (11.14), marked (ii), (iv), (v) in Figure 11.25, form invariant manifolds around which orbits rotate until they impact and slide. (The 'nonsmooth diabolo' invariant surface which we will find in Section 13.3.4 is the unfolding in Figure 11.25(v) at $\nu = 0$.)

Examples of a canard explosions of a periodic orbit or the unstable manifold to a saddle are shown in Figures 11.3 and 11.5.

11.5.4 The Omitted Singularities

The singularities not appearing in the classification above are the repelling invisible fold, the repelling visible or invisible cusp, and the invisible two-fold. It is generically impossible for a distinguished orbit to develop contact with these singularities as one parameter is varied. The reader may satisfy themselves that the unfoldings of flow topologies at these singularities create no new cases of sliding bifurcations or explosions. Theorem 11.1 tells us that all generic cases occur only at the fold, cusp, or two-fold, and the unfolding above gives a means to explore them exhaustively. A rigorous proof that the classification into eight events found here is complete would consist of a yet longer exhaustive accounting of all orbit permutations for all possible initial conditions.

11.6 Codimension Two Sliding Bifurcations and Explosions

The higher codimension singularities $S_{r\geq 4}$ include the swallowtail, the butterfly, intersections of folds with cusps/two-folds, and so on. Theorem 11.1 states that for an orbit to hit these requires at least two bifurcation parameters, in which case the second (and third ...) parameter controls the intersection of two (or more) of the elementary bifurcations from Sections 11.5.2.1 to 11.5.3.2. Let us discuss these briefly.

The results in Section 11.5 suggest how higher codimension sliding bifurcations can be derived from the k-parameter unfoldings U_k and $\tilde{U}_k$. The grazing orbit of a codimension k sliding bifurcation intersects a singular set S_k at an impact point or S_{k+1} via a sliding segment.

Consider first the cusp and two-fold sets S_2. An orbit impacting the cusp at $x_{10} = x_3 = 0$ has the form $0 = \frac{1}{3}x_1^3 + \sigma\,\text{step}\,\sigma$, $x_3 = 0$ (see (11.11)). That is, the grazing orbit has a cubic tangency to the discontinuity surface at the cusp. This orbit can be reached only by control of two parameters. Two-parameter sliding bifurcations are described in [130] at and near an attracting visible or invisible cusp, which can be reproduced from the unfolding (11.11) with ν and p as parameters (see Figures 11.22 and 11.23).

An orbit impacting the two-fold has a quadratic tangency to the discontinuity surfaceand can also be reached only by control of two parameters.

The next singularity set, S_3, is generic in $\mathbb{R}^4$. It also forms a codimension one singularity in $\mathbb{R}^3$, as considered in [206]. In the prototype for the straightened vector fields, S_3 consists of the swallowtail $\sigma = x_2 + (\frac{1}{4}x_1^4 + \frac{1}{2}x_3x_1^2 + x_4x_1)$, umbilic $\sigma = x_2 + (\frac{1}{3}x_1^3 \pm (x_3^2 + x_4)x_1)$ ('+' gives the elliptic or 'lips' case, and '−' gives the hyperbolic or 'beak-to-beak' case), and fold-cusp $\sigma = x_4 + (\frac{1}{3}x_1^3 + x_3x_1 \pm \frac{1}{2}x_2^2)$ ('±' is a Morse term similar to that in the hill/saddle of the two-folds). These topologies are those outlined in Sections 8.5.4 to 8.5.6.

Two-parameter sliding bifurcations occur where orbits impact at cusps near any of the S_3 singularities and where orbits impact at two-folds near a fold-cusp. The cusp and two-folds in $\mathbb{R}^4$ are curves in the (three-dimensional) discontinuity surface. Two-parameter bifurcations also occur where sliding orbits intersect the S_4 points themselves. Note that this intersection is often tangential—the cusp occurs where a sliding orbit grazes a fold, and a swallowtail or umbilic occurs where a sliding orbit grazes a cusp. However, a two-fold or fold-cusp occurs where the vector field is set-valued in the Filippov convention—typically an orbit enters the singularity transversally, but is not uniquely defined at the singular point itself (i.e. it is set-valued).

11.7 Maps

Through Chapters 7 to 10 on local dynamics and in classifying the mechanisms of global sliding bifurcations and explosions above, we were able to use local geometry to great effect. Working locally, we can limit the permutations of local singularities and separatrices and derive approximations for them. This is obviously not going to be effective for global behaviour in general.

One method that can be incredibly powerful for reducing global dynamics back to a local problem is to study how the flow pierces some hypersurface, effectively taking a cross-section of the flow, known commonly as a Poincaré section. An example is the line Π shown in the first portrait of Figure 11.4. Given an initial point $\mathbf{x}_\pi$ on the section Π, the flow is evolved forwards through space until it returns (if the section is well chosen) to Π. The return time is some $T(\mathbf{x})$, and the flow is denoted by Φ_t, and then the return point is $\Phi_T(\mathbf{x}_\pi) \in \Pi$, and we call $\mathbf{x}_\pi \mapsto \Phi_T(\mathbf{x}_\pi)$ the *return map* to the section Π, illustrated in Figure 11.27.

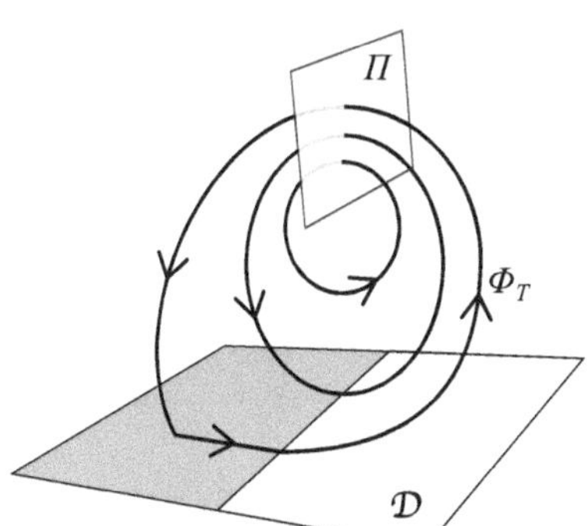

Fig. 11.27 A return map Φ_T to the section Π (often called a Poincaré section), via a smooth or piecewise-smooth flow.

Of course, Π must be chosen well to capture the parts of the flow that are of interest. In particular, it should typically lie transverse to the flow so that each trajectory pierces it in a structurally stable way. Though this is usually quite simple to apply in smooth systems, in piecewise-smooth systems it can be much more challenging. However the pay-off is clear. A periodic orbit that pierces Π is a fixed point of the map, where $\Phi_T(\mathbf{x}_\pi) = \mathbf{x}_\pi$. By studying Φ_T in the neighbourhood of $\mathbf{x}_\pi$, we thus reduce the global problem of the flow to a local problem in the map Φ_T, by characterizing the flow neighbouring the orbit through $\mathbf{x}_\pi$.

Maps are often a good way to handle discontinuities, because by integrating over an interval of the flow, they reduce the degree of any discontinuities by one. Hence just as the discontinuous vector field $\mathbf{f}$ integrates to give continuous trajectories $\mathbf{x}(t)$, so a flow that crosses transversally but non-differentiably at a discontinuity surface gives a differentiable map. Sliding dynamics is more challenging because it involves changes in the dimensionality of the flow.

It is evident that the sliding bifurcations above will lead to continuous maps. In fact it is shown in [48] that all cases lead to a map that is at least differentiable (but discontinuous in a higher derivative), except a grazing-sliding bifurcation which is discontinuous in its first derivative. The crossing-sliding, switching-sliding, and adding-sliding bifurcations are therefore governed by the restrictions of differentiable maps. The grazing-sliding bifurcation is not but can be approximated by the *border collision normal form* for a continuous piecewise-linear map [84], where a fixed point of the map hitting its discontinuity boundary (a 'border collision') corresponds to a periodic orbit undergoing a grazing-sliding bifurcation. The border collision map is itself an incredibly complex beast; see, e.g. [77, 78, 79, 84, 86, 172, 188, 189], where the reader will find hints at the true potential for complexities in the practical occurrence of grazing-sliding bifurcations in general. These include arbitrary jumps in the number, dimension, or periodicity of an orbit through the bifurcation. So complicated and incomplete is the picture that we cannot cover it here.

A useful notion for studying return maps near sliding bifurcations is that of *discontinuity mappings*. The idea is to decompose the map Φ_T into a smooth part, Φ_{glo}, which remains smooth throughout the bifurcation, and a discontinuous correction map, Φ_{DM}, determined by local geometry of the sliding bifurcation. The full return map is then

$$\Phi(\mathbf{x}_\Pi) = \Phi_{DM} \circ \Phi_{glo}(\mathbf{x}_\Pi) \,. \tag{11.17}$$

Approximating near a periodic orbit at $\mathbf{x}_\pi = \mathbf{0}$, the map Φ_{glo} can typically be expanded as

$$\Phi_{glo}(\mathbf{x}_\pi) = \mathbf{c} + \underline{\underline{C}}\mathbf{x}_\pi + h.o.t. \tag{11.18}$$

and coordinates can be found in which $\mathbf{c} = (1, 0, 0, \dots)^{\mathsf{T}}$ and $\underline{\underline{C}}$ is in some canonical form, e.g. observer canonical form (e.g. [84]). Most important is the dimension of this map: $\mathbf{x}_\pi$ is $n-1$ dimensional for grazing-sliding and crossing-sliding and $n-2$ dimensional for switching-sliding and adding-sliding, due to the way sliding dynamics contracts the flow.

Although Φ_{glo} is system dependent, the discontinuity mapping Φ_{DM} is not. Instead it is determined by the local geometry of the tangencies as studied above. The analysis is lengthy but gives the expression

$$\Phi_{DM}(\mathbf{x}) = \mathbf{x} - \mu^r \mathbf{d} + \mathsf{O}\left(\mu^{r+1/2}\right) , \qquad r = \begin{cases} 1 \text{ for grazing-sliding bif.} \\ 2 \text{ for crossing-sliding bif.} \\ 3 \text{ for switching-sliding bif.} \\ 2 \text{ for adding-sliding bif.} \end{cases} \tag{11.19}$$

and actually [48] improves the error estimate to order $\mu^{3/2}$, μ^3, μ^4, and $\mu^{5/2}$, respectively, for the four bifurcations.

The sliding explosions will instead lead to piecewise-smooth maps. These are a topic in themselves, alongside piecewise-smooth flows, with little known in general beyond the existence of multiple attracting and repelling orbits of different periods (with no upper bound) in piecewise-linear maps [107, 48], or routes to chaos in one-dimensional maps (see, e.g. [75, 76, 78, 189]), determined from the gradients of a map either side of the discontinuity.

The study of nonsmooth maps, and their applications to sliding bifurcations and explosions, remains extremely open, especially for flows of more than two dimensions and, hence, of maps with dimension more than one. The topic deserves a book in itself, indeed the book [48] does a substantial job for bifurcations, with the task remaining open for explosions.

11.8 Looking Forward

We must yet add to the global classification here the sliding bifurcations and explosions that take place at intersecting discontinuity submanifold or corners of discontinuity surfaces. The study of these has barely begun.

It is worth emphasizing that the sliding bifurcations are *global* bifurcations, in that they affect global sets (limit cycles, stable manifolds, etc.) that graze the discontinuity surface, but the bifurcation relies only on the geometry in the neighbourhood of grazing. Global conditions give orbits in the unfolding a specific identity, and a bifurcation takes place under a change of the bifurcation parameter μ (e.g. impact coordinate), with no local bifurcation of the underlying vector field. We therefore say that these are *local* mechanisms for *global* bifurcations.

Until we advance much further in discovering general features of global discontinuity-induced bifurcations, nonsmooth maps, and their interrelations, their behaviour requires case-by-case analysis. Our focus remains the general theory of how discontinuities can be modelled and analysed. To that end, it is now time to turn once again away from classifications and tackle the idealizations we have made that underlay every part of this book since Chapter 2. We are now prepared to ask more rigorously whether our discontinuities are robust models of real behaviour and whether the phenomena we have found to accompany them are robust in a practical sense.

Chapter 12
Asymptotics of Switching: Smoothing and Other Perturbations

The modest assumption made throughout this book is that a discontinuous system can be expressed in the form $\dot{\mathbf{x}} = \mathbf{f}(\mathbf{x};\boldsymbol{\lambda})$, in terms of switching multipliers $\boldsymbol{\lambda} = (\lambda_1, \dots, \lambda_m)$. It is time to ask what these assumptions involve, in theory and in application.

Intuitively, the existence of m independent switching multipliers $\lambda_j = \operatorname{sign}(\sigma_j(\mathbf{x}))$ implies that the discontinuities at the submanifolds $\mathcal{D}_j = \{\mathbf{x} \in \mathbb{R}^n : \sigma_j(\mathbf{x}) = 0\}$ behave independently. The existence of the switching layer implies that the discontinuities amount to infinitely fast evolution through a continuous interval.

We are now ready to ask to what extent these idealizations apply to the real processes that accompany discontinuity, such as multiple timescales, delays, or stochasticity. Using probabilistic logic, singular perturbation theory, and some numerical experiments, we learn under what circumstances discontinuities behave independently and thereby as a combination $\mathbf{f}(\mathbf{x};\boldsymbol{\lambda})$ and in what circumstances the switching layer dynamics is robust.

Only recently have we learned how to formulate these questions. The answers are perplexing, challenging, intriguing, and encouraging.

12.1 Probabilistic Switching

Essentially, when we write a piecewise-smooth system as some function

$$\dot{\mathbf{x}} = \mathbf{f}(\mathbf{x};\text{switch terms}) \ ,$$

we are *matching* between known regimes of behaviour $\dot{\mathbf{x}} = \mathbf{f}^{\kappa}$ on disjoint domains $\mathcal{R}_K$, across the discontinuity surface $\mathcal{D}$ between them.

M. R. Jeffrey, *Hidden Dynamics*,
https://doi.org/10.1007/978-3-030-02107-8_12

In this matching, the vector fields $\mathbf{f}^K$ exterior to $\mathcal{D}$ (what we have called the constituent fields) provide boundary conditions on what the complete function $\mathbf{f}$ might look like. It makes sense then to assume that, as a matched function, $\mathbf{f}$ looks like the hull of the $\mathbf{f}^K$'s, plus some correction terms,

$$\dot{\mathbf{x}} = \mathbf{f} = \sum_{\kappa_1=\pm} \cdots \sum_{\kappa_m=\pm} u_{\kappa_1\dots\kappa_m} \mathbf{f}^{\kappa_1\dots\kappa_m} + \text{corrections} . \tag{12.1}$$

The u_K's are coefficients that take values 1 or 0 outside $\mathcal{D}$, with $0 < u_K < 1$ inside $\mathcal{D}$, and they are normalized so $u_1 + \cdots + u_N = 1$.

If we wish to solve for dynamics in the discontinuity surface in the form of sliding modes, however, then such an expression is of limited use. If $\mathcal{D}$ consists of m transversally intersecting submanifolds $\mathcal{D}_j$, then the number of constituent fields $\mathbf{f}^K$ is $N = 2^m$, and the number of u_K's is $2^m - 1$. A sliding mode along the intersection of all m submanifolds $\mathcal{D}_j$ will follow a vector field $\mathbf{f}$ that lies in the tangent space of every $\mathcal{D}_j$. This provides m conditions to fix the $N = 2^m - 1$ independent coefficients u_i. Hence, except for the case $m = 1$, the convex hull cannot uniquely determine sliding motion (at best it provides a $2^m - 1 - m$ parameter family of *possible* sliding modes). And that is before we even consider any further ambiguity that might be introduced by the 'correction' terms in (12.1).

In Chapter 3 we set out with a refined expression (5.3), a 'canopy' rather than 'hull' of the $\mathbf{f}^K$'s, which matches across each submanifold $\mathcal{D}_j$ independently, arriving at

$$\dot{\mathbf{x}} = \mathbf{f} = \sum_{\kappa_1=\pm} \cdots \sum_{\kappa_m=\pm} \mu_1^{\kappa_1} \dots \mu_m^{\kappa_m} \mathbf{f}^{\kappa_1\dots\kappa_m} + \text{hidden} , \tag{12.2}$$

where each $\mu_j^- = 1 - \mu_j^+$. This is just a different way of writing of (5.3), with $\mu_j^\pm = \frac{1}{2}(1 \pm \lambda_j)$, but the μ_j's will be more useful for the following discussion. Ignoring the hidden terms, this contains just the right number of μ_j's (or λ_j's) to give well-defined sliding solutions, namely, m coefficients (the μ_j's) to satisfy the m sliding conditions (for $\mathbf{f}$ to lie in the tangent plane of m submanifolds $\mathcal{D}_1,\dots,\mathcal{D}_m$). Therefore, unlike the hull (12.1), the canopy (12.2) gives a well-defined continuation of the $\mathbf{f}^K$'s across $\mathcal{D}$.

The canopy (12.2) with a given hidden term is a subset of the hull (12.1), but with suitably chosen hidden terms, the canopy can reach all values in the hull (and account for any possible corrections if they have a closed analytic form). Hence the difference between (12.1) and (12.2) lies in the hidden terms whose dynamical roles have interested us throughout this book.

For a system with switching quantities $\boldsymbol{\lambda} = (\lambda_1, \dots, \lambda_m)$ to be modelled by a vector equation $\dot{\mathbf{x}} = \mathbf{f}$, there are two options:

1. the hidden term can be expressed analytically in the known variables and multipliers, i.e. as a function of $\mathbf{x}$ and $\boldsymbol{\lambda}$; hence we arrive at a *combination* $\mathbf{f} = \mathbf{f}(\mathbf{x}; \boldsymbol{\lambda})$; this book is dedicated to such systems.

2. the hidden term cannot be expressed analytically, it is a process, perhaps in variables of a higher dimensional system, perhaps a stochastic or discrete process or smoothing; this chapter is dedicated to those systems.

Our job in this chapter is then to investigate what kinds of systems fall into either of these two classes. There is no simple theorem to provide a short answer, but we shall see, via thorough investigations, that the answer *is* nonetheless somewhat simple, and Item 1 is more inclusive than a pragmatist might expect.

Item 1 leads to attractors in the discontinuity surface, making up a sliding manifold $\mathcal{M}$ which we can solve for and express analytically. Item 2 tends also to create attractors in the discontinuity surface, making up an object corresponding to a sliding set $\mathcal{M}$ which, however, will typically have no analytic expression. Nevertheless, we can evolve or simulate whatever process takes place in the switching layer, allow it to settle to its attractor $\mathcal{M}$, and extract from that the equivalent sliding dynamics. We do just this in Section 12.2.1. If we are lucky, then the resulting dynamics can be approximated by an analytic hidden term, as happens for smooth systems as we show in Section 12.3 and Section 12.4. There are hints that noisy systems or stochastic switching may also fall into this latter class, in Section 12.2.1.

First let us get some intuition for what the hidden term represents from the viewpoint of probabilities.

12.1.1 Multiplying Probabilities in the Combination

Intuitively, the existence of a combination $\mathbf{f}(\mathbf{x};\boldsymbol{\lambda})$, employing independent switching parameters $\boldsymbol{\lambda} = (\lambda_1, \ldots, \lambda_m)$, suggests an independence between the discontinuities associated with each λ_j. Interpreting the coefficients u_K and μ_j as probabilities tells us more about what this independence means.

Sliding is motion along a discontinuity surface between regions where the constituent vector fields $\mathbf{f}^K$ apply. Let us assume that sliding dynamics is made up of many small increments of the $\mathbf{f}^K$'s (with no other hidden dynamics), in a manner given by the summation terms in (12.1) or (12.2).

The coefficients u_K in (12.1) can be interpreted as occupation times or probabilities. If the proportion of time that a sliding mode follows each $\mathbf{f}^K$ is u_K, then the contribution of that mode to $\dot{\mathbf{x}}$ is $u_K\mathbf{f}^K$. We can equivalently say u_K is the probability that $\dot{\mathbf{x}}$ is following $\mathbf{f}^K$ anywhere along that sliding mode at a given instant. So $u_{++\ldots} = 1$ means the system is in mode $\mathbf{f}^{++\cdots}$, and if one of the $u_{\pm\pm\ldots}$'s in (12.1) is unity, then all others are zero (the system is not simultaneously in another mode).

The coefficients μ_j in (12.2) can also be interpreted as occupation times or probabilities, but now in relation to the submanifolds $\mathcal{D}_j$ rather than the modes themselves. So $\mu_j^+ = 1$ in (12.2) means the state lies on the positive side of $\mathcal{D}_j$, and implies $\mu_j^- = 1 - \mu_j^+ = 0$ (the system cannot also be on the negative side of $\mathcal{D}_1$), but this does not constrain the other $\mu_i^\pm$s for $i \neq j$.

So can we relate the two expressions for $\mathbf{f}$ using probabilistic logic?

The probability μ_1^+, for example, is the sum of all probabilities of being on the positive side of $\mathcal{D}_1$, i.e. the sum of all $u_{\pm\pm\ldots}$'s with a '+' in their first index, so

$$P(+\text{ side of }\mathcal{D}_1) = \sum_{\kappa_2\ldots\kappa_m=\pm} P(\text{in mode }\mathbf{f}^{+\kappa_2\ldots\kappa_m})$$

where $P(A)$ denotes the probability of event A occurring. In terms of the coefficients u_K and μ_j, this reads as

$$\mu_1^+ = u_{+++\ldots} + u_{++-\ldots} + u_{+-+\ldots} + u_{+--\ldots} + \cdots \tag{12.3}$$

with only the first index not changing. Moreover, the probability $u_{+++\ldots}$ is the probability of being on the positive side of each of $\mathcal{D}_1$, $\mathcal{D}_2$, $\mathcal{D}_3$, ..., so

$$P(\text{in mode }\mathbf{f}^{+++\cdots}) = \begin{cases} P(+\text{ side of }\mathcal{D}_1)P(+\text{ side of }\mathcal{D}_2|+\text{ side of }\mathcal{D}_1)\ldots \\ \hfill \text{if dependent} \\ P(+\text{ side of }\mathcal{D}_1)\ldots P(+\text{ side of }\mathcal{D}_m) \\ \hfill \text{if independent} \end{cases}$$

which in terms of the u's and μ's reads

$$u_{+++\ldots} = \begin{cases} \mu_1\mu_{2|1}\mu_{3|1\cap 2}\ldots & \text{if dependent} \\ \mu_1\mu_2\ldots\mu_m & \text{if independent} \end{cases} \tag{12.4}$$

if we understand $\mu_{1|2}$ to mean 'the probability of $\mu_1 = 1$ given $\mu_2 = 1$' and so on.

This means that if the probabilities $\mu_1, \ldots, \mu_m$ are mutually dependent, the probability of being on the positive side of $\mathcal{D}_1$ depends on whether the state is on the right-hand side of $\mathcal{D}_2$ (and each other $\mathcal{D}_j$), the probabilities are interdependent, and things are much more complicated.

If the probabilities $\mu_1, \ldots, \mu_m$ are independent, however, then they multiply simply and (12.4) reduce (12.1) to (12.2) by direct substitution!

Let us see if this worked out for two switches. The convex hull of modes $\mathbf{f}^{\pm\pm}$ is

$$\mathbf{f} = u_{++}\mathbf{f}^{++} + u_{-+}\mathbf{f}^{-+} + u_{+-}\mathbf{f}^{+-} + u_{--}\mathbf{f}^{--}\,, \tag{12.5}$$

where each coefficient u_K is the probability of being in the accompanying mode $\mathbf{f}^K$. The probabilities μ_1 and μ_2 (we can drop the '+' superscript) of being on the positive sides of $\mathcal{D}_1$ and $\mathcal{D}_2$, respectively, are

$$\mu_1 = u_{++} + u_{+-} \quad \& \quad \mu_2 = u_{++} + u_{-+}\,. \tag{12.6}$$

Using these to eliminate u_{+-} and u_{-+} gives

$$\mathbf{f} = u_{++}\mathbf{f}^{++} + (\mu_2 - u_{++})\mathbf{f}^{-+} + (\mu_1 - u_{++})\mathbf{f}^{+-} + (1 - \mu_1 - \mu_2 + u_{++})\mathbf{f}^{--} . \tag{12.7}$$

The probability of being in the $\mathbf{f}^{++}$ mode is

$$u_{++} = \mu_1\mu_{2|1} = \mu_{1|2}\mu_2 , \qquad \mu_{a|b} = \mu_a \text{ if independent} , \tag{12.8}$$

which allows us to eliminate u_{++},

$$\begin{aligned} \mathbf{f} = \mu_1\mu_2\mathbf{f}^{++} + \mu_2(1-\mu_1)\mathbf{f}^{-+} + \mu_1(1-\mu_2)\mathbf{f}^{+-} + (1-\mu_1)(1-\mu_2)\mathbf{f}^{--} \\ +\mu_1(\mu_{2|1} - \mu_2)\left\{\mathbf{f}^{++} - \mathbf{f}^{-+} - \mathbf{f}^{+-} + \mathbf{f}^{--}\right\} , \end{aligned} \tag{12.9}$$

which is of the form (12.2). The conditional probabilities generate a hidden term, written on the second line of (12.9). The quantity $\mu_1(\mu_{2|1} - \mu_2) = \mu_2(\mu_{1|2} - \mu_1)$ is not a simple function of the variables and can only be determined by simulation of whatever process links the probabilities during the dynamics of switching. If the probabilities are independent, however, the hidden term vanishes, and we are left with

$$\mathbf{f} = \mu_1\mu_2\mathbf{f}^{++} + \mu_2(1-\mu_1)\mathbf{f}^{-+} + \mu_1(1-\mu_2)\mathbf{f}^{+-} + (1-\mu_1)(1-\mu_2)\mathbf{f}^{--} , \tag{12.10}$$

i.e. the canopy (12.2) or (5.3) obtained by substituting $\mu_j = \frac{1}{2}(1+\lambda_j)$,

$$\begin{aligned} \mathbf{f} =& \tfrac{1}{4}(1+\lambda_1)(1+\lambda_2)\mathbf{f}^{++} + \tfrac{1}{4}(1-\lambda_1)(1+\lambda_2)\mathbf{f}^{-+} \\ &+ \tfrac{1}{4}(1+\lambda_1)(1-\lambda_2)\mathbf{f}^{+-} + \tfrac{1}{4}(1-\lambda_1)(1-\lambda_2)\mathbf{f}^{--} \end{aligned}$$

with no hidden terms.

12.1.2 The Unreasonable Effectiveness of Nonsmooth Models

Another way of characterizing the hidden terms is as representing *modelled errors*, corrections to the piecewise defined system $\dot{\mathbf{x}} = \left\{\mathbf{f}^K(\mathbf{x}) : \mathbf{x} \in \mathcal{R}_K\right\}$ required to obtain the correct dynamics—that of an actual physical system—on the discontinuity surface $\mathcal{D}$. Call the summation term in (12.2) $\mathbf{f}^{can}(\mathbf{x};\boldsymbol{\lambda})$, and rewrite (12.2) as

$$\dot{\mathbf{x}} = \mathbf{f}^{can}(\mathbf{x};\boldsymbol{\lambda}) + \underset{\text{(hidden dynamics)}}{\text{modelled errors}} . \tag{12.11}$$

The real system may contain processes or influences that we still neglect, unmodelled errors such as environmental factors,

$$\dot{\mathbf{x}} = \mathbf{f}^{can}(\mathbf{x};\boldsymbol{\lambda}) + \underset{\text{(hidden dynamics)}}{\text{modelled errors}} + \text{unmodelled errors} , \tag{12.12}$$

whose overall effect must be small for our model (12.11) to be effective.

This becomes non-trivial when discontinuities are used to approximate the complex but fleeting processes of switching. We have already seen how the hidden dynamics that we *have* modelled, while negligible almost everywhere, can nevertheless dominate at the discontinuity and massively affect the global dynamics. The assumption that any unmodelled dynamics can be neglected, even at the switch, may be in peril.

A known example is in control electronics, where unmodelled errors arise that are negligible almost everywhere, except when excited by nonlinearities, of which discontinuities are an extreme example. These excitations can lead to high-frequency chattering, resulting in mechanical wear, power loss, adverse heating, and controllability degradation (see, e.g. [129, 191, 212]).

We have some notion of how to represent the modelled errors using terms of the form $\mathbf{n}(\mathbf{x};\boldsymbol{\lambda}) = \Gamma_\varepsilon(\boldsymbol{\lambda})\mathbf{h}(\mathbf{x};\boldsymbol{\lambda})$, decomposed into a general vector field $\mathbf{h}$, and some scalar Γ_ε that vanishes outside an ε-neighbourhood of $\mathcal{D}$. For $\varepsilon \to 0$, we obtain a piecewise-smooth system with hidden term $\mathbf{n}(\mathbf{x};\boldsymbol{\lambda}) = \Gamma_0(\boldsymbol{\lambda})\mathbf{h}(\mathbf{x};\boldsymbol{\lambda})$ where $\sigma(\mathbf{x})\Gamma_0(\boldsymbol{\lambda}) = 0$ consistent with (5.4).

The unmodelled dynamics cannot be expressed in such an explicit form (otherwise we would write it as such!). All we know is that the modelled state $\mathbf{x}$ is a perturbation of the system's true state $\mathbf{x}'$. So we write the real system as

$$\dot{\mathbf{x}} = \mathbf{f}^{can}(\mathbf{x}';\boldsymbol{\lambda}) \; + \; \Gamma_\varepsilon(\mathbf{x}')\mathbf{h}(\mathbf{x}';\boldsymbol{\lambda}) \, . \tag{12.13}$$

Similarly to the interpretation of the coefficients u_i as probabilities in (12.1) and (12.2), the quantity Γ_ε can be interpreted as the probability (or *residence time* in [192]) of lying in an ε-neighbourhood of $\mathcal{D}$, where the modelled error (or hidden) term $\mathbf{h}$ is significant. If a trajectory crosses $\mathcal{D}$ transversally, then $\Gamma_\varepsilon \sim \mathsf{O}(\varepsilon)$; if a trajectory slides along $\mathcal{D}$, then Γ_ε is of order unity.

Denote the real state as a perturbation $\mathbf{x}' = \mathbf{x} + \boldsymbol{\rho}(\mathbf{x})$, where $\rho = |\boldsymbol{\rho}|$ is small. We know nothing about the form of this perturbation (e.g. it may come from a continuous function or from some distribution or external process and may be time dependent). At a point $\mathbf{x}$ when the unmodelled perturbation $\boldsymbol{\rho}$ is applied, we now have

$$\begin{aligned} \dot{\mathbf{x}} &= \mathbf{f}^{can}(\mathbf{x}';\boldsymbol{\lambda}) + \Gamma_\varepsilon(\mathbf{x}')\mathbf{h}(\mathbf{x}';\boldsymbol{\lambda}) \\ &= \mathbf{f}^{can}(\mathbf{x};\boldsymbol{\lambda}) + \Gamma_\varepsilon(\mathbf{x}+\boldsymbol{\rho}(\mathbf{x}))\mathbf{h}(\mathbf{x};\boldsymbol{\lambda}) + \mathsf{O}(\rho) \, , \end{aligned} \tag{12.14}$$

where we expand $\mathbf{f}^{can}$ and $\mathbf{h}$, which are smoothly varying in $\mathbf{x}$, for small ρ. We do not expand Γ_ε, whose dependence on $\mathbf{x}$ is not known and which requires more care.

To find Γ_ε, let the values $\mathbf{x} + \boldsymbol{\rho}(\mathbf{x})$ be distributed inside a set $\mathcal{S}_\rho(\mathbf{x})$ with a density $a(\mathbf{x})$, and then define

$$\Gamma_\varepsilon\left(\mathbf{x}+\boldsymbol{\rho}(\mathbf{x})\right) \;=\; P\left(|\sigma(\mathbf{x}+\boldsymbol{\rho}(\mathbf{x}))| < \varepsilon\right) \;=\; \frac{\int_{\mathcal{S}_\rho(\mathbf{x})\cap|\sigma|<\varepsilon} a(\mathbf{x})d\mathcal{S}}{\int_{\mathcal{S}_\rho(\mathbf{x})} a(\mathbf{x})d\mathcal{S}} \, , \tag{12.15}$$

where $P(A)$ denotes the probability of the situation A occurring. In the limit of small ρ, we have $\Gamma_\varepsilon = 1$ if $|\sigma(\mathbf{x})| < \varepsilon$, and $\Gamma_\varepsilon = 0$ if $|\sigma(\mathbf{x})| > \varepsilon$, as expected.

For example consider a point $\mathbf{x}$ on $\sigma = 0$, and assume that the size of the error, ρ, is constant, then $\mathcal{S}_\rho(\mathbf{x})$ is an (n dimensional) spherical shell of radius ρ centred on $\mathbf{x}$. Then for $\rho < \varepsilon$ we immediately have $\Gamma_\varepsilon = 1$. For $\rho > \varepsilon$, if $\mathbf{x}$ is uniformly distributed on the unit circle then $a = 1$, and we have

$$\begin{aligned}\Gamma_\varepsilon(\mathbf{x}) &= \frac{\text{surface area of } \mathcal{S}_\rho(\mathbf{x}) \cap \{\mathbf{x} : |\sigma(\mathbf{x})| < \varepsilon\}}{\text{surface area of } \mathcal{S}_\rho(\mathbf{x})} \\ &= \frac{2\arcsin(\varepsilon/\rho)}{\pi} \quad \leq \quad \frac{2\varepsilon}{\pi\rho} + \left(\frac{2\varepsilon}{\pi\rho}\right)^2\end{aligned}$$

in two dimensions. Different expressions apply in $n > 2$ dimensions but always scaling with ε/ρ. If the distribution is not uniform, then the probability that the corrected location $\mathbf{x} + \boldsymbol{\rho}(\mathbf{x})$ lies in $|\sigma| < \varepsilon$ might be greater, for example, if $\boldsymbol{\rho}$ has a preference to lie along $\sigma = 0$. As an example take a Gaussian distribution $a = e^{-\sigma^2}$, so the error tends to push $\mathbf{x}$ along the discontinuity. Again, in two dimensions but easily generalized, for $\rho > \varepsilon$, we have

$$\Gamma_\varepsilon(\mathbf{x}) = \frac{4\int_0^\varepsilon d\sigma e^{-\sigma^2}/\sqrt{\rho^2 - \sigma^2}}{4\int_0^\rho d\sigma e^{-\sigma^2}/\sqrt{\rho^2 - \sigma^2}} \leq \frac{\int_{-\varepsilon}^\varepsilon d\sigma e^{-\sigma^2}}{\int_{-\rho}^\rho d\sigma e^{-\sigma^2}} = \frac{\operatorname{Erf}\varepsilon}{\operatorname{Erf}\rho} ,$$

which is again of order ε/ρ for $\varepsilon \ll 1$ and $\rho \ll 1$, where Erf denotes the standard error function. In either case we can let $\Gamma_\varepsilon(\mathbf{x}) = \frac{\varepsilon}{\rho}\mu(\mathbf{x})$ for some $\mu \in [0, 1]$, and then (12.14) becomes

$$\dot{\mathbf{x}} = \mathbf{f}^{can}(\mathbf{x}; \boldsymbol{\lambda}) + \frac{\varepsilon}{\rho}\mu(\mathbf{x})/\varepsilon)\mathbf{h}(\mathbf{x}; \boldsymbol{\lambda}) + \mathsf{O}(\rho) , \tag{12.16}$$

and since the second term is of order ε/ρ, it vanishes as $\varepsilon \to 0$. Putting the two results for ρ greater or less than ε together, we have

$$\dot{\mathbf{x}} = \mathbf{f}^{can}(\mathbf{x}; \lambda) + \begin{cases} \Gamma_\varepsilon(\mathbf{x})\mathbf{h}(\mathbf{x}; \boldsymbol{\lambda}) + \mathsf{O}(\rho) & \text{for } \rho < \varepsilon , \\ \mathsf{O}(\rho, \varepsilon/\rho) & \text{for } \rho > \varepsilon . \end{cases} \tag{12.17}$$

If unmodelled errors $\boldsymbol{\rho}(\mathbf{x})$ are small compared to the range over which the switch occurs, $\rho < \varepsilon$, then the modelled errors $\Gamma_\varepsilon(\mathbf{x})\mathbf{h}(\mathbf{x}; \boldsymbol{\lambda})$ are significant—hidden dynamics matters in the switching layer. Unmodelled errors dominate over the modelled errors, though, for $\rho > \varepsilon$, they effectively kill off hidden dynamics. For $\varepsilon, \rho \to 0$, the $\mathsf{O}(\rho)$ terms in the first line and $\mathsf{O}(\rho, \varepsilon/\rho)$ terms in the second line vanish, but the hidden $\Gamma_\varepsilon\mathbf{h}$, as we know, does not.

This is a somewhat surprising result and overall counterintuitive. A more irregular system where unmodelled errors are more significant behaves like the simplest multilinear piecewise-smooth vector field, $\dot{\mathbf{x}} = \mathbf{f}^{can}$. The less effect there is from unmodelled errors, the more important it is that we carefully model the hidden terms $\Gamma_\varepsilon\mathbf{h}$.

To make this heuristic argument more rigorous requires choosing particular sources of unmodelled error. In Section 12.2.1, numerical experiments will investigate different switching models and show how modelled errors, due to hysteresis, time delay, or time-stepping, can be eliminated by the presence of noise or stochastic switching. As an example application, errors from external white noise affecting friction between solid bodies will be simulated in Section 14.7. The results are consistent with those above.

12.2 Convex Switching

In real-world behaviour, a discontinuity is not likely to occur at an ideal threshold $\sigma = 0$, but instead over a range $|\sigma| < \varepsilon$ for some small ε, within which there may be intermediary processes that have to complete before switching completes, resulting in smoothing, hysteresis, time delay, probabilistic switching, or overshoot due to time-stepping. We can treat these as perturbations of the idealized piecewise-smooth systems $\dot{\mathbf{x}} = \mathbf{f}(\mathbf{x}; \boldsymbol{\lambda})$ studied thus far.

We describe such nonideal processes collectively as *convex switching* if they involve abrupt switching between vector fields $\mathbf{f}^{\kappa}$ on regions $\mathcal{R}_{\kappa}$, $\kappa = 1, \ldots, N$, with zero probability of hitting the discontinuity surface $\mathcal{D}$ exactly. The description of these systems is then by means of combinations of the $\mathbf{f}^{\kappa}$'s only, as in (12.1) without the correction terms.

This is the intuition that led Filippov and others to study piecewise-smooth systems through the convex combinations of their constituent vector fields. This is not, however, a sufficient argument to say that convex combinations (the canopy for instance) are a good model of real-world discontinuity. The truth, it turns out, is much richer.

Through numerical experiments, we will investigate here how systems with nonideal switches behave, showing the freedom that convex switches have to explore Filippov's convex hull and the conditions under which our ideal piecewise-smooth models represent good approximations.

Convex switching applies when a switching process involves hysteresis, time delay, probabilistic switching, and time-stepping. It does not apply to a switch that is smoothed via sigmoid functions or intermediary variables, which we leave to Sections 12.3 to 12.5.

12.2.1 Experiments on Convex Switching

Take a three-dimensional system in coordinates $\mathbf{x} = (x_1, x_2, x_3)$, with switching submanifolds

$$\mathcal{D}_i = \left\{(x_1, x_2, x_3) \in \mathbb{R}^3 : x_i = 0\right\} \quad \text{for} \quad i = 1, 2,$$

and constituent vector fields

$$\dot{\mathbf{x}} = \mathbf{f}^{\kappa_1\kappa_2} \;, \quad \kappa_j 1 = \operatorname{sign}(x_j) \;, \quad \begin{cases} \mathbf{f}^{++} = (-1, -0.7, -0.5) \;, \\ \mathbf{f}^{-+} = (0.9 - 0.8x_3, 0.5, -0.6) \;, \\ \mathbf{f}^{+-} = (0.3, -0.9, 1) \;, \\ \mathbf{f}^{--} = (-0.6, 0.7, 0.6) \;. \end{cases} \tag{12.18}$$

The reader may perform similar experiments to those below on almost any vector field of their choosing, provided they converge on an intersection of discontinuity submanifolds $\mathcal{D}_1 \cap \mathcal{D}_2 \cap \ldots$ to ensure that sliding occurs along the intersection. (Examples where not all vector fields converge on the intersection, therefore permitting some deviation of solutions from sliding, are also tentatively explored in [121].)

To model imperfect switches, we simulate a solution $(x_1(t), x_2(t), x_3(t))$, with switching occurring approximately but not exactly at the thresholds $x_1 = 0$ and $x_2 = 0$. We employ several models:

0. *piecewise-smooth*: switching occurs at exactly $x_1(t) = 0$ and $x_2(t) = 0$, and the sliding solution on $x_1 = x_2 = 0$ is found using the canopy (5.8); this implementation is non-convex but provides our idealized 'control experiment';
1. convex hull: the theoretical hull of possible sliding solutions consists of all vectors (12.1) for $u_\kappa \in [0, 1]$ without 'corrections', and such that $\mathcal{F}$ lies in the tangent planes $x_1 = 0$ and $x_2 = 0$;
2. *hysteresis*: switching occurs at $x_j(t) = \pm\varepsilon_1$ if $\dot{x}_j(t) \gtrless 0$ for $j = 1, 2$, (and not at $x_j(t) = \mp\varepsilon_1$ if $\dot{x}_j(t) \gtrless 0$); therefore the mode inside $|x_1| < \varepsilon_1$, $|x_2| < \varepsilon_2$ depends on a solution's history;
3. *time delay*: switching occurs when $x_j(t - \varepsilon_j) = 0$ for $j = 1, 2$; hence a time ε_j after the threshold $x_j = 0$ is passed;
4. *time-stepping*: the solution is solved in discrete time increments $\Delta t = \varepsilon$, therefore overshoot of the thresholds $x_j = 0$ occurs;
5. *noise*: at each time $\Delta t = \varepsilon$, a perturbation $\Delta x_3 = \varepsilon R$ is added to the state, where R is a random number between 0 and 1;
6. *stochastic switching*: switching occurs when $x_j(t - \varepsilon R_j) = 0$ for $j = 1, 2$; hence a time εR_j after the threshold $x_j = 0$ is passed, where R_j is a random number between 0 and 1—this simulates switching occurring with some probability when the thresholds $x_j(t) = 0$ are passed;

Simulations use Mathematica™routine NDSolve in default mode, using the command WhenEvent to affect switching.

The instantaneous sliding vector field at a point $(0, 0, x_3)$ on the switching intersection is found by running a simulation with fixed x_3 for sufficient time to reach an attractor (a time $t \gtrsim 100\varepsilon$ is typically sufficient). The proportion of time u_κ spent in each mode $\mathbf{f}^\kappa$ is then found over a long time interval $\Delta t = 1000\varepsilon$, and these determine the sliding vector field via the hull (12.1).

The results of the sliding speed $\dot{x}_3$, giving the effect vector field component f_3, are plotted against x_3 in Figure 12.1 for each of the different switching models.

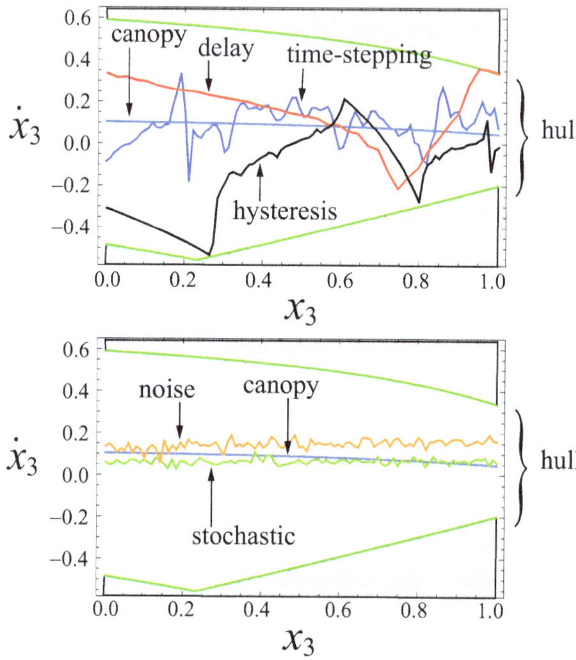

Fig. 12.1 The sliding speed $\dot{x}_3$ plotted against x_3. The hull is the unshaded region, the curves indicate results for the canopy and for hysteresis with $(\varepsilon_1, \varepsilon_2) = (0.81, 0.59) \times 10^{-2}$; time delay with $(\varepsilon_1, \varepsilon_2) = (0.59, 0.81) \times 10^{-2}$; time-stepping with $\varepsilon = 10^{-2}$; noise with $(\varepsilon_1, \varepsilon_2, \varepsilon_3) = (0.95, 0.31, 1) \times 10^{-2}$; stochastic switching with $(\varepsilon_1, \varepsilon_2) = (0.81, 0.59) \times 10^{-2}$.

The graphs show erratic jumps in the sliding speed at different x_3, which by (12.18) varies the angle of the vector field $\mathbf{f}^{-+}$. In the upper panel of Figure 12.1, the sliding solutions for hysteresis, time delay, and time-stepping explore a large range of values within the hull, seeming to ignore the canopy solution. In the lower panel of Figure 12.1, the sliding solutions with noise or stochastic switching are very irregular but remain close to the canopy solution. Similar behaviours have been observed across a great range of simulations on different vector fields (see, e.g. [121]).

Similar results are also obtained if the vector field remains constant, but instead the switching model itself varies continuously. In Figure 12.2 the effect of varying the relative sizes of the two hysteresis bounds, ε_1 and ε_2, is simulated.

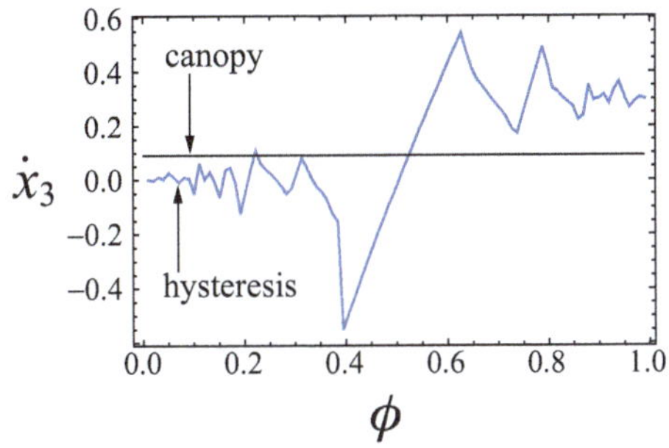

Fig. 12.2 The sliding speed $\dot{x}_3$ at fixed $x_3 = 0.25$, plotted against ϕ where $(\varepsilon_1, \varepsilon_2) = 0.01(\cos(\phi\pi/2), \sin(\phi\pi/2))$.

The graphs above show theoretical sliding speeds, giving the instantaneous vector field that propels a trajectory assuming it has reached the sliding attractor. What then is the effect of all this on the sliding trajectory $x_3(t)$? A solution from an initial condition $(x_1(0), x_2(0), x_3(0)) = (\varepsilon, \varepsilon, 0)$ with hysteresis is simulated fully in Figure 12.3(i), sliding at an irregular speed until time $t \approx 21$, where it seems to reach an equilibrium. The experimental sliding speed, calculated by taking the gradient of the graph in Figure 12.3(i), is shown by the bold curve in Figure 12.3(ii). This is overlaid by the theoretical curve (shown dashed, calculated similarly to Figure 12.1).

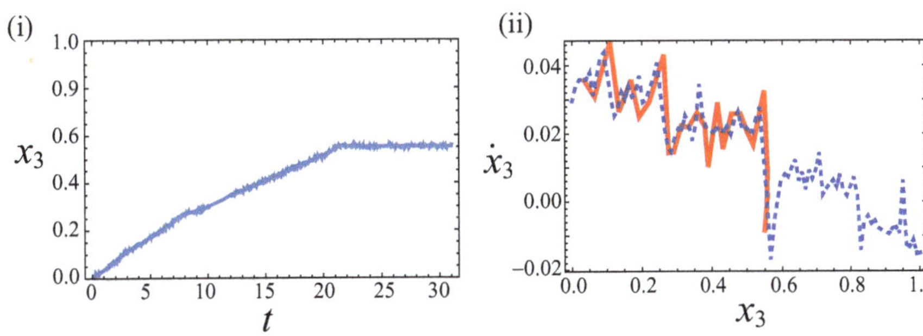

Fig. 12.3 A full simulation for a trajectory switching via hysteresis bounds $(\varepsilon_1, \varepsilon_2) = (0.92, 0.38) \times 10^{-2}$. Showing (i) the simulation of $x_3(t)$; (ii) its gradient calculated over increments $\delta t = 96 \times 10^{-2}$ (full curve) overlaid with the theoretical instantaneous sliding speed (dashed curve).

What would be the effect if we combined the different switching implementations above into a single process, as probably happens in real switching? In Figure 12.4 the effects of adding noise to hysteresis, delay, or time-stepping, are shown. The results are remarkable and yet consistent with the heuristic prediction from Section 12.1.2. When only weak noise is added, in Figure 12.4(i), the solution is only slightly perturbed from Figure 12.1, but under strong noise in Figure 12.4(ii), each solution collapses to a neighbourhood of the canopy. Again, similar behaviours have been observed across a great range of simulations on different vector fields.

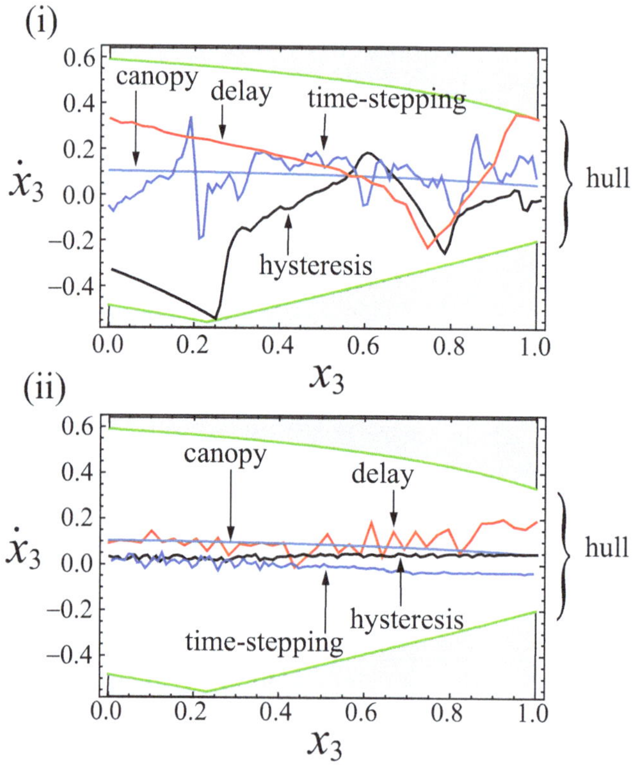

Fig. 12.4 Revisiting Figure 12.1 with stochastic perturbations. We add to hysteresis a stochastic perturbation of x_3, order 0.0001 and order 0.01; add to delay a stochastic time perturbation, order 0.01 and 0.1 of the unperturbed delay; and add to time-stepping a random perturbation of x_3, order 0.0001 and 0.01.

Examples of the attractors giving rise to these solutions are shown in Figure 12.5, at four different x_3 values for each of the sliding models from Figure 12.1. These give a sample of the many bifurcations that occur at each kink in the sliding speed graphs in Figure 12.1, between different periodic or chaotic attractors, indicating why the proportions of time u_κ spent in each mode $\mathbf{f}^\kappa$ jump between attractors, causing a requisite jump in the sliding speed.

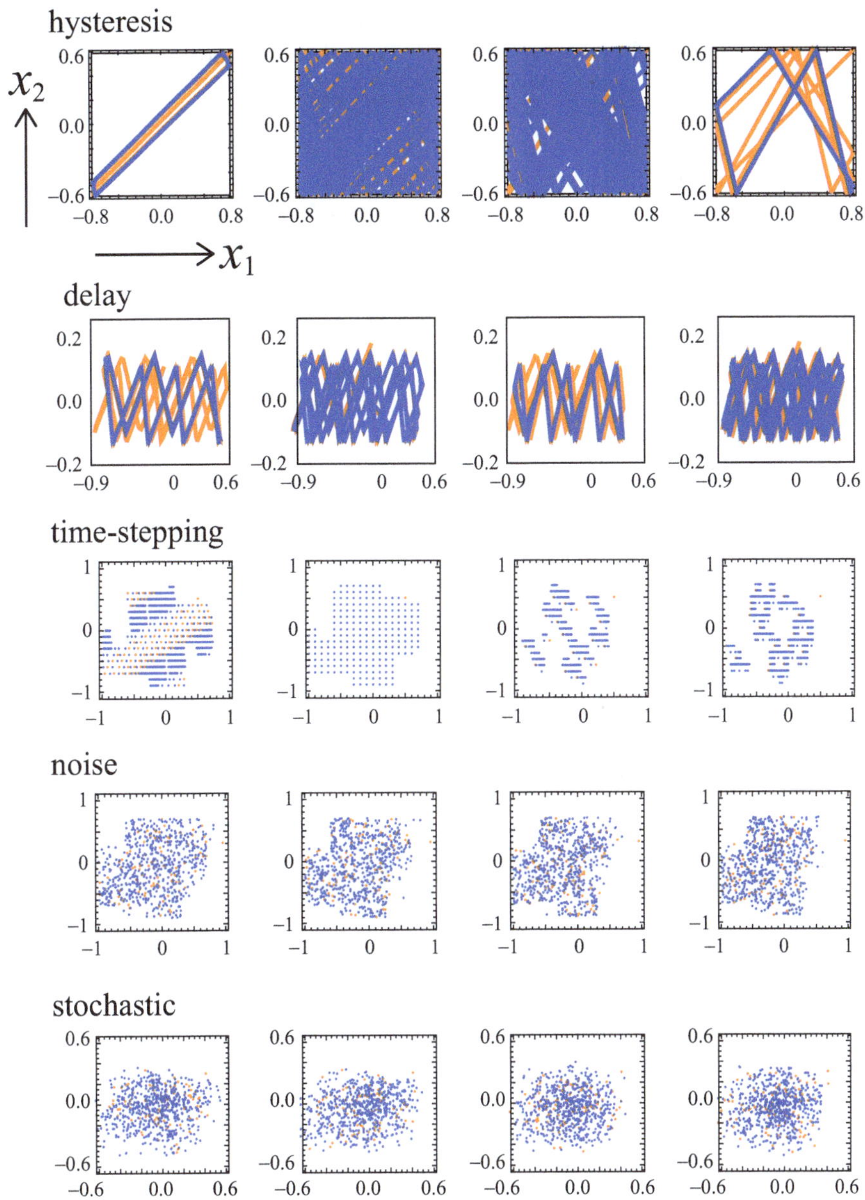

Fig. 12.5 Sliding modes attractors, showing simulations in the switching layer $|x_1| < \varepsilon_1 = 0.01\cos(\phi\pi/2)$, $|x_2| < \varepsilon_2 = 0.01\sin(\phi\pi/2)$), for different switch implementations: hull, canopy, and for $\varepsilon = 0.01$: hysteresis with $\phi = 0.4$ for $x_3 = 0.2, 0.3, 0.6, 0.62$; time delay with $\phi = 0.6$ for $x_3 = 0.25, 0.35, 0.6, 0.5$; time-stepping with $x_3 = 0.2, 0.25, 0.6, 0.65$; noise with $\phi = 0.2$ for $x_3 = 0.25, 0.3, 0.55, 0.6$; stochastic switching with $\phi = 0.4$ for $x_3 = 0.15, 0.25, 0.35, 0.42$.

For the case of hysteresis, a rigorous analysis can be carried out which was begun in [7] and continued in [121]. A hysteretic system has a definite switching layer, the hysteresis region $|\sigma_1| < \varepsilon_1$, $|\sigma_2| < \varepsilon_2$, ..., $|\sigma_m| < \varepsilon_m$, inside which we can derive a return map between exterior boundaries. The result is a map on the $(m-1)$-sphere (a circle map for two switches). The map is continuous but only piecewise differentiable, with numerous branches (e.g. eight branches for two switches). With these maps the precise attractors and bifurcations underlying the erratic behaviours above can be simulated more efficiently, for simple examples like that above, at least, but closed form solutions are still not possible in general.

For the cases of delayed, discretized, noisy, or stochastic switching, rigorous analysis of the results above remains an open problem.

Fortunately, the qualitative explanation for the behaviour in these experiments is rather simple.

12.2.2 Conclusion: Jitter Over the Convex Hull

In the course of these numerical experiments, seeking to test how closely combinations and sliding describe nonideal switching, we observe complex and erratic behaviour that is at first disheartening. This is another example of the *jitter* from Section 9.4.1. A closer look reveals intriguing structure in the complexity of jitter.

The picture we have formed throughout this book is that, as a dynamical system switches between vector fields $\mathbf{f}^\kappa$, it either crosses the discontinuity surface $\mathcal{D}$ or finds an attractor inside $\mathcal{D}$. So far this attractor has always been a sliding manifold $\mathcal{M}$, which then determines some sliding mode value for the multiplier $\lambda^\$$ and in turn prescribes the dynamics as $\dot{\mathbf{x}} = \mathbf{f}(\mathbf{x}; \boldsymbol{\lambda}^\$)$.

If the combination $\mathbf{f}(\mathbf{x}; \boldsymbol{\lambda})$ does not remain valid across the discontinuity, then something similar must still happen. Either the flow crosses the discontinuity surface $\mathcal{D}$ or finds an attractor inside $\mathcal{D}$. In convex switching, a given attractor consists of increments of evolution along each vector field $\mathbf{f}^\kappa$; therefore the sliding vector field on $\mathcal{M}$ must lie within the convex hull (12.1), except now $\mathcal{M}$ may be a complex and distributed object, possibly with periodic or chaotic properties, like those in Figure 12.5.

What we see in Figures 12.1 to 12.5 is that this complex attractor $\mathcal{M}$ may be sensitive to parameters, both of the vector fields and the switching model. Under small changes in the vector fields or the switching model, the sliding set undergoes numerous bifurcations between attractors such as those in Figure 12.5, resulting in highly erratic sliding dynamics as in Figures 12.1 to 12.4. This is another example of the jitter we observed in well-defined sliding modes in Section 9.4.1. An attractor generalizing the sliding manifold $\mathcal{M}$ exists, and sliding dynamics occurs on it, but neither can be expressed in a closed algebraic form.

In the presence of random perturbations of sufficient size, any complex attractor $\mathcal{M}$ gets washed away, in the sense that the perturbations repeatedly push the dynamics away from it. In its stead, as argued in Section 12.1.2, the system evolves towards the ideal canopy solution and the simpler, more robust attractor $\mathcal{M}$ that it provides.

12.3 Smooth Switching

We will show here that, if we smooth the discontinuity in a piecewise-smooth system, approximating the step in λ by some kind of differentiable sigmoid function, there is no direct correspondence between the resulting differentiable systems and the original piecewise-smooth system. In particular, different smoothings give systems with nonequivalent dynamics.

This means there is a fundamental problem with smoothing discontinuities that is not always obvious and, worse, can easily be missed when rigour is permitted to disguise implicit assumptions. This is so important that we shall devote a large section here to explaining it from different viewpoints. We will show explicitly, taking a piecewise-smooth system $\dot{\mathbf{x}} = \mathbf{f}(\mathbf{x}; \lambda)$, that:

S0. there is an r-family of possible smooth systems $\dot{\mathbf{x}} = \mathbf{F}_r^\varepsilon(\mathbf{x})$ obtained by replacing $\lambda \mapsto \Lambda_r^\varepsilon$, such that $\Lambda_r^\varepsilon \to \lambda$ as $\varepsilon \to 0$ for any r;
S1. those systems $\dot{\mathbf{x}} = \mathbf{F}_r^\varepsilon(\mathbf{x})$ with different r can typically lie in more than one topological class with nonequivalent dynamics;
S2. the $\varepsilon \to 0$ limit of $\dot{\mathbf{x}} = \mathbf{F}_r^\varepsilon(\mathbf{x})$ is a family of piecewise-smooth systems $\dot{\mathbf{x}} = \mathbf{f}_r(\mathbf{x}; \lambda)$ which can take more than one topological class for different r, not all equivalent to the original system $\dot{\mathbf{x}} = \mathbf{f}(\mathbf{x}; \lambda)$.

Thus the smoothing of piecewise-smooth system is not unique, and taking the limit in which the smooth system becomes piecewise-smooth gives a family of systems containing, but not restricted to, the same dynamical class as the original piecewise-smooth system. Not only is the description of the systems non-unique, but so is their dynamics, as the examples following Theorem 12.1 below show rather starkly.

12.3.1 Why Smooth?

There are three main reasons why people attempt to smooth out discontinuities in a system $\dot{\mathbf{x}} = \mathbf{f}(\mathbf{x}; \lambda)$, and these are to make the system amenable to standard numerical simulation routines for smooth systems, to make the system amenable to standard dynamical systems theory for smooth systems, or as an attempt to more closely model reality. Each of these is in some way valid and in others flawed.

Yes, smoothing a system makes it amenable to standard numerical and analytical tools; indeed we have used smoothing to make simulations throughout this book, but it depends *how* we go about smoothing. It is important to understand to what extent smoothing is or is not unique. The indeterminacies that exist in nonsmooth systems must go somewhere upon smoothing, and we must look at the conditions under which smoothing does or does not tell us anything reliable about a discontinuous system.

12.3.2 The Smoothing Tautology

Smoothing is tautological as a way of studying discontinuities, because the act of smoothing selects one member from an infinite class of possible smooth systems. Since any results we derive then apply only to that one member, not the whole class, they therefore tell us little or nothing about the piecewise-smooth system being smoothed. The principle is rather easy to show.

Theorem 12.1. *The smoothing of a piecewise-smooth dynamical system is not unique.*

Proof. Without loss of generality, take the piecewise-smooth system defined, by the convex combination

$$\dot{\mathbf{x}} = \mathbf{f}(\mathbf{x};\lambda) = \tfrac{1}{2}(1+\lambda)\mathbf{f}^+(\mathbf{x}) + \tfrac{1}{2}(1-\lambda)\mathbf{f}^-(\mathbf{x}) \quad \text{with} \quad \lambda = \operatorname{sign}\sigma\;. \tag{12.19}$$

We define a smoothing of this as the system

$$\dot{\mathbf{x}} = \mathbf{f}\left(\mathbf{x}; \Lambda_r\left(\sigma(\mathbf{x})/\varepsilon\right)\right)\;, \tag{12.20}$$

where Λ_r^ε is a smooth function, sometimes called a *smoothing* or *transfer* function (we will use the former), satisfying

$$\Lambda_r\left(\sigma/\varepsilon\right) \in \left\{\begin{matrix} \operatorname{sign}(\sigma) & \text{if } |\sigma| > \varepsilon \\ \mathcal{L}_r & \text{if } |\sigma| \le \varepsilon \end{matrix}\right\} \xrightarrow{\varepsilon\to 0} \operatorname{sign}\sigma\;, \tag{12.21}$$

where $\mathcal{L}_r \subset \mathbb{R}$ is a continuous set and $\pm 1 \in \mathcal{L}_r$. (An example of Λ_r will appear in (12.30).) We can choose the parameterization such that $\mathcal{L}_0 = (-1,+1)$ and Λ_0 is monotonic, that is,

$$\tfrac{\partial}{\partial\sigma}\Lambda_0\left(\sigma/\varepsilon\right) > 0 \qquad \text{for} \quad |\sigma| < \varepsilon\;. \tag{12.22}$$

Let

$$\Delta_r(\sigma/\varepsilon) := \Lambda_r\left(\sigma/\varepsilon\right) - \Lambda_0\left(\sigma/\varepsilon\right) \in \left\{\begin{matrix} 0 & \text{if } |\sigma| > \varepsilon\;, \\ \mathcal{L}_r & \text{if } |\sigma| \le \varepsilon\;. \end{matrix}\right. \tag{12.23}$$

Then the smoothing of (12.19) using (12.20) is

$$\dot{\mathbf{x}} = \mathbf{f}(\mathbf{x}; \Lambda_r) = \tfrac{1}{2}(1+\Lambda_r)\,\mathbf{f}(\mathbf{x};+1) + \tfrac{1}{2}(1-\Lambda_r)\,\mathbf{f}(\mathbf{x};-1) \ .$$

However, using (12.23) we can also write

$$\dot{\mathbf{x}} = \mathbf{f}(\mathbf{x}; \Lambda_r) = \mathbf{f}(\mathbf{x}; \Lambda_0) + \tfrac{1}{2}\Delta_r \{\mathbf{f}(\mathbf{x};+1) - \mathbf{f}(\mathbf{x};-1)\} \ , \qquad (12.24)$$

separating out the second term, which vanishes for $|\sigma| > \varepsilon$ but not for $|\sigma| \le \varepsilon$. It is clear that (12.24) is not one smooth system, but an r-parameterized family of smooth systems. We will show below that these are not equivalent. First let us see whether we recover (12.19) if we return to the $\varepsilon \to 0$ limit.

The function $\Lambda_0(\sigma/\varepsilon)$ is monotonic on $\sigma < \varepsilon$ and therefore has an inverse V such that $V(\Lambda_0(\sigma/\varepsilon)) = \sigma/\varepsilon$. So $\Delta_r(\sigma/\varepsilon)$ can be written in terms of Λ_0 as a function

$$\hat{\Delta}_r(\Lambda_0) := \Delta_r(V(\Lambda_0))$$

with which we have

$$\dot{\mathbf{x}} = \mathbf{f}(\mathbf{x}; \Lambda_0) + \tfrac{1}{2}\hat{\Delta}_r(\Lambda_0)\left\{\mathbf{f}^+(\mathbf{x}) - \mathbf{f}^-(\mathbf{x})\right\} \ .$$

The quantity Λ_0 is independent of ε, so let us define

$$\lambda = \Lambda_0 \ ,$$

which allows us to return to the limit $\varepsilon \to 0$. Instead of (12.19), however, we now obtain

$$\dot{\mathbf{x}} = \mathbf{f}(\mathbf{x}; \lambda) + \tfrac{1}{2}\Delta_r(\lambda)\left\{\mathbf{f}^+(\mathbf{x}) - \mathbf{f}^-(\mathbf{x})\right\} \ : \quad \lambda \in \begin{cases} \operatorname{sign}(\sigma) & \text{if } \sigma \neq 0 \ , \\ (-1,+1) & \text{if } \sigma = 0 \ . \end{cases} \qquad (12.25)$$

Hence we obtain not a unique piecewise-smooth system, but again an r-parameterized family. The systems (12.25) for different r differ from (12.19) on the discontinuity surface $\sigma = 0$, yet they are consistent for $\sigma \neq 0$.

This shows the systems are formally different. To show that they are topologically nonequivalent merely requires an example where the r-parameterized smoothings are topologically nonequivalent for different r, given by Example 12.1 which follows below. □

Example 12.1 (An r-Family of Smoothing Functions). Taking the piecewise-smooth system

$$\dot{\mathbf{x}} = \mathbf{f}(\mathbf{x}; \lambda) = \tfrac{1}{2}(1+\lambda)\mathbf{f}^+(\mathbf{x}) + \tfrac{1}{2}(1-\lambda)\mathbf{f}^-(\mathbf{x}) \ , \quad \lambda = \operatorname{sign}(\sigma) \ , \qquad (12.26)$$

we will first smooth this by just one order (making it continuous but non-differentiable) and then smooth it fully (making it infinitely differentiable).

To smooth (12.26) one order, replace the discontinuous function λ by a continuous but non-differentiable function $\Lambda_r(x/\varepsilon)$, given by

$$\Lambda_r(\sigma/\varepsilon) := \begin{cases} r\sigma^3/\varepsilon^3 + (1-r)\sigma/\varepsilon & \text{if } |\sigma| \le \varepsilon\,, \\ \text{sign}(\sigma) & \text{if } |\sigma| > \varepsilon\,. \end{cases} \tag{12.27}$$

This is monotonic if $|r| < 1$. Substituting into (12.26) for $|\sigma| \le \varepsilon$, we have

$$\begin{aligned} \dot{\mathbf{x}} = \mathbf{f}(\mathbf{x}; \Lambda_r(\sigma/\varepsilon)) &= \tfrac{1}{2}(1 + \Lambda_r(\sigma/\varepsilon))\mathbf{f}^+(\mathbf{x}) + \tfrac{1}{2}(1 - \Lambda_r(\sigma/\varepsilon))\mathbf{f}^-(\mathbf{x}) \\ &= \tfrac{1}{2}(1 + \Lambda_0(\sigma/\varepsilon))\mathbf{f}^+(\mathbf{x}) + \tfrac{1}{2}(1 - \Lambda_0(\sigma/\varepsilon))\mathbf{f}^-(\mathbf{x}) \\ &\quad + (\Lambda_0^2(\sigma/\varepsilon) - 1)\Lambda_0(\sigma/\varepsilon)(\mathbf{f}^+(\mathbf{x}) - \mathbf{f}^-(\mathbf{x}))\tfrac{r}{2}\,, \end{aligned} \tag{12.28}$$

by noticing that $\Lambda_0(\sigma/\varepsilon) = \sigma/\varepsilon$.

In the limit $\varepsilon \to 0$, the function $\Lambda_r(\sigma/\varepsilon)$ tends to sign(σ), but for each r the behaviour inside $|\sigma| \le \varepsilon$ is different, so let us take $\Lambda_0(\sigma/\varepsilon)$ as a 'root' smoothing function. This is monotonic, and in the limit $\varepsilon \to 0$, if we replace it by $\lambda = \text{sign}(\sigma)$, we obtain not one piecewise-smooth system, but an r-parameterized family of piecewise-smooth systems

$$\begin{aligned} \dot{\mathbf{x}} = \mathbf{f}_r(\mathbf{x}; \lambda) &= \tfrac{1}{2}(1 + \lambda)\mathbf{f}^+(\mathbf{x}) + \tfrac{1}{2}(1 - \lambda)\mathbf{f}^-(\mathbf{x}) \\ &\quad + \lambda(\lambda^2 - 1)(\mathbf{f}^+(\mathbf{x}) - \mathbf{f}^-(\mathbf{x}))\tfrac{r}{2} \end{aligned} \tag{12.29}$$

with $\lambda = \text{sign}(\sigma)$. These systems are not necessarily equivalent to our original system (12.26), for instance, $\lambda^\$$ is now r-dependent, e.g.

$$\lambda^\$ = (\mathbf{f}^- + \mathbf{f}^+)\cdot\delta_{\mathbf{x}}\sigma/(\mathbf{f}^- - \mathbf{f}^+)\cdot\delta_{\mathbf{x}}\sigma \qquad \text{for} \quad r = 0$$

and

$$\lambda^\$ = \left[(\mathbf{f}^- + \mathbf{f}^+)\cdot\delta_{\mathbf{x}}\sigma/(\mathbf{f}^- - \mathbf{f}^+)\cdot\delta_{\mathbf{x}}\sigma\right]^{1/3} \quad \text{for} \quad r = 1\,.$$

Even if some $\lambda^\$ = \lambda_0^\$$ is perturbed slightly to $\lambda^\$ = \lambda_0^\$ + \delta\lambda$, we have

$$\delta\lambda = \frac{r\lambda_0^\$(1 - (\lambda_0^\$)^2)}{1 + r(3(\lambda_0^\$)^2 - 1)}\,,$$

which is only small near $r = 0$ and $r = \pm 1$, implying that r-dependence persists for small perturbations. Clearly the stability, given by $d\mathbf{f}/d\boldsymbol{\xi}$ in the layer variable $\boldsymbol{\xi}$ (see Section 7.6), may also be r dependent. We illustrate the nonequivalence using example vector fields in Example 12.2, after we make the last step in our smoothing.

To fully smooth out all derivatives, we simply replace λ with $\Lambda_r(\phi(\sigma/\varepsilon))$, where Λ_r is defined in (12.27) and ϕ is a smooth-but-non-analytic function

$$\phi(v) := \begin{cases} q(-v)^{q(v)} - q(v)^{-q(v)} & \text{if } |v| \le 1\,, \\ \text{sign}(v) & \text{if } |v| > 1\,, \end{cases} \qquad q(v) = e^{2/(v-1)}\,. \tag{12.30}$$

■

There is no contradiction or paradox in the fact that infinitely many different smooth systems have the same limit for $\sigma \neq 0$; this is merely due to the non-uniqueness of the singular limits by which smooth functions may tend towards discontinuous ones.

The following two examples show nonequivalent systems obtained by smoothing.

Example 12.2 (Destruction of a Cycle by Smoothing). Taking constituent vector fields

$$\begin{aligned} \mathbf{f}^+(x_1,x_2) &= \begin{pmatrix} \frac{1}{10} & 1 \\ -1 & \frac{1}{10} \end{pmatrix} \begin{pmatrix} x_1 - 1 \\ x_2 - 1 \end{pmatrix} \quad &\text{for} \quad x_1 > 0\,, \\ \mathbf{f}^-(x_1,x_2) &= \begin{pmatrix} 1 \\ x_2 - \frac{1}{5} \end{pmatrix} \quad &\text{for} \quad x_1 < 0\,, \end{aligned} \tag{12.31}$$

start with the linear switching system

$$\dot{\mathbf{x}} = \mathbf{f}(x_1, x_2; \lambda) = \tfrac{1}{2}(1+\lambda)\mathbf{f}^+ + \tfrac{1}{2}(1-\lambda)\mathbf{f}^- \,. \tag{12.32}$$

Let us smooth this by replacing λ with Λ_r^ε as defined in (12.27). As illustrated in Figure 12.6, the system has a repelling focus at $(x_1, x_2) = (1, 1)$, a visible fold at $(x_1, x_2) = (0, 11/10)$, and a sliding saddle which lies at $(x_1, x_2) \approx (-0.19, -0.38)$ for $r = 0$. For $r = 0$ there exists a stick-slip cycle, and this persists for $r < 0$. The location of the saddle depends on r, however, and at $r \approx 0.01$ the cycle becomes a homoclinic orbit to the saddle. For $r \gtrsim 0.01$, there is no cycle, and solutions evolve on the sliding manifold towards large negative x_2.

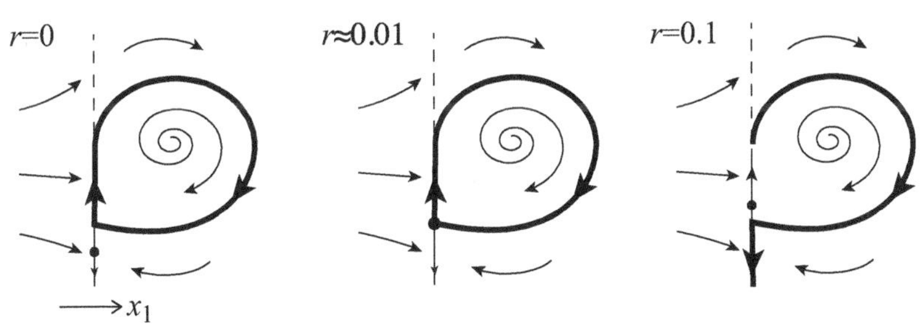

Fig. 12.6 Homoclinic bifurcation induced by smoothing. The system has (i) a stick-slip cycle for $r = 0$ which persists for $r < 0$; (ii) a homoclinic connection of a saddle via a fold at $r \approx 0.01$; (iii) the cycle ceases to exist for $r \gtrsim 0.1$ and all orbits evolve towards $(0, -\infty)$. Taken from simulations with $\varepsilon = 10^{-5}$.

■

Example 12.3 (Stabilisation of a Focus by Smoothing). The smooth function Λ_r need not even be a scalar; it could, for instance, be a matrix. Taking constituent vector fields

$$\begin{aligned}\mathbf{f}^+(x_1,x_2) &= \begin{pmatrix} -2 \\ ay+z \\ az-y \end{pmatrix} \quad \text{for} \quad x_1>0\,, \\ \mathbf{f}^-(x_1,x_2) &= \begin{pmatrix} 1 \\ 0 \\ 0 \end{pmatrix} \quad \text{for} \quad x_1<0\,, \end{aligned} \tag{12.33}$$

start with the linear switching system (12.32). Again let us smooth this by replacing λ with Λ_r^ε such that $\Lambda_r^\varepsilon = \Lambda_0^\varepsilon$ when $\Lambda_0^\varepsilon = \pm 1$ and $\Lambda_r^0 = \operatorname{sign}(\sigma)$ for $\sigma \neq 0$, but we will let

$$\Lambda_r(\Lambda_0) = \Lambda_0 - (\Lambda_0^2 - 1)r \begin{pmatrix} 1/3 & 0 & 0 \\ 0 & -a & 1 \\ 0 & -1 & -a \end{pmatrix}\,. \tag{12.34}$$

The discontinuity surface has an attracting sliding region with a sliding focus, shown in Figure 12.7. For $r=0$ and $a>0$, the focus is repelling inside $\mathcal{D}$, but for large enough r, the focus becomes attracting; this value of r is not very large, for example, if $a=0.1$, then $r \gtrsim 0.0731$ is sufficient. The constants in (12.33) can be varied without significantly changing the results.

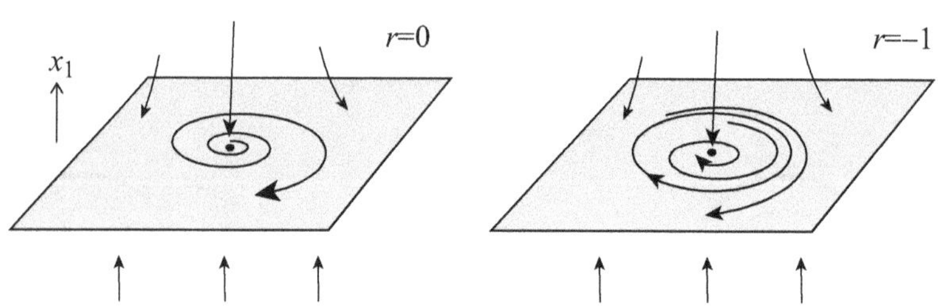

Fig. 12.7 Stability change due to smoothing. A sliding focus changes between repelling for $r=0$, to attracting for $r=a$. Simulations reveal a repelling limit cycle for $r=a$ indicating that a subcritical Hopf bifurcation has occurred.

■

Generally then, given any $\mathbf{f}^+$ and $\mathbf{f}^-$, we must consider a class of piecewise-systems of the form

$$\dot{\mathbf{x}} = \mathbf{f}(\mathbf{x};\lambda) + \Delta_r(\lambda)\,\mathbf{h}_r(\mathbf{x};\lambda)\,, \qquad \lambda = \operatorname{sign}(\sigma)\,, \tag{12.35}$$

where $\mathbf{h}_r$ is a family of finite-valued vector fields and Δ_r vanishes on $\sigma \neq 0$. We can associate with each a smoothing

$$\dot{\mathbf{x}} = \mathbf{f}\left(\mathbf{x}; \Lambda_r^\varepsilon\right) + \Delta_r\left(\Lambda_r^\varepsilon\right)\mathbf{h}_r\left(\mathbf{x}; \Lambda_r^\varepsilon\right) , \qquad \Lambda_r^\varepsilon \xrightarrow{\varepsilon\to 0} \operatorname{sign}(\sigma) , \tag{12.36}$$

for *a particular choice* of Λ_r^ε.

We can strengthen this directly so that Λ_r^ε is analytic, in which case Λ_r^ε cannot be constant for $|\sigma| > \varepsilon$ and must instead have an asymptotic form such as $\Lambda_r^\varepsilon = \operatorname{sign}(\sigma) + \mathsf{O}\left(\varepsilon/\sigma\right)$.

Smoothing can even change the number of sliding modes that exist, as the following example shows.

Example 12.4 (Numbers of Sliding Modes). Take a one-dimensional piecewise-constant system switching between $f^+ = a-1$ for $x > 0$ and $f^- = 1$ for $x < 0$; therefore the discontinuity surface $x = 0$ is attracting if $a < 1$; otherwise we have crossing. Again start with the linear switching system (12.32), in this case simply $\dot{x} = f = \frac{1}{2}(1+\lambda)(a-1) + \frac{1}{2}(1-\lambda)$. This has a single sliding mode at $\lambda^\$ = a/(a-2)$ if $a < 1$, and none otherwise. We now understand that smoothing of this, by replacing λ with for example (12.27), yields a family of systems related to the family of piecewise-smooth systems:

$$\dot{x} = f^{(r)} = \tfrac{1}{2}(1+\lambda)(a-1) + \tfrac{1}{2}(1-\lambda) + (\lambda^2-1)\lambda(a-2)\tfrac{r}{2} , \tag{12.37}$$

which has sliding modes wherever $f^{(r)} = 0$. Clearly these are r-dependent. In fact, the number of sliding modes changes, with modes being created pairwise where $f^{(r)} = f^{(r)}_{,\lambda} = 0$, which reduces to $\lambda^\$ = -3a/4\omega$ and

$$2^4w^3 - 3^3a^2w + 3^3a^2(\tfrac{a}{2}-1) = 0 , \quad w = (1-a)r + \tfrac{a}{2} - 1 , \tag{12.38}$$

solutions of which are valid if $-3a/4w \in (-1,+1)$. For example, for $a = 0$, with $r = 0$, there is one attracting sliding mode, and with $r = 2$ there are three sliding modes (one repelling surrounded by two attracting); for $a = 3/2$, with $r = 0$, there are no sliding modes, and, with $r = -8$ or $r = 11$, there are two sliding modes (one attracting and one repelling); hence crossing becomes nonlinear sliding. ■

Note the large values of r required to change the number of sliding modes in this example. We noted following (12.27) that the smoothing function Λ_r is monotonic only for $|r| < 1$, so these imply a non-monotonic smooth function.

In fact, following from Theorem 12.1, we have:

Corollary 12.1. *The monotonic smoothing of a piecewise-smooth dynamical system cannot create or destroy linear sliding modes.*

Proof. If Λ_r^ε is monotonically increasing, then the convex combination

$$\mathbf{f}\cdot\delta_{\mathbf{x}}\sigma = \left\{\tfrac{1}{2}(1+\lambda)\mathbf{f}^+ + \tfrac{1}{2}(1-\lambda)\mathbf{f}^-\right\}\cdot\delta_{\mathbf{x}}\sigma \tag{12.39}$$

interpolates monotonically between $\mathbf{f}^-\delta_{\mathbf{x}}\sigma$ and $\mathbf{f}^-\delta_{\mathbf{x}}\sigma$. Therefore if $\mathbf{f}^+\delta_{\mathbf{x}}\sigma$ and $\mathbf{f}^-\delta_{\mathbf{x}}\sigma$ have opposite signs, then $\mathbf{f}$ has exactly one zero for $\Lambda_r^\varepsilon \in (-1,+1)$

for any r and ε. Similarly if $\mathbf{f}^+\delta_{\mathbf{x}}\sigma$ and $\mathbf{f}^-\delta_{\mathbf{x}}\sigma$ have the same signs, then $\mathbf{f}$ has no zeros for $\Lambda_r^\varepsilon \in (-1,+1)$ for any r and ε. Because the number of zeros, corresponding to the number of sliding modes, is independent of r and ε, it does not vary for different monotonic smoothings.

There are two other important consequences of Theorem 12.1 that follow immediately.

Corollary 12.2. *If Λ_0 is a monotonic smoothing function and Λ_r is another (possibly non-monotonic) smoothing function, then there exists a unique vector field $\mathbf{h}$ such that the Λ_r-smoothing of $\dot{\mathbf{x}} = \frac{1}{2}(1+\lambda)\mathbf{f}^+ + \frac{1}{2}(1-\lambda)\mathbf{f}^-$ is a Λ_0-smoothing of $\dot{\mathbf{x}} = \frac{1}{2}(1+\lambda)\mathbf{f}^+ + \frac{1}{2}(1-\lambda)\mathbf{f}^- + (\lambda^2-1)\mathbf{h}$.*

Moreover the family of Λ_0-smoothed systems is larger than the family of Λ_r-smoothed linear switching ('Filippov') systems.

Corollary 12.3. *If Λ_0 is a monotonic smoothing function, then there exists a non-monotonic smoothing function Λ_r such that the Λ_0-smoothing of $\dot{\mathbf{x}} = \frac{1}{2}(1+\lambda)\mathbf{f}^+ + \frac{1}{2}(1-\lambda)\mathbf{f}^- + (\lambda^2-1)\mathbf{h}$ is a Λ_r-smoothing of $\dot{\mathbf{x}} = \frac{1}{2}(1+\lambda)\mathbf{f}^+ + \frac{1}{2}(1-\lambda)\mathbf{f}^-$, if and only if $(\lambda^2-1)\mathbf{h} = \frac{1}{2}\left(\mathbf{f}^+(\mathbf{x}) - \mathbf{f}^-(\mathbf{x})\right)\Delta(\lambda)$ such that $\sigma(\mathbf{x})\Delta(\lambda) = 0$.*

12.3.3 Deriving the Layer System via Smoothing

From the point of view of a smooth system, the origin of the layer system (3.12) is as follows. A small change in σ_j results in a change in λ_j of $\delta\lambda_j = \delta\sigma_j \frac{d}{d\sigma_j}\lambda_j$. This is zero outside $\mathcal{D}$. On $\mathcal{D}$, to find the derivative express $\lambda_j(\sigma_j)$ as the singular limit of a smooth monotonic function $\Lambda(u)$,

$$\lambda_j(\sigma_j) = \lim_{E_j\to 0} \Lambda(\sigma_j/E_j)\,, \tag{12.40}$$

where Λ satisfies

$$\begin{aligned} &\text{for } |u| \geq 1 : \Lambda(u) = (1-K_E)\text{sign}(u)\,, \quad \lim_{E\to 0} K_E = 0\,, \\ &\text{for } |u| < 1 : \Lambda(u) \in (-1,+1)\,, \quad \Lambda'(u) \in (0,1]\,, \end{aligned} \tag{12.41}$$

so that $\Lambda(\sigma/E) \xrightarrow{E\to 0} \text{sign}(\sigma)$ for $\sigma \neq 0$. Differentiating $\Lambda(\sigma_j/E_j)$ with respect to σ_j and multiplying by E_j gives

$$E_j \frac{d}{d\sigma_j}\lambda_j(\sigma_j) = E_j \lim_{E_j\to 0} \frac{d}{d\sigma_j}\Lambda(\sigma_j/E_j) = \frac{d}{du}\Lambda(u)\,. \tag{12.42}$$

Since $\Lambda(u)$ is smooth and monotonic, its derivative $\frac{d}{du}\Lambda(u)$ is just a smooth positive function $\Gamma(u) = \frac{d}{du}\Lambda(u)$. The function $\Lambda(u)$ also has an inverse $u(\Lambda)$, allowing us to define the limit

$$\gamma(\lambda_j(\sigma_j)) = \lim_{E_j \to 0} \Gamma(u(\Lambda(\sigma_j/E_j))) \in (0,1] . \tag{12.43}$$

Then $\dot{\lambda}_j = \frac{d\sigma_j}{dt}\frac{d\lambda_j}{d\sigma_j} = \dot{\sigma}_j E_j^{-1}\gamma(\lambda_j)$, or letting $\varepsilon_j = E_j/\gamma$,

$$\varepsilon_j(\lambda)\dot{\lambda}_j = \dot{\sigma}_j \quad \text{where} \quad \varepsilon_j(\lambda) = E_j/\gamma(\lambda) > 0 . \tag{12.44}$$

Since ε_j is strictly positive it represents only a time-scaling. Moreover since γ is strictly positive, for $E_j \to 0$ the factor ε_j is small, so the timescale $\tau = t/\varepsilon_j$ is fast and becomes instantaneous as $\varepsilon_j, E_j \to 0$. Putting this together with the $x_{j>r}$ dynamics, we have the layer system (3.15).

12.3.4 Equivalence of the Smoothed System?

We should recount here the major theorem behind a recent trend to study piecewise-smooth system via smoothing, in order to see how they fit with the above results. The theorem, now 20 years old, promised a more rigorous, ambiguity-slaying role for smoothing than we now know is possible. The theorem is valid, of course, but its consequences are often misconstrued as something stronger. We will sketch its proof here with references to the full analysis elsewhere.

In essence we can prove that, given a piecewise-smooth system with a sliding manifold $\mathcal{M}$, there exists a smooth system with a topologically equivalent slow manifold $\mathcal{M}^\varepsilon$ and with conjugacy between the sliding dynamics on $\mathcal{M}$ and $\mathcal{M}^\varepsilon$. Despite this conjugation, the phrase '*there exists a*' is crucial here: smoothing could give a nonequivalent system.

We first show that there exists a (smooth) singularly perturbed system whose singular limit is qualitatively equivalent to some piecewise-smooth system, before extending the result away from the critical limit.

12.3.5 Equivalence of Layer Dynamics

In this section we show that a singularly perturbed system is topologically equivalent to a piecewise-smooth system in its singular limit.

Given a piecewise-smooth system $\dot{\mathbf{x}} = \mathbf{f}(\mathbf{x};\lambda)$ with $\lambda = \operatorname{sign}\sigma$, take coordinates $\mathbf{x} = (x, \underline{y})$ where $x = \sigma$ and $\underline{y} = (x_2, \dots, x_n)$. Similarly write $\mathbf{f} = (f, \underline{g})$. The switching layer system (3.15) on $x = 0$ is

$$\begin{aligned} \varepsilon\dot{\lambda} &= f(0,\underline{y};\lambda) , \\ \dot{\underline{y}} &= \underline{g}(0,\underline{y};\lambda) , \end{aligned} \qquad \varepsilon \to 0 , \tag{12.45}$$

where $\varepsilon > 0$ is an infinitesimal singular perturbation parameter. Setting $\varepsilon \to 0$ gives a differential-algebraic system

$$\begin{aligned} 0 &= f(0, \underline{y}; \lambda) , \\ \dot{\underline{y}} &= \underline{g}(0, \underline{y}; \lambda) , \end{aligned} \tag{12.46}$$

defining dynamics on some surface where f vanishes. To understand the significance of this in the overall flow, we consider what (12.45) would look like with respect to a fast timescale $\tau = t/\varepsilon$, giving

$$\begin{aligned} \lambda' &= f(0, \underline{y}; \lambda) , \\ \underline{y}' &= \varepsilon \underline{g}(0, \underline{y}; \lambda) , \end{aligned} \tag{12.47}$$

with the prime denoting differentiation with respect to the fast time τ (instead of a dot for derivatives with respect to the normal or 'slow' time t). This becomes a one-dimensional fast subsystem in the limit $\varepsilon \to 0$,

$$\begin{aligned} \lambda' &= f(0, \underline{y}; \lambda) , \\ \underline{y}' &= 0 . \end{aligned} \tag{12.48}$$

On the fast timescale, $\underline{y}$ appears fixed, and hence behaves as a parameter, while λ evolves through $(-1, +1)$ unless it reaches some $(\lambda, \underline{y})$ where $f = 0$, whereupon the flow on $x = 0$ is attracted towards (or repelled from) a $\underline{y}$ parameterized set of equilibria of (12.47) on the fast timescale. The equilibria occupy a hypersurface

$$\mathcal{M} = \left\{(\lambda, \underline{y}) \in (-1, +1) \times \mathbb{R}^{n-1} \ : \ 0 = f(0, \underline{y}; \lambda)\right\} . \tag{12.49}$$

The manifold $\mathcal{M}$ is invariant wherever it is normally hyperbolic, that is, except at a set of points:

$$\mathcal{L} = \left\{(\lambda, \underline{y}) \in \mathcal{M} \ : \ \tfrac{\partial}{\partial \lambda} f(0, \underline{y}; \lambda) = 0\right\} . \tag{12.50}$$

We know of $\mathcal{M}$ already, of course, as the *sliding manifold*, because it is also the solution to the first row of (12.46), namely, the sliding system.

Putting everything together, in the limit $\varepsilon \to 0$ on $x = 0$, the flow outside $\mathcal{M}$ evolves on the fast timescale according to (12.48), unless it reaches a manifold $\mathcal{M}$ whereupon the fast dynamics becomes frozen, and the slow dynamics of (12.46) takes over, giving sliding motion inside $\mathcal{M}$. If a point is reached where $\mathcal{M}$ ceases to exist, because it reaches either the boundaries $\lambda = \pm 1$ or the non-hyperbolic set $\mathcal{L}$, the flow will exit back to the fast flow (12.48).

In geometric singular perturbation theory, $\mathcal{M}$ is known as the critical manifold of the system (12.45), with (12.46) and (12.48) often called the 'reduced' and 'layer' systems, respectively (the clash with our usage of 'layer' to describe the switching layer is unfortunate, but this usage need not concern us), and the existence of invariant manifolds near $\mathcal{M}\backslash\mathcal{L}$ in the $\varepsilon \neq 0$ system then follows by the theory of Fenichel [68, 124].

Now consider the smoothed system $\dot{\mathbf{x}} = \mathbf{f}(\mathbf{x}; \Lambda_r(x/\varepsilon))$, where Λ_r satisfies (12.21) and (12.22). Taking the same coordinates as above in which $\sigma = x$, let $u = x/\varepsilon$ and note by (12.21) that $\Lambda_r(x/\varepsilon) = \Lambda_r(u)$ is ε independent; then the smoothed system is

$$\begin{aligned} \varepsilon\dot{u} &= f(\varepsilon u, \underline{y}; \Lambda_r(u)) , \\ \dot{\underline{y}} &= \underline{g}(\varepsilon u, \underline{y}; \Lambda_r(u)) . \end{aligned} \tag{12.51}$$

Setting $\varepsilon \to 0$ gives the differential-algebraic system

$$\begin{aligned} 0 &= f(0, \underline{y}; \Lambda_r(u)) , \\ \dot{\underline{y}} &= \underline{g}(0, \underline{y}; \Lambda_r(u)) . \end{aligned} \tag{12.52}$$

This is evidently the sliding problem (12.46) rewritten with λ replaced by $\Lambda_r(u)$. To understand its role in the wider flow, we look on the fast timescale $\tau = t/\varepsilon$,

$$\begin{aligned} u' &= f(\varepsilon u, \underline{y}; \Lambda_r(u)) , \\ \underline{y}' &= \varepsilon \underline{g}(\varepsilon u, \underline{y}; \Lambda_r(u)) , \end{aligned} \tag{12.53}$$

which for $\varepsilon \to 0$ becomes

$$\begin{aligned} u' &= f(0, \underline{y}; \Lambda_r(u)) , \\ \underline{y}' &= 0 , \end{aligned} \tag{12.54}$$

This one-dimensional system has a $\underline{y}$-parameterized family of equilibria,

$$\mathcal{M}' = \left\{(u, \underline{y}) \in \mathbb{R}^n \; : \; 0 = f(0, \underline{y}; \Lambda_r(u))\right\} . \tag{12.55}$$

which is therefore the (critical) invariant manifold of (12.52) wherever it is normally hyperbolic, that is, excepting a set of points

$$\mathcal{L}' = \left\{(u, \underline{y}) \in \mathcal{M}' \; : \; f_{,u}(0, \underline{y}; \Lambda_r(u)) = 0\right\} . \tag{12.56}$$

This already suggests a similarity between the smooth system (12.51) and the nonsmooth system (12.45), in their critical sets at least.

However, whereas for the smoothed equation we have obtained the dynamics on $u = x/\varepsilon$, the layer equations for the nonsmooth system give dynamics on λ. The corresponding equation here would give the dynamics on $(\Lambda_r, \underline{y})$. So let us define a new variable $\lambda = \Lambda_r$, whose fast time derivative we find by the chain rule $\lambda' = \frac{du}{d\tau}\frac{d}{du}\Lambda_r(u)$. To express everything in terms of λ, we need the inverse of Λ_r, namely, a function $V(\lambda)$ such that $V(\Lambda_r(u)) = u$, which exists for $|u| < \varepsilon$ by (12.22). Write also $\gamma(\lambda) = \frac{d}{dV}\Lambda_r(V(\lambda))$. On the t timescale, we then have

$$\begin{aligned} \varepsilon\dot{\lambda} &= f(\varepsilon V(\lambda), \underline{y}; \lambda).\gamma(\lambda) , \\ \dot{\underline{y}} &= \underline{g}(\varepsilon V(\lambda), \underline{y}; \lambda) , \end{aligned} \tag{12.57}$$

which for $\varepsilon \to 0$ becomes

$$\begin{aligned} 0 &= f(\varepsilon V(\lambda), \underline{y}; \lambda).\gamma(\lambda) , \\ \underline{\dot{y}} &= \underline{g}(0, \underline{y}; \lambda) . \end{aligned} \tag{12.58}$$

On the fast $\tau = t/\varepsilon$ timescale, we have

$$\begin{aligned} \lambda' &= f(\varepsilon V(\lambda), \underline{y}; \lambda).\gamma(\lambda) , \\ \underline{y}' &= \varepsilon \underline{g}(\varepsilon V(\lambda), \underline{y}; \lambda) , \end{aligned} \tag{12.59}$$

and for $\varepsilon \to 0$ this becomes

$$\begin{aligned} \lambda' &= f(0, \underline{y}; \lambda).\gamma(\lambda) , \\ \underline{y}' &= 0 . \end{aligned} \tag{12.60}$$

Importantly, by (12.22) the quantity $\gamma(\lambda)$ is strictly positive on $|u| < \varepsilon$, so this constitutes only a time rescaling of (12.54) inside the layer. The critical sets $\mathcal{M}'$ and $\mathcal{L}'$ remain unchanged, and their dynamics appears to be that of the switching layer up to the γ fast time scaling.

Lemma 12.1. *The boundary layer dynamics of the smoothed system* (12.51) *on* $|x| < \varepsilon$ *corresponds to a switching layer system*

$$\begin{aligned} E(\lambda, \varepsilon)\dot{\lambda} &= f(0, \underline{y}; \lambda) + \mathsf{O}(\varepsilon u) , \\ \underline{\dot{y}} &= \underline{g}(0, \underline{y}; \lambda) + \mathsf{O}(\varepsilon u) , \end{aligned} \tag{12.61}$$

such that $E(\lambda, \varepsilon) \ll 1$, *where* E *denotes a continuous positive function and* ε *a small parameter, with* $0 < \varepsilon < \varepsilon^* \ll 1$ *for* $\lambda \in (\phi^-_{\varepsilon*}, \phi^+_{\varepsilon*})$, *in terms of constants* ε^* *and* $\phi^\pm_{\varepsilon*}$ *that satisfy* $\phi^\pm_{\varepsilon*} \to \pm 1$ *as* $\varepsilon^* \to 0$.

Proof. We derive from the smoothed system (12.51) the dynamics on the quantity $\lambda = \Lambda_r(x/\varepsilon)$, treated as a variable on $|x| < \varepsilon$. Differentiating $\lambda = \Lambda_r(x/\varepsilon)$ with respect to t gives $\dot{\lambda} = \frac{\dot{x}}{\varepsilon}\frac{\partial}{\partial x}\Lambda_r(x/\varepsilon)$ Defining a variable $u = x/\varepsilon$, we have from (12.21) that $\Lambda_r(x/\varepsilon) = \Lambda_r(u)$ is ε independent, with derivative $\varepsilon\frac{\partial}{\partial x}\Lambda_r(x/\varepsilon) = \frac{\partial}{\partial u}\Lambda_r(u)$. Both $\Lambda_r(u)$ and $\frac{\partial}{\partial u}\Lambda_r(u)$ are smooth with respect to u and independent of ε. Moreover $\frac{\partial}{\partial u}\Lambda_r(u)$ is strictly positive because $\Lambda_r(u)$ is strictly increasing, and $\frac{\partial}{\partial x/\varepsilon}\Lambda_r(x/\varepsilon)$ only becomes small (or vanishing) for $|x|/\varepsilon > 1$. So the quantity $\varepsilon/\left(\frac{\partial}{\partial u}\Lambda_r\right)$ is small and nonzero for $|x|/\varepsilon \leq 1$. Since $\Lambda_r(u)$ is differentiable and monotonic for $|u| < 1$, it has an inverse $V(\lambda)$ such that $V(\Lambda_r(u)) = u$, and we can define a function

$$E(\lambda, \varepsilon) := \frac{\varepsilon}{\frac{\partial}{\partial V}\Lambda_r(V(\lambda))} , \qquad \text{for} \quad |\lambda| < 1 . \tag{12.62}$$

That this quantity is small is shown as follows: the function $\Lambda_r(u)$ varies differentiably over an interval on which its extremal values are $\Lambda_r(\pm 1) = \pm 1$; therefore, there exists a point u_* where $V'(u_*) = \frac{\Lambda_r(+1) - \Lambda_r(-1)}{(+1)-(-1)} = 1$, and, by continuity since $\frac{\partial}{\partial u}\Lambda_r(\pm 1) = 0$, there exist two points $u_\varepsilon^\pm$ where $\frac{\partial}{\partial u}V(u_\varepsilon^\pm) = \pm\varepsilon$ for $0 < \varepsilon < 1$ and moreover an interval $u_\varepsilon^- < u < u_\varepsilon^+$ such that $\frac{\partial}{\partial u}\Lambda_r(u) > \varepsilon$. Fix some ε_* such that $0 < \varepsilon_* \ll 1$, then $\varepsilon/\Lambda_r(u) < 1$ for $u_{\varepsilon*}^- < u < u_{\varepsilon*}^+$, and

$$\lim_{\varepsilon\to 0} E(\lambda,\varepsilon) = 0\ ,$$

so that $E(\lambda,\varepsilon) \ll 1$ for $\varepsilon \ll \varepsilon_*$ and $u \in (u_{\varepsilon*}^-, u_{\varepsilon*}^+)$.

We therefore have the dynamical equation $E(\lambda,\varepsilon)\dot{\lambda} = \dot{x} = f(x,\underline{y};\lambda) = f(\varepsilon u,\underline{y};\lambda)$ for small E, along with $\dot{\underline{y}} = g(\varepsilon u,\underline{y};\lambda)$. Expanding in εu gives (12.61). □

This proposition identifies λ as a fast variable inside $\lambda \in (-1,+1)$ (more strictly for $\lambda \in (\phi_{\varepsilon*}^-, \phi_{\varepsilon*}^+)$ where $\phi_{\varepsilon*}^\pm = \Lambda_r(u_{\varepsilon*}^\pm)$, and $\varepsilon^* > 0$ is arbitrarily small. When λ is set-valued on $x = 0$ with $\varepsilon \to 0$, this determines the variation of λ on the timescale τ, which is instantaneous relative to the timescale t.

Standard geometrical singular perturbation theory does not apply to (12.61) when derived in this way from the smoothing (12.20), because E is a function rather than a parameter. Nevertheless we see from the foregoing analysis that the smoothed system and the original piecewise-smooth system have qualitatively similar slow timescale and fast timescale dynamics. More precisely:

Corollary 12.4. *The system* (12.47) *has equivalent slow-fast dynamics to the system* (12.51) *on the discontinuity set $x = 0$ in the critical limit $E = \varepsilon \to 0$.*

Proof. The smoothed system (12.51), and hence (12.61) derived from it, has critical manifold geometry given by (12.55) to (12.56). This is equivalent to the piecewise-smooth system (12.45) given by (12.49) to (12.50). □

For piecewise-smooth systems, the dynamics of (12.57) is of interest only in the limit $\varepsilon \to 0$, where the dynamics of λ is infinitely fast, and so further study away from this limit is beyond the interest of piecewise-smooth dynamics itself. Evidently there is more to be analysed in the relation between the nonsmooth and smoothed systems for nonzero ε. The first efforts to provide this can be found in [71, 5], showing the structural stability of piecewise-smooth systems under small perturbations by smoothing. Sotomayor-Teixeira [195] provided the first attempt to draw a rigorous equivalence between the dynamics of smooth and piecewise-smooth systems, essentially by extending the results above to $\varepsilon > 0$. The original theorem applied only to linear dependence on λ but was extended to the nonlinear case in [169].

Theorem 12.2 (Equivalence Between Sliding Dynamics and Smoothed Slow Dynamics). *Let the region $\mathcal{D}_\mathcal{M} \subset \mathcal{D}$ be expressible as a graph $x = 0$ in coordinates $\mathbf{x} = (x,\underline{y})$, and let there exist a function $\lambda^\$(\underline{y})$ such that $f(0,\underline{y};\lambda^\$(\underline{y})) = 0$ for every $(0,\underline{y}) \in \mathcal{D}_\mathcal{M}$. Then for any k-times differentiable function Λ_r, the Λ_r-smoothing contains a slow manifold* *related to $\mathcal{D}_{\mathcal{M}}$ by a k-times differentiable mapping, on which the slow dynamics is conjugated to the nonlinear sliding dynamics of (12.58) by a k-times differentiable mapping. Moreover, if $\partial f(0, \underline{y}; \lambda)/\partial\lambda \neq 0$ at $\lambda = \lambda^{\$}(\underline{y})$, then for $\delta > 0$ sufficiently small the nonlinear sliding dynamics defined on $\mathcal{M}'$ persists to order δ, on a manifold $\mathcal{M}'_\delta$ which is δ–close to $\mathcal{M}'$.*

Proof. In the $\mathbf{x} = (x, \underline{y})$ coordinate system, $\mathcal{D}_{\mathcal{M}} \subset \mathcal{M}'$ is an open subset of the hyperplane $\mathcal{D}$. The piecewise-smooth system on the switching layer is given by (12.45), becoming the differential-algebraic system (12.46) in the $\varepsilon \to 0$ limit. The Λ_r-smoothing is given by (12.51) and has a slow critical subsystem (12.52) in the $\varepsilon \to 0$ limit.

By hypothesis there exists at least one function $\lambda^{\$}(\underline{y})$ satisfying the algebraic condition in (12.46), and therefore there exists at least one slow critical manifold $\mathcal{M}'_0$ given by the restriction $\Lambda_r(u) = \lambda^{\$}(\underline{y})$. Since Λ_r has an inverse V in $(-1, +1)$ and $\lambda^{\$}(\underline{y}) \in (-1, +1)$ for every $(0, \underline{y}) \in \mathcal{D}_{\mathcal{M}}$, the manifold $\mathcal{M}'_0$ is the graph $u(\underline{y}) = V(\lambda^{\$}(\underline{y}))$. We can let $H : \mathcal{D}_{\mathcal{M}} \to \mathcal{M}'_0$ be the bijective function $H(0, \underline{y}) = (V(\lambda^{\$}(\underline{y})), \underline{y})$, for which $H(\mathcal{D}_{\mathcal{M}}) = \mathcal{M}'_0$; therefore $\mathcal{M}'_0$ is homeomorphic to $\mathcal{D}_{\mathcal{M}}$. The function H is invertible and its order of differentiability is the same as that of Λ_r.

Substituting $\Lambda_r(u) = \lambda^{\$}(\underline{y})$ into (12.52), the reduced problem on $x = 0$ becomes $\dot{\underline{y}} = \underline{g}(0, \underline{y}; \lambda^{\$}(\underline{y}))$. Taking an initial point $(0, \underline{y}_0)$ and a solution $\Phi_t(0, \underline{y}_0)$ of the piecewise-smooth system (12.46), such that $\Phi_0(0, \underline{y}_0) = (0, \underline{y}_0) \in \mathcal{S}$, then the solution $\Psi_t(H(0, \underline{y}_0))$ of the smoothed system (12.52) on the slow manifold such that $\Psi_0(H(0, \underline{y}_0)) = H(0, \underline{y}_0)$ is given by

$$\Psi_t(H(0, \underline{y}_0)) = \left(V\left(\lambda^{\$}\left(\Phi_t(0, \underline{y}_0)\Big|_{\underline{y}} \right)\right), \Phi_t(0, \underline{y}_0)\Big|_{\underline{y}} \right) = H(\Phi_t(0, \underline{y}_0)).$$

The flows of the regularized reduced (slow manifold) system and the discontinuous sliding system are therefore C^r(topologically)-conjugated.

It remains to show the persistence of the slow-fast dynamics for $\delta > 0$. On the fast timescale $\tau = t/\varepsilon$, the smooth system becomes (12.54) for $\varepsilon \to 0$. Thus $\mathcal{M}'_0$ is a manifold of critical points of the layer problem, which is normally hyperbolic if $\frac{\partial}{\partial\lambda} f(0, \underline{y}; \lambda^{\$}(\underline{y})) \neq 0$. The existence of slow manifolds $\mathcal{M}'_\delta$ that are δ-close to the slow critical manifold $\mathcal{M}'_0$, with dynamics ε-close to the reduced problem (12.53), then follows by Fenichel's theorem [68, 124]. □

This is still incomplete as a theorem equating piecewise-smooth systems and their smoothings. A result extending this equivalence to include dynamics outside the discontinuity surface, over large variations in x and not just the small scale of $x = \varepsilon u$, would require something more. It would require matching the slow and fast timescale dynamics, on the x and u and λ spatial scales, and showing equivalence of the matched dynamics in each case. This is impossible except in special cases, as the counter Example 12.2 shows. There is a more basic flaw that makes such a search perhaps futile anyway.

The analysis in Section 12.3.2 is vital to help us understand Theorem 12.2. The theorem appears to say that the smoothed system has equivalent sliding dynamics to the original piecewise-smooth system. If so, how can two different smoothings (two different r values) of the same piecewise-smooth system have different dynamics?

The resolution to the paradox is important. Each value of r denotes a *different* smoothing, and the theorem states that a particular r-smoothed system has equivalent sliding dynamics to the original piecewise-smooth system *specified by* r. A different smoothing, i.e. a different r, will correspond to a different piecewise-smooth system specified by a different r, with different nonlinear λ terms. This fact has not been clear in previous formulations of these equivalence theorems.

12.3.6 The Degeneracy of $\mathcal{L}$ Persists to $\mathcal{L}'$

Let us look at the non-hyperbolic set $\mathcal{L}$ of the sliding manifold, and its counterpart $\mathcal{L}'$ in the smoothed system. In particular it is worth noting that the degeneracy of $\mathcal{L}$ described in Section 8.8 also applies to $\mathcal{L}'$ in (12.56), because all higher derivatives of $f(0,\underline{y};\Lambda_r(u))$ with respect to u vanish, but this fact is less obvious than the vanishing of derivatives of $f(0,\underline{y};\lambda)$ with respect to λ in Section 8.8. The second derivative of f is

$$f_{,uu} = \Lambda_{r,uu}(u)f_{,\Lambda_r} + (\Lambda_{r,u}(u))^2 f_{,\Lambda_r\Lambda_r} = \frac{\Lambda_{r,uu}(u)}{\Lambda_{r,u}(u)}f_{,u} + (\Lambda_{r,u}(u))^2 f_{,\Lambda_r\Lambda_r} \,, \tag{12.63}$$

where $f_{,u}$ denotes $\partial f/\partial u$ and so on. The first term vanishes on $\mathcal{L}'$ where $f_{,u}=0$, and the second vanishes for (12.51) as $f_{,\Lambda_r\Lambda_r}=0$.

This means that, expanding the first line of (12.52) as a series in u about some $u_0 \in (-\varepsilon,+\varepsilon)$, we have

$$f(0,\underline{y};\Lambda_r(u)) = f(0,\underline{y};\Lambda_r(u_0)) + (u-u_0)f_{,u}(0,\underline{y};\Lambda_r(u_0)) + \mathsf{O}\left((u-u_0)^2\right) \,. \tag{12.64}$$

On the set $\mathcal{L}'$, the two lowest-order terms in this expression vanish, leaving

$$f(0,\underline{y};\Lambda_r(u)) = \mathsf{O}\left((u-u_0)^2\right) \,, \tag{12.65}$$

which vanishes identically if $\mathbf{f}$ has only linear dependence on Λ_r, in which case the first line of (12.52) or (12.51) is trivial on $\mathcal{L}'$, and the value of u that defines $\mathcal{L}'$ is undetermined. Geometrically this means that $\mathcal{L}'$ lies along the u coordinate direction and the fast direction of the corresponding smooth system (12.51). A typical perturbation will result in a topologically nonequivalent system. Nonlinear dependence on λ is necessary to break such degeneracies.

12.3.7 Exponential Sensitivity, Contraction, and Determinacy-Breaking

One thing that smooth systems do give us is some insight into how to interpret the dimensional changes that go on in a piecewise-smooth flow, i.e. the entry to and exit from sliding. For $\varepsilon > 0$ the smoothed system (12.51) possesses an invariant manifold $\mathcal{M}'_\varepsilon$, closely approximating a sliding manifold $\mathcal{M}$ (associated with this particular smoothing).

The manifold is an equilibrium of the one-dimensional fast system (12.53), in which $\underline{y}$ is only slow varying, so expanding about a point $(0, \underline{y}_0)$,

$$\begin{aligned} & x(t) = \varepsilon u(t) \sim \varepsilon u_0 e^{-(t-t_0)\omega/\varepsilon}\,, \quad \underline{y}(t) \sim \underline{y}_0 + (t-t_0)\underline{g}(0,\underline{y}_0; \Lambda_r(0))\,, \\ \text{or} \quad & x(\varepsilon\tau) = \varepsilon u(t) \sim \varepsilon u_0 e^{-(\tau-\tau_0)\omega}\,, \quad \underline{y}(t) \sim \underline{y}_0 + \varepsilon(\tau-\tau_0)\underline{g}(0,\underline{y}_0; \Lambda_r(0))\,, \end{aligned} \tag{12.66}$$

where $\omega = \frac{\partial}{\partial u} f(\varepsilon u, \underline{y}; \Lambda_r(u))|_{u=0, \underline{y}=\underline{y}_0}$.

We should verify that ω is not small enough to cancel the effect of ε, and in fact the derivative with respect to the first argument of f is $\mathsf{O}(\varepsilon)$, so the leading order of ω comes from differentiating the third argument of f. Writing $f = f(x, \underline{y}; \Lambda_r(u))$, in a slight abuse of notation, we can differentiate as

$$\omega = \tfrac{\partial}{\partial u} f(x, \underline{y}; \Lambda_r(u)) + \varepsilon \tfrac{\partial}{\partial x} f(x, \underline{y}; \Lambda_r(u)) = \tfrac{\partial}{\partial u} f(x, \underline{y}; \Lambda_r(u)) + \mathsf{O}(\varepsilon)$$

hence ω is not typically small in ε.

We may prefer to view this in the layer coordinates $(\lambda, \underline{y})$, treating $\lambda = \Lambda_r$ as a variable, for which

$$\lambda(\varepsilon\tau) \sim \lambda^{\$} + (\lambda_0 - \lambda^{\$})e^{-(\tau-\tau_0)\omega}\,, \quad \underline{y}(t) \sim \underline{y}_0 + E.(\tau-\tau_0)\underline{g}(0,\underline{y}_0; \Lambda_r(0))\,, \tag{12.67}$$

for small E.

The approximation (12.67) therefore tells us the $\mathsf{O}(\varepsilon)$ variation in x and $\underline{y}$ near the sliding manifold. If $\omega < 0$, the manifold is attracting; if $\omega > 0$, it is repelling. Moreover trajectories with the same $\underline{y}$ coordinate converge/diverge exponentially as they approach/depart $\mathcal{M}'_\varepsilon$, for $\omega < 0/\omega > 0$, respectively. To confirm this, take two trajectories which lie at points $\mathbf{x}_1 = (x_1, \underline{y}_1)$ and $\mathbf{x}_2 = (x_2, \underline{y}_2)$ at time $t = t_0$, such that $|x_1| < \varepsilon$, $|x_2| < \varepsilon$, and $\underline{y}_1 = \underline{y}_0 + \mathsf{O}(\varepsilon)$, $\underline{y}_2 = \underline{y}_0 + \mathsf{O}(\varepsilon)$; then the distance between them in the x direction, λ multiplier, and $\underline{y}$ subspace evolve as

$$\begin{aligned} dx &\sim \varepsilon e^{-(t-t_0)\omega/\varepsilon}\,, \quad & d\lambda &\sim e^{-(t-t_0)\omega/\varepsilon}\,, \quad & dy &\sim \mathsf{O}(\varepsilon)\,, \\ dx &\sim \varepsilon e^{-(\tau-\tau_0)\omega}\,, \quad & d\lambda &\sim e^{-(\tau-\tau_0)\omega}\,, \quad & dy &\sim \mathsf{O}(\varepsilon)\,, \end{aligned} \tag{12.68}$$

over a change in time $t = \varepsilon\tau = \mathsf{O}(\varepsilon)$.

This lends some interpretation to sliding dynamics and some of its more nefarious consequences from the point of view of slow-fast dynamics. Attracting sliding regions crush trajectories together exponentially in the direction orthogonal to the discontinuity surface, and infinitely so as $\varepsilon \to 0$. Conversely, repelling regions pull trajectories apart exponentially in the orthogonal direction, becoming explosions into infinitely many possible forward trajectories as $\varepsilon \to 0$.

At a determinacy-breaking singularity, trajectories exponentially contract as they approach the sliding manifold leading into the singularity and then exponentially diverge as they depart it, filling some region of space within a flow that is exponentially sensitive to initial conditions. To tell such trajectories apart would require precision not of order ε, but of order $e^{-\omega/\varepsilon}$, i.e. exponentially small in ε.

In fact the divergence through a determinacy-breaking singularity, or an exit point more generally, is slightly weaker than exponential and rather appears to be of order ε^p for some $p > 0$. (This does not diminish the argument of exponential sensitivity above, since trajectories are exponentially squashed elsewhere around $\mathcal{M}$.) The exponential approximation above holds only inside a sliding region, i.e. where $\mathcal{M}'$ is well defined, and requires further analysis at the boundaries where exit is possible, for example, at the non-hyperbolic set $\mathcal{L}'$, or where $\lambda^{\$} = \pm 1$. The only case that has been studied so far is, naturally, the simplest, that of a grazing, i.e. a visible fold, and assuming only linear switching. In that case, the deviation of the invariant manifold $\mathcal{M}'_\varepsilon$ from the critical manifold $\mathcal{M}'_0$ near a fold is found in [22] to be of order $\varepsilon^{(r+1)/(2r+1)}$, where the smoothing function Λ_r is r-times differentiable at $\sigma = \pm\varepsilon$.

12.3.8 The Canopy as a Series Expansion

Now that we have smoothed the discontinuity, we can apply more standard concepts to ask in what sense the canopy, as derived in Section 5.2, is a series expansion of $\mathbf{f}$.

The values $\lambda = +1$ and $\lambda = -1$ can be interpreted as 'points at infinity' with respect to a variable σ/ε for some positive scale parameter ε, meaning $\Lambda(\sigma/\varepsilon) \to \lambda = \pm 1$ corresponds to $\sigma/\varepsilon \to \pm\infty$. If we let $\varepsilon \to 0$, this means $\lambda = \pm 1$ correspond to $\sigma \gtrless 0$. So the series expansions (5.11) and (5.12) make sense if interpreted as large parameter expansions about $\sigma/\varepsilon \to +\infty$ and $\sigma/\varepsilon \to -\infty$.

To verify this, take our usual single-switch system, where $\dot{\mathbf{x}}$ takes values $\mathbf{f}^+$ above, $\mathbf{f}^-$ below, and $\mathbf{f}^{\$}$ on, a discontinuity surface. This can be expressed as the $\varepsilon \to 0$ limit of a system $\dot{\mathbf{x}} = \mathbf{F}(\mathbf{x}; \sigma/\varepsilon)$, such that

$$\lim_{\varepsilon \to 0} \mathbf{F}(\mathbf{x}, \sigma/\varepsilon) = \begin{cases} \mathbf{f}^+(\mathbf{x}) \text{ if } \sigma > 0\,, \\ \mathbf{f}^-(\mathbf{x}) \text{ if } \sigma < 0\,, \\ \mathbf{f}^{\$}(\mathbf{x}) \text{ if } \sigma = 0\,. \end{cases} \tag{12.69}$$

For small positive ε, we should then have $\mathbf{F} \sim \mathbf{f}^{+}$ for $\sigma \gg +\varepsilon$, $\mathbf{F} \sim \mathbf{f}^{-}$ for $\sigma \ll -\varepsilon$, and $\mathbf{F} \approx \mathbf{f}^{\$}$ for $\sigma \approx 0$. More precisely,

$$\begin{aligned} &\lim_{\sigma/\varepsilon \to \pm\infty} \mathbf{F}(\mathbf{x}, \sigma/\varepsilon) = \mathbf{f}^{\pm}(\mathbf{x}) \\ \& \\ &\lim_{\sigma/\varepsilon \to 0} \mathbf{F}(\mathbf{x}, \sigma/\varepsilon) = \mathbf{f}^{\$}(\mathbf{x}) \end{aligned} \tag{12.70}$$

Let us expand $\mathbf{F}$ about these different limits. One way to do this is to define $\omega = \varepsilon/\sigma$; then expanding for large σ is like expanding for small ω, requiring just a Taylor series about $\omega = 0$. Let $\mathbf{G}(\mathbf{x};\omega) = \mathbf{F}(\mathbf{x};\varepsilon/\omega)$, then

$$\mathbf{G}(\mathbf{x};\omega) = \mathbf{G}(\mathbf{x};0) + \omega\mathbf{G}_{,\omega}(\mathbf{x};0) + \omega^2\mathbf{G}_{,\omega\omega}(\mathbf{x};0) + \mathsf{O}\left(\omega^3\right) , \tag{12.71}$$

which is straighforward enough. The first term in the series is

$$\mathbf{G}(\mathbf{x};0) = \lim_{\omega\to 0} \mathbf{F}(\mathbf{x};1/\omega) = \lim_{\varepsilon\to 0} \mathbf{F}(\mathbf{x};\sigma) = \begin{cases} \mathbf{f}^{+}(\mathbf{x}) \text{ for } \sigma > 0 , \\ \mathbf{f}^{-}(\mathbf{x}) \text{ for } \sigma < 0 , \end{cases} \tag{12.72}$$

by (12.70). The derivative is given, using $\frac{\partial}{\partial\omega} = -\frac{\sigma^2}{\varepsilon}\frac{\partial}{\partial\sigma}$, by

$$\mathbf{G}_{,\omega}(\mathbf{x};0) = \lim_{\omega\to 0} \mathbf{F}_{,\omega}(\mathbf{x};1/\omega) = -\lim_{\varepsilon\to 0} \tfrac{\sigma^2}{\varepsilon}\mathbf{F}_{,\sigma}(\mathbf{x};\sigma) := \begin{cases} \mathbf{f}^{+}_{,\sigma}(\mathbf{x}) \text{ for } \sigma > 0 , \\ \mathbf{f}^{-}_{,\sigma}(\mathbf{x}) \text{ for } \sigma < 0 , \end{cases} \tag{12.73}$$

defining

$$\mathbf{f}^{+}_{,\sigma}(\mathbf{x}) = -\lim_{\sigma/\varepsilon\to+\infty} \tfrac{\sigma^2}{\varepsilon}\tfrac{\partial}{\partial\sigma}\mathbf{f}(\mathbf{x};\sigma) , \qquad \mathbf{f}^{-}_{,\sigma}(\mathbf{x}) = -\lim_{\sigma/\varepsilon\to-\infty} \tfrac{\sigma^2}{\varepsilon}\tfrac{\partial}{\partial\sigma}\mathbf{f}(\mathbf{x};\sigma) , \tag{12.74}$$

and so on for higher derivatives. We can assume that these are finite as $\varepsilon \to 0$ because $\mathbf{G}$ is ε independent. So all this gives us an expansion for large σ/ε,

$$\mathbf{F}(\mathbf{x};\sigma/\varepsilon) = \mathbf{f}^{\pm}(\mathbf{x}) + \frac{\varepsilon}{\sigma}\mathbf{f}^{\pm}_{,\sigma}(\mathbf{x}) + \frac{\varepsilon^2}{\sigma^2}\mathbf{f}^{\pm}_{,\sigma\sigma}(\mathbf{x}) + \mathsf{O}\left(\varepsilon^3/\sigma^3\right) \quad \text{for } \varepsilon \to 0 , \tag{12.75}$$

taking the $\pm$ signs for $\sigma \gtrless 0$. This is an asymptotic expansion in large σ/ε, diverging if σ is of order ε and having different forms for $\sigma > 0$ and $\sigma < 0$.

We obtain a function satisfying both $\pm$ expansions if we interpolate between them, but how? We need an interpolation parameter that runs between two values, say $+1$ and -1, between the asymptotes in which our two expansions are valid, namely, $\sigma/\varepsilon \to +\infty$ and $\sigma/\varepsilon \to -\infty$. Such a parameter therefore takes the form

$$\Lambda(\sigma/\varepsilon) = \begin{cases} +1 + \mathsf{O}\left(\varepsilon/\sigma\right) \text{ for } \sigma/\varepsilon \to +\infty , \\ -1 + \mathsf{O}\left(\varepsilon/\sigma\right) \text{ for } \sigma/\varepsilon \to -\infty , \\ \Lambda^{\$} + \mathsf{O}\left(\sigma/\varepsilon\right) \text{ for } |\sigma| < \varepsilon , \end{cases} \tag{12.76}$$

where $\Lambda^\$$ is some constant to be determined. A joint expansion of our system's dynamics is then

$$\begin{aligned}\dot{\mathbf{x}} = \mathbf{F}(\mathbf{x};\sigma/\varepsilon) &= \tfrac{1}{2}(1+\Lambda(\sigma/\varepsilon))\left\{\mathbf{f}^+(\mathbf{x}) + \tfrac{\varepsilon}{\sigma}\mathbf{f}^+_{,\sigma}(\mathbf{x}) + \tfrac{\varepsilon^2}{\sigma^2}\mathbf{f}^+_{,\sigma\sigma}(\mathbf{x}) + \mathsf{O}\left(\varepsilon^3/\sigma^3\right)\right\} \\ &\quad + \tfrac{1}{2}(1-\Lambda(\sigma/\varepsilon))\left\{\mathbf{f}^-(\mathbf{x}) + \tfrac{\varepsilon}{\sigma}\mathbf{f}^-_{,\sigma}(\mathbf{x}) + \tfrac{\varepsilon^2}{\sigma^2}\mathbf{f}^-_{,\sigma\sigma}(\mathbf{x}) + \mathsf{O}\left(\varepsilon^3/\sigma^3\right)\right\} \\ &= \tfrac{1}{2}(1+\Lambda(\sigma/\varepsilon))\mathbf{f}^+(\mathbf{x}) + \tfrac{1}{2}(1-\Lambda(\sigma/\varepsilon))\mathbf{f}^-(\mathbf{x}) + \mathbf{K}(\mathbf{x};\sigma/\varepsilon)\end{aligned} \tag{12.77}$$

where

$$\mathbf{K}(\mathbf{x};\sigma/\varepsilon) = \tfrac{\varepsilon}{2\sigma}\left\{(1+\Lambda(\sigma/\varepsilon))\mathbf{f}^+_{,\sigma}(\mathbf{x}) + (1-\Lambda(\sigma/\varepsilon))\mathbf{f}^-_{,\sigma}(\mathbf{x})\right\} + \mathsf{O}\left(\varepsilon^2/\sigma^2\right) . \tag{12.78}$$

Because we introduced it as an interpolation parameter, the function Λ must be monotonically increasing and analytic in σ. This implies it has a well-defined inverse V such that $V(\Lambda(\sigma/\varepsilon)) = \sigma/\varepsilon$, where V is both finite and non-vanishing, and in terms of which the system becomes

$$\begin{aligned}\dot{\mathbf{x}} = \mathbf{F}(\mathbf{x};V(\Lambda)) &= \tfrac{1}{2}(1+\Lambda)\mathbf{f}^+(\mathbf{x}) + \tfrac{1}{2}(1-\Lambda)\mathbf{f}^-(\mathbf{x}) + \mathbf{K}(\mathbf{x};V(\Lambda)) \\ &:= \mathbf{f}(\mathbf{x};\Lambda) ,\end{aligned} \tag{12.79}$$

giving the familiar combination vector field $\mathbf{f}(\mathbf{x};\Lambda)$. The only appearance of ε is now implicit, inside Λ, and for $\varepsilon \to 0$ we define

$$\lambda := \lim_{\varepsilon\to 0} \Lambda(\sigma/\varepsilon) = \begin{cases} +1 \text{ for } \sigma > 0 , \\ -1 \text{ for } \sigma < 0 , \\ \Lambda^\$ \text{ for } \sigma = 0 , \end{cases} \tag{12.80}$$

hence $\mathbf{x} = \mathbf{f}(\mathbf{x};\lambda)$.

Nonlinear dependence on λ is housed in the term $\mathbf{K}$, which is of order $\mathbf{K} = \mathsf{O}(\sigma/\varepsilon)$, and therefore vanishes in the limit $\sigma/\varepsilon \to \pm\infty$ where $\Lambda = \pm 1$.

An alternative expression for $\mathbf{K}$ is found if we assume we can write

$$\dot{\mathbf{x}} = \mathbf{f}\,(\mathbf{x};\ \Lambda(\sigma(\mathbf{x})/\varepsilon)) \ = \ \sum_{n=0}^{\infty} \mathbf{a}_n(\mathbf{x})\Lambda^n(\sigma(\mathbf{x})/\varepsilon) . \tag{12.81}$$

We do not need to specify whether the Λ's themselves are smooth or discontinuous. Letting $\lambda = \Lambda(\sigma(\mathbf{x})/\varepsilon)$, the derivation in Section 5.2.2 leads directly to the combination with the usual linear term,

$$\dot{\mathbf{x}} = \mathbf{f}\,(\mathbf{x};\Lambda) = \tfrac{1}{2}(1+\Lambda)\mathbf{f}^+ + \tfrac{1}{2}(1-\Lambda)\mathbf{f}^- + \mathbf{n} ,$$

plus an expression for the hidden term $\mathbf{n}$ as a series in Λ, namely,

$$\mathbf{n} = \left(\Lambda^2 - 1\right)\mathbf{h} \quad \text{with} \quad \mathbf{h} = \sum_{n=1}^{\infty} \left(\mathbf{a}_{2n} + \Lambda \mathbf{a}_{2n+1}\right) \sum_{r=0}^{n-1} \Lambda^{2r} \,.$$

Moreover this extends to multiple switches in the form (5.20).

The possibility of approximating a function of several variables $\mathbf{f}(\mathbf{x}, \Lambda)$ by a sum of functions of a single variable $\Lambda(\sigma)$ has arisen particularly in the context of universal approximation by neural networks [40, 87, 147], where the sums of interest include sigmoid functions $\Lambda(\sigma_i(\mathbf{x}))$ that may each have a different threshold $\sigma_i = \hat{\mathbf{e}}_i \cdot (\mathbf{x} - \mathbf{x}_i)$ and stiffness $|\mathbf{e}_i|$ for some vectors $\{\mathbf{e}_i, \mathbf{x}_i\}_{i=1,2,...}$. Here we are interested in series taken about a single switching threshold, say $\mathbf{x}_r = 0$, particularly in the limit of infinite stiffness $|\mathbf{e}_i| \to \infty$. An alternative is to use a series of powers of the sigmoid function $\Lambda\left(\sigma_i\right)$, and the use of such series as solutions of nonlinear differential equations has been explored in [160, 216] for $\Lambda = \tanh$.

To summarize this section, it seems that smoothing a piecewise-smooth system can be done, but is fraught with at least as many ambiguities as piecewise-smooth systems themselves. We can approximate a piecewise-smooth system by a smooth one, but *how* we smooth becomes part of the final model; it is not independent.

Can we go in the other direction? If we begin with a smooth system, can we derive from it a piecewise-smooth approximation? The rigorous method to do this is something we call *pinching.*

12.4 Pinching

Pinching can be thought of as an inverse to smoothing, a way to derive a piecewise-smooth system as an approximation to a continuous system. Given a smooth dynamical system $\dot{\mathbf{x}} = \mathbf{f}(\mathbf{x})$, a region of state space is chosen, say some $|\sigma| \leq \varepsilon$ for $\varepsilon > 0$, to be collapsed down to the zero measure set of a manifold $\mathcal{D}$ by means of a discontinuous transformation, resulting in switching of the vector field $\mathbf{f}(\mathbf{x})$ across the manifold.

The first sophisticated use of pinching was perhaps in [27], seeking to prove the structural stability of piecewise-smooth systems by studying smooth systems with regions of space 'pinched' out of existence. The concept held a lot of promise but lacked some explicit methodology which, it turned out, would require ideas from nonlinear switching. I myself made somewhat crude attempts to develop the concept of pinching to gain insight into the two-fold singularity [44, 45], but the breakthrough came in [169], with the notions of intrinsic and extrinsic pinching, and of completeness, that are due to my colleague Douglas Duarte Novaes. Since the details from [169] are constructive, we will follow them closely, adapted into the present framework.

We begin with an example.

Example 12.5 (Pinching a Hill Function). Take a system

$$(\dot{x}_1, \dot{x}_2) = (-x_1, 2\,\mathrm{H}(x_1/\alpha; p) - 1) \;, \qquad \mathrm{H}(u;p) = \frac{u^p}{1+u^p} \;. \tag{12.82}$$

The Hill function H was introduced in Table 1.1. We will consider α to be small and p to be large but finite.

There is an invariant manifold along $x_1 = 0$ with dynamics $(\dot{x}_1, \dot{x}_2) = (0, -1)$. Keeping $p \gg 1$ and $\alpha \ll 1$ fixed, we want to approximate this with a piecewise-smooth system. Of course the resulting system should be simpler than the transcendental model (12.82).

Pinching involves transforming to a coordinate $\tilde{x}_1 = \sigma - \beta\mathrm{sign}(\sigma)$ for some small $\beta \geq 0$. For (12.82), this results in a piecewise linear approximation:

$$\begin{aligned}(\dot{\tilde{x}}_1, \dot{x}_2) &= \left(-\tilde{x}_1 \mp \beta,\; 2\,\mathrm{H}\left(\tfrac{\tilde{x}_1 \pm \beta}{\alpha}; p\right) - 1\right) \\ &= \left(-\tilde{x}_1 \mp \beta,\; 2\,\mathrm{H}\left(\pm\tfrac{\beta}{\alpha}; p\right) - 1 + \mathsf{O}\left(\tilde{x}_1\right)\right) \qquad \text{for} \quad \tilde{x}_1 \gtrless 0 \;. \end{aligned} \tag{12.83}$$

There is now a choice to make about the size of β compared to the small parameters of the system, namely, α and $1/p$.

If we take β to be a small parameter β that is extrinsic to the smooth system (12.82), then expanding for small β/α gives $\mathrm{H}\left(\pm\frac{\beta}{\alpha}; p\right) = \mathsf{O}\left((\beta/\alpha)^p\right)$, which we can neglect for small enough β; hence

$$(\dot{\tilde{x}}_1, \dot{x}_2) \approx (-\tilde{x}_1 - \beta\lambda,\; -1) \;, \qquad \lambda = \mathrm{sign}(\tilde{x}_1) \;, \tag{12.84}$$

giving the system in Figure 12.8. Seeking a sliding mode, where $\dot{\tilde{x}}_1 = \tilde{x}_1 = 0$, gives $\lambda^\$ = 0$ and a sliding vector field $\dot{x}_2 = (0, -1)$, which is equivalent to the dynamics on the invariant manifold $x_1 = 0$ of (12.82).

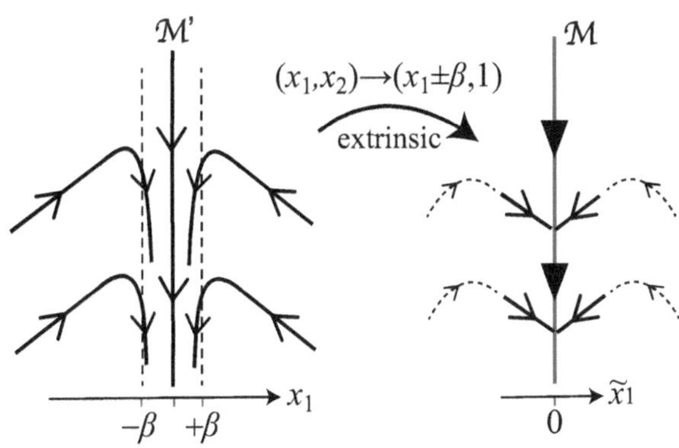

Fig. 12.8 A smooth system with an invariant manifold $x_1 = 0$ (left), pinched by removing the region $|x_1| \leq \beta$, with β a small parameter extrinsic to (i.e. not appearing in) the smooth system. This fails to capture the large $\tilde{x}_1$ behaviour (dotted).

This approximation is valid only for very small $\tilde{x}_1$ and does not capture the turning around of the flow shown dotted in the right of Figure 12.8. If we add higher-order terms, we arrive back at the exact expression in (12.83), so this approximation is quite weak, but there is a way to improve it.

The approximation is poor because it sees only the vector field for $|x_1| \approx \beta \ll \alpha$, so no matter how small α is, β is smaller and the approximation has a small range of validity. If instead we fix β against α, then we pinch with respect to a parameter that is intrinsic to the system, and the approximation improves. If we set $\beta = \alpha\sqrt{2}$, then expanding $\mathrm{H}\left(\pm\frac{\beta}{\alpha}; p\right)$ for small α/β gives $c_\pm = 1 + \mathrm{O}\left((\alpha/\beta)^p\right)$, and the piecewise linear approximation becomes

$$(\dot{\tilde{x}}_1, \dot{x}_2) \approx (-\tilde{x}_1 - \beta\,\mathrm{sign}(\tilde{x}_1),\ +1)\ , \tag{12.85}$$

shown in Figure 12.9. The behaviour at large $\tilde{x}_1$ is now improved, but what

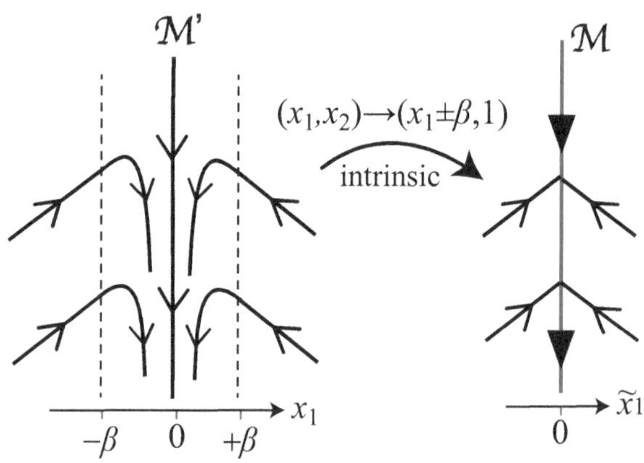

Fig. 12.9 Starting from the same smooth system (left), we pinch by removing the region $|x_1| \leq \beta$, with the parameter $\beta = \alpha\sqrt{2}$ now intrinsic to the smooth system.

happens to the dynamics for small $\tilde{x}_1$? This should be approximated by sliding dynamics. If we let $\lambda = \mathrm{sign}(\tilde{x}_1)$ in (12.85), since there is then no λ dependence in $\dot{x}_2$, we will obtain sliding dynamics where $\dot{x}_2 = +1$; hence by improving the large $\tilde{x}_1$ dynamics, we now have an incorrect representation of the dynamics of (12.82). The correct sliding dynamics must instead be given by a hidden term. Permitting a nonlinear dependence on λ, to satisfy (12.85), we write

$$(\dot{\tilde{x}}_1, \dot{x}_2) \approx (-\tilde{x}_1 - \beta\lambda,\ +1) + (\lambda^2 - 1)\mathbf{h}\ , \qquad \lambda = \mathrm{sign}(\tilde{x}_1)\ , \tag{12.86}$$

for some $\mathbf{h}$. The sliding dynamics is found by solving $\dot{\tilde{x}}_1 = \tilde{x}_1 = 0$, to give $\lambda^\$ = 0$ and hence $\dot{x}_2 = (0,1) - \mathbf{h}$. This is equivalent to the dynamics on the invariant manifold $x_1 = 0$ in (12.82) if we set $\mathbf{h} = (0,2)$. Thus the approximation (12.86) with $\mathbf{h} = (0, 2)$ correctly captures the dynamics of the smooth system, completing Figure 12.9.

■

We call (12.84) and (12.86) the *extrinsic* and *intrinsic* pinching of (12.82), respectively. In (12.84) no hidden term was necessary, but in (12.86) one was. We say in these cases that $\mathbf{h} = (0, 0)$ and $\mathbf{h} = (0, 2)$ *complete* the extrinsically and intrinsically pinched systems, (12.84) and (12.86), respectively. These ideas can be generalized as follows.

12.4.1 Extrinsic Pinching

Consider the dynamical system

$$\dot{\mathbf{x}} = \mathbf{F}(\mathbf{x}), \tag{12.87}$$

where $\mathbf{F}$ is a smooth (or at least differentiable) vector field. Assume that for $\mu = 0$, the manifold $\mathcal{D} = \{\mathbf{x} \in \mathbb{R}^n : \sigma(\mathbf{x}) = 0\}$ is invariant under the flow, that is, $\mathbf{F}(\mathbf{x}) \cdot \delta_{\mathbf{x}}\sigma(\mathbf{x}) = 0$ for every $\mathbf{x} \in \mathcal{D}$.

For small $\beta > 0$, consider the piecewise-smooth system

$$\dot{\mathbf{x}} = \begin{cases} \mathbf{F}(x + \beta\delta_{\mathbf{x}}\sigma(x)) \text{ if } \sigma(\mathbf{x}) > 0, \\ \mathbf{F}(x - \beta\delta_{\mathbf{x}}\sigma(x)) \text{ if } \sigma(\mathbf{x}) < 0, \end{cases} \tag{12.88}$$

which we call the *incomplete* extrinsically pinched system. The manifold $\mathcal{D}$ becomes a discontinuity surface as the right-hand side of (12.88) switches between some $\mathbf{f}^{\pm}(\mathbf{x}; \beta) = \mathbf{F}(\mathbf{x} \pm \beta\delta_{\mathbf{x}}\sigma(\mathbf{x}))$.

To *complete* the system (12.88), we must define the dynamics on $\sigma = 0$. To do so we form a combination $\mathbf{f}(\mathbf{x}; \lambda)$ with $\lambda = \text{sign}(\sigma)$, whose sliding dynamics agrees with the invariant dynamics of (12.87) on the manifold $\mathcal{D}$. When this is possible for some family of functions $\mathbf{h}(\mathbf{x}; \lambda, \beta)$, we say that $\mathbf{h}$ *completes* the pinched system, and using $\mathbf{f}^{\pm}(\mathbf{x}; \beta) = \mathbf{F}(\mathbf{x} \pm \beta\delta_{\mathbf{x}}\sigma(\mathbf{x}))$, we call

$$\begin{aligned} \dot{\mathbf{x}} = \mathbf{f}(\mathbf{x}; \lambda; \beta) = \tfrac{1}{2}(1 + \lambda)\, \mathbf{f}^{+}(\mathbf{x}; \lambda, \beta) + \tfrac{1}{2}(1 - \lambda)\, \mathbf{f}^{-}(\mathbf{x}; \lambda, \beta) \\ + (\lambda^2 - 1)\mathbf{h}(\mathbf{x}; \lambda, \beta) \end{aligned} \tag{12.89}$$

the *complete* extrinsically pinched system, where $\lambda \in (-1, +1)$.

Completing the pinched system in this way is possible provided that (12.87) restricted to the manifold $\mathcal{D}$ is structurally stable. The function $\mathbf{h}$ that completes the pinched system is not unique, but it can be developed in successive orders via the following.

Take coordinates $\mathbf{x} = (x, \underline{y})$ in which $\sigma = x = x_1$, $\underline{y} = (x_2, \ldots, x_n)$, and write $\mathbf{f} = (f, \underline{g})$, $\mathbf{F} = (F, \underline{G})$, $\mathbf{h} = (h, \underline{i})$, where $f = \mathbf{f} \cdot \delta_{\mathbf{x}}\sigma$, $F = \mathbf{F} \cdot \delta_{\mathbf{x}}\sigma$, $h = \mathbf{h} \cdot \delta_{\mathbf{x}}\sigma$.

Theorem 12.3. *For $\beta > 0$ sufficiently small, if there exists a continuous family $\lambda^{\$}(\underline{y};\beta) \in (-1,+1)$ of differentiable functions such that*

$$f(0,\underline{y};\lambda^{\$}(\underline{y};\beta)) = 0$$

for every $(0,\underline{y}) \in \mathcal{D}$, then the sliding mode of (12.89) satisfies

$$\dot{\underline{y}} = \underline{g}(0,\underline{y};\lambda^{\$}(\underline{y};\beta)) = \underline{G}(0,\underline{y}) + \underline{r}(\underline{y};\beta) \qquad on \quad \mathcal{M}',$$

where $\underline{r}(\underline{y};\beta)$ is continuous in β and differentiable in $\underline{y}$ and where $\underline{r}(\underline{y};\beta) \to 0$ as $\beta \to 0$. Moreover if (12.87) restricted to $\mathcal{D}$ is structurally stable, then it is topologically equivalent to the nonlinear sliding dynamics of (12.89).

Proof. Applying the sliding criteria $x = f = 0$ (i.e. $\sigma = \dot{\sigma} = 0$ in the $(x,\underline{y})$ coordinates) to (12.89) gives

$$\begin{aligned}\underline{g}(0,\underline{y}) = \tfrac{1}{2}\left(1+\lambda^{\$}_{\beta}(\underline{y},\beta)\right)\underline{G}(+\beta,\underline{y}) + \tfrac{1}{2}\left(1-\lambda^{\$}(\underline{y};\beta)\right)\underline{G}(-\beta,\underline{y})\\ + \underline{i}(0,\underline{y};\lambda^{\$}_{\beta}(\underline{y},\beta),\beta)\\ = \underline{G}(0,\underline{y}) + \underline{r}(0,\underline{y};\beta)\ ,\end{aligned}$$

the second line following because $\lambda^{\$}(\underline{y};\beta)$ is a continuous family of functions. Since the system $\dot{\underline{y}} = \underline{G}(0,\underline{y})$ is structurally stable it must therefore be topologically equivalent to $\dot{\underline{y}} = \underline{g}(0,\underline{y};\lambda^{\$}(\underline{y};\beta))$. □

Assume now that $\mathbf{F}$ is $k+1$ times differentiable and that the hidden term can be written as

$$(\lambda^2-1)\mathbf{h}(0,\underline{y};\lambda,\beta) = \sum_{j=1}^{k}\beta^j\gamma_j(\underline{y};\lambda) + \mathsf{O}\left(\beta^{k+1}\right) \tag{12.90}$$

for some functions γ_j. Expand f similarly as

$$f(0,\underline{y};\lambda,\beta) = \sum_{j=r}^{k}\beta^j\chi_j(\underline{y};\lambda)/j! \tag{12.91}$$

in terms of functions χ_j given by

$$\begin{aligned}\chi_j =& j(\underline{y};\lambda) = j!\tfrac{\partial^j}{\partial\beta^j}f(0,\underline{y};\lambda,\beta)\Big|_{\beta=0} \qquad (12.92)\\ =& \lambda^{\frac{1-(-1)^j}{2}}\left[(\delta_{\mathbf{x}}\sigma(0,\underline{y})\cdot\delta_{\mathbf{x}})^j\mathbf{F}(0,\underline{y})\right]\cdot\delta_{\mathbf{x}}\sigma(0,\underline{y}) + \gamma_j(\underline{y};\lambda)\cdot\delta_{\mathbf{x}}\sigma(0,\underline{y}),\end{aligned}$$

for $j = 1,2,\ldots,k$. Here $(\delta_{\mathbf{x}}\sigma(0,\underline{y})\cdot\delta_{\mathbf{x}})^j\mathbf{F}(0,\underline{y}) \in \mathcal{D}$ denotes the scalar derivative $(\delta_{\mathbf{x}}\sigma)\cdot\delta_{\mathbf{x}} = \sum_{k=1}^{n}\frac{\partial\sigma}{\partial x_k}\frac{\partial}{\partial x_k}$ applied j times to $\mathbf{F}$ and evaluated at $(0,\underline{y})$.

Theorem 12.4. *For $r \le k$ assume that $\chi_j = 0$ for $j = 1, 2, \ldots, r-1$ and $\chi_r \ne 0$. Suppose that there exists $\ell(\underline{y}) \in (-1, 1)$ such that $\chi_r(\underline{y}; \ell(\underline{y})) = 0$ and $\frac{\partial}{\partial \lambda}\chi_r(\underline{y}; \ell(\underline{y})) \ne 0$ for every $(0, \underline{y}) \in \mathcal{D}$. Then for $\beta > 0$ sufficiently small there exists a continuous family $\lambda^{\$}(\underline{y}; \beta) \in (-1, +1)$ of differentiable functions such that $f(0, \underline{y}; \lambda^{\$}(\underline{y}; \beta), \beta) = 0$ for every $(0, \underline{y}) \in \mathcal{D}$. Moreover if we assume that the system* (12.87) *restricted to the invariant manifold $\mathcal{D}$ is structurally stable, then on $\mathcal{D}$ it is topologically equivalent to a nonlinear sliding mode of* (12.89).

Proof. Assuming that $\chi_j = 0$ for $j = 1, 2, \ldots, r-1$, we write using (12.91)

$$f(0, \underline{y}; \lambda, \beta) = \beta^r \tfrac{\chi_r(\underline{y};\lambda)}{r!} + \mathsf{O}\left(\beta^{r+1}\right).$$

Since $\chi_r(\underline{y}; \ell(\underline{y})) = 0$ and $\frac{\partial}{\partial \lambda}\chi_r(\underline{y}; \ell(\underline{y})) \ne 0$, applying the implicit function theorem for the function $f(0, \underline{y}; \lambda, \beta)/\beta^r$, we obtain, for $\beta > 0$ sufficiently small, the existence of a differentiable family $\lambda^{\$}(\underline{y}; \beta) \in (-1, +1)$ of differentiable functions such that $f(0, \underline{y}; \lambda^{\$}(\underline{y}; \beta), \beta) = 0$ for every $(0, \underline{y}) \in \mathcal{D}$. The result follows by Theorem 12.3. □

In some cases linear switching, i.e. taking $\mathbf{h} \equiv 0$, is sufficient to complete the pinched system (12.88). The following corollary concerns cases, as in Example 12.5, which require nonlinear switching.

Corollary 12.5. *Assuming that $\mathbf{F}$ in (12.88) is at least three times differentiable, the following hold:*

(*a*) *if $[\delta_{\mathbf{x}}\sigma(0, \underline{y}) \cdot \mathbf{F}(0, \underline{y})] \cdot \delta_{\mathbf{x}}\sigma(0, \underline{y}) \ne 0$, then $\mathbf{h} \equiv 0$ is sufficient to complete the pinched system (12.89);*
(*b*) *if $[(\delta_{\mathbf{x}}\sigma(0, \underline{y}) \cdot \delta_{\mathbf{x}})\mathbf{F}(0, \underline{y})] \cdot \delta_{\mathbf{x}}\sigma(0, \underline{y}) = 0$ and $[(\delta_{\mathbf{x}}\sigma(0, \underline{y}) \cdot \delta_{\mathbf{x}})^2\mathbf{F}(0, \underline{y})] \cdot \delta_{\mathbf{x}}\sigma(0, \underline{y}) \ne 0$, then to complete the pinched system requires $\mathbf{h} = \beta^2\mathbf{c}(\underline{y})$ where $\mathbf{c}(\underline{y}) \ne 0$ and $\mathbf{c}(\underline{y}) \cdot \delta_{\mathbf{x}}\sigma(0, \underline{y}) \ne [(\delta_{\mathbf{x}}\sigma(0, \underline{y}) \cdot \delta_{\mathbf{x}})^2\mathbf{F}(0, \underline{y})] \cdot \delta_{\mathbf{x}}\sigma(0, \underline{y})$.*

Proof. Taking $\mathbf{h} \equiv 0$ we have from above that

$$\begin{aligned} f(0, \underline{y}; \lambda, \beta) = {} & \beta\lambda[(\delta_{\mathbf{x}}\sigma(0, \underline{y}) \cdot \delta_{\mathbf{x}})\mathbf{F}(0, \underline{y})] \cdot \delta_{\mathbf{x}}\sigma(0, \underline{y}) \\ & +\beta^2\tfrac{1}{2}[(\delta_{\mathbf{x}}\sigma(0, \underline{y}) \cdot \delta_{\mathbf{x}})^2\mathbf{F}(0, \underline{y})] \cdot \delta_{\mathbf{x}}\sigma(0, \underline{y}) + \mathsf{O}\left(\beta^3\right) . \end{aligned}$$

If $[\delta_{\mathbf{x}}\sigma(0, \underline{y}) \cdot \delta_{\mathbf{x}}\mathbf{F}(0, \underline{y})] \cdot \delta_{\mathbf{x}}\sigma(0, \underline{y}) \ne 0$, we can choose $\ell(\underline{y}) = 0$; thus $\chi_1(\underline{y}, 0) = 0$ and $\frac{\partial}{\partial \lambda}\chi_1(\underline{y}; \ell(\underline{y})) = [\delta_{\mathbf{x}}\sigma(0, \underline{y}) \cdot \delta_{\mathbf{x}}\mathbf{F}(0, \underline{y})] \cdot \delta_{\mathbf{x}}\sigma(0, \underline{y}) \ne 0$. Hence applying Theorem 12.4 we have statement (*a*).

If instead $[\delta_{\mathbf{x}}\sigma(0, \underline{y}) \cdot \delta_{\mathbf{x}}\mathbf{F}(0, \underline{y})] \cdot \delta_{\mathbf{x}}\sigma(0, \underline{y}) = 0$ and $[(\delta_{\mathbf{x}}\sigma(0, \underline{y}) \cdot \delta_{\mathbf{x}})^2\mathbf{F}(0, \underline{y})] \cdot \delta_{\mathbf{x}}\sigma(0, \underline{y}) \ne 0$, there is no bounded family of solutions $\lambda^{\$}(\underline{y}; \beta)$ of the equation $f(0, \underline{y}; \lambda^{\$}(\underline{y}; \beta), \beta) = 0$ for $\mathbf{h} \equiv 0$. Taking instead $\mathbf{h} = \beta^2\mathbf{c}(\underline{y})$ such that $\mathbf{c}(\underline{y}) \cdot \delta_{\mathbf{x}}\sigma(0, \underline{y}) \ne [(\delta_{\mathbf{x}}\sigma(0, \underline{y}) \cdot \delta_{\mathbf{x}})^2\mathbf{F}(0, \underline{y})] \cdot \delta_{\mathbf{x}}\sigma(0, \underline{y})$, we have that

$$f(0, \underline{y}; \lambda) = \beta^2\mathbf{c}(\underline{y}) \cdot \delta_{\mathbf{x}}\sigma(0, \underline{y}) .$$

So $\lambda^{\$}(\underline{y};\beta) = 0 \in (-1,+1)$ is a family of solutions of $f(0,\underline{y};\lambda^{\$}(\underline{y};\beta),\beta) = 0$. Applying Theorem 12.4, we then have statement (*b*). □

As we saw in Example 12.5, a simple illustration is given by $\dot{x}_1 = -x_1$ with $(\dot{x}_2,..,\dot{x}_n) = \underline{q}(x_2,\ldots,x_n)$ where $\underline{q}$ is any smooth function; this would give a complete extrinsically pinched system with linear switching (i.e. $\mathbf{h} \equiv 0$) sliding dynamics equivalent to the smooth system's invariant dynamics on $x_1 = 0$. Instead consider the following more interesting system.

Example 12.6. For $x \in \mathbb{R}$ and $\underline{y} = (x_2, x_3, \ldots, x_n) \in \mathbb{R}^{n-1}$, consider the system

$$\dot{x} = x^2, \quad \dot{\underline{y}} = \underline{q}(\underline{y}). \tag{12.93}$$

Taking $\sigma(x,\underline{y}) = x$, the manifold $\mathcal{D} = \{(x,\underline{y}) \in \mathbb{R}^n \,:\, x = 0\}$ is invariant under the flow, and the dynamics on $\mathcal{D}$ is simply $\dot{\underline{y}} = \underline{q}(\underline{y})$. The incomplete pinched system is

$$\dot{x} = \left\{\begin{array}{ll} (x+\beta)^2 & \text{if} \quad x > 0 \\ (x-\beta)^2 & \text{if} \quad x < 0 \end{array}\right\}, \qquad \dot{\underline{y}} = \underline{q}(\underline{y}). \tag{12.94}$$

Taking the combination $f(x,\underline{y};\lambda,\beta) = \frac{1}{2}(1+\lambda)(x+\beta)^2 + \frac{1}{2}(1-\lambda)(x-\beta)^2 + (\lambda^2-1)\mathbf{h}(x,\underline{y};\lambda,\beta)\cdot\delta_{\mathbf{x}}\sigma$, we have $f(0,\underline{y};\lambda,\beta) = \beta^2 + (\lambda^2-1)\mathbf{h}(0,\underline{y};\lambda,\beta)\cdot\delta_{\mathbf{x}}\sigma$. Clearly for $\mathbf{h} \equiv 0$ with $\beta > 0$, the sliding equation $f(0,\underline{y};\lambda,\beta) = 0$ has no solutions for λ; instead (12.88) has only crossing solutions, which is inconsistent with the dynamics of the smooth system (12.94). Taking instead $\mathbf{h}(x,\underline{y};\lambda,\beta) = (\beta^2,0,0,\ldots)$, we have that, for $\beta > 0$ sufficiently small, $\lambda^{\$}(\underline{y};\beta) = 0 \in (-1,+1)$ is a family of solutions of $f(0,\underline{y};\lambda,\beta) = 0$, and produces a nonlinear sliding mode with dynamics $\dot{\underline{y}} = \underline{g}(0,\underline{y};\lambda^{\$}(\underline{y};\beta),\beta) = \underline{q}(\underline{y})$. ■

In this example, therefore, we can complete the pinched system, but we cannot use Theorem 12.4 to prove equivalence between the pinched sliding dynamics and the original invariant dynamics on $\mathcal{D}$, because the term $\dot{x} = x^2$ is structurally unstable. To handle such cases, it is necessary to perturb the original system by a small quantity. It is then natural to pinch with respect to that small quantity, which is intrinsic to the system.

12.4.2 Intrinsic Pinching

In coordinates $x \in \mathbb{R}$ and $\underline{y} = (x_2, x_3, \ldots, x_n) \in \mathbb{R}^{n-1}$, consider the system

$$\dot{x} = F(x,\underline{y};\mu), \quad \dot{\underline{y}} = \mu\,\underline{G}(x,\underline{y};\mu)\,, \tag{12.95}$$

where $\mathbf{F} = (F,\underline{G})$ is a differentiable vector field and μ is a small parameter. Let $I \subset \mathbb{R}$ and $U \subset \mathbb{R}^{n-1}$ be open and bounded, and for $(x,\underline{y}) \in (I,V)$ and

$\mu = 0$, assume the graph $\mathcal{D} = \{(0, \underline{y}) : \underline{y} \in U\}$ is a critical invariant manifold of (12.95), that is, $F(0, \underline{y}; 0) = 0$ for every $\underline{y} \in U$.

Let us also assume that, for $\mu > 0$ sufficiently small, the graphs $\mathcal{D}^i_\beta = \{(m^i_\beta(\underline{y}), \underline{y}) : \underline{y} \in U\}$ for $i = 1, 2, \ldots, k$, are invariant manifolds of (12.95), where $m^i_\beta(\underline{y}) = \beta\, m_i(\underline{y}) + \mathsf{O}\left(\beta^2\right)$ for some differentiable functions $m_i : \overline{U} \to \mathbb{R}$, such that the $\mathcal{D}^i_\beta$ are order β-perturbations of $\mathcal{D}$. Finally assume $\mu = \mathsf{O}\left(\beta^r\right)$ where $r \geq 1$, so that writing $\mu = \mu(\beta)$, we have $\mu(0) = 0$. The system (12.95) induces dynamics on each $\mathcal{D}^i_\beta$, namely,

$$\underline{\dot{y}} = \mu(\beta)\underline{G}(m^i_\beta(\underline{y}), \underline{y}; \mu(\beta)) \qquad \text{on} \qquad x = m^i_\beta(\underline{y})\ . \tag{12.96}$$

Now let R be a positive real number such that $R > \max\{|m_i(\underline{y})| : \underline{y} \in \overline{U},\ i = 1, 2, \ldots, k\}$. For $\beta > 0$ sufficiently small, consider the piecewise-smooth system

$$\begin{aligned} \dot{x} &= \begin{cases} F(x + \beta R, \underline{y}; \mu(\beta)) & \text{if} \quad x > 0, \\ F(x - \beta R, \underline{y}; \mu(\beta)) & \text{if} \quad x < 0, \end{cases} \\ \underline{\dot{y}} &= \mu(\beta) \begin{cases} \underline{G}(x + \beta R, \underline{y}; \mu(\beta)) & \text{if} \quad x > 0, \\ \underline{G}(x - \beta R, \underline{y}; \mu(\beta)) & \text{if} \quad x < 0, \end{cases} \end{aligned} \tag{12.97}$$

which we call the *incomplete* intrinsically pinched system. The manifold $\mathcal{D}$ is now the discontinuity surface.

As for extrinsic pinching, to complete the system, we must now consider the dynamics on $x = 0$. In this case we must ask whether the pinched system (12.97) can be completed such that there exist k nonlinear sliding modes, each of which agrees with the dynamics of (12.96) for $i = 1, 2, \ldots, k$. When this is possible for some family of functions $\mathbf{h}(\mathbf{x}; \lambda, \beta)$, we say that $\mathbf{h}$ *completes* the pinched system, and letting $f^{\pm}(x, \underline{y}; \lambda, \beta) = F(x \pm \beta R, \underline{y}; \mu(\beta))$, $\underline{g}^{\pm}(x, \underline{y}; \lambda, \beta) = \underline{G}(x \pm \beta R, \underline{y}; \mu(\beta))$, we call

$$\begin{aligned} \dot{x} &= f(x, \underline{y}; \lambda, \beta) \\ &= \tfrac{1}{2}\left(1 + \lambda\right) f^{+}(x, \underline{y}; \mu(\beta)) + \tfrac{1}{2}\left(1 - \lambda\right) f^{-}(x, \underline{y}; \mu(\beta)) + (\lambda^2 - 1) h(x, \underline{y}; \lambda, \beta)\ , \\ \underline{\dot{y}} &= \underline{g}(x, \underline{y}; \lambda, \beta) \\ &= \tfrac{1}{2}\left(1 + \lambda\right) \underline{g}^{+}(x, \underline{y}; \mu(\beta)) + \tfrac{1}{2}\left(1 - \lambda\right) \underline{g}^{-}(x, \underline{y}; \mu(\beta)) + (\lambda^2 - 1) \underline{i}(x, \underline{y}; \lambda, \beta)\ , \end{aligned} \tag{12.98}$$

the *complete* intrinsically pinched system, where $\lambda \in (-1, +1)$. We impose $h(x, \underline{y}; \lambda, 0) = 0$.

Theorem 12.5. *Suppose that* (12.95) *has an invariant manifold defined as the graph of the function* $m_\beta(\underline{y}) = \beta m(\underline{y}) + \mathsf{O}\left(\beta^2\right)$. *If the system*

$$\begin{aligned} \underline{\dot{y}} &= \beta \mu'(0) \underline{G}(0, \underline{y}; 0) \\ &\quad + \tfrac{\beta^2}{2}\left(\mu''(0)\underline{G}(0, \underline{y}; 0) + 2\mu'(0)^2 \tfrac{\partial}{\partial \mu}\underline{G}(0, \underline{y}; 0) + 2\mu'(0) m_i(\underline{y}) \tfrac{\partial}{\partial x}\underline{G}(0, \underline{y}; 0)\right) \end{aligned}$$

is structurally stable and

$$\mu'(0)\tfrac{\partial}{\partial\mu}F(0,\underline{y};0)\tfrac{\partial}{\partial x}F(0,\underline{y};0)\neq 0,$$

then $\mathbf{h}(x,\underline{y};\lambda,\beta)=0$ *completes the intrinsically pinched system.*

Proof. The graph $\mathcal{D}_\beta=\{(m_\beta(\underline{y}),\underline{y}):\underline{y}\in U\}$ is an invariant manifold for (12.95), so if we define $s_\beta(x,\underline{y})=x-m_\beta(\underline{y})$ we have

$$\begin{aligned}0&=F(m_\beta(\underline{y}),\underline{y};\mu(\beta))\cdot\delta_{\mathbf{x}}s_\beta(m_\beta(\underline{y}),\underline{y})\\&=F(m_\beta(\underline{y}),\underline{y};\mu(\beta))-\mu(\beta)\underline{G}(\beta,m(\underline{y}),\underline{y};\mu(\beta))\cdot\tfrac{d}{d\underline{y}}m_\beta(\underline{y})\,,\end{aligned}$$

for $\beta>0$ sufficiently small. Taking the derivative at $\beta=0$, we obtain

$$\mu'(0)\tfrac{\partial}{\partial\mu}F(0,\underline{y};0)+m(\underline{y})\tfrac{\partial}{\partial x}F(0,\underline{y};0)=0. \tag{12.99}$$

For the component f, we obtain

$$\begin{aligned}f(0,\underline{y};\lambda,\beta)&=\tfrac{1+\lambda}{2}F(+\beta R,\underline{y};\mu(\beta))+\tfrac{1-\lambda}{2}F(-\beta R,\underline{y};\mu(\beta))\\&=\beta\left(\mu'(0)\tfrac{\partial}{\partial\mu}F(0,\underline{y};0)+R\lambda\tfrac{\partial}{\partial x}F(0,\underline{y};0)\right)+\mathsf{O}\left(\beta^2\right).\end{aligned}$$

Let $\mathcal{K}(\underline{y};\lambda,\beta)=f(0,\underline{y};\lambda,\beta)/\beta$. From (12.99) we have $\mathcal{K}\left(\underline{y};m(\underline{y})/R,0\right)=0$, and by hypothesis

$$\tfrac{\partial\mathcal{K}}{\partial\lambda}(\underline{y};\lambda,0)\Big|_{(\lambda,\beta)=(m_i(\underline{y})/R,0)}=R\tfrac{\partial}{\partial x}F(0,\underline{y};0)\neq 0.$$

By the implicit function theorem, for $\beta>0$ sufficiently small there now exists $\lambda(\underline{y};\beta)=\frac{m(\underline{y})}{R}+\beta\overline{\lambda}+\mathsf{O}\left(\beta^2\right)$ such that $\lambda(0,\underline{y};\beta)\in(-1,+1)$ and $f(0,\underline{y};\lambda(\underline{y};\beta),\beta)=0$ for every $\underline{y}\in U$. The expression for $\overline{\lambda}$ is easily obtained, but we do not require it here.

A nonlinear sliding mode now satisfies $f(0,\underline{y};\lambda(\underline{y};\beta),\beta)=0$ with dynamics

$$\begin{aligned}\dot{y}=&\underline{g}(0,\underline{y};\lambda(\underline{y};\beta),\beta)=\beta\mu'(0)\underline{G}(0,\underline{y};0)\;+\;\beta^2\Big(\tfrac{\mu''(0)}{2}\underline{G}(0,\underline{y};0)\\&+\mu'(0)^2\tfrac{\partial}{\partial\mu}\underline{G}(0,\underline{y};0)+\mu'(0)m(\underline{y})\tfrac{\partial}{\partial x}\underline{G}(0,\underline{y};0)\Big)+\mathsf{O}\left(\beta^3\right).\end{aligned}\tag{12.100}$$

Expanding (12.96) to second order in β, the nonlinear sliding mode (12.100) is therefore equivalent to (12.96). □

A prototype for systems satisfying the hypotheses of Theorem 12.5 is $\dot{x}=x-\mu$, $\dot{y}=\mu y$, with a slow invariant manifold $x=\mu m_\mu(y)$ that becomes the critical manifold $x=0$ when $\mu=0$.

It is clear that the function $\mathbf{h} \equiv 0$ does not complete the system if $k > 1$. In particular we have the following.

Theorem 12.6. *Suppose that* (12.95) *has two invariant manifolds defined as the graphs of the functions* $m^i_\beta(\underline{y}) = \beta m_i(\underline{y}) + \mathrm{O}(\beta^2)$ *for* $i = 1, 2$ *where* $\mu(\beta) = \mathrm{O}(\beta^2)$. *We assume that* $m_1 \neq m_2$ *and*

$$\mu''(0)\tfrac{\partial}{\partial\mu}F(0,\underline{y};0)\tfrac{\partial^2}{\partial x^2}F(0,\underline{y};0) \neq 0\,.$$

If for $\beta > 0$ *sufficiently small the system*

$$\underline{\dot{y}} = \tfrac{\mu''(0)}{2}\underline{G}(0,\underline{y};0) + \tfrac{\beta}{6}\left(\mu'''(0)\underline{G}(0,\underline{y};0) + 3\mu''(0)m_i(\underline{y})\tfrac{\partial}{\partial x}\underline{G}(0,\underline{y};0)\right)$$

is structurally stable for $i = 1, 2$, *then*

$$\mathbf{h}(x,\underline{y};\lambda,\beta) = \beta^2(\lambda^2 - 1)\left(\tfrac{R^2}{2}\tfrac{\partial^2}{\partial x^2}F(0,\underline{y};0)\,,\,\underline{0}\right)$$

completes the system.

Proof. The graph $\mathcal{D}^i_\beta = \{(m^i_\beta(\underline{y}),\underline{y}) : \underline{y} \in U\}$ is an invariant manifold for (12.95), so defining $s^i_\beta(x,\underline{y}) = x - m^i_\beta(\underline{y})$ we have

$$\begin{aligned} 0 &= F(m_i(\underline{y}),\underline{y};\mu(\beta)) \cdot \delta_{\mathbf{x}}\sigma_i(m_i(\underline{y}),\underline{y}) \\ &= F(\beta m_i(\underline{y}),\underline{y};\mu(\beta)) - \beta\underline{G}(\beta,m(\underline{y}),\underline{y};\mu) \cdot \tfrac{d}{d\underline{y}}m_i(\underline{y})\,, \end{aligned}$$

for $\beta > 0$ sufficiently small. Taking the second derivative at $\beta = 0$ gives

$$\mu''(0)\tfrac{\partial}{\partial\mu}F(0,\underline{y};0) + m_i(\underline{y})^2\tfrac{\partial^2}{\partial x^2}F(0,\underline{y};0) = 0. \tag{12.101}$$

For the component f, we then have

$$\begin{aligned} f(0,\underline{y};\lambda,\beta) &= \tfrac{1+\lambda}{2}F(+\beta R,\underline{y};\mu(\beta)) + \tfrac{1-\lambda}{2}F(-\beta R,\underline{y};\mu(\beta)) \\ &\qquad\qquad\qquad\qquad\qquad +(\lambda^2-1)h(0,\underline{y};\lambda,\beta) \\ &= \tfrac{\beta^2}{2}\left(\mu''(0)\tfrac{\partial}{\partial\mu}F(0,\underline{y};0) + R^2\lambda^2\tfrac{\partial^2}{\partial x^2}F(0,\underline{y};0)\right) + \mathrm{O}(\beta^2)\,. \end{aligned}$$

Now let $\mathcal{K}(\underline{y};\lambda,\beta) = f(0,\underline{y};\lambda,\beta)/\beta^2$. From (12.101) we have $\mathcal{K}\left(\underline{y};m_i(\underline{y})/R,0\right) = 0$, and by hypothesis

$$\tfrac{\partial}{\partial\lambda}\mathcal{K}(\underline{y};\lambda,\beta)\Big|_{(\lambda,\beta)=(m_i(\underline{y})/R,0)} = R\,m_i(\underline{y})\tfrac{\partial^2}{\partial x^2}F(0,\underline{y};0) \neq 0.$$

Hence by the implicit function theorem, for $\beta > 0$ sufficiently small there exists $\lambda_i(0,\underline{y};\beta) = \frac{m_i(\underline{y})}{R} + \mathrm{O}(\beta)$ such that $\lambda_i(0,\underline{y};\beta) \in (-1,+1)$ and $f(0,\underline{y};\lambda_i(0,\underline{y};\beta),\beta) = 0$ for every $\underline{y} \in U$ and for $i = 1, 2$.

The nonlinear sliding mode satisfies $f(0, \underline{y}; \lambda_i(0, \underline{y}; \beta), \beta) = 0$ with dynamics

$$\dot{y} = \underline{g}(0, \underline{y}; \lambda_i(0, \underline{y}; \beta), \beta) = \tfrac{\beta^2 \mu''(0)}{2} \underline{G}(0, \underline{y}; 0) + \tfrac{\beta^2}{6} \left(\mu'''(0) \underline{G}(0, \underline{y}; 0) + 3\mu''(0) m_i(\underline{y}) \tfrac{\partial}{\partial x} \underline{G}(0, \underline{y}; 0) \right) + \mathrm{O}\left(\beta^4\right) . \tag{12.102}$$

Hence expanding (12.96) to third order around $\beta = 0$, the nonlinear sliding mode (12.102) is therefore equivalent to (12.96) for each $i = 1, 2$. □

A prototype for systems satisfying the hypotheses of Theorem 12.6 is $\dot{x} = x^2 - \mu$, $\dot{y} = \mu y$, with slow manifolds $x = \pm\sqrt{\mu} m(y)$ which are normally hyperbolic for $\mu > 0$, but which coalesce onto a non-hyperbolic critical manifold $x = 0$ for $\mu = 0$.

Example 12.7 (Pinching the Canard Singularity). In Section 14.9 we apply pinching to a model inspired by neuron firing, finding that a fundamental singularity responsible for canards in slow-fast systems becomes the familiar two-fold singularity when pinched.

■

12.5 Intermediary Agents

Instead of the closed form

$$\dot{\mathbf{x}} = \mathbf{f}(\mathbf{x}; \boldsymbol{\lambda}) , \qquad \lambda_j = \mathrm{sign}(\sigma_j) ,$$

a discontinuity may be induced by an intermediary variable that shadows either σ_j's or λ_j's. This can create hysteresis or time delay, or turn the discontinuity into some distributed process, thus forming, for example, switching processes of the kind seen in our numerical experiments earlier in this chapter.

There are too many possible avenues to explore when asking how to introduce intermediary variables. These avenues seem like an important set of problems for piecewise-smooth models to begin exploring, nonetheless. An intermediary variable represents some added agent, an additional step in a switching model that introduces more precision, so whether or not its presence affects the qualitative behaviour could be vital. As we come to understand the processes behind real-world physical and biological discontinuities, intermediary variables are likely to play more of a role.

Adding intermediary variables in a discontinuity has a similar effect to adding variables to a smooth system. The behaviour of a system as it explores these extra variables may alter dramatically. Of principal interest is to determine when a system is robust to increases of dimension.

Some motivating examples are as follows.

Example 12.8 (Biological Regulation). In basic genetic models, the regulatory action of a protein species j is triggered when its concentration x_j exceeds a threshold θ_j, in the form $\dot{x}_j = B_j(Z_1, \ldots, Z_m) - \alpha_j x_j$ for $j = 1, \ldots, m$. The functions Z_j are step functions (or Hill functions in smoothed models), and B_j is a Boolean function representing a network of gene interactions. In a more precise model [57], x_i is instead the concentration of the jth species of mRNA molecule, and the protein concentration, now called y_i, is only the trigger, so

$$\left.\begin{array}{l} \dot{x}_j = B_j(Z_1, \ldots, Z_m) - \alpha_j x_j \\ Z_j = \text{step}(x_j - \theta_j) \end{array}\right\} \quad \text{becomes} \quad \left\{\begin{array}{l} \dot{x}_j = B_j(Z_1, \ldots, Z_m) - \alpha_j x_j \ , \\ \dot{y}_j = \gamma_j x_j - \beta_j y_j \ , \\ Z_j = \text{step}(y_j - \theta_j) \ , \end{array}\right. \tag{12.103}$$

with one set of equations for each species $j = 1, \ldots, m$, where $\alpha_j, \beta_j, \gamma_j$, are constants. This extended model was proposed in [57].

■

Example 12.9 (Electronic Regulation). In an electronic circuit described by equations $\dot{\mathbf{x}} = \underline{\underline{A}}.\mathbf{x} + \mathbf{b}$, where $\mathbf{x}$ is a vector of current and voltage variables, parts of a circuit may be turned on or off by adding a switching action $\underline{\underline{B}}(\mathbf{x}).\mathbf{u}$, where $\mathbf{u} = (u_1, \ldots, u_m)$, and each u_j a step function triggered as the state x_j passes a threshold θ_j. In practice, each state x_j is measured by a sensor y_j whose value triggers the control, so

$$\left.\begin{array}{l} \dot{\mathbf{x}} = \underline{\underline{A}}.\mathbf{x} + \mathbf{b} + \underline{\underline{B}}(\mathbf{x}).\mathbf{u} \\ u_j = \text{step}(x_j - \theta_j) \end{array}\right\} \quad \text{becomes} \quad \left\{\begin{array}{l} \dot{\mathbf{x}} = \underline{\underline{A}}.\mathbf{x} + \mathbf{b} + \underline{\underline{B}}(\mathbf{x}).\mathbf{u} \ , \\ \dot{u}_j = (x_j - y_j)/\mu_j \ , \\ Z_j = \text{step}(y_j - \theta_j) \ , \end{array}\right. \tag{12.104}$$

This extended model was proposed in [125].

■

Hysteresis can be added using intermediary variables too, with a system such as

$$\begin{aligned} \dot{\mathbf{x}} &= \mathbf{f}(\mathbf{x}; \mathbf{y}) \ , \qquad \lambda_j = \text{sign}(x_j + \alpha_j y_j); \\ \underline{\underline{\varepsilon}}_j \dot{y}_j &= \lambda_j - y_j \ . \end{aligned} \tag{12.105}$$

The various different effects can then be combined various different ways, for example,

$$\begin{aligned} \dot{\mathbf{x}} &= \mathbf{f}(\boldsymbol{\xi}; \boldsymbol{\eta}) \ , \qquad \lambda_j = \text{sign}(\zeta_j + \alpha \eta_j) \ , \\ \varepsilon_j \dot{\xi}_j &= x_j - \xi_j \ , \\ \mu_j \dot{\eta}_j &= \lambda_j - \eta_j \ , \\ \nu_j \dot{\zeta}_j &= \sigma_j - \zeta_j \ , \end{aligned} \tag{12.106}$$

for small constants ε_j, μ_j, ν_j, α_j, or a version proposed in [23],

$$\begin{aligned} \dot{\mathbf{x}} &= \mathbf{f}(\mathbf{s}_\mu(\mathbf{x}, \mathbf{y}); \boldsymbol{\lambda}) \ , \qquad \lambda_j = \text{sign}(S_\mu(x_j, y_j)) \\ \dot{\mathbf{y}}_j &= (x_j - y_j)/\mu_j \ , \end{aligned} \tag{12.107}$$

in coordinates where $\sigma_j = x_j$, where we define vector and scalar *shadowing* functions $\mathbf{s}_\mu(\mathbf{x}, \mathbf{y})$ and $S_\mu(x, y)$ such that $\mathbf{s}_\mu(\mathbf{x}, \mathbf{x}) = 0$ and $S_\mu(x, x) = x$, for example, $\mathbf{s}_\mu(\mathbf{x}, \mathbf{y}) = \mu\mathbf{x} + (1-\mu)\mathbf{y}$ and $S_\mu(x, y) = \mu x + (1-\mu)y$ with $\mu \in [0, 1]$.

Preliminary results on the robustness of sliding in such models were given in [23]. They were enough to show that piecewise-smooth dynamics is under some conditions robust to intermediary variables, in the sense that sliding persists. In other conditions sliding no longer occurs, but is replaced by dynamics of an oscillatory nature, whose stability and approximation to the original piecewise-smooth model are harder to ascertain.

These investigations are ongoing, and worth pursuing in as wide a range of applications as possible, particularly where sufficient theory or experiments exist allowing the switching process to be probed in detail.

12.6 Looking Forward

The most important result of this somewhat involved chapter is perhaps the result in Section 12.1.2 which, while heuristic, has been well borne out by subsequent investigations (e.g. Section 12.2.1 and Section 14.7).

The attempt to more closely model reality is where the most exciting open problems lie for nonsmooth systems. A model is not necessarily a more accurate representation of reality just because we make it more precise. Nonsmooth systems, at their most powerful, are a faithful representation not only of our knowledge but also our lack of knowledge, of physical reality. The urge to smooth comes from our urge to deal with deterministic equations, but determinism is too high an ideal for many complex systems. The non-uniqueness in nonsmooth flows represent physical motions whose trajectories are indistinguishable to any real order of precision. Hidden dynamics represents processes of switching that often lie, too, beyond any real order of empirical precision. Thus piecewise-smooth dynamical theory can help us understand the uses and interpretation of non-uniqueness in mathematical modeling of real-world dynamics.

We could have started the book with this chapter, but the search through the many complex aspects of the real physical processes involved in switching is messy. It has been a major breakthrough of only the year or so prior to publication of this book to be able to place such issues on a rigorous footing, with clear outcomes. We have seen that smoothing has its dangers, because even with monotonicity constraints it is not unique. We have seen that hysteresis, delay, and discretization, introduce irregularity, but that this is killed off by noise or stochastic processes. In all, the approach of defining a piecewise-smooth system by a combination $\dot{\mathbf{x}} = \mathbf{f}(\mathbf{x}; \boldsymbol{\lambda})$ in terms of a switching multiplier vector $\boldsymbol{\lambda}$ appears rather powerful, and provides at least a robust starting point for deeper analysis of specialized applications. Elsewhere, the central ideas of a discontinuity surface $\mathcal{D}$ and sliding modes upon it persist as strong modeling concepts, even when they are more complex objects not expressible in closed form as a combination $\dot{\mathbf{x}} = \mathbf{f}(\mathbf{x}; \boldsymbol{\lambda})$.

The deeper study of applications where piecewise-smooth theory does or does not apply, at least as presented in this book, remains an exciting challenge for dynamical systems theorists and mathematical modellers. Our final chapter glances through several applications of recent interest.

Before we get to that, however, we have unfinished business concerning the two-fold singularity.

Chapter 13
Four Obsessions of the Two-Fold Singularity

And so we return, one last time, to the *two-fold* singularity. The story of this elementary point spans the last 30 years of piecewise-smooth dynamical theory. No other aspect of nonsmooth systems has so challenged the established wisdom or tested and inspired new ideas.

There have been four obsessions in the two-fold's history, and it still makes sense to explore these in order. They are:

1. Obsession 1: a 'normal form'
2. Obsession 2: the crossing map
3. Obsession 3: the sliding flow
4. Obsession 4: blowing up the singularity

These titles reflect the state of knowledge at the inception of each obsession, below we use names that better reflect our current understanding.

The obsessions arose in attempts to apply the most fundamental notions of dynamics to nonsmooth systems: Is it structurally stable in a way that can be embodied in a 'normal form' equation? Is it asymptotically stable in a manner that organizes its phase portraits via invariants and topological singularities? And ultimately, the problem lingering for 30 years or more: what happens at the singularity itself?

The chase from one level of obsession to the next runs to the heart of the ideas of uniqueness and determinism in nonsmooth systems. This chapter brings together results that unfolded piecemeal through a long series of papers [71, 205, 206, 119, 37, 69, 38, 116]; for a much abridged introduction, see [123].

M. R. Jeffrey, *Hidden Dynamics*,
https://doi.org/10.1007/978-3-030-02107-8_13

13.1 The Generic Two-Fold: A Summary

We introduced the two-fold singularity in Definition 6.2 of Section 6.4, as a generic point $\mathbf{x}_p$ on a discontinuity surface, $\sigma(\mathbf{x}_p) = 0$, where the vector fields are tangent from both sides, $\delta_{t+}\sigma(\mathbf{x}_p) = 0$ and $\delta_{t-}\sigma(\mathbf{x}_p) = 0$.

The accompanying inequalities in Definition 6.2 specify non-degeneracy conditions: $\delta^2_{t\pm}\sigma(\mathbf{x}_p) \neq 0$ says that tangencies between the discontinuity surface and the flow are folds, while $\Delta \neq 0$ means that the pair of folds (one with respect to each of the upper and lower flows) are transversal. Examples are shown in Figure 13.1 for two invisible folds in two, three, or four dimensions.

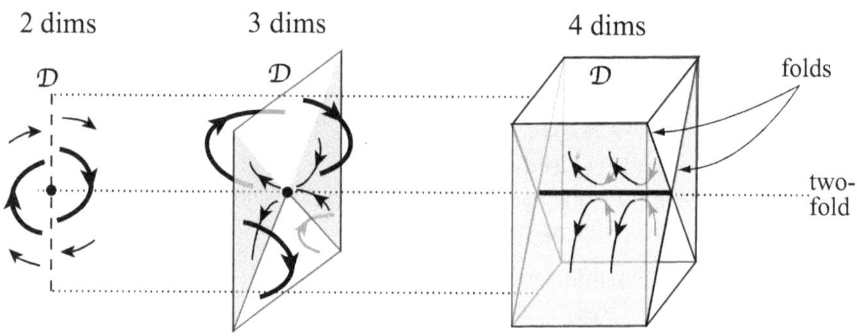

Fig. 13.1 A discontinuity hypersurface has codimension one, folds have codimension two, and the two-fold has codimension three. As a consequence, the two-fold singularity generically arises in systems with at least three dimensions. (i) In two dimensions, a pair of folds may coincide as a codimension one event (which we call a 'fold-fold'). (ii) In three dimensions, folds form curves on the discontinuity surface, which can generically intersect at a two-fold. (iii) In four dimensions the folds are surfaces and their intersection curve is a two-fold; here, we depict the three-dimensional space of the discontinuity surface. Shading indicates sliding regions.

When we fully classify the types of two-fold, we will need a further condition,

$$(\delta_{t\lambda}\mathbf{x}_p) \cdot (\partial_{\mathbf{x}}\delta_{t\pm}\sigma(\mathbf{x}_p)) \neq 0 \qquad \forall \lambda \in (-1, +1) \ , \tag{13.1}$$

which means that all values of the linear combination $\delta_{t\lambda}\mathbf{x}$ have a nonzero component in the plane of $\delta_{t+}\mathbf{x} = \mathbf{f}^+$ and $\delta_{t-}\mathbf{x} = \mathbf{f}^-$ at $\mathbf{x}_p$. An important consequence of this is

$$0 \notin \frac{d}{dt}\mathbf{x} \ , \tag{13.2}$$

which says that the flow is not stationary at the singularity. Hence the singularity is not an equilibrium or sliding equilibrium. *How* the flow passes through the singularity turns out to be non-trivial.

The flow's orientation and curvature with respect to the folds is characterized by two functions

$$v^+(\mathbf{x}) := \frac{\delta_{t+}\delta_{t-}\sigma(\mathbf{x})}{\sqrt{|\delta_{t+}^2\sigma(\mathbf{x})\delta_{t-}^2\sigma(\mathbf{x})|}} \quad \& \quad v^-(\mathbf{x}) := \frac{-\delta_{t-}\delta_{t+}\sigma(\mathbf{x})}{\sqrt{|\delta_{t+}^2\sigma(\mathbf{x})\delta_{t-}^2\sigma(\mathbf{x})|}} . \tag{13.3}$$

These are important for classifying all aspects of the local dynamics, and we define their values at the singularity as a pair of constants

$$\nu^+ := v^+(\mathbf{x}_p) \qquad \& \qquad \nu^- := v^-(\mathbf{x}_p) . \tag{13.4}$$

The local geometry of the two-fold singularity is at first glance rather simple. We have seen it previously in Figure 6.10 in coordinates where the flow is piecewise straightened and Figure 8.7 where the discontinuity surface is flat. Figure 13.2 shows it more generally. The flow folds—visibly (away from) or invisibly (towards)—the discontinuity surface $\mathcal{D}$ from either side, Figure 13.2. The two sets of folds cut $\mathcal{D}$ into four parts, an attracting sliding region $\mathcal{D}_{att}$, a repelling sliding region $\mathcal{D}_{rep}$, and crossing regions $\mathcal{D}_{C\pm}$ through which the flow threads the surface in the positive or negative σ direction.

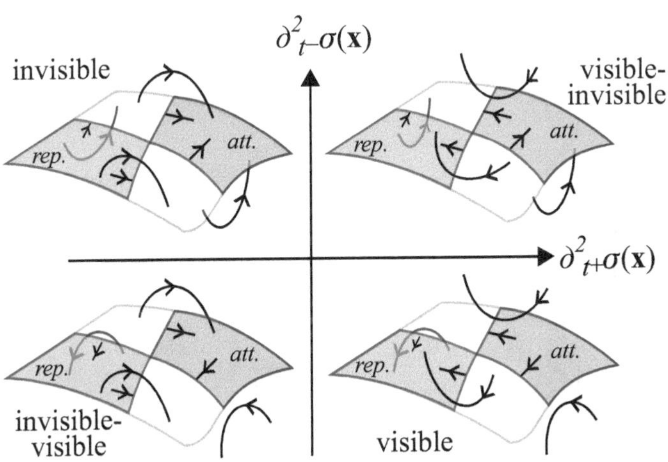

Fig. 13.2 Two-folds come in three flavours depending on whether the folds that comprise them are visible or invisible (the top-right and bottom-left cases are topologically equivalent), as determined by the signs of $\delta_{t\pm}^2\sigma$. Regions of sliding (*att*, shaded), repelling sliding (*rep*, shaded), and crossing (unshaded) all meet at the singularity.

That is where the simplicity ends, as on closer inspection our familiar intuitions of dynamical systems are betrayed, one after another.

Example 13.1. Take a look at Figure 13.3, which shows an example of each type of two-fold, simulated from the equations

(i) visible two-fold:

$$\mathbf{f}^+=(-x_2,\ \tfrac{2x_1}{5}+\tfrac{x_2}{10}-1,\ \tfrac{3x_2}{10}-\tfrac{x_2x_3+2}{5})\ ,\quad \mathbf{f}^-=(x_3,\ \tfrac{x_2x_3-3}{5},\ \tfrac{2x_3}{5}-1-x_1)\ ;$$

(ii) invisible two-fold:

$$\mathbf{f}^+=(-x_2, 1+x_1, -\tfrac{7}{5})\ ,\qquad \mathbf{f}^-=(x_3, -\tfrac{9}{10}, 1-\tfrac{3x_1}{5})\ ;$$

(iii) visible-invisible two-fold:

$$\mathbf{f}^+=(-x_2+\tfrac{1}{10}x_1, x_1-\tfrac{6}{5}, x_1-2)\ ,\quad \mathbf{f}^-=(x_3+\tfrac{x_1}{10}, x_1+\tfrac{23}{100}, 1-x_1)\ ;$$

using $(\dot{x}_1,\dot{x}_2,\dot{x}_3)=\frac{1}{2}(1+\lambda)\mathbf{f}^++\frac{1}{2}(1-\lambda)\mathbf{f}^-+\frac{1}{5}(1-\lambda^2)(1,0,0)$ with $\lambda=\mathrm{sign}(x_1)$.

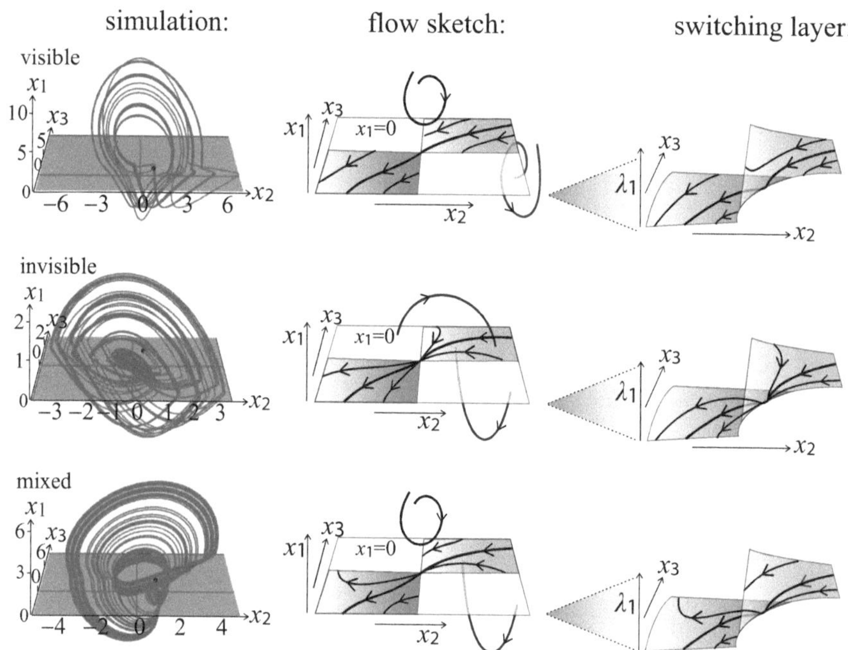

Fig. 13.3 Three examples of attractors organized around a two-fold singularity, showing (i) a simulated trajectory, (ii) a sketch of the piecewise-smooth flow inside and outside $x_1=0$ that gives rise to it, and (iii) blow-up of the switching layer on $x_1=0$.

In each case there is a two-fold singularity at the origin, through which the flow passes repeatedly as it is reinjected through some return mechanism. The flow through the singularity is determinacy-breaking and therefore set-valued. The simulations use a smoothed version of the vector field, replacing

$\lambda \mapsto \tanh(x_1/\varepsilon)$ where $\varepsilon = 10^{-7}$, and manifest the determinacy-breaking as extreme sensitivity in the simulations as the flow traverses the singularity.

■

This example is simulated by smoothing the discontinuity. What happens if instead the switch involves non-idealities as we studied in Section 12.2? What would a two-fold's effect be in a real system, and how would we recognise it? These, and all generic local possibilities for the flow local to the singularity, are what the *obssesions* of the two-fold seek to make sense of.

Evidently from Figure 13.3 we will need the concept of *canards* from Definition 4.3, trajectories evolving from attracting sliding to repelling sliding via a singularity. A simple canard is isolated, a robust canard belongs to a continuous family of canards. Faux canards are similar but evolve from repelling to attracting sliding.

We present most of the results that follow in three dimensions—the minimum required for two-folds to occur generically—though they apply in any higher number of dimensions unless stated otherwise.

13.2 Obsession 1: The Prototype in n Dimensions

Of the four obsessions, finding a local expression for the two-fold was the problem most easily resolved. It is given by

$$
\begin{aligned}
(\dot{x}_1, \dot{x}_2, \dot{x}_3) &= \left\{ \begin{matrix} \mathbf{f}^+ \text{ if } x_1 > 0 \\ \mathbf{f}^- \text{ if } x_1 < 0 \end{matrix} \right\} + \left(\mathsf{O}\left(|\mathbf{x}|^2\right), \mathsf{O}\left(|\mathbf{x}|\right), \mathsf{O}\left(|\mathbf{x}|\right) \right) \\
\text{where} \quad & \begin{matrix} \mathbf{f}^+ = (-x_2, -s^+, \nu^+) \ , \\ \mathbf{f}^- = (\ x_3\ ,\ \nu^-,\ s^-\ ) \ , \end{matrix} \\
\dot{x}_i &= \mathsf{O}\left(|\mathbf{x}|\right) \qquad \text{for } i = 4, \ldots, n \ .
\end{aligned}
\tag{13.5}
$$

for which $\sigma = x_1$, $\delta_{t\pm}\sigma(0) = 0$, and $\operatorname{sign}\left[\delta_{t\pm}^2\sigma(0)\right] = s^{\pm}$. The constants $\nu^{\pm}$ are those defined in (13.3). Local forms for the two-fold have been redefined multiple times by different authors without explicit derivation, and we have in the course of our investigations already provided two derivations, one using catastrophe theory in Section 6.1 and another by explicit choice of coordinates in Section 6.4 and Section 6.9. All are equivalent to (13.5) to leading order.

The discontinuity surface $\mathcal{D}$ is $x_1 = 0$ and is divided by the fold sets $x_1 = x_2 = 0$ and $x_1 = x_3 = 0$ into regions

$$
\begin{aligned}
\mathcal{D}_{att} &= \{\mathbf{x} \in \mathbb{R}^n : \ x_1 = 0, \ x_2 > 0, \ x_3 > 0\} \ , \\
\mathcal{D}_{rep} &= \{\mathbf{x} \in \mathbb{R}^n : \ x_1 = 0, \ x_2 < 0, \ x_3 < 0\} \ , \\
\mathcal{D}_{C^+} &= \{\mathbf{x} \in \mathbb{R}^n : \ x_1 = 0, \ x_2 < 0 < x_3\} \ , \\
\mathcal{D}_{C^-} &= \{\mathbf{x} \in \mathbb{R}^n : \ x_1 = 0, \ x_3 < 0 < x_2\} \ ,
\end{aligned}
\tag{13.6}
$$

such that the vector field outside $\mathcal{D}$ is directed into $\mathcal{D}_{att}$, out from $\mathcal{D}_{rep}$, and points through $\mathcal{D}$ in the direction of increasing/decreasing x_1 in $\mathcal{D}_{C^+}/\mathcal{D}_{C^-}$.

We can solve (13.5) to give the following as a corollary to Theorem 6.2.

Corollary 13.1. *Trajectories of* (13.5) *are arc segments of the form*

$$\left.\begin{array}{l} x_1(t) = x_1(0) - tx_2(0) + \frac{1}{2}\,\mathrm{sign}[\delta^2_{t+}\sigma(0)]t^2 + \mathsf{O}\left(t^3\right) \\ x_2(t) = x_2(0) - \mathrm{sign}[\delta^2_{t+}\sigma(0)]t + \mathsf{O}\left(t^2\right) \\ x_3(t) = x_3(0) + \nu^+ t + \mathsf{O}\left(t^2\right) \end{array}\right\} \tag{13.7a}$$

for $x_1(0) \geq 0$ *and* $x_2(0) \leq 0$ *for* t *such that* $x_1(t) \geq 0$, *and*

$$\left.\begin{array}{l} x_1(t) = x_1(0) + tx_3(0) + \frac{1}{2}\,\mathrm{sign}[\delta^2_{t-}\sigma(0)]t^2 + \mathsf{O}\left(t^3\right) \\ x_2(t) = x_2(0) + \nu^- t + \mathsf{O}\left(t^2\right) \\ x_3(t) = x_3(0) + \mathrm{sign}[\delta^2_{t-}\sigma(0)]t + \mathsf{O}\left(t^2\right) \end{array}\right\} \tag{13.7b}$$

for $x_1(0) \leq 0$ *and* $x_3(0) \leq 0$ *for* t *such that* $x_1(t) \leq 0$. *In both cases we have* $x_i(t) = x_i(0) + \mathsf{O}\left(t^2\right)$ *for* $i = 4, 5, \ldots, n$.

Proof. Expand the trajectories $(x_1(t), x_2(t), \ldots)$ of (13.5) as a power series in t, namely, $x_i(t) = x_i(0) + t\frac{dx_i(0)}{dt} + \frac{1}{2}t^2\frac{d^2x_i(0)}{dt^2} + \ldots$, substituting in $d^r x_i/dt^r$, found from (6.41), to obtain separate expansions in $x_1 > 0$ and $x_1 < 0$. These expansions yield (13.7a) and (13.7b), respectively, along with $x_i(t) = x_i(0) + \mathsf{O}\left(t^2\right)$ for $i = 4, 5, \ldots, n$. □

The expansion of $x_1(t)$ to $\mathsf{O}\left(t^2\right)$, but of $x_2(t)$ and $x_3(t)$ only to $\mathsf{O}(t)$, is sufficient to obtain the correct dependence on terms quadratic in x_2 and x_3 when both t and $\mathbf{x}$ are considered to be small.

Eliminating t from (13.7), the flow outside $x_1 = 0$ consists of arcs given by

$$(x_1, x_2, x_3) = \begin{cases} \left(\frac{1}{2s^+}(x_2^2 - x_{20}^2), x_2, x_{30} + \frac{\nu^+}{s^+}(x_{20} - x_2)\right) & \text{for } |x_2| < |x_{20}|, \\ \left(\frac{1}{2s^-}(x_3^2 - x_{30}^2), x_{20} + \frac{\nu^-}{s^-}(x_3 - x_{30}), x_3\right) & \text{for } |x_3| < |x_{30}|, \end{cases} \tag{13.8}$$

where each arc enters or leaves the discontinuity surface at coordinates $(0, x_{20}, x_{30})$ and either $\left(0, -x_{20}, x_{30} - 2\frac{\nu^+}{s^+}x_{20}\right)$ or $\left(0, x_{20} - 2\frac{\nu^-}{s^-}x_{30}, -x_{30}\right)$. These maps are oblique reflections along the directions $(0, -s^+, \nu^+)$ and $(0, \nu^-, s^-)$ about the fold lines. We return to these in Section 13.3.

The key point to note from Corollary 13.1, which completely describes the first three components of the n-dimensional system (13.5), is that the leading-order dynamics of $x_1(t), x_2(t), x_3(t)$ is independent of the variables $x_4, x_5, \ldots, x_n$. This allows us to form a classification in n dimensions based purely upon the dynamics of x_1, x_2, x_3.

In general the variables $x_4, x_5, \ldots, x_n$ do have a role to play, of course, because they enter into the values of the parameters $\nu^\pm$ and $\delta^2_{t\pm}\sigma(0)$, which depend on the point $\mathbf{x}_p$ at which the expression (13.5) is derived. In other words they are implicit in the coordinate transformation required to obtain (13.5). At any given point, the possible signs of the derivatives $\delta^2_{t\pm}\sigma(0)$ give one of three topological flavours of two-fold (Figure 13.2), and the values of the parameters $\nu^\pm$ divide these further into dynamical subclasses, explored in Sections 13.2.1–13.2.3. Different values thus determine neighbourhoods of different points $\mathbf{x} \in \mathbb{R}^n$ with different dynamics. The transition between such regions occurs where the non-degeneracy conditions in Definition 6.2 break down, at critical values of $\delta^2_{t\pm}\sigma$ and $\nu^\pm$, and this requires study of the higher-order terms labelled $\mathsf{O}(|\mathbf{x}|)$ in (13.5), something which must remain beyond our scope here.

The system outlined above comes in three main 'flavours' determined by the visibility of the folds, as shown in Figure 13.2. The details of the analysis behind these results lead us into deeper intrigues and will be the subject of Obsessions 2–4. It is useful, before proceeding, to draw together their key properties in Sections 13.2.1–13.2.3.

13.2.1 The Visible Two-Fold

If $\delta_{t+}\sigma(0) > 0 > \delta_{t-}\sigma(0)$, then the two-fold is the intersection of a pair of visible folds. The flow curves away from the discontinuity as seen in the left of Figure 13.4.

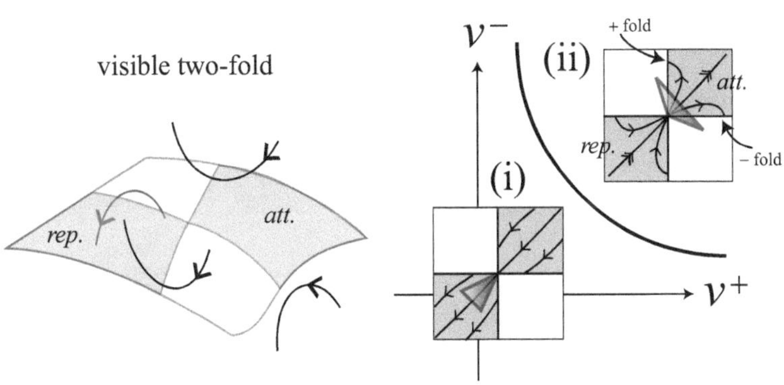

Fig. 13.4 The visible two-fold. Left: the discontinuity surface is partitioned into sliding (at) and repelling sliding (rep) regions (shaded) and crossing regions (unshaded), bounded by visible folds. Right: the $\nu^\pm$ parameter space, showing the transition curve $\nu^+\nu^- = 1$, $\nu^\pm > 0$, and inset, phase portraits of the sliding flow, with different cases: (i) simple canards if $\nu^+\nu^- < 1$ and/or $\nu^+ < 0$ and/or $\nu^- < 0$; (ii) robust faux canards if $\nu^\pm > 0$ and $\nu^+\nu^- > 1$. The set-valued velocity is illustrated by a shaded triangle at the singularity.

To leading order (13.5) becomes

$$\frac{d}{dt}\mathbf{x} = \begin{cases} (-x_2,\ -1,\ \nu^+) \text{ if } x_1 > 0, \\ (\ x_3\ ,\ \nu^-,\ -1\ ) \text{ if } x_1 < 0. \end{cases} \tag{13.9}$$

Locally trajectories will visit the discontinuity surface at most once, making the visible two-fold the simplest to understand. After any crossing or sliding contact on the discontinuity surface, any trajectory will leave the neighbourhood of the singularity. The sliding dynamics in the neighbourhood of the singularity is characterized by:

(i) if $\nu^+\nu^- > 1$ & $\nu^+ > 0$ & $\nu^- > 0$, the entire sliding flow consists of faux canards;
(ii) otherwise (if any of $\nu^+\nu^- < 1$ or $\nu^+ < 0$ or $\nu^- < 0$ hold), there exists a simple canard;

as illustrated in Figure 13.4 (right). These stem from study of the sliding flow, which follows in Obsession 3.

13.2.2 The Visible-Invisible Two-Fold

If $\delta_{t\pm}\sigma(0) > 0$, then the two-fold is the intersection of one visible fold and one invisible fold. To leading order (13.5) becomes

$$\frac{d}{dt}\mathbf{x} = \begin{cases} (-x_2, -1, \nu^+) \text{ if } x_1 > 0, \\ (\ x_3\ , \nu^-,\ 1\ ) \text{ if } x_1 < 0. \end{cases} \tag{13.10}$$

The invisible fold provides a return mechanism for the flow to the discontinuity surface, as seen in Figure 13.5(top left). The attracting and repelling sliding regions can contain canards and faux canards, robust or not, creating seven cases illustrated in the top right of Figure 13.5 and by the panels (i-vii), showing:

(i) No canards;
(ii) Robust faux canards;
(iii) Robust canards whose set-valued flow passes through only part of the crossing regions (shown dotted);
(iv) Robust canards whose set-valued flow passes through the full crossing regions (shown dotted);
(v) coexisting simple canards and faux canards, with the set-valued flow of the faux canard passing through the crossing regions;
(vi) Coexisting simple canards and faux canards, with the set-valued flow of the faux canard passing through the visible fold;
(vii) Coexisting simple canards and faux canards, with the set-valued flow of the faux canard passing through neither the crossing regions nor the visible fold.

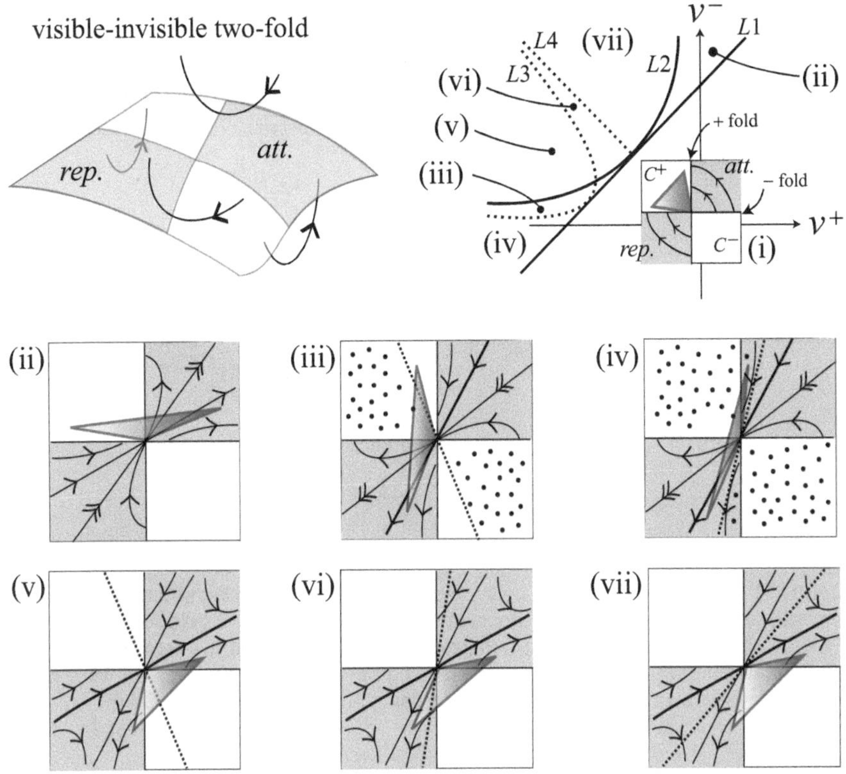

Fig. 13.5 The visible-invisible two-fold. Top-left: the sliding (*att*) and repelling sliding (*rep*) regions (shaded), and crossing regions (unshaded), are bounded by one visible and one invisible fold. Top-right: $\nu^{\pm}$ space separating regions with topologically different phase portraits of the sliding flow, shown inset/bottom in cases (i-vii).

Transitions between these topological classes occur along: the line $L1$ given by $\nu^+ - \nu^- = -2$; the curves $L2$ given by $\nu^+\nu^- = -1$ in $\nu^+ - \nu^- < -2$; the curve $L3$ given by $2\nu^-(\nu^+ + \nu^-) = -1$; and along the line $L4$ given by $\nu^+ = -\nu^-$.

These stem from study of the sliding flow which follows in Obsession 3, together with the arcs outside $\mathcal{D}$ given by (13.7b).

13.2.3 The Invisible Two-Fold

If $\delta_{t-}\sigma(0) > 0 > \delta_{t+}\sigma(0)$, then the two-fold is the intersection of a pair of invisible folds. This case so much captured the imagination that it was dubbed the *Teixeira singularity*, after M. A. Teixeira who fuelled fascination with it in [205]. To leading order (13.5) becomes

$$\frac{d}{dt}\mathbf{x} = \begin{cases} (-x_2,\ 1,\ \nu^+,\ 0) \text{ if } x_1 > 0, \\ (\ x_3\ ,\nu^-,\ 1\ ,\ 0) \text{ if } x_1 < 0. \end{cases} \tag{13.11}$$

The invisible two-fold, see Figure 13.6(left), leads to particularly interesting local dynamics, as each trajectory can return to the discontinuity repeatedly.

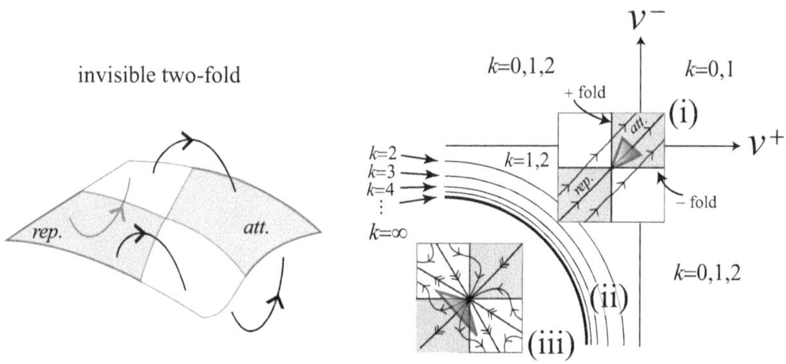

Fig. 13.6 The invisible two-fold. Left: the sliding (att) and repelling sliding (rep) regions (shaded), and crossing regions (unshaded), are bounded by invisible folds. Right: the $\nu^{\pm}$ space showing (inset) phase portraits of the sliding flow and the crossing map, with different cases: (i)-(ii) faux canards and crossing numbers k if $\nu^+\nu^- < 1$ or if ν^+ or ν^- is positive; (iii) robust canards and infinite crossing numbers if $\nu^{\pm} < 0 < \nu^+\nu^- - 1$.

In three-dimensional systems, (13.11) belongs to one of three dynamical classes. Local to the invisible two-fold:

(i) if $\nu^+ > 0$ or $\nu^- > 0$, there exists a simple faux canard, and between ejection from the repelling sliding region to impact on the sliding region, the flow crosses the discontinuity surface at most once from $\sigma < 0$ to $\sigma > 0$ if $\nu^+ > 0$, and at most once from $\sigma > 0$ to $\sigma < 0$ if $\nu^- > 0$;

(ii) if $\nu^{\pm} < 0 < 1 - \nu^+\nu^-$, there exists a simple faux canard, and between ejection from the repelling sliding region to impact on the sliding region the flow crosses the discontinuity surface at least once;

(iii) if $\nu^{\pm} < 0 < \nu^+\nu^- - 1$, the entire sliding flow consists of canards, and the crossing flow traverses the discontinuity repeatedly as it tends asymptotically towards/away from branches of a diabolical invariant manifold, whose two (piecewise smooth) cones are unstable/stable manifolds of the singularity, respectively.

Again these stem from study of the sliding flow, which follows in Obsession 3, and the now much more involved study of the crossing flow which follows in Obsession 2. We shall see in Section 13.3 that we can describe the number of crossings in case (iii) very precisely.

13.2.4 Geometry of the Angular Jump Parameter $\nu^+\nu^-$

A quantity we see recurring above, and clearly the key parameter determining the behaviour at any two-fold singularity, is the product of the parameters ν^+ and ν^- defined in (13.3). It has a simple geometrical interpretation: it quantifies the jump in the vector field at the singularity.

To show this, from the definition of $\nu^\pm$ in (13.3), we have (omitting the argument $\mathbf{x}$)

$$\nu^+\nu^- = -\frac{(\delta_{t+}\delta_{t-}\sigma)(\delta_{t-}\delta_{t+}\sigma)}{|\delta_{t+}^2\sigma\delta_{t-}^2\sigma|} .$$

Using $\mathbf{f}^\pm = \delta_{t\pm}\mathbf{x}$ and $\delta_{t\pm} = \delta_{t\pm}\mathbf{x}\cdot\delta_{\mathbf{x}} = \mathbf{f}^\pm\cdot\delta_{\mathbf{x}}$, we write

$$\begin{aligned}\nu^+\nu^- &= -\frac{(\delta_{t+}\delta_{t-}\sigma)(\delta_{t-}\delta_{t+}\sigma)}{|(\delta_{t+}\delta_{t+}\sigma)(\delta_{t-}\delta_{t-}\sigma)|}\\ &:= -Q_-^+Q_+^-/|Q_+^+Q_-^-|\end{aligned} \tag{13.12}$$

where $Q_j^i = \mathbf{f}^i\cdot(\delta_{\mathbf{x}}\delta_{tj}\sigma)$, with i and j denoting the labels $+$ or $-$. This is not a very illuminating expression yet, but let us turn these into angles expressing the directions of the constituent vector fields.

Let us measure angles from the folds to the vector directions $\mathbf{f}^\pm$, in the plane spanned by $\mathbf{f}^+$ and $\mathbf{f}^-$, and identify the angle of rotation from $\mathbf{f}^-$ to $\mathbf{f}^+$ as positive. This is possible as this plane typically intersects the folds.

We first need a reference vector that lies along each fold, while also lying in the plane spanned by $\mathbf{f}^+$ and $\mathbf{f}^-$. This is given by

$$\mathbf{e}_j = Q_j^-\mathbf{f}^+ - Q_j^+\mathbf{f}^-, \tag{13.13}$$

which lies in the fold set $\delta_{\mathbf{x}}\delta_{tj}\sigma = 0$ since a short calculation shows $\mathbf{e}_j\cdot\delta_{\mathbf{x}}\delta_{tj}\sigma = Q_j^-Q_j^+ - Q_j^+Q_j^- = 0$.

Then let θ_j^i denote the angle measured from the fold set $\delta_{\mathbf{x}}\delta_{tj}\sigma = 0$ to the vector $\mathbf{f}^i$. This satisfies

$$\cos\theta_j^i = (\mathbf{f}^i\cdot\mathbf{e}_j)/|\mathbf{f}^i||\mathbf{e}_j|. \tag{13.14}$$

In terms of these quantities θ_j^i, the angle between the folds is the difference $\phi = \theta_+^+ - \theta_-^+ = \theta_+^- - \theta_-^-$, while the angle between $\mathbf{f}^+$ and $\mathbf{f}^-$ is the difference $\psi = \theta_+^+ - \theta_+^- = \theta_-^+ - \theta_-^-$ (see Figure 13.7).

Finding the required quantities Q_j^i for (13.12) in terms of the angles θ_j^i is now just a calculation. First, take the ratio

$$\frac{\cos\theta_j^+}{\cos\theta_j^-} = \frac{(\mathbf{f}^+\cdot\mathbf{e}_j)/|\mathbf{f}^+||\mathbf{e}_j|}{(\mathbf{f}^-\cdot\mathbf{e}_j)/|\mathbf{f}^-||\mathbf{e}_j|} = \frac{Q_j^-|\mathbf{f}^+|^2 - Q_j^+\mathbf{f}^+\cdot\mathbf{f}^-}{Q_j^-\mathbf{f}^-\cdot\mathbf{f}^+ - Q_j^+|\mathbf{f}^-|^2}\cdot\frac{|\mathbf{f}^-|}{|\mathbf{f}^+|}$$

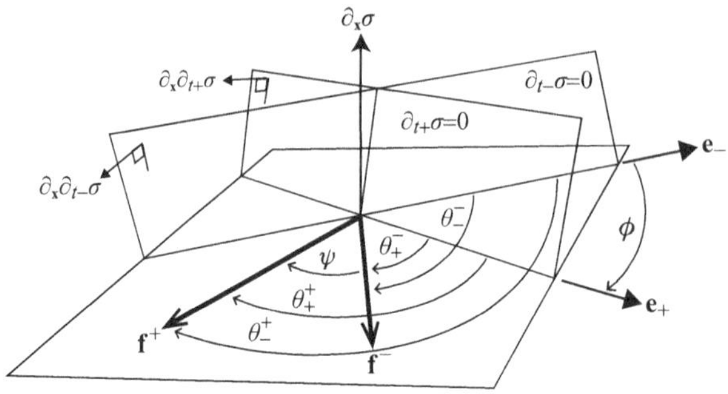

Fig. 13.7 Local geometry behind the two-fold *angular jump* parameter $\nu^+\nu^-$ as given by (13.17b). The angle θ_j^i between the vector $\mathbf{f}^i$ and the fold $\delta_{tj}\sigma = 0$, the angle ϕ between the folds, and the angle ψ between $\mathbf{f}^+$ and $\mathbf{f}^-$. All angles are measured in the plane spanned by $\mathbf{f}^\pm$, which lies inside (and in three dimensions is exactly) the tangent plane to the discontinuity surface $\sigma = 0$.

which, substituting $\mathbf{f}^+ \cdot \mathbf{f}^- = |\mathbf{f}^+||\mathbf{f}^-|\cos\psi$, rearranges to give

$$\frac{|\mathbf{f}^+|Q_j^-}{|\mathbf{f}^-|Q_j^+} = \frac{\cos\psi\cos\theta_j^- - \cos\theta_j^+}{\cos\theta_j^- - \cos\psi\cos\theta_j^+}.$$

Substituting the relations $\psi = \theta_+^+ - \theta_+^- = \theta_-^+ - \theta_-^-$ gives

$$\frac{|\mathbf{f}^+|Q_j^-}{|\mathbf{f}^-|Q_j^+} = \frac{\cos\psi\cos\theta_j^- - \cos(\psi+\theta_j^-)}{\cos(\psi-\theta_j^+) - \cos\psi\cos\theta_j^+} = \frac{\sin\theta_j^-}{\sin\theta_j^+}\,.$$

From this we can find the ratios that appear in (13.12),

$$\frac{Q_+^-Q_-^+}{Q_+^+Q_-^-} = \frac{\sin\theta_+^-\sin\theta_-^+}{\sin\theta_+^+\sin\theta_-^-}\,.$$

More usefully, through the relations $\phi = \theta_+^+ - \theta_-^+ = \theta_+^- - \theta_-^-$, we can write

$$\frac{Q_+^-Q_-^+}{Q_+^+Q_-^-} = \frac{\sin\theta_+^-\sin(\theta_+^+-\phi)}{\sin\theta_+^+\sin(\theta_+^- -\phi)} = \frac{\cot\phi - \cot\theta_+^+}{\cot\phi - \cot\theta_+^-} \tag{13.15}$$

or

$$\frac{Q_+^-Q_-^+}{Q_+^+Q_-^-} = \frac{\sin(\theta_-^- +\phi)\sin\theta_-^+}{\sin(\theta_-^+ +\phi)\sin\theta_-^-} = \frac{\cot\phi + \cot\theta_-^-}{\cot\phi + \cot\theta_-^+}\,. \tag{13.16}$$

Finally then we have

$$\nu^+\nu^- = \alpha\frac{\sin\theta_+^-\sin\theta_-^+}{\sin\theta_+^+\sin\theta_-^-}\,, \tag{13.17a}$$

or

$$\nu^+\nu^- = \alpha \begin{cases} \dfrac{\cot\phi - \cot\theta_+^+}{\cot\phi - \cot\theta_+^-} & \text{measuring angles to the `+' fold,} \\ \dfrac{\cot\phi + \cot\theta_-^-}{\cot\phi + \cot\theta_-^+} & \text{measuring angles to the `}-\text{' fold,} \end{cases} \tag{13.17b}$$

where

$$\alpha = -\operatorname{sign}[(\delta_{t+}^2\sigma)(\delta_{t-}^2\sigma)]$$

is $+1$ if both folds are visible or both are invisible and is -1 if they are mixed. The term ϕ is the angle between the folds, and θ_j^i is the angle of $\mathbf{f}^i$ from the 'j' fold, measured in the plane spanned by $\mathbf{f}^+$ and $\mathbf{f}^-$.

These are the general expressions. If the folds are orthogonal such that $\phi = \frac{\pi}{2}$ then

$$\nu^+\nu^- = \alpha\frac{\tan\theta_-^+}{\tan\theta_-^-} = \alpha\frac{\tan\theta_+^-}{\tan\theta_+^+}\,, \tag{13.18}$$

so $\nu^+\nu^-$ is simple the ratio of the tangents of the constituent vector field directions, i.e. it quantifies the angular jump in the flow that occurs between $\mathbf{f}^+$ and $\mathbf{f}^-$.

13.3 Obsession 2: The Nonsmooth Diabolo

The obsessions of the two-fold really took hold with the study of the crossing map in [205], its intricacy having been somewhat obscured in [71]. For this one obsession only, the intrigue centres squarely on the invisible two-fold, or *Teixeira singularity*. The reason is simple: in the other cases, after one or two crossings, the flow departs the local neighbourhood of the singularity, but at an invisible two-fold, it can wind around and around finitely or infinitely many times. The question is how many times does it wind around, and does it spiral into or out from the singularity as it does so?

On a sufficiently small neighbourhood, we can consider just the truncated prototype of the invisible two-fold, from (13.11), which we recall is

$$(\dot{x}_1, \dot{x}_2, \dot{x}_3) = \begin{cases} (-x_2, 1, \nu^+) & \text{if } x_1 > 0 \\ (\ x_3, \nu^-, 1\ ) & \text{if } x_1 < 0 \end{cases}.$$

13.3.1 First Return Map: The Skewed Reflection

The flow crosses through the discontinuity surface when the components of $\mathbf{f}^\pm$ normal to $x_1 = 0$ have the same sign. The crossing regions in the prototype are given by

$$\mathcal{D}_{C^+} = \left\{(x_1, x_2, x_3) \in \mathbb{R}^3 : \; x_1 = 0, \; x_2 < 0 < x_3\right\},$$
$$\mathcal{D}_{C^-} = \left\{(x_1, x_2, x_3) \in \mathbb{R}^3 : \; x_1 = 0, \; x_3 < 0 < x_2\right\},$$

where the flow crosses in the positive or negative x_1 directions, respectively.

Letting $\mathbf{y}$ denote a two-dimensional column vector with components (y, z), the arcs from Corollary 13.1 define a pair of maps, $\Phi^{\pm}$, that take any point $\mathbf{y}_m$ to its first return coordinate $\mathbf{y}_{m+1}$ on $x_1 = 0$, given by

$$\begin{aligned} \mathbf{y}_{2m+1} &= \Phi^+(\mathbf{y}_{2m}) &&= \underline{\underline{B}}^+\mathbf{y}_{2m}, && \underline{\underline{B}}^+ = \begin{pmatrix} -1 & 0 \\ -2\nu^+ & 1 \end{pmatrix}, \\ \mathbf{y}_{2m} &= \Phi^-(\mathbf{y}_{2m-1}) &&= \underline{\underline{B}}^-\mathbf{y}_{2m-1}, && \underline{\underline{B}}^- = \begin{pmatrix} 1 & -2\nu^- \\ 0 & -1 \end{pmatrix}, \end{aligned} \tag{13.19}$$

where even iterates $\mathbf{y}_{2m}$ lie in $\mathcal{D}_{C^+}$ and odd iterates $\mathbf{y}_{2m+1}$ lie in $\mathcal{D}_{C^+}$ for $m \in \mathbb{N}$. We employ this convention throughout this section, as illustrated in Figure 13.8.

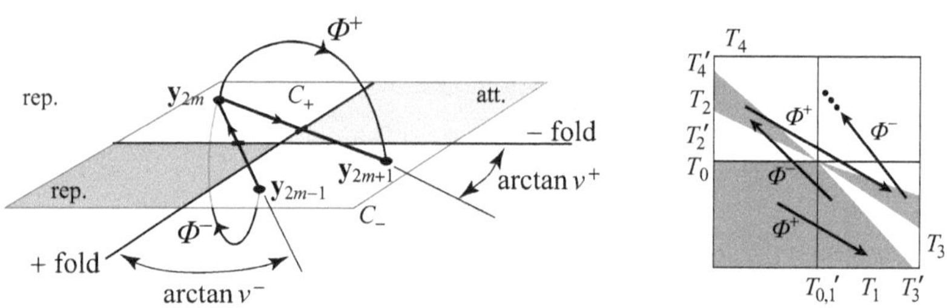

Fig. 13.8 Local mapping: In the parabolic approximation, points are reflected by the maps Φ^+ acting through $\sigma > 0$ and Φ^- acting through $\sigma < 0$. In this example the point $\mathbf{x} = (0, \mathbf{y}_{2m-1})$ is reflected obliquely by Φ^- to $\mathbf{x} = (0, \mathbf{y}_{2m})$, and then by Φ^+ to $\mathbf{x} = (0, \mathbf{y}_{2m+1})$, remaining in the crossing regions. The sliding regions (rep/att) are shaded. On the right an example of the first three mappings of the repelling sliding region are shown on $\mathcal{D}$.

The maps $\Phi^{\pm}$ obey

$$\begin{aligned} \Phi^+ &: \mathcal{D}_{att} \cup \mathcal{D}_{C^+} \mapsto \mathcal{D}_{att} \cup \mathcal{D}_{C^-}, \\ \Phi^- &: \mathcal{D}_{att} \cup \mathcal{D}_{C^-} \mapsto \mathcal{D}_{att} \cup \mathcal{D}_{C^+}. \end{aligned} \tag{13.20}$$

The domain of $\Phi^{\pm}$ and range of $\Phi^{\mp}$ overlap on $\mathcal{D}_{C^{\pm}}$, on which we can compose Φ^+ and Φ^- to define second return maps

$$\begin{aligned} \mathbf{y}_{2m+2} &= \Phi^- \circ \Phi^+(\mathbf{y}_{2m}) = \underline{\underline{A}}^+\mathbf{y}_{2m}, && \underline{\underline{A}}^+ = \underline{\underline{B}}^-\underline{\underline{B}}^+, \\ \mathbf{y}_{2m+1} &= \Phi^+ \circ \Phi^-(\mathbf{y}_{2m-1}) = \underline{\underline{A}}^-\mathbf{y}_{2m-1}, && \underline{\underline{A}}^- = \underline{\underline{B}}^+\underline{\underline{B}}^-, \end{aligned} \tag{13.21}$$

such that

$$\begin{aligned} \Phi^- \circ \Phi^+ &: \mathcal{D}_{rep} \cup \mathcal{D}_{C^+} \mapsto \mathcal{D}_{att} \cup \mathcal{D}_{C^+}, \\ \Phi^+ \circ \Phi^- &: \mathcal{D}_{rep} \cup \mathcal{D}_{C^-} \mapsto \mathcal{D}_{att} \cup \mathcal{D}_{C^-}. \end{aligned} \tag{13.22}$$

The maps $\Phi^{\pm}$ are involutions, meaning $\Phi^{+} \circ \Phi^{+}(\mathbf{x}) = \Phi^{-} \circ \Phi^{-}(\mathbf{x}) = \mathbf{x}$, and as a result

$$(\underline{\underline{B}}^{+})^2 = (\underline{\underline{B}}^{-})^2 = \underline{\underline{I}} \quad \text{and} \quad \underline{\underline{A}}^{+} = (\underline{\underline{A}}^{-})^{-1} \tag{13.23}$$

where $\underline{\underline{I}}$ denotes the 2×2 identity matrix.

The solutions to the difference equation (13.21) are obviously

$$\begin{aligned} \mathbf{y}_{2m} &= (\Phi^{-} \circ \Phi^{+})^m(\mathbf{y}_0) &= (\underline{\underline{A}}^{+})^m \mathbf{y}_0 \,, \\ \mathbf{y}_{2m+1} &= (\Phi^{+} \circ \Phi^{-})^m(\mathbf{y}_1) &= (\underline{\underline{A}}^{-})^m \mathbf{y}_1 \,, \end{aligned} \tag{13.24}$$

and a little trigonometry using a substitution $\nu^{+}\nu^{-} = \cos^2\theta$ provides

$$(\underline{\underline{A}}^{\pm})^m = \frac{\sin[2m\theta]}{\sin 2\theta}\underline{\underline{A}}^{\pm} - \frac{\sin[2(m-1)\theta]}{\sin 2\theta}\underline{\underline{I}} \,. \tag{13.25}$$

Since this is the retelling of an obsession, a brief aside on the story of its unfolding is in order. It is tempting to consider these two-dimensional maps in their Cartesian form above, as pursued in [71, 205]. It is not clear how to apply principles of asymptotic stability from smooth dynamical systems to these piecewise maps, however, and this led to some delay in their solution. For example, Teixeira [205] remarked that the second-return map $x_m \mapsto x_{m+2}$ (the map $\Phi^{+} \circ \Phi^{-}$ or $\Phi^{-} \circ \Phi^{+}$) is non-hyperbolic if $0 < \nu^{+}\nu^{-} < 1$. It is unclear whether this condition says anything about the stability of the system because orbits only cross $\mathcal{D}$ over a finite stretch of time before entering the sliding region. The two-fold's stability remained a major unsolved problem by the time an Intensive Research Program in *Complex Nonsmooth Dynamics Systems* was held at the Centre de Recerca Matemàtica in 2007. I joined the game shortly after and with Alessandro Colombo began pursuing solely the polar form outlined below. This moved things forward considerably, but turned out largely to be retracing Filippov's steps of more than 20 years earlier. It was with a young student, Soledad Fernández Garcia, that we finally derived a tractable expression for the mth iterate map leading to the trigonometric form (13.25), which fully unlocked the crossing problem, and only then could the rotation map be solved in its entirety.

13.3.2 The Rotation Map

Recalling that we use components $\mathbf{x} = (0, \mathbf{y})$ on $\mathcal{D}$ where $\mathbf{y} = (x_2, x_3)$, then given an iterate $\mathbf{y}_i$ with components $(x_{2,i}, x_{3,i})$ (this is not the cleanest notation but we need it only briefly), we define its tangent relative to the folds by

$$T_{2m} = \frac{x_{3,2m}}{x_{2,2m}}, \qquad T_{2m+1} = \frac{x_{2,2m+1}}{x_{3,2m+1}}, \qquad m \in \mathbb{N}, \tag{13.26}$$

using the convention that even iterates $\mathbf{x}_{2m}$ lie in $\mathcal{D}_{C^+}$ and odd iterates $\mathbf{x}_{2m+1}$ lie in $\mathcal{D}_{C^-}$. Then T_m is positive for points $\mathbf{y}_m$ in the sliding regions ($x_2x_3 > 0$), negative in the crossing regions ($x_2x_3 < 0$), and either zero or infinite on the folds ($x_2x_3 = 0$). We thus denote the sets of tangent values T of points in $\mathcal{D}_{rep}, \mathcal{D}_{att}$, and $\mathcal{D}_{C^\pm}$, by

$$\mathcal{T}_{s.rep} = \mathcal{T}_{s.att} = (0,\infty), \qquad \mathcal{T}_{C^+} = \mathcal{T}_{C^-} = (-\infty, 0). \tag{13.27}$$

Substituting (13.21) into (13.26), we find that the tangents T_i map as

$$T_{2m+1} = \frac{1}{2\nu^+ - T_{2m}}\,, \qquad T_{2m} = \frac{1}{2\nu^- - T_{2m-1}}\,. \tag{13.28}$$

An iterate of these maps to the crossing regions if it satisfies

$$2\nu^+ < T_{2m} < 0\,, \qquad 2\nu^- < T_{2m-1} < 0\,. \tag{13.29}$$

When these hold we can apply the alternate map ($\Phi^\mp$ if we previously applied $\Phi^\pm$) to obtain the second return. Writing

$$T_{2m} = \nu^+\tau_{2m} \qquad \text{and} \qquad T_{2m-1} = \nu^-\tau_{2m-1}\,,$$

these second return maps are a pair of Möbius transformations

$$\tau_{i+2} = \frac{\tau_i - 2}{2\nu^+\nu^-(\tau_i - 2) + 1}\,, \qquad i = 2m \text{ or } 2m-1\,, \tag{13.30}$$

which can be solved (directly or using (13.26) and (13.24)) to find

$$\tau_{\{2m,2m+1\}} = \frac{1 - \tau_{\{0,1\}}G_m(\theta)}{\nu^+\nu^-(2 - \tau_{\{0,1\}}) - G_m(\theta)}\,, \tag{13.31}$$

using again the quantity $\theta = \arccos\sqrt{\nu^+\nu^-}$ as in (13.25) and introducing the function

$$G_m(\theta) = \frac{\sin[(2m-1)\theta]}{\sin[2m\theta]\sec\theta}\,. \tag{13.32}$$

For this solution to exist, all intermediate iterates must lie in the crossing regions, satisfying (13.29). We call $\tau_i \mapsto \tau_{i+2}$ the *rotation map*, telling us how points $\mathbf{y}_i \in \mathcal{D}$ are rotated around the singularity by the flow.

By expressing the radial coordinate as $R = \sqrt{x_2^2 + x_3^2}$ and iterating twice, we find the radial maps that accompany (13.30),

$$\begin{aligned} R_{2m+2}^2 &= \tfrac{1+T_{2m+2}^2}{1+T_{2m}^2} \cdot \tfrac{2\nu^+ - T_{2m}}{T_{2m+2}} R_{2m}^2 &&:= P_{2m+2}R_{2m}^2\,, \\ R_{2m+1}^2 &= \tfrac{1+T_{2m+1}^2}{1+T_{2m-1}^2} \cdot \tfrac{2\nu^- - T_{2m-1}}{T_{2m+1}} R_{2m-1}^2 &&:= P_{2m+1}R_{2m-1}^2\,, \end{aligned} \tag{13.33}$$

and these are easily solved as

$$\begin{aligned} R^2_{2m+2} &= P_{2m+2}P_{2m}\dots P_2R_0\ , \\ R^2_{2m+1} &= P_{2m+1}P_{2m-1}\dots P_3R^2_1\ . \end{aligned} \tag{13.34}$$

Thus we have the polar maps in the plane of $\mathcal{D}$, $(R_0, T_0) \mapsto (R_{2m+2}, T_{2m+2})$ and $(R_1, T_1) \mapsto (R_{2m+1}, T_{2m+1})$.

The various expressions above make two properties of the crossing maps apparent. First, the maps are similar up to the transformations

$$(x_2, x_3, \nu^{\pm}) \leftrightarrow (x_3, x_2, \nu^{\mp}) \qquad \text{or} \qquad (T_{2m}, \nu^{\pm}) \leftrightarrow (T_{2m-1}, \nu^{\mp})\ . \tag{13.35}$$

Second, Φ^+ and Φ^- map straight lines through the singularity to each other. It is therefore sufficient to study the orientation T_i of points relative to the folds $x_2 = 0$ and $x_3 = 0$ as the matrices $\underline{\underline{B}}^{\pm}$ rotate them around the singularity. Any point on the line $\mathbf{y}_{2m} = R(\cos\theta_{2m}, \sin\theta_{2m})$ for variable $R \in \mathbb{R}$ maps to a point on another line, $\mathbf{y}_{2m+1} = R'(\cos[\theta_{2m} + \alpha(\theta_{2m})], \sin[\theta_{2m} + \alpha(\theta_{2m})])$ for some function α. So to study the images of $\mathcal{D}_{att}$ and $\mathcal{D}_{rep}$ under successive iterations of $\Phi^{\pm}$, we need only consider the rotation of their boundaries, which are straight lines through the origin as illustrated in Figure 13.8. This rotation constitutes the angular behaviour of the map.

13.3.3 Number of Crossings

The number of times any orbit winds around the invisible two-fold in its vicinity is given by the following. We state the main results and then bring together their proofs (originally from [119, 69]).

Theorem 13.1. *Number of crossings. In the system* (13.11)*, between visits to the sliding regions:*

- *(i.a) if $\nu^+ > 0$, every orbit crosses $\mathcal{D}$ at most once, and does so in $\mathcal{D}_{C+}$ from $x_1 < 0$ to $x_1 > 0$;*
- *(i.b) if $\nu^- > 0$, every orbit crosses $\mathcal{D}$ at most once, and does so in $\mathcal{D}_{C-}$ from $x_1 > 0$ to $x_1 < 0$;*
- *(i.c) If $\nu^+\nu^- < 1$ or $\nu^+ > 0$ or $\nu^- > 0$, every orbit crosses $\mathcal{D}$ a finite number of times in either direction;*
- *(ii) if $0 < \nu^+\nu^- < 1$ and $\nu^{\pm} < 0$, every orbit crosses $\mathcal{D}$ at least once;*
- *(iii) If $\nu^+\nu^- > 1$ and $\nu^{\pm} < 0$, every orbit crosses $\mathcal{D}$ an infinite number of times. There exist a pair of invariant surfaces that meet at the singularity. One of the invariant surfaces is asymptotically attractive and encloses the repelling sliding region $\mathcal{D}_{rep}$ within the domain of repulsion of the singularity; the other invariant surface is asymptotically repulsive and encloses the attracting sliding region $\mathcal{D}_{att}$ within the domain of attraction of the singularity.*

We can explore (ii) in considerably more detail too.

Lemma 13.1. *If $\nu^+\nu^- = \cos^2\frac{\pi}{k+1}$ where $k \geq 2$ is an integer and $\nu^\pm$ are negative, then between visits to the sliding regions, any orbit crosses the discontinuity surface exactly k times.*

Corollary 13.2. *If $\nu^+\nu^- = \cos^2\frac{\pi}{r+1}$ and $r > 1$ is not an integer and $\nu^\pm$ are negative, then between visits to the sliding regions:*

1. *Any orbit crosses the discontinuity surface either k or $k+1$ times, where k and $k+1$ are the integers either side of r; and*
2. *The number of crossings made by a solution passing through a point $(0, x_{20}, x_{30})$ changes across four lines, at $x_{20} = 0$ (the '+' fold), $x_{30} = 0$ (the '−' fold), $x_{30} = x_{20}\Gamma^+_{(k+1)/2}$, and $x_{20} = x_{30}\Gamma^-_{(k+1)/2}$, in terms of functions*

$$\Gamma^\pm_m = \nu^\pm / G_m(\arccos\sqrt{\nu^+\nu^-})$$

and

$$G_m(\theta) = \sin([2m-1]\theta)\cos\theta/\sin(2m\theta)\ .$$

These are illustrated in Figure 13.9.

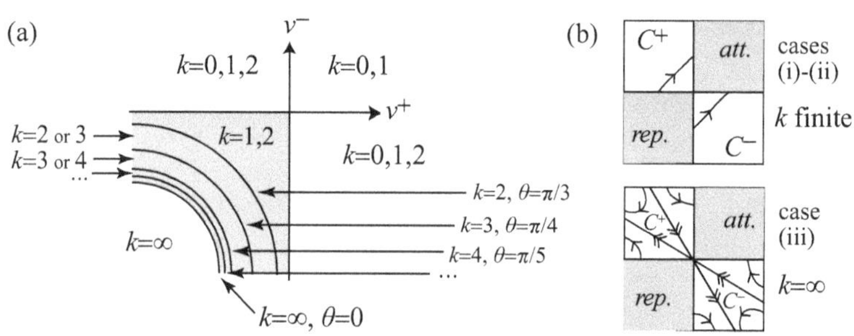

Fig. 13.9 Bifurcation diagram of the invisible two-fold, showing (a) k, the number of times any orbit crosses the discontinuity surface for different $\nu^\pm$ and (b) the corresponding phase portraits of the crossing maps (in crossing regions $\mathcal{D}_{C^\pm}$) and sliding flow (in sliding regions $\mathcal{D}_{att}$ and $\mathcal{D}_{rep}$, shaded). On each of the curves $\nu^+\nu^- = \cos^2\pi/(k+1)$ for integer k, the entire flow maps $\mathcal{D}_{rep}$ onto $\mathcal{D}_{att}$ via k crossings. In between, the number of crossings of the two bounding curves is permitted.

The lemma and its corollary mean that if $\nu^+\nu^- = \cos^2\frac{\pi}{k+1}$ where $k \geq 2$ is an integer, then the flow crosses the discontinuity exactly k times between initial ejection from the repelling sliding region to ultimate impact on the sliding region. Moreover, if $\nu^\pm < 0$ and $\nu^+\nu^- = \cos^2\frac{\pi}{r+1}$ but $r > 1$ is not an integer, then between ejection from the repelling sliding region to impact on the sliding region, different parts of the flow cross the discontinuity either k or $k+1$ times, where k and $k+1$ are the integers either side of r.

As $k \to \infty$, the transition curves $\nu^+\nu^- = \cos^2 \frac{\pi}{k+1}$ accumulate onto the particular curve $\nu^+\nu^- = 1$. On this curve the non-degeneracy conditions are violated (as in Lemma 13.2); therefore the leading-order expression (13.11) is no longer valid. The bifurcation that occurs along this line is the subject of Section 13.3.4.

Theorem 13.1 is proven using the rotation map T_{2m} (or T_{2m-1}). In its nondimensionalized form (13.30), the rotation map $\tau_i \mapsto \tau_{i+2}$ is shown in Figure 13.10. If $\nu^+\nu^- > 1$, the map has two fixed points, one at τ_* with eigenvalue $(1 - 2/\tau_*) - 2 < 1$ which is therefore asymptotically attracting and one at $1/(\tau_*\nu^+\nu^-)$ with eigenvalue $(1 - 2\nu^+\nu^-\tau_*) - 2 > 1$ which is therefore asymptotically repelling, where

$$\tau_* = 1 - \sqrt{1 - 1/\nu^+\nu^-}\ . \tag{13.36}$$

Note that $\tau_* > 0$ and $1/(\tau_*\nu^+\nu^-) > 0$. Then we have the following.

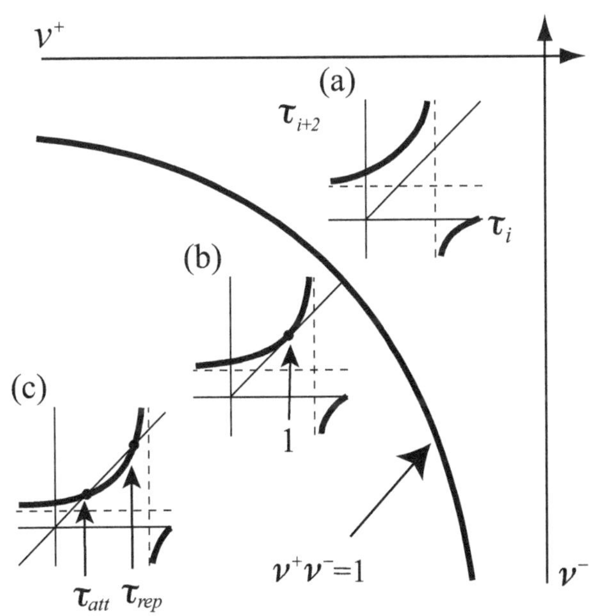

Fig. 13.10 Tangent mapping and bifurcation diagram: (a) no invariant manifolds for $\nu^+\nu^- = 1$, (b) bifurcation along $\nu^+\nu^- = 1$, (c) two fixed points $\tau_{att} = \tau_*$ and $\tau_{rep} = 1/(\tau_*\nu^+\nu^-)$ in $\nu^+\nu^- > 1$. $\tau_i = 0$ is the boundary of $\mathcal{D}_{rep}$. Lines with $\tau_i = 2 - 1/(2\nu^+\nu^-)$ map to the τ_{i+2} graph asymptotes (dashed) which are the boundaries of $\mathcal{D}_{att}$. The bound $\tau_i < 2$ ensures the existence of the intermediate step τ_{i+1}.

Proof of Theorem 13.1 [119].

(i-ii) If $\nu^+\nu^- > 1$ and $\nu^\pm > 0$, then from (13.36) the equilibria of the T_i maps are $\nu^+\tau_* > 0$ and $\nu^\pm/(\tau_*\nu^+\nu^-) > 0$, which lie in the attracting or repelling sliding regions, so the equilibria are outside of the range of the map. If $\nu^+\nu^- < 1$, there are no real-valued equilibria. In either case there are then no admissible limit points (i.e. in the crossing regions), so all trajectories intersect the attracting and repelling sliding regions after finitely many iterations.

(i.a) If $\nu^+ > 0$, (13.29) implies that any T_{2m} is mapped to $T_{2m+1} > 0$, a termination point in the sliding region. Therefore, there is at most one crossing point T_{2m} in the region $x_2 < 0 < x_3$, where orbits cross from $x_1 < 0$ to $x_1 > 0$.

(i.b) If $\nu^- > 0$, (13.29) implies that any T_{2m-1} is mapped to $T_{2m} > 0$, a termination point in the sliding region. Therefore, there is at most one crossing point T_{2m-1} in the region $x_3 < 0 < x_2$, where orbits cross from $x_1 > 0$ to $x_1 < 0$.

(ii) If $\nu^\pm < 0 < 1 - \nu^+\nu^-$, then for any $T_m > 0$ we have from the maps above that $T_{m+1} = \frac{1}{2\nu^\pm - T_m} = \frac{-1}{2|\nu^\pm|+|T_m|} < 0$ for ν^+ if m is even or ν^- if m is odd. Therefore, an iterate T_m exists for all orbits; thus there always exists at least one crossing point.

(iii) If $\nu^+\nu^- > 1$ and $\nu^\pm < 0$, the equilibria $\nu^\pm\tau_* < 0$ and $\nu^\pm/(\tau_*\nu^+\nu^-) < 0$, of the τ_m map, lie in the crossing regions. These maps are monotonic; therefore all trajectories tend asymptotically towards the equilibria either in forward or in reverse time and thus cross $\mathcal{D}$ an infinite number of times. The smooth segments of orbits starting and ending at crossing points along the $\{T_m, T_{m+1}\}$ directions thus form invariant surfaces. The surfaces intersect $\mathcal{D}$ along lines through the singularity given by $x_1 = l(1, \nu^+\tau_*)$ and $x_1 = l(1, \nu^+/(\tau_*\nu^+\nu^-))$ for constant $l \in \mathbb{R}$.

□

Lemma 13.1 is proven by considering how $\mathcal{D}_{rep}$ maps under the action of Φ^+ and Φ^-. Without loss of generality, we apply Φ^+ to a point $\mathbf{y}_0 \in \mathcal{D}_{rep}$ (corresponding results for Φ^- applied to $\mathbf{y}_0 \in \mathcal{D}_{rep}$ are given by the similarity transformation (13.35)). The subsequent orbit consists of crossing points $\mathbf{y}_1, \mathbf{y}_3, \ldots \mathbf{y}_{m-1} \in \mathcal{D}_{C^-}$ and $\mathbf{y}_2, \mathbf{y}_4, \ldots \mathbf{y}_{m-2} \in \mathcal{D}_{C^+}$ for some integer m, following the convention in [119] that even iterates $\mathbf{y}_{2m}$ lie in the domain of Φ^+, while odd iterates $\mathbf{y}_{2m+1}$ lie in the domain of Φ^-. We then have two cases: either $\mathbf{y}_{2m} \in \mathcal{D}_{att}$, giving an orbit with an odd number of crossings $k = 2m - 1$, or $\mathbf{y}_{2m} \in \mathcal{D}_{C^+}$ and $\mathbf{y}_{2m+1} \in \mathcal{D}_{att}$, giving an orbit with an even number of crossings $k = 2m$.

We find that, for certain values of the product $\nu^+\nu^-$, the boundaries of $\mathcal{D}_{rep}$ are mapped exactly onto the boundaries of $\mathcal{D}_{att}$ and take the same number of crossings (the number of iterates in $\mathcal{D}_{C^\pm}$) to do so. By implication since the maps $\Phi^{\pm}$ are linear, all points in $\mathcal{D}_{rep}$ are then mapped into $\mathcal{D}_{att}$ via the same number of crossings.

Proof of Lemma 13.1 [69]. Let $\nu^+\nu^- = \cos^2\frac{\pi}{k+1}$ where k is an integer, and let $\pi_k = \frac{\pi}{k+1}$.

If $\mathbf{y}_0$ lies on the boundary of $\mathcal{D}_{rep}$, then $T_0 = 0$ or $T_0 = \infty$, and then a little algebra gives the tangent of a subsequent point $\mathbf{y}_{2m}$ as

$$\begin{aligned} T_{2m}|_{T_0=0} &= \frac{\sin[2m\pi_k]\cos\pi_k}{\sin[(2m+1)\pi_k]\nu^-} = \frac{1}{\nu^-}G_{m+\frac{1}{2}}(\pi_k), \\ T_{2m}|_{T_0=\infty} &= \frac{\sin[(2m-1)\pi_k]\cos\pi_k}{\sin[2m\pi_k]\nu^-} = \frac{1}{\nu^-}G_m(\pi_k). \end{aligned} \tag{13.37}$$

Let us now take a point $\mathbf{y}_0 \in \mathcal{D}_{rep}$ with tangent $T_0 \in T_{rep}$ and consider the map $\mathbf{y}_0 \mapsto \mathbf{y}_{2m} = (\Phi^- \circ \Phi^+)^m(\mathbf{y}_0)$ which sends $T_0 \mapsto T_{2m}$.

Assuming $k \geq 2$, we must consider two cases:

- If k is odd, let $k = 2m-1$, and then if $\mathbf{y}_0$ lies on the boundary of $\mathcal{D}_{rep}$, by (13.37) the point $\mathbf{y}_{2m} = (\Phi^- \circ \Phi^+)^m(\mathbf{y}_0)$ has tangent $T_{2m}|_{T_0=0} = 0$ or $T_{2m}|_{T_0=\infty} = \pm\infty$ and hence lies on the boundary of $\mathcal{D}_{att}$. Since the map is linear, this implies that $\mathbf{y}_{2m} = (\Phi^- \circ \Phi^+)^m(\mathbf{y}_0) \in \mathcal{D}_{att}$ for any $\mathbf{y}_0 \in \mathcal{D}_{rep}$. By applying (13.30)), the iterate $\mathbf{y}_{2m-1}$ has tangent $T_{2m-1} \in (-\infty, 2\nu^-) \subset T_{C^-}$ and is therefore the last in a sequence of crossing points $\mathbf{y}_{1,3,\ldots,2m-1} \in \mathcal{D}_{C^-}$ and $\mathbf{y}_{2,4,\ldots,2m-2} \in \mathcal{D}_{C^+}$, which number $2m-1$ in total; see Figure 13.11(ii).
- If k is even, let $k = 2m$, and then for a point $\mathbf{y}_0$ on the boundary of $\mathcal{D}_{rep}$, (13.37) gives $T_{2m}|_{T_0=0} = \pm\infty$ and $T_{2m}|_{T_0=\infty} = 2\nu^+$, and therefore by linearity, $\mathbf{y}_{2m}$ belongs to a sequence of crossing points $\mathbf{y}_{1,3,\ldots,2m-1} \in \mathcal{D}_{C^-}$ and $\mathbf{y}_{2,4,\ldots,2m} \in \mathcal{D}_{C^+}$. One further application of Φ^+ sends $T_{2m} \in (-\infty, 2\nu^+)$ to $T_{2m+1} \in (0,\infty) = T_{att}$, so the same map sends the boundaries of $\mathcal{D}_{rep}$ to those of $\mathcal{D}_{att}$, and any $\mathbf{y}_0 \in \mathcal{D}_{rep}$ to $\mathbf{y}_{2m+1} = \Phi^+ \circ (\Phi^- \circ \Phi^+)^m(\mathbf{y}_0) \in \mathcal{D}_{att}$. Thus there are $2m$ crossing points in total; see Figure 13.11(i).

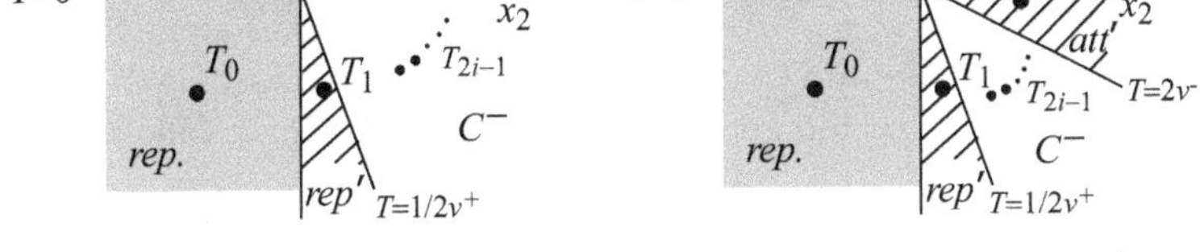

Fig. 13.11 Figure showing $\mathcal{D}_{rep}$ is the domain of T_0, and $\mathcal{D}'_{rep} = \Phi^+(\mathcal{D}_{rep}) = (-\infty, 1/2\nu^+)$ is the region where the first crossing occurs. $\mathcal{D}'_{att} = (2\nu^{\pm}, 0)$ is the region where the last crossing occurs and is given by $\Phi^+(\mathcal{D}'_{att}) = \mathcal{D}_{att}$ in the left figure and $\Phi^-(\mathcal{D}'_{att}) = \mathcal{D}_{att}$ in the right figure. There are $k = 2m$ crossings in the left figure and $k = 2m-1$ in the right, for integer m and $k \geq 2$.

In both cases the entire neighbourhood of the singularity consists of orbits that connect $\mathcal{D}_{rep}$ to $\mathcal{D}_{att}$ via a number of crossings $k = 2m$ or $k = 2m - 1$, where k is fixed by the quantity $\nu^+\nu^- = \cos^2 \frac{\pi}{k+1}$. Finally, applying the similarity transformation (13.35), which reflects the topology in Figure 13.11 in the line $x_2 = x_3$, trivially yields the same result when Φ^- instead of Φ^+ is applied first on $\mathcal{D}_{rep}$. □

Proof of Corollary 13.2 [69]. In Theorem 13.1, the boundaries of $\mathcal{D}_{rep}$ map exactly onto the boundaries of $\mathcal{D}_{att}$, and $\nu^+\nu^- = \cos^2 \frac{\pi}{k+1}$ where k is an integer. An immediate consequence of this is that if $\nu^+\nu^- = \cos^2 \frac{\pi}{r+1}$ for noninteger r, then the number of crossing points in the flow can only take the integer values immediately either side of r. To prove this explicitly consider the orbits that map from inside $\mathcal{D}_{rep}$ to the boundary of $\mathcal{D}_{att}$, and from the boundary of $\mathcal{D}_{rep}$ to the interior of $\mathcal{D}_{att}$. When these orbits are perturbed, they undergo a change in the number of crossing points they contain: if we move from a point $\mathbf{y}_i$ with $T_i > 0$ to one with $T_i < 0$ at the boundary of $\mathcal{D}_{rep}$, it changes from a starting point in $\mathcal{D}_{rep}$ to a crossing point in $\mathcal{D}_{C^+}$. Thus the number of crossing points increases by one as we go from an orbit with $T_0 \in T_{rep}$ to a nearby orbit with $T_0 \in \mathcal{D}_{C^+}$, passing through $T_0 = 0$.

It remains to consider orbits that are mapped onto the boundaries of $\mathcal{D}_{att}$. Assume the zeroth iterate $\mathbf{y}_0$ to be either the starting point of an orbit such that $\mathbf{y}_0 \in \mathcal{D}_{rep}$ or to be the first crossing point of an orbit such that $\mathbf{y}_0 = \Phi^-(\mathbf{y}_{-1}) \in \mathcal{D}_{C^+}$ for some $\mathbf{y}_{-1} \in \mathcal{D}_{rep}$. In the first case, we have $T_0 \in T_{rep}$, and in the second case, applying (13.19)) gives $T_0 \in (1/2\nu^-, 0)$, recalling that ν^- is negative. Likewise, we will assume that some later iterate $\mathbf{y}_i$ is either the end point $\mathbf{y}_i \in \mathcal{D}_{att}$ or the last crossing point $\mathbf{y}_i \in \mathcal{D}_{C^+}$ such that $\mathbf{y}_{i+1} = \Phi^+(\mathbf{y}_i) \in \mathcal{D}_{att}$. In the first case $T_i \in T_{att}$, and in the second case (13.19)) gives $T_i \in (-\infty, 2\nu^+)$, recalling that ν^+ is negative. We refer to a *complete orbit* as having a starting point $\mathbf{y}_0$ or $\mathbf{y}_{-1}$ in $\mathcal{D}_{rep}$, and an endpoint $\mathbf{y}_i$ or $\mathbf{y}_{i+1}$ in $\mathcal{D}_{att}$, for some i. This is illustrated in Figure 13.12.

Let us now find the value of T_0 for which T_i lies on the boundary of T_{att}, where $\mathbf{y}_i$ changes from a crossing point to an end point. Solving $T_{2m}|_{T_0} = 0$ and $T_{2m}|_{T_0} = \pm\infty$ from (13.31) for some integer m, we find

$$\begin{array}{ll} T_{2m}|_{T_0} = 0 & \Rightarrow T_0 = \Gamma_m, \\ T_{2m}|_{T_0} = \pm\infty & \Rightarrow T_0 = \Gamma_{m+\frac{1}{2}}, \end{array} \qquad \text{where} \qquad \Gamma_m = \frac{\nu^+}{G_m(\theta)},$$

recalling that $\theta = \arccos \sqrt{\nu^+\nu^-}$ by definition. Now let θ be equal to $\pi_r = \frac{\pi}{r+1}$ for some $r > 0$ and assume that $k < r < k + 1$. Let us assume k is odd and let $k = 2m - 1$ for some integer m. For the quantity Γ_m, it follows that

$$\frac{1}{2\nu^-} < \Gamma_m < 0 < \Gamma_{m+\frac{1}{2}}. \tag{13.38}$$

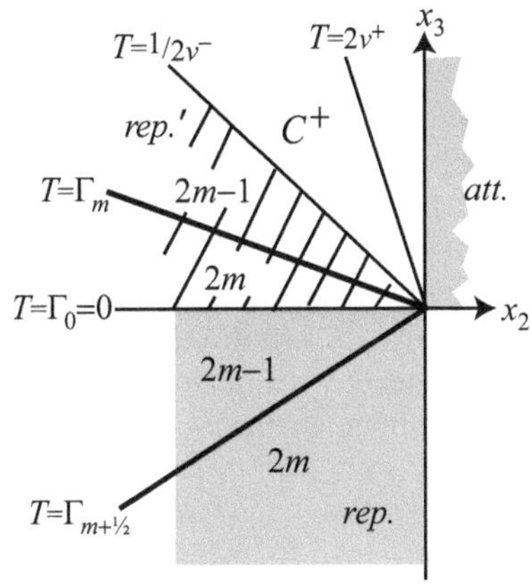

Fig. 13.12 Number of crossings for $2m-1<r<2m$ is $k=2m$ or $k=2m-1$, $m\in\mathbb{N}$, alternating as the iterate T_0 crosses the tangent values $T_{\delta k}$. To obtain the case $2m<r<2m+1$ just substitute m with $m+\frac{1}{2}$.

Recall that we allow either $T_0\in(1/2\nu^-,0)\subset\mathcal{T}_{C^+}$ or $T_0\in(0,\infty)=\mathcal{T}_{att}$. Then (13.38) partitions this range of T_0 values into four different regions:

- If $T_0\in(\frac{1}{2\nu^-},\Gamma_m)\subset\mathcal{T}_{C^+}$, then $T_{2m-2}\in(-\infty,2\nu^+)\subset\mathcal{T}_{C^+}$ from (13.31). Moreover $\mathbf{y}_0=\Phi^-(\mathbf{y}_{-1})$ for some $\mathbf{y}_{-1}\in\mathcal{D}_{rep}$, and $\Phi^+\circ(\Phi^-\circ\Phi^+)^{m-1}(\mathbf{y}_0)\in\mathcal{D}_{att}$, so the complete orbit $\mathbf{y}_{-1}\mapsto\Phi^+\circ(\Phi^-\circ\Phi^+)^{m-1}(\mathbf{y}_0)$ has $2m-1$ crossing points $\mathbf{y}_{0,1,\ldots,2m-2}\in\mathcal{D}_{C^\pm}$;
- If $T_0\in(\Gamma_{m+\frac{1}{2}},0)\subset\mathcal{T}_{C^+}$, then $T_{2m}\in(0,\infty)=\mathcal{T}_{rep}$ from (13.31). Moreover $\mathbf{y}_0=\Phi^-(\mathbf{y}_{-1})$ for some $\mathbf{y}_{-1}\in\mathcal{D}_{rep}$, so the complete orbit $\mathbf{y}_{-1}\mapsto(\Phi^-\circ\Phi^+)^{m-1}(\mathbf{y}_0)$ has $2m$ crossing points $\mathbf{y}_{0,1,\ldots,2m-1}\in\mathcal{D}_{C^\pm}$;
- If $T_0\in(0,\Gamma_{m+\frac{1}{2}})\subset\mathcal{T}_{att}$, then $T_{2m}\in(0,\infty)=\mathcal{T}_{rep}$ from (13.31), so the complete orbit $\mathbf{y}_0\mapsto(\Phi^-\circ\Phi^+)^m(\mathbf{y}_0)$ has $2m-1$ crossing points $\mathbf{y}_{1,\ldots,2m-1}\in\mathcal{D}_{C^\pm}$;
- If $T_0\in(\Gamma_{m+\frac{1}{2}},\infty)\subset\mathcal{T}_{att}$, then $T_{2m}\in(-\infty,2\nu^+)\subset\mathcal{T}_{C^+}$ from (13.31), and moreover $\Phi^+\circ(\Phi^-\circ\Phi^+)^m(\mathbf{y}_0)\in\mathcal{D}_{att}$, so the complete orbit $\mathbf{y}_0\mapsto\Phi^+\circ(\Phi^-\circ\Phi^+)^m(\mathbf{y}_0)$ has $2m$ crossing points $\mathbf{y}_{1,\ldots,2m+1}\in\mathcal{D}_{C^\pm}$.

On the bounds between these four cases, T_0 maps onto the boundary of $\mathcal{T}_{att}$ by (13.38). If k is even we instead let $k=2m$, and the corresponding result is found trivially by substituting $m\mapsto m+\frac{1}{2}$ into the cases above. Finally, we have assumed that Φ^+ is applied first, and the similarity transformation (13.35) gives the corresponding values $T=0$ and $T=\Gamma_{(k+1)/2}=\nu^-/G_{(k+1)/2}(\pi_r)$ when Φ^- is applied first. In each case the number of crossings changes between k and $k+1$, giving part 1 of the corollary, and the change takes place either at $T=0$ or at $T=\Gamma_{(k+1)/2}=\nu^\pm/G_{(k+1)/2}(\pi_r)$, giving part 2 of the corollary. □

Lemma 13.1 shows that the singularity is topologically unstable when $\nu^+\nu^- = \cos^2\frac{\pi}{k+1}$ for integer k, since the flow maps the boundaries of $\mathcal{D}_{rep}$ exactly onto to the boundaries of $\mathcal{D}_{att}$. The corollary shows that the intervening cases, when $\nu^+\nu^- = \cos^2\frac{\pi}{r+1}$ for noninteger r, are topologically stable, since the boundaries of $\mathcal{D}_{rep}$ map into the interior of $\mathcal{D}_{att}$, and the same number of crossings applies to open intervals of starting points with tangents T_0.

The equilibria of the second return rotation map (13.30) are actually a simple geometric feature forming a separatrix of the three-dimensional flow: a pair of opposing invariant cones, with apexes at the singularity, and creased where they cross the discontinuity surfaceor more concisely a *nonsmooth diabolo.*

13.3.4 The Nonsmooth Diabolo

The equilibria $\nu^+\tau_*$ and $\nu^-\tau_*$, of the maps $T_{2m} \mapsto T_{2m+2}$ and $T_{2m-1} \mapsto T_{2m+1}$, exist for $\nu^\pm < 0$ with $\nu^+\nu^- > 1$. There is an attracting equilibrium with branches $T^\pm_{att}\pm \in \mathcal{T}_{C^\pm}$ (see (13.27)) given by

$$T^+_{att}/\nu^+ = T^-_{att}/\nu^- = 1 - \sqrt{1 - 1/\nu^+\nu^-}$$

and a repelling equilibrium with branches $T^\pm_{rep} \in \mathcal{T}_{C^\pm}$ given by

$$T^+_{rep}/\nu^+ = T^-_{rep}/\nu^- = 1 + \sqrt{1 - 1/\nu^+\nu^-}\ .$$

They correspond to invariant manifolds of the second return maps (13.24) on which only radial motion occurs with respect to the singularity. From (13.29) it is clear that they divide $\mathcal{D}$ such that the attracting manifolds at angle $T^\pm_{att}$ enclose the repelling sliding region, while the repelling manifolds at angle $T^\pm$ enclose the attracting sliding region. Each pair of angles $\{T^+_{att}, T^-_{rep}\}$ and $\{T^+_{rep}, T^-_{att}\}$ forms a straight line through the origin, since $T^\pm_{att}T^\mp_{rep} = 1$.

Where the diabolo intersects $\mathcal{D}$, that is, along the directions $T^\pm_{att,rep}$ inside $\mathcal{D}_{C^\pm}$, the radial map (13.33) simplifies to

$$\frac{R_{k+2}}{R_k} = \frac{1 \pm \sqrt{1 - 1/\nu^+\nu^-}}{1 \mp \sqrt{1 - 1/\nu^+\nu^-}} \tag{13.39}$$

taking the upper signs for $T^\pm_{att}$ and the lower signs for $T^\pm_{rep}$. The index k can take even or odd values $2m$ or $2m-1$ for $m \in \mathbb{N}$. Since the diabolo exists only for $\nu^+\nu^-$, this implies

$$\begin{aligned} &R_{k+2} > R_k \ \text{ on the attracting manifolds } T^\pm_{att}, \text{ and}\\ &R_{k+2} < R_k \ \text{ on the repelling manifolds } T^\pm_{rep}, \end{aligned} \tag{13.40}$$

hence orbits move away from the singularity on the cone generated by $T^{\pm}_{att}$ (which surrounds the repelling sliding region), and orbits move towards the singularity on the cone generated by $T^{\pm}_{rep}$ (which surrounds the attracting sliding region). We might therefore describe the cones crossing $\mathcal{D}$ at angles $T^{\pm}_{att}$ and $T^{\pm}_{rep}$ as unstable and stable manifolds of the singularity, respectively.

This behaviour is illustrated in Figure 13.13.

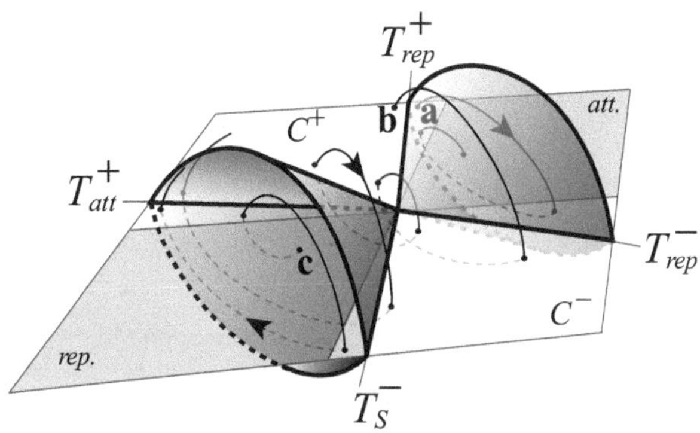

Fig. 13.13 The nonsmooth diabolo: Invariant manifolds near a two-fold singularity. The three qualitatively different types of orbit are shown. (a) An orbit starting near the inside of $\mathcal{D}_{rep}$ spirals in towards the singularity and hits the sliding region (shaded). (b) An orbit starting near the outside of $\mathcal{D}_{rep}$ initially spirals inwards towards the singularity, then spirals out away from the singularity, and tends asymptotically towards $\mathcal{D}_{att}$. (c) An orbit spirals outwards from the repelling sliding region and away from the singularity, approaching $\mathcal{D}_{att}$ asymptotically.

Though derived in three dimensions, the results above hold generally in higher dimensions because they depend solely on the leading-order part (13.11) of the full system (13.5), which depends only on the (x_1, x_2, x_3) components near the two-fold. That is, Theorem 13.1 to Corollary 13.2 applies on a neighbourhood where the conditions in Definition 6.2 are satisfied.

A trajectory may of course evolve in such a way that the higher-order terms $\mathsf{O}(|\mathbf{x}|)$ in (13.5) become significant. In the case of infinitely many crossings, the effect of higher-order terms in the vector field is to cause the invariant cones to curve such they may intersect the folds transversally regions; this effect is significant only near a bifurcation that occurs at $\nu^+\nu^- = 1$, the subject of the following section. For dynamics at the critical value $\nu^+\nu^- = 1$, we give details only for in three dimensions, and their applicability to higher dimensions is not known.

13.3.5 The Nonsmooth Diabolo Bifurcation

As $\nu^+\nu^- \to 1$ from $\nu^+\nu^- > 1$, the cones of the invariant diabolo flatten, until they at $\nu^+\nu^- = 1$ coalesce and annihilate in a *nonsmooth diabolo bifurcation* (Figure 13.14).

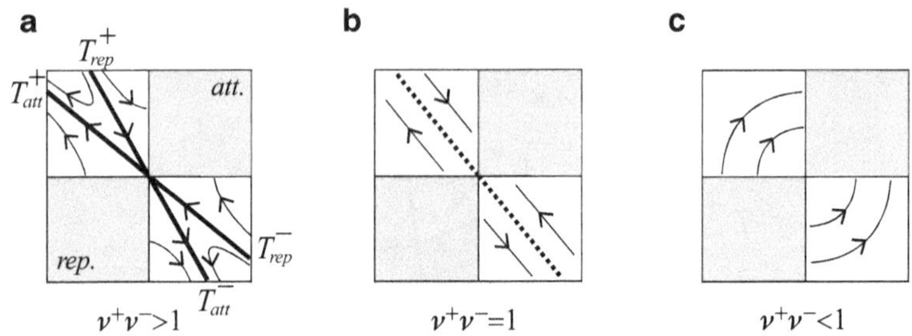

Fig. 13.14 Nonsmooth diabolo bifurcation: (a)-(c) correspond to the parameter values in the figure above. (a) $T^{\pm}_{att}$ and $T^{\pm}_{rep}$ form two stable and two unstable manifolds in the crossing regions, which are, respectively, repelling and attracting with respect to the singularity at the centre. (b) The manifolds coalesce in a line of fixed points. (c) All points map from the repelling sliding region to the sliding region in finite time.

At $\nu^+\nu^-$ the map becomes a structurally unstable shear along the direction $T_{2m} = \nu^{\pm}$. This occurs because the condition $\nu^+\nu^- = 1$ means the vector fields $\mathbf{f}^{\pm}$ are coplanar.

Away from the singularity, this coplanarity is broken by higher-order terms that we have omitted from (13.11) onwards. We can derive their effect without explicitly including them in (13.11) by introducing small perturbations of the maps (13.19) near the bifurcation, valid for $|1 - 1/\nu^+\nu^-| \ll 1$. Without loss of generality, perturb the map Φ^+. A general second-order perturbation is a two-vector of terms x_2^2, x_3^2, x_2x_3, containing six parameters, of which only two are free if the map is still locally a reflection. Thus, four parameters are set by demanding that two iterations of the map give the identity and that the fold line is invariant under the map. The result can be written

$$\mathbf{y}_{2m+1} = \left[\underline{\underline{B}}^+ + \alpha x_{2,2m}\underline{\underline{I}} - 2\beta x_{2,2m}\begin{pmatrix} 1 & 0 \\ \nu^+ & 0 \end{pmatrix}\right]\mathbf{y}_{2m} \tag{13.41}$$

in terms of parameters α and β for small x_2. The second return maps (Φ^+ followed by Φ^- and vice versa) are then

$$\begin{aligned}\mathbf{y}_{k+2} &= \underline{\underline{B}}^-\left[\underline{\underline{B}}^+ + \alpha x_{2,2m}\underline{\underline{I}} - 2\beta x_{2,2m}\begin{pmatrix} 1 & 0 \\ \nu^+ & 0 \end{pmatrix}\right]\mathbf{y}_{2m}\,,\\ \mathbf{y}_{2m+1} &= \left[\underline{\underline{B}}^+ + \alpha x_{2,2m}\underline{\underline{I}} - 2\beta x_{2,2m}\begin{pmatrix} 1 & 0 \\ \nu^+ & 0 \end{pmatrix}\right]\underline{\underline{B}}^-\mathbf{y}_{2m-1}\,.\end{aligned}$$

As well as a trivial fixed point $\mathbf{x} = 0$ that was present in the unperturbed system, a further two fixed points (one in each map) now exist near the origin when

$$(\nu^+\nu^- - 1)\left(\nu^+\nu^- - \left(\frac{\alpha}{\beta} - 1\right)^2\right) > 0. \tag{13.42}$$

Both fixed points move to the origin as $\nu^+\nu^- \to 1$ and disappear. At $\nu^+\nu^- = 1$, the eigenvalues of the Jacobian of (13.42) are both unity, satisfying the Takens-Bogdanov condition [138, 202], and we exploit this to obtain a more refined perturbation below. The eigenvalues for $\nu^+\nu^- < 1$ lie on the unit circle so the origin is non-hyperbolic, but in the nonsmooth system, this does not imply structural instability because orbits evolve under the map only for a finite time, after which they reach the sliding region.

The eigenvectors of the Jacobian of (13.42) lie in the directions given by $T^{\pm}_{rep,att}$ for $\nu^+\nu^- > 1$, so the invariant manifolds $T^{\pm}_{rep,att}$ are valid locally. For $\nu^+\nu^- < 1$ the eigenvectors are complex. As α and β vary, the fixed points move around in phase space, where their angles make tangents

$$T^{\pm} = \nu^{\pm}\left(\frac{\beta}{\alpha} + \frac{1-\frac{\beta}{\alpha}}{\nu^+\nu^-} \pm \sqrt{\left(\frac{\beta^2}{\alpha^2} + \frac{\left(1-\frac{\beta}{\alpha}\right)^2}{\nu^+\nu^-}\right)\left(1 - \frac{1}{\nu^+\nu^-}\right)}\right) \tag{13.43}$$

with the $\pm$ folds.

In the special case $\alpha = \beta$, the second return maps still preserve lines through the singularity. Then the perturbation disappears in the rotation maps, and the fixed points must lie on the invariant manifolds given by $T^{\pm}_{rep,att}$. We need then consider only the perturbation to the radial map (13.33), which takes the form

$$R_{i+2} = R_i\left(\frac{2\nu^+}{T^+_{rep,att}} - 1\right)(1 + \rho R_i) \tag{13.44}$$

for i odd or even, with the subscript *rep* or *att* chosen corresponding to the appropriate manifold direction, and where

$$\rho = \frac{T^+_{rep,att}\left(\alpha - 2\beta + 4\nu^+\nu^-\right)}{\left(2\nu^+ - T^+_{rep,att}\right)\sqrt{1 + T^+_{rep,att}}}. \tag{13.45}$$

We can determine from (13.39) that on the unstable manifold we have $2\nu^+/T^+_{rep} - 1 < 1$, while on the stable manifold, we have $2\nu^+/T^+_{att} - 1 > 1$. R_i is positive by definition, so we can determine the sign of ϵ necessary for a fixed point R_{att} to exist on $T^{\pm}_{att}$ or for a fixed point R_{rep} to exist on $T^{\pm}_{rep}$:

$$\epsilon R_{rep} = \left(\frac{2\nu^+}{T^+_{rep}} - 1\right)^{-1} - 1 > 0 \Rightarrow \epsilon > 0\,,$$
$$\epsilon R_{att} = \left(\frac{2\nu^+}{T^+_{att}} - 1\right)^{-1} - 1 < 0 \Rightarrow \epsilon < 0\,. \tag{13.46}$$

This means that a positive perturbation gives an unstable limit point on each of the unstable manifolds $T^{\pm}_{rep}$, while a negative perturbation gives a stable limit point on each of the stable manifolds $T^{\pm}_{att}$, illustrated in Figure 13.15. In the full vector space, the limit points correspond to limit cycles on the parabolic invariant cones generated by $T^{\pm}_{rep}$ and $T^{\pm}_{att}$. These limit cycles approach the singularity as $\nu^+\nu^- \to 1$ and disappear.

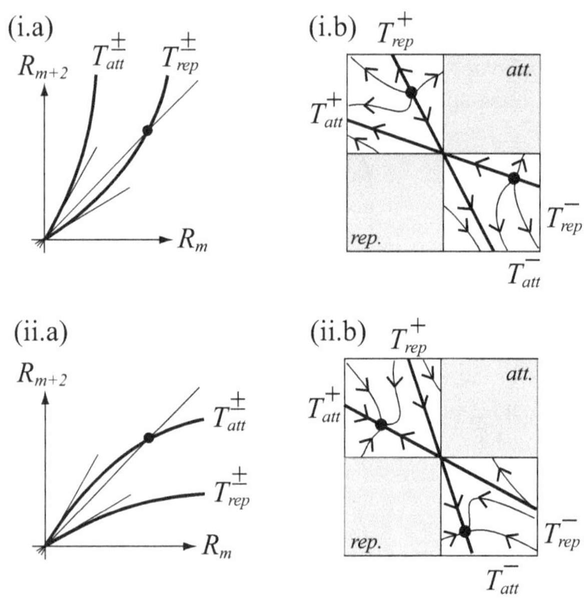

Fig. 13.15 Radial perturbation of the diabolo for $\alpha = \beta$, showing (a) the radial map on the invariant manifolds with a fixed point on $T^{\pm}_{rep}$ and (b) dynamics of the rotation map; (i) for a perturbation $\rho > 0$ and (ii) or a perturbation $\rho < 0$.

To improve on this perturbative approach, we can use the fact that, at the bifurcation, the Jacobian of the return map at the origin is nondiagonalizable and has two eigenvalues equal to one. As a consequence of Φ^+ and Φ^- being involutions, this corresponds to a degenerate $1:1$ resonance bifurcation of the map (see [138]). Proceeding in general terms in some normal form coordinates (u_2, u_3) which are a transformation of (x_2, x_3), near the bifurcation the map can be reduced to the normal form

$$\begin{aligned} u_2 &\mapsto u_2 + u_3 + \mathsf{O}\left(|\mathbf{u}|^4\right) , \\ u_3 &\mapsto (4\delta + au_2^2)u_2 + (1 + 4\delta + bu_2 + cu_2^2 + du_3^2)u_3 + \mathsf{O}\left(|\mathbf{u}|^4\right) , \end{aligned} \tag{13.47}$$

with $\delta = \nu^+\nu^- - 1$, through a lengthy series of changes of variables and parameters detailed in [37]. These transformations are not ultimately of great concern, because though (13.47) tells us what phase portraits are possible, it does not tell us in any useful way which portrait occurs in a given system (nor is this easily gleaned from the changes of variables to obtain (13.47)).

At the bifurcation and at the origin, the u_2 axis lies along the vector direction $(\nu^-, 1)$, the singular eigenvector of the Jacobian of the rotation map $\Phi^\pm \circ \Phi^\mp$, and the positive direction of u_2 points into $\mathcal{D}_{C^+}$, while the u_3 axis is tangent to the x_2 axis in the original coordinates. The normal form has a fixed point at the origin for all values of p, while two more fixed points, located at

$$u_2 = \pm 2\sqrt{-(\nu^+\nu^- - 1)/a} , \qquad u_3 = 0 ,$$

emerge when $-(\nu^+\nu^- - 1)/a$ becomes positive. The positivity of the eigenvalues $2\nu^+\nu^- - 1 \pm 2\sqrt{\nu^+\nu^-(\nu^+\nu^- - 1)}$ implies that the negative u_2 axis intersects $\mathcal{D}_{C^-}$ and lies inside the domain of the return map $\Phi^\pm \circ \Phi^\mp$, implying that the fixed point at $u_2 = -2\sqrt{-(\nu^+\nu^- - 1)/a}$ corresponds to a crossing periodic orbit near the singularity. The eigenvalues of this fixed point are

$$1 - b\sqrt{-(\nu^+\nu^- - 1)/a} \pm \sqrt{-\frac{(\nu^+\nu^- - 1)}{a}(8a + b^2)} + \mathsf{O}\left(p\right) .$$

Depending on the values of a and b the following are possible:

(c1) If $a > 0$, the two eigenvalues are real, one positive and one negative;
(c2) If $a < 0$ and $|8a| < b^2$, both eigenvalues are real, inside the unit circle if $b > 0$, outside the unit circle otherwise;
(c3) If $a < 0$ and $|8a| > b^2$, the eigenvalues are complex conjugate, inside the unit circle if $b > 0$ and outside otherwise.

To turn this into a phase portrait, we can use the fact that the orbits of the map on the (u_2, u_3) coordinates are approximated, for $\nu^+\nu^- = 1$, by the unit time shift of a flow which is equivalent to

$$\dot{v}_2 = v_2 + \mathsf{O}\left(|\mathbf{v}|^4\right) , \quad \dot{v}_3 = bv_2v_3 + av_2^3 + (\tfrac{1}{2}b^2 + c - 3a)v_2^2v_3 + \mathsf{O}\left(|\mathbf{v}|^4\right) ,$$

as shown in [37]. This is the normal form of a codimension-three Bogdanov-Takens bifurcation [139] and is unfolded in three parameters in [56]. In the situation above, changing p through 0 we explore a one-dimensional curve of parameters through the three-dimensional unfolding of cases (*c*1), (*c*2), and (*c*3), called, respectively, in [56] the saddle, elliptic, and focus cases for the topological type of the origin when $\nu^+\nu^- = 1$. Overall, cases (*c*1-*c*3) give rise to the following bifurcation scenarios of the crossing dynamics in a neighbourhood of the origin:

(c1) For $\nu^+\nu^- > 1$, the singularity is a saddle of the crossing map. For $\nu^+\nu^- < 1$, a saddle cycle emerges from the singularity, and the singularity is a centre of the crossing map.
(c2) For $\nu^+\nu^- < 1$, the singularity is a centre of the crossing map. For $\nu^+\nu^- > 1$, a node cycle emerges from the singularity, and the singularity is a saddle of the crossing map. At $\nu^+\nu^- = 1$, the crossing map at the singularity can exhibit an elliptic sector (a region within which every orbit converges on the singularity both forward and backward in time).
(c3) For $\nu^+\nu^- < 1$, the singularity is a centre of the crossing map. For $\nu^+\nu^- > 1$, a focus cycle emerges from the singularity, and the singularity is a saddle of the crossing map.

These are illustrated in Figure 13.16.

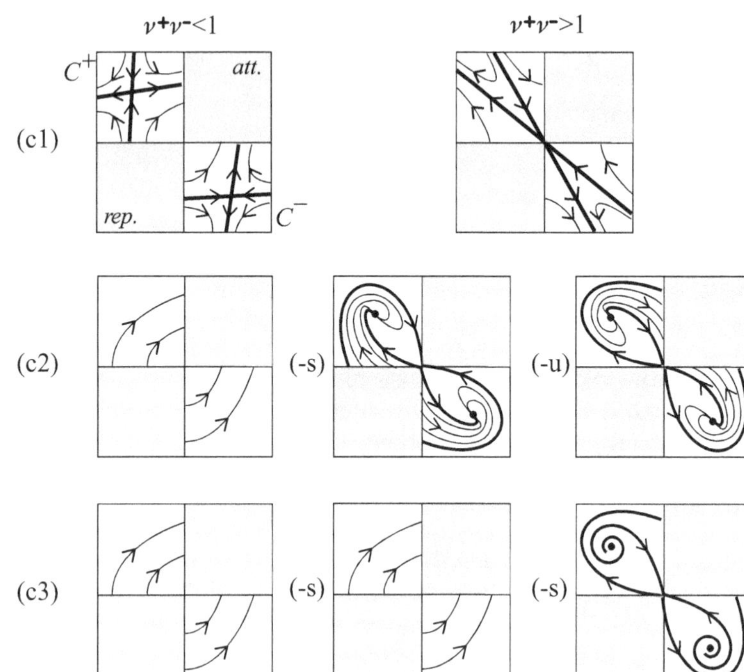

Fig. 13.16 The complete catalogue of local dynamics around an invisible two-fold when $p = \nu^+\nu^- - 1 = 0$ is obtained by composing the possible phase portraits of sliding $(s1)$, $(s2)$ and crossing $(c1)$, $(c2_{s,u})$, $(c3_{s,u})$. $(s1)$, $(s2)$ show the two possible phase portraits in the attracting and repelling sliding regions, as the bifurcation parameter p changes sign. $(c1)$, $(c2_{s,u})$, $(c3_{s,u})$ depict the intersections of orbits with the crossing regions as p changes sign (these are derived from the unfoldings in [56]). The crossing maps in $(c2_s)$ and $(c3_s)$ have a stable fixed point of node and focus type, respectively, while the dual cases in $(c2_u)$ and $(c3_u)$ have unstable fixed points. Altogether, composing cases $(s1)$, $(s2)$ with cases $(c1)$, $(c2_{s,u})$, $(c3_{s,u})$, ten qualitatively different portraits are obtained.

Caution must be taken in studying the dynamics of the crossing map because it is valid only on the subset of $\mathcal{D}_{C^-}$ whose image under Φ^- lies in $\mathcal{D}_{C^+}$, from which orbits return to $\mathcal{D}_{C^-}$ under $\Phi^- \circ \Phi^+$. Finally, some issues remain undealt with regarding the invariant sets and the structural stability of crossing dynamics. The bifurcations occurring between scenarios ($c1$)-($c3$) for $\nu^+\nu^-$ near unity suggest the existence of other structures, e.g. quasi-periodic orbits, that may emerge as p crosses 0. Also, as we have seen, when $\nu^+\nu^- < 1$, the crossing map $\Phi^\pm \circ \Phi^\mp$ is a (nonlinear) rotation from $\mathcal{D}_{rep}$ to $\mathcal{D}_{att}$. A structurally unstable scenario occurs when the image of the border of $\mathcal{D}_{rep}$ under ϕ (or multiple iterations of ϕ) is tangent to the border of $\mathcal{D}_{att}$. This is associated with the change in the number of iterations it takes to map points from the border of $\mathcal{D}_{rep}$ into $\mathcal{D}_{att}$, considered in the previous section for the local linear map.

13.4 Obsession 3: The Folded Bridge

The sliding vector field is relatively simple to understand, but just as easy to misunderstand or overlook, as happened for much of the two-fold's history. The local phase portraits take the form of folded equilibria, attracting on one side of the singularity and repelling on the other, but in some cases traversed in finite time to form a bridge between attractive and repelling regions of sliding.

Following the historical approach, we begin by studying the convex combination (substituting (13.5) into (5.7) with $\mathbf{h} \equiv 0$)

$$
\begin{aligned}
&(\dot{x}_1, \dot{x}_2, \dot{x}_3) = \tfrac{1}{2}(1+\lambda)\mathbf{f}^+ + \tfrac{1}{2}(1-\lambda)\mathbf{f}^- + \left(\mathsf{O}\left(|\mathbf{x}|^2\right), \mathsf{O}\left(|\mathbf{x}|\right), \mathsf{O}\left(|\mathbf{x}|\right)\right)\\
&\quad\text{where}\\
&\qquad \mathbf{f}^+ = \left(-x_2, -\operatorname{sign}\left[\delta^2_{t+}\sigma(0)\right], \nu^+\right),\\
&\qquad \mathbf{f}^- = \left(\ x_3\ ,\ \nu^-,\ \operatorname{sign}\left[\delta^2_{t-}\sigma(0)\right]\ \right),
\end{aligned}
\tag{13.48}
$$

with $\dot{x}_i = \mathsf{O}\left(|\mathbf{x}|\right)$ for $i = 4, \ldots, n$. The sliding vector field on $x_1 = 0$ is obtained by solving $\dot{x}_1 = x_1 = 0$ (substituting (13.5) into (7.6) for $r = 1$), giving $\lambda^\$ = (x_3 - x_2)/(x_2 + x_3)$, and

$$
\begin{aligned}
\varepsilon\dot{\lambda}_1 &= 0,\\
\dot{x}_2 &= \left(\nu^- x_2 - \operatorname{sign}\left[\delta^2_{t+}\sigma(0)\right] x_3\right) / \left(x_2 + x_3\right) + \mathsf{O}\left(|\mathbf{x}|\right),\\
\dot{x}_3 &= \left(\operatorname{sign}\left[\delta^2_{t-}\sigma(0)\right] x_2 + \nu^+ x_3\right) / \left(x_2 + x_3\right) + \mathsf{O}\left(|\mathbf{x}|\right),\\
\dot{x}_i &= \mathsf{O}\left(|\mathbf{x}|\right), \qquad i = 4, 5, \ldots, n,
\end{aligned}
\tag{13.49}
$$

on $x_1 = 0$ and $\lambda \in (-1, +1)$. The right-hand side of (13.49) is independent of the coordinates $x_4, x_5, \ldots$ to leading order. The dynamics in the x_2 and x_3 directions is thus given, to leading order, by the two-dimensional system

$$\begin{pmatrix} \dot{x}_2 \\ \dot{x}_3 \end{pmatrix} = \frac{1}{x_2+x_3} \underline{\underline{S}} \cdot \begin{pmatrix} x_2 \\ x_3 \end{pmatrix} , \qquad s_\pm = \mathrm{sign}[\delta^2_{t\pm}\sigma(0)]$$

$$\text{where} \qquad S := \begin{pmatrix} \nu^- & -s_+ \\ s_- & \nu^+ \end{pmatrix} . \tag{13.50}$$

The 2×2 matrix in (13.50) has determinant and trace

$$\det \underline{\underline{S}} = \mathrm{sign}\left[\delta^2_{t+}\sigma(0)\delta^2_{t-}\sigma(0)\right] + \nu^+\nu^- , \quad \mathrm{Tr}\, \underline{\underline{S}} = \nu^+ + \nu^- . \tag{13.51}$$

These, and the eigenvectors of the 2×2 matrix, are sufficient to determine the phase portrait of the sliding flow. The eigenvalues and eigenvectors of $\underline{\underline{S}}$, satisfying $\underline{\underline{S}}\gamma = \gamma\gamma$, are

$$\gamma_\pm = \frac{\nu^+ + \nu^- \pm \sqrt{(\nu^+ - \nu^-)^2 - 4s_+s_-}}{2} , \quad \gamma_\pm = \left(\gamma_\pm - \nu^+, s_-\right) , \tag{13.52}$$

the discriminant being $(\mathrm{Tr}\, \underline{\underline{S}})^2 - 4 \det \underline{\underline{S}}$ (the term under the square root).

The sliding vector field in the attracting sliding region has the phase portrait of a saddle if $\det \underline{\underline{S}} < 0$, of a focus if $(\mathrm{Tr}\, \underline{\underline{S}})^2 - 4 \det \underline{\underline{S}} < 0 < \det \underline{\underline{S}}$, and of a node if $4 \det \underline{\underline{S}} - (\mathrm{Tr}\, \underline{\underline{S}})^2 < 0 < \det \underline{\underline{S}}$; the focus/node are attracting if $\mathrm{Tr}\, \underline{\underline{S}} < 0$ and repelling if $\mathrm{Tr}\, \underline{\underline{S}} > 0$.

Note how we have phrased this carefully: the field '*in the attracting sliding region has the phase portrait of...*'. The prefactor $\frac{1}{x_2+x_3}$ in (13.50) amounts to a timescaling, which reverses time in the repelling sliding region and which is singular at the two-fold itself. The attracting and repelling sliding regions in the prototype occupy $x_2x_3 > 0$ on $x_1 = 0$, implying $x_2+x_3 > 0$ in the sliding region and $x_2 + x_3 < 0$ in the repelling sliding region, in both of which the system (13.49) is well-defined. To obtain the sliding vector field, we therefore neglect the prefactor and obtain the phase portrait using the matrix $\underline{\underline{S}}$ and its characteristics above. This gives the correct topological phase portrait in the attracting sliding region, and in the repelling sliding region, we just reverse time.

The sliding vector field is not well-defined at the two-fold singularity itself (nor on the line $x_2 + x_3 = 0$, but this passes through the crossing regions), where both the numerator and denominator of (13.49) vanish. So (13.49) specifies the sliding flow in the neighbourhood of the singularity, excepting the singularity itself. We will obtain a phase portrait not of an equilibrium at the singularity but of a 'folded' singularity, where trajectories enter in finite time from one side and depart it from the other. That this occurs in finite time is clear from the non-degeneracy condition (13.2): zero is not an allowed velocity at the singularity. How the sliding flow traverses the singularity is discussed in Obsession 4 below.

We can now see from the analysis above why the condition (13.1) constitutes a degeneracy of the two-fold.

Lemma 13.2. (13.1) *implies that the determinant* (13.51) *is nonzero. The parameter curve*

$$D = \nu^+\nu^- + \text{sign}[\delta_{t+}^2\sigma(0)]\,\text{sign}[\delta_{t-}^2\sigma(0)] = 0 \tag{13.53}$$

therefore constitutes a degeneracy of the two-fold, in which a sliding equilibrium lies at the singularity, subject to

$$\frac{\nu^-}{\text{sign}[\delta_{t+}^2\sigma(0)] + \nu^-} \in [0,1] \quad \& \quad \frac{\text{sign}(\delta_{t-}^2\sigma)}{\text{sign}[\delta_{t-}^2\sigma(0)] - \nu^+} \in [0,1]\,. \tag{13.54}$$

Proof. To show this we prove that the condition $0 \notin \delta_{t\lambda}\mathbf{x}$ at the singularity implied by (13.1) reduces to $D \neq 0$ in the prototype. The condition in (13.1) means that $\delta_{t\lambda}\mathbf{x}$ has a nonzero component in the plane of $\partial_{\mathbf{x}}\delta_{t+}\sigma$ and $\partial_{\mathbf{x}}\delta_{t-}\sigma$, at $\hat{\mathbf{x}}$, for all $\lambda \in [0,1]$. In the prototype (13.5) the gradient vectors $\partial_{\mathbf{x}}\delta_{t+}\sigma$ and $\partial_{\mathbf{x}}\delta_{t-}\sigma$ correspond to the x_2 and x_3 directions, so (13.1) states that the second and third components of $\delta_{t\lambda}\mathbf{x} = \lambda\delta_{t+}\mathbf{x} + (1-\lambda)\delta_{t-}\mathbf{x}$ are nonzero. Evaluating these, and eliminating λ, gives $\text{sign}[\delta_{t+}^2\sigma(0)]\,\text{sign}[\delta_{t-}^2\sigma(0)] + \nu^+\nu^- \neq 0$, which is $D \neq 0$ by (13.51). Solving for λ and substituting into the requirement $\lambda \in [0,1]$ gives (13.54). Thus $D \neq 0$ is violated on a parameter curve $\nu^+\nu^- = \pm 1$ from (13.53), with restrictions on the signs of $\nu^\pm$ as implied in (13.54). □

13.4.1 The Visible Two-Fold

The sliding vector field (13.50) for $\text{sign}[\delta_{t-}^2\sigma(0)] < 0 < \text{sign}[\delta_{t+}^2\sigma(0)]$ is

$$\begin{pmatrix} \dot{x}_2 \\ \dot{x}_3 \end{pmatrix} = \frac{1}{x_2 + x_3}\underline{\underline{S}} \cdot \begin{pmatrix} x_2 \\ x_3 \end{pmatrix}\,, \qquad \underline{\underline{S}} = \begin{pmatrix} \nu^- & -1 \\ -1 & \nu^+ \end{pmatrix}\,. \tag{13.55}$$

Very straightforward analysis of this sliding vector field gives Figure 13.4. The delimiting case is $\nu^+\nu^- = 1$ for $\nu^+, \nu^- > 0$ (from Lemma 13.2).

The first component of the eigenvector from (13.52) satisfies

$$2(\gamma_\pm - \nu^+) = \nu^- - \nu^+ \pm \sqrt{(\nu^+ - \nu^-)^2 + 4} \gtrless 0\,,$$

implying

$$\boldsymbol{\gamma}_+ = (\gamma_+ - \nu^+, -1) \in \mathcal{D}_{C^-}\,, \quad \boldsymbol{\gamma}_- = (\gamma_- - \nu^+, -1) \in \mathcal{D}_{rep}\,. \tag{13.56}$$

For $\nu^\pm > 0$ and $\nu^+\nu^- > 1$, we have $\det\underline{\underline{S}} = -1 + \nu^+\nu^- > 0$ and $\text{Tr}\,\underline{\underline{S}} = \nu^+ + \nu^- > 0$, giving a finite-time node where the flow is outwards from the singularity in $\mathcal{D}_{att}$. The eigenvalues satisfy $0 < \gamma_- < \gamma_+$, so by (13.56) the weak eigendirection $\boldsymbol{\gamma}_-$ lies in the sliding regions, and the strong eigendirection $\boldsymbol{\gamma}_+$ lies in the crossing regions, producing the phase portrait Figure 13.4(ii). The flow is outwards from the singularity in the attracting sliding region and inwards in the repelling region; thus, it consists entirely of faux canards.

For $\nu^+\nu^- < 1$, we have $\det \underline{\underline{S}} = -1 + \nu^+\nu^- < 0$, implying a finite-time saddle. The eigenvalues have signs $\gamma_- < 0 < \gamma_+$, so by (13.56) the inward eigendirection γ_- lies in the sliding regions and the outward eigendirection γ_+ lies in the crossing regions, producing the phase portrait Figure 13.4(i). Along the direction γ_- only, a single trajectory passes from attracting to repelling sliding, constituting a canard. The rest of the flow passes through the folds.

For $\nu^\pm < 0$ and $\nu^+\nu^- > 1$, we have $\det \underline{\underline{S}} = -1 + \nu^+\nu^- > 0$, implying a finite-time node where the flow is in towards the singularity in $\mathcal{D}_{att}$. The eigenvalues satisfy $\gamma_- < \gamma_+ < 0$, so by (13.56) the strong eigendirection γ_- lies in the sliding regions, and the weak eigendirection γ_+ lies in the crossing regions, again producing the phase portrait Figure 13.4(i). Again, along the direction γ_- only, a single trajectory passes from attracting to repelling sliding, constituting a canard. The rest of the flow passes through the folds.

13.4.2 The Visible-Invisible Two-Fold

To study the visible-invisible two-fold, let us take the case $\text{sign}[\delta^2_{t\pm}\sigma(0)] > 0$ (the case with $\text{sign}[\delta^2_{t\pm}\sigma(0)] < 0$ is analogous). The sliding vector field (13.50) is

$$\begin{pmatrix} \dot{x}_2 \\ \dot{x}_3 \end{pmatrix} = \frac{1}{x_2 + x_3} \underline{\underline{S}} \cdot \begin{pmatrix} x_2 \\ x_3 \end{pmatrix}, \qquad \underline{\underline{S}} = \begin{pmatrix} \nu^- & -1 \\ 1 & \nu^+ \end{pmatrix}. \tag{13.57}$$

The analysis of this is much more straightforward than for crossing and gives Figure 13.5. The delimiting case is $\nu^+\nu^- = -1$ for $\nu^+ < 0 < \nu^-$ (from Lemma 13.2, and otherwise for $\nu^- < 0 < \nu^+$ in the case $\text{sign}[\delta^2_{t\pm}\sigma(0)] < 0$).

The first component of the eigenvector from (13.52) satisfies

$$2(\gamma_\pm - \nu^+) = \nu^- - \nu^+ \pm \sqrt{(\nu^+ - \nu^-)^2 - 4} \begin{matrix} > 0 \text{ if } \nu^- > \nu^+ , \\ < 0 \text{ if } \nu^- < \nu^+ , \end{matrix}$$

implying

$$\gamma_\pm = (\gamma_\pm - \nu^+, 1) \in \mathcal{D}_{att} \qquad \text{if} \quad \nu^- > \nu^+ , \tag{13.58a}$$

and

$$\gamma_\pm = (\gamma_\pm - \nu^+, 1) \in \mathcal{D}_{C+} \qquad \text{if} \quad \nu^- < \nu^+ . \tag{13.58b}$$

For $\nu^+ < 0 < \nu^-$ and $\nu^+\nu^- < -1$, we have $\det \underline{\underline{S}} = 1 + \nu^+\nu^- < 0$, giving a finite-time saddle. The eigenvalues have signs $\gamma_- < 0 < \gamma_+$, and by (13.58a) both eigendirections lie in the sliding regions, giving the phase portraits Figure 13.5(v-vii) (the different cases are only distinguished by their intersection with the crossing dynamics). Only two trajectories pass through the singularity, a canard along the eigendirection γ_- and a faux canard along the eigendirection γ_+.

For $\nu^+ < 0 < \nu^-$, $\nu^+\nu^- > -1$, and $\nu^- - \nu^+ > 2$, we have $\det \underline{\underline{S}} = 1 + \nu^+\nu^- > 0$, implying a finite-time node. The eigenvalues are ordered as $\gamma_- < \gamma_+ < 0$ if $\nu^+ + \nu^- < 0$ and $0 < \gamma_- < \gamma_+$ if $\nu^+ + \nu^- > 0$. This creates two cases:

If $\nu^+ + \nu^- < 0$, the flow in the attracting sliding region is towards the singularity, and by (13.58a) both eigendirections lie in the sliding regions, producing the phase portraits $(iii - iv)$ in Figure 13.5. The strong eigendirection γ_- separates the sliding region into trajectories which hit the visible fold and trajectories which pass from attracting to repelling sliding, constituting a family of robust canards that includes one along the direction γ_+. We can call the canards along γ_- and γ_+ the primary strong and weak canards respectively.

If $\nu^+ + \nu^- < 0$, the flow in the attracting sliding region is away from the singularity, and by (13.58a) both eigendirections lie in the sliding regions, producing the phase portrait (ii) in Figure 13.5. The strong eigendirection γ_+ separates the sliding region into trajectories which hit the visible fold and trajectories which pass from repelling to attracting sliding, constituting a family of robust faux canards that includes one along the direction γ_-. We can call the canards along γ_+ and γ_- the primary strong and weak faux canards, respectively.

For $\nu^- - \nu^+ < 2$, we have $\det \underline{\underline{S}} = 1 + \nu^+\nu^- > 0$, but $\gamma_\pm$ are complex, implying a finite-time focus. The rotational character of the flow (lacking any real eigendirections) means all trajectories in the sliding regions rotate from the invisible to visible folds, giving portrait (i) making up the lower right majority region of Figure 13.5.

13.4.3 The Invisible Two-Fold

The sliding vector field (13.50) for $\mathrm{sign}[\delta_{t+}^2\sigma(0)] < 0 < \mathrm{sign}[\delta_{t-}^2\sigma(0)]$ is

$$\begin{pmatrix} \dot{x}_2 \\ \dot{x}_3 \end{pmatrix} = \frac{1}{x_2 + x_3}\underline{\underline{S}} \cdot \begin{pmatrix} x_2 \\ x_3 \end{pmatrix}, \qquad \underline{\underline{S}} = \begin{pmatrix} \nu^- & 1 \\ 1 & \nu^+ \end{pmatrix}. \tag{13.59}$$

Again the analysis of this is much more straightforward than for crossing, and gives Figure 13.6. The delimiting case is $\nu^+\nu^- = 1$ for $\nu^+, \nu^- < 0$ (from Lemma 13.2).

The first component of the eigenvector from (13.52) satisfies

$$2(\gamma_\pm - \nu^+) = \nu^- - \nu^+ \pm \sqrt{(\nu^+ - \nu^-)^2 - 4} \gtrless 0\,,$$

implying

$$\gamma_+ = \left(\gamma_+ - \nu^+, +1\right) \in \mathcal{D}_{att}\,, \quad \gamma_- = \left(\gamma_- - \nu^+, +1\right) \in \mathcal{D}_{C^+}\,. \tag{13.60}$$

For $\nu^\pm < 0$ and $\nu^+\nu^- > 1$, we have $\det \underline{\underline{S}} = -1 + \nu^+\nu^- > 0$ and $\operatorname{Tr} \underline{\underline{S}} = \nu^+ + \nu^- < 0$, giving a finite-time node where the flow is in towards the singularity in $\mathcal{D}_{att}$. The eigenvalues satisfy $\gamma_- < \gamma_+ < 0$, so by (13.60) the strong eigendirection $\boldsymbol{\gamma}_-$ lies in the crossing regions, and the weak eigendirection $\boldsymbol{\gamma}_+$ lies in the sliding regions, completing the phase portrait Figure 13.6(iii). The flow is outwards from the singularity in the repelling sliding region and inwards in the attracting region; thus, it consists entirely of canards.

For $\nu^+\nu^- < 1$, we have $\det \underline{\underline{S}} = -1 + \nu^+\nu^- < 0$, implying a finite-time saddle. The eigenvalues have signs $\gamma_- < 0 < \gamma_+$, so by (13.60) the inward eigendirection $\boldsymbol{\gamma}_-$ lies in the crossing regions and the outward eigendirection $\boldsymbol{\gamma}_+$ lies in the sliding regions, producing the phase portrait in Figure 13.6(i). Along the direction $\boldsymbol{\gamma}_+$ only, a single trajectory passes from repelling to attracting sliding, constituting a faux canard. The rest of the flow passes through the folds. (Cases (i–ii) have the same sliding phase portrait and are only distinguished by their crossing dynamics.)

For $\nu^\pm > 0$ and $\nu^+\nu^- > 1$, we have $\det \underline{\underline{S}} = -1 + \nu^+\nu^- > 0$, implying a finite-time node where the flow is outwards from the singularity in $\mathcal{D}_{att}$. The eigenvalues satisfy $0 < \gamma_- < \gamma_+$, so by (13.60) the strong eigendirection $\boldsymbol{\gamma}_+$ lies in the sliding regions, and the weak eigendirection $\boldsymbol{\gamma}_-$ lies in the crossing regions, again producing the phase portrait in Figure 13.6(i). Again, along the direction $\boldsymbol{\gamma}_+$ only, a single trajectory passes from repelling to attracting sliding, constituting a faux canard. The rest of the flow passes through the folds.

13.4.4 The Nonsmooth Diabolo Bifurcation: Sliding

As shown in Lemma 13.2, a bifurcation occurs for any of the two-folds when $\nu^+\nu^- = 1$ (for the visible/invisible two-folds) or $\nu^+\nu^+ = -1$ (for the mixed visibility two-fold). At the bifurcation, the condition (13.2) is violated, meaning the vector field can vanish at the singularity, and in fact we show here that it does. In the generic situation as $\nu^+\nu^-$ passes through unity, an equilibrium passes through the singularity from one sliding region to the other, changing stability in the process.

Figure 13.17 shows the sliding phase portrait for the truncated prototype, as $\nu^+\nu^-$ passes through unity in the case of the invisible two-fold. The invariant manifolds of the crossing map for the invisible two-fold are also shown, as these coalesce in the nonsmooth diabolo bifurcation at the same parameter value.

For the crossing map we investigated the nearby effect of higher-order terms at the diabolo bifurcation. To do the same for the sliding vector field is rather simpler. We will do this only for the invisible two-fold.

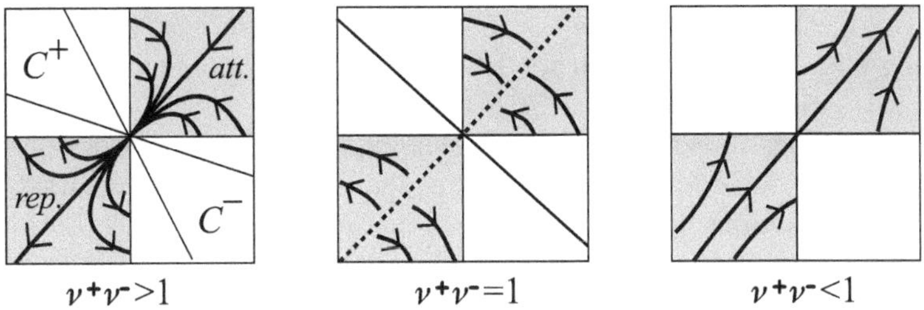

Fig. 13.17 Sliding dynamics corresponding to Figure 13.14. The sliding vector field consists of canards for $\nu^+\nu^- > 1$, a line of equilibria extending from the singularity for $\nu^+\nu^- = 1$, and a single faux canard for $\nu^+\nu^- < 1$.

Choose a coordinate u along the centre manifold—the line of equilibria in Figure 13.17 along the direction $\gamma_\pm = (-\nu^+, +1) \propto (1, -\nu^-)$. The one-dimensional sliding field along a centre manifold direction $(-\nu^+, 1)$ is

$$\begin{pmatrix} \dot{x}_2 \\ \dot{x}_3 \end{pmatrix} = \frac{1}{1-\nu^+}\underline{\underline{S}} \cdot \begin{pmatrix} -\nu^+ \\ 1 \end{pmatrix} = (1-\nu^+\nu^-)\begin{pmatrix} 1 \\ 0 \end{pmatrix} . \qquad (13.61)$$

Along this direction the vector field is constant, but vanishes identically when $\nu^+\nu^- = 1$. A lowest order perturbation (in terms of x_2 or x_3 along the centre manifold direction) gives

$$\begin{pmatrix} \dot{x}_2 \\ \dot{x}_3 \end{pmatrix} = \left(1-\nu^+\nu^- + \mu x_2\right)\begin{pmatrix} 1 \\ 0 \end{pmatrix} , \qquad (13.62)$$

illustrated in Figure 13.18. The sliding vector field is then non-vanishing along the centre manifold direction, except for an equilibrium at $x_2 = (\nu^+\nu^- - 1)/\mu$.

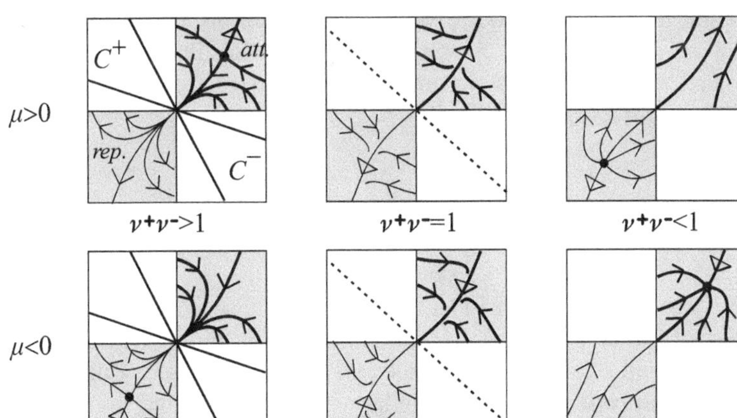

Fig. 13.18 Perturbed sliding dynamics. Triangular arrowheads indicate nonlinear drift.

If $\mu > 0$, then for $\nu^+\nu^- - 1 < 0$ this is a repelling sliding node in the repelling sliding region, which for $\nu^+\nu^- - 1 > 0$ becomes a sliding saddle in the attracting sliding region. If $\mu < 0$ then for $\nu^+\nu^- - 1 < 0$ this is an attracting sliding node in the attracting sliding region, which for $\nu^+\nu^- - 1 > 0$ becomes a sliding saddle in the repelling sliding region. This completes the picture of the sliding flow accompanying the nonsmooth diabolo bifurcation of the crossing map, as shown together in Figure 13.19.

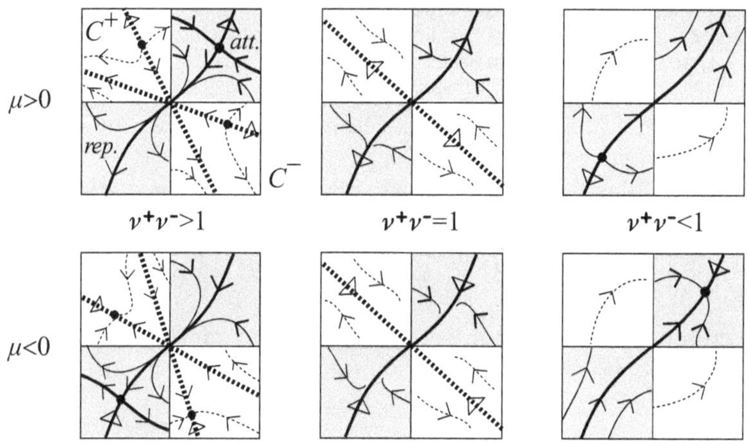

Fig. 13.19 Nonsmooth diabolo, sliding (shaded regions) and crossing (shown for the line preserving case $\alpha = \beta$).

We saw the more general treatment of the crossing maps towards the end of Section 13.3.5. The full picture near the bifurcation is therefore found by combining the different sliding flows in Figure 13.18 with all of the different crossing maps in Figure 13.16, and unfortunately for this more general analysis, we do not know, at present, which sliding flows are permitted to be matched with which crossing maps, unlike in the simpler but less general analysis of Figure 13.19.

13.5 Obsession 4: Sensitivity in the Layer

At last it seems that the crossing and sliding phase portraits in the neighbourhood of the two-fold are well understood. The sliding flow inside $\mathcal{D}$ and the crossing map outside $\mathcal{D}$ are structurally stable within certain classes, between which bifurcations occur. One last issue remains that finally brings us to the very heart of the problem: what happens at the singularity itself.

The reason for embarking upon our fourth Obsession is apparent when we combine the former three. Figure 13.20 illustrates a particular case of each

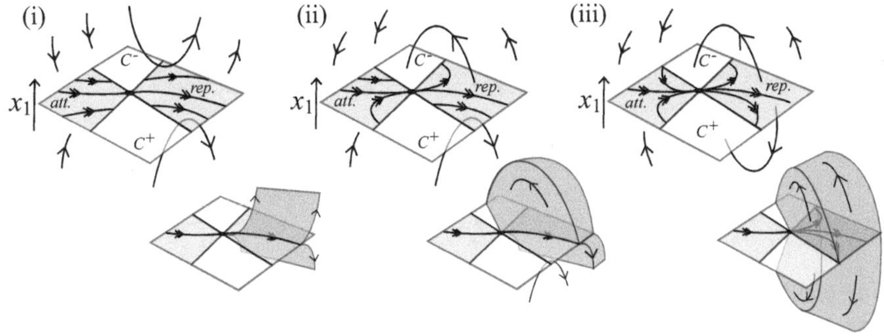

Fig. 13.20 Determinacy-breaking due to canards at each flavour of two-fold. The main figure shows the phase portrait, and the lower figure shows a single canard trajectory exploding into a flow that is set-valued, the set having two dimensions in (i) and three dimensions in (ii-iii).

flavour of two-fold that exhibits canards, that is, trajectories evolving from attractive sliding to repelling sliding.

A canard at a two-fold singularity behaves deterministically as it enters the attracting sliding region, until it hits the singularity. It then explodes into a set-valued forward evolution as it meets the repelling sliding region. At a visible two-fold, an isolated canard gives rise to a one-parameter family of outward trajectories. At a visible or mixed visibility two-fold, a robust canard gives rise to a two-parameter (or p parameter in higher dimensions with $1 < p < n - 1$) family of outward trajectories.

To inspect the flow as it actually traverses the singularity requires analysis inside the switching layer. We already showed, in Section 8.8, that at a complex tangency (of which the two-fold is the simplest example) the convex combination is structurally unstable inside the switching layer. It will therefore be necessary to perturb to a nonlinear combination, considering terms nonlinear in the switching multiplier λ. We will inspect the convex combination first, find the precise form of its degeneracy, and then perturb to obtain a structurally stable form which can be related, by explicit coordinate transforms, to a well-known smooth slow-fast dynamical system, following [117].

13.5.1 The Unperturbed System

As we did to study the sliding dynamics in Section 13.4, let us substitute the leading order of the two-fold prototype (13.5) into a convex combination, giving

$$\begin{aligned}(\dot{x}_1, \dot{x}_2, \dot{x}_3) &= \tfrac{1}{2}(1+\lambda)(-x_2, -s_+, \nu^+) + \tfrac{1}{2}(1-\lambda)(x_3, \nu^-, s_-) \qquad (13.63)\\ &= (f_1(x_1,x_2,x_3;\lambda),\ f_2(x_1,x_2,x_3;\lambda),\ f_3(x_1,x_2,x_3;\lambda)) \ .\end{aligned}$$

The dynamics on $x_1 = 0$ is given by the layer system

$$(\varepsilon\dot{\lambda}, \dot{x}_2, \dot{x}_3) = \tfrac{1}{2}(1+\lambda)(-x_2, -s_+, \nu^+) + \tfrac{1}{2}(1-\lambda)(x_3, \nu^-, s_-) . \quad (13.64)$$

The family of sliding solutions, where $\dot{\lambda} = 0$, forms a manifold

$$\mathcal{M} = \left\{ (\lambda, x_2, x_3) \in (-1, +1) \times \mathbb{R}^2 : \lambda = \frac{x_3 - x_2}{x_3 + x_2} \right\} . \quad (13.65)$$

The condition $|\lambda| < 1$ implies that $\mathcal{M}$ exists for $0 < x_2 x_3$, corresponding to attracting or repelling regions of the discontinuity surface.

The shape of $\mathcal{M}$ is perhaps easier to see expressed implicitly as $(x_3 + x_2)\lambda + (x_2 - x_3) = 0$, which describes a smooth surface that twists over near $x_2 = x_3 = 0$. As such $\mathcal{M}$ consists of two normally hyperbolic branches, one attractive in $x_2, x_3 > 0$ since $\partial f_1/\partial\lambda = -(x_3 + x_2)/2 < 0)$ and one repulsive in $x_2, x_3 < 0$ since $\partial f_1/\partial\lambda = -(x_3 + x_2)/2 > 0$, connected at $x_2 = x_3 = 0$ along a set of non-hyperbolic points

$$\mathcal{L} = \{(\lambda, x_2, x_3) \in \mathcal{M} : x_2 = x_3 = 0\} . \quad (13.66)$$

This line segment $\mathcal{L}$ constitutes the blowing up of the two-fold point $x_1 = x_2 = x_3 = 0$ into a line at $x_2 = x_3 = 0$ with $|\lambda| < 1$. Figure 13.21 shows an example of the piecewise-smooth system in (i), the switching layer showing $\mathcal{M}$ and $\mathcal{L}$ in (ii), which is then rotated in to view $\mathcal{M}$ 'face on' and show $\mathcal{L}$ more clearly in (iii).

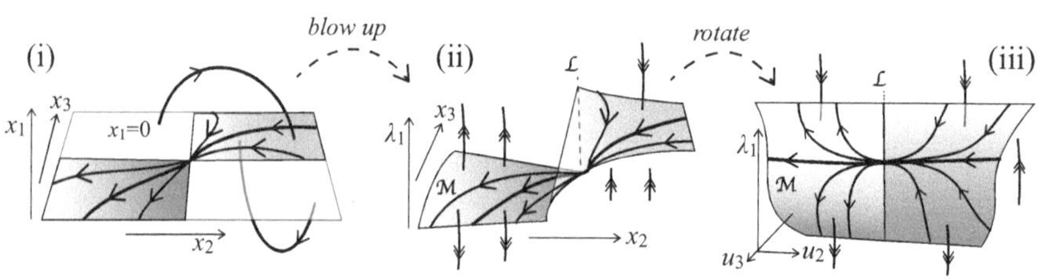

Fig. 13.21 The switching layer for the unperturbed system (13.63), for the example of an invisible two-fold. (i) The flow directions outside $x_1 = 0$ create an attracting sliding region in $x_2, x_3 > 0$ and repelling sliding region in $x_2, x_3 < 0$. (ii) The switching layer on $x_1 = 0$, where the sliding regions create a sliding manifold $\mathcal{M}$ (shaded), hyperbolic except along the vertical line $\mathcal{L}$, which aligns with the fast (double arrowed) λ dynamics. (iii) The dynamics in the manifold is best viewed along the v axis of rotated coordinates $u = x_2 + x_3$, $v = x_2 - x_3$.

By Proposition 9.1, since the prototype (13.63) depends linearly on λ and the origin is a complex tangency, the set $\mathcal{L}$ is structurally unstable in the flow. Specifically here we have the following.

Corollary 13.3. *In the switching layer system* (13.64) *of the two-fold singularity* (13.5), *the non-hyperbolic set* $\mathcal{L}$ *of the sliding manifold* $\mathcal{M}$ *lies everywhere tangent to the coordinate axis of the fast variable.*

Proof. The non-hyperbolic set $\mathcal{L}$ forms a line with tangent vector $e_{\mathcal{L}} = (1,0,0)$ in the space of (λ, x_2, x_3), which means it lies everywhere parallel to the fast λ-coordinate axis of the two-timescale system (13.64). □

As in Proposition 9.1, this degeneracy is related to the fact that all derivatives of f_1 with respect to the fast multiplier λ vanish along $\mathcal{L}$, not only the first derivative $\partial f_1/\partial\lambda = -x_3 - x_2 = 0$ which defines $\mathcal{L}$ as the set $x_2 = x_3 = 0$ but also all higher derivatives $\partial^r f_1/\partial\lambda^r = 0$ for any $r > 1$. Thus, this constitutes an infinite codimension degeneracy, as a typical perturbation may lead to any of these derivatives becoming nonzero.

What kind of perturbations will break the degeneracy? Proposition 9.1 tells us that to break the degeneracy requires terms nonlinear in λ, specifically hidden terms. It is easy to see that a term proportional to $\lambda^2 - 1$ added to to (13.63) may give $\frac{\partial^2 f_1}{\partial\lambda^2} \neq 0$. Adding constants or functions of the coordinates (x_1, x_2, x_3) to (13.63) would succeed only in moving $\mathcal{L}$ in the (x_2, x_3) plane, with the derivatives $\frac{\partial^r}{\partial\lambda^r} f_1(0, x_2, x_3; \lambda)$ for $r > 0$ still vanishing on $\mathcal{L}$. In fact we will show that perturbing $\dot{x}_1$ with a term proportional to $\lambda^2 - 1$ is sufficient for structural stability. Perturbing the $\dot{x}_2$ or $\dot{x}_3$ components of the vector field is neither necessary nor sufficient to break the degeneracy, therefore we leave them unaltered.

13.5.2 The Perturbed System

As a prototype for a structurally stable form of the two-fold, we therefore take

$$
\begin{aligned}
(\dot{x}_1, \dot{x}_2, \dot{x}_3) &= \tfrac{1}{2}(1+\lambda)\left(-x_2, -s_+, \nu^+\right) \\
&\quad + \tfrac{1}{2}(1-\lambda)\left(x_3, \nu^-, s_-\right) + (1-\lambda^2)(\alpha, 0, 0) \\
&:= \left(f_1(x_1, x_2, x_3; \lambda), f_2(x_1, x_2, x_3; \lambda), f_3(x_1, x_2, x_3; \lambda)\right) ,
\end{aligned}
\tag{13.67}
$$

where α is a constant. We will show that this does not suffer the structural stability of (13.63) and moreover relates the two-fold to an important singularity from smooth multi-timescale systems. The layer system on $x_1 = 0$ is

$$
\begin{aligned}
\left(\varepsilon\dot{\lambda}, \dot{x}_2, \dot{x}_3\right) &= \tfrac{1}{2}(1+\lambda)\left(-x_2, -s_+, \nu^+\right) \\
&\quad + \tfrac{1}{2}(1-\lambda)\left(x_3, \nu^-, s_-\right) + (1-\lambda^2)(\alpha, 0, 0) .
\end{aligned}
\tag{13.68}
$$

Proposition 13.1. *The structurally stable switching layer system* (13.68) *can be transformed into*

$$\begin{aligned} \varepsilon\dot{x} &= y + x^2 + \mathsf{O}(\varepsilon x, \varepsilon z, xz) \\ \dot{y} &= pz + qx + \mathsf{O}(z^2, xz) \\ \dot{z} &= r + \mathsf{O}(z, x) \end{aligned}$$

provided $\alpha \neq 0$ *for small* $\varepsilon > 0$, *where* p, q, r *are real constants, and provided the conditions* $\frac{1}{2}(\nu^+ - \nu^-) \leq 1 = -s_+ = s_-$ *or* $\frac{1}{2}(\nu^+ - \nu^-) \geq -1 = -s_+ = s_-$ *do not hold [38].*

The significance is that the smooth system in Proposition 13.1 is the canonical local form for a two-timescale system with transversally intersecting attracting and repelling slow invariant manifolds [218]. Such an intersection is the fundamental singularity that makes canards possible in smooth two-timescale systems. It turns out that the cases excluded by the conditions $\pm\frac{1}{2}(\nu^+ - \nu^-) \leq 1 = \mp s_+ = \mp s_-$ are those in which there are no orbits of the sliding flow passing through the singularity, i.e. where no canards exist.

The proof of the proposition involves a coordinate transformation that untwists $\mathcal{M}$ and then folds it into the parabolic surface where the $\dot{z}$ equation inProposition 13.1 vanishes. To emphasise the geometry, we proceed by way of three lemmas, establishing first the non-degeneracy of $\mathcal{L}$, second locating a new singularity that distinguishes a special point along $\mathcal{L}$, and finally morphing $\mathcal{M}$ into the $\dot{x}$ nullcline of the system in Proposition 13.1. At the end of the section, we review the basic analysis used to classify the resulting singularity.

The fixed points of the fast $\dot{\lambda}$ subsystem of (13.68) define the sliding manifold

$$\mathcal{M} = \left\{ \begin{array}{l} (\lambda, x_2, x_3) \in (-1, +1) \times \mathbb{R}^2 \, : \\ \qquad \frac{1}{2}(1 - \lambda)x_3 - \frac{1}{2}(1 + \lambda)x_2 + \alpha(1 - \lambda^2) = 0 \end{array} \right\}, \tag{13.69}$$

which is normally hyperbolic except on the set

$$\mathcal{L} = \left\{ (\lambda, x_2, x_3) \subset \mathcal{M} : \lambda = 2\frac{2\alpha + x_3 - x_2}{x_3 + x_2} = -\frac{x_3 + x_2}{4\alpha} \right\}. \tag{13.70}$$

From the two equations in (13.70), we can express $\mathcal{L}$ in parametric form as

$$(\lambda, x_2, x_3) = \mathcal{L}(\lambda) := \left(\lambda, \alpha(\lambda - 1)^2, \; -\alpha(\lambda + 1)^2\right), \tag{13.71}$$

from which we can show that $\mathcal{L}$ lies in a general position in the flow.

Lemma 13.3. *The non-hyperbolic set* $\mathcal{L}$ *is transverse to the fast direction of* (13.68).

Proof. By differentiating (13.71) with respect to λ, we find that the curve $\mathcal{L}$ has tangent vector $e_{\mathcal{L}} = (1\,,\, 2\alpha(\lambda - 1)\,,\, -2\alpha(\lambda + 1))$, which for all $|\lambda| < 1$ is transverse to the coordinate axes provided $\alpha \neq 0$. □

We said in the linear system that $\mathcal{L}$ defined in (13.66) constituted the blow-up of the two-fold singularity. If $\mathcal{L}$ is no longer degenerate, then where did the singularity go? It turns out that, since $\mathcal{L}$ is a curve, generically there may exist an isolated point along $\mathcal{L}$ where the flow's projection along the λ-direction onto the nullcline $f = 0$ is indeterminate. We shall call this the *canard* singularity, defined in the following lemma.

Lemma 13.4. *For the values of the constants s_+, s_-, ν^+, ν^-, given in Proposition 13.1, there exists an isolated singularity of the flow along the non-hyperbolic set $\mathcal{L}$, where the projection of the slow flow onto $\mathcal{M}$ lies tangent to $\mathcal{L}$.*

Proof. Consider the slow critical subsystem obtained by letting $\varepsilon \to 0$ in (13.68),

$$(0, \dot{x}_2, \dot{x}_3) = \tfrac{1}{2}(1+\lambda)\left(-x_2, -s_+, \nu^+\right) + \tfrac{1}{2}(1-\lambda)\left(x_3, \nu^-, s_-\right) + \alpha(1-\lambda^2)(1,0,0) \ . \tag{13.72}$$

$\mathcal{M}$ is the surface where $f_1 = 0$, so a solution of (13.72) that remains on $\mathcal{M}$ for an interval of time satisfies $\dot{f}_1 = 0$. We can find $\dot{\lambda}$ on $\mathcal{M}$ using the chain rule, writing

$$\begin{aligned}\dot{f}_1 &= \left(\dot{\lambda}, \dot{x}_2, \dot{x}_3\right) \cdot (\partial/\partial\lambda, \partial/\partial x_2, \partial/\partial x_3) f_1 \\ &= \dot{\lambda}\frac{\partial f}{\partial \lambda} + (f_2, f_3) \cdot \frac{\partial f_1}{\partial(x_2, x_3)} = 0 \ ,\end{aligned}$$

which rearranges to

$$\dot{\lambda} = -(f_2, f_3) \cdot \frac{\partial f_1}{\partial(x_2, x_3)} \Big/ \left(\frac{\partial f_1}{\partial \lambda}\right) \ .$$

Thus $\dot{\lambda}$ is indeterminate on $\mathcal{M}$ at points where the numerator and denominator of this vanish, or in full, where

$$0 = f_1 = \frac{\partial f_1}{\partial \lambda} = (f_2, f_3) \cdot \frac{\partial f_1}{\partial(x_2, x_3)} \ . \tag{13.73}$$

These three conditions define an isolated singularity on $\mathcal{L} \subset \mathcal{M}$. Denoting the value of f_i at the singularity as $f_{i\star}$, the equations (13.73) imply

$$\begin{aligned}0 &= \tfrac{1}{2}(f_{2\star}, f_{3\star}) \cdot (-1 - \lambda_\star, 1 - \lambda_\star) \\ &= \left(\tfrac{\nu^- - s_+}{2} - \tfrac{\nu^- + s_+}{2}\lambda_\star \, , \ \tfrac{\nu^+ + s_-}{2} + \tfrac{\nu^+ - s_-}{2}\lambda_\star\right) \cdot \left(-\tfrac{1+\lambda_\star}{2}, \tfrac{1-\lambda_\star}{2}\right) ,\end{aligned} \tag{13.74}$$

which we can solve to find that the canard singularity lies at $(\lambda, x_2, x_3) = (\lambda_\star, x_{2\star}, x_{3\star})$, where

$$\lambda_\star = \frac{\frac{s_+ - s_-}{\nu^+ - \nu^-} \pm \sqrt{1 - \frac{4s_+s_-}{(\nu^+ - \nu^-)^2}}}{1 + \frac{-s_+ s_-}{\nu^+ - \nu^-}} , \quad x_{2\star} = \alpha(\lambda_\star - 1)^2 , \quad x_{3\star} = -\alpha(\lambda_\star + 1)^2 . \tag{13.75}$$

Noting that s_+ and s_- just take values ± 1:

- In the case $-s_+ = s_- = 1$, we have $\lambda_\star = \frac{-2}{\nu^+ - \nu^-} \pm \sqrt{1 + \frac{4}{(\nu^+ - \nu^-)^2}}$, implying that there exists a unique solution $\lambda_\star \in (-1, +1)$ for any ν^+ and ν^- (the positive root for $\nu^+ > \nu^-$, the negative root for $\nu^+ < \nu^-$);
- In the case $-s_+ = s_- = -1$, we have $\lambda_\star = \frac{2}{\nu^+ - \nu^-} \pm \sqrt{1 + \frac{4}{(\nu^+ - \nu^-)^2}}$, implying that there exists a unique solution $\lambda_\star \in (-1, +1)$ for any ν^+ and ν^- (the positive root for $\nu^+ < \nu^-$, the negative root for $\nu^+ > \nu^-$);
- In the case $-s_+ = s_- = 1$, we have $\lambda_\star = \pm\sqrt{\frac{\nu^+ - \nu^- - 2}{\nu^+ - \nu^- + 2}}$, implying that there exist two solutions $\lambda_\star \in (-1, +1)$ for $\nu^+ - \nu^- > 2$, and no points otherwise.
- In the case $-s_+ = s_- = -1$, we have $\lambda_\star = \pm\sqrt{\frac{\nu^+ - \nu^- + 2}{\nu^+ - \nu^- - 2}}$, implying that there exist two solutions $\lambda_\star \in (-1, +1)$ for $\nu^+ - \nu^- < -2$, and no points otherwise.

□

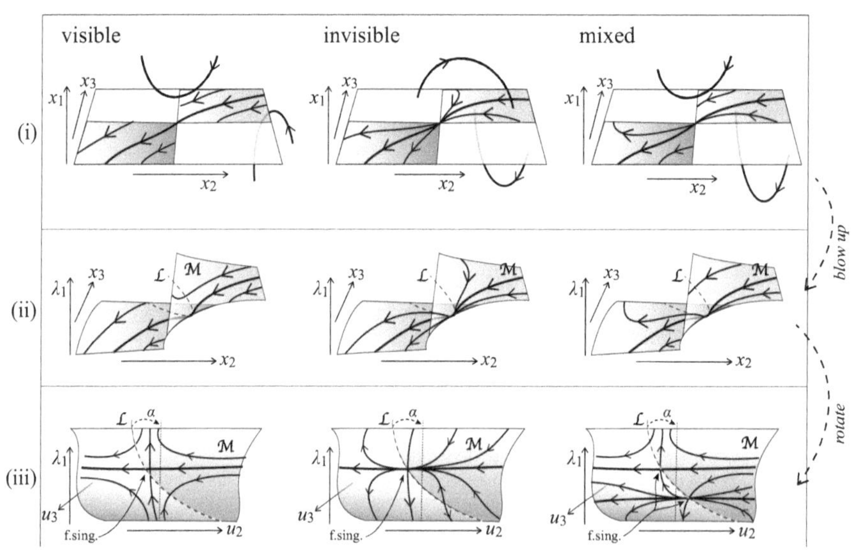

Fig. 13.22 Blowing up the perturbed ($\alpha \neq 0$) system, for examples of each flavour of two-fold. Labelling as in Figure 13.21. Note in the layer (ii) that $\mathcal{L}$ is now a curve. Rotating around the u axis in (iii), we can see the attracting branch (upper right segment) and repelling branch (lower left segment) of the sliding manifold $\mathcal{M}$ (shaded), connected by $\mathcal{L}$. The canard singularity ($\star$ sing.) appears along $\mathcal{L}$, two in the case of mixed visibility, recognized as having a phase portrait that resembles a saddle or node if we reverse time in the repelling branch of $\mathcal{M}$.

These singularities are illustrated in Figure 13.22, which shows an example of the perturbed system (13.67) and its switching layer for each flavour of two-fold in (i) (corresponding to those in Figure 13.2), followed by their switching layer (ii), and a rotation (iii) to show the phase portrait around the set $\mathcal{L}$ more clearly (similar to Figure 13.21). In the mixed visible-invisible two-fold (far right column), for example, the nonsmooth system (i) has a phase portrait with infinitely many intersecting trajectories traversing the singularity, while the layer system (ii–iv) splits these into distinguishable orbits, a finite number of which asymptote to the attracting and repelling branches of the critical manifold.

Lemma 13.4 therefore establishes the existence of at least one unique canard singularity on $\mathcal{L}$ in the cases listed in Proposition 13.1. It remains to map these onto the canonical form of the singularity within the layer. In the cases where $\lambda_\star$ has two admissible solutions, there are two canard singularities in the switching layer, and then we can proceed with the following analysis about each singularity, but we obtain different constants in the pair of final local expressions. In the cases where $\lambda_\star$ is unique, we can proceed directly with the analysis that follows. In the cases when $\lambda_\star$ does not exist, no equivalence can be formed; these are the cases when the two-fold's sliding portrait is of focal type (see [38]), and there exists no orbits passing directly between the attracting and repelling branches of sliding, since orbits wind around the two-fold but never enter or leave it. So let us exclude those cases where $-s_+ = s_- = 1$ with $\nu^+ - \nu^- \leq 2$ and $-s_+ = s_- = -1$ with $\nu^+ - \nu^- \geq -2$.

Lemma 13.5. *Coordinates can be defined in which the canard singularity of (13.68) lies at the origin, and $\mathcal{L}$ lies along a coordinate axis corresponding to a slow variable.*

Proof. Taking a valid solution of $\lambda_\star$ from (13.75) for $|\lambda| < 1$, a translation puts the singularity at the origin of the new coordinates

$$\xi_1 = \lambda - \lambda_\star\,, \qquad \xi_2 = x_2 - x_{2\star}\,, \qquad \xi_3 = x_3 - x_{3\star}\,. \tag{13.76}$$

Then f_1 becomes

$$f_1 = -\frac{1+\lambda_\star}{2}\xi_2 + \frac{1-\lambda_\star}{2}\xi_3 - \left(\frac{\xi_3+\xi_2}{2} + \alpha\xi_1\right)\xi_1\,, \tag{13.77}$$

found by using (13.74)-(13.75) to ensure that terms involving $x_{2\star}$ and $x_{3\star}$ vanish. We then need to find coordinates in which $\mathcal{L}$ lies along a coordinate axis. From (13.71) we can obtain the ξ_1-parameterized expression for $\mathcal{L}$,

$$(\xi_1, \xi_2, \xi_3) = (\xi_1, -\alpha\xi_1(2 - 2\lambda_\star - \xi_1),\ -\alpha\xi_1(2 + 2\lambda_\star + \xi_1))\ ,$$

and re-arrange this to take ξ_3 as a parameter, expressing $\mathcal{L}$ as $(\xi_1, \xi_2) = (\xi_{1L}(\xi_3), \xi_{2L}(\xi_3))$, where

$$\begin{pmatrix} \xi_{1L}(\xi_3) \\ \xi_{2L}(\xi_3) \end{pmatrix} := \begin{pmatrix} -1-\lambda_\star+\sqrt{(1+\lambda_\star)^2-\xi_3/\alpha} \\ -\xi_3-4\alpha(-1-\lambda_\star+\sqrt{(1+\lambda_\star)^2-\xi_3/\alpha}) \end{pmatrix} . \tag{13.78}$$

The derivatives of these functions are needed to evaluate the vector field components below, these are

$$\xi_{1L}'(\xi_3) = \frac{-1/2\alpha}{1+\lambda_\star+\xi_{1L}(\xi_3)} , \quad \xi_{2L}'(\xi_3) = \frac{1-\lambda_\star-\xi_{1L}(\xi_3)}{1+\lambda_\star+\xi_{1L}(\xi_3)} . \tag{13.79}$$

We can then rectify $\mathcal{L}$ to lie along some ζ_3 axis by defining new coordinates

$$\zeta_1 = \xi_1 - \xi_{1L}(\xi_3) , \qquad \zeta_2 = \xi_2 - \xi_{2L}(\xi_3) , \qquad \zeta_3 = \xi_3 . \tag{13.80}$$

The original vector field components can then be written as

$$\begin{aligned} f_1 &= -\tfrac{1+\lambda_\star+\xi_{1L}}{2}\zeta_2 - \alpha\zeta_1^2 - \zeta_2\zeta_1/2 , \\ f_2 &= f_{2\star} + (\zeta_1+\xi_{1L}(\zeta_3))\tfrac{\partial f_{2\star}}{\partial\lambda} = f_{2\star} + \mathsf{O}(\zeta_1,\zeta_3) , \\ f_3 &= f_{3\star} + (\zeta_1+\xi_{1L}(\zeta_3))\tfrac{\partial f_{3\star}}{\partial\lambda} = f_{3\star} + \mathsf{O}(\zeta_1,\zeta_3) . \end{aligned} \tag{13.81}$$

With a little algebra, we find that

$$\begin{aligned} \varepsilon\dot{\zeta}_1 &= \varepsilon\dot{\xi}_1 - \varepsilon\dot{\xi}_3\xi_{1L}'(\xi_3) \\ &= \frac{1+\lambda_\star}{2}\left(\frac{\varepsilon f_{3\star}}{\alpha(1+\lambda_\star)^2} - \zeta_2\right) - \alpha\zeta_1^2 + \mathsf{O}\left(\varepsilon\zeta_3,\varepsilon\zeta_1,\zeta_2\zeta_3,\zeta_2\zeta_1,\zeta_1^3\right) . \end{aligned}$$

A small shift $\tilde{\zeta}_2 = \zeta_2 - \frac{\varepsilon f_{3\star}}{\alpha(1+\lambda_\star)^2}$ yields, after some lengthy but straightforward algebra,

$$\dot{\tilde{\zeta}}_2 = f_2 - \dot{\zeta}_3\xi_{2L}'(\zeta_3) \;=\; \tilde{q}\zeta_1 + \frac{\tilde{p}}{\alpha}\zeta_3 + \mathsf{O}\left(\zeta_3^2,\zeta_1\zeta_3\right) ,$$

using the relations in (13.74) and (13.81) to show that any terms not proportional to ζ_1 or ζ_3 vanish, where

$$\tilde{q} = \frac{\partial f_{2\star}}{\partial\lambda} - \frac{\partial f_{3\star}}{\partial\lambda}\frac{1-\lambda_\star}{1+\lambda_\star} , \qquad \tilde{p} = -\frac{f_{3\star}}{(1+\lambda_\star)^3} - \frac{\tilde{q}}{2(1+\lambda_\star)} .$$

The last thing to do is just scaling. Collecting everything together so far we have

$$\begin{aligned} \varepsilon\dot{\zeta}_1 &= k\tilde{\zeta}_2 - \alpha\zeta_1^2 + \mathsf{O}(\varepsilon\zeta_1,\varepsilon\zeta_3,\zeta_1\zeta_3) \\ \dot{\tilde{\zeta}}_2 &= \tfrac{\tilde{p}}{\alpha}\zeta_3 + \tilde{q}\zeta_1 + \mathsf{O}\left(\zeta_3^2,\zeta_1\zeta_3\right) \\ \dot{\zeta}_3 &= f_{3\star} + \mathsf{O}(\zeta_3,\zeta_1) \end{aligned}$$

where $k = -\frac{1}{2}(1+\lambda_\star)$. Defining new variables $\eta_1 = \sqrt{|\alpha|}\zeta_1$, $\eta_2 = -k\mathrm{sign}(\alpha)\tilde{\zeta}_2$, $\eta_3 = -\mathrm{sign}(\alpha)\zeta_3$, and $\tilde{t} = -\sqrt{|\alpha|}\mathrm{sign}(\alpha)t$, gives

$$\begin{aligned}\varepsilon\dot{\eta}_1 &= \eta_2 + \eta_1^2 + \mathsf{O}\left(\varepsilon\eta_1, \varepsilon\eta_3, \eta_1\eta_3\right)\\ \dot{\eta}_2 &= p\eta_3 + q\eta_1 + \mathsf{O}\left(\eta_3^2, \eta_1\eta_3\right)\\ \dot{\eta}_3 &= r + \mathsf{O}\left(\eta_3, \eta_1\right)\end{aligned} \tag{13.82}$$

where

$$\begin{aligned} r &= \tfrac{1}{\sqrt{|\alpha|}} f_{3\star}\,, \quad p = \tfrac{1}{2\sqrt{|\alpha|}}\left\{\tfrac{q}{1+\lambda_\star} - \tfrac{f_{3\star}}{|\alpha|(1+\lambda_\star)^2}\right\}\,,\\ q &= \tfrac{1}{2|\alpha|}\left\{(1-\lambda_\star)\tfrac{\partial f_{3\star}}{\partial\lambda} - (1+\lambda_\star)\tfrac{\partial f_{2\star}}{\partial\lambda}\right\}\,. \end{aligned} \tag{13.83}$$

Replacing (η_1, η_2, η_3) with (x, y, z), this is the result in Lemma 13.5 and giving Proposition 13.1, clearly valid only for $\alpha \neq 0$ (otherwise the transformation is singular). □

The quantities

$$\begin{aligned} q &= \tfrac{\operatorname{sign}(\nu^+ - \nu^-)}{|\alpha|}\kappa \qquad \& \\ rp &= \tfrac{1}{4\alpha^2}\left(-\kappa^2 + \tfrac{|\nu^+ + \nu^-|}{2}\kappa\right)\,, \quad \kappa = \sqrt{\left(\tfrac{\nu^+ - \nu^-}{2}\right)^2 - s_+ s_-} \end{aligned} \tag{13.84}$$

will be useful below. Importantly their signs do not depend on the sign of α. Instead,

$$\operatorname{sign}(q) = \operatorname{sign}(\nu^+ - \nu^-) \quad \text{and} \quad \operatorname{sign}(rp) = \operatorname{sign}\nu^+\nu^- + s_+ s_-\,, \tag{13.85}$$

the first following since $\kappa > 0$ by definition, the latter following since also

$$-\kappa + \tfrac{|\nu^+ + \nu^-|}{2} \gtrless 0 \quad \Rightarrow \quad \tfrac{|\nu^+ + \nu^-|}{2} \gtrless \kappa \quad \Rightarrow \quad \nu^+\nu^- + s_+ s_- \gtrless 0\,.$$

In the literature on smooth two-timescale systems, the connection of attracting and repelling branches of a slow invariant manifolds has been well studied, leading to a generic canonical form (13.82) as described in [218]. In the present notation, this requires $\mathcal{M}$ and $\mathcal{L}$ to be nondegenerate, given by

$$f = 0\,, \qquad \tfrac{\partial f}{\partial\lambda} = 0\,, \qquad \tfrac{\partial f}{\partial y}, \tfrac{\partial f}{\partial z} \neq 0\,, \qquad \tfrac{\partial^2 f}{\partial\lambda^2} \neq 0\,, \tag{13.86}$$

the first three of which are satisfied on $\mathcal{L}$, while the fourth holds only for $\alpha \neq 0$. A canard singularity is a generic point along $\mathcal{L}$ where moreover $0 = f = \dot{f} = \frac{\partial f}{\partial x}$, with non-degeneracy conditions $\frac{\partial f}{\partial y}, \frac{\partial f}{\partial z}, \frac{\partial^2 f}{\partial x^2} \neq 0$.

There are different classes of the canard singularity. These have been well studied in [218], and their classification depends on the dynamics inside $\mathcal{M}$ (on the t timescale). These classes are now closely associated with the different flavours of two-fold singularity in whose blow-up they appear. From the expression (13.82) with (13.83), we see that the class depends not only on the constants s_+, s_-, ν^+, ν^- of the original piecewise-smooth system but also on the 'hidden' parameter α.

To classify the dynamics on $\mathcal{M}$ is fairly simple and confirms the different local phase portraits seen in Figure 13.22. The slow invariant manifold corresponding to $\mathcal{M}$ is now the surface $0 = \eta_2 + \eta_1^2$, and we can find the dynamics on it by projecting the system (13.82) onto $\mathcal{M}$, found by differentiating $0 = \eta_2 + \eta_1^2$ with respect to time. This gives $0 = p\eta_3 + q\eta_1 + 2\eta_1\dot{\eta}_1 + \mathsf{O}\left(\eta_3^2, \eta_1\eta_3\right)$, which we can re-arrange for $\dot{\eta}_1$, and we have $\dot{\eta}_3$ directly from (13.82), giving

$$\begin{pmatrix} \dot{\eta}_1 \\ \dot{\eta}_3 \end{pmatrix} = \frac{1}{-2\eta_1}\left\{\begin{pmatrix} q & p \\ -2r & 0 \end{pmatrix}\begin{pmatrix} \eta_1 \\ \eta_3 \end{pmatrix} + \mathsf{O}\left(\eta_3^2, \eta_1\eta_3\right)\right\} .$$

A classification then follows by neglecting the singular prefactor $1/2\eta_1$ and considering whether the phase portrait is that of a focus, a node, or a saddle. This is determined by the 2×2 matrix Jacobian, which has trace q, determinant $2pr$, and eigenvalues $(q \pm \sqrt{q^2 - 8pr})/2$. This will not be the true system's phase portrait because the timescaling from the $-1/2\eta_1$ factor is positive in the attractive branch of $\mathcal{M}$, negative (time-reversing) in the repulsive branch, and divergent at the singularity (turning infinite time convergence to the singularity into finite time passage *through* the singularity). The effect of this is to fold together attracting and repelling pairs of each equilibrium type, so each equilibrium becomes a 'folded equilibrium', forming a continuous bridge between branches of $\mathcal{M}$.

As a result, the flow on $\mathcal{M}$ is a folded saddle if $pr < 0$, a folded node if $0 < 8pr < q^2$, and a folded-focus if $q^2 < 8pr$. Canard cases occur for $q > 0$ and faux canard for $q < 0$. In the visible two-fold the singularity becomes a folded saddle, in the invisible case it becomes a folded node, while the mixed case becomes a pair consisting of one folded saddle and one folded node.

In the cases depicted in Figure 13.22 there exist one or more canard trajectories, passing from the attractive sliding region to the repelling sliding region in finite time (as we showed in Section 13.4). Thus the flow evolves deterministically until it arrives at the singularity, at which point determinacy breaking reders the flow set-valued, following a continuous family of trajectories into the repelling region and into the regions $x_1 \gtrless 0$. This is illustrated in Figure 13.20. It occurs in the invisible case when $\nu^+, \nu^- < 0$ and $\nu^+\nu^- > 1$, in the visible case when $\nu^+ < 0$ or $\nu^- < 0$ or $\nu^+\nu^- < 1$, and finally in the mixed case when $\nu^+ < 0 < \nu^-$ and $\nu^+\nu^- < -1$ or when $\nu^+ + \nu^- < 0$ and $\nu^+ - \nu^- < -2$. (The particular cases shown in Figure 13.20 are (i) $-s_+ = s_- = -1$ with $\nu^+ < 0$ or $\nu^- < 0$ or $\nu^+\nu^- < 1$; (ii) $-s_+ = s_- = 1$ with $\nu^+, \nu^- < 0$ and $\nu^+\nu^- > 1$; and (iii) $-s_+s_- = -1$ with $\nu^+ < 0 < \nu^-$ and $\nu^+\nu^- < -1$ or with $\nu^+ + \nu^- < 0$ and $\nu^+ - \nu^- < -2$.)

The phase portraits in Figure 13.22 resolve the passage through the singularity in more detail, revealing strong and weak eigendirections and showing that the determinacy-breaking trajectories persist in the switching layer.

We have referred to the origin of (13.82) as the 'canard singularity' to avoid confusion, but in smooth two-timescale systems, it is usually known as a 'folded equilibrium' (a folded node, folded saddle, etc. depending on the phase portrait inside $\mathcal{M}$) [218], occurring at the 'fold' of a slow critical manifold as observed in the η space in Figure 13.22(iv); this is an unfortunate clash of 'fold' terminologies with the piecewise-smooth literature, rooted in much older work on singularities of flows and surfaces, but it need not be a source of confusion. A qualitative association between the two-fold and the canard singularity was made in [44, 45], based on similarities between the phase portraits on $\mathcal{M}$, and the result above clarifies this relation, but requires obtaining structural stability inside the switching layer via introduction of terms nonlinear in λ.

The cases excluded by Proposition 13.1 were those in which no canards or faux canards exist in the slow-fast system. Trajectories connecting the attractive and repulsive sliding regions occur when transversal intersections exist between the attracting and repelling branches of the sliding manifolds $\mathcal{M}$. If no such intersections exist, the critical system possesses no canard singularities and hence is excluded from Proposition 13.1. Hence the omission of these cases is consistent, and ultimately their omission is obviously necessary in the equivalence sought in the proposition.

This just touches the surface of the problem, providing a classification that opens the way to deeper study of the two-fold's dynamics. The papers [218, 219, 47] delve into much more detail of the singularly perturbed canard singularity, i.e. for $\varepsilon > 0$. We obtain this, for example, if we smooth out the discontinuity, as is often done for reasons of regularization in analysis or numerical simulation. (We looked at smoothing in general in Section 12.3.) Smoothing the discontinuity in (13.63), as is done in [133, 151], for example, does not remove its degeneracy, and the authors of those papers use other blow-up techniques to establish the existence or not of canards in certain cases. Ultimately the implications for the piecewise-smooth system are of interest only in the singular limit $\varepsilon \to 0$, not, as concerns singular perturbation studies, what happens for $\varepsilon > 0$.

13.6 An Unfinished Saga

It has been nearly half a century since Filippov described how to solve a discontinuous differential equation [70], and half that since his work [71] raised the problem of the stability of the two-fold singularity. As Filippov stated in [71], there are several topological classes of the visible and mixed visibility two-fold, and infinitely many classes of the invisible ('Teixeira') two-fold. Within each class, however, the topology is structurally stable. For the invisible two-fold when $\nu^+\nu^- \lesssim 1$ and $\nu^\pm < 0$, these classes are infinitely crowded, such that a small perturbation of the system yields a class where the flow rotates a different number of times around the singularity. These bifurcations occur when $\nu^+\nu^- = \cos^2 \frac{\pi}{k+1}$ and k is an integer, whereupon the boundaries of repelling and attracting sliding are mapped exactly onto each other by the flow. Clearly this scenario is topologically unstable, and, at these values, the role of higher-order terms becomes important to break the degeneracy. For the case $k \to \infty$ this was studied via perturbations of the nonsmooth diabolo bifurcation above. For finite k, one expects that higher-order terms will perturb the bifurcation curves $\nu^+\nu^- = constant$, without altering the dynamics (in particular the range of crossing numbers exhibited by the flow for nearby parameters) significantly.

In essence there is nothing in the first 'obsession' that was not already covered by Filippov in Chap. 5 of [71] (the same cannot be said for Obsessions 2–4), though the perhaps overly concise presentation in [71] led to some results remaining obscure until their rediscovery across a series of papers [205, 206, 119, 69]. The planar fold-folds were studied in [71] too and revisited in [140] (we visited them ourselves in Section 8.6). Different forms for the three-dimensional two-folds were studied in [205, 206] which focussed on proving its structural stability, providing various necessary conditions, but running into difficulties which turned out to be due to the various bifurcations uncovered in [119, 37, 69], described in the next few sections. The n-dimensional form (13.5) was fully derived in Section 6.1 (as first done in [38])

Not wishing to extinguish the two-fold's ability to confound, there remain certain peculiarities within the classification. For instance, it was pointed out by Jesus Enrique Achire Quispe that, for certain parameter values, it would seem possible for pairs of orbits to form closed loops—pseudo-orbits—if they become connected by sliding segments in both the repelling region $\mathcal{D}_{rep}$ and the attracting region $\mathcal{D}_{att}$ (Figure 13.23(ii)). Simulations of the prototype system suggest this may occur only at $\nu^+\nu^- = 1$, but this has not been proven conclusively.

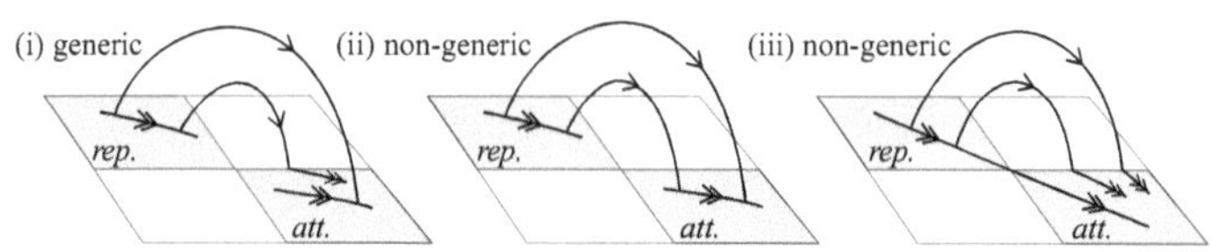

Fig. 13.23 (i) a typical stable topology, (ii) a closed loop formed by sliding and non-sliding segments, (iii) a unique sliding segment through the singularity maps onto a fold. Scenarios (ii)–(iii) represent unstable topologies but have no critical effect on dynamics.

By virtue of the local analysis presented here, the two-fold's flow intrinsically unfolds in three dimensions, with higher dimensions merely decorating the central structure, and weaving together the different dynamical substructures the flow can exhibit. In higher dimensions the two-fold singularity is a set of points, in which different subsets may exhibit different classes of the behaviours shown throughout this chapter. Transitions of the flow between these classes can be expected to result in further novel dynamics. I would think that investigations of such behaviour might lead to fascinating dynamics, a worthy course on the adventure into higher dimensions.

Highly degenerate forms of the two-folds occur in systems with symmetries. Like the two-fold, these have a long history in nonsmooth systems [4, 166] and remain a subject of current interest, particularly in electronic control. Typically, in $n > 3$ dimensions or when parameters are varied, one can expect the degeneracy conditions in Definition 6.2 to be violated at certain points, curves, surfaces, etc., depending on the value of n. For example, an equilibrium may pass through a two-fold [37, 206], a pair of two-folds may collide to form a fold-cusp (see Section 6.7, Section 8.5.6, and [206]), two-folds may occur at the intersection of multiple discontinuity submanifolds, and so on, and of such scenarios in higher dimensions, little is known.

Chapter 14
Applications from Physics, Biology, and Climate

We close this book with introductions to a few recent applications where piecewise-smooth models have arisen. They showcase many different forms that discontinuity takes and the different kinds of analysis it requires, some amenable to the methods in this book and others requiring further innovation. In particular, no good problem at the frontier of theory and experiment is complete with its paradoxes, and we will see a few examples here.

We will only sketch out these applications and suggest avenues of further study for the reader to pursue. Of the topics here, only those in climate, friction mechanics, and neuroscience are topics of major current study, and at the time of writing, none of these have had applied to them the full methodology of hidden dynamics and switching layers set out in this book, at least not beyond the preliminary steps we set out here.

14.1 In Control: Steering a Ship

The idea of automated control using switches was gaining ground through the 20th century. One of the early seminal works in dynamical systems is rich with examples for steering ships, regulating servomotors, or stabilizing steam engines [10]. Later the idea found application in aeronautics [73] and eventually in more general electronic control [211]. We shall begin with the quaint but illustrative example of a boat steered by an automated rudder.

Let φ be the deviation of a ship from its assigned course. The ship is turned by a moment $M(\varphi)$ from the ship's rudder (or other lateral propulsion) and resisted by hydrodynamic forces we write simply as $-H\frac{d}{dt}\varphi$. If the ship's moment of inertia with respect to its vertical axis is I, then the equation of rotation of the boat is

$$I\tfrac{d^2\varphi}{dt^2} + H\tfrac{d\varphi}{dt} = M\ , \qquad M = -M_0\,\mathrm{sign}(\sigma)\ . \tag{14.1}$$

M. R. Jeffrey, *Hidden Dynamics*,
https://doi.org/10.1007/978-3-030-02107-8_14

This says the rudder takes a position $\pm M_0$ at its maximal deflection, switched as some measure of the heading, σ, changes sign. The aim is to find σ to create a stable equilibrium corresponding to the desired course heading $\varphi = 0$.

The obvious choice is to switch at $\sigma = \varphi$, because for $\varphi > 0$ or $\varphi < 0$, when the boat is off course, it is obviously necessary to switch the rudder position. It will be more effective, however, to anticipate that passing $\varphi = 0$ is imminent, using the rate of course change $d\varphi/dt$. Trying a switching function

$$\sigma = \varphi + b\tfrac{d\varphi}{dt} , \tag{14.2}$$

for $b > 0$, the switch will occur before the ship passes through φ, anticipating a change in heading. It is then possible to control the ship's course. Taking $b < 0$ corresponds to a reactionary control response and is unstable. To see why we must look at the solutions of (14.2).

In dimensionless variables, letting $x = \frac{H^2}{M_0 I}\varphi$, $\beta = bH/I$, and $\tau = tH/I$, we have

$$\dot{x} = y , \qquad \dot{y} = -y - \text{sign}(\sigma) , \qquad \sigma = x + \beta y , \tag{14.3}$$

denoting the τ derivative with a dot (and rescaling σ). The phase portrait, that is, the flow made up of solutions of the differential equations plotted in the space of (x, y), is shown in Figure 14.1. Solutions spiral around, threading the discontinuity surface $x + \beta y = 0$ repeatedly, forming what is known as a *fused focus*. For the anticipatory control, $\beta > 0$, this spiral is attractive towards the desired course $x = y = 0$. For $\beta < 0$ the spiral is repulsive, so solutions diverge, meaning reactionary control is unstable.

Changes from anticipatory control to reactionary control, perhaps as a sailor becomes more fatigued or intoxicated at the helm, can also be modelled. Say at some moment β decreases with time, $\dot{\beta} = -1$, so the system transitions from the $\beta > 0$ to $\beta < 0$. The phase portrait in (x, y, β) space is shown in Figure 14.2. The phase portrait is also shown for the converse, $\dot{\beta} = +1$. On the switching threshold $\sigma = 0$, now a surface in

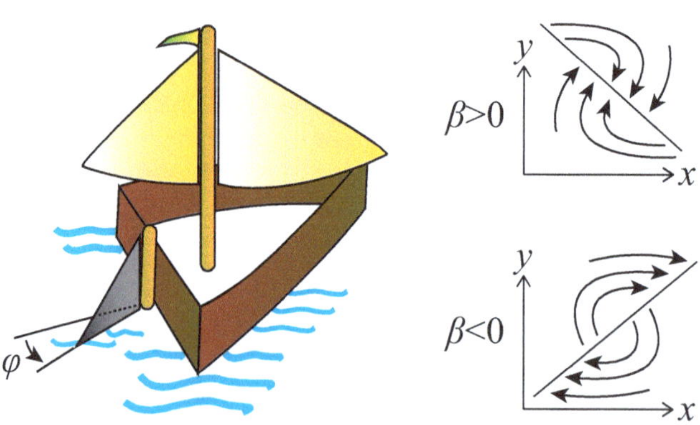

Fig. 14.1 Ship rudder control, showing the phase plane (x, y) for $\beta < 0$ (anticipatory control) and $\beta > 0$ (time lag control).

three dimensions, the shaded region shows where the surface is attractive or repulsive, with dynamics induced on it by the solutions outside. Solutions cross the surface in the unshaded regions. A two-fold singularity lies at $x = y = \beta = 0$.

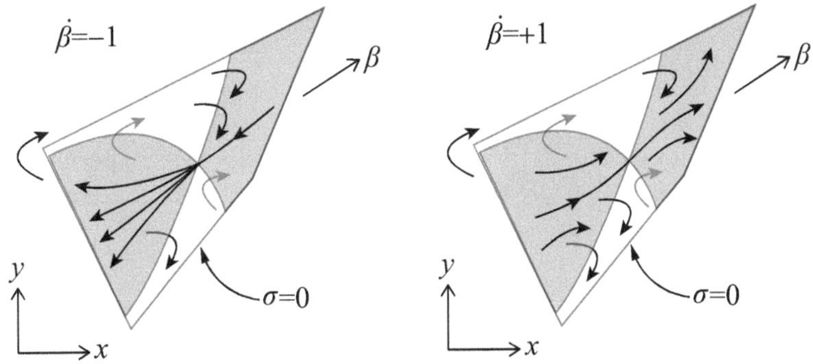

Fig. 14.2 In a change from anticipatory to reactionary control ($\dot\beta = -1$), the system becomes unstable. Changing instead from reactionary to anticipatory ($\dot\beta = +1$), the control stabilizes. The arrows show solution trajectories of the boat in the space of nondimensionalized heading x, rate of turn y, and rudder responsiveness β, relative to the switching surface $\sigma = 0$.

This sharp 'on-off' type of control may be unrealistic for a mechanical rudder, which takes time to move from one position to another, but for many other kinds of control, an abrupt model is much more accurate, for example, electrical relays or synaptic relays in the nervous system. In Chapter 12 we studied how nonidealities in the switching model can impact the dynamics.

14.2 Ocean Circulation

The Earth's climate is one of those big and complex systems that, with all our knowledge of physics, fluid mechanics, and modern computational power to help, seems like it should succumb to mathematical modelling. It has therefore become a playground for dynamicists over the past half century or so, with discontinuities increasingly playing a role, see, for example, [1, 2, 17, 32, 33, 59, 60, 109, 215].

Our everyday experience of climate is through its effect on the land and the atmosphere, but the big driving forces are beneath the surface of the seas and oceans that cover the majority of the Earth's surface, responsible for absorbing and redistributing the majority of the Sun's radiation that reaches the Earth.

The so-called general circulation models attempt to capture the fluid dynamics and thermodynamics of flows in the ocean and the atmosphere. Figure 14.3 shows two examples of jumps in physical quantities in ocean modelling that are well modelled as discontinuities. The first graph shows a *thermocline*, which is a horizontal layer in the ocean through which the temperature and salinity change abruptly, between the inertial reservoir of the deep ocean, and the more dynamic shallow surface layer sensitive to seasonal variation. The second graph shows the *albedo* of the Earth's surface—the fraction of solar energy reflected back into space—averaged across a given latitude, showing a jump between water-dominated and ice-dominated regimes around the Antarctic circle, which is sometimes modelled as a discontinuity, while the transition around the Arctic circle is more gradual.

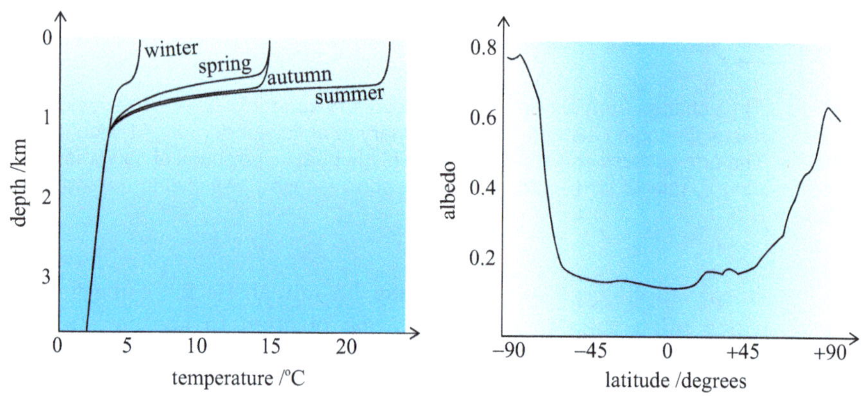

Fig. 14.3 Left: a typical thermocline, where ocean temperature jumps between a steady deep water layer and a shallow surface layer subject to seasonal variation. Right: sketch of albedo of the Earth's surface at different latitudes shows a jump at the edge of the Antarctic region but only a slow change around the Arctic.

To reduce the study of circulation to something tractable, one can use *ocean box models*. These are conceptual or 'toy' models whose modest purpose is to lay a groundwork of insight into the more detailed processes underlying circulation between the various oceans, seas, rivers, and the atmosphere.

H. Stommel [198] proposed a number of experimental set-ups upon which to base box models, including a two vessel experiment, Figure 14.4, whose equations are piecewise smooth.

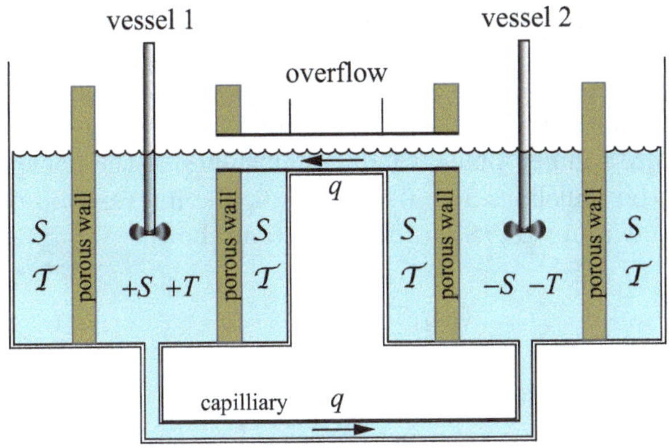

Fig. 14.4 Stommel's two-box model of ocean circulation. Two vessels contain stirred fluid with relative temperature $\pm T$ and salinity $\pm S$, linked by a capillary flow q and balancing overflow, surrounded by a reservoir of static temperature $\mathcal{T}$ and salinity $\mathcal{S}$, divided by a porous wall.

Stommel considered two vessels of stirred fluid, with temperature $\pm T$ and salinity $\pm S$ in two cavities, each relative to some mean value. There is a flow q from high pressure ρ_1 to low pressure ρ_2 via a capillary between the vessels, given by $q = (\rho_1 - \rho_2)/k$ where k is the resistance of the capillary, and an overflow is introduced to conserve fluid levels. The vessels are surrounded by a porous wall leading to a slow varying reservoir with fixed temperature $\mathcal{T}$ and salinity $\mathcal{S}$. The equations for conservation of heat and salinity are then

$$\begin{aligned} \dot{T} &= c(\mathcal{T} - T) - 2|q|T \,, \\ \dot{S} &= d(\mathcal{S} - S) - 2|q|S \,. \end{aligned} \tag{14.4}$$

The first terms on the right-hand side represent flow across the porous wall with coefficients c, d, and the second terms represent flow through the capillary. The absolute sign means that the flow of temperature and salinity across the capillary is insensitive to the direction of circulation, making these equations continuous but non-differentiable (so the discontinuity is in the derivative of the right-hand side).

We only briefly touched on such systems in Chapter 4. We could write the absolute term as $|q| = q\lambda$ where $\lambda = \text{sign}(q)$, but being continuous across the discontinuity surface $q = 0$, there is no need to form a switching layer, and we can do without introducing the switching multiplier λ explicitly.

The capillary flow rate is assumed to satisfy some equilibrium relation $q = aS - bT$ for constants a, b. Equilibria are then solutions $(T, S) = (T_*, S_*)$ of the simultaneous equations

$$0 = c(\mathcal{T} - T_*) - 2|q|T_* = d(\mathcal{S} - S_*) - 2|q|S_* = aS_* - bT_* - q \ .$$

Unfortunately these do not have a simple closed form solution, but a qualitative study is quite straightforward. Figure 14.5 shows the flow of (14.4) in the temperature-salinity plane for different values of the constants a, b, c, d. We also plot the nullclines $\dot{T} = 0$ and $\dot{S} = 0$, the intersection of which form the equilibria at roots (T_*, S_*) of the equation above.

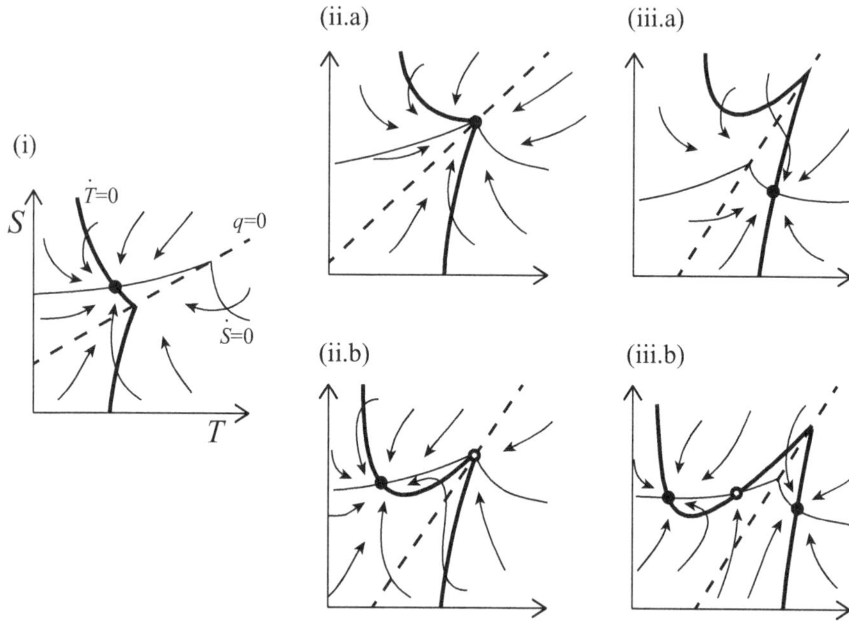

Fig. 14.5 Typical dynamics of Stommel's box model in $t$emperature-$s$alinity space, showing flows with (i) a node in $q > 0$, (ii.a) a persistence boundary equilibrium bifurcation, (iii.a) a node in $q < 0$, (ii.b) a fold boundary equilibrium bifurcation creates a new node and saddle either side of $q = 0$, and (iii.b) three equilibria with attracting nodes in both regions. The discontinuity surface $q = 0$ is shown (dashed), as are the nullclines of temperature (thick curve) and of salinity (thin curve).

Straightforward stability analysis, calculating the Jacobian $\frac{\partial(\dot{T},\dot{S})}{\partial(T,S)}$ of the vector field at each equilibrium, reveals for different parameter values either: the existence of an attracting node on one or other side of the discontinuity, or attracting nodes on both sides combined with the existence of a saddle point.

Boundary equilibrium bifurcations occur when any one of these nodes or saddle hit $q = 0$, then solving $\dot{T} = \dot{S} = 0$ gives

$$q = 0 \quad \Rightarrow \quad T_* = \mathcal{T} \ , \quad S_* = \mathcal{S} \ , \qquad a/b = \mathcal{T}/\mathcal{S} \ .$$

As the parameters are changed through these values, the system undergoes either a persistence event, in which a node crosses from $q \gtrless 0$ to $q \lessgtr 0$ (Figure 14.5 (i → ii.a → iii.a)), or a saddle-node event, in which one saddle and one node on either side of $q = 0$ are created or annihilated pairwise at $q = 0$ (Figure 14.5 (i → ii.b → iii.b)).

Saddle-node bifurcations can also occur away from $q = 0$, at parameter values where the two nullclines $\dot{T}$ and $\dot{S}$ are tangent to each other, that is, when $\left(\frac{\partial}{\partial T}, \frac{\partial}{\partial S}\right) \dot{T} = \left(\frac{\partial}{\partial T}, \frac{\partial}{\partial S}\right) \dot{S}$, which implies

$$daT_* + cbS_* + 2|aS_* - bT_*|(bS_* + aT_*) = 0 \ .$$

As the parameters are changed through values satisfying this equation, the system passes between scenarios a and b in Figure 14.5.

All of these equilibria and their bifurcations come together where the parameters satisfy two equations,

$$da\mathcal{T} + cb\mathcal{S} + 2|a\mathcal{S} - b\mathcal{T}|(b\mathcal{S} + a\mathcal{T}) = 0 \ , \qquad a/b = \mathcal{T}/\mathcal{S} \ ,$$

constituting a codimension two bifurcation.

Stommel provides the conclusion rather cogently:

The fact that even in a very simple convective system, such as here described, two distinct stable regimes can occur — one where temperature differences dominate the density differences and the flow through the capillary is from the cold to the warm vessel, and the other where salinity dominates the density difference so that the flow in the capillary is opposite, from warm to cold — suggests that a similar situation may exist somewhere in nature. One wonders whether other quite different states of flow are permissible in the ocean or some estuaries and if such a system might jump into one of these with a sufficient perturbation. If so, the system is inherently fraught with possibilities for speculation about climatic change. [198]

Stommel's model shows the ability of such a system to tend towards different, sometimes co-existing, steady states for different parameters and (during co-existence) different initial conditions.

Welander [220] proposed a variation on the model to demonstrate the origin of large scale oscillatory circulation between deep ocean, surface ocean, and the atmosphere. The deep ocean now provides a slow varying reservoir, with constant temperature T_0, salinity S_0, and density ρ_0, which we normalise to $T_0 = S_0 = \rho_0 = 0$. The atmosphere provides another reservoir with temperature T_A and salinity S_A (Figure 14.6).

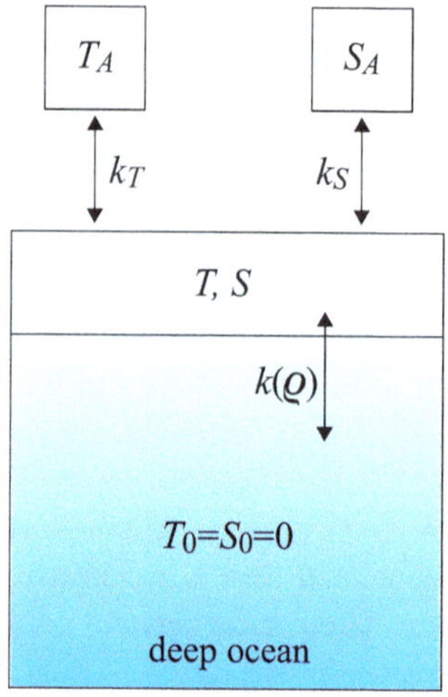

Fig. 14.6 Welander's box model for circulation between the surface layer (temperature T and salinity S) and deep ocean (temperature T_0 and salinity S_0).

The model for convection between the atmosphere, the deep ocean, and the surface with temperature T, salinity S, and density $\rho = \gamma S - \alpha T$ for some constants γ, α, is then

$$\left.\begin{array}{l} \dot{T} = (T_A - T)k_A + (T_0 - T)k(\rho) \\ \dot{S} = (S_A - S)k_B + (S_0 - S)k(\rho) \end{array}\right\} \quad : \quad k(\rho) = k_0 \,\mathrm{step}(\rho - \rho_k) \,,$$

The constants k_A and k_B are relaxation rates for atmospheric forcing. The interesting new quantity is $k(\rho)$, the rate of convective mixing between ocean layers, which switches off at low density, reflecting the fact that the ocean tends to be highly stratified into layers with little mixing between them. Welander, and later Leifeld [145, 145], considered different forms for this, including sigmoid smoothings of the step function, such as $k(\rho)/k_0 = \frac{1}{2} + \frac{1}{\pi}\arctan(\frac{S-\alpha T-\rho_k}{\varepsilon})$ for small $\varepsilon > 0$. Having studied in Section 12.3 the effect of smoothing a discontinuity, let us accept the step function as our model.

Following [146], let us make a coordinate change to $x = T$, $y = \gamma S - \alpha T - \rho_k$, and nondimensionalize so that the equations take the form

$$\left.\begin{array}{l} \dot{x} = 1 - x - x\mu \\ \dot{y} = \beta + \alpha(x - \beta x - 1) - (y + \rho_k)(\beta + \mu) \end{array}\right\} \quad : \quad \mu = \mathrm{step}(y)$$

in terms of constants given as $\alpha = 4/5$ and $\beta = 1/2$ in [220, 146], and ρ_k which can be taken as bifurcation parameter. Rewriting this as

$$\begin{pmatrix} \dot{x} \\ \dot{y} \end{pmatrix} = \begin{pmatrix} -1 - \mu & 0 \\ \alpha(1 - \beta) & -\beta - \mu \end{pmatrix} \begin{pmatrix} x - x_\mu \\ y - y_\mu \end{pmatrix} , \qquad \mu = \mathrm{step}(y) \tag{14.5}$$

where

$$(x_\mu, y_\mu) = \left(\frac{1}{1+\mu}, \frac{\beta}{\beta+\mu} - \frac{\alpha}{1+\mu} - \rho_k\right),$$

makes it clear that the system has equilibria at $(x, y) = (x_0, y_0)$ if $y_0 < 0$ and $(x, y) = (x_1, y_1)$ if $y_1 > 0$. The Jacobian is just the matrix in (14.5), with negative real eigenvalues given by the diagonals, so the equilibria are attracting nodes (since $\beta = 1/2$). The eigenvectors of the Jacobian are $(0, 1)$ and $(1, -\alpha)$.

These equilibria do not always exist. An equilibrium exists in $y > 0$ if $\rho_k < \frac{\beta}{\beta+1} - \frac{\alpha}{2} = -1/15$, in $y < 0$ if $\rho_k > 1 - \alpha = 1/5$. Thus for $\rho_k < -1/15$ there is one node above the discontinuity surface, in the regime of mixing between ocean layers. For $\rho_k > 1/5$ there is a single node in $y < 0$, in the non-mixing regime. Finally for $-1/15 < \rho_k < 1/5$ there are no nodes (at least for $y \neq 0$).

So we turn to the discontinuity surface itself, on which sliding dynamics occurs in a state of partial mixing between ocean layers. To carry out layer analysis, we can use the switching multiplier $\mu = \text{step}(y) \in [0, 1]$ directly (instead of our usual $\lambda = 2\mu - 1 = \text{sign}(y)$). The switching layer system is

$$\begin{pmatrix} \dot{x} \\ \varepsilon\dot{\mu} \end{pmatrix} = \begin{pmatrix} -1-\mu_* & -x_* \\ \alpha(1-\beta) & -\rho_k \end{pmatrix} \begin{pmatrix} x - x_* \\ \mu - \mu_* \end{pmatrix}, \qquad \mu \in (0, 1) \tag{14.6}$$

with a sliding equilibrium at $(x, \mu) = (x_*, \mu_*)$, where

$$(x_*, \mu_*) = \left(\tfrac{\alpha-\beta+R}{2\alpha(1-\beta)} - \tfrac{\rho_k}{2\alpha}, \tfrac{\alpha-\beta-R}{2\rho_k} - \tfrac{1+\beta}{2}\right),$$

and $R^2 = ((\alpha - \beta + \rho_k(\beta - 1))^2 + 4\rho_k\alpha(1-\beta)$. (There is a second point where the right-hand side of (14.6) vanishes, obtained by changing $R \mapsto -R$, but this lies outside the switching layer).

The sliding equilibrium exists for $-1/15 < \rho_k < 1/5$. The Jacobian there has determinant R/ε and trace $-\rho_k/\varepsilon$, so its attractivity depends on the sign of ρ_k, and it is a node for $\rho_k \lesssim 0.047$ and a focus for $\rho_k \gtrsim 0.047$ (the bounding value being the solution of an eighth-order polynomial in ρ_k, calculated numerically here for the given values of α and β).

These considerations lead to the flows shown in Figure 14.7. Boundary equilibrium bifurcations take place in which a node in $y \neq 0$ hits the discontinuity surface and becomes a sliding equilibrium. A node from $y > 0$ hits the discontinuity surface when $\rho_k = \frac{\beta}{\beta+1} - \frac{\alpha}{2}$, for which $R = \frac{1}{2}(4\beta - \alpha(1+\beta)^2)/(1+\beta)$, and the equilibrium lies at

$$(x_1, y_1) = (1/2, 0) \qquad \Leftrightarrow \qquad (x_*, \mu_*) = (1/2, 1), \tag{14.7}$$

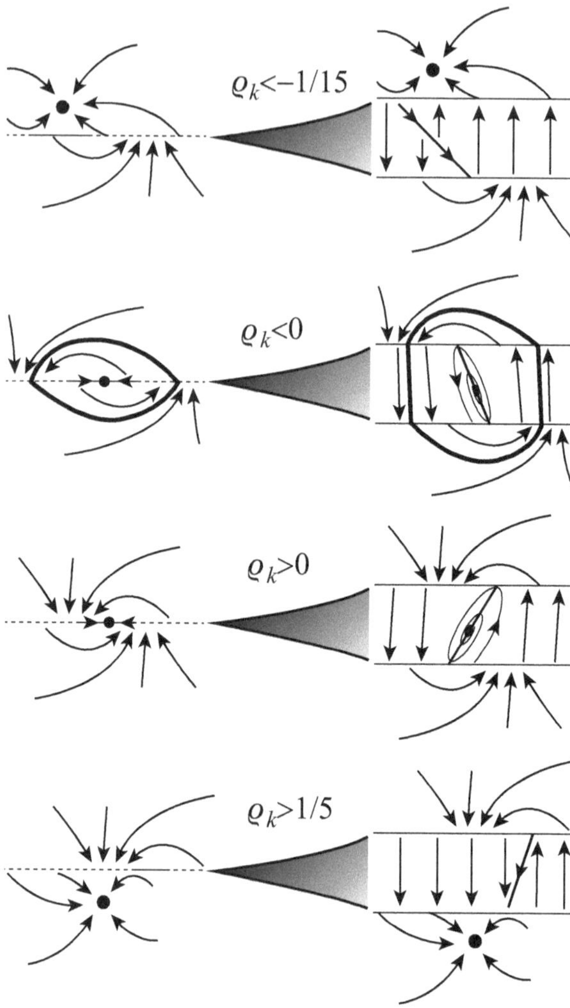

Fig. 14.7 Flows in the Welander model for different values of ρ_k, showing the piecewise-smooth system (left) blown up to reveal the switching layer (right). As ρ_k increases an attracting node above the discontinuity surface undergoes a boundary equilibrium bifurcation, becoming a repelling sliding node surrounded by an attracting limit cycle, which shrinks and changes the sliding node to attracting, before another boundary equilibrium bifurcation detaches this from the discontinuity surface to leave an attracting node below.

on the upper boundary of the switching layer. A node from $y < 0$ hits the discontinuity surface when $k = 1 - \alpha$, for which $R = 1 - \alpha\beta$ and

$$(x_0, y_0) = (1, 0) \qquad \Leftrightarrow \qquad (x_*, \mu_*) = (1, 0) \, , \tag{14.8}$$

on the lower boundary of the switching layer.

A nonsmooth analogue of a Hopf bifurcation must accompany the change of sign of ρ_k (either a super- or subcritical case, and for these particular α and β, it turns out to be supercritical), thus an attracting limit cycle shrinks down upon the sliding equilibrium as it changes from repelling to attracting.

Thus Welander's model can have stable steady states in a regime of effective mixing between ocean layers or in a regime of non-mixing, depending on how high the threshold for mixing is. It can also have a stable steady state that corresponds to partial mixing or a stable oscillatory state that makes substantial excursions into the regimes of effective mixing and non-mixing. A boundary equilibrium bifurcation and a nonsmooth analogue of the Hopf bifurcation carry the system between these states.

Turning from the ocean bodies to the Earth's surface, an example of discontinuous albedo forming a piecewise-smooth climate model, used to study the coverage of the Earth's surface by ice, can be found in [144]. The energy balance as the Sun's radiation is reflected from the Earth's surface at temperature T is given in the form

$$\begin{aligned}\dot{T} &= Q(1-\alpha(T)) - A - BT\ , \\ \dot{A} &= nT - mA + C\ ,\end{aligned} \tag{14.9}$$

where A provides a dynamics dependence on greenhouse gas concentration, with smaller A indicating more greenhouse gases, Q is the total solar radiation (treated as constant), and total outgoing long-wave radiation is the term $A + BT$. The albedo function $\alpha(T)$ measures the reflectivity of the Earth's surface, discontinuous with respect to the temperature, corresponding to a jump in reflectivity between sea water and sea ice. The analysis in [144], for example, reveals visible folds from either side of a discontinuity surface, on which lies a sliding saddle. The existence of attracting or repelling foci is possible on either side of the discontinuity surface, and as for Stommel's model but not Welander's model, for certain parameters the system exhibits bistability.

With the growing use of such models, there is growing discussion in the literature (and indeed at conferences) of how to model the discontinuity, either as a step function or as some smooth sigmoid. We have addressed the extent to which discontinuous or smoothed models can be considered approximations of one another in Section 12.3. Smoothing carries greater uncertainty and disguises that uncertainty by being 'well behaved' (i.e. smooth), and *unless* there is a clear physical law that leads to a specific smoothing function, an exploration of the uncertainties is required. A piecewise-smooth model makes any such uncertainties explicit, non-uniqueness is part of the quantitative theory, and we can study its implications for modelling directly.

There is also a discussion in climate models over the interpretation of sliding equilibria (or pseudo/virtual/zombie/singular/... equilibria or other terms that suggest something artificial), much as there is in biological models and electronic control models. In piecewise-smooth dynamics, we accept that a sliding equilibria is simply an equilibrium, as genuine a part of the system dynamics as an equilibrium outside the discontinuity surface. But a sliding equilibrium does have some special properties, in that it lies in the sliding dynamics and is approached from certain eigendirections in finite rather than infinite time. Those directions and rates of approach are obtained by the layer analysis of Chapter 7.

As new climate models are explored, the use of piecewise-smooth equations is likely to become more prevalent, and it will be interesting to see how models can be refined using switching layer analysis and consider possible nonlinear dependence on switching multipliers via hidden terms.

14.3 Chaos in a Church

In the north transept of the fifteenth century, St. Mary Redcliffe Church in Bristol, England, sits a water-driven chaotic pendulum. It consists of a horizontal crossbar with a pendulum bob hanging from its centre, so that the two rotate as a rigid body about a central pivot. Water flows into the crossbar from the pivot, running down one arm or the other depending on the crossbar's direction of tilt, before pouring out into a separate semi-circular pipe which collects and recycles the water (Figure 14.8).

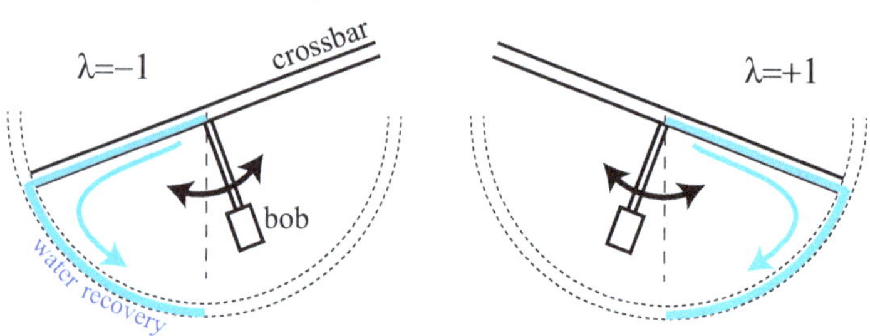

Fig. 14.8 The piecewise-smooth chaotic pendulum, driven by water flowing along the crossbar. The forcing has a discontinuity between water flowing down the left half or right half of the bar, depending on its direction of tilt, indicated by a switching multiplier $\lambda = -1$ for 'tipping left' and $\lambda = +1$ for 'tipping right'.

The pendulum is observed to swing chaotically, making large or small oscillations, sometimes rocking to and fro about a tilted position without swinging back through the central position, other times making wide sweeps from left to right and back again. Thus angular motion, fluid mechanics, and nonlinear and nonsmooth dynamics are combined to create an effect that is highly irregular and yet hypnotically peaceful.

The mathematics lying behind its unpredictable meanderings is subtle and because of the flow of the water is potentially very complex. Toy models, however, provide insight into the origin of chaos and the role of discontinuity.

We can neglect the weight of the crossbar itself. Let the bob have mass m_b and hang at a radius l. The entire apparatus has a moment of inertia I and is tilted through an angle θ at time t. There is a stop that prevents the crossbar going past the vertical, represented by an impact $\dot{\theta} \mapsto -\dot{\theta}$ if θ reaches $\pm\pi/2$. Using a switching multiplier λ, we denote a body of water of mass m_w as present in the left arm of the crossbar if $\lambda = -1$ or the right arm if $\lambda = +1$.

We will now look at three different approaches to representing the water, and they lead to three different kinds of discontinuous system: piecewise-smooth differential equations with a continuous flow (of the kind we have focussed on in this book), piecewise-smooth differential equations with a reset (sometimes called *hybrid* systems), and a piecewise-smooth discrete time map.

14.3.1 'Lumped Water' Model

Say that a fixed mass m_w of water sits in one or other arm of the crossbar, represented by a point mass a distance L from the pivot. This creates a moment $m_w g L \lambda \cos\theta$ about the pivot, with λ signifying which arm the water resides in. The obvious rule is that the water lies in the lowest arm, then $\lambda = -\operatorname{sign}\theta$. We can allow for the motion of the water complicating the precise instant at which water runs from one arm to the other by writing instead $\lambda = \operatorname{sign}(\theta - \phi(\theta, \dot{\theta}, t))$ for some function ϕ.

There is a moment $m_b g l \sin\theta$ from the bob and damping from friction in the pivot or air resistance, which we will represent as a linear retardation $\rho\dot{\theta}$. With the angular acceleration begin $I\ddot{\theta}$, Newton's second law then gives the equations of motion as

$$I\ddot{\theta} = -\rho\dot{\theta} - m_b g l \sin\theta - m_w g L \lambda \cos\theta\ , \quad \lambda = -\operatorname{sign}(\theta - \phi(\theta, \dot{\theta}, t))\ .$$

There are a lot of parameters here, but introducing $a = \rho/\sqrt{I m_b g l}$, $b = m_w L/m_b l$ and scaling time as $\tau = t\sqrt{I/m_b g l}$, the equations reduce to

$$\theta'' = -a\theta' - \sin\theta - b\lambda\cos\theta\ , \quad \lambda = \operatorname{sign}(\theta - \phi(\theta, \dot{\theta}, t))\ , \tag{14.10}$$

in terms of just two parameters, a and b.

The result is a system of piecewise-smooth differential equations with a continuous flow, with continuous dependence on a switching multiplier of the kind we have focussed on in this book. The basic phase portrait is illustrated in Figure 14.9.

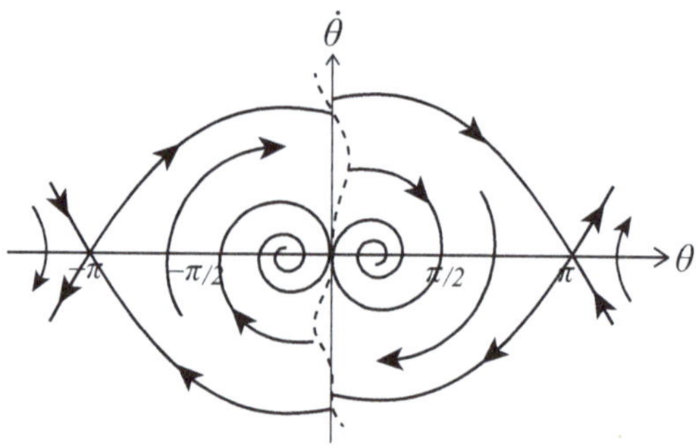

Fig. 14.9 Phase plane of the piecewise-smooth pendulum model. The dashed curve indicates the discontinuity surface.

Written as a system of first-order ordinary differential equations in terms of the angle θ and angular velocity ω, we have

$$\begin{aligned}\theta' &= \omega\ ,\\ \omega' &= -a\omega - \sin\theta - b\lambda\cos\theta\ .\end{aligned}$$

There are equilibria at $\tan\theta = -b\lambda$. For $b = 0$ (i.e. with no water, $m_w = 0$), we have a simple pendulum with foci at $\theta = 0$ and saddles at $\theta = \pi$, corresponding to hanging vertically downwards or upwards, respectively. With $m_w = \rho = 0$ the saddle has a homoclinic connection, creating potential conditions for chaos under perturbation.

Introducing $m_w \neq 0$ breaks the attracting downwards hanging equilibrium into three equilibria, two attracting foci at $\theta = \mp\arctan(b\lambda)$, and a sliding saddle on the discontinuity surface at the coincidence of two visible folds (where $\dot\theta = 0$ from either side of the discontinuity surface).

Some representative simulations of the system are shown in Figure 14.10, taking a sinusoidal discontinuity surface given by letting $\phi(\theta,\dot\theta,t) = \frac{1}{10}\sin(5t)$ for illustration. For different parameters one may observe periodic or chaotic behaviour, the simulations we have picked out show the latter.

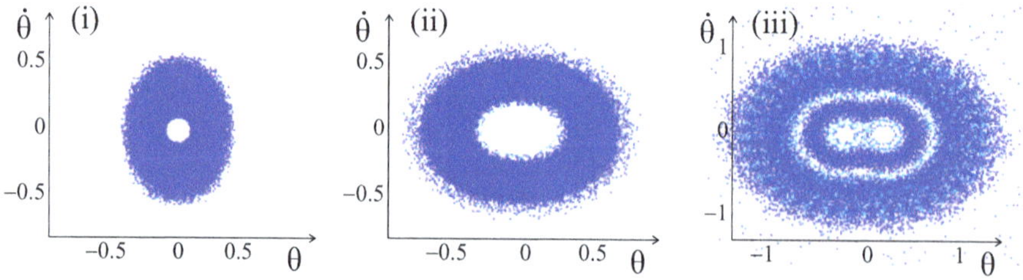

Fig. 14.10 Simulation of the lumped water model, with (i) $a = 1/3000$, $b = 7/300$; (ii) $a = 1/1000$, $b = 6/100$; (iii) $a = 1/1000$, $b = 185/1000$.

The function ϕ is a crude attempt to model the non-trivial motion of the water in the crossbar. A different approach is to actually allow the water to move in the arms, in which case we need to set our point mass in motion.

14.3.2 'Moving Point' Model

To improve on the previous model, let us picture the water as a point mass rolling down the crossbar arms, lying a changing distance x from the pivot. The moment of inertia of the apparatus will then vary as $I(x) = m_w\dot{x}^2 + m_b l^2$. If the point reaches the end of the crossbar at some $x = \pm L$, we recirculate it back to the origin by resetting x and $\dot{x}$ to zero. We will neglect any damping in this model.

With varying radii and angles, it is often better to turn to the Lagrangian approach, rather than Newton's second law as in the previous section. The kinetic energy of the system is $T = \frac{1}{2}m_w\dot{x}^2 + \frac{1}{2}I(x)\dot{\theta}^2$, with contributions from the radial motion of the water and the angular motion of the combined apparatus. The potential energy comes solely from the changing height of the centres of mass of the water and the bob, $U = m_w g x \sin\theta - m_b g l \cos\theta$.

The Lagrangian is then $L = T - U$, and the Euler-Lagrange equations

$$\frac{d}{dt}\frac{\partial L}{\partial \dot{\theta}} - \frac{\partial L}{\partial \theta} = 0 \,, \qquad \frac{d}{dt}\frac{\partial L}{\partial \dot{x}} - \frac{\partial L}{\partial x} = 0 \,,$$

yield equations of motion

$$\begin{aligned} &m_w\ddot{x} = m_w\dot{\theta}^2 x - m_w g \sin\theta \,, \\ &(m_w x^2 + m_b l^2 + \tfrac{1}{12}m_c L^2)\ddot{\theta} = -m_w g x \cos\theta - m_b g l \sin\theta - 2m_w x\dot{x}\dot{\theta} \,. \end{aligned}$$

These again depend on a number of parameters which we can reduce to just two, defining $a = \frac{l}{L} + \frac{1}{12}\frac{m_c L}{m_b l}$ and $b = \frac{m_w L}{m_b l}$, scaling time as $\tau = t\sqrt{g/L}$, giving

$$\begin{aligned} y'' &= (\theta')^2 y - \sin\theta \,, \\ (by^2 + a)\theta'' &= -by\cos\theta - \sin\theta - 2byy'\theta' \,. \end{aligned} \tag{14.11}$$

Written as a four-dimensional system of ordinary differential equations, this is

$$\begin{aligned} y' &= z \,, & z' &= \omega^2 y - \sin\theta \,, \\ \theta' &= \omega \,, & (by^2 + a)\omega' &= -by\cos\theta - \sin\theta - 2byz\omega \,. \end{aligned}$$

We must combine these with the condition that arrival of the water at $y = 1$ triggers a recirculation $(y, y') \mapsto (0, 0)$. Thus we have a system of piecewise-smooth differential equations combined with a reset, sometimes called a *hybrid* system.

A few simulations showing chaotic motion are shown in Figure 14.11. The behaviour of this is somewhat richer than the lumped model, and at different parameters exhibits a range of simple point attractors, periodic or quasi-periodic orbits, as well as chaos.

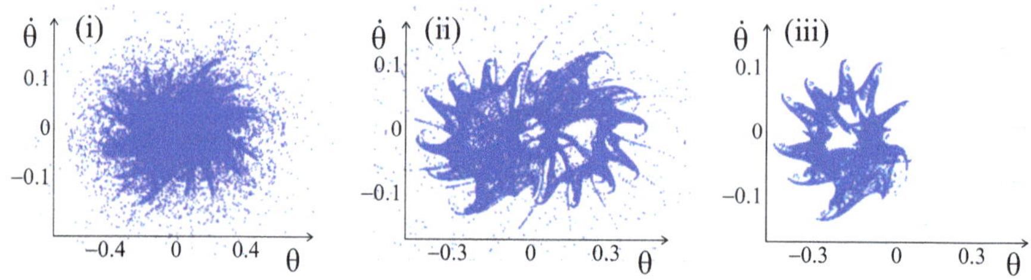

Fig. 14.11 Simulation of the moving point water model, with $b = 4.5$ and (i) $a = 0.4$; (ii) $a = 0.45$; (iii) $a = 0.46$.

We can of course try to improve the model further. Adding the weight of the crossbar would just increase I in the lumped model and a in the moving point model, so let us just focus on how we model the water. A nice improvement on the reset might be to let m_w vary, modelling the amount of fluid flowing into and out of the arm, and then consider the water as an extended body, further complicating its moment of inertia. Ultimately, of course, the water should be modelled as a continuous fluid with its own complex internal dynamics. These will all add complication to a model that already exhibits sufficient complexity in its behaviour to keep most dynamicists happy for some time. Our last model goes in the other direction, capturing the pendulum's complexity in an even simpler set of equations.

14.3.3 'Discrete Kick' Model

An alternative toy model takes the form of a piecewise-smooth map.

Let us inspect the pendulum's motion only in discrete steps, looking at how they change from some angle θ_i and angular speed ω_i, to their values θ_{i+1} and ω_{i+1} a moment later. Let us model the water by saying that the pendulum turns freely for an angle α and then receive a unit angular impulse λ, whose sign depends on the direction of tilt. Lastly we apply damping in the form of a factor F (close to unity for small damping). We obtain a piecewise-smooth discrete time map.

$$\left.\begin{aligned} \theta_{i+1} &= (\theta_i \cos\alpha + \omega_i \sin\alpha)F \\ \omega_{i+1} &= (\omega_i \cos\alpha - \theta_i \sin\alpha + \lambda)F \end{aligned}\right\} \quad : \quad \lambda = \operatorname{sign}(\theta_{i+1}) \tag{14.12}$$

A few simulations showing chaotic behaviour are shown in Figure 14.12. A particular feature of this model is that there exist multiple attractors. Which

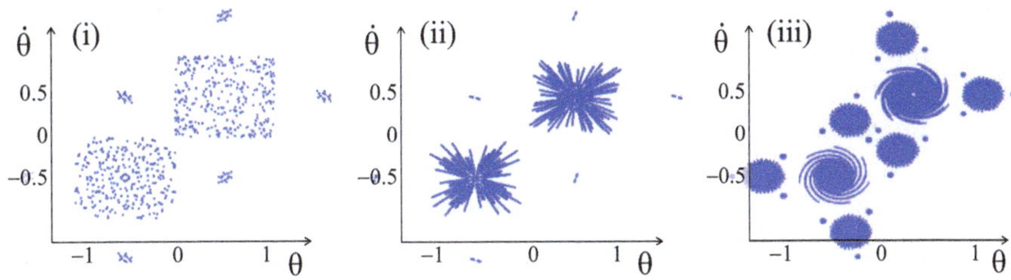

Fig. 14.12 Simulation of the discrete time model, with: (i) $a = \pi/2$, $F = 1$; (ii) $a = \pi/2$, $F = 0.999$; (iii) $a = 1.8$, $F = 0.999$.

attractor the dynamics settles into depends on its initial conditions. Each picture in Figure 14.12 shows a number of different attractors that are reached from different regions of the plane; in (iii), for example, each set of discs of a given size constitutes one attractor (so the 2 large discs are one attractor, the 6 next largest discs are another attractor, and the 18 small discs are another attractor), inside which an orbit will unpredictably jump from one disc to another.

The complexity of this model's dynamics comes partly from the deceptively simple assumption of a unit impulse λ being applied at each iteration. A constant impulse is not easily achieved and instead implies some process behind it mimicking, if not modelling directly, the complication of water in the pendulum. For further intuition we can of course change the size of the constant impulse, replacing λ with some $c\lambda$ in (14.12), for instance. As we make c smaller, the complex attractors persist but tend to become constrained towards, and eventually for $c = 0$ collapse to, either an attractor at $\theta = \omega = 0$ or a stable periodic orbit, as the reader may readily investigate.

We have not studied piecewise-smooth maps in this book, as they are a weighty and incredibly complex topic of their own, and we see how easily here they give rise to complex attractors and bifurcations. We will see another example in Section 14.6. Few entirely general results concerning singularities and bifurcations of piecewise-smooth maps are known, particularly if the maps are nonlinear (the pendulum map above being only linear). Early interest centred on their role as novel routes to chaos [156, 75, 76], while lately there have been growing attempts at a local bifurcation theory, starting with border collisions (the discrete time equivalent of a boundary equilibrium bifurcation) [172, 107, 48], which despite being the simplest normal form class for a local bifurcation, still involves incredible complication [79, 78, 188, 189], including a curse of dimensionality [84]. Work has gone on to study perturbations of ideal piecewise-smooth maps, e.g. [77, 85]. Whole communities are dedicated solely to the study of piecewise linear maps (just as entire communities dedicate themselves to piecewise linear flows). A particular challenge for their future study is to more closely understand how discontinuities and discontinuity-induced phenomena in flows are related to discontinuities in maps and the kinds of attractors and bifurcations they imply.

The St. Mary pendulum was conceived by Prof. Sir Brian Pippard and constructed by Mr. Robert Knight and introduced at the church on 25th October 1997. Visitors can freely view the pendulum in situ, or videos of it in motion are also available on the internet. Two of the models above are included thanks to a few generous individuals. The 'moving point' model above was the work of three project students I supervised in 2018, Lili Luan, Neophytos Polydorou, and Libby Walker, at the University of Bristol. The discrete model was proposed by my own former PhD supervisor, Sir Michael Berry (by private communication, 16 Oct 2013).

These ideas are not without practical potential, and one may apply similar concepts to model water wheels or other rotary systems like turbines, driven by a succession of impulses rather than perfectly smooth power input, for instance, by gusts of air flow in the case of wind turbines.

14.4 Explosion in a Superconducting Stripline Resonator

One of the first devices in which discontinuity-induced explosions were observed was a superconducting stripline resonator [15, 180, 181, 182], a measuring device conceived to experimentally observe forces arising through quantum phenomena, like quantum squeezing or the dynamic Casimir effect, and therefore requiring enormous precision. Rather than classical electronics the experimenters turned to superconductors.

The device consists of a thin circular ring of niobium nitride (NbN), whose radius is on the order of centimetres, while its cross-section is on the order of square microns (Figure 14.13).

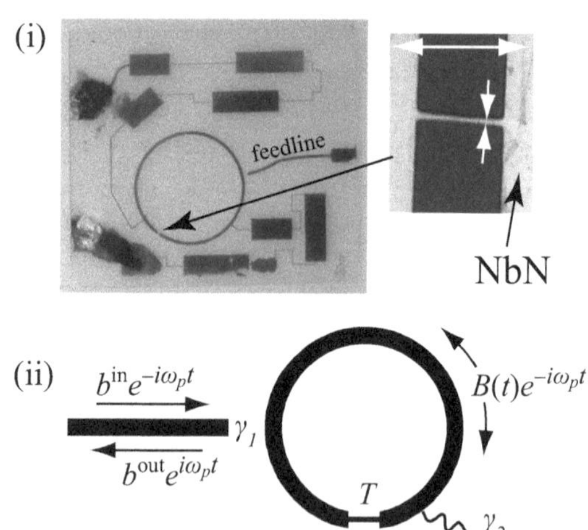

Fig. 14.13 The stripline resonator, showing (i) the device itself, consisting of a superconducting ring interrupted by a microbridge and (ii) a schematic of electromagnetic waves in the ring (Images in (i) kindly provided by the authors of [15, 180, 181, 182]).

An applied voltage excites electromagnetic waves of frequency ω_p, whose amplitude B depends on the temperature T of the microbridge. When excited by electromagnetic forcing, the device should find an equilibrium which can be 'read off' as a measurement. At least that is the intention.

The relationship between B and T is dynamic and is discontinuous across the superconducting temperature threshold $T = T_C$ (which is on the order of 10 Kelvin). In practice, self-sustaining oscillations can arise in T and B. The mechanism is simple in principle. If $T < T_C$ then the microbridge is superconducting. A strong excitation produces a large current, which heats the microbridge, and with sufficient heating the temperature grows to $T > T_C$, where the microbridge is no longer superconducting, the excitation becomes highly damped (so B shrinks), and the current decreases. However, this removes the source of heating, so the temperature falls again until $T < T_C$, the microbridge once again becomes superconducting, the damping falls, and the current grows, and so the whole process repeats in an oscillating fashion.

To explain these observations, a model was developed in the form

$$\begin{aligned} \dot{B} &= \left[i\left(\omega_{\mathrm{p}} - \omega_0\right) - \gamma_1 - \gamma_2\right] B - i\sqrt{2\gamma_1} b^{\mathrm{in}} \\ \dot{T} &= g\left[\left(T_0 - T\right) + 2\kappa\hbar\omega_0\gamma_2 \left|B\right|^2 / H\right] , \end{aligned} \tag{14.13}$$

where i is the imaginary unit $\sqrt{-1}$. We will simplify this considerably by nondimensionalizing shortly. The complex amplitude B is just a linear oscillator, while the T dynamics are driven by linear decay towards the bath temperature T_0, described by the first term, countered by the B-dependent Joule heating in the microbridge. H is a heat transfer coefficient between the resonator microbridge with thermal heat capacity C and a coolant with temperature T_0, κ measures efficiency of microbridge heating relative to power dissipated and is typically estimated to be near one, ω_0 is the angular resonance frequency, γ_1 is the coupling coefficient between the resonator and feedline, and γ_2 is the damping rate of the excitation mode in the ring. For numerical values of the various physical parameters, see [15, 118].

The ratio $g = H/C$ is a large parameter which implies fast dynamics evolving towards slow manifolds where $\dot{T} = 0$.

The crucial factor in understanding why the system oscillates is the dependence of these physical parameters on whether the microbridge is normal conducting ($T > T_C$) or superconducting ($T < T_C$). In Figure 1.6(iii) we saw how abrupt the jump is in resistivity at a typical superconducting threshold. The parameters in Section 14.4 are each piecewise constant, with a jump at $T = T_C$. The following analysis is based on work in [118, 112, 111].

Let us transform to dimensionless variables

$$\beta = \frac{B}{\sqrt{2\gamma_1} b^{\rm in}} \qquad \text{and} \qquad \theta = \frac{T - T_0}{T_C - T_0} ,$$

in which $\theta = 0$ is the bath temperature and $\theta = 1$ is the transition temperature between normal and superconducting modes, and define dimensionless parameters

$$\Lambda = \mathrm{i}\,(\omega_{\rm p} - \omega_0) - \gamma_1 - \gamma_2 , \qquad s = \frac{4\kappa\hbar\omega_0\gamma_1\gamma_2(b^{\rm in})^2}{(T_C - T_0)H} ,$$

both taking on different values for $T > T_C$ and $T < T_C$. Physically realistic values give $\mathrm{Re}\Lambda < 0$ and $s_+ > s_-$. We thus obtain a two-parameter system

$$\left.\begin{array}{r} \dot\beta = \Lambda\beta - i \\ \frac{1}{g}\dot\theta = s|\beta|^2 - \theta \end{array}\right\} \quad \text{where} \quad \left\{\begin{array}{l} \Lambda = \frac{1}{2}(1+\lambda)\Lambda_+ + \frac{1}{2}(1-\lambda)\Lambda_- , \\ s = \frac{1}{2}(1+\lambda)s_+ + \frac{1}{2}(1-\lambda)s_- , \end{array}\right. \tag{14.14}$$

with the switching multiplier

$$\lambda = \mathrm{sign}(\theta - 1) . \tag{14.15}$$

Three surfaces illustrated in Figure 14.14(i) play a vital role in the dynamics, the discontinuity surface

$$\mathcal{D} = \{(\beta, \theta) : \ \theta = 1\} , \tag{14.16}$$

and the θ-nullclines of (14.14), which are

$$\begin{array}{l} \Sigma_+ = \{(\beta, \theta) : \eta_+(\beta, \theta) = 0, \theta > 1\} , \\ \Sigma_- = \{(\beta, \theta) : \eta_-(\beta, \theta) = 0, \theta < 1\} , \end{array} \tag{14.17}$$

introducing a pair of functions

$$\eta_\pm(\beta, \theta) = s_\pm|\beta|^2 - \theta. \tag{14.18}$$

Sliding occurs on

$$\mathcal{D}_{sl} = \big\{(\beta, \theta) : \ \theta = 1, \ |\beta|^2 \in (1/s_-, 1/s_+)\big\} . \tag{14.19}$$

To find the sliding dynamics, we solve $\dot{\theta} = \theta - 1 = 0$, to find in sliding that $\lambda = \lambda^{\$} = (2/|\beta|^2 - (s_+ + s_-))/(s_+ - s_-)$, which substituted back into (14.14) gives

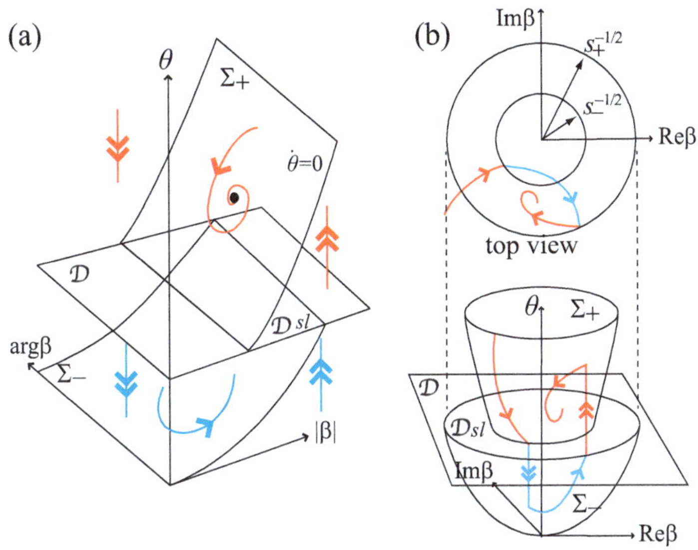

Fig. 14.14 Geometry of the resonator dynamics, consisting of fast (timescale gt) oscillation between manifolds $\Sigma_\pm$ of slow dynamics, passing through a discontinuity surface which has a repelling sliding region in the strip $\mathcal{D}_{sl}$. Shown in (a) $(|\beta|, \arg\beta, \theta)$ space and (b) $(\mathrm{Re}\,\beta, \mathrm{Im}\,\beta, \theta)$ space including the top-down view.

$$\dot{\beta} = \frac{\Lambda_+ \eta_-(\beta, 1) - \Lambda_- \eta_+(\beta, 1)}{\eta_-(\beta, 1) - \eta_+(\beta, 1)} \beta - \mathrm{i} \quad \text{on} \quad \mathcal{D}_{sl}. \tag{14.20}$$

Fixed points of the system must be contained within the surfaces $\mathcal{D}_{sl}$ or $\Sigma_\pm$, in the form of equilibria where (14.14) vanishes or sliding equilibria where (14.20) vanishes.

The zeros of (14.14) lie at $(\beta, \theta) = (\beta_\pm^{eq}, \theta_\pm^{eq}) \equiv (\mathrm{i}/\Lambda_\pm, s_\pm/|\Lambda_\pm|^2)$. Each point $(\beta_\pm^{eq}, \theta_\pm^{eq})$ is an equilibrium if it lies on its respective surface $\Sigma_\pm$, that is, (i) if $\theta_+^{eq} > 1$, implying $\eta_+(\mathrm{i}/\Lambda_+, 1) > 0$, and (ii) if $\theta_-^{eq} < 1$, implying $\eta_-(\mathrm{i}/\Lambda_-, 1) < 0$. Both of these are attracting foci, because the Jacobian matrix of (14.14) at $(\beta_\pm^{eq}, \theta_\pm^{eq})$ has one negative eigenvalue $-g$ and two complex eigenvalues with negative real parts $\mathrm{Re}\Lambda_\pm < 0$.

The number of sliding equilibria, where (14.20) vanishes, is:

- 0 if $\eta_+(\Lambda_+,1)\eta_-(\Lambda_-,1) > j^2/|\Lambda_+\Lambda_-|^2 > 0$,
- 0 if $\eta_+(\Lambda_+,1)\eta_-(\Lambda_-,1) > 0$, and $j/|\Lambda_-|^2\eta_-(\Lambda_-,1) > 0$,
- 1 if $\eta_+(\Lambda_+,1)\eta_-(\Lambda_-,1) < 0$,
- 2 if $\dfrac{j^2}{|\Lambda_+\Lambda_-|^2} > \eta_+(\Lambda_+,1)\eta_-(\Lambda_-,1) > 0$ & $\dfrac{j}{|\Lambda_-|^2\eta_-(\Lambda_-,1)} < 0$,

where $j = \frac{1}{2}(s_- + s_+) - \operatorname{Re}\left[\Lambda_+\Lambda_-^\star\right]$. To show this, instead of solving for the zeros of (14.20), it is easier to use the fact that sliding equilibria occur at points on $\mathcal{D}_{sl}$ where the constituent vector fields are anti-colinear (because dependence on λ is linear), that is when there exists $\mu < 0$ such that

$$\begin{aligned} \Lambda_+\beta - \mathrm{i} &= \mu(\Lambda_-\beta - \mathrm{i}), \\ s_+|\beta|^2 - 1 &= \mu(s_-|\beta|^2 - 1). \end{aligned} \tag{14.21}$$

implying

$$\mu = \frac{j \pm \sqrt{j^2 - |\Lambda_+\Lambda_-|^2\eta_+(\mathrm{i}/\Lambda_+,1)\eta_-(\mathrm{i}/\Lambda_-,1)}}{|\Lambda_-|^2\eta_-(\mathrm{i}/\Lambda_-,1)}. \tag{14.22}$$

Putting these together, the number of equilibria in the resonator system is either 1 or 3: if $\eta_+(\Lambda_+,1) > 0 > \eta_-(\Lambda_-,1)$, there exists one sliding equilibrium on $\mathcal{D}_{sl}$ and one equilibrium on each of Σ_+ and Σ_-; if $\eta_+(\Lambda_+,1) < 0 < \eta_-(\Lambda_-,1)$, there exists one sliding equilibrium on $\mathcal{D}_{sl}$, but there are no equilibria on Σ_- or Σ_+; and if $\eta_+(\Lambda_+,1)\eta_-(\Lambda_-,1) < 0$, there are either 2 or 0 sliding equilibria on $\mathcal{D}_{sl}$ and one equilibrium either on Σ_+ or Σ_- but not both.

This allows the number of fixed points to changes as the parameters $\Lambda_\pm$ or $s_\pm$ are varied. Since these fixed points must lie on $\mathcal{D}_{sl}$ or $\Sigma_\pm$, they can appear/disappear in only two ways.

A saddle-node bifurcation of sliding equilibria occurs when both μ values in (14.22) coincide. When $j^2 = |\Lambda_+\Lambda_-|^2\eta_+(\Lambda_+,1)\eta_-(\Lambda_-,1)$, two sliding equilibria coincide and under perturbation there exist either 0 or 2.

Boundary equilibrium bifurcations take place in which equilibria pass continuously between $\mathcal{D}_{sl}$ and either Σ_+ or Σ_-, respectively, when $\eta_+(\mathrm{i}/\Lambda_+,1) = 0$ or $\eta_-(\mathrm{i}/\Lambda_-,1) = 0$. The equilibrium's attractivity must change when it passes from $\mathcal{D}_{sl}$ to $\Sigma_\pm$, from an attracting focus on $\Sigma_\pm$, to a sliding equilibrium on $\mathcal{D}_{sl}$ that is repelling at least in the direction normal to $\mathcal{D}_{sl}$. The change in stability is another example of a nonsmooth analogy of a Hopf bifurcation: two eigenvalues of the equilibrium cross the imaginary axis, and a limit cycle is created or destroyed. In this case, when a fixed point moves from $\mathcal{D}_{sl}$ to Σ_+ or Σ_-, a limit cycle of saddle- type is created, corresponding qualitatively to a subcritical Hopf bifurcation.

Figure 14.15 shows this Hopf-like boundary equilibrium bifurcation. The top row shows a sketch in three dimensions; the bottom row shows a numerical simulation of (14.14) with $g \to \infty$ in the top-down view. A saddle sliding equilibrium on $\mathcal{D}_{sl}$ becomes an attracting focus equilibrium on Σ_+, and develops a saddle-type limit cycle.

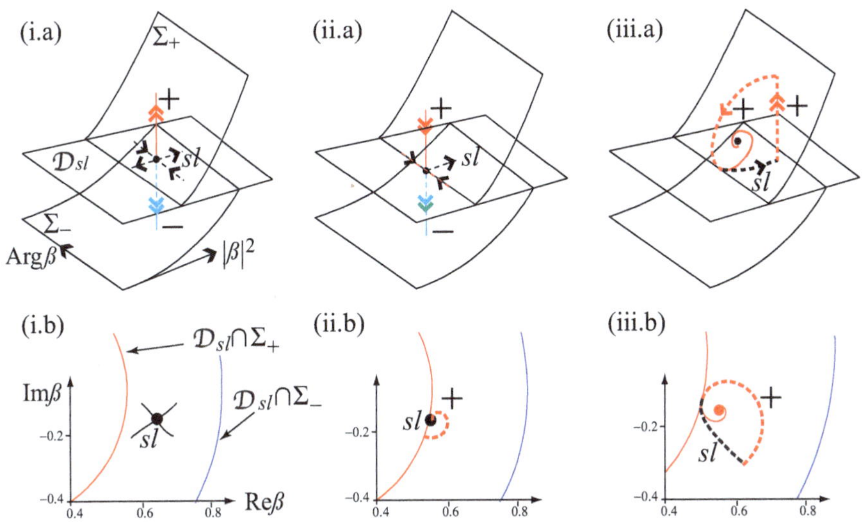

Fig. 14.15 A Hopf-like bifurcation in the resonator model, as a saddle sliding equilibrium on $\mathcal{D}_{sl}$ in (i) becomes a focus on Σ_+ in (iii). Top row shows a sketch, and bottom row shows a simulation of (14.14) with $g \to 0.$, with orbits labelled $+$, $-$, and sl, according to whether they lie on Σ_+, Σ_-, or $\mathcal{D}_{sl}$ (coloured, respectively, red, blue, black). Simulation parameters are $s_+ = 3.891$, $s_- = 1.297$, $\Lambda_- = -0.2 + \mathrm{i}$, $\mathrm{Re}\Lambda_+ = -0.5$ with (a) $\mathrm{Im}\Lambda_+ = 2.2$, (b) $\mathrm{Im}\Lambda_+ = 1.9$, (c) $\mathrm{Im}\Lambda_+ = 1.7$.

As we continue to vary $\mathrm{Im}\Lambda_+$, a switching-sliding bifurcation of a saddle-type limit cycle occurs. The cycle in Figure 14.15 grows, shown in Figure 14.16(i), eventually intersecting the boundary between $\mathcal{D}_{sl}$ and Σ_- as in Figure 14.16(ii), and in doing so it develops a segment on Σ_-, Figure 14.16(iii).

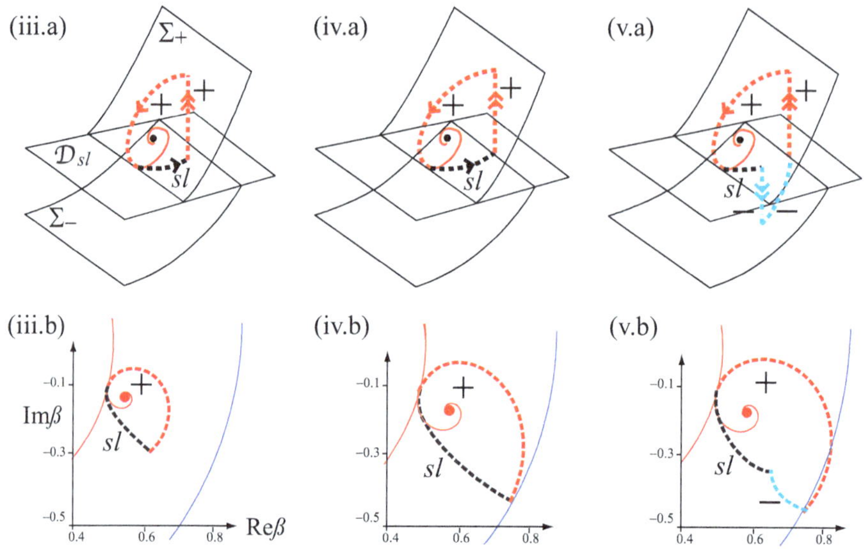

Fig. 14.16 A switching-sliding bifurcation of a piecewise-smooth saddle-type limit cycle in the resonator model. From (i) to (iii) the limit cycle (dashed) grows and develops a segment that jumps off $\mathcal{D}_{sl}$ onto Σ_-. Continued from Figure 14.15, with (a) $\mathrm{Im}\Lambda_+ = 1.6$, (b) $\mathrm{Im}\Lambda_+ = 1.58$, (c) $\mathrm{Im}\Lambda_+ = 1.573$.

A second, attracting, limit cycle exists in the system, omitted from Figures 14.15 and 14.16 for clarity but which we now show in Figure 14.17(i). Implicit formulae for the attracting limit cycle were derived in [118]. The attracting limit cycle suddenly vanishes as a parameter is varied continuously, via a grazing-sliding explosion.

The saddle-type limit cycle, born in Figure 14.15 and growing through Figure 14.16, is also involved in the explosion. As shown in Figure 14.17(i), the saddle limit cycle now visits all three regions $\mathcal{D}_{sl}$, Σ_+, and Σ_-, while the attracting limit cycle visits only the attracting regions Σ_+ and Σ_-. The attracting orbit shrinks and develops a tangency to the boundary of $\mathcal{D}_{sl}$ in Figure 14.17(ii). Meanwhile the saddle orbit grows until at least part of it coincides with the attracting orbit. The two cycles do not fully coincide when the explosions take place in Figure 14.17(ii) (as they would, for instance, in a saddle-node bifurcation of limit cycles, see, e.g. [138]). Under further parameter variation to produce Figure 14.17(iii), both limit cycles vanish.

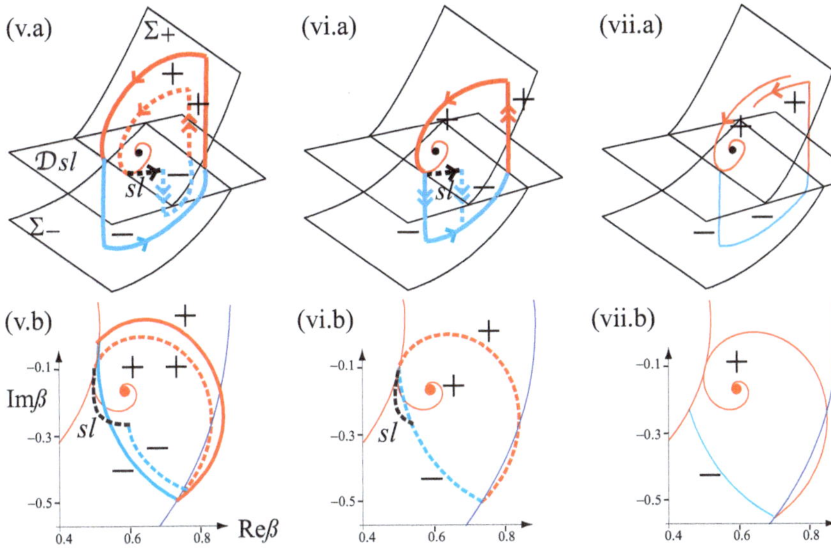

Fig. 14.17 Grazing-sliding explosion in the superconducting resonator model. In (i) an attracting limit cycle (bold curve, omitted from Figures 14.15 and 14.16 for clarity) and the smaller saddle-type limit cycle (dashed curve). In (ii) the two orbits coalesce by forming a tangency to $\mathcal{D}$, at which the forward evolution is non-unique—a solution could follow the limit cycle or evolve towards the focus. In (iii) all solutions evolve towards an attracting focus. Continued from Figure 14.16, with (a) $\mathrm{Im}\Lambda_+ = 1.572$, (b) $\mathrm{Im}\Lambda_+ = 1.57$, (c) $\mathrm{Im}\Lambda_+ = 1.56$.

Lastly we show the effect of the explosions, using a simulation of (14.14) for $g = 100$, $s^+ = 3.891$, $s^- = 1.297$, $\Lambda^- = -0.2 + i$, and $\Lambda^+ = -0.5 + ic$. For two nearby values of c (see Figure 14.18), an orbit through an initial point at $\beta = 0.8 - 0.4i$, $\theta = 3$ (labelled IP in the figure) oscillates around the cycle labelled A, crossing between $\theta > 1$ and $\theta < 1$, or finds an equilibrium labelled B in $\theta > 1$.

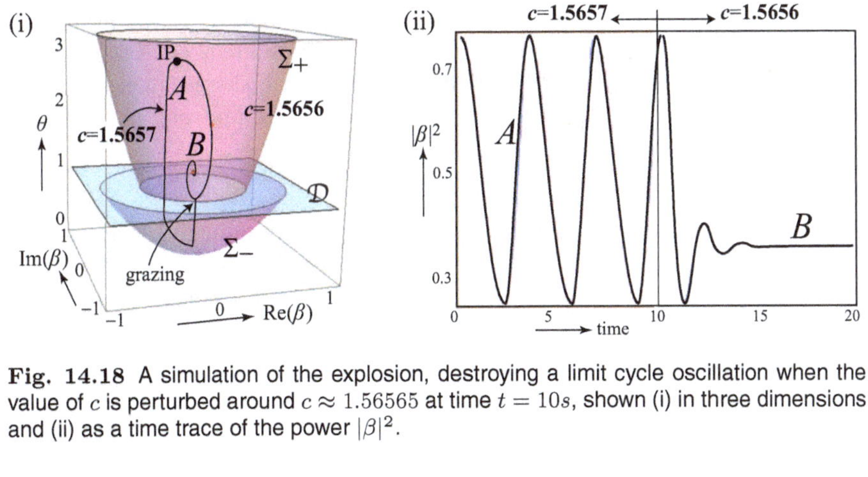

Fig. 14.18 A simulation of the explosion, destroying a limit cycle oscillation when the value of c is perturbed around $c \approx 1.56565$ at time $t = 10s$, shown (i) in three dimensions and (ii) as a time trace of the power $|\beta|^2$.

We could go further and delve into the switching layer analysis for this sequence of bifurcations, and we have some idea of what to expect in terms of the sliding equilibria being fully resolved inside the layer. The much more intriguing problem is then to look closely at how to introduce nonidealities of the model, the effects of thermal noise, for example, and how refinements might be made using nonlinear switching terms. All of this leaves interesting challenges for future authors in this or similar devices.

14.5 Conical Refraction

In Section 1.4.2 we introduced a form of discontinuity-induced determinacy-breaking that has long been known in fundamental physics, specifically in classical optics. The rays along which light travels kink when they enter a crystal. The crystal's anisotropy creates a singularity in direction space, along which one ray splits into infinitely many rays which splay out to form the surface of a cone, as depicted in Figure 1.9.

At the time when conical refraction was discovered, physics was split into two camps concerning the nature of light. Augustin-Jean Fresnel believed that light travelled in waves, Isaac Newton thought it travelled along rays. Of course they were both right. Newton's rays are the normals to the Fresnel's wave surfaces.

It was a third man, William Rowan Hamilton, who in resolving this connection discovered conical refraction.

The origin of the phenomenon is not purely dynamical (i.e. we cannot simply derive it from a switching multiplier as elsewhere in this book) but arises through the geometry of light's passage across a discontinuity in refractive index. In fact, the singularity itself is a property of any typical anisotropy tensor or, to be even less technical, any 3×3 matrix.

Light in a medium travels at a speed $v = c/n$, where n is the medium's *refractive index* and c is the speed of light in a vacuum. In an anisotropic medium, the index actually changes with direction and does not have such a simple interpretation. Instead we can then talk about a *dielectric matrix* $\underline{\underline{\mathcal{N}}}$, which relates the electric field $\mathbf{E}$ to the electric displacement field $\mathbf{D}$ (essentially an effective electric field in the presence of matter), as

$$\mathbf{E} = \tfrac{1}{\varepsilon_0}\underline{\underline{\mathcal{N}}}.\mathbf{D} \tag{14.23}$$

where ε_0 is the vacuum permittivity. The matrix components $\mathcal{N}_{ij}$ are constants $1/n_{ij}^2$, where, in an isotropic medium, the off-diagonals n_{ij} for $i \neq j$ vanish and the diagonals $n_{11} = n_{22} = n_{33}$ become the unique refractive index n. In general $\mathcal{N}_{ij}$ is a real symmetric matrix (provided the medium is transparent and optically inactive), with real eigenvalues that we call $1/n_1^2$, $1/n_2^2$, and $1/n_3^2$. We can think of n_1, n_2, and n_3 as the refractive indices along three principal directions (the eigenvectors) of $\underline{\underline{\mathcal{N}}}$. Expressed in coordinates along these directions, $\underline{\underline{\mathcal{N}}}$ is a diagonal matrix with components $1/n_{ii}^2 = 1/n_i^2$.

The phenomenon of conical refraction stems from a degeneracy of this dielectric matrix. The quadratic form

$$\sum_{ij} \mathcal{N}_{ij} x_i x_j = constant$$

defines a characteristic ellipsoid associated with the matrix, with principal radii $1/n_1^2$, $1/n_2^2$, and $1/n_3^2$, shown in Figure 14.19. For any such ellipsoid, there exists a direction (actually two directions) along which the cross-section is circular, corresponding to a singular direction of the matrix. This means that in certain coordinates (x, y, z), we can place z along the singular direction such that the degeneracy means $\mathcal{N}_{xx} = \mathcal{N}_{yy}$. If we achieve this by rotating about the principal axis associated with index n_2, so the '2' direction just becomes the 'y' direction, and then these conditions are enough to deduce uniquely that $\underline{\underline{\mathcal{N}}}$ in the (x, y, z) coordinates takes the form

$$\underline{\underline{\bar{\mathcal{N}}}} = \frac{1}{n_2^2}\underline{\underline{\mathcal{I}}} + \begin{pmatrix} 0 & 0 & \sqrt{\alpha\beta} \\ 0 & 0 & 0 \\ \sqrt{\alpha\beta} & 0 & \alpha - \beta \end{pmatrix}, \tag{14.24}$$

where $\underline{\underline{\mathcal{I}}}$ is the 3×3 identity matrix and $\alpha = \frac{1}{n_1^2} - \frac{1}{n_2^2}$, $\beta = \frac{1}{n_2^2} - \frac{1}{n_3^2}$.

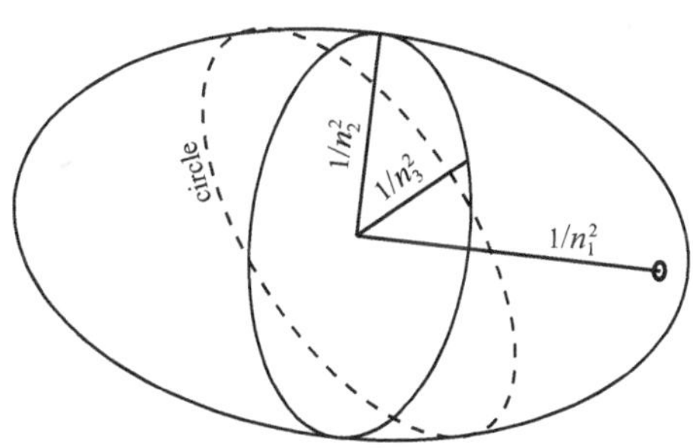

Fig. 14.19 The general ellipsoid $x_1^2/n_1^2 + x_2^2/n_2^2 + x_3^2/n_3^2 = constant.$

When applied to light entering such a medium, this degeneracy manifests itself as a breakdown in the uniqueness of the law of refraction. When light waves propagate along a wavevector $\mathbf{k}$ with frequency ω, subject to a refractive index n, they satisfy the dispersion relation

$$\omega(\mathbf{k}) = \frac{ck}{n(\mathbf{k})}. \tag{14.25}$$

Level sets of ω are called the *wave surfaces*. The waves evolve along ray paths that follow the group velocity vector $\dot{\mathbf{r}} = \frac{d}{d\mathbf{k}}\omega(\mathbf{k})$, hence normal to the wave surface.

To find the dispersion relation, we can turn to Maxwell's equations

$$\mathbf{k} \times \mathbf{E} = nck\mathbf{B}\,, \qquad nck\mathbf{D} = \mathbf{H} \times \mathbf{k}$$

where $\mathbf{B} = \mu_0 \mathbf{H}$ for the magnetic $\mathbf{H}$ and magnetic induction $\mathbf{B}$, fields, μ_0 is the vacuum permeability, and n is the refractive index along the direction of propagation. With a little matrix algebra, by taking $\mathbf{k}$ crossed with the first Maxwell equation, we obtain an eigenequation for $\mathbf{E}$,

$$\frac{1}{n^2}\underline{\underline{\mathcal{N}}}^{-1}.\mathbf{E} = -\hat{\mathbf{k}} \times \hat{\mathbf{k}} \times \mathbf{E} = \mathbf{E} - \left(\hat{\mathbf{k}} \cdot \mathbf{E}\right)\hat{\mathbf{k}}\,, \tag{14.26}$$

where $\hat{\mathbf{k}} = \mathbf{k}/|\mathbf{k}|$. We can rearrange this and write the jth component as

$$\frac{E_j}{n^2} - \frac{\hat{\mathbf{k}} \cdot \mathbf{E}}{n^2 - n_j^2}\hat{k}_j = 0\,, \tag{14.27}$$

before multiplying by $\hat{k}_j$, summing over j's, and then dividing by the first term $\hat{\mathbf{k}} \cdot \mathbf{E}/n^2$ to give, upon some rearrangement, the form of the wave surface as originally stated by Fresnel,

$$\frac{\hat{k}_1^2}{\frac{1}{n^2} - \frac{1}{n_1^2}} + \frac{\hat{k}_2^2}{\frac{1}{n^2} - \frac{1}{n_2^2}} + \frac{\hat{k}_3^2}{\frac{1}{n^2} - \frac{1}{n_3^2}} = 0\,, \tag{14.28}$$

which we can in principle solve for n. The equation defines a double-sheeted surface generated by the wavevector $\mathbf{k} = k_0 n(\mathbf{k})\hat{\mathbf{k}}$, as the unit direction vector $\hat{\mathbf{k}}$ traces out all directions on the unit sphere. Two sheets imply two sets of rays; hence typically one ray of light splits into two, $R_\pm$, with orthogonal polarisation. The two sheets meet along a singular direction—the same direction as the degeneracy of $\underline{\underline{\mathcal{N}}}$—which we now call the *optic axis* direction, where $k_2 = 0$ and $|k_1/k_3| = \tan\theta_{OA} := \sqrt{\alpha/\beta}$, as illustrated in Figure 14.20.

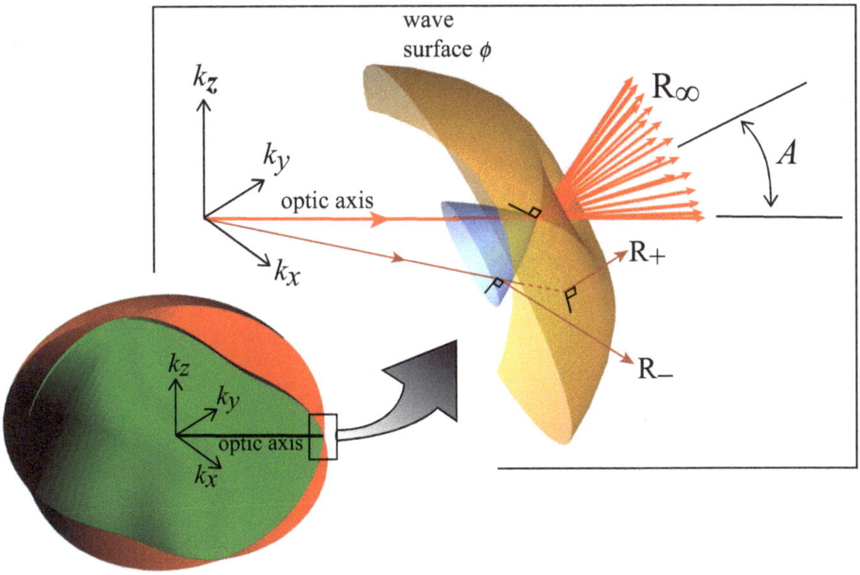

Fig. 14.20 Conical refraction: a light ray entering an anisotropic crystal (top left) splits into two rays $R_{\pm}$, except along the optic axis direction where it splits into an infinity of rays forming a hollow cone. The larger figure shows the construction of the ray directions as normals to Fresnel's wave surface, a pair of intersecting ellipsoids (of which only a small portion near the conical intersection is shown).

The wave surface near the optic axis direction has the shape of a pair of opposing cones, called a *conical* or *diabolical* point. A single ray incident upon the crystal thus degenerates into an infinite number of rays, lying in the surface of a narrow cone, and this Hamilton termed *internal conical refraction*. At the exit face, the simple laws of refraction applying, each ray leaves the crystal with the direction it entered such that the cone refracts into a cylinder. We can describe these more precisely by finding the flow of energy in the electromagnetic field, given by the so-called Poynting vector.

The z-coordinate is now distance along the optic axis and behaves like a time parameter. So let us seek the dynamical system $\dot{\mathbf{r}} \equiv d\mathbf{r}/dz = f(x, y, z)$ of light rays refracted according to Fresnel and Hamilton's theory.

Let the electric displacement vector $\mathbf{D}$, which is transverse to the wavevector, have an angle χ to the x-axis, and use the material relation (14.23) to find the electric field vector $\mathbf{E}$,

$$\mathbf{D} = D \begin{pmatrix} \cos\chi \\ \sin\chi \\ 0 \end{pmatrix} \qquad \Rightarrow \qquad \mathbf{E} = \frac{D}{n_2^2} \begin{pmatrix} \cos\chi \\ \sin\chi \\ 2A\cos\chi \end{pmatrix}, \tag{14.29}$$

defining $A = \frac{1}{2}\sqrt{\alpha\beta}$. As light is a harmonic oscillation of the electric and magnetic fields, $\mathbf{E}$ and $\mathbf{H}$, energy flows along the Poynting vector

$$\mathbf{S} = \mathbb{R}\mathrm{e}\left[\mathbf{E}^* \times \mathbf{H}\right] = \frac{D^2}{n_2^3} \begin{pmatrix} A\left(\cos 2\chi - 1\right) \\ A \sin 2\chi \\ 1 \end{pmatrix}, \tag{14.30}$$

which coincides with the ray direction in a transparent crystal and which now traces out a skewed cone, Figure 14.20. The axis of the cone is slanted such that the optic axis lies in its surface and through its apex, while the cross-section transverse to the optic axis remains circular. The half-angle of this (typically narrow) cone is A.

For our purposes we care that $d\mathbf{r}/dz \propto \mathbf{S}$, and therefore we can write a piecewise-smooth differential equation for a ray $\mathbf{r}(z)$ as

$$\frac{d}{dz}\mathbf{r}(z) = (0,0,1) + (A(\cos\phi_k - 1), A\sin\phi_k, 0)\,\mu\,, \tag{14.31}$$

where $\mu = \text{step}(z - L)$, letting refraction into the crystal medium occur at $z = L$. Solutions give the continuous but non-differentiable ϕ_k-parameterized family of rays

$$\mathbf{r}(z) = (0,0,z) + (A(\cos\phi_k - 1), A\sin\phi_k, 0)\,|z - L|\,. \tag{14.32}$$

Thus one ray $\mathbf{r}(z) = (0,0,z)$ for $z < L$ refracts into infinitely many, $\mathbf{r}(z) = (0,0,z) + (A(\cos\phi_k - 1), A\sin\phi_k, 0)$ for $z > L$, lying in the shape of a cone. The loss of determinacy here comes not from the way the flow in z-time contacts the discontinuity surface $z = L$, which must always lie transverse to the flow since z is the 'time' parameter but instead arises from the geometry that generates the vector field itself—a family of vector fields associated with different directions of travel $\mathbf{k}$, all permitted due to the singularity.

Hamilton's colleague Humphrey Lloyd verified the prediction in a set of admirable experiments using poor-quality crystals, sunlight, or candlelight and imaging the rings on his own eye, noting:

This phenomenon was exceedingly striking.
It looked like a small ring of gold viewed upon a dark background;
and the sudden and almost magical change of the appearance,
from two luminous points to a perfect luminous ring,
contributed not a little to enhance the interest. [155]

Nowadays the experiments still pose a challenge, even with the luxury of synthetic crystals, laser beams, hi-tech cameras, and digital image processing.

In fact there is more to this phenomenon, as nature conspires to obscure it. After all the mathematics we went through above, nature hides the phenomenon from view! The intensity of light that can fall on the singular point is zero, so the cone is actually dark, surrounded by inner and out bright cones from light scattered from the nearby sheets of ϕ; see [21] for further information. The effect is in some sense inverted; nevertheless the origin of the phenomenon remains. This may provide lessons for our studies of similar phenomena in dynamics.

14.6 Optical Folded Billiards

The following is more of an optical game than a serious application but follows on the theme of refraction and culminates in the same billiard-like dynamical system that we found in far different circumstances in Section 12.2.

Take four different transparent media with refractive indices n_{00}, n_{01}, and n_{10}, n_{11}, satisfying

$$n_{00}^2 + n_{11}^2 = n_{01}^2 + n_{10}^2 \,, \tag{14.33}$$

and arrange rectangular blocks of these four media in a repeating pattern as shown in Figure 14.21.

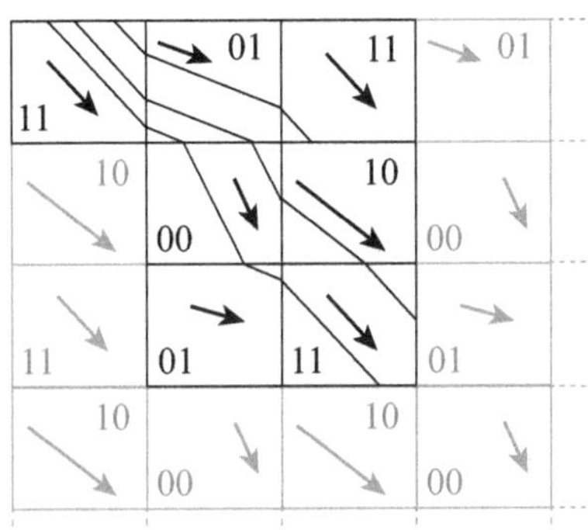

Fig. 14.21 An optical lattice of blocks with refractive indices n_{ij}, with i and j taking values 0 or 1, arranged such that i flips across vertical interfaces and j flips across horizontal interfaces. A light ray incident upon the lattice will pass through the blocks in the directions illustrated. Three sample paths are shown from block 11 to another block 11.

Let us project a ray of light through this lattice. We assume that no reflection takes place and instead at each interface the ray refracts according to Snell's law. Let θ_{ij} be the angle of a ray to the horizontal in block with index refractive n_{ij}, then Snell's law gives

$$\begin{aligned} n_{11}\sin\theta_{11} &= n_{01}\sin\theta_{01} \,, & n_{11}\cos\theta_{11} &= n_{10}\cos\theta_{10} \,, \\ n_{10}\sin\theta_{10} &= n_{00}\sin\theta_{00} \,, & n_{01}\cos\theta_{01} &= n_{00}\cos\theta_{00} \,. \end{aligned}$$

If we assume these angles are unique, in the sense that a ray returning to a block of the same refractive index does so with the same angle, these imply

$$\tan\theta_{00}\tan\theta_{11} = \tan\theta_{10}\tan\theta_{01} \,.$$

To show that the angle that light traverses each refractive index is indeed unique by virtue of (14.33), it is enough to consider three possible paths from a block of refractive index n_{11} (without loss of generality) to another of index n_{11}, writing the initial ray angle as θ_{11} and final ray angle as θ'_{11}. As illustrated in Figure 14.21, the three distinct paths are:

- via one intermediate block, e.g. $11 \mapsto 01 \mapsto 11$, then successive ray angles $\theta_{11} \mapsto \theta_{01} \mapsto \theta'_{11}$ obey

$$n_{11} \sin\theta_{11} = n_{01} \sin\theta_{01} = n_{11} \sin\theta'_{11} \quad \Rightarrow \quad \theta_{11} = \theta'_{11}$$

  hence the ray lies along angle θ_{11} in the first block and θ'_{11} in the last, and the two are equal. The alternative path $11 \mapsto 10 \mapsto 11$ is similar.
- via three intermediate blocks with no repetition, e.g. $11 \mapsto 01 \mapsto 00 \mapsto 01 \mapsto 11$, then successive ray angles obey

$$\begin{aligned} n_{11} \sin\theta_{11} &= n_{01} \sin\theta_{01} \,, & n_{01} \cos\theta_{01} &= n_{00} \cos\theta_{00} \,, \\ n_{00} \cos\theta_{00} &= n_{01} \cos\theta_{01} \,, & n_{01} \sin\theta_{01} &= n_{11} \sin\theta'_{11} \quad \Rightarrow \quad \theta_{11} = \theta'_{11} \,. \end{aligned}$$

  Alternative paths similarly give that the initial and final ray angles are equal.
- via three intermediate blocks without repetition, e.g. $11 \mapsto 01 \mapsto 00 \mapsto 10 \mapsto 11$, then successive ray angles obey

$$\begin{aligned} n_{11} \sin\theta_{11} &= n_{01} \sin\theta_{01} \,, & n_{01} \cos\theta_{01} &= n_{00} \cos\theta_{00} \,, \\ n_{00} \sin\theta_{00} &= n_{10} \sin\theta_{10} \,, & n_{10} \cos\theta_{10} &= n_{11} \cos\theta'_{11} \,, \end{aligned}$$

  which we combine successively to give

$$\begin{aligned} n_{11}^2 \sin^2\theta_{11} &= n_{01}^2 \sin^2\theta_{01} \\ &= n_{01}^2 \left[1 - \left(\tfrac{n_{00}}{n_{01}} \cos\theta_{00}\right)^2\right] \\ &= n_{01}^2 - n_{00}^2 \left[1 - \left(\tfrac{n_{10}}{n_{00}} \sin\theta_{10}\right)^2\right] \\ &= n_{01}^2 - n_{00}^2 + n_{10}^2 \left[1 - \left(\tfrac{n_{11}}{n_{10}} \cos\theta'_{11}\right)^2\right] \\ \Rightarrow \quad n_{01}^2 + n_{10}^2 &= n_{00}^2 + n_{11}^2 \left(\cos^2\theta'_{11} + \sin^2\theta_{11}\right) \\ \Rightarrow \quad \theta_{11} &= \theta'_{11} \quad \text{by (14.33)} \,; \end{aligned}$$

  thus the initial and final ray angles are equal.

As a result, our optical game is defined by four parameters. We may choose three refractive indices n_{ij} and one ray angle θ_{ij}, which fixes the remaining refractive index and ray angles. Alternatively we may choose three ray angles and one refractive index, which fixes the remaining ray angle and refractive indices.

This has the beautiful implication that the lattice dynamics is simply that of a circle map, or a billiard problem with non-conservative reflections. Each rectangular block is distinguishable only by the change of refractive index across its edges or, because of the result above, by the unique ray angle through it. With a little topology, we can transform the lattice into a single rectangular block with rays reflecting (rather than refracting) along four different angles. We do this by folding the lattice as shown in Figure 14.22, along every vertical edge (stage 1 to 2) and along every horizontal edge (stage 2 to 3), until we obtain a single multi-sheeted block, where each refractive index is associated with a specific *direction* of travel.

We obtain, as shown in stage 4 of Figure 14.22, a peculiar kind of billiard problem, where each reflection with a vertical edge permutes i and each reflection with a horizontal edge permutes j, in the refractive index n_{ij} or ray angle θ_{ij}, and the angles of *reflection* are given by the law of *refraction* by means of these indices.

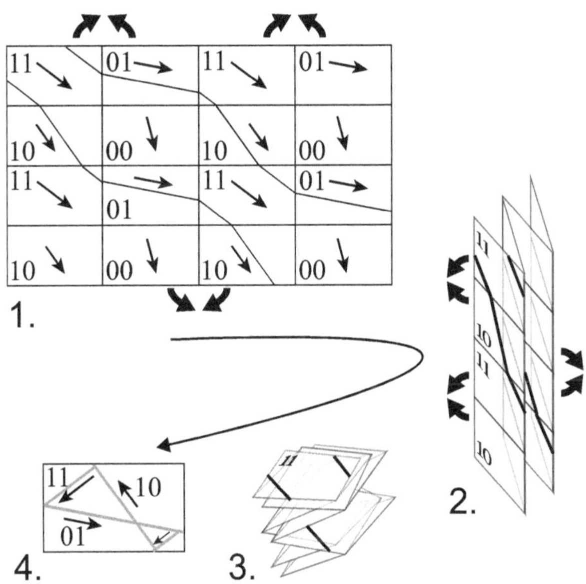

Fig. 14.22 Folding the lattice, to form a non-conservative billiard problem on a single block. The example shown has a periodic billiard trajectory.

Amusingly, the mathematical problem we obtain is identical to the dynamics inside the intersection of two discontinuity submanifolds with hysteresis, which we studied in Section 12.2.1. As we saw in that section (though the system it derives from is unrelated), the dynamics around the block can be periodic or chaotic, undergoing jumps of attractor by means of bifurcations. We omitted to study the dynamics in detail in Section 12.2.1, let us now go a little further.

First, to remind ourselves of the behaviour of this system, we simulate the dynamics on a square block in Figure 14.23 for different parameter values, showing, as θ_{11} changes, bifurcations between significantly different attractors, including what appears to be a chaotic attractor in (ii).

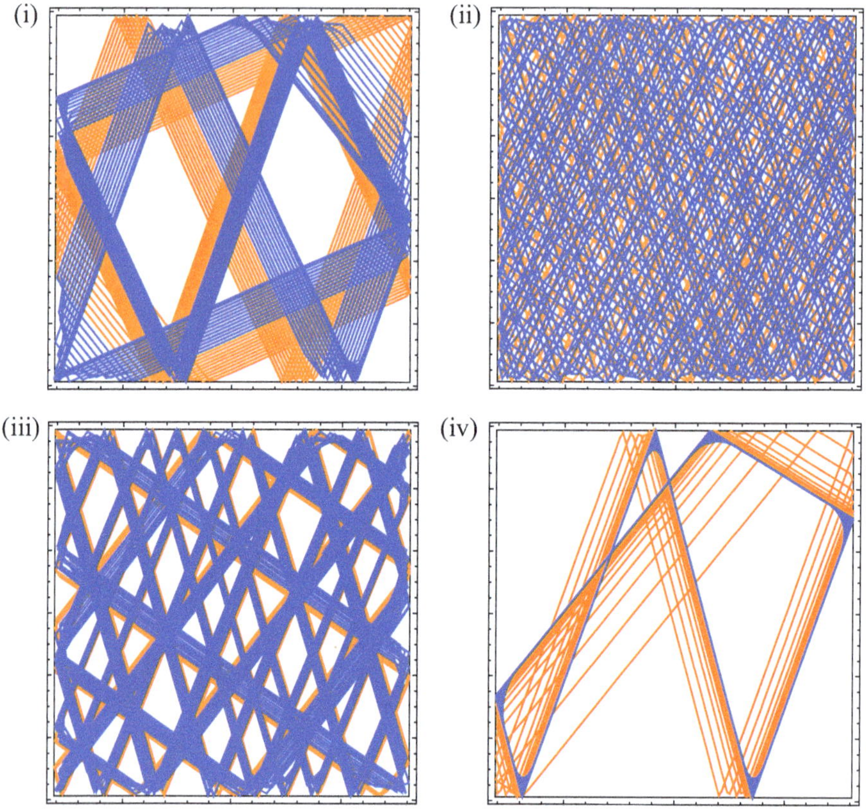

Fig. 14.23 Simulations of the folded optical billiard for $n_{11} = 0.7$, $n_{01} = 0.8$, $n_{10} = 0.9$, $n_{00} = \sqrt{n_{01}^2 + n_{10}^2 - n_{11}^2} \approx 0.98$, $\theta_{00} = \arccos(\frac{n_{01}}{n_{00}} \cos\theta_{01}) \approx 0.77$, $\theta_{01} = \arcsin(\frac{n_{11}}{n_{01}} \sin\theta_{11}) \approx 0.49$, $\theta_{10} = \arccos(\frac{n_{11}}{n_{10}} \cos\theta_{11}) \approx 1.14$, and (i) $\theta_{11} = 0.3$, (ii) $\theta_{11} = 1$, (iii) $\theta_{11} = 1.05$, (iv) $\theta_{11} = 1.1$. The first half of the solution is shown light/orange; the later half is shown dark/blue indicating the attractor.

To fully analyse the dynamics, we can study the return map to the boundary. It is not simple or instructive to write down explicitly, but we can study it qualitatively. In fact, as observed in [121], it is best to study the map obtained by return to the boundary after two reflections (the *second reflection map*), ϕ. We coordinatize the boundary with some x that runs from 0 to 1 around the rectangle, with values $x = 0, \frac{1}{4}, \frac{1}{2}, \frac{3}{4}$, at the corners, shown in Figure 14.24(left).

The second return is then a map on the unit interval, with kinks at the images or pre-images of the corners, forming an eight-branch piecewise lin-

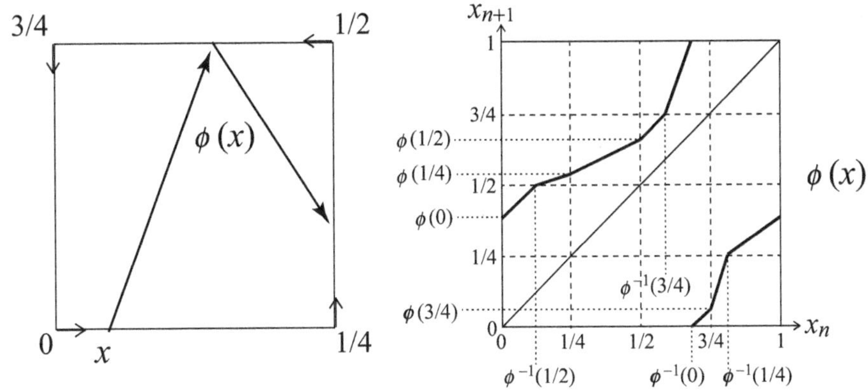

Fig. 14.24 The coordinatization of the block boundary (left) and the second reflection map ϕ. The map (right) which has kinks wherever it passes through x_n or $x_{n+1}=$ $0, \frac{1}{4}, \frac{1}{2}, \frac{3}{4}$, signifying images or pre-images of the corners.

ear map, as shown in Figure 14.24(right). This cuts the domain-range space $[0,1) \times [0,1)$ into 16 zones, the map on each of which is illustrated in Figure 14.25. For example, in the four zones lying along the diagonal $x_n = x_{n+1}$, the map is a simple translation $x_{n+1} = x_n + constant$.

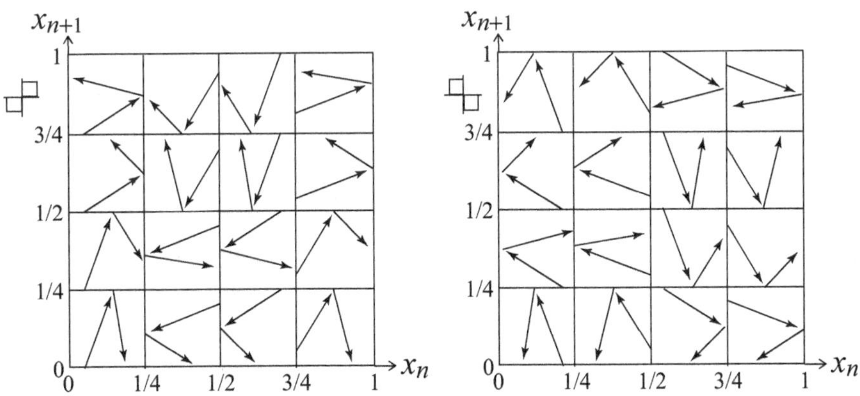

Fig. 14.25 The domain and range of the map split into 16 boxes, on each of which the map ϕ is the composition of a different pair of reflections as shown; there are two different forms of the map: in one the 00 or 11 map are always applied first (left), and in the other the 01 or 10 map are always applied first (right).

A related problem of biperiodic lattices, again from quite different motivations, was proposed in [81].

14.7 Static Versus Kinetic Friction

Friction and impact are the most obvious mechanical applications of piecewise-smooth dynamics. Let us look a little more closely at how switching layers can be put to use in the modelling of friction.

In Section 1.4.1 we discussed a simple block resting on a moving surface. In the laboratory a more controllable set-up is often favourable. Figure 14.26 shows a single degree of freedom oscillator used in some experimental studies of friction (for example in [222]; for alternatives see also [18, 105]).

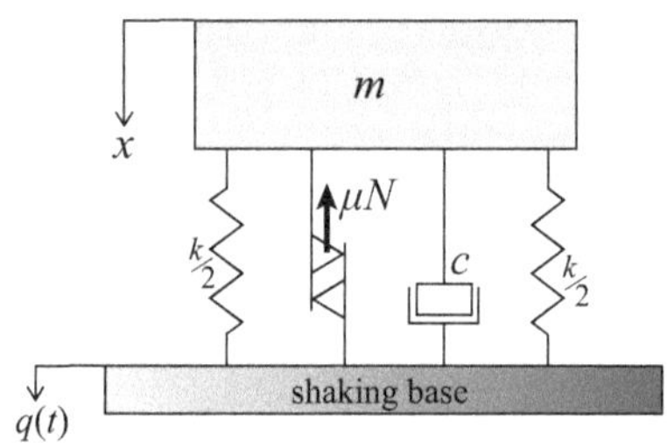

Fig. 14.26 A dry-friction oscillator. A block mass sits above a shaking base, connected by springs, a linear damper, and a dry-friction sliding contact.

A block of mass m is connected to a shaking base plate with displacement $q(t)$, via springs of overall spring coefficient k and extension x, a linear damper of coefficient c, and a sliding contact with dry-friction coefficient μ and normal reaction force N. The equation of motion is

$$m\ddot{x} = (q - x)k + (\dot{q} - \dot{x})c - N\mu \,. \tag{14.34}$$

The friction coefficient μ will be some expression of the direction of motion $\lambda = \text{sign}(v)$, where v is relative slipping speed between contact surfaces.

We saw in Chapter 2 that we can represent static friction by introducing nonlinear dependence on the switching multiplier λ. In Chapter 2 we let $\mu = \lambda + \rho(\lambda^2 - 1)$, the constant ρ giving different static and kinetic coefficients of friction, but this is not a good model because it gives some directionality to static friction. The expression for μ should be odd in λ, the simplest example therefore being $\mu = \lambda + \rho\lambda(1 - \lambda^2)$.

If we just write

$$\mu = (1 + \rho z)\lambda \,, \tag{14.35}$$

then there are other possibilities, too, for example, introducing dynamics to the nonlinear term. We can say that $z \to 1 - \lambda^2$ is only the equilibrium state of some dynamic variable z that satisfies

$$\beta\dot{z} = \gamma(\lambda) - z \,, \qquad \text{where} \quad \gamma(\lambda) = 1 - \lambda^2 \,, \tag{14.36}$$

with β, a small positive constant. More generally the function $\gamma(\lambda)$ could be any hidden function, that is, any finite-valued function satisfying $v\gamma(\lambda) = 0$.

The variable z represents some physical quantity associated with sticking, related to the degree of adhesion between the surfaces in contact. In slipping, when $v \neq 0$ (hence $\lambda = \pm 1$), z relaxes on the fast timescale t/β to the state $z = 0$. In sticking, when $v = 0$, z tends to an equilibrium state $z = \gamma(\lambda) \in (0,1)$ whose value depends on the full system's dynamics.

In this equilibrium state $z = \gamma(\lambda)$, if $\rho < 1/2$ then the friction coefficient is monotonic in λ, and if $\rho > 1/2$ it has peak values $\mu = \mu_s$ at $\lambda = \lambda_s$, defining

$$\lambda_s :=:= \pm\sqrt{\frac{1+\rho}{3\rho}} , \qquad \mu_s := \begin{cases} 2(1+\rho)^{3/2}/3\sqrt{3\rho} & \text{if } \rho > 1/2 , \\ 1 & \text{if } \rho \leq 1/2 , \end{cases} \tag{14.37}$$

The sliding region in the system, which is the region on $y = 0$ where solutions can satisfy $\dot{y} = 0$, is $\frac{1}{k}(\ddot{q} - N\mu_s) < y < \frac{1}{k}(\ddot{q} + N\mu_s)$. The bounds of this, by setting the maximum size of the region where sticking can persist, determine the maximum force of static friction $\mu_s N$.

In physical terms, ignoring the driving term $\ddot{q}$, this means that the force ky the spring must exert to overcome friction and cause breakaway from sticking is $\pm\mu_s N$.

The relative displacement between the block and the shaking base is $y = x - q$, and the slipping velocity between them is $v = \dot{y}$, in terms of which (14.34) and (14.36) become for small $\beta > 0$,

$$\left.\begin{aligned} \dot{y} &= v \\ \dot{v} &= \ddot{q} - ky - cv - N(1+\rho z)\lambda \\ \beta\dot{z} &= 1 - \lambda^2 - z \end{aligned}\right\} \quad : \quad \lambda = \operatorname{sign} v . \tag{14.38}$$

The $\dot{z}$ equation introduces a relaxation towards $z = \gamma(\lambda)$ on the fast timescale t/β, a state that is invariant for motion confined to either $v < 0$, $v > 0$, or $v = 0$, but is not invariant to motion that transitions between them, because the attractor $z = \gamma(\lambda)$ is discontinuous. To see this let $w = \gamma(\lambda) - z$, and then $\beta\dot{w} = \beta\dot{\lambda}\gamma'(\lambda) - w = w - 2\beta\dot{\lambda}\lambda$, which has a well-defined attractor at $w = 2\beta\dot{\lambda}\lambda$. For $\beta \to 0$ this attractor becomes $w = 0$, except as v approaches zero, where $\dot{\lambda}$ is infinite, and $w = 0$ could only be invariant if β were small enough that $\beta\dot{\lambda}$ were finite.

On $v = 0$ the switching layer system is

$$\left.\begin{aligned} \dot{y} &= 0 \\ \varepsilon\dot{\lambda} &= \ddot{q} - ky - N(1+\rho z)\lambda \\ \beta\dot{z} &= 1 - \lambda^2 - z \end{aligned}\right\} \quad : \quad \varepsilon \to 0 , \tag{14.39}$$

with $\lambda \in (-1, +1)$. With respect to the infinitely fast timescale t/ε, this is

$$\begin{aligned} y' &= 0 \,, \\ \lambda' &= \ddot{q} - ky - N\,(1+\rho z)\,\lambda \,, \\ \beta z' &= (\gamma(\lambda) - z)\varepsilon \;\; \rightarrow \;\; 0 \,. \end{aligned} \tag{14.40}$$

The different regimes of dynamics are sketched in Figure 14.27 for some fixed $z = \gamma(\lambda)$; the system has three dimensions plus time, so the horizontal axis represents some mixture of the t, y, z, coordinates. The dynamics in the regions of slipping in $v \neq 0$ is shown, given by (14.38), and the discontinuity surface $v = 0$ is blown up into the switching layer $\lambda \in (-1, +1)$, where the dynamics is given by (14.39).

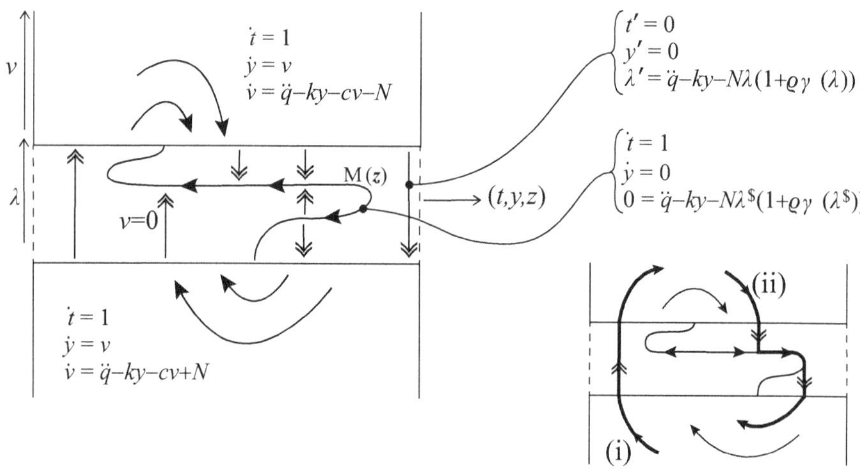

Fig. 14.27 Discontinuous model showing: slipping in the vertical (hyper)planes representing dynamics in $\{y, v, t\}$ variables (given by (14.38) for $v \neq 0$); fast transition given by (14.40) in the surface $v = 0$; sticking in the surface $\mathcal{M}(z)$ with dynamics in λ given by (14.43). The equations shown are in the steady state $z = \gamma(\lambda)$. The inset bottom-right shows two example trajectories: (i) transitions between slipping motion via the fast layer, (ii) transitions between slipping motion via an interval of sticking on the manifold $\mathcal{M}(z)$.

In the layer, the fast dynamics (14.40) is a $\{t, y, z\}$-parameterized one-dimensional system with a three-dimensional surface of fixed points

$$\mathcal{M}(z) = \left\{ (t, y, z, \lambda^{\$}) \in \mathbb{R}^3 \times (-1, +1) : \; \ddot{q} - ky = N\lambda^{\$}\,(1+\rho z) \right\} \,, \tag{14.41}$$

forming the sliding manifold. In the physical system, this manifold is where mechanical sticking occurs. $\mathcal{M}$ can be coordinatized by $\{t, y, \lambda^{\$}\}$, while z determines its shape. In particular, in $\{y, \lambda^{\$}\}$-space for fixed t, $\mathcal{M}(0)$ is a straight line with y-intercept $\ddot{q}/k$.

Let

$$\mathcal{M}_\gamma \equiv \mathcal{M}(\gamma(\lambda^{\$})) \,. \tag{14.42}$$

This is a curve with turning points at $\lambda^\$ = \pm\lambda_s$ if $|\lambda_s| < 1$. If a trajectory spends a sufficient time $t = \mathsf{O}(\beta)$ inside $v = 0$, the variable z will relax to $\gamma(\lambda^\$)$, and hence $\mathcal{M}(z)$ will relax to $\mathcal{M}_\gamma$. To find its attractivity, we calculate

$$\frac{\partial\lambda'}{\partial\lambda} = -N\left(1 + \rho\gamma(\lambda) + \rho\lambda\frac{d\gamma(\lambda)}{d\lambda}\right)$$

whose sign shows that $\mathcal{M}_\gamma$ has an attracting branch (where $\frac{\partial\lambda'}{\partial\lambda} < 0$) over $|\lambda| < \min[\lambda_s, 1]$, using λ_s defined in (14.37), onto which the system collapses on the timescale τ. If $|\lambda_s| < 1$ then $\mathcal{M}_\gamma$ also has two repelling branches over $\lambda_s < |\lambda| < 1$, with the attracting and repelling branches separated by sets of turning points at $\lambda = \pm\lambda_s$, as illustrated in Figure 14.27.

For y values inside the sliding region, the fast t/ε-timescale dynamics (instantaneous as $\varepsilon \to 0$ gives collapse to $\mathcal{M}(z)$, instigating sticking motion, while for y values outside this, the fast dynamics rapidly transitions λ between $\lambda = \pm1$, signifying the block transitioning directly between left and right slip.

The sliding dynamics inside $\mathcal{M}(z)$ is given by the $\varepsilon \to 0$ limit of (14.39), where we let $\lambda = \lambda^\$$, giving

$$\begin{aligned} \dot{t} &= 1\ , \\ 0 &= \ddot{q} - ky - N\lambda^\$\left(1 + \rho z\right)\ , \\ \beta\dot{z} &= \gamma(\lambda^\$) - z\ . \end{aligned} \tag{14.43}$$

The second line just constrains the dynamics to $\mathcal{M}(z)$, which $(\dot{t}, \beta\dot{z}) = (1, \gamma(\lambda^\$) - z)$, which as shown in Figure 14.27 give sticking dynamics either towards or away from the extrema of $\mathcal{M}(z)$.

Thus trajectories of the full system transition between (14.38), (14.40), and (14.43) and depicted in Figure 14.27. Left slip or right slip obey (14.38) with $v \neq 0$. Switches between the two are made via the fast system (14.40), as shown by trajectory (i) in Figure 14.27(bottom-right). For $\{y, t\}$ values in the sliding region, the fast system collapses to $\mathcal{M}(z)$, and sticking ensues given by (14.43). If this sticking dynamics reaches the extrema of $\mathcal{M}(z)$, at its turning points or at $\lambda^\$ = \pm1$, the system transitions via the fast system back into slipping, as shown by trajectory (ii) in Figure 14.27(bottom-right).

To see the roles of static and kinetic friction in this system, we can simulate the friction characteristic, plotting the friction force μN as the block transitions between slip and stick. The result is illustrated in Figure 14.28(i) and exhibits history-dependent (or 'hysteretic') stiction due to the way z excites or relaxes through the discontinuity surface. During slipping z approaches $\gamma = 0$, so as the block switches direction or enters into sticking, the friction coefficient μ transitions directly between ±1; therefore μN has no peaks and does not exhibit static friction. During an interval of sustained sticking on $\mathcal{M}_\gamma$, however, z again approaches z, but now this takes a value $z \in (0, 1]$, and if $\lambda_s < 1$ then the coefficient μ approaches $\mu \to \lambda(1 + \rho\gamma(\lambda))$

as $z \to \gamma$, which has peaks at $\lambda = \pm\lambda_s$. When the system subsequently exits back into slip from $v = 0$ to $v \neq 0$, it must pass through these peaks in the friction coefficient at $|\mu| = \mu_s = -\lambda_s^2 \frac{d\gamma(\lambda_s)}{dv} > 1$, forming the spikes seen in Figure 14.28(i).

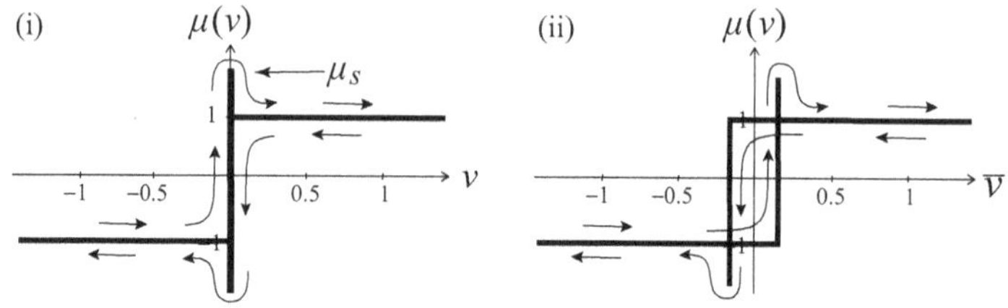

Fig. 14.28 Friction characteristic for the bouncer, discontinuous model plotted against the relative speed v in (i), and against the offset speed $\bar{v} = v - \varepsilon y$ representing compliance in (ii). Places where the graph is horizontal correspond to slip, and where it is vertical corresponds to stick.

This model treats the block as a point-like object, but we can easily consider it to be an extended body which, in resisting the contact force, suffers a small shear proportional to the spring extension y. To do this we introduce a velocity of the compliant body, some $\bar{v} = v - \varepsilon y$ for small $\varepsilon > 0$. The effect on the characteristic is shown in Figure 14.28(ii) and reveals the different directions of the hysteresis in the friction characteristic described above.

To see the motion behind these characteristics, as generated by the various dynamical elements above. We take a driving oscillation $\ddot{q}(t) = -\sigma \sin \omega t$. The parameters of the model come in three groupings: ρ, ε, and β for the friction model; k, c and N, for the bouncer; and ω and σ for the shaker.

In these simulations we will show a 'linear' sticking region, where $\mathcal{M}$ exists and sticking will occur if we assume $\gamma \equiv 0$ (a 'linear switching' or 'Filippov' system) or if we assume $\rho < 1/2$ so that the static friction coefficient μ_s is unity. We will also show the 'nonlinear' sticking region on which $\mathcal{M}$ has multiple branches (an overlapping sliding region, which can only exist in the full nonlinear system, see Definition 9.1 in Section 9.1). In each simulation one period of a repeating stick-slip oscillation is shown. For the sake of numerical computation, we replace the discontinuities with sigmoid functions

$$\lambda \mapsto \Lambda(v) = \frac{v/\varepsilon}{\sqrt{1 + (v/\varepsilon)^2}} \qquad \& \qquad \gamma \mapsto \Gamma(v) = \frac{1}{1 + (v/\varepsilon)^2}$$

for small ε.

Simulating the system (14.38), we obtain Figure 14.29. Once per period the trajectory sticks for a time interval $\delta t \approx \pi/2$, and once per period it crosses directly between right slip and left slip at around $t \approx -3\pi/4$. The relaxation time β is sufficiently fast for static friction to be observed as spikes in the friction characteristic in Figure 14.29(top left).

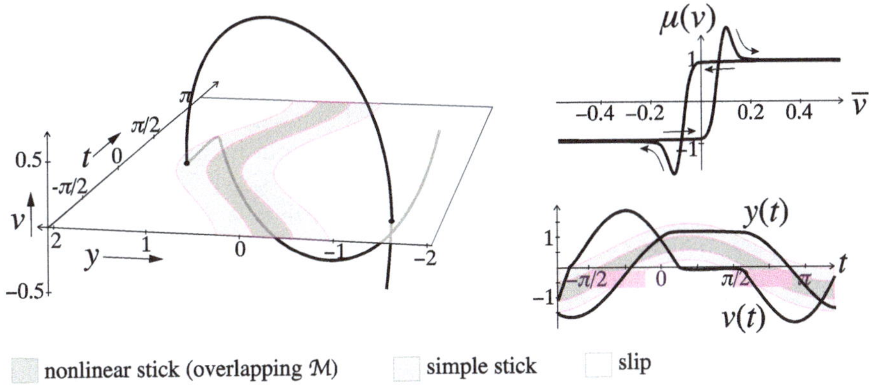

Fig. 14.29 Simulation of the oscillator with $\varepsilon = 0.1$, $\beta = 0.1$, $\lambda_s = 2/3$ ($\rho = 3$), $k = 3.1$, $c = 0.1$, $N = 1$, $\sigma = 2.4$, $\omega = 1$. Left: a trajectory of the limit cycle. Top-right: the friction characteristic μ during the simulation, as a function of the 'compliant speed' $\bar{v} = v - \varepsilon y$ introduced in Figure 14.28. Bottom-right: a plot of the y and v variables. Shading indicates the regions of simple sticking (dark grey) and overlapping sticking (light grey) corresponding to stiction. A stick-slip oscillation with sticking in the overlap region is seen.

For the stick-slip cycle shown, sticking occurs in the nonlinear sticking region. If we set $\rho < 1/2$, there should effectively be no effect from static friction, and the same orbit should pass through the discontinuity surface without sticking. As predicted, upon setting $\rho = 0$, we obtain an oscillation with left and rightward slip only, as in Figure 14.30.

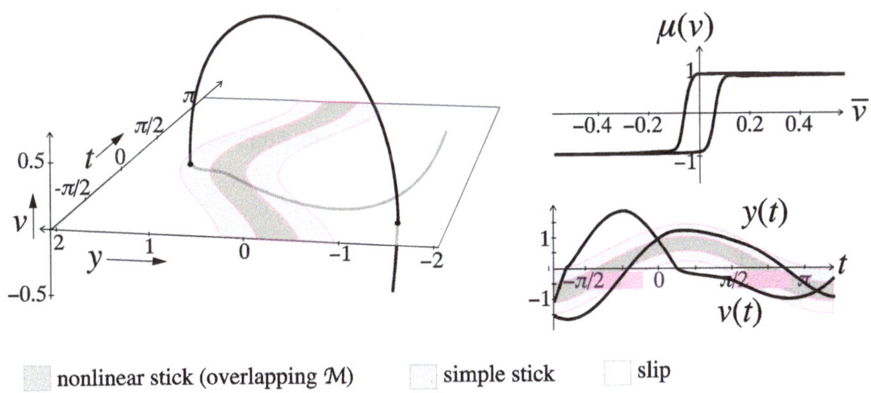

Fig. 14.30 Simulation corresponding to Figure 14.29 but with $\rho = 0$, showing an oscillation without sticking, which pierces the overlapping sticking region (light grey) without sticking.

Setting $\rho = 0$ effectively 'turns off' the nonlinear dependence of the system on the switching multiplier; hence the switching on or off of hidden terms by ρ crucially changes the global dynamics, giving a cycle with stick in Figure 14.29 but without stick in Figure 14.30.

In Section 12.1.2 we argued that unmodelled errors could kill of the effect of hidden terms, which here means they should be able to eliminate static friction. To see this in action, let us add a random perturbation of size κ to the state $\{y, v\}$ after every $1/500^{th}$ of a period (similar dynamics is obtained if perturbations are also made in z). The simulation in Figure 14.31(i) shows that nonlinear sticking persists for perturbations of size $\kappa = \varepsilon/5$ but is destroyed in Figure 14.31(ii) by perturbations of size $\kappa = 5\varepsilon$. Note that κ is small enough that the noisy and noiseless trajectories closely coincide almost everywhere, the perturbations only having a significant effect near $v = 0$, where they eliminate nonlinear sticking in the latter case. For small noise the nonlinear terms and the difference between static and kinetic friction they produce persist under what are effectively two different perturbations: the small added noise and the smoothing of the discontinuity (Figure 14.31(i)). With sufficient noise the system behaves as if $\rho = 0$ and hence neglects static friction (Figure 14.31(ii)), removing the segment of sticking.

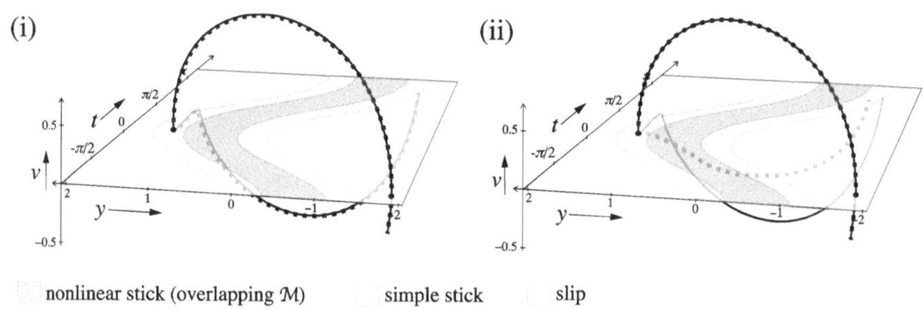

Fig. 14.31 Simulation corresponding to Figure 14.29 showing a stick-slip oscillation subject to no perturbation (full curve) or to small noise of size κ (dotted curve), parameters as in Figure 14.30 except with (i) $5\kappa = \varepsilon = \beta = 0.001$, (ii) $\kappa/5 = \varepsilon = \beta = 0.001$.

The friction characteristics for these two simulations subject to noise are given by the dotted curces in Figure 14.32.

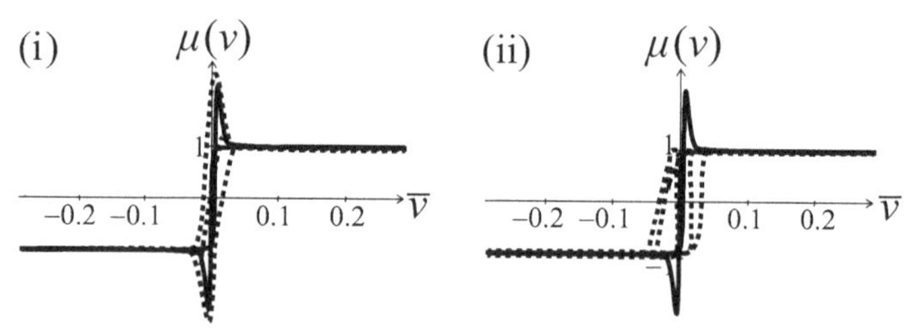

Fig. 14.32 Friction characteristics for the simulations in Figure 14.31 (i) and (ii), respectively, constituting (i) small noise and (ii) large noise. The full curve shows the unperturbed simulation. The dotted curve shows the perturbed simulation, which exhibits the static friction peaks in (i) but misses them in (ii).

They confirm that for $\kappa < \varepsilon$ that the peaks of static friction still appear in the presence of noise, but for $\kappa > \varepsilon$ the perturbations are large enough that the trajectory misses the peaks that constitute static friction. Very little compliance is visible here because ε is small.

Experimental, numerical, and analytical studies of similar oscillator models can be found in [62, 131, 89, 105, 184, 39] (though note that analytical studies usually consider $\rho = 0$ only). A similar two-dimensional problem without driving studied in [114, 122] showed that nonlinear sticking can mean the difference between decay to a steady state (with $\rho = 0$) and trapping into a stick-slip oscillation (with $\rho \neq 0$). Moreover in [122] it was shown in a stochastic setting that sticking modes can be interpreted as potential wells, escapable under sufficient perturbation due to noise.

14.8 A Paradox of Skipping Chalk

When we write on a blackboard with chalk, we tend to pull or drag the chalk along, rather than push it. Pushing seems to require more effort in terms of the force applied and can lead to skipping of the chalk along the board rather than the nice continuous line usually given by dragging. You can try it out yourself; depending on the chalk, you may just have to hold it at a steep enough angle and press hard enough to observe skipping (Figure 14.33).

Fig. 14.33 Chalk marks on a blackboard: dragging/pulling tends to give a continuous contact, while pushing can cause the chalk to skip.

The skipping arises due to a singularity in the contact dynamics of a rod (the chalk) in frictional contact with a surface (the blackboard). Since it is possible for the rod to either leave the surface or slip along it, the problem contains elements of two sources of discontinuity: dry-friction, whose force jumps from left-to-right as the rod alternates from right-to-left slip, *and* impact, whose force switches on during contact and off during free-flight. The two discontinuities intersect at a moment when the chalk appears stationary against the board, but the forces on it are changing, such that in the next instant, it might either lift-off or remain in contact and either slip left, slip right, or remain stuck in position.

Painlevé's paradox concerns a situation in which there seems to be no way to tell which of these will happen from the Newtonian equations of motion, whether it will lift-off, slip right, slip left, or stick or even follow a fifth counterintuitive possibility dubbed '*impact without collision*'.

Let us model the chalk as a simple rod of length l and mass m, with its centre of mass at coordinates (X, Y), its moment of inertia about one end I. The rod is in contact with a rough surface, to which it is inclined at an angle θ with $0 < \theta < \pi/2$, as in Figure 14.34. We can also apply an external force

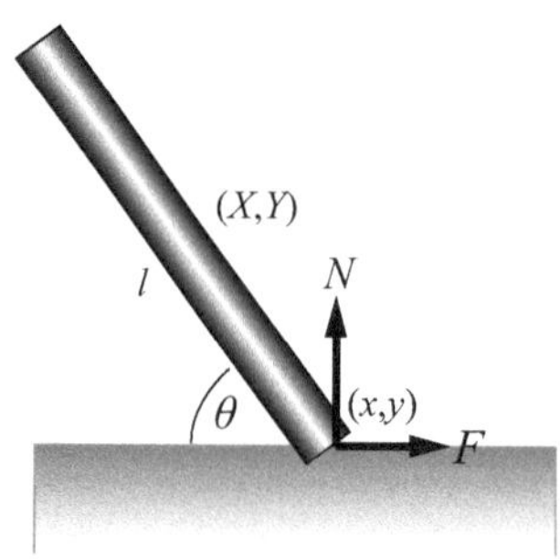

Fig. 14.34 The set-up for Painlevé's paradox, an inclined piece of chalk pushed along a surface.

(E_x, E_y) applied a distance ρ along the rod to simulate it being dragged or pushed.

If the normal reaction force on the rod is N, and the friction force is F, then the equations of motion are

$$\begin{aligned} m\ddot{X} &= F + E_x \,, \\ m\ddot{Y} &= N - mg + E_y \,, \\ I\ddot{\theta} &= \tfrac{l}{2}(F\sin\theta - N\cos\theta) - \rho(E_x\sin\theta - E_y\cos\theta) \,. \end{aligned} \tag{14.44}$$

Discontinuity enters through the friction force, which switches between left or right slip,

$$F = \mu N \,\mathrm{sign}(\dot{x}) \ , \tag{14.45}$$

and also through the normal reaction force, which is zero if the rod lifts off from the surface, for which we write

$$N = \tilde{N} \,\mathrm{step}(-y) \ , \tag{14.46}$$

for some positive $\tilde{N}$. Hence F and N are both zero during free-flight but are variable during contact.

It is really the contact point of the rod we are interested in, and this lies at

$$(x, y) = (X, Y) - \frac{l}{2}(\cos\theta, \sin\theta) \ . \tag{14.47}$$

Let $(u, v) = (\dot{x}, \dot{y})$. Coulomb's law implies $|F| \leq \mu|N|$, hence $0 \leq \mu \leq 1$, but let us say that the rod is in contact and on the point of slipping such that $F = \mu N$, then

$$\dot{y} = v \ , \tag{14.48a}$$

$$\dot{v} = b(\theta, \dot{\theta}) + \begin{cases} Na(\theta, \mu) & \text{if } y = 0 \\ 0 & \text{if } y > 0 \end{cases} \ , \tag{14.48b}$$

where

$$\begin{aligned} a(\theta, \mu) &= \tfrac{1}{m} + \tfrac{l^2}{8I}(1 + \cos 2\theta) - \tfrac{\mu l^2}{8I} \sin 2\theta \ , \\ b(\theta, \dot{\theta}) &= \tfrac{l}{2}\dot{\theta}^2 \sin\theta - g + c(\theta) \ , \\ c(\theta) &= E_x \tfrac{\rho l}{4} \sin 2\theta + E_y \left(\tfrac{1}{m} - \tfrac{\rho l}{4}(1 + \cos 2\theta) \right) \ , \end{aligned}$$

such that b represents the free vertical acceleration (i.e. when there are no contact forces).

The driving force E is really just there for generality, and we can ignore it henceforth. In fact, the whole of the following argument will be highly schematic, as this is sufficient to illustrate Painlevé's paradox, while a detailed study is very lengthy and no more illuminating.

Now let us ask whether onward motion can continue in contact, i.e. with $\dot{v} = 0$. This implies

$$N = -b/a \ . \tag{14.49}$$

Painlevé's paradox arises because a and b change sign independently, contrary to the natural intuition that N, the normal reaction force, must be positive. The cases of interest to us involve $a < 0$.

If $a < 0 < b$ then there are two possibilities for onward motion. In the absence of any contact forces, $b > 0$ implies positive vertical acceleration, i.e. the rod lifts off from the surface to $y > 0$. However, there also exists an equilibration of the normal contact force N, and a perfectly sensible one since $N = -b/a > 0$, for slipping motion in sustained contact. The onward motion is non-unique.

If $a, b < 0$ then things are even worse. This implies $N = -b/a < 0$, a normal reaction that *pulls* the rod *into* the surface! This pulls in the same direction as the downward free acceleration $b < 0$, precluding any vertical equilibrium, so the rod only accelerates into the surface at an increasing rate. If we look in free flight then $b < 0$ carries the rod onto the surface, so there is no escape there. Newton is seemingly intent on dragging the tip of the rod ever deeper into the surface, a surface which, in the rigid body problem as we have set it up, is impenetrable. Hence the onward motion appears inconsistent with the problem. Nevertheless this seems to be a real physical situation that has been dubbed *impact without collision*.

A paradox of this kind generates many an attempt at resolution or re-interpretation (see [196, 31] and references therein), as indeed it should. Paradoxes are the testing ground for our most fundamental theories. Either they reveal a mistake in our reasoning, or, frequently and often most profoundly, they reveal something true and yet counterintuitive. In this case Painlevé's paradox seems to reveal something not only true but potentially rather general and, yes, counterintuitive.

A set-up described in [31] is suggestive of this generality, and we take this as a starting point, but here we will employ layer analysis rather than rely on compliance of the surface (though the methods are analogous). Define a vector $\mathbf{q} = (x, y, \theta)$ and diagonal matrix of masses $\underline{\underline{M}} = \text{diag}[m, m, I]$; then our problem takes the form

$$\underline{\underline{M}}\ddot{\mathbf{q}} = \mathbf{f}(\mathbf{q}, \dot{\mathbf{q}}, t) + N\mathbf{p}(\mathbf{q}, t) + F\mathbf{n}(\mathbf{q}, t) \ , \tag{14.50}$$

where we have also allowed time dependence. In fact, $\underline{\underline{M}}$ can now be any positive-definite matrix and $\mathbf{p}$ and $\mathbf{n}$ any vector functions, and $\mathbf{q}$ can be any n-dimensional vector of generalized coordinates.

We will assume that we can project these equations onto a 'vertical' coordinate direction, y, whose axis lies along the direction of $\mathbf{n}$, and letting $v = \dot{y}$ as usual, we have

$$\dot{y} = v \ , \tag{14.51a}$$
$$\dot{v} = \varphi(\mathbf{q}, \dot{\mathbf{q}}, t) + N\rho(\mathbf{q}, t) + F\eta(\mathbf{q}, t) \ , \tag{14.51b}$$

for some functions φ, ρ, η, of y and the other generalized coordinates. The following analysis can be applied directly to our simple rod-surface set-up above, as just one example.

To explore the full dynamics of the problem, there are various scenarios to consider depending on the signs of various combinations of functions. We are concerned only with gaining a fundamental insight into why a paradox should occur, if it should occur at all.

It boils down to a simple local analysis. The impact surface is where $y = 0$. The layer system for (14.51a) is then

$$\left.\begin{aligned} \varepsilon\dot{\lambda} &= v \\ \dot{v} &= \varphi(\mathbf{q}, \dot{\mathbf{q}}, t) + N\rho(\mathbf{q}, t) + F\eta(\mathbf{q}, t) \end{aligned}\right\} \quad : \quad \varepsilon \to 0\,, \tag{14.52}$$

with $y = 0$ and $\lambda \in (0, 1)$; note that the layer here is over $\lambda \in (0, 1)$ rather than our usual $(-1, +1)$, as the impact force has a factor of $\lambda = \text{step}(y)$ rather than $\lambda = \text{sign}(y)$.

The simplicity of this expression suggests we introduce a scaling to remove the ε, namely, rescaling time t and speed v as $\tilde{t} = t/\sqrt{\varepsilon}$ and $\tilde{v} = v/\sqrt{\varepsilon}$, so

$$\left.\begin{aligned} \lambda' &= \tilde{v} \\ \tilde{v}' &= \varphi(\mathbf{q}, \dot{\mathbf{q}}, t) + N\rho(\mathbf{q}, t) + F\eta(\mathbf{q}, t)\,, \end{aligned}\right\} \quad : \quad \varepsilon \to 0 \tag{14.53}$$

where the prime denotes differentiation with respect to the scaled time $\tilde{t}$.

If sustained contact to the surface $y = 0$ is possible, then it corresponds to an equilibration of vertical forces where $v = 0$ and $\dot{v} = 0$. Expanding the dynamical expressions in (λ, v) layer space about such an equilibrium at $(\lambda, v) = (\lambda_0, 0)$, assuming it is nondegenerate, gives

$$\begin{pmatrix} \lambda' \\ v' \end{pmatrix} = \begin{pmatrix} 0 & 1 \\ \alpha & \beta \end{pmatrix} \begin{pmatrix} \lambda - \lambda_0 \\ v \end{pmatrix} + h.o.t. \tag{14.54}$$

for some constants α, β, and λ_0. This analysis is carried out in somewhat more detail in [31], with derivations of the initial conditions corresponding to specific dynamical scenarios and the corresponding functional forms of α and β, in particular their relation to the functions a and b above, which are not entirely simple. Nevertheless, the result is the same in any case, namely, (14.54), where λ_0 may or may not lie inside the layer $(0, 1)$ and where different signs of α and β are possible, analogous to the different signs of a and b possible above.

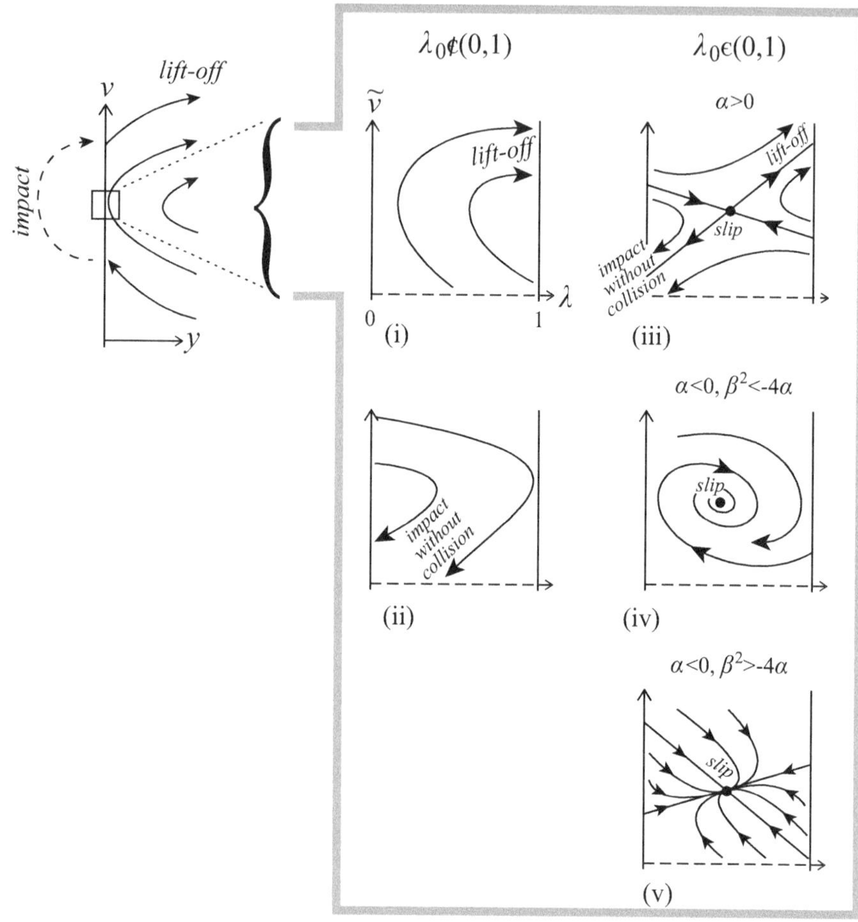

Fig. 14.35 Blow-up of the Painlevé problem. Left: the phase portrait in free-flight and nonzero speed impact. Right: the blow-up of the point $y = v = 0$ obtained by a time-rescaling inside the switching layer, with five qualitatively different phase portraits.

We have now the equations (14.48a) which describe well the dynamics outside $y = v = 0$ and (14.54) which describe the dynamics at this point using a time-rescaling inside the switching layer. The key phase portraits are illustrated in Figure 14.35, showing the free-flight and nonzero speed impact in the main picture (left) and the different possible phase portraits in the blow-up of $y = v = 0$ in the grey box (right).

Lift-off must occur $\alpha > 0$ and $\lambda_0 < 0$, or $\alpha < 0$ and $\lambda_0 > 1$, both giving portrait (i) in Figure 14.35.

Impact without collision must occur if $\alpha > 0$ and $\lambda_0 > 1$ or $\alpha < 0$ and $\lambda_0 < 0$, both giving portrait (ii) in Figure 14.35.

If $\alpha < 0$ and $0 < \lambda_0 < 1$, there is an attracting focus or node corresponding to a stable slip or stick state.

If $\alpha > 0$ and $0 < \lambda_0 < 1$, there is a saddle point, whose stable manifold separates out different initial conditions (though indistinguishable up to order ε), which can lead either to lift-off or impact without collision. The equilibrium corresponds to slip or stick, but as a saddle point, it can be reached only from a special initial condition (as [31] puts it, this would be like finding an initial condition that can make an infinitely heavy pin balance on its point). Because the limit $\varepsilon \to 0$ squeezes all of these possibilities into the impact point $y = v = 0$, as we have seen many times in preceding chapters, we have a determinacy-breaking point.

To determine whether stick or slip occurs requires looking at the tangential motion along x.

The layer analysis here is only in the impact coordinate y, with the slipping speed $\dot{x}$ taken to be fixed at the moment of impact. A more complete treatment would introduce two switching multipliers μ_1 and μ_2, one for friction and one for impact, via the tangential force

$$F = \mu N \lambda_1 , \qquad \lambda_1 = \text{sign}(\dot{x}) \tag{14.55}$$

and normal reaction

$$N = \tilde{N}\lambda_2 , \qquad \lambda_2 = \text{step}(-y) = \tfrac{1}{2}(1 - \text{sign}(y)) , \tag{14.56}$$

and explore the full five-dimensional phase space $(x, y, \theta, \dot{x}, \dot{y}, \dot{\theta})$ and associated switching layers: $(x, \lambda_2, \theta, \lambda_1, \dot{y}, \dot{\theta})$ for stick-impact and $(x, y, \theta, \lambda_1, \dot{y}, \dot{\theta})$ for slip-impact. Space permits that this, alas, must be left to future works.

14.9 Pinching Neurons

At the heart of various membrane potential models that have become popular in the mathematical biology of neurons—such as Hodgkin-Huxley, Hindmarsh-Rose, Fitzugh-Nagumo, Morris-Lecar, and so on; see, e.g. [106, 72, 163, 46, 43, 135, 171]—lie the turning points of slow manifolds. Let us see how to derive piecewise-smooth approximations of such models. Those turning points, and the novel dynamics they create, are intimately related to the singularity that has arisen more than any other throughout this book: the two-fold singularity.

A crude approach sometimes taken to studying nonlinear system is to caricature them as piecewise linear, for example, replacing x^2 by $|x|$ in the system

$$\left\{ \begin{array}{c} \varepsilon\dot{x} = y - \frac{1}{2}x^2 \\ \dot{y} = a(x,y) \end{array} \right\} \longrightarrow \left\{ \begin{array}{c} \varepsilon\dot{x} = y - |x| \\ \dot{y} = a(x,y) \end{array} \right\}$$

which has slow timescale t and fast timescale t/ε, for small positive ε. As shown in Figure 14.36, this replaces the slow critical manifold $\{(x,y) : y = \frac{1}{2}x^2\}$ and its turning point at $x = y = 0$, by a manifold $\{(x,y) : y = |x|\}$ having a corner at $x = y = 0$.

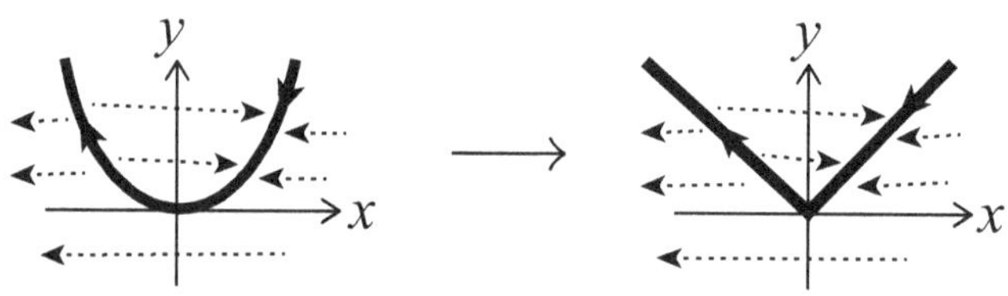

Fig. 14.36 The curve $y = \frac{1}{2}x^2$ caricatured as $y = |x|$.

Both are able to exhibit canard trajectories, which travel round from the attracting branch in $x > 0$ to the repelling branch in $x < 0$. However, the phenomena—the attractors, bifurcations, and parameter scalings—associated with the canards are different for a smooth turning point and a corner, and it is not clear in what sense the replacement above constitutes an approximation. To be candid, this is not an approach I favour.

The method of pinching from Section 12.4 gives us a more rigorous way to approximate a smooth system by a piecewise-smooth one. The constituent smooth systems can then, if appropriate, be approximated piecewise to obtain a piecewise linear system. (The following uses Section 12.4 to update unpublished preliminary work done with M. Desroches [45], at a time when we were using pinching as a crude tool to form nonsmooth analogues of canard systems, with little idea of the more rigorous theory that would develop later.)

14.9.1 Relaxation Oscillations and Canards

One of the classic problems where such turning points arise is in systems involving Lienard-like equations

$$\varepsilon\ddot{x} + f(x)\dot{x} = g(x, \dot{x}, z) \,, \tag{14.57}$$

where f and g are some smooth functions. Alfred-Marie Liénard studied the existence and uniqueness of equations of this form [148]. Here we have added a coupling to some other variable z for a little more freedom.

This can be turned into a first-order ordinary differential equation by defining $y = \dot{x}$, but the substitution $y = \varepsilon\dot{x} + \int dx f(x)$ is sometimes found to be more useful, giving

$$\begin{aligned} \varepsilon\dot{x} &= y + \textstyle\int dx f(x) \,, \\ \dot{y} &= g(x, y, z) \,, \\ \dot{z} &= \rho(x, y, z, t) \,. \end{aligned} \tag{14.58}$$

where we have given z some dynamics in terms of a smooth function $\dot{z} = \rho(x, y, z, t)$. An interesting example associated with Balthazar van der Pol and his study of relaxation oscillations [213] takes $f(x) = x^2 - 1$ and takes g as a function constant or linear in x. We will let $g(x, z) = bz - ax$, so

$$\begin{aligned} \varepsilon\dot{x} &= y + x - \tfrac{1}{3}x^3 \,, \\ \dot{y} &= bz - ax \,, \\ \dot{z} &= \rho(x, y, z, t) \,. \end{aligned} \tag{14.59}$$

On the fast timescale t/ε, this looks like

$$\begin{aligned} x' &= y + x - \tfrac{1}{3}x^3 \,, \\ y' &= \varepsilon(bz - ax) \,, \\ z' &= \varepsilon\rho(x, y, z, t) \,. \end{aligned} \tag{14.60}$$

In the limit $\varepsilon \to 0$ this has a set of equilibria forming a critical manifold $\left\{(x, y, z) : y = x - \frac{1}{3}x^3\right\}$, which is normally hyperbolic except at $x = \pm 1$ where it has turning points. The turning points separate attracting outer branches in $|x| > 1$ from a repelling inner branch in $|x| < 1$. There is an equilibrium of the (x, y) system at $x = bz/a$, $y = b^3z^3/3a^3 - bz/a$.

Ignoring any dynamics in the z-direction, the equilibrium is repelling while it lies on the inner branch with $|bz/a| < 1$, and a so-called *relaxation oscillation* forms, wrapping around the attracting branches, with fast transitions between them as shown in the first portrait of Figure 14.37.

As the equilibrium passes through either turning point, ($x = 1$ is shown in Figure 14.37), the limit cycle forming the oscillation undergoes a rapid transition. Within a parameter range $|bz/a| = 1 - \mathrm{O}\left(e^{-1/\varepsilon}\right)$, the cycle begins to shrink by travelling up along the repelling branch, until eventually when $|bz/a| > 1$, the equilibrium is an attractor and the oscillation disappears.

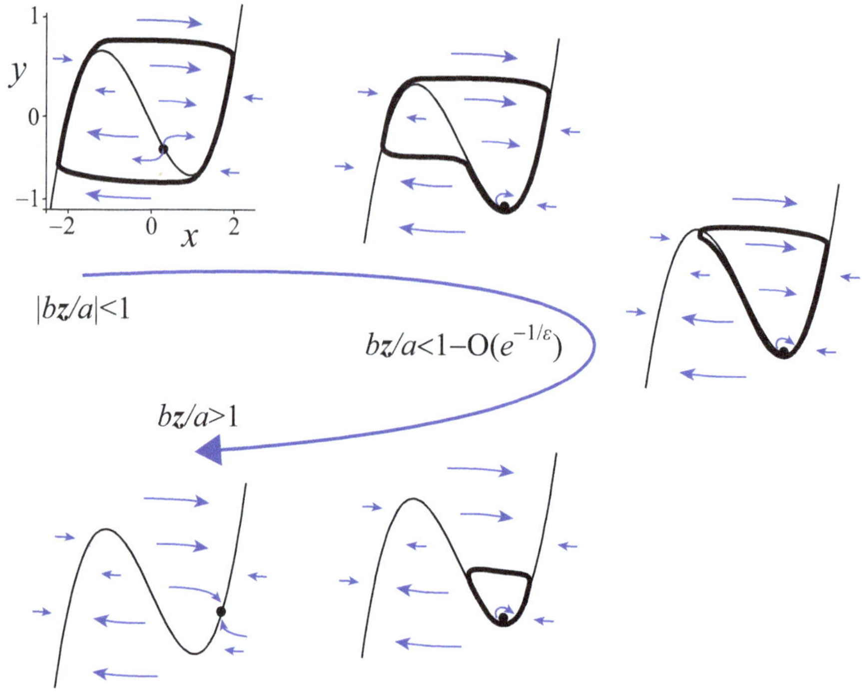

Fig. 14.37 As the equilibrium at $x = bz/a$ moves through the right-hand turning point of the critical manifold, we have a relaxation oscillation, a canard, a maximal canard, a canard without head, and an attracting fixed point.

The cycle that exists for that brief $\mathrm{O}\left(e^{-1/\varepsilon}\right)$ parameter range is a canard, travelling from an attracting manifold to a repelling manifold via a singularity—the turning point. The interesting dynamics has therefore entirely to do with what happens local to the turning points; elsewhere the flow does not change qualitatively and just provides the return mechanism to create a limit cycle *only if* the local conditions near the turning point permit it.

Let us therefore approximate about one of these turning points, say at $(x, y) = (1, -\frac{2}{3})$, in coordinates $x_1 = 1 - x$, $x_2 = -y - \frac{2}{3}$, $x_3 = (z - \frac{a}{b})$, and let $q = -a$, $p = -b$, $r = \rho(1, -\frac{2}{3}, \frac{a}{b}, t)$, giving

$$\begin{aligned} \varepsilon \dot{x}_1 &= x_2 + x_1^2 + \mathrm{O}\left(x_1^3\right) \,, \\ \dot{x}_2 &= q x_1 + p x_3 \,, \\ \dot{x}_3 &= r + \mathrm{O}\left(x_1, x_2, x_3\right) \,. \end{aligned} \tag{14.61}$$

You may recognise this as the singularity from Proposition 13.1 and (13.82). What follows therefore gives an alternative insight into our layer analysis on the two-fold in Chapter 13. From the point of view of smooth systems, what follows is an alternative insight into the simplest route to canards in two-timescale systems, via a special point at the origin of (14.61) which we shall call here the 'canard' singularity as in Section 13.5.

14.9.2 Local Geometry of the Canard Singularity

We take as our starting point the truncated local form of the singularity above,

$$\begin{aligned} \varepsilon\dot{x}_1 &= x_2 + x_1^2 , \\ \dot{x}_2 &= qx_1 + px_3 , \\ \dot{x}_3 &= r , \end{aligned} \tag{14.62}$$

in terms of the small positive parameter ε and arbitrary constants p, q, r. On the fast timescale, t/ε this reads

$$\begin{aligned} x_1' &= x_2 + x_1^2 , \\ x_2' &= \varepsilon(qx_1 + px_3) , \\ x_3' &= \varepsilon r , \end{aligned} \tag{14.63}$$

In the limit $\varepsilon \to 0$, the fast system has a family of equilibria that make up the slow *critical manifold*

$$\mathcal{C}^0 = \left\{ (x_1, x_2, x_3) \in \mathbb{R}^3 \; : \; x_2 + x_1^2 = 0 \right\} , \tag{14.64}$$

which is normally hyperbolic except along the line $x_1 = x_3 = 0$, as is illustrated in Figure 14.38. The $x_1 > 0$ branch of $\mathcal{C}^0$ is attracting and the $x_1 < 0$ branch repelling with respect to the fast flow (14.63). Once in the ε-neighbourhood of $\mathcal{C}^0$, the flow slows, and is approximated by the differential algebraic system that (14.62) becomes for $\varepsilon \to 0$, on $\mathcal{C}^0$ itself.

For $\varepsilon > 0$, Fenichel's theorem [68] tells us there exist invariant manifolds $\mathcal{C}^\varepsilon$ in the ε-neighbourhood of $\mathcal{C}^0$ wherever it is normally hyperbolic, with attracting and repelling branches corresponding to those of $\mathcal{C}^0$, on which the slow dynamics is an ε-perturbation of that on $\mathcal{C}^0$. The "normally hyperbolic" criterion is essential. Near where $\mathcal{C}^0$ ceases to be normally hyperbolic, along $x_1 = x_3 = 0$, Fenichel's manifolds $\mathcal{C}^\varepsilon$ cease to be invariant, and in the ε-neighbourhood of these points, the fast and slow flows mix in a more intricate manner.

Key to the local dynamics is now the phase portrait on $\mathcal{C}^0$. This is best found by differentiating the critical manifold's expression $x_2 + x_1^2 = 0$,

$$0 = \tfrac{d}{dt}(x_2 + x_1^2) = \dot{x}_2 + 2x_1\dot{x}_1 = (qx_1 + px_3) + 2x_1\dot{x}_1 , \tag{14.65}$$

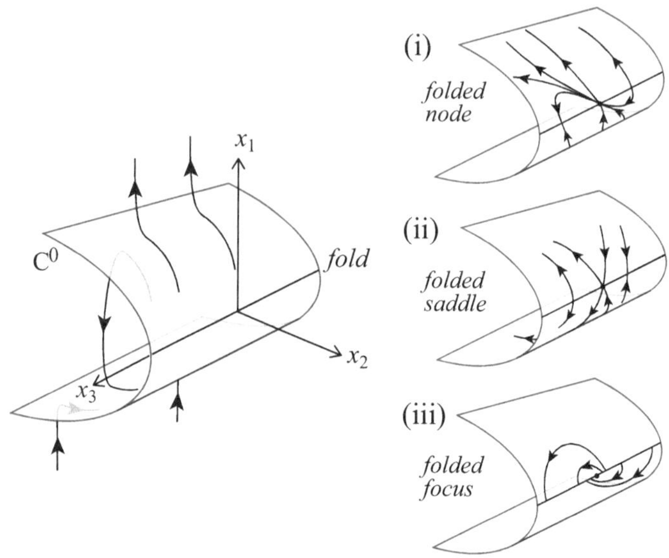

Fig. 14.38 Sketch of the slow-fast dynamics of (14.62) showing the critical manifold $\mathcal{C}^0$. Projected onto $\mathcal{C}^0$, the flow has the phase portrait of a *folded* (i) node, (ii) saddle, or (iii) focus.

substituting in $\dot{x}_2$ to obtain an expression in (x_1, x_3), which form a well-defined coordinatization of $\mathcal{C}^0$ (there is no overlap of $\mathcal{C}^0$ in these coordinates). Thus the phase portrait on the critical manifold is

$$\begin{pmatrix} \dot{x}_3 \\ \dot{x}_1 \end{pmatrix} = \frac{-1}{2x_1} \begin{pmatrix} 0 & -2r \\ p & q \end{pmatrix} \begin{pmatrix} x_3 \\ x_1 \end{pmatrix} \quad \text{on} \quad \mathcal{C}^0 . \tag{14.66}$$

This resembles a linear system with an equilibrium at $x_1 = x_3 = 0$, except for the singular timescaling $\frac{-1}{2x_1}$ at the front, which reverses time in $x_1 > 0$, and, importantly, permits the flow to traverse $x_1 = x_3 = 0$ in finite time. The phase portraits, illustrated in Figure 14.38(i-iii), are that of a node, focus, or saddle, except that the timescaling means the equilibria are 'folded'.

This point, the origin $x_1 = x_2 = x_3 = 0$ in the systems above, is the canard singularity that interests us. Thus the singularity is either a folded node, a folded saddle, or a folded focus.

The folded focus, case (iii) in Figure 14.38 when $rp > q^2/8$, does not have any trajectories passing through it; instead the flow steers around the singularity and leaves its neighbourhood while remaining close to the x_3 axis (the direction of departure depending on the sign of r.

The folded saddle, case (ii) in Figure 14.38 when $rp < 0$, has one canard trajectory, an orbit that is able to evolve counterintuitively *from* the attracting branch of C^0 *to* the repelling branch. This lies along the stable manifold of the saddle described by (14.66) when we ignore the prefactor $-1/2x_1$. In its neighbourhood, for $\varepsilon > 0$, there exist solutions that may have travelled arbitrarily far along the attracting branch and then penetrate arbitrarily far along the repelling branch, before joining the fast flow outside the neighbourhood of $\mathcal{C}^0$.

The folded node, case (i) in Figure 14.38 when $0 < rp < q^2/8$, is then the most intriguing when $q < 0$, because there exist a whole family of canards traversing the nodal point (the canard singularity). In fact every solution that passes through the singularity is a canard in this case. We can distinguish these into two *primary canards*, consisting of a *weak* and a *strong* canard, lying along the weak and strong eigendirections of the node (as described by (14.66) when we ignore the prefactor $\frac{-1}{2x_1}$). The remainder are called *secondary* canards. They play different roles in the flow for $\varepsilon > 0$, and our aim below is to understand these. If $q > 0$ then the solutions traversing the nodal points singularity are faux canard, passing from the repelling to attracting branches of $\mathcal{C}^0$, yielding much simpler dynamics.

The primary canards are two particular solutions of (14.62), which we label $\gamma^{wk,st} = \{x_1(t),\ x_2(t),\ x_3(t)\}$, that uniquely satisfy $x_1/x_3 = \mu_i/2$ for some real constant μ_i. They are

$$\gamma_i(t) = \left\{ -\tfrac{1}{2}\mu_i t,\ -\tfrac{1}{2}\mu_i\varepsilon - \tfrac{1}{4}\mu_i^2 t^2,\ rt \right\} , \qquad i = wk, st, \tag{14.67}$$

where μ_{wk} and μ_{st} are the weak and strong eigenvalues ($|\mu_{wk}| < |\mu_{st}|$) of the matrix in (14.66). These solutions are defined as canards for $\varepsilon > 0$ because they asymptote towards an ε-neighbourhood of both the attracting and repelling branches of $\mathcal{C}^0$.

The primary canards exist even for $\varepsilon > 0$. However, the singular canards—of which there are infinitely many in the limit $\varepsilon \to 0$—will degenerate into only a finite number of secondary canards for $\varepsilon > 0$, and they possess an intriguing geometry (see, e.g. [218, 47]). As illustrated in Figure 14.39, secondary canards rotate around the weak canard γ_{wk} near the singularity and asymptotically align with the strong canard γ_{st}. They are neither easy to express in closed form nor easy to simulate numerically. Their study is achieved analytically by M. Wechselberger in [218] by applying a parameter blow-up and taking cylindrical coordinates centred on the weak canard, such that the variational equation along the weak canard yields a Weber equation, whose solutions describe small oscillations that the secondary canards perform around the weak canard. Our piecewise-smooth approximation to be found here offers an alternative analytical method with consistent results.

To study the canards more closely, we will rescale the system, in preparation for pinching to obtain a piecewise-smooth approximation. It will be sufficient to let $r = 1$.

If the eigenvalues of the two-dimensional matrix in (14.66) are μ_{wk} and μ_{st}, then its trace and determinant are $q = \mu_{wk} + \mu_{st}$ and $2p = \mu_{wk}\mu_{st}$ respectively. Let us assume that μ_{wk} and μ_{st} are real and negative, giving

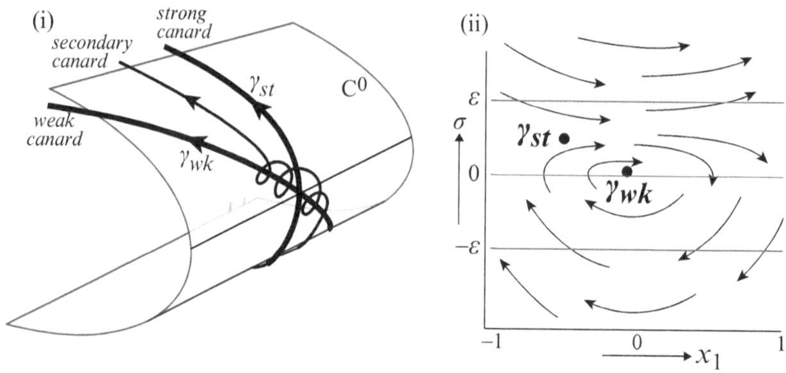

Fig. 14.39 Canards near the folded node. The weak primary canard γ_{wk} and strong primary canard γ_{st} asymptote to the ε-neighbourhood of the critical manifold C^0, shown in (i). Secondary canards rotate around the weak canard, driven by the flow which circulates around the weak canard, as simulated in (ii).

the node case. Writing the system in terms of μ_{wk} and μ_{st}, defining a ratio $\mu = \mu_{wk}/\mu_{st}$, and then introducing new coordinates and parameters $(x, y, z, t, \varepsilon)$ by the substitution

$$(x_1, x_2, x_3, t, \varepsilon) \mapsto (-x/\mu_{st}, y/\mu_{st}^2, -z/\mu_{st}^2, t/\mu_{st}^2, -\varepsilon/\mu_{st}^3) \, ,$$

we obtain

$$\begin{aligned} \varepsilon\dot{x} &= y + x^2 \, , \\ \dot{y} &= \mu z - (1+\mu)x \, , \\ \dot{z} &= 1 \, , \end{aligned} \tag{14.68}$$

in terms of only one structural parameter μ and the singular perturbation parameter ε.

The quantity $y + x^2$ will be useful to define as both a variable σ and an ε-scaled variable u,

$$\sigma := \varepsilon u := y + x^2 \, . \tag{14.69}$$

Then in the variables $\{x, u, z\}$ our system becomes

$$\begin{aligned} \dot{x} &= u \, , \\ \varepsilon\dot{u} &= \tfrac{1}{2}\mu z - (1+\mu)x + 2xu \, , \\ \dot{z} &= 1 \, . \end{aligned} \tag{14.70}$$

We now investigate canards in this system by deriving a piecewise-smooth approximation by pinching.

14.9.3 The First Pinch: A Shot in the Dark

We first make a crude attempt at pinching which serves to illustrate the concept and show how subtly (or indeed unsubtly) it can be wielded as a tool. This first pinch is extrinsic; the two that follow afterwards are intrinsic.

The pinch is simply a piecewise-smooth change of coordinates, and here we will take

$$U = u - \text{sign}(u) \;, \tag{14.71}$$

for $|u| > 1$. The interval $|u| < 1$ is then omitted from the problem, and we obtain a piecewise-smooth system

$$\left.\begin{aligned} \dot{x} &= U + \lambda \\ \varepsilon\dot{U} &= \tfrac{1}{2}\mu z - (\mu+1)x + 2x\lambda \;+\; \mathsf{O}\left(xU\right) \\ \dot{z} &= 1 \end{aligned}\right\} \quad \lambda = \text{sign}\, U \;, \tag{14.72}$$

which is piecewise linear when we omit the higher-order term $\mathsf{O}\left(xU\right)$.

The discontinuity surface $U = 0$ consists of:

- crossing in $4x^2 < \left(\frac{1}{2}\mu z - (\mu+1)x\right)^2$ (the 'bow-tie' region in Figure 14.40),
- sliding in $4x^2 > \left(\frac{1}{2}\mu z - (\mu+1)x\right)^2$, which is attracting where $x < 0$ and repelling where $x > 0$.

The sliding dynamics obeys

$$\begin{pmatrix} \dot{z} \\ \dot{x} \end{pmatrix} = \frac{-1}{2x} \begin{pmatrix} 0 & -2 \\ \frac{1}{2}\mu & -(\mu+1) \end{pmatrix} \begin{pmatrix} z \\ x \end{pmatrix} \;, \tag{14.73}$$

giving the phase portrait of a folded node as shown in Figure 14.40. The reader may easily show that this is identical to the slow manifold dynamics of (14.68) (on $y+x^2 = 0$); hence the pinched system is *complete* (see definition in Section 12.4.1).

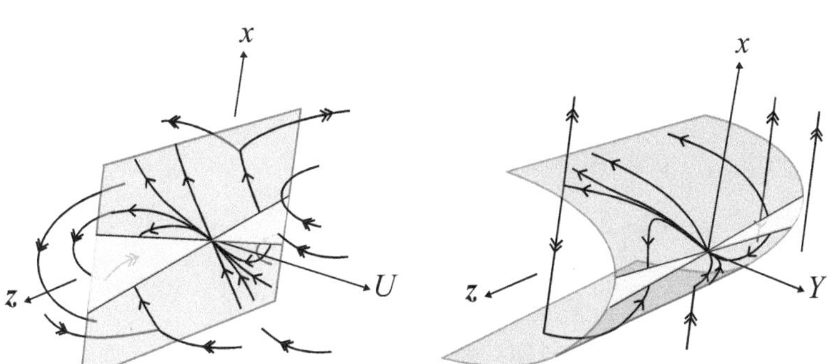

Fig. 14.40 A piecewise linear approximate of the folded node, shown in (x, U, z) space (left) and (x, Y, z) space (right) where $Y = \varepsilon U - x^2$.

The sliding dynamics contains solutions corresponding to the weak and strong canards and a continuum of other canards between them persists for $\varepsilon > 0$. This gives a crude approximation of (14.70) in which the identity of the primary and secondary canards is apparent, but there exist infinitely many other secondary canards. Qualitatively this only captures the $\varepsilon \to 0$ limit of the system's behaviour. However, we can do better.

14.9.4 The Second Pinch: Zooming in on the Manifolds

In the pinch above, there is no clear motivation for omitting the region $|u| < 1$ from our system. This motivation is provided if we first attempt to zoom in on the slow manifold.

We know the interesting dynamics takes place in an ε-neighbourhood of $u = 0$. We can magnify this by an exponential scaling, introducing a new coordinate

$$v = u^{[\varepsilon]} \ := \ |u|^{\varepsilon} \operatorname{sign}\, u \ , \tag{14.74}$$

in terms of which (14.70) becomes

$$\begin{aligned} \dot{x} &= v^{1/[\varepsilon]} \ , \\ \dot{v} &= v \left\{ 2x + \left(\tfrac{1}{2}\mu z - (\mu+1)x\right) v^{-1/[\varepsilon]} \right\} \ , \\ \dot{z} &= 1 \ . \end{aligned} \tag{14.75}$$

For $\varepsilon \ll 1$, the flow of (14.75) in the region $|v| < 1$ is dominated by the term $v^{1-\frac{1}{[\varepsilon]}}$ in the second row and so lies close to a set of fibres translating solutions between the surfaces $v = \pm 1$, while z and x are slow varying. We can approximate this by omitting the dynamics in this region and pinching the two surfaces $v = \pm 1$ together, using the transformation

$$V = v - \operatorname{sign}\, v \tag{14.76}$$

for $|v| > 1$, giving the system

$$\begin{aligned} \dot{x} &= (V + \operatorname{sign} V)^{1/[\varepsilon]} \ , \\ \dot{V} &= (V + \operatorname{sign} V) \left\{ 2x + \left(\tfrac{1}{2}\mu z - (\mu+1)x\right) (V + \operatorname{sign} V)^{-1/[\varepsilon]} \right\} \ , \\ \dot{z} &= 1 \ , \end{aligned} \tag{14.77}$$

as simulated in Figure 14.41(ii). The pinch is intrinsic because $v = u^{[\varepsilon]}$ depends on the perturbation parameter ε.

Prior to pinching, both the weak and strong canards, which now satisfy $v^{1/[\varepsilon]} = x/z = \mu/2$ and $v^{1/[\varepsilon]} = x/z = 1/2$, respectively, lie inside the region $|v| < 1$, and therefore they should become part of the sliding dynamics. At the discontinuity surface, $V = 0$, (14.77) reduces to

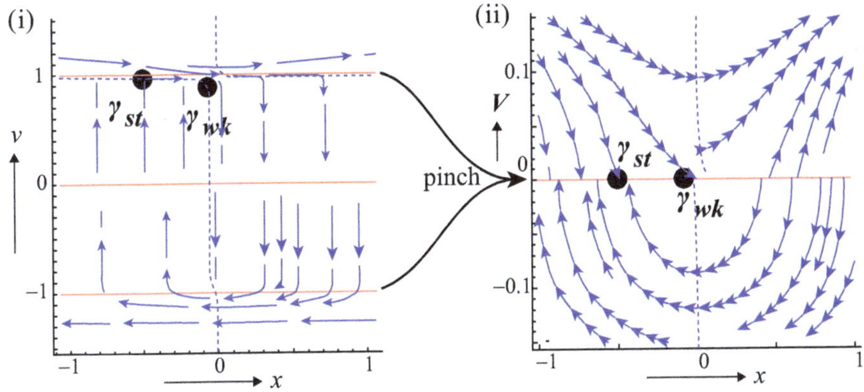

Fig. 14.41 The exponential scaling creates a trivial strip $|v| < 1$ shown in (i), in which the weak and strong canard are found, and under pinching these become part of the sliding dynamics. Simulated in the plane $z = -1$ with $\varepsilon = 0.05$ and $\mu = 1/8.5$.

$$
\begin{aligned}
\dot{x} &= \operatorname{sign} V \ , \\
\dot{V} &= \tfrac{1}{2}\mu z - (\mu + 1)x + 2x \ \operatorname{sign} V \ , \qquad (14.78)\\
\dot{z} &= 1 \ ,
\end{aligned}
$$

which is equivalent to (14.72) up to an ε scaling in x, z, t. The crossing and sliding dynamics on the discontinuity surface are therefore exactly as

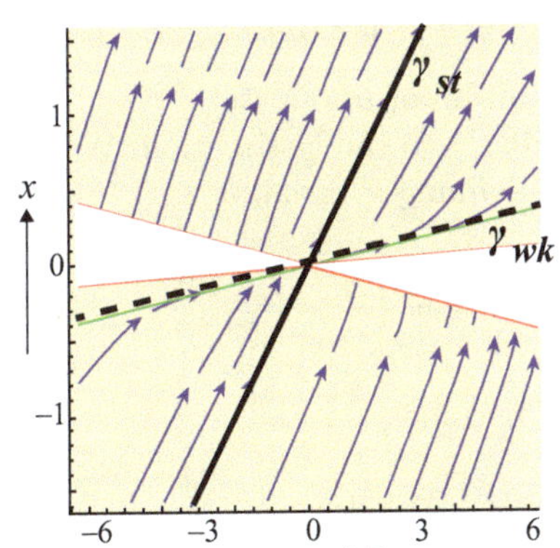

Fig. 14.42 Sliding dynamics of the pinched system (14.78), in the regions $4x^2 > (\frac{1}{2}\mu z - (\mu + 1)x)^2$ on $V = 0$. This shows a folded node, with a continuum of secondary canards between the weak and strong canards.

described in (14.73) and as depicted in Figure 14.42, and this implies the pinched system is *complete* (see definition in Section 12.4.2).

So here we have gained a motivation for pinching, but the result is similar to the previous section, as pinching again gives a continuum of canards. To resolve the different canards, we will have to refine the preparatory scaling $u \mapsto v$.

14.9.5 The Third Pinch: Catching the Canards

We have failed to resolve the canards above because we have not honed in very precisely on the slow manifolds. The first approximation for the slow invariant manifolds is that they lie in an ε-neighbourhood of $\mathcal{C}^0$, where $u = 0$. The slow dynamics actually lies not on, but is stationary with respect to, the critical manifold $\mathcal{C}^0$, and so a better approximation is that the slow manifolds lie close to (actually in an ε^2-neighbourhood of) the nullcline $\dot{u} = 0$.

The surface

$$\left\{(x, u, z) \in \mathbb{R}^3 : \; u = \frac{\mu + 1}{2} - \frac{\mu z}{4x}\right\} , \tag{14.79}$$

where $\dot{u} = 0$, can only approximate a slow manifold where it lies in the ε-neighbourhood of the critical manifold $u = 0$. This evidently cannot hold near $x = 0$ (because $-\frac{\mu z}{4x}$ diverges). Instead for large x, the nullcline is approximated by

$$\mathcal{N} = \left\{(x, u, z) \in \mathbb{R}^3 : u = \frac{\mu + 1}{2}\right\} , \tag{14.80}$$

and for this to lie within ε of the critical manifold, which corresponds to the region $|u| < 1$, we must have $|(\mu + 2)/2| < 1$. Combining this with the folded node condition $\mu > 0$ gives the range of permitted values for the parameter μ as

$$0 < \mu < 1 , \tag{14.81}$$

consistent with our definition of μ.

We now proceed similarly to the last section. We first zoom in on $\mathcal{N}$ with an exponential scaling, defining a variable

$$w = \left(\frac{u}{\varepsilon} - \frac{1 + \mu}{2\varepsilon}\right)^{[\varepsilon]} , \tag{14.82}$$

and the system (14.70) becomes

$$\begin{aligned} \dot{x} &= \varepsilon w^{1/[\varepsilon]} + \frac{1 + \mu}{2} , \\ \dot{w} &= w \left\{2x + \frac{\mu z}{2\varepsilon} w^{-1/[\varepsilon]}\right\} , \\ \dot{z} &= 1 , \end{aligned} \tag{14.83}$$

illustrated in Figure 14.43(i). The weak and strong canards now satisfy $\{w,x\} = \{-1/(2\varepsilon)^{\varepsilon}, \mu z/2\}$ and $\{w,x\} = \{-(\mu/2\varepsilon)^{\varepsilon}, z/2\}$, respectively.

The nullcline $\dot{w} = 0$ is the curve

$$\mathcal{Q} = \left\{(w,x) \in \mathbb{R}^2 : \; w = -\left(\frac{\mu z}{4x\varepsilon}\right)^{[\varepsilon]}\right\} . \tag{14.84}$$

The flow projection onto $w = 0$ is again (14.73), but the fact that the flow organises around the nullcline $\mathcal{Q}$ as apposed to $\mathcal{N}$ gives different dynamics, as will be revealed when we pinch.

The pinch is now the transformation

$$W = w - \text{sign } w \,, \tag{14.85}$$

which is intrinsic as w depends on the perturbation parameter ε and which yields

$$\begin{aligned} \dot{x} &= \varepsilon(W + \text{sign}\, W)^{1/[\varepsilon]} + \frac{1+\mu}{2} \,, \\ \dot{W} &= (W + \text{sign}\, W)\left\{2x + \frac{\mu z}{2\varepsilon}(W + \text{sign}\, W)^{-1/[\varepsilon]}\right\} \,, \\ \dot{z} &= 1 \,, \end{aligned} \tag{14.86}$$

as illustrated in Figure 14.43(ii).

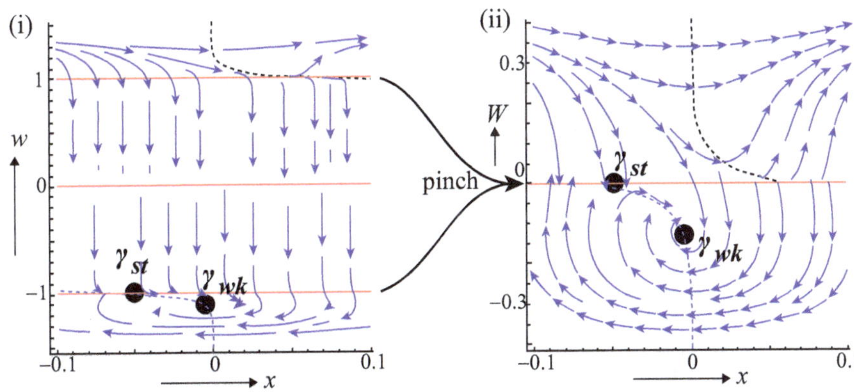

Fig. 14.43 The zoom around the surface $\mathcal{N}$ captures the strong canard in the region $|w| < 1$ but not the weak canard, as shown in (i), and therefore under pinching only the strong canard becomes part of the sliding dynamics. Simulated in the plane $z = -1$ with $\varepsilon = 0.05$ and $\mu = 1/8.5$.

The sliding flow is superficially given again by (14.73), as is found by solving for $\dot{W} = 0$ on $W = 0$. The $W > 0$ and $W < 0$ subsystems in (14.86) are tangent to the discontinuity surface where $\dot{W}$ reaches zero as

W approaches zero from above or below. These tangencies lie along $\frac{x}{z} = -\frac{\mu}{4\varepsilon}\,\text{sign}\,W$; hence the sliding regions are where

$$|x/z| > \mu/4\varepsilon \quad \text{on} \quad W = 0\,, \tag{14.87}$$

illustrated in Figure 14.44 for different values of μ.

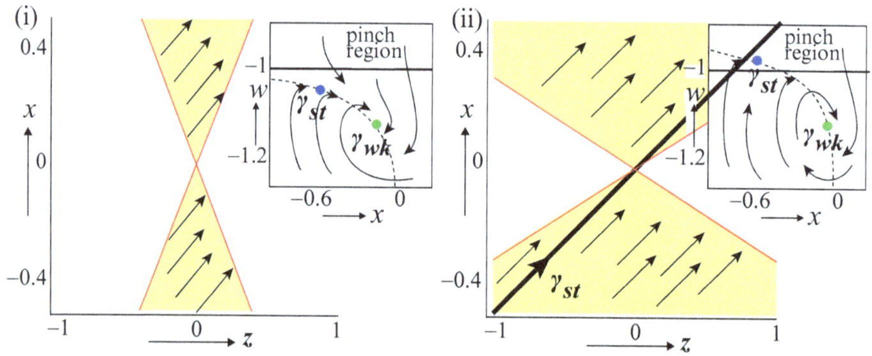

Fig. 14.44 Sliding dynamics on $W = 0$ occurs in the regions $|x/z| > \mu/4\varepsilon$. In (i) there are no canards for $\mu = 1/4$, and in (ii) the strong canard can be seen in the sliding flow for $\mu = 1/16$. Simulated for $\varepsilon = 0.05$.

The curvature of the flow is specified by the second derivative, $\ddot{W} = (1 \pm 2\varepsilon)(\mu \pm 2\varepsilon)/2\varepsilon$. Since we have $0 < \mu < 1$ and $0 < \varepsilon \ll 1$, the flow in $W > 0$ satisfies $\ddot{W} = (1+2\varepsilon)(\mu+2\varepsilon)/2\varepsilon > 0$ on its line of tangency with the discontinuity surface, $x/z = \mu/4\varepsilon$, and hence forms a visible fold. The flow in $W < 0$ satisfies $\ddot{W} = (1-2\varepsilon)(\mu-2\varepsilon)/2\varepsilon$ on its tangency line $x/z = -\mu/4\varepsilon$ and hence forms an invisible fold if $\mu < 2\varepsilon$ and a visible fold if $\mu > 2\varepsilon$. Although we are interested in arbitrarily small ε, either of these can be satisfied for small enough μ.

This happens because the weak eigendirection of (14.73) lies outside the sliding region. The strong eigendirection lies inside the sliding regions if $\mu > 2\varepsilon$ or outside the sliding regions if $\mu < 2\varepsilon$, meaning that in the former case the sliding dynamics captures no canards and in the latter captures a single canard. (This is immediately in contrast to the continuum of sliding canards in (14.78).) The two different cases are shown in Figure 14.44.

The weak and strong canards in the unpinched system lie at $\{x, w\} = \{\mu z/2, -1/(2\varepsilon)^{\varepsilon}\}$ and $\{x, w\} = \{z/2, -(\mu/2\varepsilon)^{\varepsilon}\}$, respectively. The weak canard clearly avoids the pinch region $|w| < 1$ for $\varepsilon < 1/2$ and then lies at $\{W, x\} = \{1 - 1/(2\varepsilon)^{\varepsilon}, \mu z/2\}$. The strong canard also avoids the pinch region if $\mu > 2\varepsilon$ and is given by $\{W, x\} = \{1 - 1/(2\varepsilon)^{\varepsilon}, z/2\}$. If $\mu < 2\varepsilon$ the strong canard falls inside the pinch region and is not part of (14.86) for $W \neq 0$; if it exists it is part of the sliding dynamics on $W = 0$. Indeed we see that is exactly the case in Figure 14.44.

It is known from [218] that the folded node possesses $\langle\frac{1-\mu}{2\mu}\rangle$ secondary canards, where $\langle n\rangle$ denotes the largest integer less than n. These canards rotate around the weak canard near $z=0$, with rotation numbers taking all integers from 1 to $\langle\frac{1-\mu}{2\mu}\rangle$, and then connect to sliding solutions in the attracting and repelling sliding regions that take them to $z\to\pm\infty$, as sketched in Figure 14.45. Note that the rotation takes place in $W<0$, therefore secondary canards satisfy $W\leq 0$.

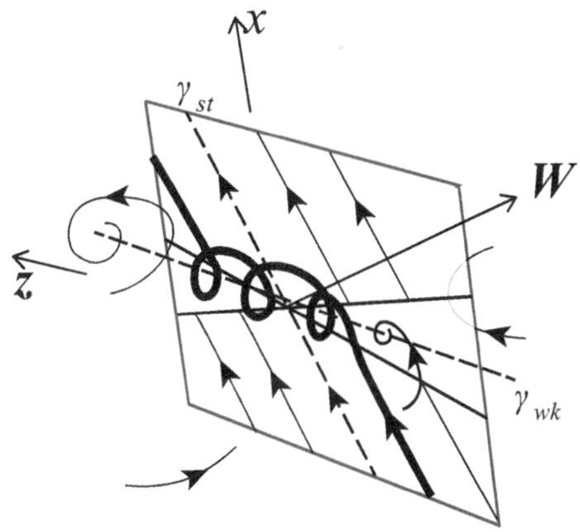

Fig. 14.45 Secondary canards are solutions that detach from the attracting sliding region, rotate around the weak canard, and then reattach to the repelling sliding region.

We can retrieve this same result using the piecewise-smooth system. Let us make a further approximation.

Treat the $W>0$ and $W<0$ systems separately, making independent approximations in the two regions about the dominant singularities. In the region $W<0$, we linearize about the weak canard at $\{W,x\}=\{1-(2\varepsilon)^{-\varepsilon},\mu z/2\}$. In the region $W>0$, we expand about the tangency to the discontinuity surface at $(W,x)=(0,-\mu z/4\varepsilon)$. To leading order these give

$$\text{in } W>0:\left\{\begin{array}{ll}\dot{x} &= \varepsilon+\frac{1+\mu}{2}+\mathsf{O}\left(W\right),\\ \dot{W} &= \frac{\mu z+4x\varepsilon}{2\varepsilon}+\mathsf{O}\left(W\right),\\ \dot{z} &= 1,\end{array}\right.$$
$$\text{in } W<0:\left\{\begin{array}{ll}\dot{x} &= \frac{1}{2}\mu+(2\varepsilon)^{\varepsilon-1}(W-1+\frac{1}{(2\varepsilon)^{\varepsilon}})+\mathsf{O}\left(\delta W^2\right),\\ \dot{W} &= \frac{\mu z-2x}{(2\varepsilon)^{\varepsilon}}+\frac{\mu z}{\varepsilon}(W-1+\frac{1}{(2\varepsilon)^{\varepsilon}})+\mathsf{O}\left(\delta W^2,\varepsilon\delta x\delta W\right),\\ \dot{z} &= 1,\end{array}\right. \tag{14.88}$$

where $\delta W=W-1+(1/2\varepsilon)^{\varepsilon}$ and $\delta x=x-\mu z/2$. These are shown in Figure 14.46.

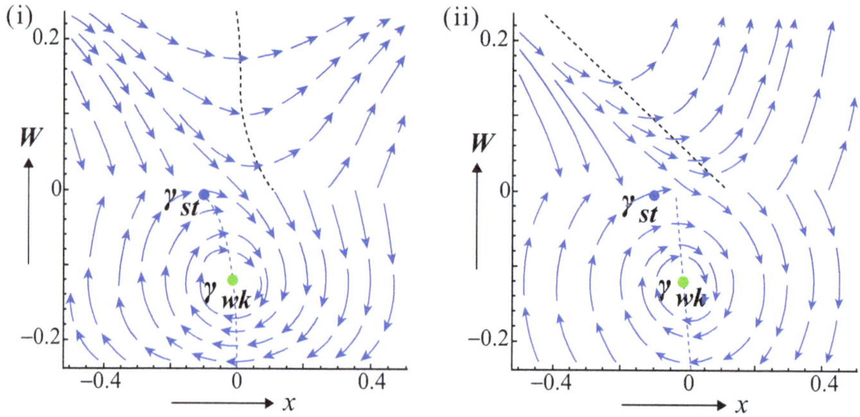

Fig. 14.46 The piecewise-smooth approximation (14.88), simulated in the plane $z = -1$ for $\varepsilon = 0.05$, $\mu = 1/8.5$.

We now omit the higher-order corrections and find solutions to the truncated local equations. Note that $1/(2\varepsilon)^{\varepsilon} \approx 1 - \varepsilon \log(2\varepsilon) + \mathsf{O}\left((\varepsilon \log 2\varepsilon)^2\right)$ deviates quickly from unity as ε increases from zero, so we cannot approximate it by unity. Approximating around the weak canard in $W < 0$ leads to a slight shift in the sliding region (14.87). The boundary where the $W > 0$ system is tangent to $W = 0$ is given, as before, by $x/z = -\mu/4\varepsilon$. The boundary where the $W < 0$ system is tangent to $W = 0$ is now given by

$$x/z = \mu(\varepsilon - 1 + (2\varepsilon)^{\varepsilon})/2\varepsilon \,. \tag{14.89}$$

Let us now find the secondary canards. We will assume that they satisfy two conditions. Firstly, since (14.88) is symmetric under the substitution $\{z, x, t\} \mapsto \{-z, -x, -t\}$, canards are expected to inherit this symmetry, implying that $z = x = 0$ at $t = 0$. Secondly, the canards pass through the boundary of the sliding region (14.89) and lie in $W \leq 0$, implying that they curve into ($\ddot{W} < 0$) the region $W < 0$. These give secondary canards the following properties:

$C1$. $z(0) = x(0) = 0$.
$C2$. $\ddot{W}(t_c) < \dot{W}(t_c) = W(t_c) = 0$ at some $t = t_c \neq 0$.

So we seek solutions of the $W < 0$ system that satisfy these two conditions. Considering (14.88) in $W < 0$ we have, noting $\dot{x} = dx/dz = dx/dt$,

$$\ddot{z} = \frac{1}{(2\varepsilon)^{1-\varepsilon}} \dot{W} = \frac{\mu z - 2x}{2\varepsilon} + \frac{\mu z}{\varepsilon}(\dot{x} - \tfrac{1}{2}\mu) \,.$$

Letting $x=\zeta+\mu z/2$ and $x=\tau\sqrt{\varepsilon/\mu}$, this rearranges to the Hermite equation

$$\zeta''-\tau\zeta'+\frac{1}{\mu}\zeta=0\,, \tag{14.90}$$

whose general solution can be written

$$\zeta(\tau)=\tau\zeta'(0)\,{}_1F_1\left(\frac{\mu-1}{2\mu},\frac{3}{2},\frac{\tau^2}{2}\right)+\zeta(0)\,{}_1F_1\left(-\frac{1}{2\mu},\frac{1}{2},\frac{\tau^2}{2}\right)\,,$$

with derivative

$$\zeta'(\tau)=\zeta'(0)\,{}_1F_1\left(\frac{\mu-1}{2\mu},\frac{1}{2},\frac{\tau^2}{2}\right)-\frac{\tau\zeta(0)}{\mu}\,{}_1F_1\left(1-\frac{1}{2\mu},\frac{3}{2},\frac{\tau^2}{2}\right)\,,$$

in terms of the confluent hypergeometric function ${}_1F_1$ (also known as Kummer's function M where ${}_1F_1(\alpha,\beta;\gamma)=M(\alpha,\beta,\gamma)$, see [3]).

Applying condition $C1$, at a point where $z(0)=x(0)=0$, we have $\dot{W}(0)=0$, hence $\ddot{x}(0)=0$. In the transformed coordinates this gives initial conditions

$$\zeta(0)=0,\qquad \zeta''(0)=0\,,$$

the former of which simplifies the solution above to

$$\frac{\zeta(\tau)}{\zeta'(0)}=\tau\,{}_1F_1\left(\frac{\mu-1}{2\mu},\frac{3}{2},\frac{\tau^2}{2}\right)\,,\qquad \frac{\zeta'(\tau)}{\zeta'(0)}={}_1F_1\left(\frac{\mu-1}{2\mu},\frac{1}{2},\frac{\tau^2}{2}\right)\,, \tag{14.91}$$

or in terms of the gamma function Γ and Hermite polynomials H_n [3],

$$\zeta(\tau)=\frac{i\zeta'(0)e^{\tau^2/4}\Gamma\left(\frac{\mu+1}{2\mu}\right)}{\sqrt{\pi}\left(i\frac{\Gamma\left(\frac{\mu+1}{2\mu}\right)}{\Gamma\left(-\frac{1}{2\mu}\right)}+\frac{\Gamma\left(\frac{2\mu+1}{2\mu}\right)}{\Gamma\left(\frac{\mu-1}{2\mu}\right)}\right)}\left\{\frac{\Gamma\left(\frac{2\mu+1}{2\mu}\right)}{\Gamma\left(\frac{\mu-1}{2\mu}\right)}D^+\left(1,\mu,i\tau\right)-D^+\left(0,\mu,\tau\right)\right\}\,,$$

where

$$D^\pm\left[m,\mu,\tau\right]=2^{\pm(\mu+2)/2\mu}e^{-\tau^2/4}H_{m\mp1/\mu}\left(\frac{\tau}{\sqrt{2}}\right)\,.$$

Substituting back in $\tau=t\sqrt{\mu/\varepsilon}$, we find that in $W<0$ there exist solutions given by

$$\begin{aligned}
x(t)&=\frac{\mu t}{2}+t\dot{x}(0)\,{}_1F_1\left[\frac{\mu-1}{2\mu},\frac{3}{2},\frac{\mu t^2}{2\varepsilon}\right]\,,\\
W(t)&=1-(2\varepsilon)^{-\varepsilon}(1+\varepsilon\mu-2\varepsilon\dot{x}(0){}_1F_1\left(\frac{\mu-1}{2\mu},\frac{1}{2},\frac{\mu t^2}{2\varepsilon}\right))\,,\\
z(t)&=t\,,
\end{aligned} \tag{14.92}$$

and these form the portions lying outside the discontinuity surface. To find the secondary canards we must then apply the second condition, $\mathcal{C}2$, to pick out solutions in (14.92) that tangentially contact (or *graze*) the boundaries of the sliding regions on $W = 0$, where they connect to sliding solutions that form the tails of the secondary canards. Substituting the conditions $W(t_c) = \dot{W}(t_c) = 0$ at some $t = t_c \neq 0$ into (14.88) for $W < 0$, gives $x(t_c)$ and $\dot{x}(t_c)$ as

$$x(t_c)/\mu t_c = \dot{x}(t_c) + \varepsilon(1-\mu)/2\varepsilon = (1 + \varepsilon - (2\varepsilon)^\varepsilon)/2\varepsilon\ . \tag{14.93}$$

Figure 14.47 shows a simulation of solutions given by (14.92) subject to the boundary conditions (14.93).

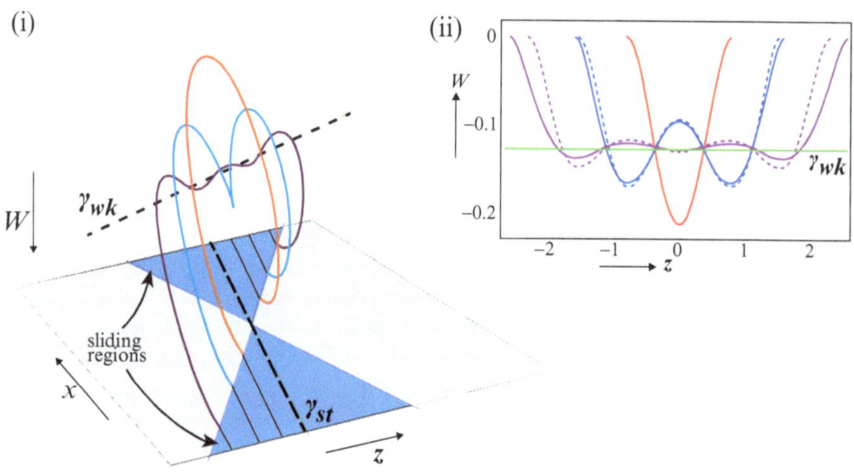

Fig. 14.47 The secondary canard, calculated by finding non-sliding solutions that graze and thereby enter/leave the sliding regions. Simulated for $\varepsilon = 0.05$ and $\mu = 1/8.5$, in (i) the (x, W, z) space, and (ii) as a graph of $W(z)$.

The number of rotations the secondary canards make around the weak canard is now easily found. The confluent hypergeometric function ${}_1F_1(-a, b, c)$ has $2\langle a+1\rangle$ real zeros [3] (again with $\langle n\rangle$ denoting the largest integer smaller than n), between which the function oscillates through $2\langle a+1\rangle - 1$ maxima/minima, and between these the functions makes $\langle a\rangle$ complete oscillations. These oscillations form the rotations of the secondary canards.

Using the boundary conditions $W(t_c) = \dot{W}(t_c) = 0$, a given value of $t_c > 0$ picks out one of the solutions (14.92), with $\langle\frac{1-\mu}{2\mu}\rangle$ rotations. Any of the rotation numbers from 1 up to $\langle\frac{1-\mu}{2\mu}\rangle$ are permitted, because the boundary conditions may be satisfied at any one of the maxima for different solutions with unique values of t_c. Hence there exist $\langle\frac{1-\mu}{2\mu}\rangle$ secondary canards with rotation numbers $1, 2, 3, \ldots, \langle\frac{1-\mu}{2\mu}\rangle$.

A more explicit form than the standard function above can be obtained using the large parameter asymptotic approximation

$$ {}_1F_1\left(a,b,y\right) = \frac{\Gamma(b)}{\sqrt{\pi}}\left(\tfrac{y}{2}(b-2a)\right)^{(1-2b)/4} e^{y/2}\cos\left(\sqrt{2y(b-2a)}+\tfrac{\pi}{4}(1-2b)\right) \tag{14.94} $$

(see, e.g. Eq. 13.5.14 of [3]). This further approximation is sufficient to capture solutions for small $y=\tau^2/2$, but due to lack of accuracy in how they contact the sliding boundary, they are not accurate enough to correctly give the number of secondary canards or the number of rotations they make.

14.10 Looking Forward

There is plenty left to be discovered in all of these applications and many more besides. Our aim here has been to indicate some of the areas of current interest and suggest avenues of further study, rather than review or solve such a wide range of problems. Piecewise-smooth models are increasingly found in climate, neuroscience, robotics and automatic control, mechanics and novel materials, chemical reactions, social or ecological modelling, ... the list is ever-growing. The biggest gains in future study are likely to be in close modelling of hidden terms, nonlinearity of switching, and perturbations from idealizations of discontinuities.

Discontinuity is often itself hidden in the way a system is modelled, particularly in the discrete models that are very common particularly in economics, in political or social modelling, and also in neuroscience. If a model proceeds in discrete steps, then with evolution itself occurring in jumps, one might easily expect it not to matter if a discontinuous change occurs in the dynamical laws from one step to the next. In a computer, a discontinuous change in a discrete model is so easily implemented that the conceptual issues of discontinuity in a differential equation might seem not to arise at all, far from it. The discontinuity has only been overlooked, of course, and is still there. The discretized simulations from Section 1.6 and Section 12.2 should cause some concern about what truly lies beneath, and closer study of the way such systems are modelled, with closer understanding of the connections between piecewise-smooth flows and piecewise-smooth maps, is certainly warranted.

Appendix A
Discontinuity as an Asymptotic Phenomenon: Examples

In Section 1.4.4, we outlined how discontinuities arise in asymptotics, using examples of discontinuities appearing in the asymptotic expansion of sigmoid transition functions. Here we show also how they arise in asymptotic expansions of solutions of ordinary and partial differential equations and integrals.

A.1 Changes of Stability

Let a transition occur between dynamical regimes $\dot{x} = a(x)$ and $\dot{x} = b(x)$, regulated by a function $y(\sigma)$ as σ changes sign. Assume we can write $\dot{x} = f(x, y)$, where $f(x, -1) \approx a(x)$ and $f(x, +1) \approx b(x)$, for some function $f(x, y)$, where y is a sigmoid function. We assume that it is known how f depends on y, and we then seek an asymptotic expression for y. We shall look at three cases exemplifying ordinary differential equations, partial differential equations, and integral equations.

A.1.1 Large-Scale Bistability, Small-Scale Decay

Say that y represents a population that has two steady values, normalized to $y \approx +1$ and $y \approx -1$ when some quantity σ is positive or negative, respectively. As it transitions between them, let y grow or decay exponentially as $\dot{y} \sim -y$. Let us assume that σ depends only on the state x of the main system and not on y directly.

We need to assign a scale to the transition, so let us refine these values such that $y \sim \pm 1$ for $|\sigma| \gg \varepsilon$, and $\dot{y} \sim -y$ for $|\sigma| \ll \varepsilon$, for some small ε. A possible model is then a system of ordinary differential equations:

M. R. Jeffrey, *Hidden Dynamics*,
https://doi.org/10.1007/978-3-030-02107-8

$$\begin{aligned} \dot{x} &= f(x,y) , \\ \varepsilon\dot{y} &= (1-y^2)\sigma(x) - \varepsilon y . \end{aligned} \tag{A.1}$$

The small ε leading the second line implies that y varies fast compared to x, so if we look first at the dynamics of y, we can treat x as slow varying. The solution to the y system in (A.1) treating $\sigma = \sigma(x)$ as a fixed parameter is easily found to be

$$y(t,\sigma) = -(\varepsilon/2\sigma) + \omega \tanh\left(\omega t\sigma/\varepsilon + k_0\right) , \tag{A.2}$$

where $\omega = \sqrt{1+(\varepsilon/2\sigma)^2}$ and $k_0 = \operatorname{arctanh}\left(\frac{(\varepsilon/2\sigma)+y(0,\sigma)}{\omega}\right)$. As t evolves this rushes towards a stationary state (where $\dot{y} = 0 > \partial\dot{y}/\partial y$), given by

$$y_*(\sigma) = -\varepsilon/2\sigma + \operatorname{sign}(\sigma)\sqrt{1+(\varepsilon/2\sigma)^2} . \tag{A.3}$$

For large σ the attractor sits close to either $+1$ or -1 depending the sign of σ. If we consider what happens as σ changes sign between these states, $y_*(\sigma)$ jumps between ± 1, and a series expansion for large σ reveals

$$y_*(\sigma) = \operatorname{sign}(\sigma)\left\{1 - \varepsilon/2\sigma + \varepsilon^2/8\sigma^2\right\} + \mathsf{O}\left((\varepsilon/\sigma)^3\right) . \tag{A.4}$$

The asymptotic terms in the tail mitigate the transition through $|\sigma| < 1$.

A more precise approximation of the full system (A.1) can be found, but this is sufficient. We now have a main system $\dot{x} = f(x,y)$, where y is seen to exhibit abrupt switching. Asymptotic analysis shows that the variable y relaxes to y_* on a timescale $t = \mathsf{O}(\varepsilon)$, giving an approximation $y \approx y_* \approx \operatorname{sign}(\sigma) + \mathsf{O}(\varepsilon/\sigma)$.

Looking more closely at the expansion of y, we see that it is of the form (1.17). The question asked in Chapter 1 and Chapter 12 is then, if f depends nonlinearly on y, and given the abruptness of the switch in $y \approx \operatorname{sign}(\sigma)$ as $\sigma(x)$ changes sign in the limit $\varepsilon \to 0$, should we model (A.1) as $\dot{x} \approx \frac{1}{2}(1+y)f(x,+1) + \frac{1}{2}(1-y)f(x,-1)$ (a linear switching system) or as $\dot{x} \approx f(x,y)$ (a nonlinear switching system)? Framed in this way, assuming such a function $f(x,y)$ exists, the nonlinear system may seem obvious, but it carries certain consequences. If we know only the x-dependence of the functions $f(x,\pm 1)$, say as a pair of functions $f^{\pm}(x)$, then the linear combination may seem obvious, but that too has consequences and limitations in terms of modelling. All of these consequences and limitations are the motivation for this book.

There are endless other forms of ordinary differential equations that create asymptotic sigmoid behaviour, for example, adding more dimensions,

$$\begin{aligned} \dot{x} &= f(x,y) , \\ \varepsilon\dot{y} &= y^2\sigma(x) + \varepsilon z , \\ \mu\dot{z} &= y - y^3 , \end{aligned} \tag{A.5}$$

which has equilibria at $(y, z) = (\text{sign}(\sigma), -\sigma/\varepsilon)$ that exchange stability when σ changes sign (as well as a saddle point at $(0, 0)$).

The example above shows how easily ordinary differential equations like (A.1) lead to switching with an asymptotic form $y \sim \text{sign}(\sigma)$, but they arise just as easily in partial differential equations.

A.1.2 Large-Scale Bistability, Small-Scale Dissipation

If y represents a physical property, like temperature, for example, it might have both spatial and temporal variations that become significant during transition.

Proceeding similarly to the previous example, we will consider a switch between two behaviours where $y \approx \pm 1$, as some quantity σ changes sign across a transition region $|\sigma| < \varepsilon$, for small $\varepsilon > 0$.

For $|\sigma| \ll \varepsilon$ assume y satisfies the heat equation $y_{\sigma\sigma} \sim \mu \dot{y}$ for some small positive μ, where y_σ denotes $\partial y/\partial \sigma$. For small enough μ, this gives $y_{\sigma\sigma} \sim 0$, consistent with the $|\sigma| \gg \varepsilon$ limit where $y \sim \pm 1$ implies $y_\sigma \sim 0$. This asymptotic character is satisfied by the system of partial differential equations:

$$\begin{aligned} \dot{x} &= f(x, y)\;, \\ \mu \dot{y} &= \sigma(x) y_\sigma + \varepsilon^2 y_{\sigma\sigma}\;. \end{aligned} \tag{A.6}$$

Again let us treat x, and hence $\sigma(x)$, as slow varying compared to y. The y system moreover evolves on a fast timescale t/μ to a slow subsystem $\sigma(x) y_\sigma + \varepsilon^2 y_{\sigma\sigma} = 0$, which has solutions $y = y_*(\sigma)$ given by

$$y_*(\sigma) = y_*(0) + \varepsilon y_{\sigma *}(0) \sqrt{\pi/2}\, \text{Erf}\left[\sigma/\sqrt{2}\varepsilon\right]\;, \tag{A.7}$$

where Erf denotes the standard error function [3]. For large $|\sigma|$ this asymptotes to $y \sim \pm 1$, implying $y_*(0) = 0$ and $y_{\sigma *}(0) = \varepsilon^{-1}\sqrt{2/\pi}$.

Solutions of the full system around these slow states can be found in the form

$$y(t, \sigma) = y_*(\sigma) + e^{-t/\sqrt{\mu}} Y(\sigma)\;.$$

Substituting this into the partial differential equation for y in (A.6) yields

$$0 = \left\{\sigma(x) y_{\sigma *} + \varepsilon^2 y_{\sigma\sigma *}\right\} + e^{-t/\sqrt{\mu}} \left\{\sqrt{\mu} Y + \sigma(x) Y_\sigma + \varepsilon^2 Y_{\sigma\sigma}\right\}\;.$$

The first bracket vanishes by the definition of y_*; the second gives an ordinary differential equation for Y with solution

$$Y(\sigma) = e^{-\frac{\sigma^2}{2\varepsilon^2}} \left\{ y(0)\, {}_1F_1\left[\tfrac{1-\sqrt{\mu}}{2}, \frac{1}{2}; \tfrac{\sigma^2}{2\varepsilon^2}\right] + \mu^{-\frac{1}{4}} \sigma y_\sigma(0)\, {}_1F_1\left[\tfrac{2-\sqrt{\mu}}{2}, \tfrac{3}{2}; \tfrac{\sigma^2}{2\varepsilon^2}\right] \right\}, \tag{A.8}$$

where ${}_1F_1$ is the Kummer confluent hypergeometric function [3, 88]. The exact functions are less interesting to us than their large variable asymptotics. Our interest lies in slow behaviour that y collapses to on the $t/\sqrt{\mu}$ timescale, given by

$$y_*(\sigma) \sim \operatorname{sign}(\sigma) - \tfrac{1}{\sigma}\sqrt{\tfrac{2\varepsilon}{\pi}}e^{-\frac{\sigma^2}{2\varepsilon^2}}\left(1 - \varepsilon/\sigma + \mathsf{O}\left(\varepsilon^2/\sigma^2\right)\right) , \tag{A.9}$$

and for completeness to obtain the full solution for y, we also have

$$Y(\sigma) \sim \sqrt{\pi}\left(\frac{y(0)}{\Gamma\left[\frac{1-\sqrt{\mu}}{2}\right]} + \frac{\varepsilon\operatorname{sign}(\sigma)y_\sigma(0)}{\sqrt{2\sqrt{\mu}}\Gamma\left[1-\frac{\sqrt{\mu}}{2}\right]}\right)\left|\frac{\sqrt{2}\varepsilon}{\sigma}\right|^{\sqrt{\mu}} + \mathsf{O}\left(\varepsilon^2/\sigma^2\right) .$$

The function $Y(\sigma)$ deviates from the sigmoid of $y_*(\sigma)$ by an amount greatest near $\sigma \approx 0$ and decreasing inversely with $\sigma^{\sqrt{\mu}}$. Moreover this deviation disappears with time t, so we approximate $y \approx y_* \approx \operatorname{sign}(\sigma)$. The asymptotic form (A.9) is again consistent with (1.17), and we can again ask how to include this in the full (x, y) space, either as linear or nonlinear switching.

Finally, our exploration of sources of sigmoids and discontinuous sign functions in asympotic expansions would not be complete without a brief look at integral equations.

A.2 In Integrals: Stokes' Phenomenon

Say that y is known to be an integral of oscillations $e^{i\tau z}$ under some envelope $a(z)$, in a system

$$\begin{aligned} \dot{x} &= f(x, u) , \\ u &= \int_{-z-}^{z+} dz a(z) e^{itz} . \end{aligned} \tag{A.10}$$

Such integrals typically arise when solving a problem using Fourier or Laplace methods, where typical forms for $a(z)$ might be $e^{Np(z)}\cos\theta(z)$ or $[p(z)]^N$ for large N and polynomials p, θ, so that there are an exponential variation (from e^{Np} or $e^{N\mathbb{R}\mathrm{e}[\log p]}$) and an oscillation (from $\cos\theta$ or $e^{iN\mathbb{I}\mathrm{m}[\log p]}$).

Integrals of this form can be analysed using stationary phase or steepest descent methods. If we let z take complex values, the integration path $z \in [z_-, z_+]$ can be deformed to a curve in the complex z plane, chosen to pass through maxima of the integrand where $|ae^{i\tau z}|$ is stationary, Figure A.1, ultimately allowing us to approximate the integrand near those maxima [19, 54, 61, 99]. A subtle but powerful geometrical game ensues to determine which stationary points contribute to the integral, depending on how the integration contour deforms within the complex plane.

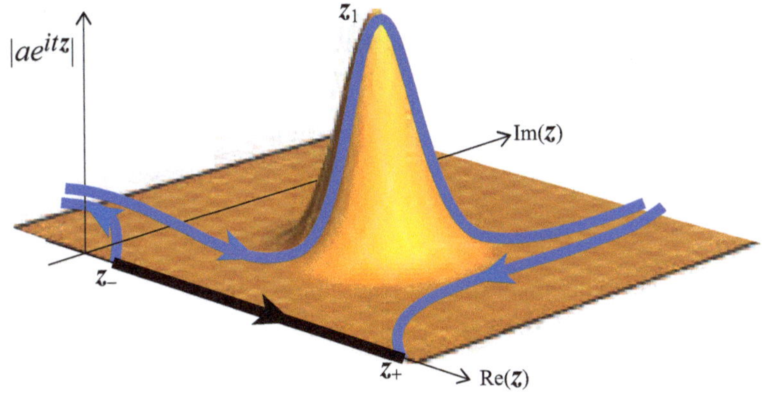

Fig. A.1 An integration contour is deformed from the real line interval $[z_-, z_+]$, to a curve in the complex plane, passing through a maximum z_1 of the integrand (the contour can pass through infinity, as this example does twice, if the integrand vanishes there).

The result is an asymptotic series composed by summing approximations near the stationary points, some $z_1, z_2, \ldots, z_r$, of the integrand in (A.10). If the end points $z_{1,2}$ are also maxima of the integrand along the contour, the approximation around these may contribute too. Each contribution from z_i is multiplied by a factor

$$\text{step}\left[\sigma_i(z_+, z_-, z_1, \ldots)\right]$$

for some function σ, which turns that contribution to the integral on or off as σ_i changes sign, and these form a set of switching functions $u_1, u_2, \ldots, u_r$ (the functions σ_i compare the imaginary parts of the exponent of the integrand at each z_i to the other z_j's, as this quantity controls switching).

To illustrate, let us say the integrand of u has just a single stationary point. We can place this at $z_1 = 0$ by a coordinate transformation. The integrand looks near $z = 0$ like $e^{-z^2/2}$ (which has a simple maximum at $z = 0$), and the integral locally resembles the error function, a sigmoid function we have seen already. Take end points $[-\infty, \sigma/\varepsilon]$ where ε is some small non-negative parameter; then

$$\begin{aligned} u &= \tfrac{1}{\sqrt{2\pi}} \int_{-\infty}^{\sigma/\varepsilon} dz\, e^{-z^2/2} = \tfrac{1}{2}\left\{1 + \tfrac{1}{2}\,\text{Erf}(\sigma/\sqrt{2}\varepsilon)\right\} \\ &\sim \text{step}(\sigma) - \tfrac{1}{\sqrt{2\pi}} e^{-\sigma^2/2\varepsilon^2}\left\{\tfrac{\varepsilon}{\sigma} + \mathsf{O}\left(\varepsilon^3/\sigma^3\right)\right\}\,, \end{aligned} \tag{A.11}$$

(with a normalization by $\sqrt{2\pi}$ being introduced for convenience). The step function is the contribution from the stationary point, which switches on/off discontinuously as σ passes through zero (step$(\sigma) = 1$ for $\sigma > 0$ and step$(\sigma) = 0$ for $\sigma < 0$). The second term, the asymptotic tail, comes from the end point $z = \sigma/\varepsilon$.

To make the example a little harder, and more typical of physical problems, introduce a steady oscillation of wavelength $2\pi/\rho$ inside the integrand, and then we have instead

$$u = \frac{1}{\sqrt{2\pi}}\int_{-\infty}^{\sigma/\varepsilon} dz\, e^{-z^2/2}\cos(\rho z) = \tfrac{1}{2}\left\{1 + \mathbb{R}\mathrm{e}\left[\mathrm{Erf}\left(\frac{\sigma/\varepsilon - i\rho}{\sqrt{2}}\right)\right]\right\}$$
$$\sim \mathsf{step}(\sigma) - \frac{1}{\sqrt{2\pi}} e^{(\rho^2-\sigma^2/\varepsilon^2)/2}\cos(\rho\sigma/\varepsilon)\left\{\tfrac{\varepsilon}{\sigma} + \mathsf{O}\left(\sigma^{-3}\right)\right\}\,. \tag{A.12}$$

There is still one maximum of the integrand, but the cosine term shifts it from $z = 0$ to $z = i\rho$. Its contribution to the integral is the same, $\mathsf{step}(\sigma)$, but the contribution from the end point $z = s$ changes, giving a different, oscillatory, asymptotic tail. Now y, or $y = 1 + 2u$, is not a simple sigmoid, having peaks near the transition of height $|y| \approx 1 + \sqrt{\frac{2}{\pi}\frac{4\rho^3}{\pi^2}}e^{-\pi^2/8\rho^2}$ at $y \approx \pm\pi/2\rho$.

As in our differential equations above, in either case y is of the form (1.17), and only exponentially small tails distinguish the function y from a discontinuous sign function. We can again ask how to include this in the full (x, y) space, either as linear or nonlinear switching, each with different consequences and limitations for modelling of physical and biological problems. Hopefully this book serves as a first step in elucidating this step into the study of nonlinear switching and hidden dynamics in practice.

A.3 In Closing

Above we see various classes of dynamical system $\dot{x} = f(x, y)$, in terms of some parameter y that jumps rapidly between steady values, where the model for y reveals an asymptotic form $y \sim \mathrm{sign}(\sigma) +$ tail, be it an ad hoc sigmoid, or a set of ordinary and partial differential equations, or an integral equation.

When we approximate or take the limit such that $y = \mathrm{sign}(\sigma)$, what happens to the tails of the asymptotic expansions? They might vanish in a simple manner and have no non-trivial effect near the discontinuity, or else they might become part of the hidden dynamics through nonlinear dependence on the switching multiplier y.

Appendix B
A Few Words from Filippov and Others, Moscow 1960

For the most comprehensive account of Filippov's approach to discontinuities in dynamical systems, one must of course refer to his 1988 book [71] (translated from the 1985 edition in Russian). But a quick and eye-opening insight into the state of the theory is given by a paper from the Proceedings of the First International Congress of the International Federation of Automatic Control, Moscow 1960, which we reprint below for the reader's interest.[1]

The paper comes complete with comments from colleagues at the Congress. The contents list reveals a number of articles on discontinuities in control applications.

[1] Reproduced with permission of the publishers. Inconsistent spelling of A. F. Filippov's name as A. G. Filippov on the paper itself is most likely a translation error.

M. R. Jeffrey, *Hidden Dynamics*,
https://doi.org/10.1007/978-3-030-02107-8

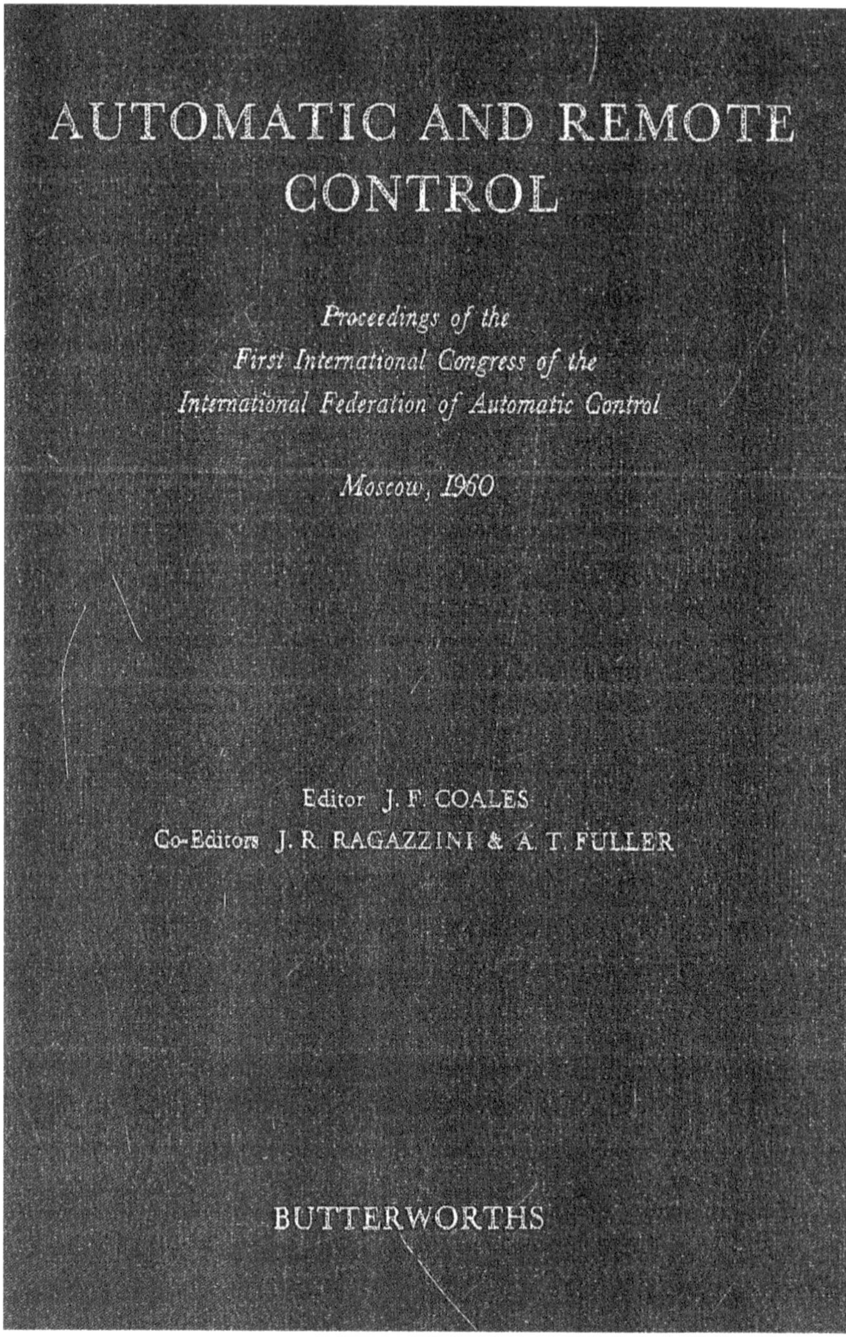

AUTOMATIC AND REMOTE CONTROL

Proceedings of the
First International Congress of the
International Federation of Automatic Control

Moscow, 1960

Editor J. F. COALES
Co-Editors J. R. RAGAZZINI & A. T. FULLER

BUTTERWORTHS

CONTENTS

CONTENTS

Application of the Theory of Differential Equations with Discontinuous Right-hand Sides to Non-linear Problems in Automatic Control

A. G. FILIPPOV

In the study of non-linear oscillations, described by the equation $\ddot{x} + f(\dot{x}) + g(x) = e(t)$, where the friction force $f(\dot{x})$ and the restoring force $g(x)$ are discontinuous functions, difficulties arise in those cases where, in a finite interval of time, the solution crosses the lines of discontinuity of the functions f and g in the plane $(x, \dot{x})$ an infinite number of times. In such cases the solution cannot consist of a finite number of smooth arcs, and there arises the problem of the existence of a solution. In automatic control theory, similar difficulties may appear in the study of relay systems, for example, if slip can occur in the system. The development of the theory of differential equations with discontinuous right-hand sides makes it possible to overcome these difficulties and to establish a basis for the application of many well known methods in the qualitative theory of differential equations to such equations.

In the domain G of space $x_1, x_2 \ldots x_n$ we consider the system of equations in vector form

$$\frac{dx}{dt} = f(t, x) \tag{1}$$

where $x = (x_1, x_2 \ldots x_n)$, $f = f(f_1, f_2 \ldots f_n)$. Let the domain G be divided by a continuous differentiable surface S into two domains G^+ and G^-, and in the domains $G^+ + S$ and $G^- + S$ let there be given vector functions f^+ and f^- which are continuous in t and continuously differentiable with respect to $x_1, x_2 \ldots x_n$; for equation 1, $f = f^+$ in G^+ and $f = f^-$ in G^-. In this case the solution of equation 1 passing through domain G is continuous in a well-defined manner until it reaches the surface S. If the vector f in domain G^- is directed towards

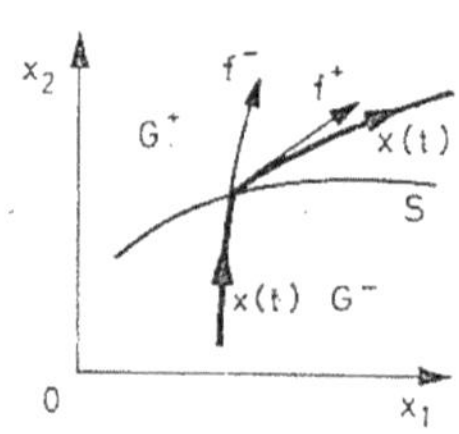

Figure 1

surface S, and in the domain G^+ from surface S, then the solution passes from G^- to G^+, intersecting surface S in one point (*Figure* 1). If the vector f is directed towards the surface S in both the domains G^- and G^+, then the solution, on reaching the surface S, cannot be continued in either G^- or G^+, and the problem arises of how to continue the solution. In automatic control relay systems, slip occurs in such cases. This feature is studied under the assumption that the transition from equation

$dx/dt = f^-$ to $dx/dt = f^+$ does not take place at the moment of reaching surface S but after some delay, owing to which the solution $x(t)$ oscillates about the surface S, being displaced along it at the same time with some mean velocity. It can be shown that as the delay tends to zero, limit motion and velocity along surface S are

$$\frac{dx}{dt} = f_0(t, x) \equiv \alpha f^+(t, x) + (1 - \alpha) f^-(t, x) \qquad (2)$$

where $x \in S$, and the number $\alpha (0 \leqslant \alpha \leqslant 1)$ such that the vector $f_0(t, x)$ is tangential to the surface S at point x. This was

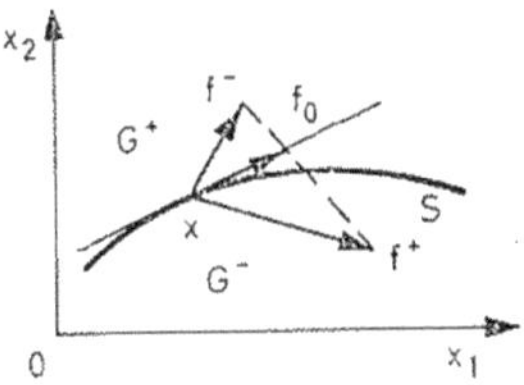

Figure 2

proved[1] under less general assumptions. The vector f_0 in equation 2 can be constructed as follows. At the point $x \in S$ we construct the vectors $f^+(t, x)$ and $f^-(t, x)$ and join their ends by a segment, The point of intersection with the tangent plane to S at point x is the end of the required vector f_0 (*Figure* 2). The function $x(t)$ which satisfies equation 1 in the domains G^- and G^+ and equation 2 on surface S can be called the solution of equation 1 in a generalized sense.

Equations with Discontinuous Right-hand Side

Cases can be imagined when the solution $x(t)$ cannot consist of a finite or even a countable number of arcs, each of which passes through domain G^- or G^+ and satisfies equation 1, or lies on the surface S and satisfies equation 2. To cover such cases, and also those when the solution falls on the line of intersection of surfaces of discontinuity, it is necessary to give a more general definition of the solution of equation 1 with a discontinuous right-hand side, which will be a direct generalization of the above. Let the vector function $f(t, x)$, where f and x are n-dimensional vectors, be measurable in domain D of space (t, x), and for any closed bounded domain $D \subset G$ let there exist a summable function $M(t)$ such that almost everywhere in D we have $|f(t, x)| \leqslant M(t)$. The absolutely continuous vector function $x(t)$ is called the solution of equation $dx/dt = f(t, x)$ if for almost all t the vector dx/dt belongs to the least convex closed set containing all the limiting values of vector

$f(t, x^\star)$, where $x^\star$ tends towards x in an arbitrary manner, and the values of function $f(t, x^\star)$ on a set of zero measure in the x space are ignored.

In particular, for the case shown in *Figure* 2, the values of function $f(t, x)$ on surface S must be ignored, then the limiting values of vector $f(t, x^\star)$ as $x^\star \to x \in S$ are simply the vectors $f^+(t, x)$ and $f^-(t, x)$, while the least convex closed set containing them is the segment joining their ends (*Figure* 2). Since the solution on *Figure* 2 cannot leave S, the vector $\mathrm{d}x/\mathrm{d}t$ must be tangential to S. Hence its end lies at the point of intersection of the tangent to S and the above segment, i.e. $\mathrm{d}x/\mathrm{d}t = f_0$.

Similarly the vector $\mathrm{d}x/\mathrm{d}t$ can also be defined in cases when there are several intersecting surfaces of discontinuity of the function f (but then it is not always defined uniquely).

Further Considerations

Apart from what has just been said, such a definition of the solution is justified by the following considerations.

(*a*) The requirement that the very concept of a solution should not depend on the choice of a coordinate system in the x space leads to such a definition. More exactly, let the right-hand side of equation 1 be bounded, and let the absolutely continuous vector function $x(t)$ have the following property. With any choice of the coordinate system in the x space and for almost all t and any $\delta > 0$, for each of its components $x_i(t)$ the derivative $\mathrm{d}x_i/\mathrm{d}t$ lies between the lower and upper bounds of the function $f_i(t, x)$ in a δ neighbourhood of point $x(t)$ of the x space (ignoring the values of this function on a set of zero measure); $f_i(t, x)$ be the right-hand side of the ith equation of system 1 in the chosen coordinate system. Then $x(t)$ is the solution of equation 1 in the sense of the above definition. Conversely each solution has the indicated property.

(*b*) The limit transition from an equation with a continuous right-hand side leads to the same definition. Let $\delta_k > 0$ and $f_{\delta_k}(t, x)$ be the mean value (with any positive weight) of function $f(t, x)$ in an n dimensional δ_k neighbourhood of point x. Then the limit $\delta_k \to 0$ of a uniformly convergent sequence of solutions of the equations $\mathrm{d}x/\mathrm{d}t = f_{\delta_k}(t, x)$ is a solution of equation 1 in the sense of the preceding section.

(*c*) The limit $k \to \infty$ of the uniformly convergent sequence of solutions of the equations with time lags

$$\frac{\mathrm{d}x}{\mathrm{d}t} = f[t, x\{t - \tau_k(t)\}]$$

where $0 \leqslant \tau_k(t) \leqslant \delta_k \to 0$, is a solution of equation 1. Moreover, the time lag can be constant or variable, depending on any factors whatever.

Solutions in the sense of the above-mentioned definition have many of the properties of the usual solutions[2]. If the conditions defining the solution are satisfied, then at least one solution of equation 1 passes through any point $(t_0, x_0) \in G$. The solution can be continued as long as it does not tend to infinity or come arbitrarily close to the boundary of domain G.

If for any $(t, x') \in G$ and $(t, x'') \in G$

$$(x'' - x')(f(t, x'') - f(t, x')) \leqslant K(t)(x'' - x')^2$$

where the function $K(t)$ is summable and the products of the vectors are scalar, then the solution passing through point $(t_0, x_0) \in G$ is unique (for $t \geqslant t_0$) and depends continuously on the initial conditions and on the right-hand side of the equation.

For solutions in the above-mentioned sense many theorems of the qualitative theory of differential equations hold true.

Equations with Piecewise Continuous Right-hand Sides

From the general theorems formulated above, new results[2] can be obtained for equations with piecewise continuous right-hand sides. Let equation 1 satisfy the conditions mentioned. Suppose that at each point of the surface S at least one of the vectors f^+ and f^- is directed from the domain G^+ (or G^-) towards surface S. (This assumption is sufficiently general, since if from both sides of surface S the vector field defined by equation 1 is directed from the surface, the solution $x(t)$ does not reach the surface.) Then for any initial condition $x(t_0) = x_0$ equation 1 gives a solution. For $t \geqslant t_0$ this solution is unique and depends continuously on the initial conditions and the right-hand side of the equation. For all t, apart from, at most, a countable set of moments at which the surface S is reached, the solution has the derivative dx/dt and satisfies equation 1 in the domains G^- and G^+, while at surface S it satisfies equation 2. Further, in this case, the solution of the mean equations and the equations with time lags (see above, b and c) for initial conditions $x(t_0) = x_0$ converge uniformly to the solution of equation 1 with the same initial conditions.

Discontinuities

If the right-hand side of equation 1 has a discontinuity, not on one surface S but on several intersecting surfaces $S_1, S_2 \ldots S_m$, then the investigation becomes more complicated. We shall assume that the right-hand side of equation 1 is the sum of m functions each of which is discontinuous on only one of the surfaces $S_1, S_2 \ldots S_m$, and satisfies the conditions mentioned, and that these surfaces form non-zero angles with each other and with vector f on the right-hand side of equation 1.

The conditions were indicated under which the solution can be continued uniquely after reaching one of the surfaces $S_1, S_2 \ldots S_m$. Consider now the $(n-2)$-dimensional manifold S_{12} which is the intersection of surfaces S_1 and S_2. Let the solution be continued uniquely after reaching surfaces S_1 and S_2.

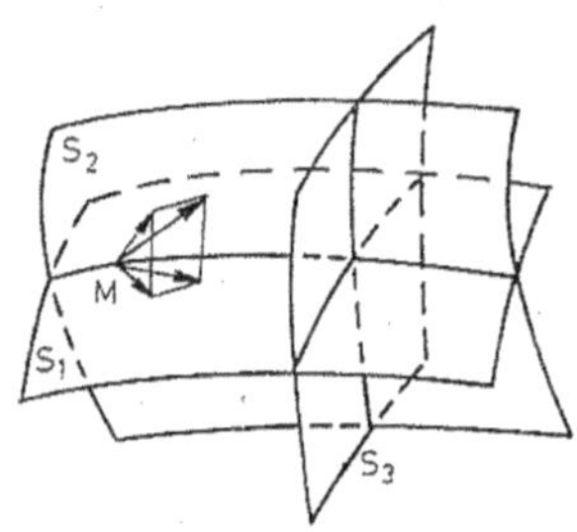

Figure 3

At each point M of S_{12} we construct four vectors, i.e. the limiting values of vector $f(t, x)$ in equation 1, as $x \to M$ (*Figure* 3). Their ends lie in one plane and form the vertices of a parallelogram. If the parallelogram does not intersect the tangent to S_{12} at point M, then the solution on reaching point M must immediately leave the manifold S_{12}, but if it does intersect the tangent, then the point of intersection determines a vector of possible velocity of motion along manifold S_{12} (corresponding to vector f_0 in *Figure* 2). By studying the directions of the above-mentioned four vectors and also vector f_0 constructed for surfaces S_1 and S_2, we can indicate the cases in which, on reaching S_{12}, the solution can be continued uniquely and those when it cannot. It may happen that in the neighbourhood of point M the solution intersects the surfaces S_1 and S_2 an infinite number of times in a finite time interval. This case is in some respects similar to a focus type singularity for a system of differential equations, and its investigation is more complicated.

Having investigated whether, on reaching the $(n-2)$-dimensional manifold S_{12}, the solution can or cannot be continued uniquely, and similarly for S_{13} and S_{23}, we consider by the same method the $(n-3)$-dimensional manifold S_{123}, which

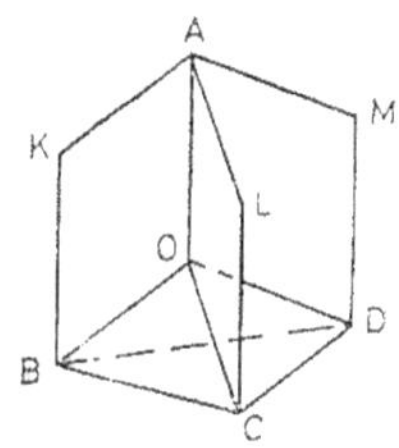

Figure 4. S_1 AOBK, S_2 AOCL, S_3 AODM, S_4 BOC, S_5 COD, S_6 DOB

is the intersection of the three surfaces S_1, S_2, S_3, etc. If, on reaching the intersection of any k of the surfaces $S_1, S_2 \ldots S_m$ $(k = 1, 2 \ldots m)$ the solution can be continued uniquely for $t \geqslant t_0$, then it depends continuously on the initial conditions and the right-hand side of the equation and the theorems about convergence hold true.

We can similarly consider cases of intersection of surfaces $S_1, S_2 \ldots S_m$ such as, for example, that of *Figure* 4.

Wider Applications

Many methods of the qualitative theory of differential equations hold true also for equations with discontinuous right-hand sides. For example, if the total derivative of the function $F(t, x_1, x_2 \ldots x_n)$ in equation 1 is non-positive, then the solution of this system does not leave the domain $F(t, x_1, x_2 \ldots x_n) < C$. If, as t increases, not one solution of system 1 leaves a closed domain in x space, which is the homeomorph of an n-dimensional sphere, if the solution for any initial conditions (t_0, x_0) in this domain is unique for $t \geqslant t_0$, and if $f(t + T, x) \equiv f(t, x)$, then in this domain there exists a solution of period T. The majority of the well-known theorems of sufficient conditions for the existence of periodic solutions of the equation for non-linear oscillations

$$\ddot{x} + f(\dot{x}) + g(x) = \mathrm{e}(t) \qquad \mathrm{e}(t + T) \equiv \mathrm{e}(t) \tag{3}$$

holds true also for the case of piecewise continuous functions f, g, e. In particular, problems of non-linear mechanical oscillations with dry friction and certain problems in electrical networks with rectifiers or relays lead to such an equation. Using the theory of differential equations with discontinuous right-hand sides more general results[2] than hitherto[3,4] are obtained for equation 3. This also applies in the case $\mathrm{e}(t) = 0$ (self-oscillating system), which is usually investigated using Bendixson's theorems on the behaviour of the solutions of a system of two equations in the phase plane, since they are only insignificantly changed for equations with discontinuous right-hand sides.

Lastly, using methods of the theory of differential equations with discontinuous right-hand sides leads to an easy proof of the existence of solutions of certain problems in the theory of optimum control[5].

References

1 André, J. and Seibert, P. Über stückweise lineare Differentialgleichungen, die bei Regelungsproblemen auftreten. II. *Arch. Math. Karlsruhe* 7 No. 3 (1956)

2 Filippov, A. F. Differential equations with discontinuous right-hand sides. *Rec. Math., Moscow* 51, No. 1 (1960)

3 Zheleztsov, N. A. The method of point transformations and the problem of forced oscillations of an oscillator with 'combined' friction. *Appl. Math. Mech., Leningr.* 13, No. 1 (1949) 3

4 Reissig, R. *Math. Nachr.* 1954–1956, in particular 11 Nos. 4–6, 13, No. 5, 14, No. 1, 15, No. 3

5 Filippov, A. F. On some problems in the theory of optimum control. *Vestn. Moscow State University* No. 2 (1959)

DISCUSSION

F. R. Gantmacher (*U.S.S.R.*)

Can the solution of the equation with delay, when the delay tends to zero, tend *not* to the solution of the equation with a discontinuous right-hand side, in the sense defined in the paper, but to some other function? Will this not depend on the way in which the delay tends to zero?

A. G. Filippov, *in reply.* If the delay (irrespectively of whether it is constant or variable) is the same for all the components of the vector, then at least one will always obtain the solution defined in the paper. If the delay is different for the different components, then the solution of the equation with delay may not tend to the above solution.

S. Ziemba (*Poland*)

What conditions are imposed on the function f in order that the solution may be extended to infinity?

A. G. Filippov, *in reply.* The conditions are the same as in the case of equations with continuous right-hand sides.

Yu. I. Alimov (*U.S.S.R.*)

In the papers read by Filippov, and also those by André and Seibert, the equations of the dynamics of relay systems, namely,

$$\dot{\bar{x}} = \bar{f}(t,\bar{x}) \tag{1}$$

are treated as equations with discontinuous right-hand sides. In our opinion this approach is not in accordance with the physical meaning of the problem, and leads to an unwelcome distinction between classical and 'relay' differential equations as objects for mathematical study. It is probably simpler and more correct to look upon the right-hand sides of these equations as continuous but not single-valued.

In fact, both the input and output quantities $\sigma(t)$ and $\phi[\sigma(t)]$ of any real element, which is approximated to by a non-inertial relay element, are in reality continuous functions of time. Only when the function $\sigma(t)$ passes rapidly through threshold values lying on the switching surface

$$\sigma = \sigma_K(t,\bar{x}) \quad K = 1, 2 \ldots \tag{2}$$

in which case $\phi = \phi[\sigma(t)]$ will also vary very rapidly, can the function $\phi(\sigma)$ be looked upon as discontinuous. However, the discontinuous characteristics of $\phi(\sigma)$ can be justifiably neglected when one is concerned with the description of those motions of a real system for which

$\sigma(t)$ remains in the neighbourhood of switching thresholds for long periods of time. The idealization involved in the introduction of discontinuous right-hand sides leads, in this case, to the fact that vector lines of the directional field corresponding to equation 1 cannot, in general, be extended to $t \to \infty$. In order to set up the correct model within the framework of the chosen structural scheme it is necessary to provide a more accurate description of $\phi(\sigma)$ in the neighbourhood of equation 2. Before we pursue this any further, it is tempting to try the following simple continuous approximation to a relay characteristic at the switching points: $\phi(\sigma_K) = \xi_K$ where ξ_K is any number lying between $\phi(\sigma_K - 0)$ and $\phi(\sigma_K + 0)$. This approximation was successfully employed by Coulomb in the case of dry friction.

In this approach, the directional field for the system given by equation 1 becomes many-valued on the surfaces given by equation 2, but remains continuous, since for any sufficiently close points $(t_1, \bar{x}_1)$ and $(t_2, \bar{x}_2)$ of the space $t.\bar{x}$ one can find values of the infinitely-valued function $\bar{f}(t,\bar{x})$ which are sufficiently close to each other. From the point of view of the determination of the velocity as a function of coordinates, we obtain a system with incomplete information.

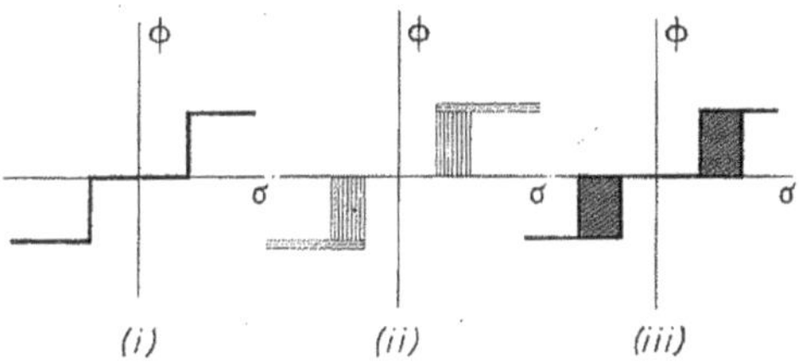

Figure A

The continuous integral curve $\bar{x}(t)$ of equation 1 can then be naturally defined and thus the motion at each point in the space $t.\bar{x}$ should, roughly speaking, proceed along one of the directions determined by the many-valued function $\bar{f}(t,\bar{x})$ at this point. Many-valuedness will then be the price which one has to pay in order to retain a simple model.

This approach to the determination of the solution of equation 1 is equivalent to treating such equations as differential in contingencies with infinitely-valued right-hand sides, or quasi-continuous in the terminology of Myshkis.[1] The fundamentals of the general theory of such equations, which is very similar to the theory of ordinary differential equations, were developed by Zaremba[2] and Marchand[3]. Solutions of contingent equations satisfy the axioms of the generalized dynamic system of Barbashin[4], therefore many of the methods of the qualitative theory of ordinary differential equations can be applied to them, including the direct method of Lyapunov.

Determination of the solutions of the 'relay' system (1) put forward by Filippov is, in fact, a modification of the Zaremba–Marchand approach[1–3] and does not appear to have any advantages over the latter as far as applications are concerned.

If one looks upon the equations of motion of relay systems in control applications as contingent equations, then it is possible, using general considerations, to obtain information on the dynamics of the system without referring in each case to a detailed description of the phase-space of the model (1).

Moreover, it also appears possible to approach, from a single point of view, those cases where $\bar{f}(t,\bar{x})$ is many-valued on subsets of the space $t.\bar{x}$ with non-zero volume. Strictly speaking, none of the elements of a real system can be characterized by a single-valued input-output dependence over a long period of time. Thus, for example, repeated determinations of the static characteristics of a relay lead to results which are more similar to *Figure A*(*ii*) than to *Figure A*(*i*). The switching of a relay in different experiments occurs for different input values (the shading density in *Figure A*(*ii*) is proportional to the relative switching frequency for the corresponding values of σ). On the first approximation attention can be confined to the strictly determined characteristics of *Figure A*(*i*), which represents some kind of average of real characteristic. The contingent equations may serve as the second approximation and correspond to the state of affairs represented by *Figure A(ii)*, since the aggregate of their trajectories is known to include all the motions of the exact model shown in *Figure A(ii)*. The latter considerations may be useful in studying the systems with hysteretic elements.

References

[1] MYSHKIS, A. D. Unique Solutions of Systems of Linear Partial Differential Equations. *Matem. sb.*, 19 (61); 3 (1946) 489

[2] ZAREMBA, S. C. Sur les equations au paratingent. *Bull. Sci. math.* 2s, LX (1936) 139

[3] MARCHAND, A. *Sur les champs continus de demicônes convexes et leurs intégrales*

[4] BARBASHIN, YE. A. On the Theory of Generalized Dynamic Systems. *Uchen. Zap. MGU*, vyp. 135, *Matem.*, t. 11 (1949), 110

J. ANDRÉ (*U.S.A.*)

Earlier, we were concerned with the different kinds of change in the right-hand sides of equations on passing through the switching surface. In all cases it was assumed that there exists a band G_0 in the neighbourhood of the surface S which is such that $f(x,t) = f^{\pm}(x,t)$ when $x \varepsilon G^{\pm} - G_0$ and assumes a value equal to some intermediate value between f^+ and f^- or (coincides with one of them) when $x \varepsilon G_0$.

This was followed by a consideration of the limit transition $G_0 \to S$.

In particular, the following two types of change were considered:

Discrete changes ('*switching*'). Here one can have a switching, from f^+ to f^- and back, on two different surfaces (on elements of the boundary G_0). In this case the function f is two-valued in G_0, and its values depend on the direction in which the transition takes place (the function is 2^k-valued at the point of intersection of k surfaces $S_1, \ldots, S_k$).

Continuous changes. Here the function f is looked upon as continuous. The limit transition $G_0 \to S$ was carried out by us for the first type of change (cf. *Compte Rendu* 245 (1957) 625–627 and *Boll. Mat. Mexicano*. In press).

The results obtained by Filippov represent, in our opinion, a significant contribution to the theory, since he showed that the limit transition always leads to the same results, irrespective of the fact whether the changes are discrete or continuous.

The result that the characteristic of the changes has the form of a hysteresis loop in which f is continuous and depends only on whether the trajectory enters G_0 from G^+ or G^-, is of particular practical interest. One would expect that in this case, limit curves are also solutions in Filippov's sense, as in the above two types of change. Another interesting problem is: can any solution in Filippov's sense (even when it lies on the intersection of more than one surface S) be obtained, by a limit transition $G_0 \to S$, from a continuous system.

YU. I. NEYMARK (*U.S.S.R.*)

Before I formulate a theorem concerned with the definition of solutions of systems of differential equations with discontinuous right-hand sides, I should like to make one general remark. From the mathematical point of view, any method of definition of solutions of equations with discontinuous right-hand sides is equally acceptable. The mathematical problem consists of the comparison of the methods and, in particular, in determination of conditions under which they are equivalent. However, if one bears in mind the fact that in the present context the differential equations are being used to describe physical systems, then the different results obtained by different methods suggest that none of them can be accepted unconditionally. It is therefore useful to try and analyse specific physical systems. In this connection one may mention the concept of constraints in classical mechanics, the well-known Prandtl paradox associated with the discontinuity and many-valuedness of the dry friction characteristic, the multivibrator, and the theory of discontinuous vibrations. I shall now formulate a result which is, to some extent, capable of unifying the mathematical and the physical points of view. A system of n equations with discontinuous (on some smooth surface $S = 0$) right-hand sides, may be written in the form

$$\frac{dx_i}{dt} = \phi_i(x_1, \ldots, x_n) + \sum_{j=1}^{n} \gamma_{ij}\psi_j(x_1, \ldots, x_n;\ \operatorname{sgn} S) \qquad (1)$$

where $\phi_i(x_1, \ldots, x_n)$ and $\psi_j(x_1, \ldots, x_n)$ are smooth functions of $x_1, \ldots, x_n$, and γ_{ij} are constants. In the neighbourhood of the surface $S = 0$, $\dot{S}S < 0$. We then have the following results: if the rank of the matrix of the coefficients γ_{ij} is m, then independently of the type of limit transition from smooth functions into discontinuous functions in the right-hand sides of equation 1, we have the following $n - m$ equations on the surface $S = 0$:

$$\sum_{\beta=1}^{n} C_{\alpha\beta}\dot{x}_\beta = F_\alpha(x_1, \ldots, x_n)$$

The missing $m - 1$th equation will depend on the method whereby the limit transition is carried out. On the other hand, when $m = 1$, the limit equations of motion on the surface $S = 0$ do not depend on the method whereby the limit transition is carried out. We note that when $m = 1$ the introduction of delays, with subsequent limit transitions, leads to the same equations, while when $m > 1$ this is not the case. In accordance with this theorem, and the considerations expressed above, it follows that when $m > 1$, the equations given by equation 1 are insufficient to describe the motions. On the other hand, when $m = 1$, equations 1 define the motion of the system on the surface $S = 0$ as well.

Discontinuities in relay systems correspond to the case $m = 1$, and hence the determination of slip ('sliding motion') may be carried out using the usual differential equations of motion.

P. Seibert and J. André (*U.S.A.*)

(*a*) Theorem 1 was established by us also for some cases where the delay is not constant. For example, this was done in the case where the switching occurs when the trajectory leaves a given open set containing the surface S (cf. our paper: Differential equations with piece-wise discontinuous right-hand sides. *Boll. Mat. Mexicano*).

(*b*) Theorem 1 may be applied directly to non-autonomous systems $dx/dt = f(t,x)$ after transforming them into $(n + 1)$ dimensional (autonomous) systems

$$\frac{dX}{dt} = F(X)$$

where $X = X(x,y)$, $y = t$, $F = (f,1)$. It follows that Theorem 2 also holds for non-autonomous systems.

(*c*) Filippov's suggestion, that the condition $\|z(t,x)\| < h$ imposed on the perturbing forces in Theorem 2 should be replaced by a less stringent limiting integral condition, is in our opinion very useful.

A. G. Filippov, *in reply*. The solution of a differential equation in the sense defined in my paper is at the same time a solution of a contingent equation. The work of Polish mathematicians in the 'thirties was concerned with the case where the field of directions is quasi-continuous with respect to the variables t and x, while in my paper it is quasi-continuous with respect to x only. However, even in the more general case, the principal properties of contingent equations remain the same. In particular, the solutions form a dispersed dynamic system in the sense of B. M. Budak, and this is discussed in my paper published in *Matematicheskom sbornike*. **51**, No. 1 (1960).

Thus the definitions of the solution of an equation with a discontinuous right-hand side, which were given in my paper and in the contribution by Yu. I. Alimov, are, in fact, entirely equivalent.

It was assumed in my paper that all the coordinates of the vector $(x_1, \ldots, x_n)$ are interdependent. This leads to the definition of the solution as given in my paper. If these coordinates are not interdependent, for example, if they include coordinates of different physical bodies, then the solution may be different.

Whatever the form of the hysteresis loop, the solution of the equation will tend to the solution of the limit equation with a discontinuous right-hand side if the quantity whose graph forms a hysteresis loop is one-dimensional.

A limit transition from equations with continuous right-hand sides will lead to any solution of an equation with a discontinuous right-hand side. This may be achieved by averaging the discontinuous right-hand side with a suitably chosen weight.

Exercises

Exercises for Chapter 1 (Origins)

1. For the equation (1.3):
   a. Sketch the one-dimensional phase portrait of (1.3) (i.e. represent the dynamics of $x(t)$ by drawing arrows on the real line).
   b. Sketch the solution of (1.3), given by (1.4), as a graph $x(t)$ for $\alpha = 0, \frac{1}{2}, 1, \frac{3}{2}$. Try sketching it from $t = 0$ to $t = 2$ with $x_0 = -1$ or from $t = 0$ to $t = -2$ with $x_0 = 4$.
   c. Calculate the time taken to traverse a distance $|x| \leq \varepsilon$ (i.e. travel from $x(0) = -\varepsilon$ to $x(t) = +\varepsilon$). Show that if we let $\varepsilon \to 0$, this time is infinite if $\alpha \geq 1$ and zero if $0 \leq \alpha < 1$.
   d. [Harder] Show that the time taken to traverse the distance $|x| \leq \varepsilon$ is actually arbitrary for α close to 1. [Hint: show that the time is $\Delta t = 2\varepsilon^{1-\alpha}/(1-\alpha)$, and show that for any positive number B, we can find α such that $\Delta t = B$ as $\varepsilon \to 0$.]
2. In the scenario shown in Figure 1.3, show (rather trivially) that any subset of points in the inset is connected by the flow to any subset of points in the outset. While trivial, this means, for instance, that there exists a Smale horseshoe (among infinitely many other chaotic and non-chaotic topologies) in the picture shown on the right of Figure 1.3.
3. Derive the series expansion in (1.6), (1.7), and (1.11), and find their second-order terms.
4. To obtain an example of Figure 1.5, consider the function

$$f(x) = -x - 2\pi\left(\alpha e^{-x^2/\varepsilon^2} + \mathrm{sign}(x)(1 - e^{-|x|/\varepsilon})\right) .$$

   Graph this for $\varepsilon = 0.1$ and $\alpha = 0, -1, +1$, showing it takes the three forms in the left-hand side of the figure. Thus show that $\dot{x} = -df/dx$ has no attractor if $\alpha = 0$, an attractor at $x \approx -0.17$ with peak $f \approx 5.7$ if $\alpha = -1$, and an attractor at $x \approx 0.04$ with peak $f \approx -7.5$ if $\alpha = +1$.
5. Derive the series expansions in (1.16). In some you form a Taylor expansion in $w = \varepsilon/x$; in others you need to expand in a different quantity, like $w = e^{-2|x|/\varepsilon}$. For the error function, split the integral (see Table 1.1) into two half-infinite integrals, one from $u = 0$ to infinity and one from $u = x/\varepsilon$ to infinity.

M. R. Jeffrey, *Hidden Dynamics*,
https://doi.org/10.1007/978-3-030-02107-8

6. Simulate the systems in examples 1.2 to 1.4 by different methods. These will give you some insight into the systems and simulations that follow throughout the book. You will have to decide how to handle the discontinuity, and there are lots of options! Try ...

   a. smoothing: replace $\lambda = \text{sign}(\sigma)$ with $\lambda = \tanh(\sigma/\varepsilon)$ (or any of the other sigmoids in Table 1.1) for small $\varepsilon > 0$;
   b. discretizing: evaluate the right-hand sides of (1.22) to (1.24) in discrete time steps $t = 0, \delta t, 2\delta t, 3\delta t, \ldots,$ for some small δt;
   c. smoothing and discretizing: combine the last two, and compare results for $\delta t \ll \varepsilon$ versus $\varepsilon \ll \delta t$ (e.g. $\varepsilon = 0.1$ and $\delta t = 0.01$ or $\delta t = 10$;
   d. introducing a spatial (Δx) or temporal (Δt) delay in evaluating the switching condition $\text{sign}(\sigma)$;
   e. simulating noise by discretizing and introducing random kicks at each time step.

Exercises for Chapter 2 (Primer)

1. In Example 2.1, expand the exponential in (2.7) to obtain the second row. Show that either expression on the right-hand side is a solution of (2.4).
2. In Example 2.2, show that (2.10) is a solution to (2.4) to order t^2.
3. In Example 2.3, fix $|bv + ax| < 1$. Show that the vector field points towards the discontinuity surface $y = v$ in this region. Show that $\lambda = \lambda^{\$} = -bv - ax$ on the discontinuity surfacein (2.6) gives sticking with dynamics $(\dot{x}, \dot{y}) = (v, 0)$, with solution $x(t) = \dot{x}(0) + vt$. Finally show that (2.12) and (2.13) are solutions to (2.4) to order t^2 and t^3, respectively.
4. In Example 2.4, looking at (2.20):

   a. show for $s > 1/2$ that $-1 < \lambda_-^{\$} < -1/2s$ and $\lambda_-^{\$}$ exists only for $1 < bv + ax < 1/4s + s$;
   b. show also for $s > 1/2$ that $-1/2s < \lambda_+^{\$} < +1$ and $\lambda_+^{\$}$ exists for $|bv + ax| < 1$;
   c. what happens for $0 < s < 1/2$?
   d. show that on $\lambda = \lambda_+^{\$}$ the dynamics is given by $(\dot{x}, \dot{\lambda}) = (v, 0)$;
   e. from (2.20) derive the approximation (2.21).

5. In Section 2.5.1, show that (2.24) is a solution of (2.22). Also show that (2.25) is a solution of (2.22).
6. In Example 2.5, show that for $c > 0$, a saddle and a sliding node exist if $b > 0$, and only a saddle exists if $b < 0$. Investigate what happens as c decreases through zero, showing the saddle-node pair coalesces and disappears if $b > 0$, and the saddle becomes a sliding saddle if $b < 0$.
7. Investigate what happens in Figure 2.14 as you move the equilibria and/or the discontinuity surface with respect to each other. You may like to consider explicit expressions, for example:

   - (i). $(\dot{x}_1, \dot{x}_2) = (x_2 - \alpha, \sqrt{2}x_1 + x_1^2 + (1 - \lambda))$,
   - (ii). $(\dot{x}_1, \dot{x}_2) = (x_2 - \alpha, -\sqrt{2}x_1 - x_1^2 + (1 - \lambda))$,
   - (iii). $(\dot{x}_1, \dot{x}_2) = (x_2 + \alpha, -\lambda x_1 + \frac{1}{2}(1 + \lambda))$,
   - (iv). $(\dot{x}_1, \dot{x}_2) = (x_2 - x_1^2 - \alpha, x_1 - (x_2 - 1)^2) + 10(1 - \lambda)(1, 1)$,
   - (v). $(\dot{x}_1, \dot{x}_2) = (x_2 - \alpha, 2x_1 + \frac{1}{5}x_2 + x_1^2 + (1 - \lambda))$,
   - (vi). $(\dot{x}_1, \dot{x}_2) = (x_2 - \alpha\lambda, \frac{1}{6}x_2 - x_1 - \lambda)$,

each with $\lambda = \text{sign}(x_2)$, for each of which you should find the bifurcation occuring close to $\alpha = 1$. You will probably need to plot the vector fields or flows of these using a computer, but for a tougher problem study them analytically (find any equilibria, and find expressions for the stable/unstable manifolds from the saddles, and the orbits through the tangencies, proving that a connection occurs for a special value of α).

8. Show that the bifurcations in Figures 2.15 and 2.16 take place in the systems:

   - Figure 2.15(i). $(\dot{x}_1, \dot{x}_2) = (px_1 + x_2, px_2 - x_1 + (1 - \lambda))$,
   - Figure 2.15(ii). $(\dot{x}_1, \dot{x}_2) = (x_2 - px_1, -x_1 - px_2 + (1 - \lambda))$,
   - Figure 2.16. $(\dot{x}_1, \dot{x}_2) = (\frac{1}{2}x_1 + x_2, -x_1 - \lambda - \tilde{\lambda})$,

   each with $\lambda = \text{sign}(x_2 + 1)$, $\tilde{\lambda} = \text{sign}(x_2 - 1)$, and $p = \frac{1}{10}(\alpha - x_1^2 - x_2^2)$. For each of these, you should find the bifurcation occurring close to $\alpha = 1$ (and exactly at $\alpha = 1$ in the first two). Again you will probably need to plot the vector fields or flows of these using a computer or study them analytically by finding expressions for any separatrices and limit cycles.

Exercises for Chapter 3 (Vector Field)

1. For Figure 3.1, show that the discontinuity surface $\sigma = x_1^2 - x_2^2$ exhibits a transversal intersection, and $\sigma = x_1^4 - x_2^2$ exhibits a non-transversal intersection, as depicted.
2. For Example 3.1 (after you've read Section 3.4), find the switching layer systems on the discontinuity submanifolds and their intersections. [Hint: for the last example it is best to change to coordinates $(y_1, y_2) = (-x_1, x_1 + x_2^2 - 1)$, noting there is no discontinuity and hence no layer system on $\mathcal{D}_3$].
3. For Figure 3.5, let $\mathbf{f}(\lambda) = (1, \lambda - 1)$ with $\lambda = \text{sign}(x_1)$, and consider

$$\mathcal{F} = \{ \ (2r\cos\theta, r\sin\theta) + (1, -1) \ : \ r \in (0,1), \ \theta \in [0, 2\pi) \ \} \ .$$

   a. Derive $\mathbf{f}^{\pm}$ and show $\mathbf{f}^{\pm} \in \mathcal{F}$ for some r and θ.
   b. Show that there exist a family of vectors $\mathbf{v}(r,\theta) \in \mathcal{F}$ such that $\mathbf{v}$ lies in the tangent plane of the discontinuity surface $\mathcal{D}$. (The (r,θ)-parameterized set of all such vectors $\mathbf{v}$ forms the sliding set $\mathcal{F}_C$.)

4. For Figure 3.6, let $\mathbf{f}(\lambda) = \left(1 - 2\cos\frac{\pi\lambda}{2}, -1 + \sin\frac{\pi\lambda}{2}\right)$, and show that there exist two values $\lambda = \lambda_{\pm}^{\$}$ such that $\mathbf{f}(\lambda_{\pm}^{\$})$ lie in the tangent plane of the discontinuity surface $\mathcal{D}$.

Exercises for Chapter 4 (Flow)

1. For Figure 4.1, show (analytically or by numerical simulation) that the system

$$(\dot{x}_1, \dot{x}_2, \dot{x}_3) = \left(\ (\tfrac{1}{2} - x_3)(1 - \lambda_2) - \lambda_1, \ -\lambda_2 - 2x_1, \ 1 \ \right)$$

   with $\lambda_1 = \text{sign}(x_1)$, $\lambda_2 = \text{sign}(x_2)$ exhibits a trajectory as depicted in the figure.
2. Show that the trajectories:

   a. $\Gamma_a = (t, \mathbf{0})$;
   b. $\Gamma_b = (\cos t, \sin t, \mathbf{0})$;

c. $\Gamma_c = (1 + ke^{-t})\,(\cos t, \sin t, \mathbf{0})$ for any $k > -1$;
d. $\Gamma_d = (2(t-1) + |t-1|, t, \mathbf{0})$;
e. $\Gamma_e = (1-t, 0, \mathbf{0})\,\mathrm{step}(-t) + \Gamma_b\,\mathrm{step}(t)$;
f. $\Gamma_f = (t-1, k(1-t)\,\mathrm{step}(1-t), \mathbf{0})$ for any $k \in \mathbb{R}$;
g. $\Gamma_g = (t, (k - t\,\mathrm{sign}(k))\,\mathrm{step}(|k|-t), \mathbf{0})$ for any $k \in \mathbb{R}$;

give examples of the flows depicted in Figure 4.3.

3. For Example 4.2, sketch the phase portraits of the perturbed and unperturbed systems described in (i)–(iii).
4. For Example 4.3 and Figure 4.7(i), given an explicit example $(\dot{x}_1, \dot{x}_2) = (-\lambda - cx_1,\ x_1)$, show this is a fused centre for $c > 0$ where all orbits form closed loops and is a fused focus for small constant $c \neq 0$, where all orbits converge on the focal point (is this in finite or infinite time?).

Exercises for Chapter 5 (Canopy)

1. Using the canopy formula (5.3), show that substituting in the constituent vector fields:

   a. $\mathbf{f}^{++} = (-1,-1)$, $\mathbf{f}^{+-} = (-1,+1)$, $\mathbf{f}^{-+} = (+1,-1)$, $\mathbf{f}^{--} = (+1,+1)$,
   b. $\mathbf{f}^{++} = (+1,-1)$, $\mathbf{f}^{+-} = (-1,-1)$, $\mathbf{f}^{-+} = (+1,+1)$, $\mathbf{f}^{--} = (-1,+1)$,

   gives a canopy combination that simplifies to

   a. $\mathbf{f}(\lambda_1,\lambda_2) = (-\lambda_1, -\lambda_2)$, b. $\mathbf{f}(\lambda_1,\lambda_2) = (\lambda_2, -\lambda_1)$.

2. In Section 5.2.1, use the formulae (5.11) to (5.13) to expand a vector field $\mathbf{f}(\mathbf{x};\lambda)$ with $\lambda = \mathrm{sign}(x_1)$, given that

$$\left.\frac{\partial^k}{\partial\lambda^k}\mathbf{f}(\mathbf{x};\lambda)\right|_{\lambda=\pm 1} = \begin{cases} (1,\ -1 \pm 1) & k = 0 \\ (\pm(-1)^{(k-1)/2}\pi^k/2^{k-1},\ 0) & k > 0 \text{ odd} \\ (0,\ \pm(-1)^{k/2}\pi^k/2^k) & k > 0 \text{ even} \end{cases}$$

   Expand $\mathbf{f}^{\pm}$ to order $(\lambda \mp 1)^p$ and derive the combination $\mathbf{f}(\mathbf{x};\lambda)$ from (5.14) assuming $\mathbf{h} \equiv 0$, and seek solutions to $f_1(\mathbf{x};\lambda) = 0$ (where f_1 is the first component of $\mathbf{f}$). Truncating the expansion at order $(\lambda \mp 1)^p$, show that:

   - for $p = 0$, $f_1(\mathbf{x};\lambda) = 0$ has no solutions on $\lambda \in (-1,+1)$;
   - for $p = 1, 2$, $f_1(\mathbf{x};\lambda) = 0$ has solutions $\lambda \approx \pm 0.83$;
   - for $p = 3, 4$, $f_1(\mathbf{x};\lambda) = 0$ has solutions $\lambda \approx \pm 0.56$;
   - for $p = 5, 6$, $f_1(\mathbf{x};\lambda) = 0$ has solutions $\lambda \approx \pm 0.69$; ...

   Show that this expansion is consistent with (and indeed converges on as p increases) the vector field given in *Exercises for Chapter 3: Ex 4*.

3. In Section 5.2.1, take the vector field $\mathbf{f}(\lambda) = \left(1 - 2\cos\frac{\pi\lambda}{2}, -1 + \sin\frac{\pi\lambda}{2}\right)$ from *Exercises for Chapter 3: Ex 4*, and carry out the expansion as a series of signs. First note that $\mathbf{f}^{\pm} = \left(1 - 2\cos\frac{\pi}{2}, -1 \pm \sin\frac{\pi}{2}\right)$, and

$$\mathbf{a}_k = \begin{cases} (-1,-1) & k = 0 \\ \left(0,\ (-1)^{(k-1)/2}\frac{\pi^k}{2^k k!}\right) & k > 0 \text{ odd} \\ \left((-1)^{k/2}\frac{\pi^k}{2^{k-1}k!},\ 0\right) & k > 0 \text{ even} \end{cases}$$

   Derive the linear combination $\frac{1}{2}(1+\lambda)\mathbf{f}^+ + \frac{1}{2}(1-\lambda)\mathbf{f}^-$ and the hidden term $\mathbf{h}$.

Exercises for Chapter 6 (Shape)

The text in this chapter, brought together, says to take constituent vector fields $\mathbf{f}^+ = (1, 0, 0, 0, \dots)$ and $\mathbf{f}^- = (0, 1, 0, 0, \dots)$ in coordinates $\mathbf{x} = (y_1, y_2, y_3, y_4, \dots)$, with:

1. $\sigma = y_2 \pm \frac{1}{2}y_1^2$ for the fold;
2. $\sigma = y_3 + \frac{\alpha}{2}y_1^2 - \frac{\beta}{2}y_2^2 + y_1 y_2$ for the two-fold;
3. $\sigma = y_2 \pm (\frac{1}{3}y_1^3 + y_3 y_1)$ for the cusp;
4. $\sigma = y_2 \pm (\frac{1}{4}y_1^4 + \frac{1}{2}y_3 y_1^2 + y_4 y_1)$ for the swallowtail;
5. $\sigma = y_2 \pm (\frac{1}{3}y_1^3 \pm (y_3^2 + y_4)y_1))$ for the umbilics;
6. $\sigma = y_2 \pm (\frac{1}{3}y_1^3 + y_4 y_1 + y_2^2)$ for the fold-cusp;

For these systems find:

a. the sliding regions and their boundaries;
b. the attractivity of the sliding region(s);
c. the lower-order singularities (e.g. each swallowtail sits at the meeting point of cusps, which sits at the meeting point of folds, while the fold-cusp sits at the meeting point of a fold and cusp, so find these lower-order sets emanating from each singularity, and classify their 'visibility').

I highly recommend taking a look at Teixeira's paper [206] which deals with this restricted to three dimensions.

Exercises for Chapter 7 (Layer Analysis)

1. In Example 7.1, for the two systems given, find the sliding manifold $\mathcal{M}$ in the switching layer, and find its region of existence on the discontinuity submanifolds.
2. In Example 7.1 and Example 7.2, find the end points of $\mathcal{M}$ for the three systems and the matrix $\underline{\underline{B}}$ given by (7.10) for these systems. For the system in Example 7.2, show that the attracting and repelling branches of $\mathcal{M}$ meet at TP.
3. For the two-switch example from Example 7.1, find the eigenvalues and eigenvectors of the matrix $\underline{\underline{B}}$ given by (7.10). What do they tell you about the local dynamics?
4. In Example 7.3, solve the sliding conditions $\dot{\lambda}_1 = 0$ on $x_1 = 0$, $\dot{\lambda}_2 = 0$ on $x_2 = 0$, and $\dot{\lambda}_1 = \dot{\lambda}_1 = 0$ on $x_1 = x_2 = 0$, to verify the sliding modes stated in the example. Find the eigenvalues and eigenvectors of $\underline{\underline{B}}$, showing how they relate to the phase portraits depicted in Figure 7.3.
5. In Example 7.5, show that a Hopf bifurcation occurs at $\mu = 0$ [hint: find the eigenvalues of the matrix $\underline{\underline{B}}$ as defined by (7.10), and show that they pass through the imaginary axis at μ changes sign]. Let $x_3 = \mathsf{const} := \mu$ to obtain Figure 7.5.
6. In Example 7.5, simulate the system for $\mu \gtrless 0$, and simulate a third variable x_3 with dynamics $\dot{x}_3 = k(\lambda_1 - x_3)$ for a large constant value of k. Investigate the resulting behaviour in $x_3(t)$.
7. Taking system (7.20), by finding the different branches of the sliding manifold $\mathcal{M}$ and their stability, show that for $p = 1$ $\mathcal{M}$ undergoes a transcritical bifurcation, and for $p = 2$ $\mathcal{M}$ undergoes a pitchfork bifurcation, at $x_2 = 0$ [you may need to look up the definition of these bifurcations, as we do not consider them in this book]. Verify the symmetry conditions stated in the paragraph below (7.20).
8. Derive the sliding modes for the systems in Example 7.8.

9. In Example 7.14, carry out the rotation from (x_1, x_2) to (y_1, y_2) coordinates to verify the expression given for the vector field $(\dot{y}_1, \dot{y}_2)$.
10. In Example 7.15, by finding any codimension 1 and 2 sliding equilibria, show that for different values of the constants a_i, b_{ij}, c_i, there are cases where a codimension 2 focus or node becomes a codimension 1 node, or a codimension 2 saddle becomes a codimension 1 saddle. To do this, either find examples of constants for which these occur [easier], or find general ranges of the constants for which the two behaviours occur [harder].

Exercises for Chapter 8 (Linear Switching)

1. Using Section 8.2, show that the combination $\mathbf{f}(\mathbf{x}; \boldsymbol{\lambda})$ defined in (8.1) is convex.
2. For Filippov's problem given in (1.22) of Example 1.2, take the piecewise-smooth system with $\lambda = \text{sign}(x_2)$. Prove that the linear switching system (replacing the λ^3 term with λ) is convex, while the nonlinear system (including the λ^3 term) is non-convex. Show that a solution evolving in infinitesimal increments along the constituent vector fields (i.e. with x_2 arbitrarily close to but never exactly on zero) evolves according to the linear switching system, while a solution evolving exactly along $x_2 = 0$ evolves along the nonlinear system (you might want to look at Filippov's book [71] for his take on this, in the example on p79–80).
3. Show that Theorem 8.1 holds for the following three systems:

$$\begin{aligned}
(\dot{x}_1, \dot{x}_2) &= (-\lambda,\ -\lambda + (1+\lambda)x_2)\ , \\
(\dot{x}_1, \dot{x}_2, \dot{x}_3) &= (-\lambda,\ -\lambda + (1+\lambda)(x_2 + x_3),\ (1+\lambda)x_2)\ , \\
(\dot{x}_1, \dot{x}_2, \dot{x}_3) &= (-\lambda_1,\ -\lambda_2,\ (\lambda_1 + \lambda_2)(x_3 - 2) + (1 + \lambda_1\lambda_2)x_3)\ ,
\end{aligned}$$

where $\lambda = \text{sign}(x_1)$, as depicted in Figure 8.2. First show the existence of a sliding equilibrium and then show that the neighbouring fields are anti-colinear (i.e. have exactly opposing directions) at the equilibrium. The equilibria exist in sliding of codimension 1, 1, and 2, respectively, for the three systems, while the 'neighbouring fields' are non-sliding, and codimension 1 is sliding, respectively.
4. Show that (8.9) produces the 13 different singularities in Figures 8.3 to 8.5, finding either specific values [easier] or the complete ranges [harder] of the constants a, b, c, d, that produce each portrait. Find the sliding regions, their attractivity, and any equilibria or sliding equilibria that exist through the boundary equilibrium bifurcation as μ changes sign.
5. Similar to the previous exercise, show that (8.29) produces the 12 different singularities in Figure 8.18, finding either specific values [easier] or the complete ranges [harder] of the constants a, b, c, d, that produce each portrait. Find the sliding regions, their attractivity, and any sliding equilibria that exist through the sliding boundary bifurcation as μ changes sign.
6. Show that Theorem 8.2 holds for the system

$$\begin{pmatrix} \dot{x}_1 \\ \dot{x}_2 \\ \dot{x}_3 \\ \dot{x}_4 \end{pmatrix} = \begin{pmatrix} x_3(1+\lambda_1)(1+\lambda_2) + 4\lambda_2 - 2(1+\lambda_1\lambda_2) \\ x_4(1+\lambda_1)(1+\lambda_2) - 4\lambda_2 + 2(1+\lambda_1\lambda_2) \\ 2(\lambda_1 + \lambda_2) \\ 2(1+\lambda_1\lambda_2) \end{pmatrix}$$

where $\lambda_j = \text{sign}(x_j)$ for $j = 1, 2$, with a tangency between the $\mathbf{f}^{++}$ constituent vector field and the intersection $x_1 = x_2 = 0$ at $(x_1, x_2, x_3, x_4) = (0, 0, 0, 0)$.

7. Show that Theorem 8.2 holds for the system

$$\begin{pmatrix} \dot{x}_1 \\ \dot{x}_2 \\ \dot{x}_3 \\ \dot{x}_4 \end{pmatrix} = \begin{pmatrix} 2(3+\lambda_1+(\lambda_1+\lambda_2)(x_3-3)) \\ -2(5+2\lambda_1-5\lambda_2) \\ 2(\lambda_1-1)+(1+\lambda_1)(1+\lambda_2+2x_4) \\ 2(\lambda_1+\lambda_2) \end{pmatrix}$$

where $\lambda_j = \mathrm{sign}(x_j)$ for $j = 1, 2, 3$, with a tangency between the intersection $x_1 = x_2 = x_3 = 0$ and the $\mathbf{f}^{\$+}$ sliding vector field on $\{x_1 = 0 < x_2\} \subset \mathcal{D}_1$, at $(x_1, x_2, x_3, x_4) = (0, 0, 0, 0)$.

Exercises for Chapter 9 (Nonlinear Switching)

1. In Section 9.3.1, we describe a saddle-node bifurcation that takes place within codimension 2 sliding, inside the intersection of $x_1 = 0$ and $x_2 = 0$. Find any equilibria of codimension one sliding on the discontinuity submanifolds $x_1 = 0$ or $x_2 = 0$. Boundary equilibrium bifurcations can occur where any of the equilibria (in codimension 1 or 2 sliding) leave their switching layers, moving onto/off the intersection or moving onto/off the discontinuity surface. Find all boundary equilibrium bifurcations that occur in this system.
2. In Section 9.6.1, find the sliding equilibrium of (9.43) without using switching layer methods. [Hint: there are two ways to do this since the system is linear in λ; either find the sliding dynamics by looking for solutions that evolve along the discontinuity surface (so $\dot{x}_1 = 0$) and seek its equilibria or simply find where the constituent vector fields are anti-colinear.]
3. In Section 9.6.2, similarly to the previous exercise, find the sliding equilibrium of (9.49) without using switching layer methods.
4. In Section 9.6.1 and Section 9.6.2, investigate alternative perturbations to study whether they would break the degeneracy. You may consider perturbing the constituent vector fields or just the hidden term, perturbing the x_2 components rather than the x_1 component, perturbing with terms proportional to higher orders than λ^2.

Exercises for Chapter 10 (Breaking Determinacy)

1. In Section 10.2.2, derive the layer and sliding dynamics for the systems (10.1a), showing how they are consistent with the phase portraits in Figure 10.1.
2. In Section 10.3.1, show that the sliding manifold given by (10.12) is attracting in $x_1 < 0$ and repelling in $x_1 > 0$. Without the x_3 term, the system (10.8) would be degenerate, with the connection between the attracting and repelling branches of $\mathcal{M}$ at $x_3 = 0$ extending along the intersection. Omit the x_3 term from (10.8), and propose a hidden term that would replace it, i.e. break the degeneracy for $x_3 = 0$ while leaving a connection at $x_3 = 0$. Try to find two alternatives: a hidden term that perturbs the system only on $x_2 = 0$ and a hidden term that perturbs the system only on $x_1 = x_2 = 0$.
3. This is more of an open-ended problem. The simulations in Figures 10.15 and 10.16 all show orbits that seem to enter and exit the intersection at approximately the same distance from the origin. This happens because there is a symmetry of the smoothed systems about $x_3 = 0$. Investigate alternative smoothings, or alternative

simulation methods (e.g. changing the integration time step or introducing noise), to show that such symmetric orbits are not typical.

4. In Section 10.3.4 for the system (10.30), verify the attractivity of the two branches as claimed in the text, and find an expression for orbits of the sliding dynamics or otherwise, and show that a canard exists when $c = 0$.
5. Section 10.4 introduces the notion of a 'stranger attractor'. By showing that any two points can be connected by infinitely many different trajectories, show that there exist stranger attractors of the types shown in Figure 10.18 in the systems

$$(\dot{x}_1, \dot{x}_2) = (\ -1 + \alpha x_1 + \lambda_2 x_2,\ -\lambda_2 x_1\ )\ ,$$
$$(\dot{x}_1, \dot{x}_2) = (\ 1 - \lambda_2 x_2 + \alpha(\lambda_1 - 1),\ \lambda_1\lambda_2\ )\ ,$$

for small enough $\alpha > 0$, where $\lambda_1 = \text{sign}(x_1)$ and $\lambda_2 = \text{sign}(x_2)$.

Exercises for Chapter 11 (Global Phenomena)

1. Concerning the global bifurcations depicted in Figure 11.1, do either of the following:

   - [harder] by identifying the essential singularities and separatrices needed to produce these bifurcations, propose equations for systems that exhibit them.
   - [easier] by deriving the equilibria and separatrices, prove that these bifurcations occur in the systems

   $$\begin{pmatrix} \dot{x}_1 \\ \dot{x}_2 \end{pmatrix} = (1+\lambda)\begin{pmatrix} x_1(1-x_1^2-x_2^2)+x_2 \\ x_2(1-x_1^2-x_2^2)-x_1 \end{pmatrix} + (1-\lambda)\begin{pmatrix} 1 \\ 0 \end{pmatrix}$$

   and

   $$\begin{pmatrix} \dot{x}_1 \\ \dot{x}_2 \end{pmatrix} = (1+\lambda)\begin{pmatrix} r+x_2-x_1^2 \\ x_2-x_1 \end{pmatrix} + (1-\lambda)\begin{pmatrix} 1 \\ 0 \end{pmatrix}$$

   at $r \approx 1.097$.
   - [easiest] simulate the equations above to verify that their phase portraits resemble those in Figure 11.1.

   The 'harder' problem above is the most instructive—identifying the essential characteristics of the phenomenon and testing your intuition by constructing an explicit example. To this end you could also try finding examples of systems exhibiting the bifurcations depicted in Figures 11.2 to 11.8.
2. In Example 11.1, prove that the system has a two-fold singularity. Find the sliding regions and the sliding dynamics on them, investigating the phase portrait particularly near the two-fold. We will see how to classify these in terms of two key quantities $\nu^{\pm}$ in Chapter 13.

Exercises for Chapter 12 (Asymptotics)

1. In Section 12.1, consider a vector field

$$\mathbf{f}(\mathbf{x};\boldsymbol{\lambda}) = \mathbf{a}(\mathbf{x}) + \mathbf{b}(\mathbf{x})\lambda_1 + \mathbf{c}(\mathbf{x})\lambda_2 + \mathbf{d}(\mathbf{x})\lambda_1\lambda_2\ .$$

Put this is the form of (12.1) and (12.2), expressing the constituent vector fields $\mathbf{f}^{\pm\pm}$ in terms of the fields $\mathbf{a}, \mathbf{b}, \mathbf{c}, \mathbf{d}$ and the coefficients $u_{\pm\pm}$ and $\mu_j^{\pm_j}$ in terms of λ_1 and λ_2.
Repeat this exercise with a hidden term $\mathbf{e}(\mathbf{x})(\lambda_1^2 - 1)$ added to the expression above.

2. In Section 12.3.2, follow through the calculations in Example 12.1 for an analytic sigmoid function $\Lambda_r(\sigma/\varepsilon)$, for example,

$$\Lambda_r(\sigma/\varepsilon) = \tanh(\sigma/\varepsilon) - r\,\mathrm{sech}^2(\sigma/\varepsilon)\ .$$

You might like to try instead constructing other functions using the sigmoids from Table 1.1. What you are showing is in principle that the results of this section extend more widely than functions satisfying (12.21). An analytic function necessarily cannot satisfy (12.21) exactly, because analyticity prevents a function being constant for some values ($|\sigma| > \varepsilon$) and varying for others ($|\sigma| \le \varepsilon$). But an analytic function can lie close to (12.21) to order ε or even $e^{-1/\varepsilon}$, and analytic functions are easier to use for smoothing; indeed we use them in many of our simulations in this book.

3. Derive the stability of the sliding focus in Example 12.3, and show that it reverses (by undergoing a Hopf bifurcation) at a critical value of r.
4. Follow through Section 12.3.3 for a single switch (so $j = 1$ only), letting $\Lambda(\sigma_1/E_1)$ be the smoothing function you used in exercise 2 above (with E_1 replacing ε). Find the functions γ and ε_1, and the fast system of the layer dynamics given by (12.44).
5. Carry out the series expansions described in Section 12.3.8 for the functions

   a. $\mathbf{f}(\mathbf{x}; \lambda) = \sin(a\lambda)$,
   b. $\mathbf{f}(\mathbf{x}; \lambda) = e^{a\lambda}$,

   deriving their constitutent vector fields and hidden terms.

Exercises for Chapter 13 (Two-Fold)

1. Two-fold singularities occur numerous times throughout the book, sometimes explicitly as part of classifications and sometimes less obviously but playing a nonetheless important role in systems exhibiting other phenomena.
   Identify and classify the two-fold singularities in the following systems:

   a. in Figure 2.11 if we replace the constant c with a third variable $\dot{x}_3 = 1$;
   b. in Section 9.4.3 each time the fold curves $x_2 = -\sin\left[(1 \pm \frac{1}{2})\pi t\right]$ on the discontinuity surface intersect;
   c. in the sliding dynamics in system (10.18), described by (10.24)—these are not strictly two-folds according to our definition, but extend the notion of two-folds to higher codimension sliding;
   d. the last three cases in Table 11.2, basing your classification solely on inspection of the phase portraits;
   e. Example 11.1 of Section 11.3, in the system defined by (11.1);
   f. the ship problem depicted in Figure 14.2;
   g. the climate problem depicted in Figure 14.7 if we replace the constant ρ_k with a third variable $\dot{x}_3 = 1$;
   h. the slow-fast problems in Section 14.9 obtained after the three pinches given in Section 14.9.3, Section 14.9.4, and Section 14.9.5.

Exercises for Chapter 14 (Applications)

We will not set explicit examples for this chapter, but let the reader explore putting the various concepts in this book to use. While all of these problems are inspired by (or directly review) the existing literature, there are numerous open avenues that could be investigated, both in theoretical study and in improvement of the models, not only as an exercise but for meaningful research. In particular one may choose to investigate perturbations of the models shown or extensions of them using hidden terms.

References

1. D. S. Abbot, M. Silber, and R. T. Pierrehumbert. Bifurcations leading to summer arctic sea ice loss. *J. Geophys. Res.*, 116(D19):120(1–11), 116 (2011), p. D19120.
2. D. S. Abbot, A. Voigt, and D. Koll. The Jormungand global climate state and implications for Neoproterozoic glaciations. *J. Geophys. Res.*, 116(D18):103(1–14), 2011.
3. M. Abramowitz and I. Stegun. *Handbook of Mathematical Functions.* Dover, 1964.
4. M. A. Aizerman and F. R. Gantmakher. On the stability of equilibrium positions in discontinuous systems. *Prikl. Mat. i Mekh.*, 24:283–93, 1960.
5. M. A. Aizerman and E. S. Pyatnitskii. Fundamentals of the theory of discontinuous systems I,II. *Automation and Remote Control*, 35:1066–79, 1242–92, 1974.
6. J. C. Alexander and T. I. Seidman. Sliding modes in intersecting switching surfaces, i: Blending. *Houston J. Math*, 24(3):545–569, 1998.
7. J. C. Alexander and T. I. Seidman. Sliding modes in intersecting switching surfaces, ii: Hysteresis. *Houston J. Math*, 25(1):185–211, 1999.
8. G. Amontons. De la resistance causée dans les machines, tant par les frottemens des parties qui les composent, que par roideur des cordes qu'on y employe, et la maniere de calculer l'un et l'autre. *Histoire de l'Académie royale des sciences*, 1699.
9. A. A. Andronov and S. E. Khaikin. *Theory of Oscillations, Part I.* Moscow: ONTI NKTN (in Russian), 1937.
10. A. A. Andronov, A. A. Vitt, and S. E. Khaikin. *Theory of oscillations.* Moscow: Fizmatgiz (in Russian), 1959.
11. Aristotle. *The Physics.* 1837 Bekker edition of Aristotle's Works, 3rd century BC.
12. Robin Waterfield (Translator)] Aristotle [David Bostock (Editor). *Physics.* Oxford World's Classics, 2008.
13. V. I. Arnold, V. V. Goryunov, O. V. Lyashko, and V. A. Vasiliev. *Dynamical Systems VI: Singularity Theory I. Local and global theory*, volume 6 of *Encyclopedia of Mathematical Sciences.* Springer Verlag, 1993.
14. V. I. Arnold, V. V. Goryunov, O. V. Lyashko, and V. A. Vasiliev. *Dynamical Systems VIII: Singularity Theory I. Classification and Applications*, volume 39 of *Encyclopedia of Mathematical Sciences.* Springer Verlag, 1993.

M. R. Jeffrey, *Hidden Dynamics*,
https://doi.org/10.1007/978-3-030-02107-8

15. G. Bachar, E. Segev, O. Shtempluck, E. Buks, and S. W. Shaw. Noise induced intermittency in a superconducting microwave resonator. *EPL*, 89(1):17003, 2010.
16. J. B. Barbour. *The Discovery of Dynamics.* Oxford University Press, 2001.
17. A. M. Barry, R. McGehee, and E. Widiasih. A Filippov framework for a conceptual climate model. *arXiv:1406.6028*, 2014.
18. J. Bastien, G. Michon, L. Manin, and R. Dufour. An analysis of the modified Dahl and Masing models: Application to a belt tensioner. *Journal of Sound and Vibration*, 302(4–5):841–864, 2007.
19. C. M. Bender and S. A. Orszag. *Advanced mathematical methods for scientists and engineers I. Asymptotic methods and perturbation theory.* Springer-Verlag, New York, 1999.
20. E. Benoit, J. L. Callot, F. Diener, and M. Diener. Chasse au canard. *Collect. Math.*, 31–32:37–119, 1981.
21. M. V. Berry, M. R. Jeffrey, and J. G. Lunney. Conical diffraction: observations and theory. *Proc. R. Soc. A*, 462:1629–1642, 2006.
22. C. Bonet-Revés and T. M. Seara. Regularization of sliding global bifurcations derived from the local fold singularity of Filippov systems. *arXiv:1402.5237*, 2014.
23. E. Bossolini, R. Edwards, P. Glendinning, M. R. Jeffrey, and S. Webber. Regularization by external variables. *Trends in Mathematics: Research Perspectives CRM Barcelona (Birkhauser)*, 8:19–24, 2017.
24. F P Bowden and D Tabor. *The friction and Lubrication of Solids.* Oxford University Press, 1950.
25. B. Brogliato. *Nonsmooth mechanics – models, dynamics and control.* Springer-Verlag (New York), 1999.
26. M. Brons, M. Krupa, and M. Wechselberger. Mixed mode oscillations due to the generalized canard phenomenon. *Fields Institute Communications*, 49:39–63, 2006.
27. M. E. Broucke, C. Pugh, and S. Simic. Structural stability of piecewise smooth systems. *Computational and Applied Mathematics*, 20(1–2):51–90, 2001.
28. J. A. bu Qahouq, H. Mao, H. J. Al-Atrash, and I. Batarseh. Maximum efficiency point tracking (mept) method and digital dead time control implementation. *IEEE Transactions on Power Electronics*, 21(5):1273–1281, 2006.
29. C. A. Buzzi, T. de Carvalho, and M. A. Teixeira. Birth of limit cycles bifurcating from a nonsmooth center. *J. Math. Pure Appl.*, 102:36–47, 2014.
30. J. Carr. *Applications of center manifold theory.* Springer-Verlag, 1981.
31. A. R. Champneys and P. L. Varkonyi. The Painleve paradox in contact mechanics. *IMA Journal of Applied Mathematics*, 81(3):538–88, 2016.
32. A. Colin de Verdière. A simple sodel of millennial oscillations of the thermohaline circulation. *J. Phys. Oceanogr.*, 37:1142–55, 2007.
33. A. Colin de Verdière. The instability of the thermohaline circulation in a low-order model. *J. Phys. Oceanogr.*, 40:757–73, 2010.
34. P. Collett and J-P. Eckmann. *Iterated Maps on the Interval as Dynamical Systems.* Boston: Birkhauser, 1980.
35. A. Colombo, M. di Bernardo, E. Fossas, and M. R. Jeffrey. Teixeira singularities in 3D switched feedback control systems. *Systems & Control Letters*, 59(10):615–622, 2010.
36. A. Colombo, M. di Bernardo, S. J. Hogan, and M. R. Jeffrey. Bifurcations of piecewise smooth flows: perspectives, methodologies and open problems. *Physica D: Special Issue on Dynamics and Bifurcations of Nonsmooth Dynamical Systems*, 241(22):1845–1860, 2012.

37. A. Colombo and M. R. Jeffrey. Non-deterministic chaos, and the two-fold singularity in piecewise smooth flows. *SIAM J. App. Dyn. Sys.*, 10:423–451, 2011.
38. A. Colombo and M. R. Jeffrey. The two-fold singularity: leading order dynamics in n-dimensions. *Physica D*, 263:1–10, 2013.
39. G. Csern/'ak and G. Stépán. On the periodic response of a harmonically excited dry friction oscillator. *Journal of Sound and Vibration*, 295:649–658, 2006.
40. G. Cybenko. Approximation by superpositions of a sigmoidal function. *Math. Control Signals Systems*, 2:303–314, 1989.
41. Charles Darwin. *Origin of Species.* John Murray, London, 1859.
42. R. Descartes. *Traité du monde et de la lumière.* modern reprint: Abaris Books, 1979, 1677.
43. M. Desroches, J. Guckenheimer, B. Krauskopf, C. Kuehn, H. M. Osinga, and M. Wechselberger. Mixed-mode oscillations with multiple time scales. *SIAM Rev.*, 54(2):211–288, 2012.
44. M. Desroches and M. R. Jeffrey. Canards and curvature: nonsmooth approximation by pinching. *Nonlinearity*, 24:1655–1682, 2011.
45. M. Desroches and M. R. Jeffrey. Nonsmooth analogues of slow-fast dynamics: pinching at a folded node. *arXiv:1506.00831*, 2013.
46. M. Desroches, B. Krauskopf, and H. M. Osinga. Mixed-mode oscillations and slow manifolds in the self-coupled Fitzhugh-Nagumo system. *Chaos*, 18(1):015107, 2008.
47. M. Desroches, B. Krauskopf, and H. M. Osinga. Numerical continuation of canard orbits in slow-fast dynamical systems. *Nonlinearity*, 23(3):739–765, 2010.
48. M. di Bernardo, C. J. Budd, A. R. Champneys, and P. Kowalczyk. *Piecewise-Smooth Dynamical Systems: Theory and Applications.* Springer, 2008.
49. M. di Bernardo, P. Kowalczyk, and A. Nordmark. Bifurcations of dynamical systems with sliding: derivation of normal-form mappings. *Physica D*, 170:175–205, 2002.
50. M. di Bernardo, P. Kowalczyk, and A. Nordmark. Sliding bifurcations: a novel mechanism for the sudden onset of chaos in dry friction oscillators. *Int.J.Bif.Chaos*, 13:2935–2948, 2003.
51. L. Dieci. Sliding motion on the intersection of two surfaces: spirally attractive case. *Commun. Nonlin. Sci. Numer. Simul.*, 26, 65–74, 2015.
52. L. Dieci, C. Elia, and L. Lopez. A Filippov sliding vector field on an attracting co-dimension 2 discontinuity surface, and a limited loss-of-attractivity analysis. *J. Differential Equations*, 254:1800–1832, 2013.
53. L. Dieci and L. Lopez. Sliding motion on discontinuity surfaces of high codimension. a construction for selecting a Filippov vector field. *Numer. Math.*, 117:779–811, 2011.
54. R. B. Dingle. *Asymptotic Expansions: their derivation and interpretation.* Academic Press London, 1973.
55. D. Dowson. *History of Tribology.* Professional Engineer Pulbishing, London, 1998.
56. F. Dumortier, R. Roussarie, J. Sotomayor, and H. Zoladek. *Bifurcations of Planar Vector Fields: Nilpotent Singularities and Abelian Integrals.* Springer-Verlag, 1991.
57. R. Edwards, A. Machina, G. McGregor, and P. van den Driessche. A modelling framework for gene regulatory networks including transcription and translation. *Bull. Math. Biol.*, 77:953–983, 2015.
58. J. Eggers. Singularities in droplet pinching with vanishing viscocity. *SIAM J. Appl. Math.*, 60(6):1997–2008, 2000.
59. I. Eisenman. Factors controlling the bifurcation structure of sea ice retreat. *J. Geophys. Res.*, 117:D01111, 2012.

60. I. Eisenman and J. S. Wettlaufer. Nonlinear threshold behavior during the loss of Arctic sea ice. *Proc. Natl. Acad. Sci. U.S.A.*, 106:28–32, 2009.
61. A. Erdelyi. *Asymptotic Expansions.* Dover, 2003.
62. B. Feeny and F.C. Moon. Chaos in a forced dry-friction oscillator: Experiments and numerical modelling. *J. Sound Vib.*, 170(3):303–323, 1994.
63. M. I. Feigin. Doubling of the oscillation period with C-bifurcations in piecewise continuous systems. *Journal of Applied Mathematics and Mechanics (Prikladnaya Matematika i Mechanika)*, 34:861–869, 1970.
64. M. I. Feigin. On the generation of sets of subharmonic modes in a piecewise continuous systemq. *Journal of Applied Mathematics and Mechanics (Prikladnaya Matematika i Mechanika)*, 38:810–818, 1974.
65. M. I. Feigin. On the structure of C-bifurcation boundaries of piecewise continuous systems. *Journal of Applied Mathematics and Mechanics (Prikladnaya Matematika i Mechanika)*, 42:820–829, 1978.
66. M. I. Feigin. *Forced Oscillations in Systems with Discontinuous Nonlinearities.* Nauka, Moscow. (In Russian), 1994.
67. H. Feinberg. *Physicists: Epoch And Personalities*, volume 4 of *History of Modern Physical Sciences.* World Scientific, Singapore, 2011.
68. N. Fenichel. Geometric singular perturbation theory. *J. Differ. Equ.*, 31:53–98, 1979.
69. S. Fernández-Garcia, D. Angulo-Garcia, G. Olivar-Tost, M. di Bernardo, and M. R. Jeffrey. Structural stability of the two-fold singularity. *SIAM J. App. Dyn. Sys.*, 11(4):1215–1230, 2012.
70. A. F. Filippov. Differential equations with discontinuous right-hand side. *American Mathematical Society Translations, Series 2*, 42:19–231, 1964.
71. A. F. Filippov. *Differential Equations with Discontinuous Righthand Sides.* Kluwer Academic Publ. Dortrecht, 1988 (Russian 1985).
72. R. FitzHugh. Impulses and physiological states in theoretical models of nerve membranes. *Biophysics J.*, 1:445–466, 1961.
73. I. Flügge-Lotz. *Discontinuous Automatic Control.* Princeton University Press, 1953.
74. A. Kh. Gelig, G. A. Leonov, and V. A. Iakubovich. Stability of nonlinear systems with nonunique equilibrium position. *Moscow, Izdatel'stvo Nauka (in Russian)*, page 400, 1978.
75. P. Glendinning. Robust new routes to chaos in differential equations. *Physics Letters A*, 168:40–46, 1992.
76. P. Glendinning. The anharmonic route to chaos: kneading theory. *Nonlinearity*, 6:349–367, 1993.
77. P. Glendinning. The border collision normal form with stochastic switching surface. *SIADS*, 13(1):181–93, 2014.
78. P. Glendinning. Bifurcation from stable fixed point to n-dimensional attractor in the border collision normal form. *Nonlinearity*, 28(10):3457–3464, 2015.
79. P. Glendinning. Bifurcation from stable fixed point to two-dimensional attractor in the border collision normal form. *IMA Journal of Applied Mathematics (accepted)*, 81(4):699–710, 2016.
80. P. Glendinning. Classification of boundary equilibrium bifurcations of planar Filippov systems. *Chaos*, 26(013108):1–12, 2016.
81. P. Glendinning. Geometry of refractions and reflections through a biperiodic medium. *SIAM J. Appl. Math.*, 76(4):1219–38, 2016.
82. P. Glendinning, S. J. Hogan, M. E. Homer, M. R. Jeffrey, and R. Szalai. Uncountably many cases of Filippov's sewed focus. *submitted*, 2016.
83. P. Glendinning, S. J. Hogan, M. E. Homer, M. R. Jeffrey, and R. Szalai. Thirty-nine steps: Classication of Filippov's type 3 singularities in piecewise smooth systems. *in preparation*, 2018.

84. P. Glendinning and M. R. Jeffrey. Grazing-sliding bifurcations, the border collision normal form, and the curse of dimensionality for nonsmooth bifurcation theory. *Nonlinearity*, 28:263–283, 2015.
85. P. Glendinning and P. Kowalczyk. Micro-chaotic dynamics due to digital sampling in hybrid systems of Filippov type. *Physica D*, 239(1–2):58–71, 2010.
86. P. Glendinning, P. Kowalczyk, and A. Nordmark. Multiple attractors in grazing-sliding bifurcations in an explicit example of Filippov type. *IMA Journal of Applied Mathematics*, 81(4):711–22, 2016.
87. A. N. Gorban. Approximation of continuous functions of several variables by an arbitrary nonlinear continuous function of one variable, linear functions, and their superpositions. *Appl. Math. Lett.*, 11(3):45–49, 1998.
88. I. S. Gradshtein and I. M. Ryzhik. *Tables of integrals, series and products.* Boston, Mass.: Academic Press, 1994.
89. M. Guardia, S. J. Hogan, and T. M. Seara. Sliding bifurcations of periodic orbits in the dry friction oscillator. *SIADS*, 9:769–98, 2010.
90. M. Guardia, T. M. Seara, and M. A. Teixeira. Generic bifurcations of low codimension of planar Filippov systems. *J. Differ. Equ.*, pages 1967–2023, 2011.
91. J. Guckenheimer, K. Hoffman, and W. Weckesser. The forced van der Pol equation I: the slow flow and its bifurcations. *SIAM J. App. Dyn. Sys.*, 2:1–35, 2002.
92. J. Guckenheimer and P. Holmes. *Nonlinear oscillations, dynamical systems, and bifurcations of vector fields.* Applied Mathematical Sciences 42. Springer, 2002.
93. N. Guglielmi and E. Hairer. Classification of hidden dynamics in discontinuous dynamical systems. *SIADS*, 14(3):1454–1477, 2015.
94. A. Guran, F. Pfeiffer, and K. Popp, editors. *Dynamics with Friction: Modeling, Analysis and Experiment I & II*, volume 7 of *Series B*. World Scientific, 1996.
95. O. Hájek. Discontinuous differential equations, I. *J. Differential Equations*, 32(2):149–170, 1979.
96. O. Hájek. Discontinuous differential equations, II. *J. Differential Equations*, 32(2):171–185, 1979.
97. W. R. Hamilton. Third supplement to an essay on the theory of systems of rays. *Transactions of the Royal Irish Academy, presented in 1832*, 17:1–144, 1837.
98. P. Hartman. *Ordinary differential equations.* Wiley: New York, 1964.
99. J. Heading. *An introduction to phase-integral methods.* Methuen, 1962.
100. D. Hilbert. Mathematische probleme. *Göttinger Nachrichten*, pages 253–297, 1900.
101. D. Hilbert. Mathematische Probleme. *Archiv der Mathematik und Physik*, 3(1):44–6, 213–237, 1901.
102. D. Hilbert. Mathematical problems. *Bulletin of the American Mathematical Society*, 8(10):437–479, 1902.
103. A. V. Hill. The possible effects of the aggregation of the molecules of haemoglobin on its dissociation curves. *Proc. Physiol. Soc.*, 40:iv–vii, 1910.
104. E. J. Hinch. *Perturbation Methods.* Cambridge University Press, 1991.
105. N. Hinrichs, M. Oestreich, and K. Popp. On the modelling of friction oscillators. *J. Sound Vib.*, 216(3):435–459, 1998.
106. A.L. Hodgkin and A.F. Huxley. A quantitative description of membrane current and its application to conduction and excitation in nerve. *J Physiol.*, 117(4):500–544, 1952.
107. S. J. Hogan, L. Higham, and T. C. L. Griffin. Dynamics of a piecewise linear map with a gap. *Proc. R. Soc. A*, 463:49–65, 2007.

108. S. J. Hogan, M. E. Homer, M. R. Jeffrey, and R. Szalai. Piecewise smooth dynamical systems theory: the case of the missing boundary equilibrium bifurcations. *Journal of Nonlinear Science*, 26(5):1161–73, 2016.
109. A. M. Hogg. Glacial cycles and carbon dioxide: A conceptual model. *Geophys. Res. Lett.*, 35:L01701, 2008.
110. I. M. Hutchings. Leonardo da vinci's studies of friction. *Wear*, 360–361:51–66, 2016.
111. M. R. Jeffrey. Non-determinism in the limit of nonsmooth dynamics. *Physical Review Letters*, 106(25):254103, 2011.
112. M. R. Jeffrey. Three discontinuity-induced bifurcations to destroy self-sustained oscillations in a superconducting resonator. *Physica D: Special Issue*, 241(22):2077–2082, 2011.
113. M. R. Jeffrey. Dynamics at a switching intersection: hierarchy, isonomy, and multiple-sliding. *SIADS*, 13(3):1082–1105, 2014.
114. M. R. Jeffrey. Hidden dynamics in models of discontinuity and switching. *Physica D*, 273–274:34–45, 2014.
115. M. R. Jeffrey. Exit from sliding in piecewise-smooth flows: deterministic vs. determinacy-breaking. *Chaos*, 26(3):033108:1–20, 2016.
116. M. R. Jeffrey. Hidden degeneracies in piecewise smooth dynamical systems. *Int. J. Bif. Chaos*, 26(5):1650087(1–18), 2016.
117. M. R. Jeffrey. The ghosts of departed quantities in switches and transitions. *SIAM Review*, 60(1):116–36, 2017.
118. M. R. Jeffrey, A. R. Champneys, M. di Bernardo, and S. W. Shaw. Catastrophic sliding bifurcations and onset of oscillations in a superconducting resonator. *Phys. Rev. E*, 81(1):016213–22, 2010.
119. M. R. Jeffrey and A. Colombo. The two-fold singularity of discontinuous vector fields. *SIAM Journal on Applied Dynamical Systems*, 8(2):624–640, 2009.
120. M. R. Jeffrey and S. J. Hogan. The geometry of generic sliding bifurcations. *SIAM Review*, 53(3):505–525, 2011.
121. M. R. Jeffrey, G. Kafanas, and D. J. W. Simpson. Jitter in dynamical systems with intersecting discontinuity surfaces. *IJBC*, 28(6):1–22, 2018.
122. M. R. Jeffrey and D. J. W. Simpson. Non-Filippov dynamics arising from the smoothing of nonsmooth systems, and its robustness to noise. *Nonlinear Dynamics*, 76(2):1395–1410, 2014.
123. Mike R. Jeffrey. An update on that singularity. *Trends in Mathematics: Research Perspectives CRM Barcelona (Birkhauser)*, 8:107–122, 2017.
124. C. K. R. T. Jones. *Geometric singular perturbation theory*, volume 1609 of *Lecture Notes in Math. pp. 44–120.* Springer-Verlag (New York), 1995.
125. G. Kafanas. Sensor effects in sliding mode control of power conversion cells. *Trends in Mathematics: Research Perspectives CRM Barcelona (Birkhauser)*, 8:113–117, 2017.
126. A. Kelley. The stable, center stable, center, center unstable and unstable manifolds. *J. Diff. Eqns.*, 3:546–570, 1967.
127. M. A. Kiseleva and N. V. Kuznetsov. Coincidence of the Gelig-Leonov-Yakubovich, Filippov, and Aizerman-Pyatnitskiy definitions. *Vestnik St. Petersburg University: Mathematics*, 48(2):66–71, 2015.
128. Klymkowsky and Cooper. Biofundamentals, 2016.
129. R. L. Kosut, B. D. O. Anderson, and I. Mareels. Stability theory for adaptive systems: methods of averaging and persistency of excitation. *Proc. of the 24th Conf. on Decision and Control*, 24:278–483, 1985.
130. P. Kowalczyk and M. di Bernardo. Two parameter degenerate sliding bifurcations in Filippov systems. *Physica D*, 204:204–229, 2005.

131. P. Kowalczyk and P.T. Piiroinen. Two-parameter sliding bifurcations of periodic solutions in a dry-friction oscillator. *Physica D: Nonlinear Phenomena*, 237(8):1053–1073, 2008.
132. J. Krim. Friction at macroscopic and microscopic length scales. *Am. J. Phys.*, 70:890–897, 2002.
133. K. U. Kristiansen and S. J. Hogan. On the use of blowup to study regularization of singularities of piecewise smooth dynamical systems in R3. *SIADS*, 14(1):382–422, 2015.
134. M. Krupa and P. Szmolyan. Relaxation oscillation and canard explosion. *J. Differ. Equ.*, 174:312–368, 2001.
135. C. Kuehn. *Multiple time scale dynamics*. Springer, 2015.
136. V. Kulebakin. On theory of vibration controller for electric machines. *Theor. Exp. Electon (in Russian)*, 4, 1932.
137. M. Kunze. *Non-Smooth Dynamical Systems*. Springer, 2000.
138. Y. A. Kuznetsov. *Elements of Applied Bifurcation Theory*. Springer, 3rd Ed., 2004.
139. Y. A. Kuznetsov. Practical computation of normal forms on center manifolds at degenerate Bogdanov-Takens bifurcations. *Int. J. Bif. Chaos*, 15:3535–46, 2005.
140. Yu. A. Kuznetsov, S. Rinaldi, and A. Gragnani. One-parameter bifurcations in planar Filippov systems. *Int. J. Bif. Chaos*, 13:2157–2188, 2003.
141. J. Larmor. *Sir George Gabriel Stokes: Memoir and Scientific Correspondence*, volume 1. Cambridge University Press, 1907.
142. A. Le Bot and E. Bou Chakra. Measurement of friction noise versus contact area of rough surfaces weakly loaded. *Tribology Letters*, 37:273–281, 2010.
143. G. W. (Translated Leibniz, edited by P., and J.) Bennett. *New Essays on Human Understanding c.1704*, volume IV:16. New York: Cambridge University Press, 1981.
144. J. Leifeld. Persistence of saddle behavior in the nonsmooth limit of smooth dynamical systems. *arxiv*, pages 1–16, 2015.
145. J. Leifeld. Nonsmooth homoclinic bifurcation in a conceptual climate model. *arxiv*, pages 1–14, 2016.
146. J. Leifeld. Perturbation of a nonsmooth supercritical Hopf bifurcation. *arxiv*, pages 1–12, 2016.
147. G. Lewicki and G. Marino. Approximation of functions of finite variation by superpositions of a sigmoidal function. *Applied Mathematics Letters*, 17:1147–1152, 2004.
148. A. Liénard. Etude des oscillations entretenues. *Revue générale de l'électricité*, 23:901–912 and 946–954, 1928.
149. E. Lindelöf. Sur l'application de la méthode des approximations successives aux équations différentielles ordinaires du premier ordre. *C. R. Hebd. Seances Acad. Sci*, 114:454–457, 1894.
150. C. Linnaeus. *Philosophia Botanica*, chapter III, 77. Typis Joannis Thomae de Trattner, 1st edition, 1751.
151. J. Llibre, P. R. da Silva, and M. A. Teixeira. Sliding vector fields via slow-fast systems. *Bull. Belg. Math. Soc. Simon Stevin*, 15(5):851–869, 2008.
152. J. Llibre, D. N. Novaes, and M. A. Teixeira. Maximum number of limit cycles for certain piecewise linear dynamical systems. *Nonlinear Dynamics*, 82(3):1159–1175, 2015.
153. H Lloyd. Further experiments on the phaenomena presented by light in its passage along the axes of biaxial crystals. *Lond Ed Phil Mag*, 2:207–210, 1833.
154. H Lloyd. On the phaenomena presented by light in its passage along the axes of biaxial crystals. *Lond Ed Phil Mag*, 2:112–120, 1833.

155. H Lloyd. On the phenomena presented by light in its passage along the axes of biaxial crystals. *Trans Roy Irish Acad*, 17:145–158, 1837.
156. R. Lozi. Un attracteur étrange (?) du type attracteur de Hénon. *Journal de Physique*, 39:C5–9, 1978.
157. A.J.S. Machado, S.T. Renosto, C.A.M. dos Santos, L.M.S. Alves, and Z. Fisk. *Superconductors - Materials, Properties and Applications*, chapter Defect Structure Versus Superconductivity in MeB2 Compounds (Me = Refractory Metals) and One-Dimensional Superconductors. InTech, 2012.
158. A. Machina, R. Edwards, and P. van den Dreissche. Sensitive dependence on initial conditions in gene networks. *Chaos*, 23(025101):1–9, 2013.
159. A. Machina, R. Edwards, and P. van den Dreissche. Singular dynamics in gene network models. *SIADS*, 12(1):95–125, 2013.
160. W. Malfliet. The tanh method: a tool for solving certain classes of nonlinear evolution and wave equations. *J. Comput. and Appl. Math.*, 164–165:529–541, 2004.
161. J. E. Marsden and M. McCracken. *The Hopf bifurcation and its applications.* Springer-Verlag, 1976.
162. J. D. Meiss. *Differential Dynamical Systems.* SIAM, 2007.
163. J. Nagumo, S. Arimoto, and S. Yoshizawa. An active pulse transmission line simulating nerve axon. *Proc. IRE*, 1962:2061–70, 50.
164. NASA. Voyager recent history, http://voyager.gsfc.nasa.gov/heliopause/recenthist.html, last viewed Aug 2016, 2012.
165. J. Nash. The imbedding problem for Riemannian manifolds. *Annals of Mathematics*, 63(1):20–63, 1956.
166. Yu. I. Neimark and S. D. Kinyapin. On the equilibrium state on a surface of discontinuity. *Izv. VUZ. Radiofizika*, 3:694–705, 1960.
167. I. Newton. *PhilosophiæNaturalis Principia Mathematica.* 1729.
168. G. Nikolsky. On automatic stability of a ship on a given course. *Proc Central Commun Lab (in Russian)*, 1:34–75, 1934.
169. D. N. Novaes and M. R. Jeffrey. Regularization of hidden dynamics in piecewise smooth flow. *J. Differ. Equ.*, 259:4615–4633, 2015.
170. D. N. Novaes and E. Ponce. A simple solution to the braga–mello conjecture. *IJBC*, 25(1):1550009:1–7, 2015.
171. J. Nowacki, J. T. Osinga, H. O. Brown, and K Randall, A. D. Tsaneva-Atanasova. Sensitivity analysis to explain the excitability in a pyramidal neuron with application to alzheimer's disease. *BMC Neuroscience*, 12(Suppl 1):P342, 2011.
172. H. E. Nusse, E. Ott, and J. A. Yorke. Border-collision bifurcations: An explanation for observed bifurcation phenomena. *Phys. Rev. E*, 49(2):1073–1076, 1994.
173. D. Paillard and F. Parrenin. The antarctic ice sheet and the triggering of deglaciations. *Earth and Planetary Science Letters*, 227:263–271, 2004.
174. B. N. J. Persson. *Sliding Friction: Physical Principles and Applications.* Springer, 1998.
175. E. Plahte and S Kjøglum. Analysis and generic properties of gene regulatory networks with graded response functions. *Physica D*, 201:150–176, 2005.
176. T. Poston and I. N. Stewart. *Catastrophe theory and its applications.* Dover, 1996.
177. C. Pugh and M. M. Peixoto. Structural stability. *Scholarpedia*, 3(9):4008, 2008.
178. T. Putelat, J. H. P. Dawes, and J. R. Willis. On the microphysical foundations of rate-and-state friction. *Journal of the Mechanics and Physics of Solids*, 59(5):1062–1075, 2011.

179. M. Radiguet, D. S. Kammer, P. Gillet, and J.-F. Molinari. Survival of heterogeneous stress distributions created by precursory slip at frictional interfaces. *PRL*, 111(164302):1–4, 2013.
180. E. Segev, B. Abdo, O. Shtempluck, and E. Buks. Extreme nonlinear phenomena in NbN superconducting stripline resonators. *Physics Letters A*, 366(1–2):160–164, 2007.
181. E. Segev, B. Abdo, O. Shtempluck, and E. Buks. Novel self-sustained modulation in superconducting stripline resonators. *EPL*, 78(5):57002 (5pp), 2007.
182. E. Segev, B. Abdo, O. Shtempluck, and E. Buks. Thermal instability and self-sustained modulation in superconducting NbN stripline resonators. *J. Phys: Condens. Matter*, 19(9):096206(14pp), 2007.
183. T. I. Seidman. Existence of generalized solutions for ordinary differential equations in Banach spaces. *Int. J. Evolution Equations*, 1:107–119, 2005.
184. S. W. Shaw. On the dynamics response of a system with dry friction. *J. Sound Vib.*, 108(2):305–325, 1986.
185. L. Sherwood. *Human Physiology, From Cells to Systems.* Cengage Learning, 2012.
186. J. Shi, J. Guldner, and V. I. Utkin. *Sliding mode control in electro-mechanical systems.* CRC Press, 1999.
187. J. Sijbrand. *Studies in nonlinear stability and bifurcation theory.* Rijksuniversiet Utrecht, 1981.
188. D. J. W. Simpson. Scaling laws for large numbers of coexisting attracting periodic solutions in the border-collision normal form. *Int. J. Bif. Chaos*, 24(9):1–28, 2014.
189. D. J. W. Simpson. Border-collision bifurcations in R^n. *SIAM Review*, 58(2):177–226, 2016.
190. D. J. W. Simpson and R. Kuske. Stochastically perturbed sliding motion in piecewise-smooth systems. *Discrete Contin. Dyn. Syst. Ser. B*, 19(9):2889–2913, 2014.
191. W. Singhose and K. Grosser. Limiting excitation of unmodeled high modes with negative input shapers. *Proceedings of the 1999 IEEE Int. Conf. on Control Applications*, 3(40):545–550, 1999.
192. J-J. E. Slotine and W. Li. *Applied Nonlinear Control.* Prentice Hall, 1991.
193. M. Sorensen, S. DeWeerth, G Cymbalyuk, and R. L. Calabrese. Using a hybrid neural system to reveal regulation of neuronal network activity by an intrinsic current. *Journal of Neuroscience*, 24(23):5427–5438, 2004.
194. J. Sotomayor and M. A. Teixeira. Vector fields near the boundary of a 3-manifold. *Lecture Notes in Mathematics*, 1331:169–195, 1988.
195. J. Sotomayor and M. A. Teixeira. Regularization of discontinuous vector fields. *Proceedings of the International Conference on Differential Equations, Lisboa*, pages 207–223, 1996.
196. D. E. Stewart. Existence of solutions to rigid body dynamics and the Painlevé paradoxes. *Comptes Rendus de l'Acadámie des Sciences - Series I - Mathematics*, 325(6):689–693, 1997.
197. G G Stokes. On the discontinuity of arbitrary constants which appear in divergent developments. *Trans Camb Phil Soc*, 10:106–128, 1864.
198. H. Stommel. Thermohaline convection with two stable regimes of flow. *Tellus*, 13:224–230, 1961.
199. S. Strogatz. *Nonlinear Dynamics and Chaos.* Westview Press, 2nd edition, 2014.
200. C. Studer. *Numerics of Unilateral Contacts and Friction*, volume 47 of *Lecture Notes in Applied and Computational Mechanics.* Springer Verlag, 2009.
201. D. Tabor. Triobology - the last 25 years. a personal view. *Tribology International*, 28(1):7–10, 1995.

202. F. Takens. Forced oscillations and bifurcations. *Comm. Math. Inst. Rijksuniv. Utrecht 2*, pages 1–111, 1974.
203. M. A. Teixeira. Structural stability of pairings of vector fields and functions. *Boletim da S.B.M.*, 9(2):63–82, 1978.
204. M. A. Teixeira. On topological stability of divergent diagrams of folds. *Math. Z.*, 180:361–371, 1982.
205. M. A. Teixeira. Stability conditions for discontinuous vector fields. *J. Differ. Equ.*, 88:15–29, 1990.
206. M. A. Teixeira. Generic bifurcation of sliding vector fields. *J. Math. Anal. Appl.*, 176:436–457, 1993.
207. R. Thom. *Structural stability and morphogenesis.* Benjamin-Addison Wesley, translation by D.H.Fowler edition, 1975.
208. G.A. Tomlinson. A molecular theory of friction. *Philos. Mag.*, 7(7):905–939, 1929.
209. Y. Tsypkin. *Theory of Relay Control Systems.* (in Russian), Moscow: Gostechizdat, 1955.
210. Y. Ueda. Strange attractors and the origin of chaos. *Nonlinear Science Today*, 2:1–16, 1992.
211. V. I. Utkin. Variable structure systems with sliding modes. *IEEE Trans. Automat. Contr.*, 22:212–222, 1977.
212. V. I. Utkin. *Sliding modes in control and optimization.* Springer-Verlag, 1992.
213. B. Van der Pol. On relaxation-oscillations. *The London, Edinburgh and Dublin Phil. Mag. & J. of Sci.*, 2(7):978–992, 1926.
214. A. J. Vander, D. Luciano, and J. Sherman. *Movement of Molecules Across Cell Membranes.* Human Physiology: The Mechanism of Body Function. McGraw-Hill Higher Education, 2001.
215. J. Walsh and E. Widiasih. A dynamics approach to a low-order climate model. *Discret. Contin. Dyn. S.*, pages 257–79, 19 2014.
216. A-M. Wazwaz. The tanh method for traveling wave solutions of nonlinear equations. *Appl. Math. Comput.*, 154:713–723, 2004.
217. W. R. Webber and F. B. McDonald. Recent voyager 1 data indicate that on 25 august 2012 at a distance of 121.7 au from the sun, sudden and unprecedented intensity changes were observed in anomalous and galactic cosmic rays. *Geophysical Research Letters*, 40(9):1665–1668, 2013.
218. M. Wechselberger. Existence and bifurcation of canards in $\mathbb{R}^3$ in the case of a folded node. *SIAM J. App. Dyn. Sys.*, 4(1):101–139, 2005.
219. M. Wechselberger. A propos de canards (apropos canards). *Trans. Amer. Math. Soc*, 364:3289–3309, 2012.
220. P. Welander. A simple heat-salt oscillator. *Dyn. Atmos. Oceans*, 6(4):233–242, 1982.
221. A. J. Weymouth, D. Meuer, P. Mutombo, T. Wutscher, M. Ondracek, P. Jelinek, and F. J. Giessibl. Atomic structure affects the directional dependence of friction. *PRL*, 111(126103):1–4, 2013.
222. J. Wojewoda, S. Andrzej, M. Wiercigroch, and T. Kapitaniak. Hysteretic effects of dry friction: modelling and experimental studies. *Phil. Trans. R. Soc. A*, 366:747–765, 2008.
223. J. Woodhouse, T. Putelat, and A. McKay. Are there reliable constitutive laws for dynamic friction? *Phil. Trans. R. Soc. A*, 373(20140401):1–21, 2015.
224. V. A. Yakubovich, G. A. Leonov, and A. K. Gelig. Stability of stationary sets in control systems with discontinuous nonlinearities. *Singapore: World Scientific*, 2004.

Glossary

Terminology

switching function $\sigma = \sigma_1 \ldots \sigma_m$

switching multiplier $\lambda_j = \operatorname{sign} \sigma_j$ forming a vector $\boldsymbol{\lambda} = (\lambda_1, \ldots, \lambda_m)$

combination vector field $\mathbf{f}(\mathbf{x}; \boldsymbol{\lambda})$

constituent vector field $\mathbf{f}^{\pm\pm\cdots}(\mathbf{x}) \equiv \mathbf{f}(\mathbf{x}; \pm 1, \pm 1, \ldots)$

discontinuity submanifold $\mathcal{D}_j = \{\mathbf{x} \,:\, \sigma_j = 0\}$

discontinuity surface $\mathcal{D} = \mathcal{D}_1 \cup \cdots \cup \mathcal{D}_m$

switching layer is a region of space that 'blows up' the discontinuity surface

sliding manifold $\mathcal{M}$ is the set of points in the switching layer where sliding occurs

invariant manifold is a general term for a subspace made up of solutions of a dynamical system, and locally equivalent to Euclidean space of a given dimension. Usually one singles out invariant manifolds made up of special solutions (e.g. sliding orbits, or trajectories separating different orbit topologies).

Indices Related to Switching

n = dimension of the state space, $\mathbf{x} \in \mathbb{R}^n$

m = number of switching submanifolds

r = codimension of sliding (number of intersecting switching submanifolds)

N = number of constitutent vector fields $\mathbf{f}^{\pm\pm\cdots}$ on regions $\mathcal{R}_{\pm\pm\ldots}$

j = index running from 1 to m (usually associated with switching submanifolds)

K $= \kappa_1 \ldots \kappa_m =$ binary index where each $\kappa_j = \pm$

i is reserved for indices running from 1 to n where possible

M. R. Jeffrey, *Hidden Dynamics*,
https://doi.org/10.1007/978-3-030-02107-8

Standard Functions

sign(x) has value $+1$ if $x > 0$ and -1 if $x < 0$, with $\operatorname{sign} x \in (-1, +1)$ for $x = 0$

step(x) has value 1 if $x > 0$ and 0 if $x < 0$, with $\operatorname{step} x \in (0, 1)$ for $x = 0$

Erf(x) is the standard error function $\operatorname{Erf}(x) = \frac{2}{\sqrt{\pi}} \int_0^x e^{-t^2} dt$.

Index

M. R. Jeffrey, *Hidden Dynamics*,
https://doi.org/10.1007/978-3-030-02107-8